DICTIONARY OF CHEMICAL TERMINOLOGY

edited by Dobromiła Kryt

DICTIONARY OF CHEMICAL TERMINOLOGY

IN FIVE LANGUAGES

English
German
French
Polish
Russian

ELSEVIER SCIENTIFIC PUBLISHING COMPANY
Amsterdam—Oxford—New York   1980

Editorial Staff: D. KRYT M. Sc.(Chem.),
G. MALUDZIŃSKA M. Sc.(Chem.), M. WERNER M. Sc.(Chem.),
J. KARPOWICZ, E. SOZAŃSKA M. A., M. SZPAKOWSKA

Graphic Design A. KOWALEWSKI

Published in coedition with Wydawnictwa Naukowo-Techniczne,
Warszawa

Distribution of this book is handled by the following publishers:
   for the U.S.A. and Canada
Elsevier/North-Holland, Inc.
52 Vanderbilt Avenue
New York, N.Y. 10017
   for the East European Countries, China, Northern Korea, Cuba,
   Vietnam and Mongolia
Wydawnictwa Naukowo-Techniczne
ul. Mazowiecka 2/4,
00-048 Warsaw, Poland
   for all remaining areas
Elsevier Scientific Publishing Company
335 Jan van Galenstraat
P.O. Box 211, 1000 AE Amsterdam, The Netherlands

Library of Congress Cataloging in Publication Data
Main entry under title:
Dictionary of chemical terminology.
   Based on the earlier Polish dictionary Słownik terminologii chemicznej
polsko-niemiecko-angielsko-francusko-rosyjski published in 1974.
   Includes indexes.
   1. Chemistry—Dictionaries—Polyglot.
2. Dictionaries, Polyglot. I. Kryt, Dobromiła.
QD5.D49   540'.3   79-20852

ISBN 0-444-99788-1

PREFACE

The Technical Terminology Division of Wydawnictwa Naukowo-
Techniczne, Publishers, has been concerned for over twenty five years
with the compiling and editing of bi- and multilingual dictionaries
in various branches of science and technology. The present dictionary
is the first multilingual terminological dictionary to cope with modern
chemistry. The dictionary comprises mutually correlated terminology
from the basic fields of chemistry, taking due account of its most
recent areas of development, as well as the terminology from related
disciplines which are entering modern chemistry in ever increasing
amounts. It does not, however, include the nomenclature of chemical
compounds and terms from chemical engineering and technology.
The dictionary covers the following subject fields:
Atomic and atomic nucleus structure, Chemical bond, Chemical
elements, Chemical kinetics, Chemical reactions, Chemical
thermodynamics, Chromatography, Colloid chemistry, Conformation,
Coordination chemistry, Electrical properties of molecules,
Electroanalytical methods, Electrochemistry, Electronic and steric
effects of atoms and atomic groups, Fundamental particles,
Gravimetric methods, Groups of elements, Isomerism, Isotopes,
Macromolecular compounds, Magnetochemistry, Mechanisms of
reactions, Molecular structure, Nuclear reactions, Optical methods of
analysis, Periodic table, Phase changes, Phase systems, Photochemistry,
Quantum mechanics, Radiation chemistry, Radioactivity, Radiochemical
methods of analysis, Radiochemistry, Spectroscopy, States of matter,
Statistical mechanics, Statistics and error estimation, Surface
chemistry, Thermochemistry, Types of reagents, Volumetric analysis.
The dictionary was compiled on the basis of the contemporary literature
in chemistry and related disciplines in the relevant languages
(handbooks, monographs, encyclopaedias, scientific journals, various
IUPAC and ISO publications, etc.).
The symbols (a list is given on a separate page) and terminology
of physicochemical quantities and units are given in this dictionary
in accordance with the recommendations of IUPAC and ISO.
The dictionary has been prepared on the basis of the earlier published
Polish dictionary "Słownik terminologii chemicznej", which was
compiled by a team of outstanding Polish scientists headed by an
Editorial Board appointed by The Polish Chemical Society.
The present dictionary is intended for scientists, researchers,
engineers, students and all those who in their work are concerned with
modern chemistry and scientific literature.
Any comments on the usefulness of the dictionary and on the gaps and
errors noticed will be gratefully acknowledged by the publisher,
serving to improve the future editions.

THE EDITOR

EXPLANATORY NOTES

The dictionary contains 3805 English entries in alphabetical order with their definitions followed by equivalents in German (D), French (F), Polish (P) and Russian (R). Efforts were made to formulate the English definitions as concisely as possible, preferably in a single sentence, but so as to explain the semantic range of the entry. Care has been taken to provide the English entries and the equivalents in the other languages as fully as possible with synonyms. English synonyms are given in parentheses, synonyms of German, French, Polish and Russian terms are separated by commas. Where there are two synonymous forms of a term, a full and an abbreviated one, the part of the composite term which can be omitted is given in parentheses, for instance the term "(ко)валентные кристаллы", contains two synonymous forms: "ковалентные кристаллы", and "валентные кристаллы", similarly the term "równowaga (promieniotwórcza) przejściowa" also has two synonymous forms: "równowaga promieniotwórcza przejściowa" and "równowaga przejściowa". A term in German, French, Polish and Russian can be easily found by using the alphabetical indexes placed at the end of the volume. The gender of nouns is indicated (only in the indexes) by the generally used abbreviations: *m* for masculine, *f* for feminine, *n* for neuter. The abbreviation *pl* stands for plural.

SYMBOLS USED IN THE DICTIONARY

A — area
A — mass number
A_r — relative atomic mass
a — acceleration
a_B — relative activity of substance B
Ar — aryl
C — capacitance
C — heat capacity
c — concentration
c — speed of light
d — deuteron
d — diameter
d — relative density
div — divergence
E — chemical element
E — electromotive force
E — energy
E_k — kinetic energy
E_p — potential energy
e — base of natural logarithm
e — electron
e — elementary charge
exp — exponential of e
F — Faraday
F — force
f — fission (*in nuclear reactions*)
G — thermodynamic potential
G — weight
g — acceleration of gravity
H — enthalpy
H — Hamiltonian function
h — height
h — Planck constant
$\hbar$ — Planck constant divided by 2π
I — electric current
i — imaginary unit
K — kelvin
k — Boltzmann constant
l — length
lg — decadic logarithm
ln — natural logarithm
M — metal
M — molar mass
M_r — relative molecular mass
m — mass
N — number of molecules

N_A — Avogadro constant
n — amount of substance
n — neutron
P — power
p — pressure
p — proton
Q — heat
Q — quantity of electricity
R — gas constant
R — organic radical
R — resistance
r — radius
S — entropy
s — path; length of arc
T — period
T — thermodynamic temperature
t — empirical temperature
t — time
t — triton
U — internal energy
U — voltage
V — electric potential
V — volume
v — velocity
W — work
X — halogen
Z — atomic number
z — charge number
α — α-particle
Γ — Γ function
γ — photon
η — viscosity
λ — wavelength
μ — reduced mass
μ_B — Bohr magneton
μ_B — chemical potential of substance B
ν — frequency
ν — kinematic viscosity
π — 3.14...
ρ — density
ρ — resistivity
Σ — sum
σ — wavenumber
τ — characteristic time interval; relaxation time
ω — angular velocity

SUPERSCRIPTS

* — pure substance
$\ominus$ — standard state
$\ddagger$ — activated complex

GREEK ALPHABET

A α — alpha	N ν — nu
B β — beta	Ξ ξ — xi
Γ γ — gamma	O o — omicron
Δ δ — delta	Π π — pi
E ε — epsilon	P ρ — rho
Z ζ — zeta	Σ σ — sigma
H η — eta	T τ — tau
Θ ϑ — theta	Υ υ — upsilon
ι — iota	Φ φ — phi
K ϰ — kappa	X χ — chi
Λ λ — lambda	Ψ ψ — psi
M μ — mu	Ω ω — omega

Main Text

1

ABEGG RULE (octet rule).
Most stable electronic configuration in atoms is that with completely filled s and p subshells of valence shell (noble gas configuration).

D Oktettprinzip
F règle de l'octet, règle des gaz rares
P reguła oktetu, reguła gazu szlachetnego
R принцип октета, правило октета

2

ABSOLUTE ACTIVITY
(*of component*), λ_B.
Dimensionless quantity

$$\lambda_B = \exp\left(\frac{\mu_B}{RT}\right) \quad \text{or} \quad \mu_B = RT \ln \lambda_B$$

where μ_B = chemical potential of B component, R = molar gas constant, and T = thermodynamic temperature.

D absolute Aktivität
F activité absolue, facteur de dégénérescence
P aktywność bezwzględna
R абсолютная активность

ABSOLUTE ACTIVITY MEASUREMENT.
See ABSOLUTE COUNTING.

3

ABSOLUTE CONFIGURATION.
Determination of real arrangement of atoms (or groups of atoms) around chiral centre. Absolute configuration is described according to R-S nomenclature. It has been established that absolute configuration of enantiomers of glyceraldehyde is in accordance with their arbitrarily adopted configuration.

D absolute Konfiguration
F configuration absolue
P konfiguracja absolutna, konfiguracja bezwzględna
R абсолютная конфигурация

ABSOLUTE CONSTANTS. *See*
UNIVERSAL CONSTANTS.

4

ABSOLUTE COUNTING
(absolute activity measurement).
Direct measurement of total rate of radioactive decay, i.e. measurement of number of disintegrations per unit time in the whole mass of preparation examined.

D absolute Aktivitätsbestimmung
F mesure absolue de l'activité
P pomiar aktywności bezwzględny
R измерение абсолютной активности

5

ABSOLUTE ELECTRODE POTENTIAL.
Difference of inner potentials of electrode and electrolyte; non-measurable quantity (*cf.* relative electrode potential).

D absolutes Elektrodenpotential, Galvanispannung
F potentiel absolu d'électrode, tension absolue d'électrode
P potencjał elektrody bezwzględny
R электродный потенциал, потенциал электрода

6

ABSOLUTE ERROR.
Difference between result of measurement (analytical determination) and value assumed as true.

D Absolutfehler
F erreur absolue
P błąd bezwzględny, błąd absolutny
R абсолютная ошибка

ABSOLUTE IONIC MOBILITY. *See*
IONIC MOBILITY.

7

ABSOLUTE REACTION RATE.
Reaction-rate constant of elementary reaction previously calculated on basis of properties of reacting molecules and activated complex, nowadays determined also experimentally.

D absolute Reaktionsgeschwindigkeit
F vitesse absolue de réaction
P szybkość reakcji bezwzględna
R абсолютная скорость реакции

8
ABSOLUTE REACTION RATE THEORY
(activated-complex theory, transition-state theory).
Reaction rate theory based on model of elementary reaction proceeding over activated complex.
D Theorie des aktivierten Komplexes, Theorie des Übergangszustandes
F théorie des vitesses absolues, théorie du complexe activé, théorie de l'état de transition
P teoria bezwzględnej szybkości reakcji, teoria kompleksu aktywnego, teoria stanu przejściowego
R теория абсолютных скоростей реакций, теория активного комплекса, теория переходного состояния

9
ABSOLUTE REFRACTIVE INDEX.
Ratio of velocities of light in vacuum and in a given medium.
D Brechungsindex, Brechungskoeffizient, Brechzahl
F indice de réfraction absolu
P współczynnik refrakcji bezwzględny, współczynnik załamania światła bezwzględny
R абсолютный показатель преломления

ABSOLUTE TEMPERATURE. *See*
THERMODYNAMIC TEMPERATURE.

ABSOLUTE ZERO OF TEMPERATURE.
See ZERO TEMPERATURE.

10
ABSORBANCE (absorbancy, optical density).
Logarithm to base 10 of ratio of intensity of radiation passing through reference, I_0, and through investigated sample, I, $A = \lg I_0/I$. Absorbance corresponds to loss of energy absorbed in a given medium in comparison with loss in reference. It is assumed that losses for reflection, solvent absorption, and refraction are compensated and that there is no interference due to scattered radiation.
D Extinktion
F absorbance
P absorbancja, wartość absorpcji
R поглощение, оптическая плотность, экстинкция

ABSORBANCY. *See* ABSORBANCE.

ABSORBANCY INDEX. *See*
ABSORPTIVITY.

11
ABSORBED DOSE, D.
Amount of energy imported to matter by ionizing particles per unit mass

of irradiated material at place of interest.
In SI system it is expressed in grays.
D Energiedosis, absorbierte Dosis
F dose absorbée
P dawka pochłonięta, dawka zaabsorbowana
R поглощённая доза

ABSORBING CENTRES. *See* COLOUR
CENTRES.

12
ABSORPTION.
Penetration of substance (absorbate) into body of another (absorbent).
D Absorption
F absorption
P absorpcja, sorpcja wgłębna
R абсорбция

13
ABSORPTION CELL (cuvette, sample cell).
In studies of radiation absorption small vessel containing investigated sample (liquid or gaseous). Distance between the two parallel propagation windows, perpendicular to direction of radiation determines thickness of investigated sample.
D Absorptionsküvette, Küvette
F cuve absorbante
P kiuweta absorpcyjna, naczynko (pomiarowe), kuweta
R кювет(к)а

ABSORPTION COEFFICIENT. *See*
ABSORPTIVITY.

14
ABSORPTION COEFFICIENT (*of gas*).
Coefficient describing volume of gas (reduced to normal conditions) which dissolves in unit volume of liquid at the given temperature and under normal pressure.
D Absorptionskoeffizient
F coefficient d'absorption
P współczynnik absorpcji Ostwalda
R коэффициент поглощения

15
ABSORPTION CROSS-SECTION.
Cross-section of nucleus of atom for absorption of bombarding particles.
D Absorptionsquerschnitt
F section efficace d'absorption
P przekrój czynny na absorpcję, przekrój czynny absorpcji
R сечение поглощения

16
ABSORPTION CURVE.
Dependence of absorption or of any of its functions on wavelength, wavenumber or frequency of electromagnetic radiation.

D Absorptionskurve
F courbe d'absorption
P krzywa absorpcji, krzywa
spektrofotometryczna
R спектральная кривая

17
ABSORPTION CURVE (absorption signal).
In magnetic resonance, curve representing
dependence of energy absorbed by a given
substance from alternating high-frequency
magnetic field on strength of static
magnetic field (at fixed frequency) or on
frequency of field (at constant strength
of static magnetic field). It represents
changes of imaginary part χ'' of magnetic
susceptibility.

D Absorptionskurve, Absorptionssignal
F courbe d'absorption, signal d'absorption
P krzywa absorpcji, sygnał absorpcji
R кривая (резонансного) поглощения,
сигнал поглощения

18
ABSORPTION FILTER (colour filter).
Optical filter absorbing radiation in some
region of frequencies.

D Absorptionsfilter
F filtre coloré
P filtr absorpcyjny
R абсорбционный светофильтр

19
ABSORPTION OF IONIZING
RADIATION.
Whole of processes occurring when
ionizing radiation passes through matter;
in result intensity of radiation is reduced
and its composition changed.

D Absorption ionisierender Strahlen
F absorption de rayonnement ionisant
P absorpcja promieniowania jonizującego
R поглощение ионизирующего излучения

20
ABSORPTION OF LIGHT.
Transformation of light energy passing
through matter into energy of different
type.

D Absorption des Lichtes, Lichtabsorption
F absorption de la lumière, absorption
lumineuse
P absorpcja światła, pochłanianie światła
R поглощение света

ABSORPTION SIGNAL. See
ABSORPTION CURVE.

21
ABSORPTION SPECTROANALYSIS.
Spectroanalysis based on investigation
of absorption spectra.

D Absorptionsspektralanalyse
F analyse par spectrophotométrie
d'absorption, spectranalyse d'absorption
P analiza spektralna absorpcyjna
R абсорбционный спектральный анализ

22
ABSORPTION SPECTROPHOTOMETRY
(absorption spectroscopy).
Branch of spectrophotometry dealing with
spectra of electromagnetic radiation
absorbed by matter.

D Absorptionsspektralphotometrie,
Absorptionsspektroskopie
F spectrophotométrie d'absorption,
absorptiométrie
P spektrofotometria absorpcyjna
R абсорпционная спектрофотометрия

23
ABSORPTION SPECTROSCOPY.
Branch of spectroscopy comprising
methods of determination of composition
and structure of chemicals by study of
absorption spectra.

D Absorptionsspektroskopie
F spectroscopie d'absorption
P spektroskopia absorpcyjna
R абсорбционная спектроскопия

ABSORPTION SPECTROSCOPY. See
ABSORPTION SPECTROPHOTOMETRY.

24
ABSORPTION SPECTRUM.
Spectrum arising as result of propagation
of electromagnetic radiation through
absorbing medium.

D Absorptionsspektrum
F spectre d'absorption
P widmo absorpcyjne
R спектр поглощения

25
ABSORPTIVITY (absorption coefficient,
specific absorbance, absorbancy index), α.
Proportionality constant in Lambert-Beer
law equation, equal numerically to
absorbance of solution of unit
concentration and unit path length.

D Absorptionskoeffizient, spezieller
dekadischer Extinktionskoeffizient,
Extinktionsmodul
F pouvoir absorbant specifique, coefficient
d'absorption
P współczynnik absorpcji, absorpcyjność,
współczynnik absorbancji,
absorbowalność
R коэффициент ослабления,
коэффициент поглощения,
коэффициент погашения

26
ABUNDANCE (abundance ratio).
Content of particular isotope in given
sample of the natural element, expressed
as percentage, e.g. 99.76% of ^{16}O in natural
oxygen.
D Isotopenhäufigkeit
F teneur isotopique, abondance isotopique
P rozpowszechnienie izotopu, obfitość
izotopu
R относительная распространённость
изотопа

ABUNDANCE RATIO. See ABUNDANCE.

27
ACCELERATED ELECTRONS.
Electrons whose kinetic energy is
increased in an electric potential.
D beschleunigte Elektronen
F électrons accélérés
P elektrony przyspieszone
R ускоренные электроны

ACCELERATOR. See CHARGED
PARTICLE ACCELERATOR.

28
ACCEPTOR (*in induced reaction*).
Substance which reacts with actor in
system of coupled reactions.
D Akzeptor (der induzierten Reaktion)
F ...
P akceptor (reakcji sprzężonej)
R акцептор (сопряжённой реакции)

ACCEPTOR. See ACCEPTOR CENTRE.

29
ACCEPTOR CENTRE (acceptor, acceptor
impurity).
Atom of valency lower than normal
constituent of crystal (e.g. admixture of
boron in germanium crystal), the presence
of which gives rise to new energy band
not occupied by electron and located
just above ground state (in range of
forbidden energy band).
D Akzeptor, Akzeptor-Zentrum
F accepteur
P centrum akceptorowe, akceptor
R акцептор

ACCEPTOR IMPURITY. See ACCEPTOR
CENTRE.

30
ACCEPTOR LEVEL (acceptor state *in
band model of solid state*).
Local, allowed, unoccupied energy level
situated in area of forbidden band,
connected with presence of acceptor
centre in crystal.
D Akzeptorniveau, Akzeptorenterm
F niveau accepteur
P poziom akceptorowy, stan akceptorowy

R уровень акцептора, акцепторный
уровень

ACCEPTOR STATE. See ACCEPTOR
LEVEL.

ACCIDENTAL ERROR. See RANDOM
ERROR.

31
ACCUMULATOR (secondary cell, storage
cell).
Cell in which processes occurring during
its work (discharge of accumulator),
can be carried out in reverse direction
by electrolysis (charging of accumulator).
D Akkumulator, Sammler,
Sekundärelement
F accumulateur, élément secondaire
P akumulator, ogniwo galwaniczne wtórne
R аккумулятор, вторичный элемент

32
ACCURACY.
Measure of agreement between result
of measurement (determination) or mean
in series of measurements and value
of measured quantity assumed as true.
D Richtigkeit, Genauigkeit
F justesse
P dokładność
R точность, правильность

33
ACCURACY OF METHOD.
Quantitative (absolute or relative)
measure of agreement of mean value of
measurement results obtained by a given
method and value of measured property
assumed as true.
D Richtigkeit, Genauigkeit (*der Methode*)
F justesse d'une méthode
P dokładność metody
R точность метода

34
ACETOLYSIS.
Metathesis between organic compound
and acetic acid, the latter acting
additionally as solvent, e.g.

$$RCOOR' + CH_3COOH \rightarrow CH_3COOR' + RCOOH$$

D Acetolyse
F acétolyse
P acetoliza
R ацетолиз

35
ACETYLATION.
Substitution of hydrogen atom by acetyl
group CH_3CO- in chemical molecule, e.g.

$$(CH_3CO)_2O + NH_3 \rightarrow CH_3CONH_2 + CH_3COOH$$

$$CH_3COCl + C_6H_5OH \rightarrow CH_3COOC_6H_5 + HCl$$

D Acetylierung
F acétylation

P acetylowanie
R ацетилирование

36
ACETYLENIC BOND.
Triple bond between two carbon atoms.
D Acetylenbindung
F liaison acétylénique
P wiązanie acetylenowe
R ацетиленовая связь

37
ACHIRALITY.
Lack of chirality; identity of configuration
of molecule with its mirror image, caused
by presence of plane of symmetry, or
centre of symmetry, or four-fold
inversion (symmetry) axis in molecule.
D Achiralität, Unchiralität
F achiralité
P achiralność
R ахиральность

38
ACID.
Substance which exhibits characteristic
properties predicted by acid-base theories
from which following are the most
important: ability to neutralize base,
displacement of weaker acids from their
compounds, definite influence on colour
of acid-base indicators, catalytic
properties.
D Säure
F acide
P kwas
R кислота

39
ACID-BASE CATALYSIS.
Catalysis due to substances of acid or base
type. In this type of catalysis, reaction
of catalyst with substrate consists in
proton transfer.
D Säure-Base-Katalyse
F catalyse acido-basique
P kataliza kwasowo-zasadowa
R кислотно-основный катализ

40
ACID-BASE INDICATOR.
Weak acid or organic base which on
reacting with water gives conjugated
acid-base system. Acid-base indicators
are used for measurement of pH of
solution and determination of end-point
of acid-base titration by visual method.
D pH-Indikator, Säure-Basen-Indikator
F indicateur de pH, indicateur coloré,
   indicateur acido-basique
P wskaźnik pH, wskaźnik alkacymetryczny,
   wskaźnik kwasowo-zasadowy

R pH-индикатор, кислотно-основный
   индикатор, кислотно-щелочной
   индикатор

41
ACID-BASE TITRATION.
Common name for acidimetry and
alkalimetry.
D Säure-Base-Titration,
   Neutralisationstitration,
   Neutralisationsanalyse
F acido-alcalimétrie
P alkacymetria, miareczkowanie
   alkacymetryczne
R алкалиметрия и ацидиметрия, метод
   кислотно-основного титрования, метод
   нейтрализации

42
ACID HYDROLYSIS.
Decomposition of β-ketoacid ester by
concentrated base, leading to formation
of corresponding carboxylic acid salts
and an alcohol, e.g.

$$CH_3\overset{O}{\overset{\|}{C}}CHR\overset{O}{\overset{\|}{C}}OC_2H_5 \xrightarrow[H_2O]{NaOH} CH_3\overset{O}{\overset{\|}{C}}ONa +$$

$$+ RCH_2\overset{O}{\overset{\|}{C}}ONa + C_2H_5OH$$

D Säurespaltung
F hydrolyse acide
P rozpad kwasowy
R кислотное расщепление

43
ACIDIC GROUP.
Part of chelating agent containing
hydrogen ion which can be replaced by
metal ion, e.g. hydroxyl group —OH,
carboxyl group —COOH, oxime group
=NOH, thiol group —SH, arsonium group
—AsO(OH)$_2$.
D saure Gruppe
F groupement acide
P grupa kwasowa, grupa solotwórcza
R солеобразующая группа

ACIDIC SOLVENT. *See* PROTOGENIC
SOLVENT.

ACIDIMETRIC TITRATION. *See*
ACIDIMETRY.

44
ACIDIMETRY (acidimetric titration).
Determination of substance by acid
titration.
Remark: Some authors use in this case
incorrectly the therm alkalimetry.
D Acidimetrie
F acidimétrie
P acydymetria, miareczkowanie
   acydymetryczne
R ацидиметрия, титрование кислотами

45
ACIDITY.
Characteristic feature of electrolyte solutions in which concentration of so-called "hydrogen ions" H^+ (in fact oxonium ions, ROH_2^+, or hydroxonium or hydronium ions, H_3O^+, present in corresponding protophilic solvents in their solvated forms) is greater than that of hydroxylic ions OH^-.
D Azidität
F acidité
P kwasowość
R кислотность

46
ACIDITY FUNCTION H (Hammett acidity function), H_0.
Conventional, numerical approximate scale of acidity of solutions

$$H_0 = -\lg a_{H^+} - \lg \frac{f_B}{f_{BH^+}}$$

where a_{H^+} = hydrogen ion activity, f_B and f_{BH^+} = activity coefficients of basic, electrically neutral and acidic forms of substance respectively.
D Hammettsche Acidatäts-Funktion
F fonction d'acidité de Hammett
P funkcja kwasowa Hammetta
R функция кислотности Гаммета

47
ACIDITY OF BASE.
Number of hydroxylic groups in molecule of base which may be exchanged by various acid radicals during neutralization process.
D Azidität der Base
F ...
P kwasowość zasady
R кислотность основания

48
ACIDOLYSIS.
Metathesis between organic compound and acid, the latter serving additionally as solvent, e.g.

RCO—O—COR + R'COOH → RCO—O—COR' +
+ RCOOH

D Acidolyse
F acidolyse
P acydoliza
R ацидолиз

49
ACID RADICAL.
Atom or group of atoms (uni- or multivalent) that remain after splitting off one or more hydrogen atoms from molecule of acid.
D Säure-Rest
F radical acide

P reszta kwasowa
R кислотный остаток

50
ACI FORM (aci-nitro form).
Tautomeric form of formula:

existing in equilibrium with primary- or secondary nitro compound. Aci form is less stable than nitro compound.
D aci-Form
F forme aci
P odmiana aci(-nitrowa)
R аци-форма, кислотная форма

ACI-NITRO FORM. See ACI FORM.

ACTINIDE ELEMENTS. See ACTINOIDS.

ACTINIDES. See ACTINOIDS.

ACTINIDE SERIES. See ACTINOIDS.

51
ACTINIUM, Ac.
Atomic number 89.
D Actinium, Aktinium
F actinium
P aktyn
R актиний

52
ACTINIUM D, AcD.
Isotope of lead. Atomic number 82. Mass number 207.
D Actiniumblei, Actinium D, Aktiniumblei
F actino-plomb, plomb de l'actinium
P aktynoołów, aktyn D
R актиний D

53
ACTINIUM SERIES.
Naturally occurring series with first and last members uranium, ^{235}U, and lead, ^{207}Pb, respectively. The mass number of each member of this series is $A = 4n+3$, where n = natural number.
D Actinium-Reihe
F famille de l'actinium, famille de l'actino-uranium
P szereg urano-aktynowy
R ряд актиноурана

54
ACTINOIDS (actinides, actinide series, actinide elements).
Series of radioactive elements in seventh period of periodic system: actinium, thorium, protactinium, uranium, neptunium, plutonium, americium, curium, berkelium, californium, einsteinium, fermium, mendelevium, nobelium, lawrencium. Electronic

structure of actinoids is not accurately
established as yet. These elements
are transition elements and form second
series of inner transition elements
starting with thorium (Seaborg) or
neptunium (Dawson) filling of
subshell 5f.

D Actin(o)ide, Aktinide
F actinides, série de l'actinium
P aktynowce
R актин(о)иды, семейство актинидов

55
ACTINON (actinium emanation), An.
Isotope of radon. Atomic number 86. Mass
number 219.

D Actinon, Actinium-Emanation, Aktinon,
  Aktinium-Emanation
F actinon
P aktynon, emanacja aktynowa
R актинон, эманация актиния

56
ACTINOURANIUM, AcU.
Isotope of uranium. Atomic number 92.
Mass number 235.

D Actinouran, Aktinouran
F actino-uranium
P aktynouran
R актиноуран

57
ACTIVATED ADSORPTION.
Chemisorption characterized by fairly
high energy of activation.

D aktivierte Adsorption
F adsorption activée
P chemisorpcja powolna, adsorpcja
  aktywowana
R активированная адсорбция

58
ACTIVATED COMPLEX (transition state,
transition complex, critical complex).
Unstable configuration of reactant atoms
to which corresponds maximum of
potential energy in reaction path.

D aktiv(iert)er Komplex,
  Übergangszustand, aktiver
  Zwischenzustand, kritischer Komplex
F complexe activé, état de transition
P kompleks aktywny, stan przejściowy,
  stan pośredni, zespół aktywny
R актив(ирован)ный комплекс,
  переходное состояние

ACTIVATED-COMPLEX THEORY. *See*
ABSOLUTE REACTION RATE THEORY.

59
ACTIVATED MOLECULE (active
molecule).
Molecule which is able to react as result
of activation.

D aktiv(iert)es Molekül
F molécule activée
P cząsteczka aktyw(owa)na
R актив(ирован)ная молекула

60
ACTIVATION (radioactivation).
Process of inducing radioactivity by
irradiation with neutrons, charged
particles or photons.

D Aktivierung, Radioaktivierung
F (radio)activation
P (radio)aktywacja
R активация

61
ACTIVATION ANALYSIS
(radioactivation analysis).
Method of chemical analysis based on
identification and measurement of
characteristic radiation from radioactive
nuclides formed in a sample by
irradiation.

D Aktivierungsanalyse,
  Radioaktivierungsanalyse
F analyse par (radio)activation
P analiza aktywacyjna
R активационный анализ

62
ACTIVATION CROSS-SECTION.
Measure of probability of formation of
specified radioactive product in nuclear
reaction.

D Aktivierungsquerschnitt
F section efficace d'activation
P przekrój czynny na aktywację, przekrój
  czynny aktywacji
R сечение активации

63
ACTIVATION ENERGY.
Excess energy over non-reacting state
which must be acquired by atomic system
in order that a particular process may
occur.

D Aktivierungsenergie
F énergie d'activation
P energia aktywacji
R энергия активации

ACTIVATION ENTROPY. *See* ENTROPY
OF ACTIVATION.

64
ACTIVATION INDUCED BY
ELECTRONS.
Activation by electrons with energy of at
least 15 MeV.

D Aktivierung mit Elektronen
F activation par les électrons
P aktywacja elektronowa
R активация (произведенная) быстрыми
  электронами

65
ACTIVATION OF MOLECULE.
Uptake of activation energy by molecule.
D Aktivierung des Moleküls
F activation de la molécule
P aktywacja cząsteczki
R активация молекулы

66
ACTIVATION POLARIZATION
(in electrochemistry).
Sum of transition and reaction
polarizations. Activation polarization
is expressed by activation overvoltage.
D Aktivierungspolarisation
F polarisation d'activation
P polaryzacja aktywacyjna
R активационная поляризация

67
ACTIVE CENTRES (active sites).
Active sites on surface of solid catalyst
where catalytic reaction may proceed
due to concentrated adsorption.
D aktive Zentren, aktive Stellen
F centres actifs, centres d'activité
P centra aktywne
R активные центры

68
ACTIVE DEPOSIT (in radiochemistry).
Any deposit containing radioactive
substances.
D aktiver Niederschlag
F dépôt actif
P osad aktywny
R активный осадок

ACTIVE MOLECULE. See ACTIVATED
MOLECULE.

ACTIVE SITES. See ACTIVE CENTRES.

69
ACTIVE SURFACE OF ADSORBENT.
Surface area at which various unbalanced
forces act, resulting in adsorption of
adsorbate by adsorbent.
D ...
F ...
P powierzchnia aktywna (adsorbenta)
R активная поверхность (адсорбента)

70
ACTIVITY (nuclear activity).
Number of disintegrations of radioactive
nuclei of a given quantity of a substance
per unit time.
D Aktivität
F activité (nucléaire)
P aktywność (promieniotwórcza)
R активность

71
ACTIVITY (thermodynamic activity of
component), a_B.

Effective concentration of a given
component in two- or multi-component
system, defined by

$$\mu_B = \mu_B^{\ominus} + RT \ln a_B$$

where μ_B = B component's chemical
potential in mixture, $\mu_B^{\ominus}$ = B component's
chemical potential in standard state,
R = molar gas constant and
T = thermodynamic temperature;
concentration is usually given in mole
fractions, molalities or as molar
concentration.
D Aktivität
F activité
P aktywność (stężeniowa), aktywność
termodynamiczna
R (термодинамическая) активность

72
ACTIVITY COEFFICIENT, f_B.
Measure of deviation of thermodynamic
properties of constituent in a given
solution from its properties in ideal
solution, defined by ratio

$$f_B = \frac{a_B}{x_B}$$

where a_B = activity of solution
constituent B, x_B = mole fraction of this
constituent; activity coefficient depends
on pressure, temperature and other
components of solution.
D Aktivitätskoeffizient
F coefficient d'activité
P współczynnik aktywności
R коэффициент активности

73
ACTIVITY COEFFICIENT OF SINGLE
ION, f_+ or f_-.
Quantity defined by formula

$$f_+ = \frac{a_+}{c_+} \quad \text{or} \quad f_- = \frac{a_-}{c_-}$$

where a_+ and a_- = ion activities and
c_+ and c_- = their concentrations; this
quantity is not measurable.
D Ionenaktivitätskoeffizient
F coefficient d'activité ionique
P współczynnik aktywności jonu
R коэффициент активности иона

ACTIVITY SOLUBILITY PRODUCT.
See SOLUBILITY PRODUCT.

74
ACTOR (in coupled reactions).
Substance which is common substrate of
both primary and secondary reactions in
system of coupled reactions.
D Aktor (der induzierten Reaktionen)
F ...
P aktor (reakcji sprzężonych)
R актор (сопряжённых реакций)

75
ACYLATION.
Substitution of hydrogen atom by acyl group RCO— in organic molecule, e.g.

$(RCO)_2O + C_2H_5OH \rightarrow RCOOC_2H_5 + RCOOH$

$RCOCl + C_6H_6 \xrightarrow{AlCl_3} C_6H_5COR + HCl$

D Acylierung
F acylation
P acylowanie
R ацилирование

76
ACYLOIN CONDENSATION.
Linking together of two molecules of carboxylic ester under action of sodium; also intramolecular reaction of dicarboxylic acid ester, resulting in formation of acyloin, e.g.

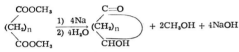

D Acyloinkondensation
F condensation des acyloïnes
P kondensacja acyloinowa
R ацилоиновая конденсация

77
ADDITION, A.
Reaction of chemical compounds involving decrease of unsaturation of molecule. The following types of addition are known (depending on reaction mechanism): electrophilic, nucleophilic and free-radical, e.g. electrophilic addition to double bond

$RCH=CH_2 + HBr \rightarrow RCHBr-CH_3$

D Addition
F addition
P przyłączenie, addycja
R присоединение

78
ADDITION COMPOUNDS (molecular compounds).
Compounds formed by linking of molecules of substances capable of independent existence.
D Molekülverbindungen, Verbindungen höherer Ordnung
F combinaisons moléculaires
P związki addycyjne, związki cząsteczkowe, związki wyższego rzędu
R координационные соединения

79
ADDITION POLYMER.
Polymer obtained from addition polymerization.
D Additionspolymere
F polymère d'addition
P polimer addycyjny
R полимеризационный полимер

80
ADDITION POLYMERIZATION.
Polymerization by repeated addition process.
D Additionspolymerisation, Polyaddition
F polymérisation par addition
P polimeryzacja addycyjna, poliaddycja
R полимеризация

81
ADDITIVE PROPERTIES.
Properties which, for a given system, are sum of corresponding properties of constituents.
D additive Eigenschaften
F propriétés additives
P własności addytywne
R аддитивные свойства

82
ADDITIVITY (additivity law).
Law according to which some definite property of a given set of elements is sum of corresponding properties of constituent elements, e.g. molecular mass is sum of all masses of constituent atoms.
D Additivität
F additivité
P addytywność, prawo addytywności
R аддитивность

ADDITIVITY LAW. See ADDITIVITY.

83
ADHESION.
Bonding occuring between surface layers of two different adhering bodies or liquids due to mutual intermolecular interactions.
D Adhäsion, Haftvermögen
F adhésion, adhérence
P adhezja, przyleganie, przyczepność
R адгезия, прилипание

84
ADIABATIC (reversible adiabatic).
Set of thermodynamic equilibrium states of the same system with identical entropy.
D Adiabate
F adiabatique (réversible)
P adiabata, izentropa
R (обратимая) адиабата

85
ADIABATIC CALORIMETER.
Type of calorimeter in which heat exchange between calorimeter vessel and its outer jacket is practically negligible.
D adiabatisches Kalorimeter
F calorimètre adiabatique
P kalorymetr adiabatyczny
R адиабатический калориметр

86
ADIABATIC PARTITION.
Partition which does not allow mass
or energy transfer in form of heat.
D adiabatische Wand, thermisch
   isolierende Wand
F paroi adiabatique
P odgraniczenie adiabatyczne, ścianka
   adiabatyczna, osłona adiabatyczna
R адиабатная оболочка, адиабатная
   диафрагма

87
ADIABATIC PROCESS (adiabatic
transition).
Thermodynamic process without any
exchange of matter and heat with
surroundings.
D adiabatische Zustandsänderung
F transformation adiabatique
P przemiana adiabatyczna, proces
   adiabatyczny
R адиабатный процесс, адиабатический
   процесс

ADIABATIC TRANSITION. *See*
ADIABATIC PROCESS.

88
ADSORBATE (adsorptive).
Substance adsorbed at surface in process
of adsorption.
Remark: In some cases term adsorbate refers
a substance which has already been adsorbed,
whereas term adsorptive is restricted to
substance (present in gaseous or liquid bulk
phases) which can be adsorbed under given
conditions.
D Adsorbat, Adsorptiv
F adsorbat
P adsorbat, adsorptyw
R адсорбат, адсорбтив

ADSORBED FILM. *See* SURFACE-
ACTIVE FILM.

89
ADSORBENT.
Crystalline, amorphous or liquid substance
or gel with greatly increased outer or
inner surface area at which adsorption
takes place.
D Adsorbens, Adsorptionsmittel
F adsorbant
P adsorbent
R адсорбент

90
ADSORPTION.
Existence of higher concentration of any
particular substance at interface between
two phases.
D Adsorption
F adsorption
P adsorpcja, sorpcja powierzchniowa
R адсорбция

91
ADSORPTION CENTRE (adsorption site).
Site on surface of adsorbent to which
adsorbate molecule is bound.
D Adsorptionszentrum, Adsorptionsstelle
F centre d'adsorption
P centrum adsorpcyjne, centrum adsorpcji
R адсорбционный центр

92
ADSORPTION CHROMATOGRAPHY.
Chromatography in which solid that can
adsorb components being separated is
stationary phase.
D Adsorptions-Chromatographie
F chromatographie d'adsorption
P chromatografia adsorpcyjna
R адсорбционная хроматография

93
ADSORPTION COEFFICIENT, K.
Constant in Langmuir equation equal
to ratio of adsorption rate constant to
corresponding desorption rate constant
and which depends on kind of system and
on temperature.
D Adsorptionskoeffizient
F coefficient d'adsorption
P współczynnik adsorpcji
R коэффициент адсорбции

94
ADSORPTION CURRENT.
In polarography and related methods,
current which is observed when substrate
or product of electrode reaction is
adsorbed on electrode surface.
D Adsorptionsstrom
F courant d'adsorption
P prąd adsorpcyjny
R адсорбционный ток

95
ADSORPTION FORCES.
Forces responsible for adsorption. They
may be of different natures, depending
on kinds of adsorption interactions,
e.g. physical adsorption or chemisorption.
D Adsorptionskräfte
F forces d'adsorption
P siły adsorpcyjne
R адсорбционные силы

96
ADSORPTION HYSTERESIS.
Non-coincidence of adsorption and
desorption isotherms.
D Adsorptionshysterese
F hystérésis d'adsorption
P histereza adsorpcyjna, histereza
   adsorpcji
R сорбционный гистерезис,
   капиллярно-конденсационный
   гистерезис

97
ABSORPTION INDICATOR.
Indicator used in precipitation titration, which at equivalence point is adsorbed or desorbed, with change of colour or fluorescence due to change of charge on precipitate surface.
D Adsorptionsindikator
F indicateur d'adsorption
P wskaźnik adsorpcyjny
R адсорбционный индикатор

98
ADSORPTION ISOSTERE.
Curve which shows dependence of gas equilibrium pressure on its temperature determined for constant amount of gas adsorbed at surface of unit mass of adsorbent.
D Adsorptionsisostere
F isostère d'adsorption
P izostera adsorpcji
R изостера адсорбции

99
ADSORPTION ISOTHERM.
Curve which shows dependence of amount of gas adsorbed on its equilibrium pressure (or equilibrium concentration of adsorbate) determined at constant temperature.
D Adsorptionsisotherme
F isotherme d'adsorption
P izoterma adsorpcji
R изотерма адсорбции

100
ADSORPTION POTENTIAL.
Change in energy of particle due to its migration from bulk of liquid (or gas) to adsorbent surface.
D Adsorptionspotential
F potentiel d'adsorption
P potencjał adsorpcyjny
R адсорбционный потенциал

ADSORPTION SITE. *See* ADSORPTION CENTRE.

ADSORPTIVE. *See* ADSORBATE.

101
ADSORPTIVE CAPACITY (*of surface*).
Quantity which describes tendency of surface (most commonly referred to defined unit of its area) of a given adsorbent to adsorb various substances.
D Adsorptionskapazität
F ...
P pojemność adsorpcyjna
R адсорбционная ёмкость

102
ADSORPTIVITY (concentrating power of adsorbent).

Ability of a given adsorbent to adsorb certain substance.
D Adsorptionsfähigkeit, Adsorptionskraft
F pouvoir adsorptif, capacité d'adsorption
P adsorpcyjność
R адсорбционная способность

103
AEROSOL.
Colloidal system in which dispersion medium is gaseous and solid or liquid is dispersed phase.
D Aerosol
F aérosol
P gazozol
R аэрозоль

A FAMILY. *See* MAIN GROUP.

AFFINITIES. *See* THERMODYNAMIC FORCES.

104
AFFINITY (*of chemical reaction*), *A*.
Intensive quantity being a force of chemical reaction

$$A = - \sum_{B} \nu_B \mu_B$$

where ν_B = stoichiometric coefficient of reagent B (positive for products, negative for substrates), μ_B = chemical potential of reagent B, and $\sum_{B}$ denotes summation over all components.
D Affinität
F affinité chimique
P powinowactwo chemiczne
R химическое сродство

105
AFTEREFFECT (post-effect, post-irradiation effect).
Chemical consequences of ionizing radiation absorption, occurring much later after irradiation.
D Nach(wirkungs)effekt
F rémanence, post-effet
P efekt następczy, postefekt
R пост-эффект, эффект последействия, послерадиационный эффект

106
AGEING OF CATALYST.
In heterogeneous catalysis, gradual loss of activity by catalyst increasing during exploitation. Ageing of catalyst is most often caused by its recrystallization.
D Alterung des Katalysators, Erschöpfung des Katalysators
F vieillissement du catalyseur
P zmęczenie katalizatora, starzenie się katalizatora
R старение катализатора

107
AGEING OF PRECIPITATE.
Change of properties of precipitate with time.
D Alterung, Altern
F vieillissement
P starzenie się osadu
R созревание осадка

108
AGEING OF SOLS.
Change of physical and chemical properties of colloidal solution with time.
D Alterung der Sole
F vieillissement des colloïdes
P starzenie się roztworu koloidalnego
R старение коллоидных растворов

109
AGGREGATION.
In colloidal chemistry, the gathering of single colloidal particles into larger entities or aggregates. Process may apply either to dispersed phase in colloids or to macromolecules in molecular colloids.
D Aggregation
F agrégation
P agregacja
R агрегация

110
ALBEDO.
Ability of given surface to reflect electromagnetic radiation or beam of particles, expressed as ratio of intensity of reflected radiation to intensity of incident radiation.
D Albedo
F albédo
P albedo
R альбедо

111
ALCOHOLIC FERMENTATION.
Anaerobic process of hexose decomposition, e.g. of glucose to ethanol and carbon dioxide, caused by zymase produced by yeast

$$C_6H_{12}O_6 \rightarrow 2C_2H_5OH + 2CO_2$$

D alkoholische Gärung
F fermentation alcoolique
P fermentacja alkoholowa
R алкогольное брожение

112
ALCOHOLOMETRY.
Determination of ethanol content in water-ethanol solutions in weight or volume per cent by means of alcoholometer.
D Alkoholometrie
F alcoo(lo)métrie
P alkoholometria
R алкоголиметрия

113
ALCOHOLYSIS.
Metathesis between organic compound and alcohol, the latter serving additionally as solvent, e.g.

$$RCOCl + R'OH \rightarrow RCOOR' + HCl$$

D Alkoholyse
F alcoolyse
P alkoholiza
R алкоголиз

114
ALDEHYDE-KETONE REARRANGEMENT.
Acid-catalyzed conversion of aldehydes with no hydrogen atoms at α-position into isomeric ketones, involving rearrangement-1,2 of alkyl or aryl groups, e.g.

$$(CH_3)_3CCHO \xrightarrow{H^{\oplus}} (CH_3)_2CHCOCH_3$$

D Aldehyd→Keton-Umlagerung
F transformation d'aldéhyde en cétone, isomérisation d'aldéhyde en cétone
P przegrupowanie aldehydowo-ketonowe
R альдегидо-кетонная перегруппировка

115
ALDOL CONDENSATION (aldolization).
Linking together of aldehydes and ketones (usually base-catalyzed) containing hydrogen atom linked with α-carbon atom, resulting in formation of aldols, e.g.

D Aldolkondensation, Aldoladdition, Aldolisation
F condensation aldolique, aldolisation
P kondensacja aldolowa
R альдольная конденсация

ALDOLIZATION. See ALDOL CONDENSATION.

116
ALKALI METALS.
Elements in first main group of periodic system with structure of outer electronic sub-shells of atoms ns^1: lithium, sodium, potassium, rubidium, caesium, francium.
D Alkalimetalle
F métaux alcalins
P litowce, metale alkaliczne, pierwiastki alkaliczne
R щёлочные металлы

ALKALIMETRIC TITRATION. See ALKALIMETRY.

117
ALKALIMETRY (alkalimetric titration).
Determination of substance by base titration.
Remark: Some authors use in this case incorrectly the therm acidimetry.

D Alkalimetrie
F alcalimétrie
P alkalimetria, miareczkowanie alkalimetryczne
R алкалиметрия, титрование основаниями

118
ALKALINE-EARTH METALS.
Elements of second main group in periodic system, except for beryllium and magnesium (calcium, strontium, barium, radium).

D Erdalkalimetalle, Erdalkalien
F métaux alcalino-terreux
P wapniowce, metale ziem alkalicznych
R элементы подгруппы кальция, щёлочноземельные металлы, щёлочноземельные элементы

ALKALINITY. *See* BASICITY.

119
ALKYLATION.
Introduction of alkyl radical into a molecule in place of hydrogen, halogen or metal atoms, e.g.

$$CH_3I + NH_3 \rightarrow CH_3\overset{\oplus}{N}H_3\overset{\ominus}{I}$$

$$C_6H_5ONa + (CH_3)_2SO_4 \xrightarrow[\Delta]{H_2O}$$

$$C_6H_5OCH_3 + CH_3\overset{\ominus}{S}O_4\overset{\oplus}{N}a$$

D Alkylierung, Alkylation
F alcoylation
P alkilowanie
R алкилирование

120
ALLINGER'S RULE.
Alicyclic conformer of small steric bulk (equatorial-axial conformer) has higher enthalpy, higher boiling point, higher density and higher refractive index than conformer having large steric bulk (equatorial-equatorial conformer).

D Allingersche Regel
F règle d'Allinger
P reguła Allingera
R правило Аллингера

121
ALLOTROPY.
Occurrence of simple substance (chemical element) in two or more forms (at the same state of aggregation).

D Allotropie
F allotropie
P alotropia
R аллотропия

122
ALLOWED ENERGY BAND (energy band *in band theory of solids*).
Quasi-continuous range of energy states in crystal, which may be occupied by electrons according to Pauli exclusion principle.

D (erlaubtes) Energieband
F bande d'énergie (permise)
P pasmo energetyczne (dozwolone)
R (разрешённая) энергетическая зона, полоса разрешённых энергий

123
ALLOWED TRANSITION.
Transition between two energy levels whose probability, according to selection rules, is different from zero.

D erlaubter Übergang
F transition permise
P przejście dozwolone
R разрешённый переход, допустимый переход

124
ALLOY.
Substance obtained by melting of two or more constituents.

D Legierung
F alliage
P stop
R сплав

125
ALLYLIC REARRANGEMENT.
Rearrangement of compound containing allyl group, involving migration of an atom (or group of atoms) from position 3 to 1, with simultaneous migration of double bond, e.g.

$$\begin{array}{ccc} \underset{R'}{\overset{R}{C}}\underset{X}{\overset{3}{-}}\underset{}{\overset{2}{C}}=\underset{R'''}{\overset{1}{C}}\overset{R''}{} & \longrightarrow & \underset{R'}{\overset{R}{C}}=\underset{}{\overset{2}{C}}\underset{X}{\overset{1}{-}}\underset{R'''}{C}\overset{R''}{} \end{array}$$

D Allylumlagerung
F transposition allylique
P przegrupowanie allilowe
R аллильная перегруппировка

126
ALPHA DECAY (alpha disintegration).
Radioactive decay in which an alpha particle is emitted by nucleus.

D α-Zerfall
F désintégration α
P przemiana α, rozpad α
R α-распад

ALPHA DISINTEGRATION. *See* ALPHA DECAY.

127
ALPHA-ELIMINATION MECHANISM.
This mechanism involves proton removal caused by base; deprotonation takes place at the same carbon atom from which leaving group X subsequently moves away. The latter stage is rate determining. α-elimination product is labile and undergoes further transformation, e.g.

$$CHCl_3 + OH^{\ominus} \rightarrow \overset{\ominus}{C}Cl_3 + H_2O$$

$$\overset{\ominus}{C}Cl_3 \rightarrow Cl_2C: + Cl^{\ominus}$$

$$Cl_2C: + H_2O \rightarrow Cl_2\overset{\ominus}{C}-\overset{\oplus}{O}H_2 \rightarrow CO + 2HCl$$

D Mechanismus der α-Eliminierung
F mécanisme de l'élimination α
P mechanizm eliminacji α
R механизм α-элиминирования

128
ALPHA PARTICLE (α-particle).
Helium nucleus ${}^4_2$He emitted in alpha radioactive decay.
D α-Teilchen
F particule α
P cząstka α
R α-частица

129
ALPHA-PARTICLE MODEL.
Model of atomic nucleus, for which basic structural unit are α-particles (strongly bound internally). Alpha-particle model is particular case of cluster model; it explains certain properties of light nuclei structure.
D Alpha-Teilchen-Modell
F modèle nucléaire alpha
P model alfowy (jądra)
R альфа-частичная модель (ядра)

130
ALPHA RAYS.
Radiation composed of alpha particles.
D α-Strahlen, α-Teilchen
F rayons α, particules α
P promie(niowa)nie α
R α-лучи, α-частицы

131
ALTERNATING COPOLYMER.
Copolymer formed of two kinds of monomeric units alternatively arranged.
D alternierendes Copolymer
F ...
P kopolimer przemienny
R регулярно чередующийся сополимер

132
ALTERNATING-CURRENT POLAROGRAPHY.
Method based on recording and analysis of alternating current which flows in polarographic circuit when voltage linearly increasing with time applied to electrodes is modulated with sinusoidal alternating voltage of frequency 50—60 Hz and amplitude 5—50 mV.
D Wechselstrompolarographie
F polarographie à tension sinusoïdale surimposée
P polarografia zmiennoprądowa (sinusoidalna)
R полярография с наложением синусоидального напряжения

133
ALUMINIUM (aluminum *US*), Al.
Atomic number 13.
D Aluminium
F aluminium
P glin
R алюминий

134
ALUMINIUM GROUP.
Elements of third main group in periodic system, except boron (aluminium, gallium, indium, thallium).
D ...
F ...
P glinowce
R ...

ALUMINUM. *See* ALUMINIUM.

AMADORI N-GLYCOSIDE TRANSFORMATION. *See* AMADORI REACTION.

135
AMADORI REACTION (Amadori rearrangement, Amadori N-glycoside transformation).
Transformation of aldose *N*-glycosides obtained from primary aromatic amines, into 1-amino-1-desoxyketoses under action of acid catalysis, e.g.

$$CH_3C_6H_4NH-CH \quad \xrightarrow[O]{H^{\oplus}} \quad CH_3C_6H_4NH-CH_2$$

(with structure: H—C—OH, H—C—OH on left; HO—C, HO—C—H O on right)

D Amadori (Aldose-N-Glykosid→ Isoglykosamin)-Umlagerung
F transposition d'Amadori
P reakcja Amadori
R перегруппировка Амадори

AMADORI REARRANGEMENT. *See* AMADORI REACTION.

136
AMAGAT'S LAW.
Total volume of mixture of different ideal gases V, is equal to sum V_i of volumes

of constituents before mixing, at the same temperature and under the same pressure, equal to pressure of mixture.

$$V = \sum_{i=1}^{n} V_i$$

Amagat's law is empirical law.

D Amagatsches Gesetz
F loi d'Amagat
P prawo Amagata
R закон Амага

137
AMALGAM POLAROGRAPHY.
Variation of polarography where instead of dropping mercury electrode, dropping amalgam electrode containing investigated metal is used.

D Amalgampolarographie
F polarographie par redissolution anodique
P polarografia amalgamatów, polarografia amalgamatowa
R амальгамная полярография

138
$A_{Ac}1$ MECHANISM.
Mechanism of acid-catalyzed monomolecular esterification or hydrolysis of esters in which cleavage of acyl-oxygen bond occurs. The following reaction steps take place: protonation of acid or ester; cleavage of acyl-oxygen bond with formation of acylium ion; reaction of the latter with alcohol or water, e.g. $A_{Ac}1$ mechanism in hydrolysis

$$C_6H_5C{-}OR + H^{\oplus} \xrightleftharpoons{\text{fast}}$$
$$\quad \overset{\parallel}{O}$$

$$C_6H_5\overset{\oplus}{C}{=}O + ROH$$

$$C_6H_5\overset{\oplus}{C}{=}O + H_2O \xrightleftharpoons{\text{fast}} C_6H_5COOH + H^{\oplus}$$

D $A_{Ac}1$-Mechanismus
F mécanisme $A_{Ac}1$
P mechanizm $A_{Ac}1$
R механизм $A_{Ac}1$

139
$A_{Ac}2$ MECHANISM.
Mechanism of acid-catalyzed bimolecular esterification or hydrolysis of esters in which cleavage of acyl-oxygen bond occurs. The following reaction steps take place: protonation of acid or ester; addition of alcohol (or water) molecule

to carbonium ion; cleavage of acyl-oxygen bond with elimination of water (or alcohol) molecule, e.g. $A_{Ac}2$ mechanism in hydrolysis

$$RC{-}OR' + H^{\oplus} \rightleftharpoons R\overset{\oplus}{C}{-}OR' \xrightleftharpoons[-H_2O]{+H_2O}$$

$$\overset{OH}{\underset{OH}{R\overset{\oplus}{C}}} \rightleftharpoons RCOOH + H^{\oplus}$$

D $A_{Ac}2$-Mechanismus
F mécanisme $A_{Ac}2$
P mechanizm $A_{Ac}2$
R механизм $A_{Ac}2$

140
$A_{Al}1$ MECHANISM.
Mechanism of acid-catalyzed monomolecular esterification or hydrolysis of esters in which cleavage of alkyl-oxygen bond occurs. The following reaction steps are postulated for case of hydrolysis: protonation of ester; cleavage of alkyl-oxygen bond with formation of carbonium ion; reaction of the latter with water, e.g.

$$\xrightleftharpoons{\text{slow}} CH_3COOH + CH_3{-}\overset{CH_3}{\underset{CH_3}{\overset{\oplus}{C}}}$$

$$CH_3{-}\overset{CH_3}{\underset{CH_3}{\overset{\oplus}{C}}} + H_2O \rightleftharpoons CH_3{-}\overset{CH_3}{\underset{CH_3}{C}}{-}OH + H^{\oplus}$$

D $A_{Al}1$-Mechanismus
F mécanisme $A_{Al}1$
P mechanizm $A_{Al}1$
R механизм $A_{Al}1$

141
AMERICIUM, Am.
Atomic number 95.

D Americium, Amerizium
F américium
P ameryk
R америций

142

AMIDE-IMIDOL TAUTOMERISM.

Type of prototropic tautomerism connected with reversible rearrangement of amids into imine

$$-\overset{\underset{\displaystyle O}{\|}}{C}-NH- \; \rightleftarrows \; -\overset{\underset{\displaystyle OH}{|}}{C}=N-$$

D Säureamid-Imid-Tautomerie, Laktam-Laktim-Tautomerie
F tautomérie amide-isoamide, tautomérie lactame-lactime
P tautomeria amido-iminowa
R лактим-лактамная таутомерия

143

α-AMIDOALKYLATION.

Introduction of amidoalkyl group RCONHCHR′ into molecule in place of hydrogen atom, with formation of new C—C bond; this results from reaction of N-substituted amide with nucleophile, e.g.

$$C_6H_5CONHCH_2Cl + :\overset{\underset{\displaystyle CH_3}{|}}{C}(COOC_2H_5)_2 \xrightarrow{PCl_3}$$

$$C_6H_5CONHCH_2\overset{\underset{\displaystyle CH_3}{|}}{C}(COOC_2H_5)_2 + Cl^{\ominus}$$

D α-Amidoalkylierung
F α-amidoalcoylation
P α-amidoalkilowanie
R α-амидоалкилирование

144

AMINATION.

Substitution of hydrogen atom in molecule by amino group —NH₂, —NHR or NRR′, e.g.

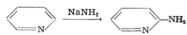

D Aminierung
F amination
P aminowanie
R аминирование

145

AMINOLYSIS.

Metathesis between organic compound and amine, e.g.

$$RCOOR' + NH_2R'' \to RCONHR'' + R'OH$$

D Aminolyse
F aminolyse
P aminoliza
R аминолиз

14

AMINOMETHYLATION (Mannich reaction).

Introduction of aminomethyl group —CH₂NRR′ into molecule of compound containing acidic C—H bond (ketone) in place of active hydrogen, due to

condensation of that compound with formaldehyde and ammonia (or primary and secondary amine), e.g.

$$C_6H_5COCH_3 + HCHO + (C_2H_5)_2NH\cdot HCl \xrightarrow{-H_2O}$$
$$C_6H_5COCH_2CH_2N(C_2H_5)_2\cdot HCl$$

D Aminomethylierung, Mannich-Reaktion
F aminométhylation, réaction de Mannich
P aminometylowanie, reakcja Mannicha
R аминометилирование, реакция Манниха

147

AMMINE COMPLEX (metal ammine).

Complex in which molecules of ammonia are ligands.

D Amminokomplex
F complexe amminé
P am(min)okompleks
R амминокомплекс

148

AMMONOLYSIS.

Metathesis between chemical compound and ammonia, e.g.

$$RCOCl + NH_3 \to RCONH_2 + HCl$$

D Ammonolyse
F ammonolyse
P amonoliza
R аммонолиз

149

AMORPHOUS BODY (amorphous solid).

Substance in condensed phase, the atoms or molecules of which are disoriented.

D amorpher Stoff, amorpher Körper
F corps amorphe
P ciało bezpostaciowe, ciało amorficzne
R аморфное тело

AMORPHOUS SOLID. See **AMORPHOUS BODY.**

150

AMORPHOUS STATE.

State characterized by random, spatial distribution of macromolecules; three kinds of amorphous state of polymers are distinguished: vitreous, high-elastic and plastic state.

D amorpher Zustand
F état amorphe des polymères
P stan bezpostaciowy, stan amorficzny
R аморфное строение полимеров

151

AMOUNT OF SUBSTANCE, *n.*

Quantity proportional to number of corresponding particles, *viz.*, atoms, molecules, ions, free radicals, elementary particles or groups of these particles. Proportionality constant is reciprocal Avogadro constant.

D Stoffmenge, Teilchenmenge
F quantité de matière

P ilość substancji, ilość materii, liczność materii
R количество вещества

152
AMPEROMETRIC TITRATION (amperometry).
Method of analysis based on measurement of limiting current flowing in polarographic circuit during titration of analysed substance with reagent at chosen constant potential.
D amperometrische Titration, Amperometrie
F titrage ampérométrique, ampérométrie
P miareczkowanie amperometryczne, amperometria, analiza amperometryczna, miareczkowanie polarometryczne
R амперометрическое титрование, амперометрия

AMPEROMETRY. *See* AMPEROMETRIC TITRATION.

AMPEROMETRY WITH TWO POLARIZED ELECTRODES. *See* BI-AMPEROMETRIC TITRATION.

153
AMPHIPROTIC SOLVENT (amphoteric solvent).
Solvent whose molecules may be both donors and acceptors of protons, i.e. solvent which undergoes autoprotolysis reaction, e.g. water, ammonia, acetic acid.
D ampholytisches Lösungsmittel
F ampholyte, solvant à la fois acide et basique
P rozpuszczalnik amfiprotonowy, rozpuszczalnik amfoteryczny
R амфипротный растворитель, амфотерный растворитель

AMPHOLYTE. *See* AMPHOTERIC ELECTROLYTE.

AMPHOTERIC CHARACTER. *See* AMPHOTERISM.

154
AMPHOTERIC ELECTROLYTE (ampholyte).
Electrolyte in which intramolecular exchange of proton may occur, e.g.

$$H_2N \cdot R \cdot COOH \rightleftarrows {}^+H_3N \cdot R \cdot COO^-$$

i.e. in acidic solution $^+H_3N \cdot R \cdot COOH$ positive ions and in basic solution $NH_2N \cdot R \cdot COO^-$ negative ions are formed.
D amphoterer Elektrolyt, Ampholyt
F ampholyte
P elektrolit amfoteryczny, amfolit
R амфотерный электролит

155
AMPHOTERIC ION (zwitterion).
Ion with both positive and negative charge; it may be formed as a result of intramolecular exchange of protons in amphoteric electrolytes, e.g.

$$H_2NRCOOH \rightleftarrows {}^+H_3NRCOO^-$$

D Zwitterion
F ion mixte
P jon obojnaczy, jon dwubiegunowy
R амфион

156
AMPHOTERIC ION EXCHANGER.
Ion exchanger simultaneously containing both cation and anion exchange groups.
D amphoterer Ionenaustauscher
F résine amphotère
P jonit amfoteryczny
R амфотерный ионит

AMPHOTERIC SOLVENT. *See* AMPHIPROTIC SOLVENT.

157
AMPHOTERISM (amphoteric character).
Ability of some complex and simple substances to exhibit both acidic and basic functions.
D Amphoterie
F amphotérie
P amfoteryczność
R амфотерность

158
AMPLIFICATION PROCEDURE.
Analytical methods with amplification of measured physical quantities with respect to concentration of investigated component by using convenient and suitable stoichiometric reactions. Amplification procedure is frequently used in colorimetric analysis.
D Verstärkungsverfahren
F méthodes d'amplification
P metody amplifikowane
R ...

ANALYSIS BY ELECTRODEPOSITION. *See* ELECTROGRAVIMETRY.

159
ANALYSIS BY FLAME SPECTROPHOTOMETRY (flame spectrophotometry).
Flame photometric analysis with radiation measured after its splitting in monochromator.
D flammenspektrophotometrische Analyse, Flammenspektrophotometrie
F (analyse par) spectrophotométrie de flamme
P analiza spektrofotometryczna płomieniowa, spektrofotometria płomieniowa
R анализ по спектрофотометрии пламени, спектрофотометрия пламени

ANALYSIS ERROR. *See* ANALYTICAL ERROR.

160
ANALYSIS OF VARIANCE.
Method of statistical investigation
(verification of hypotheses) by analysis
of possible components of variance of a
given feature. Analysis of variance allows
determination of source and magnitude
of errors of measurement and evaluation
of significance of differences between
investigated materials, procedures or
methods.
D Varianzanalyse
F analyse de variance
P analiza wariancji
R дисперсионный анализ

ANALYSIS VIA FUNCTIONAL GROUPS.
See FUNCTIONAL GROUP ANALYSIS.

161
ANALYTICAL ABSORPTION
SPECTROSCOPY.
Analytical methods based on the
measurement of radiation absorption by
given medium.
D Absorptionsspektralanalyse
F analyse spectrophotométrique
P analiza spektrofotometryczna,
  spektrofotometria absorpcyjna
R спектрофотометрический анализ

162
ANALYTICAL CHEMISTRY.
Branch of chemistry dealing with
qualitative and quantitative analysis
of natural and artificial (man-made)
products.
D analytische Chemie
F chimie analytique
P chemia analityczna
R аналитическая химия

163
ANALYTICAL CURVE.
Plot of dependence of measured quantity
on concentration of analysed component.
D Eichkurve
F courbe d'étalonnage
P krzywa wzorcowa, krzywa analityczna
R калибровочный график,
  калибровочная кривая

164
ANALYTICAL ERROR (analysis error).
Difference between real and determined
content of a given component in sample.
D Analysenfehler
F erreur d'analyse
P błąd analizy
R ошибка аналитического определения

165
ANALYTICAL FACTOR (stoichiometric
factor, chemical factor).
Ratio of relative atomic mass of

determined atoms or atom groups to
relative molecular mass weighed
compound.
D analytischer Faktor, stöchiometrischer
  Faktor
F coefficient d'analyse, facteur d'analyse
P mnożnik analityczny
R аналитический фактор, аналитический
  множитель

166
ANALYTICAL GAP (electrode gap).
In emission spectral analysis region
between two electrodes in which sample
is excited either by arc or by electric
spark.
D Elektrodenabstand (*Funkenstrecke oder
  Bogenstrecke*)
F entrode, intervalle analytique
P przerwa analityczna
R аналитический промежуток

ANALYTICAL GRADE REAGENT. *See*
ANALYTICAL REAGENT.

167
ANALYTICAL GROUP.
Group of cations or anions which exhibit
identical reactions, e.g. precipitation with
group reagent under specified conditions.
D Gruppe (von Elementen)
F groupe (analytique)
P grupa analityczna
R (аналитическая) группа

168
ANALYTICAL LINE (*in emission spectral
analysis*).
Spectral line used for analytical purposes.
D Analysenlinie
F raie d'analyse, raie analytique
P linia analityczna
R аналитическая линия

ANALYTICALLY PURE REAGENT. *See*
ANALYTICAL REAGENT.

ANALYTICAL RADIOCHEMISTRY. *See*
RADIOCHEMICAL ANALYSIS.

169
ANALYTICAL REACTION.
Chemical reaction which may be used
for analytical purposes.
D ...
F réaction analytique
P reakcja analityczna
R аналитическая реакция

170
ANALYTICAL REAGENT (analytical
grade reagent, analytically pure reagent).
Reagent with producer's certificate
guaranteeing content of specified
impurities below given limits.

D Reagens zur Analyse, pro-analytisches Reagens
F ...
P odczynnik czysty do analizy
R реактив чистый для анализа

171
ANAPHORESIS.
Process of electrophoresis in which particles of dispersed phase migrate to anode.
D Anaphorese
F anaphorèse
P anaforeza
R анафорез

172
ANCHIMERIC ASSISTANCE.
Rise of reaction rate caused by intervention of adjacent groups, e.g.

D anchimere Beschleunigung
F accélération anchimérique
P przyspieszenie anchimeryczne
R анхимерное ускорение

173
ANCHIMERIC ASSISTANCE EFFECT.
In S_N2 reactions, effect exerted by nucleophile on reaction centre located in the same molecule not further than 5 to 6 atoms away. Such an effect may increase reaction rate.
D Effekt der anchimeren Beschleunigung
F ...
P efekt przyspieszenia anchimerycznego
R ...

174
ANGULAR MOMENTUM (orbital angular momentum).
Vector product of particle momentum and its position vector relative to assumed origin of angular momentum.
D Drehimpuls
F moment cinétique orbital, moment angulaire orbital
P moment pędu orbitalny, kręt
R момент импульса, орбитальный момент, момент количества движения

175
ANGULAR MOMENTUM CONSERVATION LAW.
In isolated mechanical system or in system in external spherically symmetric potential field, if no friction forces are present, angular momentum of centre of mass is constant.
D Drehimpulssatz
F loi de conservation du moment cinétique, loi de conservation de l'impulsion, loi de conservation de la quantité de mouvement
P zasada zachowania momentu pędu
R закон сохранения момента количества движения, закон сохранения углового момента

176
ANHARMONIC OSCILLATOR.
Oscillator in which acting force directed towards equilibrium position is not proportional to displacement from equilibrium position.
D anharmonischer Oszillator
F oscillateur anharmonique
P oscylator anharmoniczny
R ангармонический осциллятор

177
ANION.
Ion with negative charge.
D Anion
F anion
P anion
R анион

178
ANION EXCHANGER.
Ion exchanger containing anions which can be exchanged for other anions from solution.
D Anionenaustauscher
F échangeur d'anions
P anionit, wymieniacz anionów
R анионит, анионообменник

179
ANIONOTROPY.
Reversible migration of anion from one position to another within the same molecule, e.g.

$$CH_3—CH=CH—CH_2—Cl \rightleftarrows CH_3—CH(Cl)—CH=CH_2$$

D Anionotropie
F anionotropie
P anionotropia
R анионотропия

180
ANISOTROPY.
Dependence of a given physical property on direction (most frequently crystallographic direction); for instance refractive index, elasticity coefficient, thermoconductivity, and dielectric constant may be anisotropic.
D Anisotropie
F anisotropie
P anizotropia
R анизотропия

181
ANNEALING.
Removal, mainly by thermal treatement, of unstable energetic or structural inhomogeneities in solids, especially of radiation-induced damage.
D Ausheilung, Anlassen
F recuit
P odprężanie, anniling
R отжиг

182
ANNIHILATION.
Process in which particle-antiparticle pair meets and converts spontaneously into one or more photons and/or mesons. Both particles disappear.
D Annihilation, Zerstrahlung, Paarzerstrahlung, Paarvernichtung
F annihilation
P anihilacja
R аннигиляция

183
ANNIHILATION PHOTON.
Photon with energy 0.511 MeV formed in collision of a positron and an electron.
D Zerstrahlung-Gammaquant
F photon d'annihilation
P foton anihilacyjny, kwant anihilacyjny
R аннигиляционный фотон

184
ANNIHILATION RADIATION.
Radiation resulting from annihilation of a particle and its antiparticle, e.g. electron and positron.
D Annihilationsstrahlung, Vernichtungsstrahlung
F rayonnement d'annihilation
P promieniowanie anihilacyjne
R аннигиляционное излучение

185
ANODE.
Electrode at which electrolytic oxidation occurs: (a) negative electrode in galvanic cell, (b) electrode connected to positive pole of external current source in electrolytic cell.
D Anode
F anode
P anoda
R анод

186
ANODIC CURRENT DENSITY (anodic partial current density).
Current density which flows across the metallic conductor — electrolyte phase boundary, equal to rate of oxidation at anode.
D anodische Teilstromdichte
F densité du courant anodique
P gęstość prądu anodowego (cząstkowego)
R плотность (частного) анодного тока

ANODIC OXIDATION. See ELECTROLYTIC OXIDATION.

ANODIC PARTIAL CURRENT DENSITY. See ANODIC CURRENT DENSITY.

187
ANOLYTE.
Part of electrolyte in vicinity of anode in which changes of physico-chemical properties occur during electrolysis or discharge of galvanic cell; if solutions in vicinity of both electrodes of cell are separated, e.g. by membranes, it is that part of electrolyte surrounding anode.
D Anolyt, Anodenlösung, Anodenflüssigkeit
F solution anodique, anolyte
P anolit
R анолит

188
ANOMERS.
Diastereoisomers of cyclic forms of sugars and their derivatives (e.g. glycosides), differing in configuration at chiral carbon atom, C_1 (in aldoses) or C_2 (in ketoses). Anomers are marked by symbols α or β, e.g.

α-D-(+)-glucopyranose

β-D-(+)-glucopyranose

D Anomere
F anomères
P anomery
R аномеры

189
ANSA COMPOUNDS.
Macro-ring compounds composed of aromatic system and polymethylene bridge, linking directly (or by heteroatoms) meta- or para-positions of ring; most often they are hydroquinone or resorcinol derivatives, e.g.

D Ansa-Verbindungen
F composés ansa
P ansa-związki
R анса-соединения

190
ANTAGONISM OF IONS.
Effect of increasing coagulation value
of one electrolyte due to addition of
another electrolyte it occurs, e.g. during
coagulation of negative sols with mixture
of electrolytes: $NaCl + CaCl_2$, $KCl + FeCl_3$.
D antagonistischer Effekt
F antagonisme de deux électrolytes
P antagonizm jonowy
R антагонизм электролитов

191
ANTI-AUXOCHROME (anti-auxochromic
group).
Electron-accepting group which shifts
absorption spectrum towards longer
wavelengths in dye molecule with
system of conjugated π-bonds, e.g.

$-SO_2NH_2 < -C\equiv N <$

(where groups are arranged according
to their increasing power of spectrum
shift).
D Antiauxochrom, antiauxochrome Gruppe
F anti-auxochrome, acichromophore
P antyauksochrom, grupa
   antyauksochromowa
R антиауксохромная группа

ANTI-AUXOCHROMIC GROUP. See
ANTI-AUXOCHROME.

192
ANTIBONDING ELECTRONS.
Electrons occupying antibonding molecular
orbitals.
D antibindende Elektronen, lockernde
   Elektronen
F électrons antiliants
P elektrony antywiążące
R антисвязывающие электроны,
   разрыхляющие электроны

193
ANTIBONDING MOLECULAR ORBITAL.
Molecular orbital expressed as linear
combination of atomic orbitals with
energy higher than that of any valence
atomic orbital contributing to it.
Electrons occupying antibonding orbitals
lead to decrease of chemical bond
strength.
D antibindendes Molekülorbital,
   antibindender Zustand
F orbitale antiliante
P orbital antywiążący
R разрыхляющая молекулярная
   орбиталь

ANTI-CLINAL CONFORMATION. See
PARTIALLY ECLIPSED
CONFORMATION.

ANTI CONFORMATION. See FULLY
STAGGERED CONFORMATION.

ANTIFERROMAGNETIC CURIE POINT.
See NÉEL TEMPERATURE.

ANTIFERROMAGNETIC SUBSTANCES.
See ANTIFERROMAGNETS.

194
ANTIFERROMAGNETISM.
Physical phenomena in and properties
of antiferromagnets.
D Antiferromagnetismus
F antiferromagnétisme
P antyferromagnetyzm
R антиферромагнетизм

195
ANTIFERROMAGNETS
(antiferromagnetic substances).
Crystals built of two ferromagnetic
sublattices compensating each other.
Magnetic ordering in antiferromagnets
arises below some characteristic
temperature, called Néel temperature.
D Antiferromagnetika,
   antiferromagnetische Stoffe
F antiferromagnétiques
P antyferromagnetyki
R антиферромагнетики

196
ANTIMATTER.
Matter composed of antiparticles, e.g.
antinuclei (composed of antiprotons and
antineutrons), antiatoms (composed of
antinuclei and positrons), antimolecules
(composed of antiatoms).
D Antimaterie
F antimatière
P antymateria
R антивещество

ANTIMERS. See ENANTIOMERS.

197
ANTIMONY, Sb.
Atomic number 51.
D Antimon
F antimoine
P antymon
R сурьма

198
ANTINEUTRINO, v̄.
General term for antiparticles
corresponding to electronic and muonic
neutrinos.
D Antineutrino
F antineutrino
P antyneutrino
R антинейтрино

199
ANTINEUTRON, $\bar{n}$.
Antiparticle corresponding to neutron,
of baryon number $B = -1$ and third
component of isospin $I_3 = 1/2$, electrically
neutral, unstable.
D Antineutron
F antineutron
P antyneutron
R антинейтрон

200
ANTIPARTICLE.
Elementary particle defined by
corresponding particle but having opposite
sign: electrical charge $Q \to -Q$, baryon
number $B \to -B$, lepton number $L \to -L$,
strangeness $S \to -S$, third component
of isospin $I_3 \to -I_3$, and magnetic moment
$\mu \to -\mu$. Mass m, spin quantum number J,
isospin I, mean lifetime and in case of
bosons parity P of antiparticle are
identical as for corresponding particle.
Annihilation may occur in the event
of interaction of antiparticle with particle.
D Antiteilchen, Antipartikel
F antiparticule
P antycząstka
R античастица

ANTI-PERIPLANAR CONFORMATION.
See FULLY STAGGERED
CONFORMATION.

201
ANTIPROTON, $\bar{p}$.
Antiparticle corresponding to proton, of
negative elementary electric charge,
baryon number $B = -1$ and third
component of isospin $I_3 = -1/2$, stable.
D Antiproton
F antiproton
P antyproton
R антипротон

202
ANTI-STOKES LINE.
Raman line of frequency higher than
that of excitation line.
D Antistokessche Linie
F ...
P linia antystokesowska
R антистоксова линия

203
APPARENT ACTIVATION ENERGY.
Activation energy of reaction taking place
in multiphase system, determined from
Arrhenius equation. Apparent activation
energy is smaller than activation energy
of the same reaction in one-phase
system by value of heat of adsorption
of activated complex on catalyst.
D scheinbare Aktivierungsenergie
F énergie d'activation apparente

P energia aktywacji pozorna
R кажущаяся энергия активации

204
APPARENT HALF WIDTH (apparent
width).
Experimentally observed line width which
is larger than true width because
of instrument's imperfection (optical,
electric and mechanical errors); it is
usually given as apparent half-width
$\Delta\nu_{1/2}^{a}$.
D scheinbare Halbwertsbreite $\Delta\nu_{1/2}^{s}$
F largeur apparente (à mi-absorption)
P szerokość linii widmowej pozorna
R кажущаяся ширина спектральной
линии

205
APPARENT MOLAL HEAT CAPACITY
(for substance in solution), Φ_s.
Intensive quantity which expresses excess
of heat capacity of solution above that
of pure solvent alone, calculated for one
mole of particular solute, viz.

$$\Phi_s = \frac{C_p - n_{solv.} C_{p, solv.}^{0}}{n_s}$$

where C_p = heat capacity of solution,
$C_{p, solv.}^{0}$ = molal heat capacity of pure
solvent, $n_{solv.}$ = number of moles of
solvent, n_s = number of moles of solute.
D scheinbare Molwärme
F chaleur spécifique molaire apparente
P ciepło molowe pozorne
R кажущаяся молярная теплоёмкость

206
APPARENT ORDER OF REACTION
(pseudo-order of reaction).
Order of reaction in cases where it is not
equal to molecularity of reaction.
D scheinbare Reaktionsordnung,
Quasi-Ordnung der Reaktion
F ordre apparent de la réaction
P rząd reakcji pozorny
R кажущийся порядок реакции,
псевдопорядок реакции

APPARENT WIDTH. See APPARENT
HALF WIDTH.

207
APPLIED CHEMISTRY.
Branch of chemistry dealing with studies
on manufacture of chemical products on
industrial scale as well as with
applications of chemical research in other
branches of science and technique.
D angewandte Chemie
F chimie appliquée
P chemia stosowana
R прикладная химия

APPROXIMATE EQUILIBRIUM
CONSTANT. *See* EQUILIBRIUM
CONSTANT.

208
APROTIC SOLVENT (indifferent solvent).
Solvent whose molecules are neither
donors nor acceptors of protons, e.g. liquid
hydrocarbons.
D aprotisches Lösungsmittel, indifferentes
   Lösungsmittel
F solvant aprotique, solvant inerte
P rozpuszczalnik aprotonowy,
   rozpuszczalnik obojętny
R апротонный растворитель

209
AQUAMETRY.
Determination of water in substances
by means of Karl Fischer reagent and
similar reagents.
D Aquametrie
F aquamétrie
P akwametria
R акваметрия

210
AQUO-COMPLEX.
Complex in which ligands are water
molecules.
D Aquokomplex
F complexe aquo
P akwokompleks
R аквокомплекс

211
ARBUSOV REARRANGEMENT.
In general sense, rearrangement of
trialkylphosphites into dialkyl
phosphonate taking place on heating the
former with alkyl halides, e.g.

$$P(OR)_3 + R'X \xrightarrow{\Delta} [R'\overset{\oplus}{P}(OR)_3]^{\oplus}X^{\ominus} \rightarrow$$

$$R'-P(O)(OR)_2 + RX$$

D Arbusow-Michaelis
   (Trialkylphosphit)-Umwandlung
F réaction d'Arbousov
P przegrupowanie Arbuzowa
R реакция Арбузова

ARC LINE. *See* ATOM LINE.

212
ARC SPECTRUM.
Emission spectrum of element obtained
by its excitation in electric arc; it contains
chiefly emission lines of neutral atoms
but, in the case of atoms with low
ionization potentials, also spectral lines
of singly ionized species.
D Bogenspektrum
F spectre d'arc

P widmo łukowe
R дуговой спектр

213
ARGON, Ar.
Atomic number 18.
D Argon
F argon
P argon
R аргон

214
ARITHMETIC MEAN (*in series*), $\bar{x}$.
Statistic given by equation

$$\bar{x} = \frac{1}{n} \sum_{i=1}^{n} x_i$$

where x_i = value of element of series
(e.g. result of measurement or
determination), n = number of elements
in series.
D arithmetisches Mittel
F moyenne arithmétique
P średnia arytmetyczna
R среднее арифметическое

215
ARITHMETIC MEAN VELOCITY
(average velocity *of gas molecules*), $\bar{c}$.
Velocity which is arithmetic mean of
velocities of all molecules in particular
population

$$\bar{c} = \sqrt{\frac{8RT}{\pi M}}$$

where R = universal gas constant,
T = thermodynamic temperature,
M = molar mass.
D mittlere Geschwindigkeit
F vitesse moyenne
P prędkość średnia
R средняя (арифметическая) скорость

216
ARNDT-EISTERT REACTION.
Preparation of higher carboxylic acid
homologue by treating acid chloride with
diazomethane. Resulting diazoketone
undergoes rearrangement with elimination
of nitrogen, in presence of catalyst (e.g.
colloidal silver), with formation of higher
acid homologue

$$RCOCl + 2CH_2N_2 \longrightarrow RCOCHN_2 + CH_3Cl + N_2$$

$$RCOCHN_2 + R'OH \xrightarrow{Ag} RCH_2COOR' + N_2$$

D Arndt-Eistert-Wolff-Reaktion,
   Arndt-Eistert Carbonsäure-Aufbau
F méthode d'Arndt-Eistert
P synteza Arndta i Eisterta
R синтез Арндта-Эйстерта

217
AROMATIC BOND.
System of σ-bonds and delocalized
π-bonds present in molecules of aromatic
compounds, e.g. in benzene molecule.
D aromatischer Bindungszustand
F liaison aromatique
P wiązanie aromatyczne
R ароматическая связь

218
AROMATIC CHARACTER (aromaticity).
Characteristic features of aromatic
molecules: ring-like, sterically plane
molecule containing $(4n+2)\pi$-electrons
in conjugated system; relatively high
resonance energy ensuring stability of
system; tendency to undergo electrophilic
substitution (rather than addition).
D aromatischer Charakter, Aromatizität
F caractère aromatique, aromaticité
P charakter aromatyczny
R ароматический характер,
  ароматичность

AROMATICITY. *See* AROMATIC
CHARACTER.

AROMATIC NUCLEUS. *See* AROMATIC
RING.

219
AROMATIC RING (aromatic nucleus).
Polyenic plane ring containing (according
to Hückel rule) $(4n+2)\pi$-electrons in
conjugated system ($n = 0, 1, 2, 3...$).
Generally, it is a six- or five-membered
ring with a sextet of π-electrons, e.g.
benzene, thiophene or pyrrol ring

D aromatischer Ring
F noyau aromatique
P pierścień aromatyczny, układ
  aromatyczny jednopierścieniowy
R ароматическое ядро

220
AROMATIC SEXTET.
Set of six π-electrons in aromatic
molecules or ions.
D Elektronensextett
F sextet d'électrons, sextet électronique,
  sextet aromatique
P sekstet elektronowy, sekstet aromatyczny
R секстет электронов

221
AROMATIZATION.
Formation of aromatic rings from organic
compounds as a result of dehydration or
elimination, e.g.

$$CH_3(CH_2)_5CH_3 \longrightarrow \text{[CH}_3\text{-ring]} + 4H_2$$

$$3\,\text{[ring]} \xrightarrow{Pt} 2\,\text{[ring]} + \text{[ring]}$$

D Aromatisierung
F aromatisation
P aromatyzacja
R ароматизация

222
ARRHENIUS COMPLEX.
In homogeneous catalysis,
substrate-catalyst complex has rate
constant of decay into starting compounds
substantially higher than rate of decay
into reaction products.
D Arrheniusscher Zwischenkörper,
  Arrheniusscher Komplex
F complexe d'Arrhenius, intermédiaire
  d'Arrhenius
P substancja przejściowa typu Arrheniusa,
  kompleks Arrheniusa
R промежуточное соединение Аррениуса,
  промежуточное вещество Аррениуса

223
ARRHENIUS EQUATION.
Equation describing dependence of
reaction-rate constant on temperature

$$\ln k = \ln A - \frac{E}{RT}$$

where k = rate constant, A = Arrhenius
coefficient (specific for given reaction),
T = thermodynamic temperature,
E = activation energy, R = universal gas
constant.
D Arrhenius(sche) Gleichung
F équation d'Arrhenius, formule
  d'Arrhenius
P równanie Arrheniusa
R уравнение Аррениуса

224
ARRHENIUS-OSTWALD'S THEORY OF
ELECTROLYTIC DISSOCIATION.
Theory basing on assumption that part of
electrolyte dissociates into free ions;
moreover solution is diluted and hence
one may apply classical law of mass action
to electrolytic dissociation and assume
that ionic mobilities are independent of
concentration.
D Arrhenius-Ostwald-Theorie der
  elektrolytischen Dissoziation
F théorie de la dissociation électrolytique
  d'Arrhenius et Ostwald
P teoria dysocjacji elektrolitycznej
  Arrheniusa i Ostwalda
R теория электролитической диссоциации
  Аррениуса и Оствальда

225
ARRHENIUS THEORY OF ACIDS AND BASES.
Theory according to which acidic properties of substance are connected with presence in substance of hydrogen which may dissociate as H^+ ion; basic properties are connected with presence of hydroxyl groups which may dissociate to form OH^- ions.

D Arrheniussche Säuren-Basen-Theorie
F théorie ionique des acides et des bases, théorie des acides et des bases d'Arrhenius
P teoria kwasów i zasad jonowa, teoria kwasów i zasad Arrheniusa
R теория кислот и оснований Аррениуса

226
ARSENIC, As.
Atomic number 33.

D Arsen
F arsenic
P arsen
R мышьяк

227
ARSONATION.
Substitution of hydrogen atoms of aromatic nucleus, e.g. in phenols or aromatic amines by arsonic group $-AsO_3H_2$.

D Arsonierung
F arsonation
P arsonowanie
R арсенирование

ARTIFICIAL RADIOACTIVITY. *See* INDUCED RADIOACTIVITY.

228
ARYLATION.
Substitution of hydrogen atom by aryl group in molecule, e.g.

Br—⟨⟩—$\overset{\oplus}{N}$≡N$\overset{\ominus}{Cl}$ + C_6H_6 + NaOH →

Br—⟨⟩-⟨⟩ + N_2 + NaCl + H_2O

D Arylierung
F arylation
P arylowanie
R арилирование

229
ASSOCIATED MOLECULES.
Unstable set of two or more identical molecules, kept together by hydrogen bonds or by dipole interaction.

D Assoziat
F associate
P asocjat
R ассоциат

230
ASSOCIATED PRODUCTION.
Simultaneous production of two strange particles in strong interactions (in accordance with strangeness conservation law valid in these interactions).

D gemeinsame Erzeugung
F production associée
P produkcja łączna
R парное рождение

231
ASSOCIATION.
Formation of associate of two or more identical molecules A: $nA \rightleftarrows (A)_n$. Equilibrium between molecules A and associates $(A)_n$ is influenced by temperature and often by concentration.

D Assoziation
F association
P asocjacja
R ассоциация

232
ASSOCIATION COLLOID (semicolloid).
Colloid composed of micelles formed by association of particles of dispersed substance under influence of van der Waals forces.

D Assoziationskolloid, Micellarkolloid, Semikolloid
F colloïde micellaire
P koloid asocjacyjny, koloid micelarny, semikoloid
R семиколлоид, полуколлоид

233
ASSOCIATION DEGREE OF IONS.
In Bjerrum's theory of ionic association, fraction of number of ions in solution which associate into stable ion pairs or higher ionic associates.

D Assoziationsgrad der Ionen
F coefficient d'association, degré d'association
P stopień asocjacji jonów
R степень ассоциации ионов

234
ASTATINE, At.
Atomic number 85.

D Astat(in)
F astate
P astat
R астатин

ASTROCHEMISTRY. *See* COSMOCHEMISTRY.

235
ASYMMETRIC CARBON ATOM.
Carbon atom bound to four different
atoms (or groups of atoms). Molecule
containing one asymmetric carbon atom
is said to be asymmetric and exists in
two enantiomeric forms, e.g.

D-(+)-glyceric aldehyde L-(−)-glyceric aldehyde

D asymmetrisches Kohlenstoffatom
F carbone asymétrique
P atom węgla asymetryczny
R асимметрический атом углерода

236
ASYMMETRIC SYNTHESIS.
Synthesis of optically active compound,
resulting in formation of new chiral
centre in molecule with one configuration
(of the two possible) predominating, e.g.

$$C_6H_5CHO + HCN \xrightarrow{\text{quinine*}} C_6H_5\overset{*}{C}H(OH)CN$$

D asymmetrische Synthese
F synthèse asymétrique
P synteza asymetryczna
R асимметрический синтез

237
ASYMMETRIC TOP MOLECULE.
Molecule, whose three principal momenta
of inertia are different, e.g. CH_2BrCl.

D asymmetrisches Kreiselmolekül
F molécule cuspidale asymétrique
P cząsteczka typu bąka niesymetrycznego,
cząsteczka-rotator asymetryczny
R молекула типа асимметричного волчка

ASYMPTOTIC CURIE POINT. *See*
PARAMAGNETIC CURIE POINT.

238
ATACTIC POLYMER.
Regular polymer in whose macromolecules
there is essential randomness with regard
to configurations at all main-chain sites
of steric isomerism

```
     H  R  H  H  H  R  H  H  H  R  H  H  H  H  R
     |  |  |  |  |  |  |  |  |  |  |  |  |  |  |
···—C—C—C—C—C—C—C—C—C—C—C—C—C—C—C—···
     |  |  |  |  |  |  |  |  |  |  |  |  |  |  |
     R  H  H  H  H  R  H  R  H  H  H  H  R  H  H
```

D ataktisches Polymer
F polymère atactique
P polimer ataktyczny
R атактический полимер

239
ATHERMAL SOLUTION.
Solution characterized by enthalpy of
mixing H^M (*see* functions of mixing)
equal to zero, i.e. it has the same value
as in ideal solution and entropy of
mixing S^M is not equal to entropy of
mixing of ideal solution

$$S^M \neq -R \sum_i x_i \ln x_i$$

where R = universal gas constant,
x_i = molar fraction of solution constituent,
$\sum_i$ = summation of all constituents of
solution.

D athermische Mischung
F solution athermique
P roztwór atermiczny
R атермический раствор

240
ATOM.
Electrically neutral system composed
of single nucleus and electrons.

D Atom
F atome
P atom
R атом

241
ATOMIC ABSORPTION ANALYSIS
(atomic absorption spectrophotometry).
Spectroanalysis based on investigation
of absorption spectrum of gaseous
(atomized) sample.

D Atomabsorptionsanalyse,
Atomabsorptionsspektro(photo)metrie
F analyse par absorption atomique,
spectrophotométrie d'absorption
atomique
P analiza absorpcyjna atomowa,
spektrofotometria absorpcyjna atomowa
R атомный абсорбционный спектральный
анализ, атомная абсорбционная
спектрофотометрия

ATOMIC ABSORPTION
SPECTROPHOTOMETRY. *See* ATOMIC
ABSORPTION ANALYSIS.

242
ATOMIC BOMB (fission bomb).
Explosion system based on chain reaction
of uranium $^{235}_{92}U$ or plutonium $^{239}_{64}Pu$
nuclides with slow neutrons.

D Kernbombe, Atombombe
F bombe atomique
P bomba jądrowa, bomba atomowa
R ядерная бомба, атомная бомба

ATOMIC BOND. *See* COVALENT BOND.

243
ATOMIC CORE.
Atom with valence electrons removed.

D Atomrumpf
F tronc d'atome
P rdzeń atomowy, zrąb atomowy
R корпус атома, атомный остов

244
ATOMIC CROSS-SECTION.
Product of isotopic cross-section and relative abundance of a given isotope in a natural element.
D atomarer Wirkungsquerschnitt
F section efficace atomique
P przekrój czynny atomowy
R атомное (эффективное) сечение

245
ATOMIC CRYSTALS (covalent crystals).
Crystals with covalent bond between atoms predominant, they are characterized by high hardness, low ionic conductance and low electron conductivity.
D Atomkristalle
F cristaux covalents
P kryształy atomowe, kryształy (ko)walencyjne
R атомные кристаллы, (ко)валентные кристаллы

246
ATOMIC FLUORESCENCE ANALYSIS (atomic fluorescence spectrometry).
Emission spectral analysis based on investigation of fluorescence spectrum of gaseous (atomized) sample.
D Atomfluoreszenzanalyse, Atomfluoreszenzspektrometrie
F analyse par fluorescence atomique, spectrométrie de fluorescence atomique
P analiza fluorescencyjna atomowa, spektrometria fluorescencyjna atomowa
R атомный флуоресцентный анализ, атомная флуоресцентная спектрометрия

ATOMIC FLUORESCENCE SPECTROMETRY. *See* ATOMIC FLUORESCENCE ANALYSIS.

ATOMIC HEAT. *See* ATOMIC HEAT CAPACITY.

247
ATOMIC HEAT CAPACITY (atomic heat, atomic specific heat).
Product of atomic mass and heat capacity measured either at constant pressure or at constant volume.
D Atomwärme
F chaleur (spécifique) atomique
P ciepło atomowe
R атомная теплоёмкость

ATOMIC LINE. *See* ATOM LINE.

ATOMIC MAGNETIC MOMENT. *See* MAGNETIC MOMENT OF ATOM.

248
ATOMIC MASS SCALE (scale of atomic masses, scale of atomic weights).
Scale of relative atomic and nuclear masses. *See also*: chemical scale of atomic weights, physical scale of atomic weights, carbon-12 atomic mass scale.
D Atommassenskala, Atomgewichtsskala
F échelle des masses atomiques, échelle des poids atomiques
P skala mas atomowych
R шкала атомных масс, шкала атомных весов

249
ATOMIC NUCLEUS.
System of nucleons (protons and neutrons) bound by nuclear forces; central part of atom with which almost its entire mass (about 99,9%) is associated. Atomic nucleus is defined usually by two numbers: atomic number Z (number of protons in nucleus), and mass number A (total number of nucleons).
D Kern, Atomkern
F noyau atomique
P jądro (atomowe)
R (атомное) ядро

250
ATOMIC NUMBER, Z.
Number of protons in given nucleus. Atomic number is also equal to element number in periodic table.
D Atomnummer, Ordnungszahl
F nombre atomique
P liczba atomowa, liczba porządkowa
R атомное число, атомный номер

251
ATOMIC ORBITAL.
Wave function of single electron in atom.
D Atomorbital, Atombahn
F orbitale atomique
P orbital atomowy
R атомная орбиталь

252
ATOMIC PARACHOR (parachor equivalent).
Contributions of atomic parachors determined experimentally in parachor of chemical compound; e.g. parachor of chloroform is equal to sum of atomic parachors

$$P_{CHCl_3} = P_C + P_H + 3P_{Cl}$$

D Atomparachor
F parachor atomique
P parachora atomowa
R атомный парахор, парахор атома

253
ATOMIC POLARIZABILITY.
Dielectric polarization (per molecule) arising in unit local field due to shift of atomic nuclei.
D atomische Polarisierbarkeit
F polarisabilité atomique
P polaryzowalność atomowa
R атомная поляризуемость

254
ATOMIC POLARIZATION.
Dielectric polarization due to translation
or rotation of atoms or ions.
D Atomverschiebungspolarisation
F polarisation atomique
P polaryzacja atomowa
R атомная поляризация

255
ATOMIC POLARIZATION, P_A.
Measure of atomic polarization, defined
by

$$P_A = \frac{1}{3\varepsilon_0} N_A \alpha_A \quad \text{(in SI units)}$$

where N_A = Avogadro constant,
α_A = atomic polarizability,
ε_0 = permittivity of free space.
D Atomverschiebungspolarisation
F polarisation atomique
P polaryzacja atomowa (molowa)
R атомная поляризация

256
ATOMIC REFRACTION.
Part of experimental value of molar
refraction computed for atom in molecule
by using additive scheme; atomic
refraction computed in this way is
different from atomic refraction of
elementary substances.
D Atomrefraktion
F réfraction atomique
P refrakcja atomowa
R атомная рефракция

ATOMIC SPECIFIC HEAT. See ATOMIC
HEAT CAPACITY.

257
ATOMIC SPECTROANALYSIS.
Spectroanalysis based on investigation
of atomic spectra.
D Atomspektralanalyse
F analyse par spectroscopie atomique
P analiza spektralna atomowa
R атомный спектральный анализ

258
ATOMIC SPECTROSCOPY.
Branch of spectroscopy dealing with
investigation of atomic emission and
absorption spectra.
D Atomspektroskopie
F spectroscopie atomique
P spektroskopia atomowa
R атомная спектроскопия

259
ATOMIC SPECTRUM.
Emission spectrum with line structure
corresponding to transitions between
different atomic energy states.
D Atomspektrum

F spectre atomique
P widmo atomowe
R атомный спектр

260
ATOMIC TERM.
Set of quantum mechanical states of
atomic electrons with defined total
orbital and spin angular momentum.
D atomarer Term
F terme atomique
P term atomowy
R атомный терм

261
ATOMIC UNITS.
System of units with $h/2\pi$ (h = Planck
constant), electron mass, and absolute
value of the electron charge all set equal
to unit value.
D atomare Einheiten
F unités atomiques
P jednostki atomowe
R атомные единицы

262
ATOMIC VOLUME, V_a.
Volume of one gram-atom of given
element, i.e. ratio of relative atomic mass
(in grams) and element density.
D Atomvolumen
F volume atomique
P objętość atomowa
R атомный объём

ATOMIC WEIGHT. See RELATIVE
ATOMIC MASS.

263
ATOM LINE (atomic line, arc line *in
emission spectral analysis*).
Spectral line in emission spectrum
of excited atoms.
D Atomlinie, Bogenlinie
F raie d'atome, raie atomique, raie d'arc
P linia atomowa
R атомная линия, дуговая линия

264
ATROPISOMERISM.
Type of stereoisomerism caused by
restricted rotation about single bond. High
value of rotational barrier
(ca. 20—35 kcal/mole) enables isolation
of enantiomeric atropisomers, e.g.
6,6'-dinitrodiphenyl acid

D Atrop-Isomerie
F atropo-isomérie
P atropoizomeria
R атропоизомерия

265
AUGER EFFECT.
Transition of atom from excited electronic energy state to lower state with emission of electron instead of photon; it is analogous to internal conversion in nuclear transitions.
D Auger-Effekt
F effet Auger
P efekt Augera
R эффект Оже

266
AUGER ELECTRONS.
Electrons ejected in Auger effect, differing from β^--radiation in spectrum and generally having lower energy.
D Auger-Elektronen
F électrons Auger
P elektrony Augera
R электроны Оже

267
AUTOCATALYSIS.
Catalytic activity of reactants or products of reaction.
D Autokatalyse
F autocatalyse
P autokataliza
R автокатализ, аутокатализ

268
AUTOCATALYST.
Product or reactant acting catalytically.
D Autokatalysator
F autocatalyseur
P autokatalizator
R автокатализатор, аутокатализатор

269
AUTOCATALYTIC REACTION.
Reaction catalyzed by its own substrate or product, e.g. aqueous hydrolysis of amyl acetate is catalyzed by acetic acid present in system as a product
$$CH_3COOC_5H_{11} + H_2O \rightarrow CH_3COOH + C_5H_{11}OH$$
D autokatalytische Reaktion
F réaction autocatalytique
P reakcja autokatalityczna
R аутокаталитическая реакция

270
AUTOCOMPLEX.
Complex compound formed in salt solution through association of simple primary molecules, e.g. in solutions of cadmium chloride following autocomplexation reaction occurs
$$2CdI_2 = Cd[CdI_4] \quad or \quad 3CdI_2 = Cd[CdI_3]_2$$
D Autokomplex
F auto-complexe
P autokompleks
R автокомплекс

271
AUTOIGNITION.
Ignition of substance due to exothermal process taking place in it, with no apparent energy impulse from outside.
D Selbst(ent)zündung
F auto-allumage, auto-ignition
P samozapłon
R самовоспламенение, самовозгорание

272
AUTOIGNITION POINT.
Lowest temperature necessary for autoignition of substance in some precisely defined, normalized conditions.
D Selbstentzündungstemperatur
F température d'auto-ignition
P temperatura samozapłonu
R температура самовоспламенения

273
AUTOINDUCTION.
Chemical induction in cases when concentration of inductor increases as function of time.
D ...
F ...
P samoindukcja reakcji
R самоиндукция (реакции)

274
AUTOIONIZATION (preionization).
Electron detachment from excited atom or molecule with simultaneous transition of another excited electron to ground state. Autoionization involves at least doubly excited states.
D Präionisation, Autoionisation
F préionisation
P prejonizacja, autojonizacja
R преионизация

275
AUTOPROTOLYSIS (*of solvent*).
Electrolytic dissociation of solvent only, as a result of proton exchange between its molecules, e.g.
$$2H_2O \rightleftarrows H_3O^+ + OH^-$$
$$2CH_3COOH \rightleftarrows CH_3COOH_2^+ + CH_3COO^-$$
Autoprotolysis occurs between amphiprotic solvents, e.g. liquids: NH_3, H_2O, CH_3COOH, H_2SO_4
D Autoprotolyse
F autoprotolyse
P autoprotoliza, autojonizacja
R автопротолиз

276
AUTORADIOGRAPH.
Result of autoradiography.
D Autoradiogramm
F autoradiogramme
P autoradiogram
R радиоавтограмма

277
AUTORADIOGRAPHY (radioautography).
Determination of distribution and content
of radioactive substances in object
examined by means of a photographic
emulsion placed in close contact with
substances emitting radiation.
D Autoradiographie
F autoradiographie
P autoradiografia
R радиоавтография

278
AUTORADIOLYSIS.
Radiolysis caused by ionizing radiation
from an internal source.
D strahleninduzierte Selbstzersetzung,
Autoradiolyse
F autoradiolyse
P autoradioliza
R авторадиолиз

279
AUTOXIDATION.
Oxidation of substance by compound
(usually peroxide or hydroperoxide)
formed in that reaction, or present as
contaminant; autoxidation is caused by
radical reaction induced by atmospheric
oxygen, e.g.

$A + O_2 \rightarrow AO_2$

$AO_2 + B \rightarrow AO + BO$

A and B substances can be identical.
D Autoxydation
F auto-oxydation
P samoutlenianie, autooksydacja
R автоокисление

280
AUXOCHROME (auxochromic group).
Electron-donating group which shifts
absorption spectrum towards longer
wavelenghts in dye molecule with system
of conjugated π-bonds, e.g.

$-CH_3 < -OH < -OCH_3 < -NH_2 < -NHCH_3 <$
$< -N(CH_3)_2$

(where groups are arranged according to
their approximate increasing power of
spectrum shift).
D Auxochrom, auxochrome Gruppe
F auxochrome
P auksochrom, grupa auksochromowa
R ауксохромная группа

AUXOCHROMIC GROUP. See
AUXOCHROME.

AVAILABLE ENERGY. See
HELMHOLTZ FREE ENERGY.

AVAILABLE NET WORK. See
NET MAXIMUM WORK.

AVAILABLE WORK AT CONSTANT
TEMPERATURE. See HELMHOLTZ
FREE ENERGY.

AVERAGE DENSITY. See PARTICLE
NUMBER DENSITY.

AVERAGE LIFE. See MEAN LIFE.

AVERAGE LIGAND NUMBER. See
LIGAND NUMBER.

281
AVERAGE OCCUPATION NUMBER,
$\langle N_k \rangle$.
Statistically averaged number of particles
in k-th (single-particle) state or energy
level.
D mittlere Besetzungszahl
F nombre (moyen) d'occupation
P liczba obsadzenia średnia
R среднее число заполнения,
заселённость состояния

AVERAGE SAMPLE. See LABORATORY
SAMPLE.

AVERAGE VELOCITY. See ARITHMETIC
MEAN VELOCITY.

282
AVOGADRO'S CONSTANT (Avogadro's
number), N_A.
Number of molecules in one mole of
substance: $N_A = 6.02252 \cdot 10^{23} \text{ mol}^{-1}$.
D Avogadro-Konstante, Avogadro-Zahl
F constante d'Avogadro, nombre
d'Avogadro
P stała Avogadry, liczba Avogadry
R постоянная Авогадро, число Авогадро

AVOGADRO'S HYPOTHESIS. See
AVOGADRO'S LAW.

283
AVOGADRO'S LAW (Avogadro's
hypothesis).
Equal volumes of different ideal gases
at the same pressure and temperature
contain the same number of molecules.
Molar volumes of ideal gases are the same
at identical pressures and temperatures.
D Avogadrosches Gesetz
F loi d'Avogadro, hypothèse d'Avogadro
P prawo Avogadry
R закон Авогадро

AVOGADRO'S NUMBER. See
AVOGADRO'S CONSTANT.

284
AXIAL BOND.
One of the six bonds parallel to three-fold
axis of symmetry in chair form of
cyclohexane. Axial bond directions are
alternately upwards and downwards,
parallel to each other and perpendicular

with respect to "mean" plane of ring,
i.e. plane bisecting all C—C ring-forming
bonds

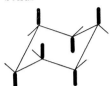

D axiale Bindung
F liaison axiale
P wiązanie aksjalne, wiązanie a
R аксиальная связь

285
AXIAL HALOKETONE RULE.
Particular case of octant rule; according
to this introduction of chlorine, bromine
or iodine atoms into axial position
adjacent to carbonyl group in cyclohexene
derivatives influences in defined way the
sign of Cotton effect in initial ketone.
Introduction of halogen atom into
equatorial position in ketone does not
change sign of Cotton effect.

D Djerassi-Klyne-Regel
F règle de Djerassi et Klyne
P reguła aksjalnych chlorowcoketonów,
      reguła Djerassiego i Klyne'a
R правило Джерасси и Клейна,
      аксиальное правило галогенкетонов

286
AZEOTROPE.
Mixture of liquids distilling at constant
temperature without change of
composition.

D Azeotrop, azeotrope Mischung
F azéotrope
P azeotrop, mieszanina azeotropowa
R азеотропная смесь

287
AZEOTROPIC AGENT.
Invariable component in series of two-
component azeotropic systems in which
second component is compound belonging
to defined homologous series.

D azeotrop(isch)er Faktor
F agent azéotropique
P czynnik azeotropowy
R азеотропный агент

288
AZEOTROPIC POINT.
Point in phase diagram which defines
composition and boiling temperature,
or composition and vapour pressure of
azeotrope.

D azeotroper Punkt
F point azéotropique
P punkt azeotropowy
R азеотропная точка

289
AZEOTROPIC RANGE.
Range of temperatures covering boiling
points of homologues which form
azeotropic system with azeotropic agent.

D azeotroper Bereich
F domaine azéotropique
P zasięg azeotropowy
R азеотропный предел

290
AZEOTROPIC SYSTEM.
System of constituents forming azeotrope.

D azeotropes System
F système azéotrope
P układ azeotropowy
R азеотропная система

291
AZEOTROPY.
Boiling (distillation) of mixture of liquids
without change of composition.

D Azeotropie
F azéotropisme
P azeotropia
R азеотропия

AZIMUTHAL QUANTUM NUMBER. *See*
ORBITAL QUANTUM NUMBER.

BACKBONE CHAIN. *See* MAIN CHAIN.

292
BACKGROUND.
Ionizing radiation coming from sources
other than the radiation to be detected or
measured, e.g. from naturally occurring
radioactive nuclides, cosmic radiation,
or radioactive contamination in laboratory.

D Nulleffekt, Nullwert
F bruit de fond
P tło promieniowania
R фон излучения

293
BACKGROUND (*in spectrophotometry*).
Absorption due to all components except
for the one to be determined.

D Untergrund
F fond spectral
P podłoże
R (спектральный) фон

294
BACKGROUND (background counting
rate).
Counting rate recorded by a counter in
absence of a radioactive sample.

D Nulleffekt, Nullwert, Untergrund
F mouvement propre, bruit de fond
P tło licznika, bieg własny licznika
R фон детектора

BACKGROUND COUNTING RATE.
See. BACKGROUND (294).

295
BACKSCATTERING.
Scattering of radiation at angle greater than 90°.
D Rückstreuung
F rétrodiffusion
P rozproszenie wsteczne, rozproszenie zwrotne
R обратное рассеяние

296
BACK-SCATTER PEAK.
Peak in gamma radiation spectrum formed by photoelectric absorption of radiation scattered at angle of 180°.
D Rückstreuspitze, Rückstreumaximum, Rückstreu-Peak
F pic de rétrodiffusion
P pik rozproszenia wstecznego, pik rozproszenia zwrotnego
R пик обратного рассеяния

297
BACK STRAIN EFFECT (B-strain effect).
Effect displayed by B-strain — one of the steric strains which are attributed to difference in molecular compactness between initial and final stages of reaction. E.g., formation of ammonium compound from amine R—N—R causes compression of R—N—R bond angle due to change from planar to pyramidal arrangement of nitrogen substituents.
D Biegespannungseffekt, B-Spannung-Effekt
F ...
P efekt napięcia B
R ...

298
BAEYER STRAIN THEORY.
Former theoretical explanation of differences in chemical and physical properties of cyclic compounds. Baeyer theory assumes that alicyclic compounds are planar and that their stability depends on angles of valency bonds. Cyclic compound should be the more stable, the more nearly the angles of valency bonds in the cycle approximate to 109°28'. Later it was shown that six-membered and more rings are not planar and the Baeyer theory partly lost its importance, reserved only for cycles containing less than 6 carbon atoms.
D Baeyersche Spannungstheorie
F théorie de Baeyer
P teoria napięć Baeyera
R теория напряжения Байера

299
BAEYER-VILLIGER OXIDATION (Baeyer-Villiger peracid rearrangement).
Oxidation of aldehydes and ketones to esters, effected with hydrogen peroxide or peracids, e.g.

D Baeyer-Villiger Keton→Ester-Oxydation, Baeyer-Villiger Umlagerung
F ...
P utlenianie Baeyera i Villigera
R окисление по Байеру-Виллигеру

BAEYER-VILLIGER PERACID REARRANGEMENT. See BAEYER-VILLIGER OXIDATION.

BALANCED REACTION. See REVERSIBLE REACTION.

300
BALANCE EQUATION FOR THE MOMENTUM DENSITY (conservation equation for the momentum).
Equation

$$\frac{\partial}{\partial t}(\rho v) = -\mathrm{div}(\rho v v + P) + \rho F$$

where t = time, v = velocity of centre of mass, ρ = density, P = pressure tensor, and F = external force per unit mass.
D Bilanzgleichung für die Impulsdichte
F équation de bilan pour l'impulsion dans les systèmes continus
P zasada zachowania pędu w układach ciągłych
R закон сохранения импульса в непрерывных системах

301
BALANDIN MULTIPLET (multiplet).
Active centre composed of several atoms or ions of catalyst, distributed in regular manner and remaining in geometrical consistence with distribution of ions or atoms in molecules of reactants which participate in catalytical reaction.
D ...
F multiplet (de Balandin)
P multiplet Bałandina
R мультиплет Баландина

302
BALANDIN THEORY OF MULTIPLETS (theory of multiplets).
Theory of heterogeneous catalysis which assumes progress of catalytic reaction on

active centres of multiplet nature.
D (Balandin-) Multiplett-Theorie
F théorie des multiplets (de Balandin)
P teoria multipletowa (Bałandina)
R мультиплетная теория (Баландина), мультиплетная теория катализа

BANANA BOND. See BENT-BOND.

BAND EDGE. See BAND HEAD.

BAND GAP. See FORBIDDEN BAND.

303
BAND HEAD (band edge).
In rotation-oscillation-electronic spectrum, direction of convergence limit of group of rotation lines.
D Bandenkopf, Bandenkante
F tête de bande
P głowica pasma
R голова полосы, кант полосы

304
BAND SPECTRUM.
Type of spectrum characteristic for molecules. Band spectrum consists of several closely spaced groups of lines corresponding to rotation-vibration-electronic transitions.
D Bandenspektrum
F spectre de bande
P widmo pasmowe
R полосатый спектр

305
BAND THEORY OF SOLIDS.
Quantum mechanical theory connecting physical and chemical properties of solid state with electronic structure of atoms in substance, and particularly successfully describing electrical properties of metals and semiconductors. According to this theory, particular energetic states of electrons belonging to isolated atoms (ions) have corresponding energy bands in solid, which result from splitting of these states due to atomic interactions, and are extended throughout the whole crystal.
D Bändermodell der Festkörpertheorie, Bändertheorie von Festkörper
F théorie des bandes d'état solide, théorie des bandes des solides
P teoria pasmowa ciała stałego
R зонная теория твёрдого тела

306
BARBIER-WIELAND DEGRADATION.
Transformation of carboxylic acid into its lower homologue by treatment with Grignard reagent, followed by oxidation of thus formed tertiary alcohol or alkene, e.g.

$$RCH_2CH_2COOC_2H_5 \xrightarrow[\text{acid hydrolysis}]{2C_6H_5MgBr}$$

$$RCH_2CH_2C(C_6H_5)_2 \xrightarrow[-H_2O]{} RCH_2CH=C(C_6H_5)_2$$
$$\overset{|}{OH}$$

$$\xrightarrow{CrO_3} RCH_2COOH + (C_6H_5)_2C=O$$

D Barbier-Wieland-Reaktion, Barbier-Wieland Carbonsäure-Abbau
F dégradation de Barbier-Wieland
P degradacja Barbiera (, Locquina) i Wielanda
R расщепление по Барбье-Виланду

307
BARIUM, Ba.
Atomic number 56.
D Barium
F baryum
P bar
R барий

308
BARKHAUSEN EFFECT.
Sequence of small discontinuous changes in magnetization of ferromagnets at continuously increasing external magnetic field.
D Barkhausen-Effekt
F phénomène de Barkhausen, effet de Barkhausen
P efekt Barkhausena
R эффект Баркгаузена

309
BARN, b.
Obsolete unit of surface area used in nuclear physics to define values of cross-sections. In SI system 1 b = 100 fm² (see femtometre).
D Barn
F barn
P barn
R барн

310
BAROMETRIC FORMULA.
In colloid chemistry, relation between height of rise and concentration of sol at sedimentation equilibrium (analogous to relation between altitude and air pressure). In case of monodisperse sol with spherical particles barometric formula can be expressed as

$$\frac{RT}{N_A} \ln \frac{n_0}{n} = \frac{4}{3} \pi r^3 (\gamma - \gamma_0) gh$$

where R = gas constant, N_A = Avogadro constant, n_0 = number of particles at zero level, n = number of particles at height h, r = radius of particle, γ and γ_0 = densities of dispersion medium and dispersed phases respectively.
D hypsometrische Formel
F équation de Perrin
P wzór barometryczny
R гипсометрический закон

BART ARYLARSONIC ACID SYNTHESIS. See BART REACTION.

311
BART REACTION (Bart arylarsonic acid synthesis).
Transformation of aromatic diazonium salt into corresponding arylarsonic acid, effected by sodium arsenite in presence of catalyst (copper or its salts), e.g.

$$C_6H_5N_2Cl + Na_3AsO_3 \xrightarrow{Cu}$$

$$C_6H_5AsO(ONa)_2 + NaCl + N_2$$

D Bart-Reaktion, (Bart) Arsenit-Arylierung
F réaction de Bart
P reakcja Barta
R реакция Барта

312
BARYON NUMBER, B.
Additive quantum number attributed to every particle and antiparticle and their systems. For proton and particles decaying to proton, $B = 1$, for corresponding antiparticles $B = -1$. Other particles and antiparticles have $B = 0$. Baryon number is conserved in all interactions.

D Baryonenzahl
F nombre baryonique
P liczba barionowa
R барионное число, барионный заряд

313
BARYON NUMBER CONSERVATION LAW.
Sums of baryon numbers of initial and final particles in given process are equal. Baryon number conservation law is observed in all interactions.

D Erhaltungssatz der Baryonenzahl
F loi de conservation du nombre baryonique
P prawo zachowania liczby barionowej
R закон сохранения барионного числа

314
BARYON RESONANCES.
Resonances with baryon number $B = 1$.

D Baryonenresonanzen
F résonances des baryons
P rezonanse barionowe
R барионные резонансы

315
BASE.
Substance which exhibits characteristic properties predicted by acid-base theories from which following are the most important: ability to neutralize acid, displacement of weaker bases from their compounds, definite influence on colour of acid-base indicators, catalytic properties.

D Base
F base

P zasada
R основание

316
BASE LINE METHOD.
Technique in quantitative spectrophotometric analysis. In order to set up reference point for absorption evaluation of the determined component and to eliminate base influence, supplementary line is drawn on absorption curve.

D Basis-Linien-Verfahren, Methode graphischer Ausschaltung von Störstoffen
F méthode de correction géométrique
P metoda linii podstawowej
R метод базисных линий

317
BASICITY (alkalinity).
Characteristic feature of electrolyte solutions in which concentration of hydroxylic ions OH^- is greater than that of hydrogen ions H^+.

D Basizität, Alkalität
F basicité
P zasadowość, alkaliczność
R основность

318
BASICITY OF ACID.
Number of hydrogen atoms in molecule of acid which are capable to be removed.

D Basizität der Säure
F basicité d'un acide
P zasadowość kwasu
R основность кислоты

BASIC SOLVENT. See PROTOPHILIC SOLVENT.

319
BATHOCHROME (bathochromic group).
Atom or group of atoms which, when substituted into molecule of organic compound, shifts its absorption spectrum towards longer wavelenghts.

D bath(m)ochrome Gruppe
F groupement bathochrome
P batochrom, grupa batochromowa
R батохромная группа

BATHOCHROMIC GROUP. See BATHOCHROME.

320
BATHOCHROMIC SHIFT.
Shift of light absorption towards lower frequencies.

D bathochrome Absorptionsverschiebung
F effet bathochrome
P przesunięcie batochromowe, efekt batochromowy, przesunięcie czerwone
R батохромное смещение

321
B BAND.
In electronic spectrum of benzene derivatives and its homologues, band corresponding to π-π^* electronic transition and analogous to 254 nm band of benzene, i.e. band of rather low intensity with well-developed vibrational structure.
D B-Band
F bande B
P pasmo B
R полоса B

322
BEAM HOLE (reactor channel for experiments).
Space accessible from outside, which is close to, or inside core of nuclear reactor, and enables utilization of produced radiation.
D Experimentierkanal
F canal d'expérimentation
P kanał reaktora doświadczalny
R экспериментальный канал

323
BEAM OF RADIATION.
Ionizing radiation (corpuscular or electromagnetic) propagating through space at a specified solid angle.
D Strahl
F faisceau (de rayonnement)
P wiązka promieniowania
R пучок (излучения)

324
BEATTIE-BRIDGEMAN EQUATION OF STATE.
Thermal equation of state of real gas.

$$pV_\mathrm{m}^2 = RT\left[V_\mathrm{m}+B_0\left(1-\frac{b}{V_\mathrm{m}}\right)\right]\left(1-\frac{c}{V_\mathrm{m}T^3}\right)-$$
$$+A_0\left(1-\frac{a}{V_\mathrm{m}}\right)$$

where p = pressure, V_m = molar gas volume, T = thermodynamic temperature, R = universal gas constant, A_0, B_0, a, b, c = constants characteristic for the given gas.
D Zustandsgleichung von Beattie-Bridgeman
F équation d'état de Beattie-Bridgeman
P równanie stanu Beattie'go i Bridgemana
R уравнение состояния Битти-Бриджмена

325
BÉCHAMP REDUCTION.
Reduction of nitro compounds to amines with iron (or ferrous salt) and dilute acid, e.g.

$R\!-\!NO_2 + 2Fe + 6HCl \rightarrow R\!-\!NH_2 + 2H_2O + 2FeCl_3$

D Béchamp (Nitroaryl)-Reduktion
F réduction de Béchamp
P redukcja Béchampa
R восстановление по Бешану

326
BECKMANN REARRANGEMENT.
Conversion of ketoximes (or aldoximes) into substituted amides of carboxylic acids, catalyzed by PCl_5, H_2SO_4, BF_3, CH_3COCl, polyphosphoric acid etc., e.g.

$$\begin{array}{c} C_6H_5 \quad CH_3 \\ \diagdown\,C\,\diagup \\ \parallel \\ N \\ \diagdown \\ OH \end{array} \xrightarrow{PCl_5} CH_3CONHC_6H_5$$

D Beckmann (Oxim → Amid)-Umlagerung
F transposition de Beckmann
P przegrupowanie Beckmanna
R перегруппировка Бекмана

327
BECQUEREL, Bq.
Unit or nuclear activity giving one disintegration per second.
D ...
F ...
P bekerel
R беккерель

328
BENDING VIBRATIONS.
Vibrations which lead to changes of valence angles.
D Deformationsschwingungen
F vibrations de déformation
P drgania zginające, drgania deformacyjne
R деформационные колебания

329
BENT-BOND (banana bond).
Chemical bond occurring in strained ring systems, when axes of bonding orbitals do not correlate with bond directions, e.g. in cyclopropane

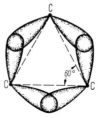

D gebogene Bindung, Bananen-Bindung
F liaison banane
P wiązanie zgięte, wiązanie τ, wiązanie bananowe
R изогнутая связь, банановая связь

330
BENZIDINE REARRANGEMENT.
Acid-catalyzed conversion of hydrazobenzene and its derivatives into benzidine (and its derivatives), e.g.

D Benzidinumlagerung
F transposition benzidinique
P przegrupowanie benzydynowe
R бензидиновая перегруппировка

331
BENZILIC ACID REARRANGEMENT.
Base-catalyzed conversion of aromatic α-diketones into corresponding α-hydroxy acids, involving formation of carboxylic group and migration of one of aryl groups (with rearrangement-1,2), e.g.

D Benzilsäureumlagerung
F transposition benzilique
P przegrupowanie benzilowe
R бензильная перегруппировка

332
BENZOIN CONDENSATION.
Specific reaction of aromatic or heterocyclic aldehydes, catalyzed by cyanide ions, resulting in formation of benzoins, e.g.

D Benzoinkondensation
F condensation des benzoïnes
P kondensacja benzoinowa
R бензоиновая конденсация

333
BENZOYLATION.
Substitution of hydrogen atom in molecule by benzoyl group C_6H_5CO-, e.g.

$$C_6H_5COCl + (CH_3)_2NH \rightarrow C_6H_5CON(CH_3)_2 + HCl$$

D Benzoylierung
F benzoylation
P benzoilowanie
R бензоилирование

334
BENZYLATION.
Substitution of hydrogen atom in molecule by benzyl group $C_6H_5CH_2-$.
D Benzylierung
F ...
P benzylowanie
R ...

335
BERKELIUM, Bk.
Atomic number 97.
D Berkelium
F berkélium
P berkel
R беркелий

336
BERTHELOT EQUATION OF STATE.
Empirical (thermal) equation of state of real gas.

$$\left(p + \frac{a}{V_m^2 T}\right)(V_m - b) = RT$$

where p = pressure, V_m = molar gas volume, T = thermodynamic temperature, R = universal gas constant, a, b = constants characteristic for the given gas.
D Berthelotsche Zustandsgleichung
F équation d'état de Berthelot
P równanie stanu Berthelota
R уравнение состояния Бертло

337
BERTHELOT-NERNST DISTRIBUTION LAW (homogeneous distribution law).
Distribution of a micro- and macrocomponent between crystals and solution, for two substances forming mixed crystals at real termodynamic equilibrium, is expressed by the equation

$$\frac{x}{y} : \frac{a-x}{b-y} = D$$

where x, y = quantities of micro- and macrocomponent, respectively, which passed to crystalline phase, a, b = initial quantities of micro- and macrocomponent in system, D = crystallization coefficient.
D Berthelot-Nernst-Verteilungsgesetz
F loi de Berthelot-Nernst
P prawo Chłopina
R закон Хлопина

BERTHOLIDES. *See* NON-STOICHIOMETRIC COMPOUNDS.

338
BERYLLIUM, Be.
Atomic number 4.
D Beryllium
F béryllium, glucinium

P beryl
R бериллий

339
BERYLLIUM, MAGNESIUM AND ALKALINE EARTH METALS.
Elements of second main group in periodic system with structure of outer electronic shells of atoms ns^2: beryllium, magnesium, calcium, strontium, barium, radium.

D Erdalkaligruppe
F glucinium, magnésium et métaux alcalino-terreux
P berylowce
R элементы группы бериллия, щёлочноземельные металлы

340
BETA DECAY (beta disintegration).
Radioactive decay in which an electron (β^--decay) or a positron (β^+-decay) is emitted and, respectively, an antineutrino $\bar{\nu}$ or a neutrino ν.

D β-Zerfall
F désintégration β
P przemiana β, rozpad β
R β-распад

BETA DISINTEGRATION. See BETA DECAY.

BET ADSORPTION ISOTHERM. See BET ISOTHERM.

341
BETA PARTICLE (β-particle).
Electron or positron emitted in beta decay.

D β-Teilchen
F particule β
P cząstka β
R β-частица

342
BETA PARTICLE ABSORPTION TECHNIQUE.
Method of analysis based on measurement of attenuation of a beta particle stream in layer of material tested. It is used for determining composition of mixtures, mainly for hydrogen content.

D Methode der Beta-Durchstrahlung
F analyse par absorption des rayons bêta
P analiza absorpcyjna beta
R метод анализа по поглощению β-лучей

343
BETA PARTICLE BACK-SCATTERING TECHNIQUE.
Method of quantitative analysis in which use is made of the relation between intensity and energy of reflected β radiation, and atomic number of elements occurring in material examined.

D Beta-Rückstreuverfahren
F analyse par rétrodiffusion du rayonnement bêta

P analiza metodą wstecznego rozpraszania cząstek β
R анализ веществ по отражению β-частиц

BETA PARTICLES. See BETA RAYS.

344
BETA RAYS (beta particles, β-particles).
Radiation composed of electrons (β^--particles) or positrons (β^+-particles) with continuous spectrum.

D β-Strahlen, β-Teilchen
F rayons β, particules β
P promie(niowa)nie β
R β-лучи, β-частицы

345
BETATRON.
Cyclic accelerator in which electrons are accelerated to energies up to 100 MeV.

D Betatron
F bêtatron
P betatron
R бетатрон

346
BET EQUATION (derived by Brunauer, Emmett and Teller).
Equation of the adsorption isotherm for multilayer adsorption of vapour:

$$\frac{p}{x(p_0-p)} = \frac{1}{x_m c} + \frac{c-1}{x_m c}\frac{p}{p_0}$$

where x = amount of gas adsorbed by unit mass of adsorbent, x_m = amount of gas necessary for formation of monomolecular layer at surface of unit mass of adsorbent, p = equilibrium pressure, p_0 = saturated vapour pressure for pure adsorbate at temperature of isotherm, c = constant characteristic for given system and temperature.

D BET-Gleichung
F équation BET
P równanie BET, równanie Brunauera, Emmetta i Tellera
R уравнение БЭТ

BETHE CYCLE. See CARBON-NITROGEN CYCLE.

347
BET ISOTHERM (BET adsorption isotherm).
Adsorption isotherm corresponding to BET (Brunauer, Emmett and Teller) equation for multilayer adsorption of vapours.

D BET-Adsorptionsisotherme
F courbe BET
P izoterma (adsorpcji) BET
R ...

B FAMILY. See SUB-GROUP.

348
BI-AMPEROMETRIC TITRATION
(amperometry with two polarized electrodes).
"Dead stop" or "dead stop end point" titration in which to two strictly identical electrodes, usually platinum, immersed in the titration cell a constant voltage (50—500 mV) is applied and current passing through cell is measured as a titrant is added.
D amperometrische dead-stop-Methode
F ampérométrie avec deux électrodes indicatrices
P miareczkowanie amperometryczne z dwiema polaryzowanymi elektrodami, miareczkowanie do punktu martwego, miareczkowanie „dead-stop"
R амперометрическое титрование с двумя поляризованными электродами

BIAS. *See* SYSTEMATIC ERROR.

349
BIMOLECULAR REACTION.
Elementary unit reaction of molecularity equal to two.
D bimolekulare Reaktion, dimolekulare Reaktion
F réaction bimoléculaire
P reakcja dwucząsteczkowa, reakcja bimolekularna
R бимолекулярная реакция, двухмолекулярная реакция

350
BINARY EUTECTIC.
Eutectic composed of two constituents.
D doppeltes Eutektikum
F eutectique binaire
P eutektyk podwójny, eutektyk dwuskładnikowy
R двойная эвтектика

351
BINARY SYSTEM (two-component system).
System composed of two independent constituents.
D Zweikomponentensystem, binäres System
F système binaire
P układ dwuskładnikowy, układ binarny
R бинарная система, двухкомпонентная система

352
BINOMIAL DISTRIBUTION.
Probability distribution of step-like random variable X with probability function given by

$$P(X = k) = \binom{n}{k} p^k (1-p)^{n-k}$$

for $k = 0, 1, 2, ..., n$ $(0 < p < 1)$

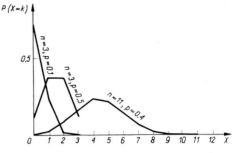

D Binomial-Verteilung
F distribution binomiale, répartition du binôme
P rozkład dwumianowy
R биноминальное распределение

353
BIOCATALYST.
Organic substance, usually of complicated structure which acts catalytically on chemical reactions proceeding in living organisms, e.g. enzyme, hormone, vitamin.
D Biokatalysator
F biocatalyseur
P biokatalizator, katalizator biochemiczny
R биокатализатор

354
BIOCHEMISTRY.
Science dealing with chemistry of compounds present in living organisms as well as with their transformations in these systems.
D Biochemie, biologische Chemie
F biochimie, chimie biologique
P biochemia
R биохимия, биологическая химия

355
BIOLOGICAL OXIDATION.
Oxidation of chemical compounds caused by enzymes, which act as electron or hydrogen atom carriers.
D biologische Oxydation
F oxydation biologique
P utlenianie biologiczne
R биологическое окисление

356
BIOLOGICAL SHIELD.
Layer of material of suitable absorption characteristics and thickness, placed between source of ionizing radiation and operator, securing attenuation of radiation to admissible level.
D biologischer Schild
F bouclier biologique
P osłona biologiczna
R биологическая защита

357
BIOLUMINESCENCE.
Chemiluminescence accompanying certain chemical reactions in living organisms.
D Biolumineszenz
F bioluminescence
P bioluminescencja
R биолюминесценция

358
BIOSYNTHESIS.
Synthesis of organic compounds, e.g. alkaloids, proteins, antibiotics, terpenes, steroids, effected by living organisms.
D Biosynthese
F biosynthèse
P synteza biologiczna, biosynteza
R биосинтез

359
BIPOLYMER.
Product of copolymerization of mixture of two different monomers.
D Bipolymer
F bipolymère
P bipolimer
R биполимер

360
BIRADICALS (diradicals).
Free radicals possessing unpaired electrons at each of two different atoms, e.g. propene biradical $CH_3\dot{C}H—\dot{C}H_2$, oxygen biradical $\cdot\ddot{O}—\ddot{O}\cdot$
D Diradikale, Biradikale
F diradicaux, biradicaux
P dwurodniki
R бирадикалы

361
BIRCH REDUCTION.
Hydrogenation of arenes with metallic sodium (or sodium and alcohol) in liquid ammonia, e.g.

D Birch-Hückel-Hydrierung
F réduction de Birch
P redukcja Bircha
R восстановление по Берчу

362
BIREFRINGENCE.
Measure of optical anisotropy of crystals, most frequently arising from difference of principal refractive indices.
D Doppelbrechung
F biréfringence
P dwójłomność
R двупреломление

363
BISCHLER-NAPIERALSKI REACTION.
Preparation of 3,4-dihydroisoquinoline from acyl derivatives of β-phenylethylamine under action of dehydrating agents, e.g.

D Bischler-Napieralski-Reaktion, Bischler-Napieralski Isochinolin-Ringschluß
F réaction de Bischler et Napieralski
P reakcja Bischlera i Napieralskiego
R реакция Бишлера-Наперальского

364
BISMUTH, Bi.
Atomic number 83.
D Bismut, Wismut
F bismuth
P bizmut
R висмут

365
BIURET REACTION.
Reaction of peptides and proteins with copper sulfate in alkaline medium, resulting in formation of violet-coloured complex.
D Biuretreaktion
F réaction du biuret
P reakcja biuretowa
R биуретовая реакция

366
BIVARIANT PHASE SYSTEM (bivariant System).
System with two thermodynamic degrees of freedom.
D divariantes Phasensystem, divariantes System
F système divariant (de phases)
P układ (fazowy) dwuzmienny
R двухвариантная (фазовая) система

BIVARIANT SYSTEM. *See* BIVARIANT PHASE SYSTEM.

367
BJERRUM ASSOCIATION THEORY
OF IONS.
Theory which predicts formation of pair
from ions of opposite charges if distance
separating them l is

$$a \leqslant l \leqslant \frac{|z_i z_j| e^2}{2\varepsilon kT}$$

Life-time of this pair as a result of
coulombic interactions is sufficiently long
to consider it as unit resistant against
thermic motions in solution; a = effective
ionic diameter, for other symbols *see*
Debye-Hückel's limiting law.
D Assoziationstheorie von Bjerrum
F théorie de l'association des ions de
Bjerrum
P teoria asocjacji jonów Bjerruma
R теория ассоциации ионов Семенченки
и Бьеррума

368
BJERRUM THEORY OF ACIDS AND
BASES.
Attempt at combination of Brönsted and
Lewis theories of acids and bases; Bjerrum
defined acids as substances having
tendency to lose protons, antibases — with
tendency to attach electron pairs, and
bases as species having tendency to add
on protons or to lose electron pairs to
form bond.
D Bjerrumsche Säuren-Basen-Theorie
F théorie des acides et des bases de
Bjerrum
P teoria kwasów i zasad Bjerruma
R теория кислот и оснований Бьеррума

369
BLACK BODY.
Body which absorbs completely all
incident radiation falling upon it, no
matter what the wavelength of radiation
and temperature is.
D schwarzer Körper
F corps (parfaitement) noir
P ciało doskonale czarne
R абсолютно чёрное тело

370
BLACKENING (optical density
of photographic emulsion), *S*.
Quantity given by function $S = \lg \dfrac{I_0}{I}$,
where I_0 = intensity of light passing
through non-exposed part of emulsion,
I = intensity of light passing through
exposed part.
D (photographische) Schwärzung
F noircissement, densité optique
P zaczernienie
R почернение

371
BLANC REACTION.
Preparation of cyclic ketones by heating
1,4- or 1,5-dicarboxylic acid with acetic
anhydride, e.g.

D Blanc (Dicarbonsäure→Keton)-
Cyclisierung
F réaction de Blanc
P reakcja Blanca, zamknięcie pierścienia
Blanca
R реакция Бланка

BLANC REACTION. *See*
CHLOROMETHYLATION.

372
BLANC RULE.
Conversion of dicarboxylic acid into cyclic
ketone (upon heating with acetic
anhydride) proceeds with good yield
when resulting ketone is 5- or
6-membered.
D Blancsche Regel
F règle de Blanc
P reguła Blanca
R правило Бланка

373
BLANK (blank test, blank determination).
Test carried out under specified conditions
with all reagents with exception of
determined substance to account for
content of this substance in reagents
used.
D Blindprobe, Leerversuch
F essai à blanc
P próba ślepa, próba zerowa
R холостой опыт, слепой опыт

BLANK DETERMINATION. *See* BLANK.

BLANK TEST. *See* BLANK.

374
BLOCK.
Portion of polymer macromolecule
comprising many constitutional units,
that portion having constitutional or
configurational feature not present in
adjacent multiple-unit portions.
D Block
F bloc
P blok
R блок

BLOCKING GROUP. *See* PROTECTING
GROUP.

BLOCKING GROUP. *See* STERIC
HINDRANCE.

375
BLOCK POLYMER.
Linear polymer whose macromolecules contain two or more species of blocks attached linearly, e.g.

—AAAAAAA—BBBBBB—AAAAAAA—

D Block(co)polymer
F (co)polymère séquencé, copolymère bloc
P (ko)polimer blokowy
R блоксополимер

376
BLOCK POLYMERIZATION.
Process in which one or more monomers and a polymer are converted together into block polymer.
Remark: Term block copolymerization has been reserved for case of block polimerization with two or more participant monomers.

D Blockpolymerisation
F polymérisation en bloc
P polimeryzacja blokowa
R блоксополимеризация

377
$B_{Ac}2$ MECHANISM.
Mechanism of bimolecular hydrolysis reaction of esters in basic medium, involving cleavage of acyl-oxygen bond. The following reaction steps are postulated: addition of $HO^{\ominus}$ ion to carbonyl carbon atom of ester, cleavage of acyl-oxygen bond with formation of acid and alcoholate ion, proton exchange between acid and alcoholate ion, e.g.

$$HO^{\ominus} + \overset{\overset{O}{\|}}{\underset{\underset{R}{|}}{C}}-OR' \xrightarrow{slow} \left[HO-\overset{\overset{O^{\ominus}}{|}}{\underset{\underset{R}{|}}{C}}-OR' \right] \xrightarrow{fast}$$

$$HO-\overset{\overset{O}{\|}}{\underset{\underset{R}{|}}{C}} + R'O^{\ominus} \to RCOO^{\ominus} + R'OH$$

D $B_{Ac}2$-Mechanismus
F mécanisme $B_{Ac}2$
P mechanizm $B_{Ac}2$
R механизм $B_{Ac}2$

378
$B_{Al}1$ MECHANISM.
Mechanism of monomolecular hydrolysis of esters in basic medium involving cleavage of alkyl-oxygen bond. Two reaction steps are postulated: cleavage of alkyl-oxygen bond with formation of carbonium ion and acid anion, and reaction of the former with $HO^{\ominus}$ ion, e.g.

$$R-\overset{\overset{O}{\|}}{C}-O-R' \to R\overset{\overset{O}{\|}}{C}-O^{\ominus} + R'^{\oplus}$$

$$R'^{\oplus} + OH^{\ominus} \to R'OH$$

D $B_{Al}1$-Mechanismus
F mécanisme $B_{Al}1$
P mechanizm $B_{Al}1$
R механизм $B_{Al}1$

379
BOAT FORM.
Labile (flexible) conformation in saturated six-membered carbocyclic and heterocyclic compounds, e.g. cyclohexane. In this conformation carbon bond angles maintain value of 109.5°; however, C—C bond system at two pairs of adjacent carbon atoms is ecliptic and repulsion between substituents is observed at 1,4-flag-pole position

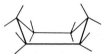

D Bootsform, Wannenform
F forme bateau, bateau
P konformacja łódkowa
R конформация ванны, ванна

380
BODROUX-CHICHIBABIN ALDEHYDE SYNTHESIS (Bodroux-Chichibabin synthesis).
Preparation of aldehydes by treating ethyl orthoformate with Grignard reagent, e.g.

$$RMgX + HC(OC_2H_5)_3 \longrightarrow RHC(OC_2H_5)_2 + XMgC_2H_5$$

$$RHC(OC_2H_5)_2 \xrightarrow[H_2O]{H^{\oplus}} RCHO + 2C_2H_5OH$$

D Bodroux-Tschitschibabin-Formylierung
F méthode de Bodroux, réaction de Tschichibabine
P synteza Bodrouxa i Cziczibabina
R реакция Бодру-Чичибабина

BODROUX-CHICHIBABIN SYNTHESIS.
See BODROUX-CHICHIBABIN ALDEHYDE SYNTHESIS.

BODY. *See* SUBSTANCE.

381
BOHR MAGNETON, μ_B (β).
Unit of atomic (electronic) magnetic moment; 1 $\mu_B = 9.273 \cdot 10^{-24}$ A m^2
D Bohrsches Magneton
F magnéton de Bohr
P magneton Bohra
R магнетон Бора

382
BOHR ORBIT.
Circular or elliptic orbit around nucleus. In old quantum theory this was electron path in atoms.
D Bohrsche Bahn
F orbite de Bohr
P orbita Bohra
R боровская орбита

383
BOHR POSTULATES.
Postulates of old quantum theory. They
refer to existence of stationary states
of electrons in atoms and to transitions
between these states.
D Bohrsche Postulate
F postulats de Bohr
P postulaty Bohra
R постулаты Бора

384
BOHR RADIUS, a_0.
Radius of lowest energy Bohr orbit of
hydrogen atom (1s). For hydrogen atom
in 1s state distance of maximum electron
density is

$$a_0 = \frac{\hbar^2}{2me^2} = 0.529167 \cdot 10^{-10} \text{ m}$$

where m = electron mass, e = elementary
charge, $\hbar$ = Planck constant divided
by 2π.
D Bohr-Radius, Wasserstoffradius
F rayon de Bohr
P promień Bohra
R радиус Бора, боровский радиус

385
BOILING.
Production of vapour in bulk of liquid
appearing when saturated vapour pressure
is equal to external pressure.
D Sieden
F ébullition
P wrzenie
R кипение

386
BOILING POINT.
Temperature at which pressure of
saturated vapour above liquid is equal
to external pressure.
D Siedetemperatur
F température d'ébullition
P temperatura wrzenia
R температура кипения

387
BOLTZMANN CONSTANT, k.
Quotient of universal gas constant R
and Avogadro number N_A

$$k = \frac{R}{N_A} = 1.38054 \cdot 10^{-23} \text{ J K}^{-1}$$

D Boltzmann-Konstante
F constante de Boltzmann
P stała Boltzmanna
R постоянная Больцмана

BOLTZMANN DISTRIBUTION
FUNCTION. *See* MAXWELL-
BOLTZMANN DISTRIBUTION.

388
BOLTZMANN EQUATION (Boltzmann
transport equation).
Conservation equation for unimolecular
partition function

$$\frac{\partial \rho_1}{\partial t} + v \frac{\partial \rho_1}{\partial x} + \frac{F}{m} \frac{\partial \rho_1}{\partial v} = \left(\frac{\partial \rho_1}{\partial t}\right)_{coll}$$

where $\rho_1 (t, v, x)$ = unimolecular partition
function, t = time, x = molecule position
vector and v = molecule velocity,
F = external forces and m = molecule

mass, $\left(\dfrac{\partial \rho_1}{\partial t}\right)_{coll}$ = collision integrals.

Boltzmann equation describes processes
in rarified gases.
D Boltzmannsche Stoßgleichung
F équation de Boltzmann
P równanie Boltzmanna
R (газо)кинетическое уравнение
  Больцмана, уравнение Больцмана

389
BOLTZMANN FACTOR.
Exp $(-E/kT)$ where E = energy of
  molecule,
k = Boltzmann constant,
T = thermodynamic temperature.
Boltzmann factor occurs in formula for
Maxwell-Boltzmann distribution, etc.
D Boltzmann-Faktor
F facteur de Boltzmann
P czynnik Boltzmanna
R фактор Больцмана

390
BOLTZMANN-MAXWELL STATISTICS.
Statistics of systems of classical particles.
D Boltzmann-Maxwell-Statistik
F statistique de Maxwell-Boltzmann
P statystyka Maxwella i Boltzmanna
R статистика Максвелла-Больцмана

BOLTZMANN TRANSPORT EQUATION.
See BOLTZMANN EQUATION.

391
BOMB CALORIMETER (oxygen-bomb
calorimeter).
Calorimeter for determining combustion
heats of organic solids and liquids.
D kalorimetrische Bombe
F bombe calorimétrique
P bomba kalorymetryczna
R калориметрическая бомба

392
BOND DIPOLE MOMENT (bond moment).
Dipole moment previously ascribed to
given chemical bond according to
additivity concepts. Bond dipole moment
was assumed to be parallel to bond axis;
however, because of asymmetry of charge
distribution in atoms with lone electron

pairs, its direction need not have been parallel to that of total molecular dipole moment.

Remark: Term group dipole moment recommended instead of this term.

D Bindungsmoment
F moment de liaison
P moment (dipolowy) wiązania
R дипольный момент связи

BOND ENERGY. *See* CHEMICAL BOND ENERGY.

393
BOND FORCE CONSTANT (stretching force constant), k (F_r).
Constant characterizing harmonic forces due to changes of bond length during vibrations of atoms in molecules. It is equal to second derivative of total molecular energy with respect to a given bond length and computed at position of nuclear equilibrium.

D Bindungskraftkonstante
F constante des forces de liaison
P stała siłowa wiązania
R силовая постоянная связи

394
BONDING ELECTRONS.
Electrons assumed to occupy bonding molecular orbitals. They play important role in formation of chemical bonds.

D bindende Elektronen
F électrons liants
P elektrony wiążące
R связывающие электроны, связующие электроны

395
BONDING ORBITAL.
Molecular orbital represented as linear combination of atomic orbitals; its energy is lower than that of any atomic orbital contributing to it. Electrons occupying bonding orbitals lead either to formation of chemical bond or to strengthening of already existing bond.

D bindendes Molekülorbital, bindender Zustand
F orbitale liante
P orbital wiążący
R связывающая молекулярная орбиталь

396
BOND LENGTH (internuclear distance).
Distance between nuclei of two atoms forming chemical bond, measured in molecular ground state.

D Bindungslänge, Atomabstand, Kernabstand
F longueur de liaison, distance interatomique
P długość wiązania
R длина связи

BOND MOMENT. *See* BOND DIPOLE MOMENT.

397
BOND ORDER.
Number denoting bond multiplicity; for conjugated double bonds rational number between 1 and 2.

D Bindungsordnung
F indice de liaison
P rząd wiązania
R порядок связи

398
BOND PAIR.
Electron pair taking part in formation of both chemical bond and system of hybridized orbitals.

D bindendes Elektronenpaar
F doublet liant
P para elektronowa wiążąca
R связывающая электронная пара

399
BOND POLARITY.
Effect of non-uniform distribution of electronic charge along bond. Electron density is shifted towards one atom and leads to bond dipole moment; characteristic for bonds between atoms of different electronegativities.

D Bindungspolarität
F polarité de liaison
P polarność wiązania
R полярность связи

400
BOND STRENGTH.
Measure of stability of chemical bond. In molecular orbital method bond strength is determined by overlap of atomic orbitals used to construct molecular orbitals and indirectly by degree of hybridization of atomic orbitals.

D Bindungsfestigkeit
F force de liaison
P moc wiązania, siła wiązania
R прочность связи

401
BORN EQUATION.
Equation which describes Gibbs free energy of solvation ΔG of one mole of ions derived from assumption that it is equivalent to difference of electrical field energy of ions in vacuum and in solvent

$$\Delta G = -\frac{N_A z_B^2 e^2}{2a}\left(1 - \frac{1}{\varepsilon_r}\right)$$

where a = ionic radius and ε_r = relative permittivity.

D Bornsche Gleichung
F équation de Born
P równanie Borna
R формула Борна

402
BORN-HABER CYCLE.
Imaginary thermodynamic cyclic process
used for calculating energy of ionic crystal
lattice.
D Born-Haberscher Kreisprozeß
F cycle de Born-Haber
P cykl Borna i Habera
R цикл Борна-Габера

403
BORN-OPPENHEIMER
APPROXIMATION.
In molecular electronic state lying far
enough energetically from other electronic
states nuclei move in effective potential
field resulting from electron density
distribution in this state.
D Born-Oppenheimer Näherung
F approximation de Born-Oppenheimer
P przybliżenie Borna i Oppenheimera,
   rozdzielność ruchów elektronów i jąder
R приближение Борна-Оппенгеймера

404
BORN THEORY OF SOLVATION.
Theory which considers solvation process
as ordering action of electrical field of ion
on polar molecules of solvent.
D Bornsche Solvatation-Theorie
F théorie de la solvatation de Born
P teoria solwatacji Borna
R теория сольвации Борна

405
BORON, B.
Atomic number 5.
D Bor
F bore
P bor
R бор

BORON FAMILY. See BORON GROUP.

406
BORON GROUP (boron family).
Elements of third main group in periodic
system with structure of outer electronic
shells of atoms ns^2np^1 : boron, aluminium,
gallium, indium, thallium.
D Borgruppe, Bor-Aluminium-Gruppe
F groupe du bore
P borowce
R группа бора, элементы группы бора

BOSE DISTRIBUTION FUNCTION. See
BOSE-EINSTEIN DISTRIBUTION.

407
BOSE-EINSTEIN CONDENSATION.
State of system of identical noninteracting
bosons at low temperatures. It corresponds
to occupation number of the lowest
single-particle energy level comparable
with total number of particles in system.

D Einsteinkondensation
F condensation d'Einstein
P kondensacja (Bosego i) Einsteina
R конденсация Бозе-Эйнштейна

408
BOSE-EINSTEIN DISTRIBUTION (Bose
distribution function).
Equilibrium distribution of particles
among single-particle energy levels

$$\langle N_\varepsilon \rangle = \frac{g_\varepsilon}{\frac{1}{\lambda} \exp \frac{\varepsilon}{kT} - 1}$$

where k = Boltzmann constant,
T = thermodynamic temperature,
λ = absolute activity and g_ε = degeneracy
of energy level ε, $\langle N_\varepsilon \rangle$ = average
occupation number of this level.
Bose-Einstein distribution can be used
for weakly interacting systems
of identical bosons.
D Bose-Einsteinsches Verteilungsgesetz
F répartition de Bose-Einstein
P rozkład Bosego i Einsteina
R распределение Бозе

409
BOSE-EINSTEIN STATISTICS.
Statistics applicable to systems of quantum
particles whose wave function is
symmetric with respect to particle
permutations (cf. Bose-Einstein
distribution).
D Bose-Einstein-Statistik
F statistique de Bose-Einstein
P statystyka Bosego i Einsteina
R статистика Бозе-Эйнштейна

410
BOSONS.
Elementary particles or complex systems
with integral spin quantum number, e.g.
mesons, photons, nuclei of even number
of nucleons. Bosons are described by
Bose-Einstein statistics.
D Bosonen
F bosons
P bozony
R бозоны

BOUGUER-BEER LAW. See
LAMBERT-BEER LAW.

411
BOUGUER'S LAW (Lambert's law).
For monochromatic light passing through
homogeneous medium absorbance A
is proportional to path length b, $A = ab$,
where a = absorptivity.
D Bouguer-Lambert-Gesetz
F loi de Lambert

P prawo Lamberta, prawo Bouguera
i Lamberta
R закон Бугера-Ламберта

BOUNDARY BETWEEN PHASES. *See*
INTERFACE.

412
BOUND ENERGY.
Extensive thermodynamic function TS,
where T = thermodynamic temperature
and S = entropy.
D gebundene Energie
F énergie liée
P energia związana
R связанная энергия

413
BOUVEAULT ALDEHYDE SYNTHESIS.
Preparation of aldehydes by treating
Grignard reagents with disubstituted
formamide derivatives, e.g.

$$RMgX + HC-N\underset{O}{\overset{R'}{\|}}\diagdown_{R''} \rightarrow RCH-N\underset{OMgX}{\overset{R'}{|}}\diagdown_{R''} \xrightarrow{HX}$$

$$RCH + R'NHR'' + MgX_2$$
$$\underset{O}{\|}$$

D Bouveault-Formylierung
F méthode de Bouveault
P synteza aldehydów Bouveaulta
R синтез альдегидов Буво

414
BOUVEAULT-BLANC REDUCTION.
Reduction of esters to alcohols with
sodium and ethanol, e.g.

$$RCOOC_2H_5 \xrightarrow[Na]{C_2H_5OH} RCH_2OH + C_2H_5OH$$

D Bouveault-Blanc (Ester)-Reduktion
F réaction de Bouveault et Blanc
P redukcja Bouveaulta i Blanca
R восстановление по Буво-Блану

415
BOYLE'S LAW (Mariotte's law).
Product of pressure p and volume V of
the given mass of gas is constant at
constant temperature; pV = const.
Boyle's law is empirical.
D Boyle-Mariottesches Gesetz
F loi de Boyle et Mariotte
P prawo Boyle'a i Mariotte'a
R закон Бойля и Мариотта

BOYLE'S POINT. *See* BOYLE'S
TEMPERATURE.

416
BOYLE'S TEMPERATURE (Boyle's
point), T_B.

Temperature at which, in range of low
pressures, perfect gas satisfies
Boyle-Mariotte's law, i.e.

$$\left(\frac{\partial(pV)}{\partial p}\right)_{T_B, p \to 0} = 0$$

where p = pressure, V = gas volume; for
gas satisfying van der Waals equation
$T_B = a/Rb$, where a and b = constants
entering this equation, R = molar gas
constant.
D Boyle-Temperatur, Boyle-Punkt
F température de Boyle, point de Boyle
P temperatura Boyle'a
R температура Бойля

417
BRAGG CURVE.
Graph for average number of ions per
unit distance along beam of initially
monoenergetic α-particles passing through
gas.
Remark: This original definition is sometimes
extended to other ionizing particles and other
media.
D Braggsche Kurve
F courbe de Bragg
P krzywa Bragga
R кривая Брэгга

418
BRAGG-GRAY CAVITY CHAMBER.
Hollow chamber filled with gas inside
solid medium, used for absorbed dose
measurements, mainly by ionization
methods.
D Bragg-Gray-Krammer
F chambre à cavité Bragg-Gray
P komora wnękowa Bragga i Graya
R кавитационная камера Брэгг-Грея

419
BRAGG IONIZATION SPECTROMETER
(Bragg spectrometer).
Device for direct measurement of
reflection angle for X-ray radiation
reflected by set of parallel
crystallographic planes; for appropriately
chosen position of crystal (*see* Bragg
equation) reflected ray appears, which
is recorded either by Geiger-Müller
or by scintillation counter.
D Bragg-Spektrometer
F spectromètre de Bragg
P spektrometr Bragga
R спектрометр Брэгга

420
BRAGG'S EQUATION (Bragg's law).
Condition under which monochromatic
X-ray beam of wavelength λ is reflected
from set of parallel planes in crystal of
indices h, k, l: $2d_{hkl} \sin \Theta = n\lambda$,
where d_{hkl} = distance between
neighbouring lattice planes, Θ = angle
between hkl plane and reflected beam,
n = order of interference ($n = 1$, 2, ...).

D Braggsche Gleichung
F équation de Bragg, loi de Bragg
P równanie Braggów, prawo Braggów
R уравнение Брэгга-Вульфа, закон
  Вульфа-Брэгга

BRAGG'S LAW. *See* BRAGG'S
EQUATION.

BRAGG SPECTROMETER. *See* BRAGG
IONIZATION SPECTROMETER.

BRANCH. *See* SIDE CHAIN.

BRANCHED CHAIN. *See* FORKED
CHAIN.

421
BRANCHED POLYMER.
Polymer composed of branched
macromolecules; sectors between
branches and those between chain
terminals and branches are linear.

D verzweigtes Polymer
F polymère multibranche, polymère
  ramifié
P polimer rozgałęziony
R разветвлённый полимер

422
BRANCH HEAD.
In rotation-oscillation spectrum of
molecules, convergence limit for
rotational lines of a given branch.

D Zweigkopf
F tête de branche
P głowica gałęzi
R голова ветви

423
BRANCHING.
In radiochemistry, existence of at least
two competitive modes of decay of
radioactive nuclide.

D Verzweigung
F embranchement
P rozpad rozgałęziony, rozpad złożony
R параллельный распад, разветвлённый
  распад

424
BRANCHING CHAIN REACTION.
Chain reaction which proceeds with chain
branching.

D verzweigte Kettenreaktion
F réaction en chaîne ramifiée

P reakcja łańcuchowa rozgałęziona
R разветвлённая цепная реакция

425
BRANCHING RATIO.
In radiochemistry, ratio of number of
radioactive disintegrations of specified
mode to number of remaining
disintegrations of the same nuclide.

D Verzweigungsverhältnis
F rapport d'embranchement
P stosunek rozgałęzień
R отношение разветвления

426
BRAVAIS LATTICE.
14 basic types of crystal lattice,
of different symmetry, built by simple
repetition of 14 types of Bravais cells
in direction of their three edges on
distances equal to length of edges.

D Bravaisgitter, Translationsgitter
F réseau de Bravais
P sieć (przestrzenna) Bravais'go, sieć
  translacyjna
R решётка Браве, трансляционная
  решётка

BREADTH OF SPECTRAL LINE. *See*
LINE WIDTH.

BREAKTHROUGH. *See* COLUMN
BREAKTHROUGH.

427
BREAKTHROUGH CAPACITY (*of
column*), Q_B.
In ion exchange chromatography number
of miliequivalents of given ion retained
by column up to moment of breakthrough.

D Durchbruchskapazität (der Säule)
F capacité utile, capacité dynamique
  limitée à la percée
P pojemność wymienna robocza,
  pojemność wymienna do przebicia
R полная ёмкость колонки, ёмкость
  (колонки) до проскока

428
BREAKTHROUGH CURVE (effluent
concentration history, exchange isoplane).
In frontal analysis, diagram showing
dependence of substance concentration
in effluent, C, (or concentration ratio
in effluent and original solution
respectively, C/C_0) upon effluent volume
or time.

D Durchbruchskurve
F courbe de répartition des ions dans
  l'effluent, isoplane
P krzywa przebicia
R выходная кривая

429
BREDT'S RULE.
In polycyclic compounds (excluding very
large cycles), C=C and C=N double

bond systems cannot exist at nodal carbon atoms.
D Bredtsche Regel
F règle de Bredt
P reguła Bredta
R правило Бредта

430
BREEDER REACTOR.
Nuclear reactor which produces more fissionable material than is present in original batch of fuel.
D Brutreaktor, Brüter
F réacteur surrégénérateur, breeder
P reaktor mnożący, reaktor powielający
R реактор-размножитель

431
BREIT-WIGNER FORMULA.
Approximate formula determining cross-section for given nuclear reaction when energies of incident particles are close to resonance level.
D Breit-Wigner-Formel
F formule de Breit-Wigner
P wzór Breita i Wignera
R формула Брейта-Вигнера

BREMSSTRAHLUNG. *See* X-RAY BACKGROUND RADIATION.

BRIDGED COMPLEX. *See* POLYNUCLEAR COMPLEX.

432
BRIDGING ATOM.
Atom of bridging group linked with two central atoms of polynuclear complex.
D Brückenatom
F atome pont
P atom mostkowy
R мостиковый атом

433
BRIDGING GROUP.
Group attached to two central atoms of polynuclear complex.
D brückenbildende Gruppe, Brücken-Gruppe
F groupe pont
P grupa mostkowa
R мостиковая группа

434
BRILLOUIN ZONES.
Zones in space of wave vector limited by such values of vector for which waves coupled with electron (or phonon) movement in crystal are standing waves (Bragg condition is fullfilled, *see* Bragg's equation).

D Brillouin-Zonen
F zones de Brillouin
P strefy Brillouina
R зоны Бриллюэна

435
BROMINATION.
Substitution of hydrogen atom in molecule by bromine atom, e.g.

$$C_6H_6 + Br_2 \xrightarrow{Fe} C_6H_5Br + HBr$$

D Bromierung
F brom(ur)ation
P bromowanie
R бромирование

436
BROMINE, Br.
Atomic number 35.
D Brom
F brome
P brom
R бром

437
BRÖNSTED-BJERRUM EQUATION.
Equation describing dependence of reaction-rate constant on activity coefficient and activated complex

$$k' = k_0 \frac{f_A f_B}{f^+}$$

where k' = reaction-rate constant in real solution, k_0 = reaction-rate constant in ideal solution, f_A, f_B = activity coefficients of substrates A and B, f = activity coefficient of activated complex.
D Brönsted-Bjerrum Gleichung
F équation de Brönsted-Bjerrum
P równanie Brönsteda i Bjerruma
R уравнение Бренстеда-Бьеррума

BRÖNSTED CATALYSIS LAW. *See* BRÖNSTED RELATIONSHIP.

438
BRÖNSTED RELATIONSHIP (Brönsted catalysis law).
Empiric law according to which catalytic reaction-rate constant k of acid- or base-catalyzed reaction is linearly proportional to acidity (alkalinity) coefficient of catalyst K to αth power

$$k = GK^\alpha$$

where G and α are constants for given reaction.
D Brönsted(sche) Beziehung
F relation de Brönsted
P prawo katalizy Brönsteda
R ...

439
BRÖNSTED THEORY OF ACIDS AND BASES.
Theory according to which acid is substance which is able to lose proton, base is substance which binds proton (see protolysis), irrespective of whether these substances exist as neutral molecules or ions, e.g.

acid   base

$H_2SO_4 \rightleftarrows HSO_4^- + H^+$

$HSO_4^- \rightleftarrows SO_4^{2-} + H^+$

D Brönstedsche Säuren-Basen-Theorie
F théorie des acides et des bases de Brönsted
P teoria kwasów i zasad protonowa, teoria kwasów i zasad Brönsteda i Lowry'ego
R теория кислот и оснований Бренстеда, протолитическая теория кислот и оснований

440
BROWNIAN MOTION (Brownian movement).
Continuous, erratic movement, similar to vibrations and hops, of particles of dispersed phase in disperse system due to continual impacts with molecules of dispersion medium which undergo thermal agitation.
D Brownsche Bewegung
F mouvement brownien
P ruchy Browna
R Броуновское движение

BROWNIAN MOVEMENT. See BROWNIAN MOTION.

B-STRAIN EFFECT. See BACK STRAIN EFFECT.

441
BUCHERER-BERGS HYDANTOIN SYNTHESIS.
Preparation of hydantoin derivatives from carbonyl compounds, hydrogen cyanide and ammonium carbonate, e.g.

$$\underset{R}{\overset{R}{>}}CO \xrightarrow[HCN]{(NH_4)_2CO_3} \underset{R}{\overset{R}{>}}C\underset{CO-NH}{\overset{NH-CO}{<}}$$

D Bucherer-Bergs Hydantoin-Ringschluß
F synthèse de Bucherer
P synteza (pochodnych) hydantoiny Bucherera
R синтез гидантоинов Бухерера(-Берга)

442
BUCHERER REACTION.
Substitution of amino group in primary aromatic amines (mainly naphthylamines) by hydroxyl group, effected by sodium hydrogen sulfite and sodium hydroxide; also reverse process: substitution of hydroxyl group in naphthols by amino group effected by ammonium sulfite and ammonia

D Bucherer-Lepetit-Reaktion, (Bucherer-Lepetit) Naphthol→ Naphthylamin-Umwandlung
F réaction de Bucherer
P reakcja Bucherera
R реакция Бухерера

443
BUFFER (buffer solution).
Solution of weak acid (base) and its salt with strong base (acid) exhibiting constant concentration of hydrogen ions, which remains practically constant in time of dilution or after addition of small quantities of strong acids (bases).
D Puffer, Pufferlösung, Puffergemisch
F solution tampon
P bufor pH, roztwór buforowy pH
R буферный раствор

BUFFER SOLUTION. See BUFFER.

444
BUILD-UP FACTOR.
Ratio of intensity of observed effect of absorption of ionizing radiation to effect of absorption of particles or photons of the primary radiation.
D Zuwachsfaktor
F facteur de correction
P współczynnik narostu
R фактор накопления

445
BULK OF MATERIAL.
Total quantity of material which is estimated on basis of analysis of mean sample.
D Partie
F partie de matériau
P partia materiału
R партия материала

446
BULK POLYMERIZATION.
Polymerization of pure monomer effected in liquid phase with no solvent or diluent added, in presence of catalyst.
D Massenpolymerisation
F polymérisation en masse
P polimeryzacja w masie
R полимеризация в массе, блочная полимеризация

447
BUNSEN ICE-CALORIMETER.
Isothermal calorimeter in which amount of heat evolved or absorbed is determined by change in volumes of water and ice which are in equilibrium inside calorimeter jacket.

D Bunsensches Eiskalorimeter
F calorimètre à glace de Bunsen
P kalorymetr lodowy Bunsena
R ледяной калориметр Бунзена

BUNSEN-ROSCOE LAW. *See* RECIPROCITY LAW.

BURNING. *See* COMBUSTION.

448
BURNT-UP FUEL ELEMENTS (spent fuel elements).
Nuclear fuel elements which have reached final degree of burn-up, as defined for the particular type or batch.

D abgestellte Brennelemente, ausgebrannte Brennelemente
F combustible nucléaire usé
P elementy (paliwowe) wypalone
R отработанные тепловыделяющие элементы, отработанные твэлы

449
BURN-UP.
In radiochemistry, destruction or transformation of specified nuclei as a result of neutron capture reaction.

D Abbrand, Ausbrand
F combustion, consommation
P wypalenie
R выгорание

450
BURN-UP.
In reactor technique, degree of consumption of initial load of reactor fuel as a result of reactor operation.

D Abbrand, Ausbrand
F combustion, consommation
P wypalenie
R выгорание

451
BURN-UP FRACTION.
In radiochemistry fraction, usually expressed as percentage, of initial quantity of nuclei of a given type which has undergone fission.

D Abbrand
F taux d'épuisement
P stopień wypalenia
R степень выгорания

452
BUTTRESSING EFFECT.
Effect typical of atropoisomers —
derivatives of biphenyl and similar systems. Substitution of 3,3′ — and 5,5′ — hydrogen atoms increases stability of atropoisomers (by decreasing racemization rate); e.g. in 2′-methoxy-3′-methyl- -2-nitrodiphenyl-6-carboxylic acid, buttressing effect is exhibited by methyl group

D ...
F ...
P efekt podpory
R поддерживающее влияние

453
BUTYRIC FERMENTATION.
Type of oxidative fermentation of hexoses, e.g. of glucose, caused by *Bacillus butylicus*, leading to decomposition of carbohydrate with formation of *n*-butanol, butyric acid, ethanol, formic acid, acetic acid, lactic acid and carbon dioxide.

D Buttersäuregärung
F fermentation butyrique
P fermentacja masłowa
R маслянокислое брожение

454
CADIOT-CHODKIEWICZ REACTION.
Formation of conjugated acetylenic system by coupling of two asymmetric acetylene derivatives (ethylenic and 1-bromoacetylenic derivatives) in presence of $Cu^{\oplus}$ ions

D Cadiot-Chodkiewicz-Reaktion
F réaction de Cadiot-Chodkiewicz
P reakcja Cadiota i Chodkiewicza
R реакция Кадио-Ходкевича

455
CADMIUM, Cd.
Atomic number 48.
D Cadmium, Kadmium
F cadmium
P kadm
R кадмий

456
CAESIUM (cesium *US*), Cs.
Atomic number 55.
D Caesium, Cäsium, Zäsium
F caesium, césium
P cez
R цезий

457
CAGE EFFECT.
Increased (higher than anticipated from diffusion) recombination of free radicals, originating from the same molecule.
D Käfigeffekt
F effet de cage
P efekt klatkowy, efekt Francka i Rabinovitcha
R эффект „клетки", эффект ячейки

458
CAILLETET AND MATHIAS LAW (law of the rectilinear diameter).
Average density of liquid and its saturated vapour is linear function of temperature; intersection of this straight line with curve describing dependence of density of liquid and saturated vapour on temperature determines density of this substance in critical state.
D Satz von Cailletet und Mathias, Gesetz der Mittellinie
F loi de Cailletet et Mathias, loi du diamètre rectiligne
P reguła Cailleteta i Mathiasa, reguła średniej gęstości
R правило Кальете и Матиаса, правило прямолинейного диаметра

459
CALCIUM, Ca.
Atomic number 20.
D Calcium, Kalzium
F calcium
P wapń
R кальций

CALCIUM-45. *See* RADIOCALCIUM.

460
CALIBRATION ERROR.
Error due to incorrect calibration of instruments.
D Kalibrierungsfehler, Eichfehler
F erreur de justesse
P błąd wzorcowania
R ...

461
CALIFORNIUM, Cf.
Atomic number 98.
D Californium, Kalifornium
F californium

P kaliforn
R калифорний

462
CALOMEL ELECTRODE.
Electrode of second kind, Hg, $Hg_2Cl|Cl^-$ used as reference electrode; electrode reaction is

$$^1/_2 Hg_2Cl_2 + e \rightleftarrows Hg + Cl^-$$

and potential of electrode,

$$\varepsilon = \varepsilon^0 - \frac{RT}{F}\ln a_{Cl^-}$$

where ε^0 = standard electrode potential.
D Kalomelelektrode
F électrode au calomel
P elektroda kalomelowa
R каломельный электрод

463
CALORIC EQUATION OF STATE.
Equation which gives phase internal energy U dependence on its temperature T, volume V and number of moles of its constituents $n_1, n_2, ..., n_c$

$$U = U(T, V, n_1, n_2, ..., n_c)$$

D calorische Zustandsgleichung
F équation calorique d'état
P równanie stanu kaloryczne
R калорическое уравнение состояния

464
15° CALORIE, cal_{15}.
Obsolete energy unit; in SI system $1\ cal_{15} = 4.1855$ J.
D 15-Grad-Kalorie
F calorie de 14,5 à 15,5°C
P kaloria piętnastostopniowa
R пятнадцатиградусная калория

465
CALORIMETER.
Device for measuring amount of heat evolved or absorbed during carrying out a given chemical reaction or physical process inside it.
D Kalorimeter
F calorimètre
P kalorymetr
R калориметр

466
CALORIMETRY.
Set of methods of measurement applied for determination of heat effects accompanying various chemical reactions and physical processes.
D Kalorimetrie
F calorimétrie
P kalorymetria
R калориметрия

467
CANNIZZARO REACTION.
Disproportionation of two molecules of
aldehyde devoid of hydrogen atoms at
α-position (usually aromatic aldehyde),
resulting in formation of mixture of an
appropriate alcohol and acid by action
of concentrated sodium hydroxide
solution, e.g.

$$2C_6H_5CHO \xrightarrow{HO^{\ominus}} C_6H_5COOH + C_6H_5CH_2OH$$

D Cannizzaro-Reaktion, (Cannizzaro)
    Aldehyd-Dismutation
F réaction de Cannizzaro, dismutation
    selon Cannizzaro
P reakcja Cannizzaro
R реакция Канниццаро

CANONICAL ASSEMBLY. *See*
CANONICAL ENSEMBLE.

CANONICAL DISTRIBUTION. *See*
CANONICAL ENSEMBLE.

468
CANONICAL ENSEMBLE (canonical
assembly, canonical distribution).
Statistical ensemble given by following
probability distribution function

$$P(Q) = \exp\left[\frac{F - E(Q)}{kT}\right]$$

where $E(Q)$ = energy of system in state Q,
k = Boltzmann constant,
T = thermodynamic temperature,
and F = free energy. Canonical ensemble
is utilized for statistical description of
thermodynamic equilibrium in
conservative systems.

D kanonische Gesamtheit, kanonische
    Verteilung
F ensemble canonique de Gibbs
P zespół kanoniczny, rozkład kanoniczny
R канонический ансамбль Гиббса,
    каноническое распределение (Гиббса)

CANONICAL STRUCTURES. *See*
RESONANCE STRUCTURES.

469
CAPILLARITY (capillary action).
Rise or fall of meniscus of liquid in
capillary tubes.
D Kapillarität
F capillarité
P włoskowatość, kapilarność
R капиллярность

CAPILLARY ACTION. *See*
CAPILLARITY.

470
CAPILLARY COLUMN (Golay column).
In gas chromatography, column with small
diameter, without packing in which layer
of liquid held at inner walls of column
serves as stationary phase.
D Kapillarsäule, Golay-Säule
F colonne capillaire
P kolumna kapilarna, kolumna Golaya
R капиллярная колонка, колонка Голея

471
CAPILLARY CONDENSATION.
Condensation of vapours in small pores
of adsorbent (preferentially in its
capillaries) due to fact that saturated
vapour pressure of liquid adsorbate
in capillary tube is lower than that above
its plane surface.
D Kapillarkondensation
F condensation capillaire
P kondensacja kapilarna
R капиллярная конденсация

472
CAPTURE CROSS-SECTION.
Cross-section for radiative capture of
nucleon by nucleus.
D Einfangsquerschnitt
F section efficace de capture
P przekrój czynny na wychwyt
R сечение захвата

473
CARATHÉODORY'S PRINCIPLE
(principle of inaccessibility).
For every state there are other states
which cannot be reached from this state
via adiabatic process.
D Carathéodory-Unerreichbarkeits-Axiom,
    Prinzip von Carathéodory,
    Carathéodorysches Postulat
F principe de Carathéodory
P zasada Carathéodory'ego
R принцип адиабатической
    недостижимости Каратеодори

474
CARBANION.
Anion with negatively charged carbon
atom

D Carbeniat-Ion, Carbanion
F carbanion
P karboanion
R карбанион

475
CARBENE.
Electrically neutral molecule of general
formula :CX₂, i.e. containing carbon atom
with six electrons in outer sphere, two
of them unshared. Chemical properties
of a carbene depend on its electronic
structure. When spins of non-bonding
electrons are anti-parallel, electronic
structure of carbene corresponds to less
stable but more reactive singlet state

. When spins of electrons are

parallel, electronic structure corresponds
to more stable but less reactive triplet

state ↑C⟨

D Carben
F carbène
P karben
R карбен

476
CARBENOID.
Transient intermediate reacting under
certain conditions in a way typical for
a carbene.
D ...
F ...
P karbenoid
R ...

477
CARBO-CATION (carbonium ion).
Cation with positively charged carbon
atom

R′ R″
⟍⊕⟋
C
│
R‴

D Carbeniumion, Carboniumion
F carbocation
P karbokation, jon karboniowy
R ион карбония, карбониевый ион

478
CARBOCYCLIC RING.
Isocyclic ring composed of carbon atoms.
D carbocyclischer Ring
F noyau carbocyclique
P pierścień karbocykliczny
R карбоциклическое кольцо

479
CARBON, C.
Atomic number 6.
D Kohlenstoff
F carbone
P węgiel
R углерод

480
CARBON-12 ATOMIC MASS SCALE
(carbon-12 scale of atomic weights).
Scale of atomic masses in use since 1961
with 1/12 of the ¹²C nuclide mass as unit
of mass.
D vereinheitlichte relative
Atommassenskala, Kohlenstoff-12
Atomgewichtsskala
F échelle des masses atomiques de
carbone-12, échelle des poids atomiques
de carbone-12
P skala mas atomowych węglowa
R углеродная атомная шкала

481
CARBON CHAIN.
System of mutually linked carbon atoms
forming skeleton of compound, e.g.

D Kohlenstoffkette
F chaîne de carbone, chaîne carbonée
P łańcuch węglowy
R углеродная цепь

CARBON FAMILY. *See* CARBON
GROUP.

482
CARBON GROUP (carbon family).
Elements of fourth main group in
periodic system with structure of outer
electronic shells of atoms ns^2np^2: carbon,
silicon, germanium, tin, lead.
D Kohlenstoff-Silicium-Gruppe
F groupe du carbone, famille du carbone
P węglowce
R группа углерода, элементы
(под)группы углерода

CARBONIUM ION. *See* CARBO-CATION.

483
CARBONIUM-ION REARRANGEMENT.
Rearrangement with carbonium ion as an
intermediate.
D ...
F transposition carbo-cationique
P przegrupowanie karbokationowe
R секстетная перегруппировка

484
CARBON-NITROGEN CYCLE (Bethe
cycle).
Series of thermonuclear reactions leading
to synthesis of helium from hydrogen.
In the cycle, nuclei of carbon and nitrogen
are involved.

$$^{12}C(p, \gamma) \ ^{13}N(\beta^+v) \ ^{13}C(p, \gamma) \ ^{14}N(p, \gamma) \rightarrow$$
$$^{15}O(\beta^+v) \ ^{15}N(p, \alpha) \ ^{12}C$$

According to Bethe, carbon-nitrogen chain
is source of energy of stars.

D CN-Zyklus, Bethe-(Weizsäcker-)Zyklus
F cycle de Bethe, cycle du carbone
P cykl węglowo-azotowy
R углеродно-азотный цикл,
   углеродный цикл

CARBON-12 SCALE OF ATOMIC
WEIGHTS. *See* CARBON-12 ATOMIC
MASS SCALE.

485
CARBONYLATION.
Addition of carbon monoxide molecule
to molecule of organic or inorganic
compound, e.g.

$$\begin{array}{c}\diagup\\C=C\\\diagdown\end{array} + CO + H_2 \xrightarrow[\substack{30 \text{ atm}\\ \text{catalyst}}]{200°} \begin{array}{c}\diagup\\CH-\overset{|}{\underset{|}{C}}-C\diagup^O_H\\\diagdown\end{array}$$

D Carbonylierung
F ...
P karbonylowanie
R карбонилирование

486
CARBOXYLATION.
Introduction of carboxylic group —COOH
into molecule of organic compound usually
effected by action of carbon dioxide on
metalorganic molecule with subsequent
decomposition of formed salt, e.g.

$$RMgX \xrightarrow{CO_2} RCOOMgX \xrightarrow{H_2O, H^{\oplus}} RCOOH$$

$$RLi \xrightarrow{CO_2} RCOOLi \xrightarrow{H_2O, H^{\oplus}} RCOOH$$

D Carboxylierung
F carboxylation
P karboksylowanie
R карбоксилирование

487
CARBOXYMETHYLATION.
Substitution of hydrogen atom (bound
with carbon, nitrogen or oxygen atom)
of organic compound molecule by
carboxymethyl group —CH₂COOH, e.g.

$$\underset{\overset{|}{O}}{RCCH_3} + ClCH_2COOR' \xrightarrow{NaNH_2} \underset{\overset{|}{O}}{RCCH_2CH_2COOR'}$$

$$\underset{CH_2NH_2}{CH_2NH_2} \xrightarrow[H_2O/NaOH]{HCHO + NaCN} \underset{CH_2N(CH_2COOH)_2}{CH_2N(CH_2COOH)_2}$$

D *(Einführung der Carboxymethylgruppe)*
F ...
P karboksymetylowanie
R ...

488
CARBYLAMINE REACTION (isocyanide
reaction).
Reaction of primary amines with
chloroform and sodium (or potassium)

hydroxide, involving isonitrile formation
(manifested by very unpleasant smell);
applied for detection of amino group, e.g.

$$RNH_2 + CHCl_3 + 3NaOH \longrightarrow$$

$$R-\overset{\oplus}{N}\equiv\overset{\ominus}{C} + 3NaCl + 3H_2O$$

D Isonitrilreaktion
F ...
P reakcja izonitrylowa
R карбиламинная реакция

489
CARNOT CYCLE.
Cyclic process proceeding through four
subsequent steps: (1) adiabatic expansion
leading to lowering of temperature,
(2) isothermal compression at this lower
temperature, (3) adiabatic compression
leading to initial temperature, and
(4) isothermal expansion leading to initial
state.

D Carnot Prozeß, Carnotscher Kreisprozeß
F cycle de Carnot
P obieg Carnota, cykl Carnota
R цикл Карно

490
CARNOT'S PRINCIPLE.
Yield of Carnot cycle is the same for all
substances. Carnot's principle follows from
second law of thermodynamics.

D Carnotscher Satz
F théorème de Carnot
P teoremat Carnota i Clausiusa
R теорема Карно-Клаузиуса, первая
   теорема Карно

491
CARRIER.
Substance introduced into system in
macro-amounts in order to separate
or retain quantitatively a microcomponent
(e.g. impurities, radioactive substances)
which undergoes the same transformations
in the system as the macrocomponent.

D Träger
F entraîneur
P nośnik
R носитель

CARRIER. *See* ELECTRIC CURRENT
CARRIER.

492
CARRIER DISTILLATION.
In spectroanalysis fractional distillation in
presence of spectroscopic carrier.

D Trägerdestillation
F distillation avec un entraîneur
P destylacja nośnikowa
R фракционная дистилляция с носителем

493
CARRIER-FREE RADIOACTIVE
PREPARATION.
Radioactive preparation which does not
contain non-radioactive isotopes. It usually
has a high specific activity.
D trägerfreies radioaktives Präparat
F préparation radioactive sans entraîneur
P preparat promieniotwórczy
  beznośnikowy
R радиоактивный препарат без носителя

494
CARRIER GAS (in chromatography).
Gas used to elute sample from column.
D Trägergas
F gaz porteur
P gaz nośny, gaz wymywający
R газ-носитель

CASCADE PARTICLE. See XI HYPERON.

495
CATALYSIS.
Phenomenon of acceleration or retardation
of chemical reaction due to presence of
catalyst.
D Katalyse
F catalyse
P kataliza
R катализ

496
CATALYST.
Element or compound that changes rate
of chemical reaction but remains
unchanged in chemical composition after
reaction.
D Katalysator
F catalyseur
P katalizator
R катализатор

497
CATALYST ACTIVITY (catalytic
activity).
Quantitative expression of ability of
catalyst to change rate of reaction.
D Aktivität des Katalysators, katalytische
  Aktivität, katalytische Wirksamkeit
F activité du catalyseur, activité
  catalytique
P aktywność katalizatora, aktywność
  katalityczna
R активность катализатора,
  каталитическая активность

CATALYST CARRIER. See CATALYST
SUPPORT.

498
CATALYST POISON (catalytic poison).
In heterogeneous catalysis, substance
adsorbed at surface of catalyst, blocking
access of substrates and therefore

diminishing, or even destroying activity
of catalyst.
D Katalysatorgift, Kontaktgift
F poison du catalyseur
P trucizna katalizatora
R катализаторный яд, контактный яд

499
CATALYST POISONING.
In heterogeneous catalysis, loss of catalyst
activity, due to action of catalytic poison.
D Katalysatorvergiftung
F empoisonnement du catalyseur
P zatrucie katalizatora
R отравление катализатора

500
CATALYST PROMOTER.
In heterogeneous catalysis a substance
which is not a catalyst or has weak
catalytic properties, but which in mixture
with a catalyst causes growth of its
catalytic activity (sometimes also its
selectivity or persistence).
D Aktivator, Verstärker, Promotor
F promoteur (en catalyse)
P promotor, aktywator katalizatora
R промотор (катализатора), активатор

501
CATALYST SUPPORT (catalyst carrier).
Substance which is not active catalytically,
usually of large specific surface, acting as
support for catalyst.
D Katalysatorträger
F support de catalyseur
P nośnik katalizatora
R носитель катализатора

CATALYTIC ACTIVITY. See CATALYST
ACTIVITY.

502
CATALYTIC COEFFICIENT (catalytic
constant).
Quotient of rate constant and
concentration of catalyst.
D Katalysekoeffizient
F ...
P stała katalityczna (szybkości reakcji)
R каталитический коэффициент,
  константа скорости каталитической
  реакции

CATALYTIC CONSTANT. See
CATALYTIC COEFFICIENT.

503
CATALYTIC CURRENT.
In polarography and related methods,
current which results from catalytic
chemical reaction which transforms
product of electrode reaction into
substrate.
D katalytischer Strom
F courant catalytique

P prąd katalityczny
R каталитический ток

CATALYTIC POISON. *See* CATALYST
POISON.

504
CATALYTIC REACTION (catalyzed
reaction).
Reaction proceeding with participation
of catalyst.
D katalytische Reaktion
F réaction catalytique
P reakcja katalityczna, reakcja
katalizowana
R каталитическая реакция

CATALYZED REACTION. *See*
CATALYTIC REACTION.

505
CATAPHORESIS.
Electrophoresis in which particles of
dispersed phase migrate to cathode.
Remark: Term "cataphoresis" is sometimes
incorrectly used in wider sense of
"electrophoresis".
D Kataphorese
F cataphorèse
P kataforeza
R катафорез

506
CATENANES.
Compounds whose molecules are
constructed of two or more large rings
(usually saturated) having no common
carbon atoms (or atoms of any other
element); such rings form chain-link
structures, e.g.

D Catenane
F caténanes
P katenany
R катенаны

507
CATHODE.
Electrode at which reduction process take
place; (a) in galvanic cell — positive
electrode, (b) in electrolytic cell —
electrode connected to negative pole of
external current source.
D Kathode
F cathode
P katoda
R катод

CATHODE-RAY POLAROGRAPHY. *See*
OSCILLOPOLAROGRAPHY.

508
CATHODIC CURRENT ⌐ENSITY
(cathodic partial current density).
Current density which flows across
metallic conductor-electrolyte phase
boundary, equal to rate of reduction at
cathode.
D kathodische Teilstromdichte
F densité du courant cathodique
P gęstość prądu katodowego (cząstkowego)
R плотность (частного) катодного тока

CATHODIC PARTIAL CURRENT
DENSITY. *See* CATHODIC CURRENT
DENSITY.

509
CATHOLYTE.
Part of electrolyte in vicinity of cathode
in which changes of physico-chemical
properties occur during electrolysis or
discharge of galvanic cell; if solutions
in vicinity of both electrodes of a cell are
separated, e.g. by membranes it is that
part of electrolyte surrounding cathode.
D Katholyt, Kathodenlösung,
Kathodenflüssigkeit
F solution cathodique, catholyte
P katolit
R католит

510
CATION.
Ion with positive charge.
D Kation
F cation
P kation
R катион

511
CATION EXCHANGER.
Ion exchanger containing cations that can
be exchanged for other cations from
solution.
D Kationenaustauscher
F échangeur de cations
P kationit, wymieniacz kationów
R катионит, катионообменник

512
CATIONOTROPY.
Reversible migration of cation from one
position to another in molecule.
D Kationotropie
F cationotropie
P kationotropia
R катионотропия

513
CAVITY FIELD.
Part of local electric field in cavity in
dielectric (obtained by removing
a molecule) due to influence of external
electric field.
D Richtungsfeld
F champ de cavité
P pole wnękowe (elektryczne)
R поле полости

514
CELL CONSTANT OF
CONDUCTOMETRIC VESSEL (vessel
constant), k.
Characteristic parameter for vessels used
in comparative measurements, of specific
conductivity of electrolyte solutions:
$k = R\varkappa$, where R and $\varkappa$ = resistance and
specific conductivity of standard solution
of KCl respectively.
D Zellkonstante des Leitfähigkeitgefäßes
F constante de la cellule de conductivité
P pojemność oporowa naczynka, stała
naczynka
R постоянная сосуда, ёмкость
сопротивления сосуда, постоянная
ячейки, константа ячейки

515
CENTIGRAM METHOD (semimicro
method).
Method of analysis where mass of
investigated sample is $x \cdot 10^{-2}$ g and in the
case of gases volume is $x \cdot 10^0$ cm³.
D Centigramm-Methode,
Halbmikro-Methode
F méthode centigrammique,
semi-microméthode
P metoda centygramowa, półmikrometoda
R полумикрометод

516
CENTRAL ATOM (nuclear atom).
In coordination chemistry, atom or ion
in complex around which ligands are
coordinated.
D Zentralatom, Kernatom
F atome central
P atom centralny
R центральный атом

517
CENTRAL FORCE FIELD.
Force field acting in direction of single
fixed point, e.g. gravitational field.
D Zentral(kraft)feld
F champ des forces centrales
P pole sił centralnych
R поле центральных сил, центральное
поле

518
CENTRAL ION (central metal ion).
Ion around which ligands are coordinated.

D Zentralion
F ion central
P jon centralny
R центральный ион

CENTRAL METAL ION. See CENTRAL
ION.

519
CENTRE FOR COORDINATION.
Ion or atom that coordinates ligands.
D Koordinationszentrum
F centre coordinateur
P ośrodek koordynacji, centrum
koordynacji, rdzeń kompleksu
R комплексообразователь

520
CENTRIFUGAL BARRIER.
Part of kinetic energy in system of two
particles, related to their relative
circulation (non-zero angular momentum).
Centrifugal barrier is a potential barrier
preventing mutual approach of particles.
D Zentrifugalpotentialwall
F barrière centrifuge
P bariera odśrodkowa, bariera energii
odśrodkowej
R барьер центробежной энергии

521
CERENKOV COUNTER.
Particle counter in which Cerenkov light
is observed.
D Čerenkov-Zähler, Tscherenkow-Zähler
F compteur de Tcherenkov
P licznik Czerenkowa
R счётчик Черенкова

CERENKOV EFFECT. See CERENKOV
RADIATION.

522
CERENKOV RADIATION (Cerenkov
effect).
Polarized electromagnetic (visible)
radiation produced when charged particle
traverse a transparent medium with
velocity exceeding velocity of light in the
medium.
D Čerenkov-Strahlung,
Tscherenkow-Strahlung
F rayonnement de Mallet-Tcherenkov
P promieniowanie Czerenkowa
R излучение Вавилова-Черенкова

523
CERIUM, Ce.
Atomic number 58.
D Cer(ium), Zer(ium)
F cérium
P cer
R церий

524
CERIUM EARTHS.
Oxides of lanthanium, cerium,
praseodymium, neodymium and samarium.
D Ceriterden
F terres cériques
P ziemie cerytowe
R цериевые земли

CESIUM. *See* CAESIUM.

525
CHAIN BRANCHING.
Formation of two or more chain carriers
in reaction of chain initiation or in
elementary reaction of chain propagation.
D Kettenverzweigung
F branchement de chaîne, ramification de
la chaîne
P rozgałęzienie łańcucha reakcji
R разветвление цепи

CHAIN BREAKING REACTION. *See*
CHAIN TERMINATION.

526
CHAIN CARRIER (chain propagator).
Active variety of molecule (e.g. free
radical) which is recreated many times in
subsequent links of chain and secures
progression of chain reaction.
D Kettenreaktionsträger
F propagateur de chaîne, porteur de
chaîne
P nośnik łańcucha reakcji, propagator
łańcucha, kontynuator łańcucha
R активная частица, переносчик реакции

527
CHAIN EXPLOSION.
Explosion caused by chain-like process.
D ...
F explosion en chaîne
P wybuch łańcuchowy
R цепной взрыв

CHAIN INDUCTOR. *See* CHAIN
INITIATOR.

528
CHAIN INITIATION.
Elementary reaction resulting in formation
of first carrier of reaction chain.
D Kettenstartreaktion, Startreaktion,
Kettenstart
F initiation de la chaîne, réaction
inductrice (de la chaîne)
P inicjowanie łańcucha reakcji,
zapoczątkowanie łańcucha reakcji
R зарождение цепи, реакция зарождения
цепи, инициирование

529
CHAIN INITIATOR (chain inductor).
Substance which enters into elementary
reaction with substrate, thus initiating
chain.
D Initiator der Kettenreaktion
F initiateur d'une réaction en chaîne
P inicjator reakcji łańcuchowej
R инициатор цепной реакции

530
CHAIN ISOMERISM.
Type of structural isomerism caused
by different structure of carbon chain
in isomeric compounds, e.g. butane and
isobutane.
D Kettenisomerie
F isomérie de la chaîne
P izomeria łańcuchowa
R изомерия скелета

531
CHAIN LENGTH.
Number of unit steps in reaction initiated
by one carrier of chain of reactions,
formed during initiation stage.
D Kettenlänge
F longueur de la chaîne cinétique
P długość łańcucha reakcji, długość
łańcucha kinetycznego
R длина цепи

532
CHAIN LENGTH OF POLYMER.
Total chain length of linear
macromolecule, measured from one atom
to another along chain.
D Kettenlänge (des Polymers)
F longueur de chaîne (du polymère)
P długość łańcucha (polimeru)
R длина (полимерной) цепи

533
CHAIN OF REACTIONS.
Train of elementary unit reactions
initiated by one carrier of reaction chain.
D Reaktionskette
F chaîne (cinétique) de réaction
P łańcuch reakcji, łańcuch kinetyczny
R реакционная цепь

534
CHAIN PROPAGATION.
Multiple reproduction of chain carrier.
D Kettenentwicklung
F propagation de la chaîne
P rozwijanie łańcucha reakcji
R развитие цепи

CHAIN PROPAGATOR. *See* CHAIN
CARRIER.

535
CHAIN REACTION.
Train of elementary unit reactions
comprising initiation, propagation and
termination of chain of reactions.
D Kettenreaktion
F réaction en chaîne
P reakcja łańcuchowa
R цепная реакция

536
CHAIN TERMINATION (chain breaking
reaction).
Reaction resulting in chemical change of
propagation carrier of chain into
non-reactive product.
D Kettenabbruch(reaktion)
F terminaison de chaîne, rupture de
  chaîne
P zakończenie łańcucha reakcji
R реакция обрыва цепи, обрыв цепи

537
CHAIR CONFORMATION (chair form).
Privileged conformation in saturated
six-membered carbocyclic and heterocyclic
compounds, e.g. cyclohexane. In chair
conformation carbon bond angles maintain
value of 109.5°, C—C bond system at
adjacent carbon atoms is staggered. In this
conformation each carbon atom has one
axial (a) and one equatorial (e) bond

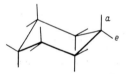

D Sesselform
F conformation chaise, forme chaise
P konformacja krzesłowa
R конформация кресла, форма кресла

CHAIR FORM. See CHAIR
CONFORMATION.

538
CHALCOGENS.
Elements in sixth main group of periodic
system with structure of outer electronic
shells of atoms ns^2np^4: oxygen, sulfur,
selenium, tellurium, polonium.
D Sauerstoff-Schwefel-Gruppe
F famille de l'oxygène, chalcogènes
P tlenowce
R группа кислорода, элементы
  (под)группы кислорода

539
CHAMIÉ AGGREGATES.
Aggregates of radioactive substance
indicated by unequally distributed
darkenings of the radiograms of gases
or liquids probably formed due to

adsorption of radioactive substance on
surface of a photographic emulsion.
D radioaktive Aggregate
F agrégats de Chamié
P agregaty Chamié
R агрегаты Шамье

540
CHAPMAN-ENSKOG METHOD.
Method for solution of Boltzmann
equation based on assumption that
unimolecular partition function depends
on time only through time dependence of
temperature, density and local velocity.
D Chapman-Enskog-Methode
F méthode d'Enskog et Chapman
P metoda Chapmana i Enskoga
R метод Энскога-Чепмена

541
CHAPMAN REARRANGEMENT.
Intramolecular rearrangement of
iminoesters into N-disubstituted amides,
effected by heating the former to 200°C

$$Ar-C=N-Ar' \xrightarrow{\Delta} Ar-C-N-Ar'$$
$$\underset{O-R}{|} \qquad\qquad \overset{\|}{O}\ \overset{|}{R}$$

D Chapman (Imidoester→Amid)-
  Umlagerung
F transposition de Chapman
P przegrupowanie Chapmana
R перегруппировка Чэпмена

542
CHARACTERISTIC COORDINATION
NUMBER.
Coordination number characteristic for
given central atom, generally equal to
maximum coordination number with some
exceptions, e.g. for Cu^{2+} ion it is 4, and
maximum coordination number is 6.
D charakteristische Koordinationszahl
F nombre de coordination caractéristique
P liczba koordynacyjna charakterystyczna
R ...

CHARACTERISTIC CURVE. See
EMULSION CALIBRATION CURVE.

543
CHARACTERISTIC FREQUENCY (Debye
frequency).
Maximum frequency of elastic wave in
crystal at defined temperature.
D charakteristische Frequenz
F fréquence caractéristique
P częstość charakterystyczna (Debye'a)
R характеристическая частота

544
CHARACTERISTIC FREQUENCY (in
spectroscopy).
Oscillation frequency for a part of
molecule relatively independent of
vibrations of other fragments of molecule.
D charakteristische Absorptionsfrequenz

F fréquence caractéristique
P częstość charakterystyczna (układu drgającego)
R характеристическая частота

545
CHARACTERISTIC FUNCTIONS (thermodynamic potentials, thermodynamic functions).
Extensive functions of state parameters such that all thermodynamic properties can be expressed in terms of them and their derivatives with respect to state parameters, which are arguments of these functions.
D charakteristische Funktionen, thermodynamische Potentiale
F fonctions caractéristiques, potentiels thermodynamiques
P funkcje charakterystyczne, potencjały termodynamiczne
R характеристические функции, термодинамические потенциалы

CHARACTERISTIC PHOTOGRAPHIC CURVE. *See* EMULSION CALIBRATION CURVE.

546
CHARACTERISTIC TEMPERATURE, Θ.
Temperature at which energy $(k\Theta)$ is equal to energy of elastic wave in crystal with frequency maximum ω_D at a given temperature, $k\Theta = h\omega_D$, where k = Boltzmann constant, and h = Planck constant.
D charakteristische Temperatur
F température caractéristique
P temperatura charakterystyczna (Debye'a)
R характеристическая температура (Дебая)

547
CHARACTERISTIC X-RADIATION.
Discrete electromagnetic radiation of spectrum characteristic for the element that emits it. It is connected with rearrangement of electrons in inner shell of atoms.
D charakteristische Röntgenstrahlung
F rayonnement X caractéristique
P promieniowanie rentgenowskie charakterystyczne
R характеристическое рентгеновское излучение, характеристические X-лучи

548
CHARACTERISTIC X-RAY SPECTRUM.
Emission spectrum with lines corresponding to transitions between inner shell atomic energy levels.
D charakteristisches Röntgenspektrum
F spectre discontinu des rayons X
P widmo rentgenowskie charakterystyczne
R рентгеновский характеристический спектр

549
CHARGE CLOUD.
Spatial distribution of electric charge of particle.
D Ladungswolke, Ladungsverteilung
F nuage de charge
P chmura ładunku
R зарядное облако

550
CHARGE DENSITY (*of particle*).
Product of particle charge and corresponding probability density.
D Ladungsdichte
F densité de charge
P gęstość ładunku
R плотность заряда

551
CHARGED PARTICLE ACCELERATOR (accelerator).
Device for acceleration of electrons or other charged particles in electric or electromagnetic fields.
D Beschleuniger, Teilchenbeschleuniger
F accélérateur (de particules chargées)
P akcelerator
R ускоритель (заряженных частиц)

552
CHARGED-PARTICLE ACTIVATION ANALYSIS.
Activation method of analysis in which a stream of charged particles, e.g. protons, deuterons, α-particles is used for activation.
D Aktivierungsanalyse mit geladenen Kernteilchen, Aktivierungsanalyse mit schnellen Ionen
F analyse par activation aux particules chargées
P analiza aktywacyjna z zastosowaniem cząstek naładowanych
R активационный анализ с применением заряженных частиц

553
CHARGE MULTIPLETS (isomultiplets).
Families of elementary particles having close values of mass, identical baryon number B, strangeness S, lepton number L, quantum spin number J, but differing in electrical charge. Value of isospin I of particles belonging to given charge multiplet is defined by formula $n = 2I+1$, where n = number of particles belonging to given charge multiplet (e.g. nucleons have $I = 1/2$; there exists an isospin proton-neutron doublet).
D Isospin-Multipletts
F multiplets de charge
P multiplety izospinowe, multiplety ładunkowe
R зарядовые мультиплеты

554
CHARGE-TRANSFER BAND (CT-band).
In electronic spectra of molecules and intermolecular complexes band corresponding to electronic transition with substantial transfer of charge between molecular fragments or complex components; e.g. in spectra of transition metal complexes transfer of charge from ligands to central ion or vice versa.
D Ladungsaustauschband, CT-Band
F bande de transfert de charge
P pasmo przeniesienia ładunku, pasmo CT
R полоса переноса заряда

CHARGE-TRANSFER COMPLEX. See DONOR-ACCEPTOR COMPLEX.

555
CHARGING CURRENT.
Current arising from charging of electrical double layer to a given potential.
D Kapazitätsstrom
F courant capacitif
P prąd pojemnościowy
R ток заряжения

556
CHARLES'S LAW (Gay-Lussac's law).
At constant pressure, volume V_t of a given mass of gas depends on temperature according to formula

$$V_t = V_0(1+\alpha t) \quad \text{or} \quad V_t = \frac{V_0}{273.15 \text{ K}} T$$

where V_0 = volume of gas at the given pressure at temperature 273.15 K ($= 0°C$),
t = (empirical) temperature in °C,
T = thermodynamic temperature,
α = expansion coefficient of gas.
Gay-Lussac's law is empirical.
D Gay-Lussacsches Gesetz
F loi de Gay-Lussac
P prawo Gay-Lussaca
R закон Гей-Люссака

557
CHELATE (chelate complex, chelate compound).
Complex in which metal is incorporated in ring formed by coordination of multidentate ligand.
D Chelatkomplex, Chelatverbindung, Metallchelat
F (composé) chélate
P kompleks chelatowy, związek chelatowy, związek kleszczowy
R циклический комплекс, хелат

558
CHELATE BONDS.
System of bonds in complex compounds in which central atom is linked with ligands and forms chelate rings, most often five- or six-membered.
D Chelatbindungen
F ...
P wiązania chelatowe, wiązania kleszczowe
R клещевые связи

CHELATE COMPLEX. See CHELATE.

CHELATE COMPOUND. See CHELATE.

559
CHELATE EFFECT.
Increase in stability of chelate complex, compared with corresponding non-chelate complex, as result of chelate ring formation.
D Chelateffekt
F effet de la chélation
P efekt chelatowy
R циклический эффект

560
CHELATE LIGAND.
Ligand attached to one central atom through two or more coordinating atoms.
D ...
F ...
P ligand chelatowy
R ...

561
CHELATE RING.
Heterocyclic ring formed by linking central atom with multidentate ligand.
D Chelatring
F cycle chélaté
P pierścień chelatowy
R хелатное кольцо

562
CHELATE RING FORMATION.
Formation of rings by coordination of multidentate ligand with central atom.
D Chelatbildungsreaktion
F chélation
P chelatowanie, chelatacja
R циклообразование

563
CHELATING AGENT.
Organic compound forming chelate complexes with cations.
D Chelatligand
F agent chélatant
P (od)czynnik chelatujący
R циклообразующий агент, хелатообразующий лиганд

CHELATING RESIN. See SPECIFIC ION EXCHANGER.

CHELATOMETRIC TITRATION. See CHELATOMETRY.

564
CHELATOMETRY (chelatometric titration).

Branch of compleximetry with compounds (usually aminopolycarboxylic acids or salts) forming water-soluble chelates with titrated ion.
D Chelatometrie
F chélatométrie
P chelatometria
R хелатометрия

CHEMICAL ADSORPTION. See CHEMISORPTION.

565
CHEMICAL ANALYSIS.
Identification and determination of components (elements, ions, radicals, molecules) in analysed sample.
D chemische Analyse
F analyse chimique
P analiza chemiczna
R химический анализ

566
CHEMICAL BOND.
Bond between atoms in a molecule.
D chemische Bindung
F liaison chimique
P wiązanie chemiczne
R химическая связь

567
CHEMICAL BOND ENERGY (bond energy).
For diatomic molecule bond energy is equal to dissociation energy into atoms in their ground states. In polyatomic molecule bond energy cannot be defined in unique way and is defined as part of total binding energy corresponding to given bond. Chemical bond energy is obtained experimentally by using indirect procedures.
D (chemische) Bindungsenergie
F énergie de liaison chimique
P energia wiązania chemicznego
R энергия химической связи

568
CHEMICAL CELL.
Galvanic cell in which electrical work results from chemical work.
D chemische Zelle, chemische Kette
F pile chimique, cellule chimique
P ogniwo chemiczne
R химическая цепь

569
CHEMICAL COMPOUND.
Homogeneous substance consisting of two or more chemical elements combined with one another in some ordered way and usually in definite quantitative proportions.
D chemische Verbindung
F composé

P związek chemiczny, połączenie chemiczne
R химическое соединение

570
CHEMICAL CONSTITUTION (constitution).
Arrangement of atoms or groups of atoms and determination of types of bonds present in a given molecule.
D Konstitution
F constitution
P budowa (chemiczna)
R строение

571
CHEMICAL DOSIMETRY.
Branch of radiation technique dealing with determination of absorbed doses of ionizing radiation on basis of chemical changes in system.
D chemische Dosimetrie
F dosimétrie chimique
P dozymetria chemiczna
R химическая дозиметрия

572
CHEMICAL EFFECTS OF NUCLEAR TRANSFORMATIONS.
Chemical reactions occurring with participation of hot atoms formed as a result of nuclear reactions or radioactive decay.
D radiochemische Rückstoßeffekte
F effets chimiques associés aux transformations nucléaires
P efekty chemiczne przemian jądrowych
R химические последствия ядерных превращений

573
CHEMICAL ELEMENT (element).
Set of atoms of the same atomic number.
D (chemisches) Element
F élément
P pierwiastek (chemiczny)
R (химический) элемент

574
CHEMICAL EQUATION (stoichiometric equation).
Shortened form of notation of chemical reaction course by placing substrate formulae on left and product formulae on right. Each formula is supplied by numerical factor determining number of reacting particles.
D chemische Gleichung, Reaktionsgleichung, stöchiometrische Gleichung
F équation chimique, équation stœchiométrique
P równanie chemiczne, równanie reakcji, równanie stechiometryczne
R химическое уравнение, уравнение реакции, стехиометрическое уравнение

575
CHEMICAL EQUILIBRIUM (reaction equilibrium).
Thermodynamic state with vanishing affinity of reaction.
D chemisches Gleichgewicht, Reaktionsgleichgewicht
F équilibre chimique
P równowaga chemiczna, równowaga reakcji (trwała)
R химическое равновесие

576
CHEMICAL EQUIVALENT.
Ratio of relative atomic mass to valence of a given element.
D chemisches Äquivalent
F équivalent (chimique)
P równoważnik chemiczny
R химический эквивалент

CHEMICAL FACTOR. See ANALYTICAL FACTOR.

577
CHEMICAL FORMULA.
Abbreviated notation which shows quantitatively and qualitatively composition of chemical compound and sometimes also its electronic structure and spatial arrangement.
D chemische Formel
F formule chimique
P wzór chemiczny
R химическая формула

CHEMICAL IMPERFECTIONS. See IMPURITIES.

CHEMICAL INDIVIDUAL. See SUBSTANCE.

578
CHEMICAL INDUCTION.
Incitement or acceleration of induced (secondary) reaction by primary reaction in system of coupled reactions.
D chemische Induktion
F induction chimique
P indukcja chemiczna
R химическая индукция

579
CHEMICAL KINETICS (reaction kinetics).
Branch of physical chemistry dealing with investigation of chemical reaction rates.
D Reaktionskinetik, chemische Kinetik
F cinétique chimique, cinétique des réactions
P kinetyka chemiczna
R химическая кинетика

580
CHEMICALLY PURE REAGENT.
D chemisch reines Reagens
F ...

P odczynnik chemicznie czysty
R химически чистый реактив

581
CHEMICAL PHYSICS.
Branch of physics dealing with experimental and theoretical research of chemical species with emphasis put on physical aspects of all problems under consideration.
D chemische Physik
F physique chimique
P fizyka chemiczna
R химическая физика

CHEMICAL POLARIZATION. See REACTION POLARIZATION.

582
CHEMICAL POTENTIAL (of component in phase), μ_B.
Intensive quantity

$$\mu_B = \left(\frac{\partial U}{\partial n_B}\right)_{S,V,n_c} = \left(\frac{\partial H}{\partial n_B}\right)_{S,p,n_c} =$$

$$= \left(\frac{\partial F}{\partial n_B}\right)_{T,V,n_c} = \left(\frac{\partial G}{\partial n_B}\right)_{T,p,n_c}$$

where n_B = number of moles of B component in phase, n_C = number of moles of other components in this phase, U, H, F, and G = internal energy, enthalpy, Helmholtz free energy and Gibbs free energy, respectively, S = entropy, V = volume of phase, T = its thermodynamic temperature and p = pressure.
D chemisches Potential
F potentiel chimique
P potencjał chemiczny
R химический потенциал

583
CHEMICAL REACTION.
Thermodynamic process with some of the components vanishing and replaced by formation of new ones.
D chemische Reaktion
F réaction chimique
P reakcja chemiczna
R химическая реакция

CHEMICAL REACTION RATE. See REACTION RATE.

CHEMICAL REACTIVITY INDICES. See REACTIVITY INDICES.

584
CHEMICAL RELAXATION.
Establishment of chemical equilibrium previously perturbed e.g. by change of temperature, pressure, electric field.
D chemische Relaxation
F relaxation chimique

P relaksacja chemiczna
R химическая релаксация

585
CHEMICAL RHEOLOGY.
Branch of rheology concerned with study of changes in viscoelastic properties brought about by chemical reactions.
D Chemorheologie
F rhéologie chimique
P chemoreologia
R химическая реология

586
CHEMICAL SCALE OF ATOMIC WEIGHTS.
Previously used scale of atomic masses with 1/16 of mean mass of oxygen atom (natural mixture of nuclides) as mass unit.
D chemische Massenskala
F échelle chimique des poids atomiques
P skala mas atomowych chemiczna
R химическая шкала атомных масс, химическая шкала атомных весов

587
CHEMICAL SHIFT.
Shift of resonance frequency ν_0 or of resonance strength of static magnetic field H_0 in nuclear magnetic resonance; caused by magnetic shielding of nucleus by surrounding electrons; in practice chemical shift denotes relative chemical shift

$$\frac{\nu_0 - \nu_0'}{\nu_0'} = \frac{H_0 - H_0'}{H_0'}$$

where ν_0 and H_0 refer to investigated resonance signal, and ν_0' and H_0' correspond to resonance signal of suitably chosen standard (reference substance).
D chemische Verschiebung
F déplacement chimique
P przesunięcie chemiczne
R химический сдвиг

588
CHEMICAL SPECIES.
Atom, molecule, ion, free radical, ion radical, cluster, solvated ion etc. which may be identified as exhibiting specific properties notwithstanding short time of life.
D Individuum
F espèce chimique
P indywiduum chemiczne
R индивидуум

589
CHEMICAL STAGE OF RADIOLYSIS.
Final phase of changes due to absorption of ionizing radiation conventionally taken to be longer than 10^{-8} s from moment of passage of ionizing particle or quantum.

D ...
F ...
P stadium radiolizy chemiczne
R ...

590
CHEMICAL SYMBOL (of element).
Arbitrary, international abbreviation applied for each of the known elements which generally has only first or first and one of following letters taken from their Latin names.
D chemisches Symbol
F symbole de l'élément, symbole chimique
P symbol chemiczny
R символ химического элемента

591
CHEMILUMINESCENCE.
Type of luminescence which depends upon emission of light by atoms or molecules previously excited by energy of chemical reaction.
D Chemilumineszenz
F chimiluminescence
P chemiluminescencja
R хемилюминесценция

592
CHEMISORPTION (chemical adsorption).
Irreversible adsorption which involves forces of chemical nature between adsorbate and adsorbent.
D Chemosorption, Chemisorption, chemische Adsorption
F chimisorption, adsorption chimique
P chemisorpcja, adsorpcja chemiczna
R хемисорбция, химическая адсорбция

593
CHEMISTRY.
Science dealing with studies on formation, properties, constitution and structure of substances as well as on their transformations and conditions which determine directions and rates of these transformations.
D Chemie
F chimie
P chemia
R химия

594
CHEMONUCLEAR REACTOR.
Nuclear reactor in which moderator and/or coolant is medium of useful chemical reactions initiated by ionizing radiation.
D Chemiekernreaktor
F réacteur de radiochimie
P reaktor chemojądrowy, reaktor chemonuklearny
R хемоядерный реактор

595
CHICHIBABIN AMINATION REACTION.
Process of amination of heterocyclic
systems, e.g. pyridine, quinoline and their
derivatives, with metal amides

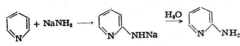

D Tschitschibabin Pyridin-Aminierung
F réaction de Tschichibabine
P reakcja aminowania Cziczibabina
R реакция Чичибабина

596
CHIRAL CENTRE (chirality centre).
The smallest grouping of atoms (usually
fragment of molecule) responsible for
existence of chirality in molecule; such a
molecule is non-superimposable on its
mirror image and is said to be optically
active (with the exception of the *mezo*
form). Asymmetric atom is the simplest
chiral centre. More complex chiral centres
are present in specially substituted allene
derivatives, alkylidenecycloalkanes,
spiranes, biphenyl and deformed aromatic
systems.
D Chiralitätszentrum
F centre de chiralité
P centrum chiralności
R центр хиральности

597
CHIRALITY.
Non-identity in steric structure
(configuration) of molecule with its mirror
image, caused by total lack of symmetry
elements or presence only of simple
symmetry axes.
D Chiralität, Händigkeit
F chiralité
P chiralność
R хиральность

CHIRALITY CENTRE. *See* CHIRAL
CENTRE.

598
CHI-SQUARE DISTRIBUTION
(χ^2-distribution).
Distribution of χ^2 statistic with probability
density function given by

$$f(\chi^2) = \frac{1}{2^{n/2}\Gamma\left(\dfrac{n}{2}\right)} \exp(-\chi^{n/2})\chi^{n/2-1}$$

$(0 < \chi^2 < \infty)$ and with $n-1$ degrees of
freedom

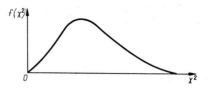

D χ^2-Verteilung
F distribution de la variable aléatoire χ^2
P rozkład χ^2
R χ^2-распределение

599
CHI-SQUARE STATISTIC (χ^2-statistic).
Statistic which is measure of disagreement
between hypothetical and empirical
distribution; given by

$$\chi^2(n) = \sum_{i=1}^{n} [(x_i - \mu)/\sigma_i]^2$$

where independent random variables x_i
have normal distribution with average
value μ and variances σ_i^2; χ^2-statistic has
χ^2-distribution.
D ...
F variable aléatoire χ^2
P statystyka χ^2, parametr χ^2
R χ^2-статистика, χ^2-параметр

600
CHI-SQUARE TEST OF GOODNESS OF
FIT (χ^2-test of goodness of fit).
Test of goodness of fit used for verification
of hypothesis about agreement between
series of results and a given distribution
of population. Results of series are divided
into classes of values and for each class
theoretical quantities are determined from
hypothetical distribution and compared
with empirical ones by using appropriate
χ^2 statistic

$$\chi^2 = \sum_{i=1}^{r} (n_i - np_i)^2/np_i$$

where n_i = empirical quantities in each
class, $n = \Sigma n_i$, p_i = probability that
random variable X of a given distribution
will have values belonging to i-th class
$(i = 1, 2, ..., r)$; calculated value of χ^2
is compared with those from tables.
D Chi-Quadrat-Test
F test du χ^2
P test χ^2
R критерий χ^2

601
CHLORINATION.
Substitution of hydrogen atom in molecule
of organic compound by chlorine atom, e.g.

$$CH_4 + Cl_2 \xrightarrow{h\nu} CH_3Cl + HCl$$

$$C_6H_6 + Cl_2 \xrightarrow{Fe} C_6H_5Cl + HCl$$

D Chlorierung
F chlor(ur)ation
P chlorowanie
R хлорирование

602
CHLORINE, Cl.
Atomic number 17.
D Chlor
F chlore
P chlor
R хлор

603
CHLORINOLYSIS.
Decomposition of molecule of chemical compound by chlorine, e.g.

$$ClCH_2CH_2SSCH_2CH_2Cl + Cl_2 \rightarrow 2ClCH_2CH_2SCl$$

D Chlorolyse
F ...
P chloroliza
R ...

604
CHLOROMETHYLATION (Blanc reaction).
Substitution of hydrogen atom of ring by chloromethyl group —CH_2Cl, effected by formaldehyde and hydrogen chloride in presence of $ZnCl_2$, e.g.

D (Blanc) Chlormethylierung, Blancsche Reaktion
F réaction de Blanc
P chlorometylowanie, reakcja Blanca (i Queleta)
R хлорометилирование, реакция Бланка

605
CHLOROSULFONATION (sulfochlorination).
Substitution of hydrogen atom in molecule of organic compound by chlorosulfonic group —SO_2Cl, usually effected by chlorosulfonic acid $HOSO_2Cl$ or sulfur dioxide and chlorine, e.g.

$$C_6H_6 + 2HOSO_2Cl \rightarrow C_6H_5SO_2Cl + H_2SO_4 + HCl$$

$$RH + SO_2 + Cl_2 \xrightarrow{h\nu} RSO_2Cl + HCl$$

D Sulfochlorierung
F chlorosulfonation
P chlorosulfonowanie, sulfochlorowanie
R сульфохлорирование

606
CHROMATOGRAM.
Fixed image of results of chromatographic separation in form of: (a) sheet or disk of filter paper or plate with adsorbent, showing after development spots due to substances being separated; (b) column with sorbent after development and detection of separated substances; (c) registration obtained from detector that recorded concentrations of substances coming out of chromatographic column or separated on filter paper or chromatographic plate.

D Chromatogramm
F chromatogramme
P chromatogram
R хроматограмма

607
CHROMATOGRAPHIC ANALYSIS.
Determination of composition of bi- and multicomponent mixtures with aid of chromatography.

D chromatographische Analyse
F analyse chromatographique
P analiza chromatograficzna
R хроматографический анализ

608
CHROMATOGRAPHIC BAND.
Distribution of adsorbate on surface or within volume of stationary phase (on filter paper, in column etc.).

D chromatographische Zone
F zone chromatographique
P pasmo chromatograficzne
R хроматографическая полоса

609
CHROMATOGRAPHIC COLUMN.
Tube (most often cylindrical) filled with stationary phase and equipped with inlet and outlet for mobile phase.

D chromatographische Säule
F colonne chromatographique
P kolumna chromatograficzna
R хроматографическая колонка

610
CHROMATOGRAPHY.
Set of separation methods that make use of different rates of migration of components being separated in porous media, in which one phase is immobile layer of substance of developed surface and the other is stream of gas or liquid flowing through it.

D Chromatographie
F chromatographie
P chromatografia
R хроматография

611
CHROMATOPOLAROGRAPHY.
Electroanalytical method which combines chromatographic separation of mixture of ions or molecules with their polarographic determination in effluent.
D Chromatopolarographie
F chromatopolarographie
P chromatopolarografia, analiza chromatopolarograficzna
R хроматополярография, хроматополярографический анализ, хроматополярографический метод

612
CHROMIUM, Cr.
Atomic number 24.
D Chrom
F chrome
P chrom
R хром

CHROMIUM FAMILY. See CHROMIUM GROUP.

613
CHROMIUM GROUP (chromium family).
Elements of sixth sub-group in periodic system with following structures of outer electronic shells of atoms: chromium $3d^54s^1$, molybdenum $4d^55s^1$ and tungsten $5d^46s^2$.
D Gruppe des Chroms
F groupe du chrome, métaux de la famille du chrome
P chromowce
R подгруппа хрома, элементы подгруппы хрома

614
CHROMOGEN.
Molecule of chemical compound containing one or more chromophores.
D Chromogen
F chromogène
P chromogen
R хромоген

615
CHROMOPHORE (chromophoric group).
Group of atoms in molecules of organic compounds which gives them their characteristic, light-absorbing properties within visible range or in its close vicinity.
D Chromophor, chromophore Gruppe
F chromophore
P chromofor, grupa chromoforowa
R хромофор

CHROMORPHORIC GROUP. See CHROMOPHORE.

616
CHRONOAMPEROMETRY.
Electroanalytical method consisting of analysis of dependence of current on time constant potential of indicator electrode placed in non-stirred solution or in molten salt.
D . . .
F chronoampérométrie
P chronoamperometria
R хроноамперометрия

617
CHRONOCOULOMETRY.
Method of investigation of kinetics of electrode reactions and adsorption of electroactive substances consisting in analysis of dependence of charge flowing in electrolytic cell in time of electrolysis carried out at constant potential of indicator electrode.
D . . .
F . . .
P chronokulometria
R хронокулонометрия

618
CHRONOPOTENTIOMETRIC ANALYSIS.
Analytical method that measures time of transition in a programmed-current (usually constant) electrolysis.
D . . .
F analyse chronopotentiométrique
P analiza chronopotencjometryczna
R хронопотенциометрический анализ

619
CHRONOPOTENTIOMETRY.
Electrochemical method which consists in recording and analysis of changes of potentials of indicator electrode in time of electrolysis carried out at programmed current intensity (usually current intensity is constant).
D Chronopotentiometrie
F chronopotentiométrie
P chronopotencjometria
R хронопотенциометрия

620
CHUGAEV REACTION (Tschugaeff dehydration).
Transformation of alcohol to alkene involving formation of xanthogenate (first stage) and then thermal decomposition of the latter, e.g.

$$RCH_2CH_2OH \xrightarrow[-H_2O]{NaOH,\ CS_2} S=C\begin{smallmatrix}OCH_2CH_2R\\ \\SNa\end{smallmatrix} \xrightarrow{\Delta}$$

$$RCH=CH_2 + HS-\overset{O}{C}-SNa$$

D Tschugaeff-Reaktion, Tschugaeff Xanthogenat-Spaltung
F réaction de Tschugaeff
P reakcja Czugajewa
R реакция Чугаева, ксантогеновая реакция

621
CIRCULAR DICHROISM (Cotton effect), CD.
Dichroism caused by different absorption of left and right circularly polarized light by optically active samples. Circular dichroism is a result of elliptical polarization of transmitted light.
D Zirkulardichroismus, Cotton-Effekt
F dichroïsme circulaire, effet Cotton
P dichroizm kołowy, efekt Cottona
R круговой дихронизм

622
CIRCULAR POLARIZATION OF LIGHT.
Fixing of direction of rotation of light wave electric vector E.
D zirkulare Polarisation des Lichtes
F polarisation circulaire de la lumière
P polaryzacja światła kołowa
R круговая поляризация света

CIRCULAR TECHNIQUE. *See* RADIAL CHROMATOGRAPHY.

CIS FORM. *See* CIS ISOMER.

623
CIS ISOMER (*cis* form).
Diastereoisomer in which identical or similar groups are situated on the same side of plane passing through double bond or through ring surface, e.g. *cis*-2-butene

D *cis*-Isomer, *cis*-Form
F isomère *cis*
P izomer *cis*, odmiana *cis*
R *цис*-изомер, *цис*-форма

624
CIS ISOMER (*cis* form).
In coordination chemistry, isomer in which two similar ligands are near each other, i.e. in square complexes [Ma$_2$b$_2$] the ligands occupy adjacent vertices of square, and in octahedral complexes [Ma$_4$b$_2$] adjacent corners of octahedron.
D *cis*-Form
F isomère *cis*
P izomer *cis*
R изомер *цис*

625
CIS-TRANS ISOMERISM (geometric isomerism).
Type of diastereoisomerism caused by restricted rotation about double bonds or considerable restriction in rotation about single bonds in ring systems. *Cis-trans*

isomerism is observed in ethylene-1,2-dicarboxylic acids

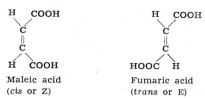

Maleic acid   Fumaric acid
(*cis* or Z)   (*trans* or E)
Remark: Abbreviations Z-E from German zusammen and entgegen.

D *cis-trans*-Isomerie, geometrische Isomerie
F isomérie *cis-trans*, isomérie géométrique
P izomeria *cis-trans*, izomeria Z-E, izomeria geometryczna
R *цис-транс* изомерия, геометрическая изомерия

626
CIS-TRANS ISOMERISM (geometrical isomerism).
In coordination chemistry, type of isomerism based on different arrangements around central ion by two similar ligands. This isomerism occurs in planar square and octahedral complexes, but does not occur in tetrahedral complexes and in complexes of coordination number 2 or 3.
D *cis-trans*-Isomerie, geometrische Isomerie
F isomérie géométrique
P izomeria *cis-trans*, izomeria geometryczna
R геометрическая изомерия

627
CIS-TRANS ISOMERIZATION.
Establishment of thermodynamic equilibrium between *cis* and *trans* isomers, caused by conversion of one isomer into another.
D *cis-trans*-Isomerisierung
F isomérisation *cis-trans*
P stereomutacja
R *цис-транс*-изомеризация

628
CITRIC ACID FERMENTATION.
Type of oxidative fermentation of hexoses, e.g. of glucose, caused by Citromicetes, Aspergilla and Penicillia resulting in decomposition of carbohydrates to citric acid.
D Zitronensäuregärung
F fermentation citrique
P fermentacja cytrynowa
R лимоннокислое брожение

629
CLAISEN CONDENSATION (ester condensation).
Condensation of ester with compound containing active methylene group (e.g. ester, ketone or nitrile), effected by basic catalyst (sodium alcoholate, sodium amide, triphenylmethylsodium etc.) resulting in formation of β-ketoester, β-diketone or β-ketonitrile, respectively, e.g.

$$2CH_3COOC_2H_5 \xrightarrow{C_2H_5ONa}$$
$$CH_3COCH_2COOC_2H_5 + C_2H_5OH$$

$$CH_3COOC_2H_5 + CH_3COCH_3 \xrightarrow{C_2H_5ONa}$$
$$CH_3COCH_2COCH_3 + C_2H_5OH$$

D Claisen-Kondensation, Esterkondensation
F condensation de Claisen, condensation par estérification
P kondensacja Claisena, kondensacja estrowa
R конденсация Клайзена, сложноэфирная конденсация

630
CLAISEN REARRANGEMENT.
Intramolecular rearrangement of allyloarylic ethers and ethers of allylic species of enolic ketones, effected by heating into allylic derivatives of phenols or carbonyl compounds, e.g.

$$CH_3-C=CH-COOC_2H_5 \rightarrow CH_3-C-CH-COOC_2H_5$$
$$O-CH_2-CH=CH_2 \qquad O \quad CH_2-CH=CH_2$$

D Claisen $(O\text{-Allyl}\rightarrow C\text{-Allyl})$-Umlagerung
F transposition de Claisen
P przegrupowanie Claisena
R перегруппировка Клайзена

631
CLAISEN-SCHMIDT CONDENSATION.
Condensation of aromatic aldehyde with aliphatic aldehyde or ketone in presence of aqueous base solution, resulting in formation of α,β-unsaturated aldehyde or ketone, e.g.

$$C_6H_5CHO + CH_3COC_6H_5 \xrightarrow{NaOH}$$
$$C_6H_5CH=CH-COC_6H_5 + H_2O$$

D Claisen-Schmidt (Chalkon)-Kondensation
F synthèse de Claisen-Schmidt
P kondensacja Claisena i Schmidta
R конденсация Клайзена-Шмидта

CLAPEYRON'S EQUATION. *See* CLAUSIUS-CLAPEYRON EQUATION.

CLAPEYRON'S RELATION. *See* CLAUSIUS-CLAPEYRON EQUATION.

CLASSICAL GAS. *See* IDEAL GAS.

CLASSICAL HARMONIC OSCILLATOR. *See* PLANCK'S OSCILLATOR.

CLASSICAL METHOD. *See* GRAM METHOD.

CLASSICAL PHASE SPACE. *See* PHASE SPACE.

632
CLASSICAL THERMODYNAMICS (thermostatics).
Theory of equilibrium states in macroscopic systems based on a few phenomenological laws called laws of thermodynamics.

D klassische Thermodynamik
F thermostatique, thermodynamique classique, thermodynamique des phénomènes réversibles
P termodynamika klasyczna, termostatyka
R классическая термодинамика

633
CLATHRATE COMPOUNDS.
Inclusion compounds which are formed during recrystallization (from solution or in gaseous atmosphere). Certain ("host") compounds are capable of enclosing other foreign ("guest") molecules or atoms in free spaces of their crystal lattice.

D Clathrate, Käfigeinschlußverbindungen
F clathrates, composés en cage
P klatraty
R клатраты, клатратные соединения

634
CLAUSIUS-CLAPEYRON EQUATION (Clapeyron's equation, Clapeyron's relation).
Equation which determines relation between thermodynamic temperature T and pressure p in one-component two-phase system in thermodynamic equilibrium

$$\frac{\partial p}{\partial T} = \frac{H_m^\alpha - H_m^\beta}{T(V_m^\alpha - V_m^\beta)}$$

where H_m = molar enthalpy, V_m = molar volume and α and β denote phases.

D Clausius-Clapeyronsche Formel
F relation de Clapeyron
P równanie Clausiusa i Clapeyrona
R формула Клапейрона, формула Клапейрона-Клаузиуса, уравнение Клапейрона-Клаузиуса

635
CLAUSIUS EQUATION OF STATE.
Empirical (thermal) equation of state of real gas

$$\left[p + \frac{a}{T(V_m + c)^2}\right](V_m - b) = RT$$

where p = pressure, V_m = molar gas volume, T = thermodynamic temperature, R = universal gas constant, a, b, c = constants characteristic for the given gas.
D Clausiussche Zustandsgleichung
F équation d'état de Clausius
P równanie stanu Clausiusa
R уравнение состояния Клаузиуса

636
CLAUSIUS-MOSSOTTI EQUATION.
Equation relating permittivity ε to polarizability α of molecule

$$\frac{\varepsilon - 1}{\varepsilon + 2}\frac{M}{\rho} = \frac{4}{3}\pi N_A \alpha$$

where M = molar mass, ρ = density, N_A = Avogadro number.
D Clausius-Mossotti Gleichung
F formule de Clausius-Mossotti
P równanie Clausiusa i Mossottiego
R уравнение Клаузиуса-Моссотти

637
CLEMMENSEN REDUCTION.
Reduction of carbonyl group to methylene group by zinc amalgam and hydrochloric acid, e.g.

$$RCOR' + 4(H) \xrightarrow{\frac{Zn(Hg)}{HCl}} RCH_2R' + H_2O$$

D Clemmensen (Carbonyl→Methylen)-Reduktion
F réduction selon Clemmensen
P redukcja Clemmensena
R восстановление по Клемменсену

638
CLOSED SYSTEM.
System which can exchange energy but not mass with surroundings.
D geschlossenes System
F système fermé
P układ zamknięty
R закрытая система

639
CLUSTER (in nuclear physics).
Group of nucleons inside atomic nucleus forming structure similar to light nuclei; it is treated as a particular substructure of the whole nucleus.
D Cluster, Nukleonengruppe
F faisceau

P klaster, grono
R ассоциация, кластер

640
CLUSTER MODEL.
Model of atomic nucleus which describes nucleus as system of comparatively weakly bound lighter atomic nuclei, e.g. helium nuclei in alpha model. Cluster model describes properties of excited nucleus in terms of nucleus-cluster properties.
D Büschelmodell, Cluster-Modell
F modèle de faisceaux
P model gronowy, model klastrowy
R модель ассоциаций

641
COACERVATION.
Incomplete coagulation, separation of lyophilic sol, e.g. during salting out with electrolytes. One constituent of separated phase is a concentrated sol, and the other, a dilute sol.
D Koazervation
F coacervation
P koacerwacja
R коацервация

642
COAGEL.
Gel formed in process of incomplete coagulation of sol when precipitate forms porous structure filled with solvent.
D Koagel
F coagel, précipité gélatineux
P koażel
R коагель, студенистый осадок

643
COAGULATION.
Process of decreasing number of particles of dispersed phase in unit volume of colloidal system by gathering of single particles into larger species (aggregates).
D Koagulation
F coagulation
P koagulacja
R коагуляция

644
COAGULATION CONCENTRATION (flocculation value).
Smallest concentration of electrolyte, expressed most often as mmoles/dm³, causing quick visible coagulation of sol.
D Flockungs(schwellen)wert, Flockungskonzentration
F valeur de coagulation
P wartość koagulacyjna, wartość progowa, stężenie koagulujące
R порог коагуляции, число коагуляции

645
COALESCENCE.
Separation of disperse system, usually emulsion, into two continuous phases, caused by gathering of particles of dispersed phase.
D Koaleszenz
F coalescence
P koalescencja
R коалесценция

646
COBALT, Co.
Atomic number 27.
D Kobalt
F cobalt
P kobalt
R кобальт

COBALT-60. See RADIOCOBALT.

647
COBALT BOMB.
Common name for $^{60}_{27}$Co source of ionizing radiation; used for therapeutical purposes and for irradiation of materials and reacting systems.
D Kobalteinheit
F bombe au cobalt
P bomba kobaltowa
R кобальтовая пушка

648
COBALT GROUP.
Three elements in eighth sub-group of periodic system with structure of outer electronic shells of atoms: cobalt — $3d^7 4s^2$, rhodium — $4d^8 5s^1$, iridium — $5d^7 6s^2$.
D ...
F ...
P kobaltowce
R ...

649
COEFFICIENT OF COMPRESSIBILITY (compressibility, isothermal compressibility), $\varkappa$.
Relative change of gas volume V caused by unit change of pressure p at constant thermodynamic temperature T taken with opposite sign

$$\varkappa = -\frac{1}{V}\left(\frac{\partial V}{\partial p}\right)_T \text{ [atm}^{-1}]$$

or [Pa^{-1}] in SI system.
Remark: Coefficient of compressibility is used sometimes also in meaning of adiabatic coefficient of compressibility, which describes relative change of volume due to pressure, but in adiabatic conditions.
D Kompressibilitätskoeffizient, Kompressibilität
F coefficient de compressibilité
P współczynnik ściśliwości izotermiczny
R коэффициент изотермического сжатия

650
COEFFICIENT OF THERMAL EXPANSION (of gas), α.
Relative change of volume V of gas, caused by change of its thermodynamic temperature T by 1 K at constant pressure p

$$\alpha = \frac{1}{V}\left(\frac{\partial V}{\partial T}\right)_p \text{ [K}^{-1}]$$

Coefficient of expansion of ideal gas is

$$\alpha = \frac{1}{273.15} \text{ K}^{-1}$$

D (thermischer) Ausdehnungskoeffizient
F coefficient de dilatation thermique
P współczynnik rozszerzalności
R коэффициент (теплового) расширения

651
COEFFICIENT OF VARIATION.
Relative standard deviation in per cent.
D relative Standardabweichung in Prozenten
F coefficient de variation
P współczynnik zmienności
R коэффициент вариации

652
COERCIVE FORCE.
Magnetic field strength corresponding to zero value of magnetic induction and magnetization in ferro- or ferrimagnets during their overmagnetization.
D Koerzitivkraft
F champ coercitif
P koercja, pole koercji, siła koercji
R коэрцитивная сила

653
COHERENT SCATTERING.
Scattering due to simultaneous interaction with more than one scattering centre, resulting in interference of waves scattered on particular centres.
D kohärente Streuung
F diffusion cohérente
P rozpraszanie spójne, rozpraszanie koherentne
R когерентное рассеяние

654
COHESION.
Effect of mutual attraction of molecules of a given substance due to short-range forces of intermolecular interaction.
D Kohäsion
F cohésion
P kohezja, spójność
R когезия, сцепление

COHESION PRESSURE. See INTERNAL PRESSURE.

655

CO-IONS.

Mobile ions (able to diffuse freely) of the same charge as that of ion exchanger's skeleton.

D Coionen
F co-ions
P kojony, współjony
R коионы

656

COLD PLASMA.

Plasma of temperature of order of tens of thousands kelvins.

D kaltes Plasma
F plasma froid
P plazma zimna
R холодная плазма

657

COLLECTIVE MODELS.

Models of atomic nucleus describing it as a whole and not as system of nucleons and taking into account strong, multinucleonic correlations. The prototype collective model is liquid drop model.

D kollektive Kernmodelle
F modèles collectifs
P modele kolektywne
R коллективные модели

658

COLLIGATIVE PROPERTIES.

Properties which depend primarily on the number of molecules concerned and not on their nature, e.g. volume of gas, at constant temperature and pressure. These properties are chiefly encountered in study of solutions.

D ...
F propriétés colligatives
P własności koligatywne
R коллигативные свойства

659

COLLISION INTEGRAL.

Integral term in Boltzmann equation

$$\left(\frac{\partial \rho_1}{\partial t}\right)_{\text{coll.}} = n \int d^3v \int d\Omega \sigma(\Omega)|v-v_I| \times$$

$$\times [\rho_1(v')\rho_1(v_I') - \rho_1(v)\rho_1(v_I)]$$

where t = time, n = particle density, ρ_1 = unimolecular partition function, Ω = angle between vectors $v-v_I$ and $v'-v_I'$, $\sigma(\Omega)$ = differential cross-section for two molecule collision with initial velocities v, v_I and final velocities v', v_I'. Collision integral takes into account influence of intermolecular collisions on partition function and their effect on gas phase processes.

D Stoßintegral
F intégrale des chocs
P całka zderzeniowa
R интеграл столкновений

660

COLLISION NUMBER, $\bar{z}$.

Average number of two-molecular collisions in unit time interval which take place in unit volume of rarified gas in equilibrium conditions

$$\bar{z} = n^2 \frac{\pi}{4}\left(\frac{m}{\pi kT}\right)^{3/2} \int\limits_0^\infty \sigma v^3 \exp\left[-\frac{mv^2}{4kT}\right] dv$$

where k = Boltzmann constant, T = thermodynamic temperature, n = molecule density, v = relative velocity of pair of molecules, and σ = cross-section for two-molecular collision. In hard sphere core model σ is constant and then $\bar{z} = 2n^2d^2\left(\frac{\pi kT}{m}\right)^{1/2}$, where d = molecule diameter.

D Stoßzahl, Zahl der Zweierstöße
F fréquence moyenne de collision
P liczba zderzeń
R (газокинетическое) число столкновений, частота столкновений

661

COLLISION THEORY (in chemical kinetics).

Reaction rate theory based on model of elementary chemical reaction as due to collisions of molecules possessing defined surplus of energy (activation energy).

D Stoßtheorie, Kollisionstheorie, Theorie der erfolgreichen Stöße
F théorie des collisions, théorie de l'activation par collision
P teoria zderzeń aktywnych
R теория активных столкновений

662

COLLOID (colloidal system).

Disperse system of high degree of dispersion and developed surface of separation of phases in which dispersed phase consists of colloidal particles (colloid) or system physically homogeneous containing macromolecules as one of components (molecular colloid).

D Kolloid, kolloides System
F colloïde, système colloïdal
P koloid, układ koloidalny
R коллоид, коллоидная система

663
COLLOIDAL PARTICLES.
Particles of dispersed phase of dimensions
from 1 nm to 500 nm which can be
observed under ultramicroscope, including
particles of two or only one dimension
of this order (laminar and thread colloidal
particles).
D Kolloidteilchen, kolloide Partikeln
F particules colloïdales
P cząstki koloidalne
R коллоидные частицы

664
COLLOIDAL STATE.
State of matter characterized by
appropriate degree of colloidal dispersion
(see colloid, colloidal particles).
D kolloider Zustand
F état colloïdal
P stan koloidalny
R коллоидное состояние

COLLOIDAL SYSTEM. See COLLOID.

665
COLLOID CHEMISTRY.
Branch of chemistry dealing with studies
on colloidal systems.
D Kolloidchemie
F chimie des colloïdes
P chemia koloidów
R коллоидная химия

666
COLLOID MILL.
Device in which direct mechanical
disintegration of substance suspended in
dispersion medium takes place until
particles of colloidal size are obtained.
D Kolloidmühle
F moulin colloïdal, broyeur colloïdal
P młyn koloidalny
R коллоидная мельница

667
COLORIMETER.
Device used to perform colorimetric
analyses either by absorption
measurement or by comparison of
absorption of solutions. Measurements are
carried out with white light or simple
monochromatization is used.
Remark: This term is also used to describe
photocolorimeter.
D Kolorimeter
F colorimètre
P kolorymetr (wizualny)
R колориметр

668
COLORIMETRIC ANALYSIS
(colorimetry).
Analytical method utilizing measurement
of colour intensity of solution for
determination of amount of substance
which is coloured or which forms
a coloured compound in chemical reaction.
D Kolorimetrie
F analyse colorimétrique, colorimétrie
P analiza kolorymetryczna, kolorymetria
R колориметрический анализ

669
COLORIMETRIC TITRATION
(duplication method).
Method of colorimetric analysis:
a coloured reaction in investigated
solution is carried out in one of the
absorption cells; in other cell all the
reagents are mixed in the same quantities
and then solution of component to be
determined is added from microburette
until solutions in both cells become
identically coloured.
D kolorimetrische Titration
F titrage colorimétrique
P miareczkowanie kolorymetryczne
R колориметрическое титрование,
дублирование

COLORIMETRY. See COLORIMETRIC
ANALYSIS.

670
COLORIMETRY.
Branch of optics dealing with quantitative
determination of colours based on
interpretation of absorption or re-emission
spectrum by using specially devised and
normalized measurement system which
enables all colours to be synthesized from
the three primary ones.
D Farbmetrik
F colorimétrie
P barwometria, kolorymetria
R колориметрия

671
COLOUR CENTRES (absorbing centres).
Crystal lattice imperfections which cause
selective absorption of light by crystal.
D Farbzentren
F centres de couleur
P centra barwne
R центры окрашивания

COLOUR FILTER. See ABSORPTION
FILTER.

672
COLOUR TEMPERATURE.
Temperature at which black body
radiation spectrum is identical with
radiation spectrum of grey body at a given
temperature.
D Farbtemperatur
F température de couleur
P temperatura barwy, temperatura
barwna
R цветовая температура

673
COLUMN BREAKTHROUGH
(breakthrough *in chromatography*).
Appearance of given component in
effluent from column (in practice moment
when concentration of this component
exceeds certain conventional value).
D Durchbruch (der Säule)
F percée
P przebicie (kolumny)
R проскок

COLUMN EFFICIENCY. *See* COLUMN
PERFORMANCE.

674
COLUMN PERFORMANCE (column
efficiency).
Measure of chromatographic band
spreading expressed in terms of number
of theoretical plates.
D Trennschärfe der Kolonne
F efficacité de la colonne
P sprawność kolumny
R эффективность колонки

675
COMBES QUINOLINE SYNTHESIS.
Condensation of arylamine with
β-diketone followed by cyclization of
resulting anil under action of concentrated
sulfuric acid on 2,4-disubstituted quinoline
derivative, e.g.

D Combes-Chinolin-Synthese
F méthode de Combes
P synteza (pochodnych chinoliny) Combesa
R синтез хинолина Комба

676
COMBINATION BAND (*summation or
difference band*).
In vibration spectrum band with
frequency equal to sum or difference of
two or more fundamental vibrations;
it appears due to anharmonicity of
vibrations.
D Kombinationsband
F bande de combinaison
P pasmo kombinacyjne
R составная частота

COMBINED SAMPLE. *See* GROSS
SAMPLE.

677
COMBUSTION (burning).
Rapid exothermal reaction accompanied
usually by phenomena of light emission;
most often reaction with oxygen.
D Verbrennung
F combustion, ignition
P spalanie, palenie
R горение

678
COMPACT DOUBLE LAYER (Helmholtz
double layer).
Part of electrical double layer at
metal—electrolyte solution interphase,
it directly adheres to metal surface and
is built from adsorbed or electrostatically
bound ions; identical with Helmholtz
model (*see* theories of double layer).
D Helmholtzsche Doppelschicht,
Helmholtz-Schicht
F couche (rigide) d'Helmholtz
P warstwa zwarta, warstwa Helmholtza
R конденсированный двойной слой,
гельмгольцевский двойной слой

679
COMPARATOR.
In colorimetric analysis, instrument for
determination of concentration of coloured
substances by visual comparison with
standard scale.
D Komparator
F comparateur
P komparator
R компаратор

COMPARATOR. *See* DOUBLE
COMPARATOR.

680
COMPARISON SOLUTION.
Solution with colour identical with that
which titrated solution should have just
before or after end-point of titration.
D Vergleichslösung
F . . .
P świadek miareczkowania
R . . .

COMPETING REACTIONS. *See*
PARALLEL REACTIONS.

681
COMPLETE ANALYSIS (total analysis).
Determination of all constituents in
analysed sample.
D Gesamtanalyse
F analyse totale
P analiza całkowita
R полный анализ

COMPLETE INTERNAL EQUILIBRIUM.
See THERMODYNAMIC EQUILIBRIUM.

682
COMPLETELY POLARIZABLE
ELECTRODE.
Electrode (half-cell) with no flow of
charged particles between bordering
phases of metallic conductor and
electrolyte, e.g. mercury in solutions of
alkali metal chlorides in a certain
potential range.

D vollständig polarisierbare Elektrode
F électrode idéalement polarisée
P elektroda doskonale polaryzowalna
R идеально поляризуемый электрод

683
COMPLETE MISCIBILITY (unlimited
compatibility).
Ability of mixture to undergo mixing in
any proportion.

D unbegrenzte Mischbarkeit, vollständige
  Mischbarkeit
F miscibilité illimitée
P mieszalność nieograniczona
R полная смешиваемость,
  неограниченная взаимная
  растворимость

COMPLEX. *See* COORDINATION
COMPOUND.

COMPLEX COMPOUND. *See*
COORDINATION COMPOUND.

684
COMPLEX DIELECTRIC
POLARIZATION (MOLAR), P_m^*.
Quantity defined by

$$P_m^* = \frac{\varepsilon^*-1}{\varepsilon^*+2}\ \frac{M_r}{\rho}$$

where ε^* = complex permittivity,
M_r = relative molecular mass, = density

D komplexe dielektrische Polarisation,
  komplexe Molpolarisation
F polarisation diélectrique complexe
P polaryzacja dielektryczna zespolona
  (molowa)
R комплексная диэлектрическая
  поляризация

COMPLEX FORMATION REACTION. *See*
COORDINATION.

COMPLEXIMETRIC TITRATION. *See*
COMPLEXIMETRY.

685
COMPLEXIMETRY (complexometry,
compleximetric titration, complexometric
titration).
Titration based on reactions of formation
of soluble complexes of determined ion
with titrant.

D Komplexometrie, Compleximetrie,
  komplexometrische Titration

F complexométrie
P kompleksometria, miareczkowanie
  kompleksometryczne
R комплексометрия,
  комплексометрическое титрование

686
COMPLEX ION.
Ion composed of central ion and ligands
attached to it.

D Komplexion
F ion complexe
P jon kompleksowy
R комплексный ион

COMPLEXOMETRIC TITRATION. *See*
COMPLEXIMETRY.

COMPLEXOMETRY. *See*
COMPLEXIMETRY.

687
COMPLEX PERMITTIVITY, ε^*.
Quantity defined by $\varepsilon^* = \varepsilon - i\varepsilon'$,
where ε = permittivity, $i = \sqrt{-1}$,
ε' = dielectric losses.

D komplexe Dielektrizitätskonstante,
  komplexe Dielektrizitätszahl
F permittivité complexe
P przenikalność elektryczna zespolona
  (bezwzględna)
R комплексная диэлектрическая
  проницаемость

688
COMPLEX REACTION.
System of two or more mutually
connected elementary reactions.

D zusammengesetzte Reaktion, komplexe
  Reaktion
F réaction complexe, réaction composée
P reakcja złożona
R сложная реакция

689
COMPLEX SALT.
Salt which contains complex ion.

D Komplexsalz
F sel complexe
P sól kompleksowa
R комплексная соль

690
COMPLEX SUBSTANCE.
Chemical compound, i.e. substance which
may be decomposed, giving simpler
chemical compounds or elements.

D . . .
F corps composé
P substancja złożona
R сложное вещество

COMPOUND CATALYST. *See* MIXED
CATALYST.

691
COMPOUND CHROMOPHORES
(conjugated chromophores).
Two or more chromophores within one
molecule with common conjugated bond
system; they exhibit fairly strong light
absorption usually shifted strongly
bathochromically and hyperchromically as
compared with those of separate
component chromophores.

D konjugierte Chromophoren
F chromophores conjugués
P chromofory sprzężone
R сопряжённые хромофоры

692
COMPOUND NUCLEUS.
Excited nucleus, in state of complex
structure, constituting intermediate stage
in nuclear reactions. Final state of such
reactions is reached independently of
initial state, as result of decay of
compound nucleus.

D Zwischenkern, Compoundkern
F noyau composé
P jądro złożone
R составное ядро, промежуточное ядро

693
COMPRESSIBILITY.
Dependence of volume of substance on
pressure.

D Kompressibilität
F compressibilité
P ściśliwość
R сжимаемость

COMPRESSIBILITY. *See* COEFFICIENT
OF COMPRESSIBILITY.

694
COMPRESSIBILITY FACTOR, z.
Empirical coefficient depending on nature
of substance, pressure p and
thermodynamic temperature T, applied in
description of properties of real gas by
equation of state in the form

$$z = \frac{pV_m}{RT}$$

where V_m = molar volume of real gas,
R = universal gas constant.

D Kompressibilitätsfaktor, Realfaktor
F facteur de compressibilité
P współczynnik ściśliwości
R коэффициент сжимаемости, фактор
сжимаемости

695
COMPTON CONTINUUM.
Energy spectrum of electrons liberated in
Compton effect.

D Comptonverteilung
F fond Compton, distribution Compton

P kontinuum komptonowskie
R комптоновское распределение

696
COMPTON EDGE.
Maximum energy which can be gained by
electron in Compton effect.

D Comptonkante
F front Compton
P krawędź komptonowska
R комптоновский край

697
COMPTON EFFECT (Compton scattering).
Scattering of gamma radiation, the
wavelength of which is increased as
a result of energy transfer to orbital
electrons.

D Compton-Effekt, Compton-Streuung
F effet Compton, diffusion Compton
P efekt Comptona, rozpraszanie
komptonowskie
R комптон-эффект, комптоновское
рассеяние

COMPTON SCATTERING. *See*
COMPTON EFFECT.

CONCENTRATING POWER OF
ADSORBENT. *See* ADSORPTIVITY.

698
CONCENTRATION CELL.
Cell in which electrical work results from
equalization of concentrations of
electrolyte, or composition of metallic
phase of electrodes, or pressures of gases
which form given electrode systems.

D Konzentrationszelle,
Konzentrationskette,
Konzentrationselement
F pile de concentration, chaîne de
concentration
P ogniwo stężeniowe
R концентрационная цепь,
концентрационный элемент

699
CONCENTRATION CELL WITHOUT
TRANSFERENCE.
Concentration cell in which electrolytes
of different concentrations m^I and m^{II} are
not in direct contact, e.g.

$$\text{Ag, AgCl}\,|\,\text{NaCl}(m^I)\,|\,\text{Na}_x\text{Hg}-\text{NaHg}_x|$$

$$|\,\text{NaCl}(m^{II})\,|\,\text{AgCl, Ag}$$

or concentration cell with one electrolyte
and different partial pressures p^I and p^{II}
of gases in electrode systems, e.g.

$$\text{Pt, H}_2(p^I)\,|\,\text{HCl}\,|\,\text{Pt, H}_2(p^{II})$$

D Konzentrationszelle ohne Überführung
F pile de concentration sans transport
d'ions
P ogniwo stężeniowe bez przenoszenia
R концентрационная цепь без переноса

700
CONCENTRATION CELL WITH TRANSFERENCE.
Concentration cell with two different concentrations m^I and m^{II} of electrolyte which are in contact (for example through diaphragm), e.g.

Pt, H_2 | $HCl(m^I)$ | $HCl(m^{II})$ | Pt, H_2

D Konzentrationszelle mit Überführung
F pile de concentration à transport d'ions
P ogniwo stężeniowe z przenoszeniem
R концентрационная цепь с переносом

701
CONCENTRATION OF SOLUTION.
Quantitative description of solution composition.

D Konzentration (der Lösung)
F concentration
P stężenie roztworu
R концентрация раствора

702
CONCENTRATION POLARIZATION.
Type of electrolytic polarization consisting in changes of electrolyte concentration at electrode as result of flow of polarizing current. Concentration polarization is expressed by concentration overvoltage.

D Konzentrationspolarisation
F polarisation de concentration
P polaryzacja stężeniowa
R концентрационная поляризация

703
CONDENSATION.
Linking together of two or more molecules, resulting in formation of bigger molecule, usually with elimination of some simple substance, e.g. water, ammonia or hydrogen halide, e.g.

$$2C_6H_5Cl + Cl_3CCHO \xrightarrow{H_2SO_4}$$

$$\xrightarrow{\hspace{1cm}} \begin{array}{c} ClC_6H_4 \\ ClC_6H_4 \end{array}\!\!\!\!> CHCCl_3 + 2H_2O$$

$$2H_3PO_4 \rightarrow H_4P_2O_7 + H_2O$$

D Kondensation
F condensation
P kondensacja
R конденсация

704
CONDENSATION (of vapour).
Transition from vapour to liquid or solid state of a substance.

D Kondensation
F condensation
P kondensacja (pary)
R конденсация

705
CONDENSATION METHOD.
In colloid chemistry, methods of obtaining colloids from systems of molecular

disintegration by aggregation, polymerization or polycondensation of ions, atoms or molecules.

D Kondensationsmethoden
F méthodes de condensation
P metody kondensacyjne
R конденсационные методы

CONDENSATION POINT. See CONDENSATION TEMPERATURE.

706
CONDENSATION POLYMER (polycondensat).
Polymer formed by condensation polymerization.

D Polykondensat
F polycondensat
P polimer kondensacyjny, polikondensat
R поликонденсационный полимер

707
CONDENSATION POLYMERIZATION (polycondensation).
Polymerization by repeated condensation, i.e. with simultaneous elimination of some simple substance as by-product.

D Polykondensation
F polycondensation
P polimeryzacja kondensacyjna, polikondensacja
R поликонденсация

708
CONDENSATION TEMPERATURE (condensation point).
Temperature at which vaporized substance is converted into liquid or solid state.

D Kondensationstemperatur
F température de condensation
P temperatura kondensacji
R температура конденсации

CONDENSED NUCLEI. See CONDENSED RING SYSTEM.

709
CONDENSED PHASES.
Liquids and solid phases.

D kondensierte Phasen
F phases condensées, corps condensés
P fazy skondensowane
R конденсированные фазы

CONDENSED POLYCYCLIC COMPOUND. See CONDENSED RING SYSTEM.

CONDENSED RINGS. See CONDENSED RING SYSTEM.

710
CONDENSED RING SYSTEM (condensed rings, condensed polycyclic compound, condensed nuclei).
System of two or more rings fused

together, having common pairs of atoms, e.g. naphthalene

D kondensiertes Ringsystem
F noyaux condensés
P pierścienie skondensowane
R конденсированные циклы

CONDENSED SURFACE FILM. *See*
SURFACE FILM.

711
CONDENSED SYSTEM.
System composed of condensed phases only.
D kondensiertes System
F système condensé
P układ skondensowany
R конденсированная система

712
CONDITION OF ISOTHERMAL MECHANICAL STABILITY.
Condition $\varkappa > 0$, where $\varkappa$ = isothermal compressibility coefficient of phase. It ensures that phase cannot decompose irreversibly into parts with different pressure.
D Bedingung für die mechanische Stabilität
F condition de stabilité mécanique
P warunek stabilności mechanicznej (fazy)
R условие механической устойчивости, условие механической стабильности

713
CONDITION OF MATERIAL STABILITY (with respect to diffusion).
Condition according to which phase cannot decompose irreversibly into parts with different values of chemical potential of the same constituent if matrix $[\mu_{ij}]$ is positive definite, where $\mu_{ij} = \dfrac{\partial \mu_i}{\partial x_j}$,
μ_i = chemical potential of i-th constituent in a given phase, x = mole fraction and i, j refer to given phase constituents.
D Bedingung für die Stabilität in Bezug auf die Diffusion
F condition de stabilité diffusion
P warunek stabilności dyfuzyjnej (fazy)
R условие диффузионной устойчивости, условие диффузионной стабильности

714
CONDITION OF THERMAL STABILITY.
Condition $C_V > 0$, where C_V = phase molar heat capacity at constant volume; it ensures that phase cannot decompose irreversibly into parts with different temperature.
D Bedingung für die thermische Stabilität
F condition de stabilité thermique
P warunek stabilności termicznej (fazy)
R условие термической устойчивости, условие термической стабильности

715
CONDUCTION BAND (*in band theory of solids*).
Energy band derived from overlap of lowest unoccupied energy levels of all atoms or ions forming crystal lattice.
D Leitungsband, Leitfähigkeitsband
F bande de conductivité, bande de conduction
P pasmo przewodnictwa
R зона проводимости, проводящая зона, полоса проводимости

716
CONDUCTOMETER.
Apparatus used for measurements of conductivity of electrolytes.
D Leitfähigkeitsmesser
F conductimètre
P konduktometr
R мостик для измерения электропроводности

CONDUCTOMETRIC ANALYSIS. *See*
CONDUCTOMETRIC TITRATION.

717
CONDUCTOMETRIC TITRATION (conductometric analysis).
Method of quantitative analysis consisting in measurements of changes of conductivity of solution during titration of analysed substance with suitable reagent.
D konduktometrische Titration, Leitfähigkeitstitration
F titrage conductimétrique
P miareczkowanie konduktometryczne, analiza konduktometryczna
R кондуктометрическое титрование

718
CONDUCTOMETRY.
Method of determination of material present in mixture based on measurement of its influence on electrolytic conductivity of mixture.
D Konduktometrie, Leitfähigkeitsmessungen
F conductimétrie
P konduktometria
R кондуктометрия

719
CONDUCTOR (*of electricity*).
Material of low specific resistance (conventionally lower than $10^{-4}\,\Omega$ cm); often used as general term for metallic conductors characterized by negative temperature coefficient of conductance. According to band model of solid state, crystal is conductor with continuous zone of incompletely occupied quantum states, accesible to electrons.
D Leiter
F conducteur
P przewodnik
R проводник

720
CONFIDENCE INTERVAL.
Random interval determined by distribution of estimator and having feature that it covers range of values of estimated parameter Θ with high pre-determined probability.
D Konfidenzintervall, Vertrauensintervall
F intervalle de confiance, intervalle de sécurité
P przedział ufności
R доверительный предел

721
CONFIDENCE LEVEL (probability level).
Probability that random variable (result of measurement or mean value in series) assumes value within confidence interval.
D Konfidenzkoeffizient
F coefficient de sécurité, seuil de confiance
P poziom ufności
R доверительная вероятность

722
CONFIGURATION.
Steric arrangement of atoms (or groups of atoms) within the chiral or rigid (double bond or ring) part of molecule, disregarding changes caused by rotation about simple bonds. Configuration of molecule possessing a few chiral centres is fully determined when configuration of each centre is known.
D Konfiguration
F configuration
P konfiguracja
R конфигурация

723
CONFIGURATIONAL BASE UNIT (*of polymer or oligomer*).
Constitutional base unit of which configuration is defined at least at one site of steric isomerism in main chain. In regular polymer, configurational base units must correspond to constitutional repeating unit.
D konfigurative Grundeinheit
F ...
P jednostka podstawowa konfiguracyjna (*polimeru lub oligomeru*)
R ...

724
CONFIGURATIONAL BLOCK.
Block containing only one species of configurational base unit.
D ...
F ...
P blok konfiguracyjny
R ...

CONFIGURATIONAL INTEGRAL. *See* PARTITION FUNCTION.

CONFIGURATIONAL INTERACTION. *See* CONFIGURATION INTERACTION.

CONFIGURATIONAL PARTITION FUNCTION. *See* PARTITION FUNCTION.

CONFIGURATIONAL PROPERTIES. *See* CONFIGURATIONAL THERMODYNAMIC FUNCTIONS.

725
CONFIGURATIONAL THERMODYNAMIC FUNCTIONS (configurational properties).
Those parts of thermodynamic quantities which can be expressed in terms of configurational integral; other parts do not depend on intermolecular interactions.
D Konfigurationsgrößen
F propriétés configurationnelles
P wielkości konfiguracyjne
R конфигурационные величины

726
CONFIGURATION INTERACTION (configurational interaction).
Interaction between ground and excited states of atom, ion or molecule assumed for interpretation of hyperfine structure of spectra of electron spin resonance.
D Konfigurations-Wechselwirkung
F interaction de configuration
P oddziaływanie konfiguracyjne
R конфигурационное взаимодействие

727
CONFORMATION.
One of the non-identical spatial arrangements of atoms in a given molecule, formed by rotation about single bond (or bonds) without its (their) rupture. From an infinite number of possible conformations, the most stable (privileged) is the one in which all interactions between mutually non-bound substituents are the weakest.
D Konformation, Konstellation
F conformation
P konformacja
R конформация, констелляция

728
CONFORMATIONAL ANALYSIS.
Analysis of physical and chemical properties with respect to conformation of molecule in ground, transition or excited states.
D Konformationsanalyse
F analyse conformationnelle
P analiza konformacyjna
R конформационный анализ

CONFORMATIONAL ISOMERS. *See* CONFORMERS.

729
CONFORMERS (conformational isomers).
Conformations of molecule in equilibrium
and impossible to be isolated (rotational
barrier is approximately 2—10 kcal/mol).
D Konformere, Konformationsisomere,
  Rotamere, Konstellationsisomere
F isomères rotationnels
P konformery
R конформеры, конформационные
  изомеры

CONGLOMERATE. See RACEMIC
MIXTURE.

CONGO RUBIN NUMBER. See RUBIN
NUMBER.

730
CONGRUENT MELTING POINT.
Point on phase diagram in which solid
phase melts and changes into liquid phase
of the same composition.
D kongruenter Punkt
F point congruent
P punkt kongruentny
R конгруэнтная точка

CONJUGATED CHROMOPHORES. See
COMPOUND CHROMOPHORES.

731
CONJUGATED DOUBLE BONDS.
Double bonds isolated one from another
by single bond, e.g. butadiene
$CH_2=CH-CH=CH_2$
D konjugierte Doppelbindungen
F liaisons doubles conjuguées
P wiązania podwójne sprzężone
R конъюгированные двойные связи,
  сопряжённые двойные связи

732
CONRAD-LIMPACH SYNTHESIS OF
QUINOLINES.
Preparation of 4-hydroxyquinoline
derivatives by condensation of arylamine
with β-ketoester, followed by cyclization
of condensation product at 250°C, e.g.

D Conrad-Limpach
  (4-Hydroxy-)Chinolin-Synthese
F synthèse de Conrad et Limpach

P synteza (pochodnych chinoliny) Conrada
  i Limpacha
R синтез (хинолина) Конрада-Лимпаха

733
CONSECUTIVE INSTABILITY
CONSTANT (stepwise dissociation
constant).
Reciprocal of consecutive stability
constant.
D individuelle Dissoziationskonstante (des
  Komplexes)
F constante d'instabilité successive
P stała nietrwałości kolejna, stała
  nietrwałości stopniowa
R ступенчатая константа нестойкости

734
CONSECUTIVE REACTIONS.
Complex reaction in which product of
proceeding reaction is substrate of
subsequent one: A→B→C etc.
D Folgenreaktionen, Folgenreaktionssystem
F réactions consécutives
P reakcje następcze
R последовательные реакции,
  консекутивные реакции

735
CONSECUTIVE STABILITY CONSTANT
(stepwise formation constant), K_n.
Equilibrium constant of consecutive
formation reaction of complex
$ML_{n-1}+L \rightleftarrows ML_n$; under conditions
guaranteeing constancy of activation
coefficients

$$K_n = \frac{[ML_n]}{[ML_{n-1}][L]}$$

where $[ML_n]$ and $[ML_{n-1}]$ = equilibrium
concentrations of complexes,
$[L]$ = equilibrium concentration of ligand.
D individuelle Stabilitätskonstante,
  individuelle Bildungskonstante
F constante de stabilité succesive,
  constante consécutive
P stała trwałości kolejna, stała trwałości
  stopniowa, stała tworzenia
R ступенчатая константа устойчивости

CONSERVATION EQUATION FOR
MASS. See MASS CONSERVATION LAW
IN CONTINUOUS SYSTEMS.

CONSERVATION EQUATION FOR THE
MOMENTUM. See BALANCE EQUATION
FOR THE MOMENTUM DENSITY.

CONSOLUTE TEMPERATURE. See
CRITICAL SOLUTION TEMPERATURE.

736
CONSTANT-CURRENT COULOMETRY.
Electroanalytical method of determination
of substance from charge which flows in
time taken to reduce or oxidize this
substance completely at constant current.
D Coulometrie bei konstanter Stromstärke
F coulométrie à intensité constante
P kulometria przy stałym prądzie
R кулонометрия при контролируемой
силе тока

CONSTITUTION. *See* CHEMICAL
CONSTITUTION.

737
CONSTITUTIONAL BASE UNIT.
Smallest possible regularly repeated
constitutional unit of regular polymer.
D...
F...
P jednostka powtarzalna konstytucyjna,
jednostka podstawowa konstytucyjna
R...

738
CONSTITUTIONAL BLOCK.
Block containing only one species of
constitutional base unit.
D...
F...
P blok konstytucyjny
R...

739
CONSTITUTIONAL UNIT (mer *of
polymer or oligomer*).
Atoms or groups of atoms present in
macromolecules of polymer or oligomer.
D Grundbaustein (*des Polymers*), Mer
F mer
P jednostka konstytucyjna (*polimeru lub
oligomeru*), mer
R элементарное (повторяющееся) звено
(*полимера или олигомера*)

740
CONSTITUTIVE PROPERTIES.
Properties which depend on constitution
or structure of molecule.
D konstitutive Eigenschaften
F propriétés constitutives
P własności konstytucyjne
R конститутивные свойства

741
CONTACT ANGLE (wetting angle), Θ.
In system of coexisting liquid-solid-gas
phases, angle between flat surface of solid
body and tangent to liquid surface at
common point of three interfaces; in
equilibrium state

$$\cos \Theta = \frac{\sigma_{sg}-\sigma_{sl}}{\sigma_{lg}}$$

where σ_{sg} = solid-gas surface tension,
σ_{sl} = solid-liquid surface tension,
and σ_{lg} = liquid-gas surface tension.
D Kontaktwinkel, Randwinkel
F angle de contact, angle de mouillage
P kąt zwilżania
R краевой угол

742
CONTACT POTENTIAL DIFFERENCE
(Volta potential difference).
Difference of outer potentials of two
phases.
D Voltaspannung
F tension Volta
P napięcie Volty, potencjał Volty, różnica
potencjałów kontaktowa, potencjał
kontaktowy
R вольта-потенциал

743
CONTAINER (*in radiation technique*).
Device for storage and transportation
of radioactive material.
D Behälter, Kontainer
F conteneur, container
P pojemnik (ochronny), zasobnik, kontener
R контейнер

744
CONTAMINATION (*radioactive*).
Pollution of environment of any objects,
plants, animals, or human environment
by substances controlled and removed by
methods of radiological protection.
D Kontamination
F contamination
P skażenie promieniotwórcze,
kontaminacja
R радиоактивное загрязнение

CONTINUOUS PHASE. *See* DISPERSION
MEDIUM.

745
CONTINUOUS SOLID SOLUTION.
Solid solution of unlimited miscibility
of constituents.
D...
F...
P roztwór stały ciągły
R непрерывный твёрдый раствор

746
CONTINUOUS SPECTRUM.
Spectrum whose spectroscopic analysis
indicates presence of all possible
frequencies.
D kontinuierliches Spektrum
F spectre continu
P widmo ciągłe
R сплошной спектр

747
CONTINUOUS SYSTEM.
Macroscopic system with properties such
as temperature, pressure, concentration
of constituents, etc., changing continuously
in space.

D kontinuierliches System
F système continuel
P układ ciągły
R непрерывная система, сплошная
система

748
CONTRAST (*of photographic emulsion*).
Tangent of angle between linear part of
emulsion calibration curve and abscissa
axis (radiation intensity axis).

D Gradation
F gamma, contraste
P współczynnik kontrastowości,
współczynnik γ
R коэффициент контрастности

CONTRIBUTING STRUCTURES. *See*
RESONANCE STRUCTURES.

749
CONTROL CHART.
Graphical method of time variation
control of parameter; usually represented
by diagram with central line and two
parallel lines which determine tolerance
limits. Numbers of subsequent
measurements are given on abscissa and
values of parameter or deviations from
its nominal value are presented on
ordinate axis. Control chart is used for
production quality control, investigation
of conditions of processes and for
accuracy and precision control in
analytical methods.

D Kontrollkarte
F carte de contrôle
P karta kontrolna
R контрольная диаграмма

750
CONTROLLED POTENTIAL
ELECTROLYSIS.
Electrolysis carried out in conditions with
potential of working electrode having
constant value or changing only within
small limits as a result of changes of total
potential applied to electrodes.

D potentiostatische Elektrolyse
F électrolyse à potentiel contrôlé
P elektroliza z kontrolowanym
potencjałem
R электролиз при контролируемом
потенциале

751
CONTROL LINE PAIR (fixation pair).
In emission spectral analysis, pair of lines
of approximately equal intensity. Ratio of
their intensities depends crucially on
excitation conditions; it is used to control
stability of excitation source.

D Fixierungspaar
F raies de paire de fixage
P para linii kontrolna
R фикспара линий

752
CONVENTIONAL ANALYSIS.
Determination of component or properties
of investigated sample under conventional
but strictly defined conditions. As a result
one obtains not absolute value but one
dependent on method of analysis.

D ...
F ...
P analiza umowna, analiza
konwencjonalna
R ...

753
CONVERSE PIEZOELECTRIC EFFECT
(piezoelectricity).
Mechanical deformation of crystal due to
polarizing influence of external electric
field.

D umgekehrter piezoelektrischer Effekt,
Piezoelektrizität
F effet piézoélectrique inverse,
piézo-électricité
P efekt piezoelektryczny (odwrotny)
R пьезоэлектрический эффект,
пьезоэлектричество

754
CONVERSION (percentage conversion), p.
Fraction (or per cent) of initial substrate
concentration undergoing conversion;
$p = x/a$ where $x =$ concentration of
substrate after definite time, $a =$ initial
substrate concentration.

D ...
F taux de transformation
P stopień przemiany, stopień
przereagowania
R степень превращения

755
CONVERSION ELECTRONS.
Electrons emitted by atoms in internal
conversion. As opposed to β^- rays, they
possess a linear energy spectrum.

D Konversionselektronen
F électrons de conversion
P elektrony konwersji
R электроны конверсии

756
CONVERSION OF CHAIR FORMS
(interconversion of chair forms).
Conversion of one chair conformation of
a six-membered ring into another chair
form with intermediate formation of skew
boat form. In the majority of cyclohexane
derivatives, conversion of chair forms is
responsible for establishing equilibrium
between conformers, differing in axial or
equatorial location of substituent, e.g.

D Konformationsumwandlung der
  Sesselformen
F transformation de conformation en
  chaise, inversion du cycle
P inwersja pierścienia, konwersja
  pierścienia
R инверсия цикла

757
COOLING (cooling time).
Time allowed to elapse between removal
of radioactive product from reactor and
its chemical treatment, to reduce its
activity by decay.
D Abkühlung
F refroidissement
P schładzanie, studzenie
R остывание, охлаждение

COOLING TIME. See COOLING.

758
COORDINATE BOND (semipolar bond,
dative bond).
Particular case of polarized covalent bond,
when both electrons forming doublet are
supplied by one atom (donor), e.g.
in NH_4^+ ion.
D koordinative Bindung, halbpolare
  Bindung, semipolare Bindung
F liaison de coordination, liaison
  coordinative, liaison semipolaire, liaison
  (de covalence) dative
P wiązanie koordynacyjne, wiązanie
  półbiegunowe, wiązanie semipolarne
R координационная связь, семиполярная
  связь, донорно-акцепторная связь

COORDINATE OF THERMODYNAMIC
SYSTEM. See STATE VARIABLE.

759
COORDINATING ATOM (ligating atom,
donor atom).
Atom of ligand linked directly with
central atom.
D Ligandatom
F atome (donneur) du coordinat
P atom donorowy, atom ligandowy, atom
  koordynujący
R место присоединения лиганда

760
COORDINATING GROUP.
Part of chelating agent containing donor
atom, e.g. amine group —NH_2, —NHR or
—NR_2, azo group —N=N—, carbonyl
group =CO, ether group —O—.
D koordinierende Gruppe
F second groupe fonctionnel
P grupa koordynująca
R координирующая группа

761
COORDINATION (complex formation
reaction).
Process leading to formation of complex.
D Koordination, Komplexbildungsreaktion
F coordination, réaction de complexation
P koordynacja
R координация, реакция
  комплексообразования

762
COORDINATION CHEMISTRY.
Chemistry of complex compounds dealing
chiefly with complexes of metals.
D Koordinationschemie, Komplexchemie,
  chemische Koordinationslehre
F chimie de coordination
P chemia koordynacyjna
R координационная химия

763
COORDINATION COMPOUND (complex,
complex compound).
Compound containing complex ion or
compound which is neutral complex.
D Komplexverbindung,
  Koordinationsverbindung, Komplex
F complexe, composé de coordination
P związek kompleksowy, kompleks,
  związek koordynacyjny, związek
  zespolony
R комплексное соединение, комплекс

764
COORDINATION ISOMERISM.
Type of isomerism of complex salts
containing complex cation and complex
anion, depending on different distribution
of particular ligands in these complex
ions, e.g. [Co(NH₃)₆][Cr(CN)₆] and
[Cr(NH₃)₆][Co(CN)₆].
D Koordinationsisomerie
F isomérie de coordination
P izomeria koordynacyjna
R координационная изомерия

765
COORDINATION NUMBER (of atom or
ion in crystal).
Number of closest neighbours of atoms
or ions in crystal lattice, characteristic for
particular type of lattice.

D Koordinationszahl
F indice de coordination, chiffre de coordination, coordinence, nombre de coordination
P liczba koordynacyjna
R координационное число

766
COORDINATION NUMBER (ligancy).
In coordination chemistry, number of coordinate bonds formed by central atom.
D Koordinationszahl
F nombre de coordination
P liczba koordynacyjna
R координационное число

767
COORDINATION POSITION ISOMERISM.
Isomerism of polynuclear complexes consisting of different distribution of ligands around central atoms

$$\left[(H_3N)_3ClCo\underset{\underset{H}{O}}{\overset{\overset{H}{O}}{<}}Co(NH_3)_3Cl\right]SO_4$$

and

$$\left[(H_3N)_2Cl_2Co\underset{\underset{H}{O}}{\overset{\overset{H}{O}}{<}}Co(NH_3)_4\right]SO_4$$

D koordinative Stellungsisomerie
F ...
P izomeria pozycji koordynujących, izomeria rozmieszczenia
R изомерия положения

768
COORDINATION SPHERE (first coordination sphere, inner coordination sphere).
Space around central atom occupied by coordinated ligands.
D Koordinationssphäre, erste Sphäre
F sphère de coordination
P sfera koordynacyjna (pierwsza)
R внутренняя (координационная) сфера

769
COORDINATIVELY SATURATED COMPLEX.
Complex in which number of coordinate bonds is equal to maximum coordination number of central atom.
D koordinativ abgesättigter Komplex
F ...
P kompleks koordynacyjnie nasycony
R координационно насыщенный комплекс

770
COPE REARRANGEMENT.
Intramolecular thermal rearrangement of 1,5-dienes involving migration of substituents from allylic to vinylic position, e.g.

X i Y = $-COOR$, $-COOH$, $-CN$, $-C_6H_5$.

D Cope (Allyl→Vinyl)-Umlagerung
F transposition selon Cope
P przegrupowanie Cope'a
R перегруппировка Копа

771
COPLANARITY.
Arrangement of two or more planar groups such as: $-OH$, $>C=O$, $-NO_2$, $-CH=CH_2$, aromatic rings, etc. in one plane common for them.
D koplanare Einstellung
F coplanarité
P koplanarność
R копланарность

772
COPOLYMER.
Product of polymerization of mixture of more than one species of monomer.
D Copolymer
F copolymère
P kopolimer
R сополимер

773
COPOLYMERIZATION.
Polymerization of mixture containing at least two monomers.
D Copolymerisation, Multipolymerisation, Mischpolymerisation
F copolymérisation
P kopolimeryzacja
R сополимеризация, кополимеризация

774
COPPER, Cu.
Atomic number 29.
D Kupfer
F cuivre
P miedź
R медь

COPPER FAMILY. *See* COPPER GROUP.

775
COPPER GROUP (copper family).
Elements in first sub-group of periodic system with structure of outer electronic shells of atoms $(n-1)d^{10}ns^1$: copper, silver, gold.
D Kupfergruppe
F famille du cuivre
P miedziowce
R подгруппа меди, элементы подгруппы меди

776
CO-PRECIPITATION.
Precipitation of required substance
together with precipitates of other
substances present in solution, e.g.
impurities or very small amount of
radioactive substance with precipitate of
carrier.
D Mitfällung
F coprécipitation
P współstrącanie, koprecypitacja
R соосаждение

777
CORPUSCULAR RADIATION (particle
radiation).
Radiation composed of particles of rest
mass greater than zero, e.g. α-, β-,
neutron, proton, fission fragment
radiation.
D korpuskulare Strahlung
F rayonnement de particules, rayonnement
   corpusculaire
P promieniowanie korpuskularne
R корпускулярное излучение

778
CORRECTED SELECTIVITY
COEFFICIENT, $k_{A,B}^{a}$.
Selectivity coefficient of ion exchange
reaction calculated taking into account
activity coefficients of ions in solution.
D korrigierter Selektivitätskoeffizient
F constante apparente d'échange corrigée
P współczynnik selektywności poprawiony
R исправленный коэффициент
   селективности

779
CORRELATION.
Quantity which determines degree of
dependence between a given variable and
another one or between a given variable
and several others.
D Korrelation
F corrélation
P korelacja
R корреляция

780
CORRELATION OF CONFIGURATIONS.
Specification of mutual correlation
between configuration of an examined
compound and that of a reference
compound having relative configuration,
or a compound having established absolute
configuration.
D Korrelation der Konfigurationen
F corrélation des configurations
P korelacja konfiguracji
R корреляция конфигураций

CORRELATION OF ELECTRONS. See
ELECTRON CORRELATION.

781
CORRESPONDING STATES.
States of two substances characterized by
the same values of reduced parameters,
i.e. by the same reduced pressure p_r,
reduced temperature T_r and reduced
volume V_r.
D übereinstimmende Zustände
F états correspondants
P stany odpowiadające sobie
R соответственные состояния

COSMIC CHEMISTRY. See
COSMOCHEMISTRY.

782
COSMOCHEMISTRY (cosmic chemistry,
astrochemistry, space chemistry).
Branch of chemistry dealing with studies
on composition, abundance, formation and
exchange of substances present in celestial
bodies, cosmic dust and interstellar
matter.
D Kosmochemie, Astrochemie
F astrochimie
P kosmochemia, astrochemia
R космохимия, астрохимия

COTTON EFFECT. See CIRCULAR
DICHROISM.

783
COTTON-MOUTON EFFECT.
Double refraction of light in liquids in
magnetic field perpendicular to incident
direction of light.
D Cotton-Mouton-Effekt
F effet Cotton-Mouton
P efekt Cottona i Moutona
R эффект Коттон-Мутона

784
COULOMB ENERGY.
That part of electrostatic energy of
electron interaction which does not take
into account electron indistinguishability.
D Coulombwechselwirkungenergie
F énergie coulombienne
P energia kulombowska
R кулоновская энергия

785
COULOMB FORCES (electrostatic forces).
Interaction forces F between electric
charges. For two point charges q_1 and q_2
value of Coulomb forces is given by the
Coulomb law

$$F = \frac{q_1 q_2}{\varepsilon r^2} \quad \text{(in unrationalized electrostatic CGS system)}$$

or

$$F = \frac{q_1 q_2}{4\pi\varepsilon r^2} \quad \text{(in SI units)}$$

where: r = distance between charges,
ε = permittivity of medium.
D Coulomb-Kräfte
F forces de Coulomb
P siły kulombowskie, siły elektrostatyczne
R кулоновские силы

786
COULOMB POTENTIAL BARRIER.
Potential barrier of Coulomb (electrostatic) repulsive forces. In nuclear physics: barrier surrounding atomic nucleus and preventing penetration of positively charged particles, e.g. protons or α particles, into nucleus or escaping from nucleus.
D Coulomb-Barriere, Coulombscher Potentialwall
F barrière (de potentiel) coulombienne
P bariera potencjału kulombowskiego, bariera kulombowska
R потенциальный кулоновский барьер

787
COULOMETER (voltameter).
Apparatus used for measurement of electrical charge flowing in circuit; based on Faraday's laws of electrolysis.
D Coulometer, Voltameter
F coulomètre, voltamètre
P kulometr, woltametr
R кулонометр, вольтаметр

788
COULOMETRIC TITRATION.
Determination of substance by its reaction with reagent generated electrolytically with constant current. Final result is calculated from product of current intensity and time of generation of reagent.
D coulometrische Titration
F titrage coulométrique
P miareczkowanie kulometryczne
R кулонометрическое титрование

789
COULOMETRY.
Group of methods based on Faraday's laws of electrolysis used mainly in measurements of quantity of electricity and analytical determinations.
D Coulometrie
F coulométrie
P kulometria
R кулонометрия

790
COULOMETRY AT CONTROLLED POTENTIAL.
Electroanalytical method of determination of substance from charge which flows

in time taken to reduce or oxidize this substance completely at constant potential of electrode.
D Coulometrie bei konstantem Potential
F coulométrie à potentiel contrôlé
P kulometria przy stałym potencjale
R кулонометрия при контролируемом потенциале

791
COULOSTATIC METHOD.
Method of investigation of kinetics of electrode reactions from changes of potential of indicator electrode after injection of known electric charge to electrolytic system.
D coulostatische Impulsmethode
F méthode coulostatique
P metoda kulostatyczna
R кулоностатический метод

COUNTER. *See* RADIATION COUNTER.

792
2π-COUNTER.
Windowless radiation counter with semi-geometrical counting efficiency.
D 2π-Zähler
F compteur 2π
P licznik 2π
R счётчик с телесным углом 2π

793
4π-COUNTER.
Radiation counter with active volume completely surrounding source of radioactivity, and geometrical counting efficiency unity.
D 4π-Zähler
F compteur 4π
P licznik 4π
R счётчик с геометрией 4π, 4π-счётчик

794
COUNTER CHARACTERISTIC.
Curve of counting rate of a counter, against applied voltage.
D Zählrohrcharakteristik eines Auslösezählrohres
F courbe caractéristique du compteur
P charakterystyka licznika
R счётная характеристика счётчика, рабочая характеристика счётчика

795
COUNTER ELECTRODE.
Electrode placed in front of electrode containing sample.
D Gegenelektrode
F contre-électrode
P przeciwelektroda (spektralna)
R подставной электрод, противэлектрод

7 Dictionary of Chemical Terminology

796
COUNTER GEOMETRY.
Mutual spatial location of sample and detector.
D Meß-Geometrie
F géometrie de comptage
P geometria pomiaru aktywności (promieniotwórczej)
R геометрия измерения

797
COUNTER-IONS.
Mobile (able to diffuse freely) ions of opposite charge to that of ion exchanger skeleton.
D Gegenionen
F contre-ions
P przeciwjony
R противоионы

COUNTING EFFICIENCY. See
COUNTING YIELD.

798
COUNTING RATE.
In radiometry, number of counts recorded per unit time.
D Zählrate
F taux de comptage, vitesse de comptage
P szybkość liczenia
R скорость счёта

799
COUNTING YIELD (counting efficiency).
Ratio of observed counting rate to disintegration rate.
D Zählausbeute, Wirkungsgrad
F efficacité du comptage, rendement du comptage
P wydajność liczenia
R эффективность счёта

800
COUPLED REACTIONS (induced reactions).
System of two parallel reactions in which path of secondary reaction (induced reaction) is determined by result of primary reaction

A+B → M primary reaction

A+C → N induced (secondary) reaction

where A = actor in coupled reactions,
B = inductor in coupled (primary) reaction,
C = acceptor in coupled (secondary) reaction.
D induzierte Reaktionen, gekoppelte Reaktionen
F réactions de l'induction mutuelle
P reakcje sprzężone
R сопряжённые реакции, индуцируемые реакции

801
COUPLING.
Reaction of diazonium salt with phenol or aromatic amine, resulting in formation of coloured azo compound, e.g.

$(C_6H_5-N\equiv N)^{\oplus}X^{\ominus} + \langle\rangle-N(CH_3)_2 \longrightarrow$

$C_6H_5-N=N-\langle\rangle-N(CH_3)_2 + HX$

D Kupplung
F copulation
P sprzęganie
R азосочетание

802
COUPLING CONSTANT.
Value characterizing force of interaction between elementary particles; occurs as coefficient in interaction hamiltonian.
D Kopplungskonstante
F constante de couplage
P stała sprzężenia
R постоянная связи, константа связи

803
COVALENCE.
Number of covalent bonds formed by atom of given element.
D Kovalenz
F covalence
P wartościowość kowalencyjna, kowalencyjność
R ковалентность

804
COVALENT BOND (atomic bond, homopolar bond).
Chemical bond formed by sharing valence electrons in pairs, each electron originating from a different atom, e.g. in molecules of H_2, N_2 and Cl_2.
D kovalente Bindung, Atombindung, homöopolare Bindung, unpolare Bindung, unitarische Bindung, Elektronenpaarbindung
F liaison covalente, liaison atomique, liaison homopolaire, liaison de covalence
P wiązanie kowalencyjne, wiązanie atomowe, wiązanie homeopolarne, wiązanie niebiegunowe
R ковалентная связь, атомная связь, гомеополярная связь

COVALENT CRYSTALS. See ATOMIC CRYSTALS.

805
COVALENT RADIUS.
One half of bond length in homonuclear diatomic molecule. Sum of covalent radii of two different atoms determines to good approximation covalent bond length in corresponding molecule.
D kovalenter Radius
F rayon de covalence, rayon covalent

P promień kowalencyjny atomu
R ковалентный радиус

COVERING GROUP. *See* PROTECTING GROUP.

806
CRAM'S RULE.
Rule describing steric addition in asymmetric synthesis of organometallic compound to acyclic carbonyl compounds with chirality centre in α-position. If substituents are denoted by: L > M > S (depending on their steric bulks), nucleophilic attack on carbonyl group in compound with energetically privileged conformation (see example below) will take place from the most uncovered side

D Regel von Cram
F règle de l'orientation stérique
P reguła Crama
R правило Крама

CRITICAL COMPLEX. *See* ACTIVATED COMPLEX.

CRITICAL CONSTANTS. *See* CRITICAL PROPERTIES.

807
CRITICAL INDICES, χ.
Numbers characterizing speed of approaching zero by a given thermodynamic quantity X when reaching critical state

$$\chi^{\pm} = \lim_{T \to T_c^{\pm}} \frac{\ln X(V, T)}{\ln[\pm(T-T_c)]}$$

where V = system's volume, T = thermodynamic temperature, T_c = critical temperature; way of approaching critical state is separately defined for each quantity; most frequently $V = V_c$.

D kritische Indexe
F indices critiques
P wykładniki krytyczne
R критические показатели

808
CRITICAL ISOTHERM.
p-V isotherm of real gas at critical temperature. Inflection point of critical isotherm has coordinates p_c, V_c.

D kritische Isotherme
F isotherme critique
P izoterma krytyczna
R критическая изотерма

809
CRITICAL MASS.
Smallest mass of fissionable material in which chain fission reaction can take place.

D kritische Masse
F masse critique
P masa krytyczna
R критическая масса

810
CRITICAL OPALESCENCE.
Very strong light scattering in critical state or in its neighbourhood. Critical opalescence is caused by large density fluctuations.

D kritische Opaleszenz
F opalescence critique
P opalescencja krytyczna
R критическая опалесценция

811
CRITICAL PHASE.
Phase in critical state.

D kritische Phase
F phase critique
P faza krytyczna
R критическая фаза

812
CRITICAL PRESSURE, P_c.
Pressure in critical state of system.

D kritischer Druck
F pression critique
P ciśnienie krytyczne
R критическое давление

813
CRITICAL PROPERTIES (critical constants).
Values of intensive quantities in critical state of system.

D kritische Daten, kritische Größen
F grandeurs critiques, constantes critiques
P parametry krytyczne, wielkości krytyczne
R критические параметры, критические константы

814
CRITICAL SOLUTION TEMPERATURE (consolute temperature).
Temperature at which interface vanishes in system of two liquids of limited mutual solubility. System has one phase in whole range of concentrations above upper or below lower critical solution temperature.

D kritische Mischungstemperatur, kritische Lösungstemperatur
F température critique de solubilité
P temperatura mieszalności krytyczna, temperatura rozpuszczania krytyczna
R критическая температура растворения

815
CRITICAL STATE.
State in which all properties of two
coexisting phases become identical.
D kritischer Zustand
F état critique
P stan krytyczny
R критическое состояние

816
CRITICAL TEMPERATURE, T_c.
Temperature of system in critical state.
D kritische Temperatur
F température critique
P temperatura krytyczna
R критическая температура

817
CRITICAL VOLUME, V_c.
Molar volume of system in critical state.
D kritische Volumen
F volume critique
P objętość krytyczna
R критический объём

818
CROSS BOMBARDMENT METHOD.
Method for assigning atomic masses to
radionuclides by producing a radionuclide
of a given half-life in different nuclear
reactions.
D Kreuzfeuermethode
F recoupement par bombardement
P metoda reakcji krzyżowych
R метод перекрёстных реакций

819
CROSSLINKING.
Process of formation of numerous
intermolecular covalent bonds between
polymer chains.
D Vernetzung
F réticulation
P sieciowanie
R сшивание

820
CROSS-SECTION.
Expression for probability of defined
elementary process (e.g. elastic collision,
chemical reaction, nuclear reaction,
recombination of free radicals) between
incident particles and particles constituing
target.
D Wirkungsquerschnitt
F section efficace
P przekrój czynny
R эффективное сечение

821
CROWN CONFORMATION (crown form).
Fairly stable conformation of cyclooctane
and its derivatives

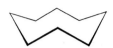

D Krone-Konformation
F conformation de couronne
P konformacja koronowa
R конформация короны

CROWN FORM. See CROWN
CONFORMATION.

822
CRYOHYDRATE.
Eutectic system formed from salt and
water.
D Kryohydrate
F cryohydrate
P kriohydrat
R криогидрат

823
CRYOHYDRATE POINT.
Eutectic point in system in which water
is one of constituents.
D Kryohydratpunkt, kryohydratischer
  Punkt
F point de cryohydrate
P punkt kriohydratu
R криогидратная точка

824
CRYOMETRY (cryoscopy).
Methods for measurement of freezing
temperature of uni- or multi-component
systems. Cryometric methods are used for
determination of molar mass,
determination of purity of substances,
for investigation of liquid-solid phase
equilibrium, etc.
D Kryometrie, Kryoskopie
F cryométrie, cryoscopie
P kriometria
R криометрия, криоскопия

825
CRYOSCOPIC CONSTANT, K.
Quantity given by equation

$$K = \frac{RT^2}{\Delta H} M$$

where R = molar gas constant,
T = thermodynamic temperature of
freezing of solvent at pressure
$0.980\,665 \cdot 10^5$ Pa, ΔH = molar enthalpy
(heat) of fusion of solvent at this
temperature, and M = molar mass of
solvent.
D kryoskopische Konstante, molare
  Gefrierpunktserniedrigung
F constante cryoscopique
P stała krioskopowa, molalne obniżenie
  temperatury krzepnięcia
  (rozpuszczalnika)
R криоскопическая константа

CRYOSCOPY. See CRYOMETRY.

826
CRYOSOL.
Colloid stable only at relatively low
temperatures, e.g. ice in chloroform,
stable at $\sim -20°C$.
D Kryosol
F cryosol
P kriozol
R криозоль

827
CRYSTAL.
Solid body characterized by periodic
distribution of particles (molecules, atoms
or ions) in space.
D Kristall
F cristal
P kryształ
R кристалл

828
CRYSTAL AXES (crystallographic axes).
System of three (sometimes four) axes
intersecting at one point and connected
with crystal in such a way that symmetry
operations mutually transform axes. Edges
of unit cell are parallel to axes, according
to accepted convention.
D kristallographische Achsen,
Kristallachsen
F axes cristallographiques, axes cristallins
P osie krystalograficzne
R кристаллографические оси

829
CRYSTAL CHEMISTRY.
Branch of chemistry dealing with
establishment of dependence of chemical
composition on structure and properties
of crystals.
D Kristallchemie
F cristallochimie
P krystalochemia
R кристаллохимия

830
CRYSTAL CLASSES.
Thirty two classes among infinite number
of symmetry classes, occurring in ideal
crystals.
D (natürliche) Kristallklassen
F classes cristallographiques
P klasy krystalograficzne
R кристалл(ограф)ические классы

CRYSTAL DEFECTS. See LATTICE
DEFECTS.

831
CRYSTAL FIELD (crystalline field).
Electric field at a given point inside
crystal produced by surrounding ions or
dipoles.
D Kristallfeld
F champ cristallin
P pole krystaliczne
R кристаллическое поле

832
CRYSTAL FIELD SPLITTING ENERGY,
Δ (10 Dq).
Difference in energies of orbitals in crystal
field theory, e.g. between e_g and t_{2g}
of central atom in the case of octahedral
complex.
D Aufspaltungsenergie
F énergie de séparation
P energia rozszczepienia polem
(krystalicznym) ligandów, parametr
rozszczepienia polem (krystalicznym)
ligandów
R энергия расщепления

833
CRYSTAL FIELD STABILIZATION
ENERGY (ligand field stabilization
energy).
Difference between average and true
orbital energy of d electrons of central
atom in complex.
D Kristallfeldstabilisierungsenergie,
Stabilisierungsenergie
F énergie de stabilisation
P energia stabilizacji polem
(krystalicznym) ligandów
R энергия стабилизации
кристаллическим полем

834
CRYSTAL FIELD THEORY.
Theory assuming bond between central ion
of transition metal and ligand to be of
ionic character, and considering influence
of field of negative ligands surrounding
the cation on energy of its d orbitals
(cf. crystal field splitting energy).
D Kristallfeldtheorie
F théorie du champ cristallin
P teoria pola krystalicznego
R теория кристаллического поля

835
CRYSTAL LATTICE (crystal space
lattice).
Regular set of geometrical points (nodes)
of ideal periodicity in three directions in
space. Lattice is geometrical concept; there
are 14 basic types of lattice (see Bravais
lattice).
D Kristallgitter, Raumgitter
F réseau cristallin
P sieć przestrzenna kryształu, sieć
krystaliczna
R решётка кристалла, кристаллическая
решётка, пространственная решётка

CRYSTALLINE FIELD. See CRYSTAL
FIELD.

836
CRYSTALLINE POLYMER.
Polymer containing considerable regions of crystalline species.
D krystallines Polymer
F polymère cristallin
P polimer krystaliczny
R кристаллический полимер

837
CRYSTALLINE STATE.
Stable state of matter characterized by periodic distribution of atoms, ions and molecules in space.
D kristallin(isch)er Zustand
F état cristallin
P stan krystaliczny
R кристаллическое состояние

838
CRYSTALLITE.
A part (grain) of solid with high degree of order (arrangement) of dimensions of some scores of nanometres.
D Kristallit
F cristallite
P krystalit
R кристаллит

839
CRYSTALLIZATION.
Process of crystal formation and growth from melt, solution or gaseous phase.
D Kristallisation
F cristallisation
P krystalizacja
R кристаллизация

840
CRYSTALLIZATION FROM GASEOUS PHASE.
Process of formation and growth of crystals from gaseous phase.
D Kristallisation aus der Gasphase
F cristallisation de la phase gazeuse
P krystalizacja z fazy gazowej, kondensacja sublimacyjna
R кристаллизация газовой фазы

841
CRYSTALLIZATION POLARIZATION.
Type of electrolytic polarization consisting in slow removal or accumulation of ion in lattice of electrode metal in comparison to other steps of electrode process. Crystallization polarization is expressed by crystallization overvoltage.
D Kristallisationspolarisation
F polarisation de cristallisation
P polaryzacja krystalizacyjna
R кристаллизационная поляризация

CRYSTALLOGRAPHIC AXES. See
CRYSTAL AXES.

842
CRYSTALLOGRAPHIC POINT GROUPS.
32 groups among infinite number of point groups, constructed from common axes or one-, two-, three-, four- and six-fold axes which may occur in crystal lattices.
D ...
F groupes ponctuels cristallins
P grupy punktowe krystalograficzne
R точечные кристаллографические группы

843
CRYSTALLOGRAPHIC SYSTEMS (crystal systems).
General mode of repartition of crystal substances in respect to geometrical relations between lengths of elementary cell a, b, c and angles between them α, β, γ. There are seven crystal systems:
triclinic
$(a \neq b \neq c, \alpha \neq \beta \neq \gamma \neq 90°)$,
monoclinic
$(a \neq b \neq c, \beta \neq \alpha = \gamma = 90°)$,
orthorhombic
$(a \neq b \neq c, \alpha = \beta = \gamma = 90°)$,
trigonal (rhombohedral)
$(a = b = c, \alpha = \beta = \gamma < 120°)$,
hexagonal
$a = b \neq c, \alpha = \beta = 90°, \gamma = 120°)$,
tetragonal
$(a = b \neq c, \alpha = \beta = \gamma = 90°)$,
cubic
$(a = b = c, \alpha = \beta = \gamma = 90°)$.
D Kristallsysteme
F systèmes cristallins
P układy krystalograficzne
R кристаллические системы, сингонии кристаллов

CRYSTALLOGRAPHIC ZONE. See
CRYSTAL ZONE.

844
CRYSTALLOID.
Historical term denoting all crystalline substances (in contrast to colloidal substances).
D Kristalloid
F cristalloïde
P krystaloid
R кристаллоид

845
CRYSTALLOLUMINESCENCE (luminescence of crystallization).
Emission of electromagnetic radiation during crystallization from solutions.

D Kristall-Lumineszenz
F cristalloluminescence, luminescence
de cristallisation
P krystaloluminescencja, luminescencja
krystalizacji
R кристаллолюминесценция,
люминесценция кристаллизации

846
CRYSTAL PHOSPHORS.
Inorganic crystalline luminophors mostly
prepared by sintering some appropriate
solids (e.g. sulfides or silicates of alkaline
earth metals) with a little admixture of
activator (e.g. copper, manganese or silver)
and flux.

D Kristallphosphoren
F cristaux phosphorescents
P fosfory krystaliczne (aktywowane)
R кристаллофосфоры

CRYSTAL SPACE LATTICE. *See*
CRYSTAL LATTICE.

847
CRYSTAL STRUCTURE.
Real population of particles (atoms, ions,
molecules) positioned regularly in space.
There are 14 crystal lattices, but crystal
structures are infinite in number.

D Kristallstruktur
F structure cristalline
P struktura krystaliczna, struktura
kryształu
R кристаллическая структура,
структура кристалла

CRYSTAL SYSTEMS. *See*
CRYSTALLOGRAPHIC SYSTEMS.

848
CRYSTAL ZONE (crystallographic zone).
Set of planes (not necessarily crystal faces)
which have one common direction in space
(axis of zone).

D Kristallzone
F zone (cristallographique)
P pas krystalograficzny
R кристаллическая зона,
кристаллический пояс

C-T BAND. *See* CHARGE-TRANSFER
BAND.

849
CUMULATED DOUBLE BONDS.
Double bonds located in molecule next to
each other, e.g. $R_2C=C=CR_2$.

D kumulierte Doppelbindungen
F liaisons (doubles) cumulées
P wiązania podwójne skumulowane
R кумулированные двойные связи

CUMULATIVE FORMATION CONSTANT.
See STABILITY CONSTANT.

CUMULATIVE INSTABILITY
CONSTANT. *See* OVERALL
INSTABILITY CONSTANT.

850
CUPELLATION.
Separation of noble metals from other
components by melting with other metals
(e.g. lead) and heating resulting alloy in
porous crucible in presence of air. Melted
oxide of non-noble metal is absorbed by
walls of crucible leaving noble metals.

D Kupellation
F coupellation
P analiza kupelacyjna, kupelacja
R купеляция, купелирование

851
CURIE, Ci.
Obsolete unit of radioactivity giving
$3.7 \cdot 10^{10}$ disintegrations per second;
in SI system 1 Ci = 37 GBq.

D Curie
F curie
P curie, kiur
R кюри

852
CURIE CONSTANT, C.
Constant entering Curie law and Curie-
Weiss law, $C = N_A \mu^2/3k$, where N_A =
Avogadro constant, μ = atomic or
molecular magnetic moment, k =
Boltzmann constant.

D Curie-Konstante
F constante de Curie
P stała Curie
R константа Кюри

853
CURIE LAW.
Physical law which determines
temperature dependence of paramagnet's
magnetic susceptibility χ, $\chi = C/T$, where C
denotes Curie constant.

D Curie-Gesetz
F loi de Curie
P prawo Curie
R закон Кюри

CURIE POINT. *See* CURIE
TEMPERATURE.

854
CURIE TEMPERATURE (Curie point).
Critical temperature at which ferro- and
ferrimagnets lose their spontaneous
magnetization and become paramagnets or
critical temperature at which ferroelectrics
lose their spontaneous polarization.

D Curie-Temperatur, Curie-Punkt
F température de Curie, point de Curie
P temperatura Curie, punkt Curie
R температура Кюри, точка Кюри

855
CURIE-WEISS LAW.
Physical law which determines
temperature dependence of normal
paramagnet magnetic susceptibility χ

$$\chi = \frac{C}{T-\Theta}$$

where C = Curie constant, Θ = Weiss
constant. Curie-Weiss law is satisfied by
susceptibilities of ferro- and
antiferromagnets above Curie and Néel
temperatures, respectively.
D Curie-Weiss-Gesetz
F loi de Curie-Weiss
P prawo Curie i Weissa, prawo Curie
uogólnione
R закон Кюри-Вейсса

856
CURIUM, Cm.
Atomic number 96.
D Curium
F curium
P kiur
R кюрий

857
CURIUM SERIES.
Radioactive series with first member of
reasonably long half-life curium, ^{245}Cm,
and last member bismuth, ^{209}Bi,
respectively. The mass number of each
member of this series is $A = 4n+1$,
where n = natural number.
Remark: Previously this series was called the
neptunium series.

D Curium-Reihe
F famille de curium
P szereg kiurowy
R ряд кюрия

858
CUROIDS.
Inner transition elements belonging to
actinide series of atomic number
$Z = 96—103$: curium, berkelium,
californium, einsteinium, fermium,
mendelevium, nobelium, lawrencium.
D Curoide
F curides
P kiurowce
R кюриды

859
CURRENT DENSITY, j.
Ratio of current intensity I in electrolytic
circuit and of geometrical electrode
surface A which is in contact with
electrolyte: $j = I/A$ (in the case of anode,
j = anodic current density, for cathode,
j = cathodic current density).
D Stromdichte
F densité du courant
P gęstość prądu
R плотность тока

860
CURRENT EFFICIENCY.
Ratio of charge used by a given electrode
reaction to total charge which flows
across circuit during electrolysis; if only
one process takes place on electrode then
current efficiency is 100 per cent.
D Stromausbeute
F rendement en courant
P wydajność prądowa
R выход по току

CURRENTS. *See* THERMODYNAMIC
FLUXES.

861
CURTIN-HAMMETT PRINCIPLE.
Composition of *cis-* and *trans*-alkenes in
elimination product depends on activation
energy of process (not on presence of
conformers satisfying requirements for
trans-elimination).
D Curtin-Hammettsche Regel
F règle de Curtin et Hammett
P reguła Curtina i Hammetta
R правило Куртина и Гаммета

CURTIUS DEGRADATION. *See* CURTIUS
REACTION.

862
CURTIUS REACTION (Curtius
degradation).
Transformation of carboxylic acid azides
into primary amines containing one carbon
atom less in molecule; reaction follows
route: decomposition of azide with
rearrangement to isocyanate; hydrolysis of
the latter with formation of amine, e.g.

$$RC\overset{O}{\underset{N_3}{\diagdown}} \xrightarrow{-N_2} R-N=C=O \xrightarrow{H_2O} RNH_2 + CO_2$$

D Curtius (Carbonsäureazid→Amin)-
-Abbau
F transposition de Curtius, décomposition
selon Curtius
P reakcja Curtiusa
R реакция Курциуса

CUVETTE. *See* ABSORPTION CELL.

863
CYANOETHYLATION.
Substitution of acidic hydrogen atom in
molecule of chemical compound by
cyanoethyl group $-CH_2CH_2CN$, effected
by vinyl cyanide in presence of basic
catalyst, e.g.

$$H_2S + CH_2=CHCN \xrightarrow{base}$$

$$HSCH_2CH_2CN \xrightarrow{CH_2=CHCN} S\overset{CH_2CH_2CN}{\underset{CH_2CH_2CN}{\diagup}}$$

$$C_6H_5COCH_3 + 3CH_2=CHCN \xrightarrow{base}$$
$$C_6H_5COC(CH_2CH_2CN)_3$$

D Cyanethylierung
F cyanoéthylation
P cyjanoetylowanie
R цианэтилирование

864
CYANOMETHYLATION.
Substitution of nitrogen-bound hydrogen in molecule of chemical compound such as amine by cyanomethyl group —CH_2CN, effected by sodium cyanide, formaldehyde and sodium hydrogen sulfite

$$R_2NH + HCHO + NaCN \xrightarrow[-Na_2SO_3]{NaHSO_3}$$

$$R_2NCH_2CN + H_2O$$

D Cyanmethylierung
F cyanométhylation
P cyjanometylowanie
R цианметилирование

865
CYCLE.
Thermodynamic process with final and initial state identical.
D Kreisprozeß
F cycle fermé, transformation cyclique
P przemiana cykliczna, proces cykliczny, obieg termodynamiczny
R круговой процесс, цикл

866
CYCLIC FORM.
Tautomeric form of cyclic structure, existing in equilibrium with its open-chain form. Tautomeric equilibrium often exists in several cyclic forms, differing in size, e.g. open-chain D-ribose exists in tautomeric equilibrium with α- and β-D-ribofuranose and with α- and β-D-ribopyranose.
D Ringform
F forme cyclique
P odmiana cykliczna, odmiana pierścieniowa
R циклическая форма

CYCLIC TRIANGULAR WAVE METHOD.
See CYCLIC VOLTAMMETRY.

867
CYCLIC VOLTAMMETRY (cyclic triangular wave method).
Method of study of mechanism of electrode processes based on electrolysis carried out under conditions of linearly changing potential in cyclic manner between two previously prescribed potentials. Product of reduction formed in cathodic half-cycle may be oxidized in anodic one.
D ...
F voltammétrie cyclique
P chronowoltamperometria cykliczna
R циклическая хроновольтамперометрия

868
CYCLIZATION (ring closure, ring formation).
Formation of cyclic molecules, e.g.

D Cyclisierung, Ringschlußreaktion
F cyclisation
P cyklizacja, zamykanie pierścienia
R циклизация

869
1,1-CYCLOADDITION (*of carbenes and nitrenes*).
Addition of carbene or nitrene in singlet state to unsaturated system with formation of three-membered ring; reaction is stereospecific, e.g.

D (*Addition von Carben oder Nitren an Olefine*)
F addition 1,1 (*du carbène ou nitrène sur les éthyléniques*)
P przyłączenie wielocentrowe 1,1, cykloaddycja 1,1 (*karbenów i nitrenów*)
R (*присоединение карбенов или нитренов к этиленовым связям*)

870
1,2-CYCLOADDITION (2+2 cycloaddition).
Dimerization of unsaturated compounds containing electron-withdrawing substituents, leading generally to formation of four-membered ring, e.g.

D 1,2-cyclische Addition
F cycloaddition 1,2
P przyłączenie wielocentrowe 1,2, cykloaddycja 1,2
R циклоприсоединение в положение 1,2

2+2 CYCLOADDITION. *See* 1,2-CYCLOADDITION.

871
CYCLOTRON.
Magnetic resonance accelerator of charged particles (protons, deuterons and heavy ions).
D Zyklotron
F cyclotron
P cyklotron
R циклотрон

872
DAKIN HYDROXYLATION REACTION.
Substitution of formyl or keto group in aromatic o- or p-hydroxyaldehydes and ketones by hydroxy group, effected with hydrogen peroxide in alkaline medium, e.g.

D Dakin-Reaktion, (Dakin) Phenolaldehyd-
-Oxydation
F réaction de Dakin
P reakcja Dakina
R реакция Дакина

DAKIN-WEST ACYLAMINO-KETONE
SYNTHESIS. *See* DAKIN-WEST
REACTION.

873
DAKIN-WEST REACTION (Dakin-West acylamino-ketone synthesis).
Formation of α-(acylamino)-methylo-ketones from α-amino acids on heating with acetic anhydride in presence of base (usually pyridine), e.g.

$C_6H_5CH_2-CH-COOH + (CH_3CO)_2O \longrightarrow$
            |
            NH_2

$C_6H_5CH_2-CH-COCH_3 + CO_2 + H_2O$
            |
            $NHCOCH_3$

D Dakin-West (α-Acylaminoketon)-
-Synthese
F réaction de Dakin et West
P reakcja Dakina i Westa
R ...

DALTONIDES. *See* STOICHIOMETRIC
COMPOUNDS.

874
DALTON'S LAW (*of partial pressures*).
Total pressure p of mixture of gases is equal to sum of partial pressures p_i of constituents of mixture

$$p = \sum_{i=1}^{n} p_i$$

Dalton's law is empirical.
D Daltonsches Gesetz
F loi de Dalton

P prawo Daltona
R закон Дальтона

875
DANIELL CELL.
Cell of $Cu|CuSO_4aq|ZnSO_4aq|Zn$ type.
D Daniell-Element
F pile Daniell, élément Daniell
P ogniwo Daniella
R элемент Даниеля-Якоби

876
DARZENS-CLAISEN GLYCIDIC ESTER
CONDENSATION.
Reaction of aldehyde or ketone with α-haloester in presence of basic condensation agent (e.g. sodium alcoholate, sodium amide), resulting in formation of glycidic ester, e.g.

$RR'CO + XCHR''COOC_2H_5 \xrightarrow{NaNH_2}$

D Darzens-Erlenmeyer-Claisen-
-Kondensation, Glycidester-Kondensation
(nach Darzens)
F condensation de Darzens
P kondensacja (estrów glicydowych)
Darzensa
R синтез глицидных эфиров по Дарзану

877
DARZENS SYNTHESIS OF
UNSATURATED KETONES.
Preparation of unsaturated cyclic ketones from cycloalkenes by action of acetyl chloride, in presence of aluminium chloride, e.g.

D Darzens Cycloolefin-Acylierung
F synthèse de Darzens
P synteza nienasyconych ketonów
Darzensa
R синтез непредельных кетонов Дарзана

DATIVE BOND. *See* COORDINATE
BOND.

878
DAUGHTER (daughter nuclide).
Nuclide formed as a result of decay of another nuclide.
D Tochternuklid
F descendant (radioactif), produit
de filiation
P nuklid pochodny
R дочерний нуклид

DAUGHTER NUCLIDE. *See* DAUGHTER.

879
DEACTIVATION OF CATALYST.
Loss of ability to catalyze by catalyst,
e.g. occurring because of ageing, sintering
or poisoning.
D Aktivitätsverlust des Kontaktes
F réduction de l'activité du catalyseur
P dezaktywacja kontaktu
R дезактивация катализатора

880
DEACTIVATION OF EXCITED STATE
(*of atom or molecule*).
Loss of excitation energy of atom or
molecule due either to non-radiative or
radiative transitions or to chemical
reaction.
D Desaktivierung des angeregten
Zustandes
F désactivation d'état excité
P dezaktywacja stanu wzbudzonego
R дезактивация возбуждённого
состояния

881
DEACTIVATION OF MOLECULE.
Loss of energy by molecule which
previously was able to react.
D Desaktivierung der Molekel,
Stabilisierung des Moleküls
F désactivation de la molécule
P dezaktywacja cząsteczki, stabilizacja
cząsteczki
R дезактивация молекулы

882
DEAD TIME (insensitive time).
Time immediately after receiving a
stimulus during which counter is
insensitive to another impulse or stimulus.
D Totzeit
F temps mort
P czas martwy (licznika)
R мёртвое время

883
DEAD VOLUME (*of chromatographic
column*).
Volume of mobile phase contained
between lower end of bed of stationary
phase and outlet of column.
D Totvolumen
F espace mort
P objętość martwa
R мёртвый объём, внеколоночный объём

884
DEALKYLATION.
Elimination of alkyl group from molecule
of organic compound.
D Entalkylierung
F dé(s)alcoylation

P dezalkilowanie, odalkilowanie
R дезалкилирование

885
DEAMINATION (desamination).
Elimination of amino group from molecule,
e.g.

$$2CH_3(CH_2)_2NH_2 \xrightarrow[-N_2]{HNO_2} CH_3(CH_2)_2OH + \\ + CH_3CH=CH_2$$

D Desaminierung
F dé(s)amination
P dezaminowanie, odaminowanie
R дезаминирование

886
DE BROGLIE WAVE.
Matter wave ascribed to moving particle.
De Broglie wavelength $\lambda = h/p$, frequency
$\nu = E/h$, where h = Planck constant,
E = particle energy and p = particle
momentum.
D Materiewelle
F onde de de Broglie
P fala de Broglie'a, fala materii
R волна материи

DEBYE-FALKENHAGEN'S EFFECT. *See*
DISPERSION OF CONDUCTANCE.

DEBYE FREQUENCY. *See*
CHARACTERISTIC FREQUENCY.

887
DEBYE-HÜCKEL-ONSAGER'S
LIMITING LAW FOR CONDUCTIVITY.
Formula for equivalent conductivity Λ_c of
diluted solution of strong electrolyte:

$$\Lambda_c = \Lambda_0 - \left[\frac{2.801 \cdot 10^6 |z_+ z_-| q}{(\varepsilon_r T)^{3/2} (1 + \sqrt{q})} \Lambda_0 + \right. \\ \left. + \frac{41.25 (|z_+| + |z_-|)}{\eta (\varepsilon_r T)^{1/2}} \right] \sqrt{J}$$

where:

$$q \equiv \frac{|z_+ z_-|}{|z_+| + |z_-|} \frac{\lambda_{0+} + \lambda_{0-}}{|z_+| \lambda_{0-} + |z_-| \lambda_{0+}}$$

and η = viscosity of solution, Λ_0 = limiting
equivalent ionic conductivity, ε_r = relative
permittivity of solution, J = ionic
strength; in aqueous diluted solutions of
1.1-electrolytes ($c \leqslant 10^{-3}$ mole/dm³), this
law gives results which agree with
experimental results.
D Debye-Hückel-Onsagersches
Grenzgesetz
F loi limite de Debye, Hückel et Onsager
P równanie Debye'a, Hückela i Onsagera
graniczne
R предельное уравнение
электропроводности
Дебая-Гюккеля-Онзагера

888
DEBYE-HÜCKEL'S EQUATION.
Semiempirical formula which describes mean activity coefficient $f_\pm$ of strong electrolyte

$$\lg f_\pm = -\frac{A|z_+ z_-|\sqrt{J}}{1+Ba\sqrt{J}}$$

$$B \equiv \left(\frac{8\pi N_A e^2}{1000k}\right)^{1/2} \frac{10^{-8}}{(\varepsilon_r T)^{1/2}} = \frac{50.29}{(\varepsilon_r T)^{1/2}}$$

where a = effective diameter of ion; other symbols *see* Debye-Hückel's limiting law.

D Debye-Hückelsche Gleichung
F formule de Debye et Hückel
P równanie Debye'a i Hückela
R уравнение Дебая-Гюккеля

889
DEBYE-HÜCKEL'S LIMITING LAW.
Formula which describes mean activity coefficient $f_\pm$ of strong electrolyte in diluted solution:

$$\lg f_\pm = -A|z_1 z_2|\sqrt{J}$$

$$A \equiv \sqrt{\frac{2\pi N_A}{1000}}\ \frac{e^3}{2.303(k\varepsilon_r T)^{3/2}} = \frac{1.8246\cdot 10^6}{(\varepsilon_r T)^{3/2}}$$

where e = elementary charge, ε_r = relative permittivity, k = Boltzmann constant, J = ionic strength; it gives results which agree with experiment if J does not exceed 0.02—0.03 (in aqueous solution).

D Debye-Hückelsches Grenzgesetz, Debye-Hückelsche Grenzgleichung
F loi limite de Debye et Hückel
P równanie Debye'a i Hückela graniczne, prawo Debye'a i Hückela graniczne
R предельный закон Дебая-Гюккеля

890
DEBYE-HÜCKEL'S THEORY OF STRONG ELECTROLYTES.
Theory which describes simplified model of diluted solution of strong electrolyte: (a) interionic interactions are exclusively coulombic in nature and are small in comparison to energy of thermal agitation, (b) ions are non-polarized spherical charges, and dielectric constant of solution is equal to that of pure solvent, (c) electrolyte is completely dissociated; theory introduces concept of ionic atmosphere and describes model mathematically.

D Debye-Hückel-Theorie der starken Elektrolyte
F théorie des électrolytes forts de Debye et Hückel
P teoria elektrolitów mocnych Debye'a i Hückela
R теория (сильных электролитов) Дебая-Гюккеля

891
DEBYE-SCHERRER METHOD (powder diffraction method).
X-ray diffraction technique for investigation of crystal structure. X-ray beam is scattered by powdered micro-crystalline sample and diffracted rays are recorded.

D Debye-Scherrer-Verfahren, Pulverbeugungsverfahren
F procédé de Debye et Scherrer, méthode de Hull-Debye-Scherrer
P metoda Debye'a, Scherrera i Hulla, metoda proszków krystalicznych, metoda proszkowa
R метод Дебая-Шеррера, метод поликристалла

892
DEBYE THEORY.
Theory of dielectric polarization which was the first to take into account orientation polarization. Assuming that local fields is given by Lorenz field, Debye theory expresses relative permittivity ε_r, and density of medium ρ in terms of induction polarizability α_I and permanent dipole moment μ

$$P_m = \frac{\varepsilon_r - 1}{\varepsilon_r + 2}\ \frac{M_r}{\rho} = \frac{N_A}{3\varepsilon_0}\left(\alpha_I + \frac{\mu^2}{3kT}\right)$$

(in SI units)
where: P_m = (molar) polarization, M_r = relative molecular mass, N_A = Avogadro constant, ε_0 = permittivity of free space, k = Boltzmann constant, T = thermodynamic temperature.

D Debyesche Theorie
F théorie de Debye
P teoria Debye'a
R теория Дебая

893
DEBYE T³ LAW.
At very low temperatures, i.e. those below $\Theta/10$, where Θ = characteristic Debye temperature, atomic heat capacities of crystalline simple substances are approximately proportional to cube of thermodynamic temperature.

D Debyesches T³-Gesetz
F approximation en T³ de Debye
P prawo Debye'a, prawo trzeciej potęgi
R закон кубов

894
DEBYE UNIT, D.
Unit of dipole moment; $1\,D = 1\cdot 10^{-18}$ e.s.u. of electric dipole moment; in SI units $1\,D = 3.33\cdot 10^{-30}$ C m.

D D(ebye)-Einheit, Debye
F debye

P debaj
R дебай

895
DECARBONYLATION.
Elimination of carbon monoxide molecule from molecule of organic compound, e.g.

$$RCOCH_2COCOOR \rightarrow RCOCH_2COOR + CO$$

D Decarbonylierung
F décarbonylation
P dekarbonylacja
R декарбонилирование

896
DECARBOXYLATION.
Elimination of carbon dioxide molecule from organic acid molecule (or its salt) with introduction of hydrogen atom at place of former carboxylic group, e.g.

$$\underset{\underset{COOH}{|}}{\overset{\overset{COOH}{|}}{CH_2}} \xrightarrow{\Delta} CH_3COOH + CO_2$$

D Decarboxylierung
F décarboxylation
P dekarboksylacja
R декарбоксилирование

DECAY LAW. *See* LAW OF RADIOACTIVE DECAY.

897
DECAY SCHEME (disintegration scheme).
Graphical presentation of decay modes of radionuclide.

D Zerfallsschema
F schéma de désintégration
P schemat rozpadu
R схема распада

898
DECOMPOSITION (decomposition reaction).
Reaction of chemical separation of a given compound into two or more simple compounds or substances, e.g.

$$2H_2O = 2H_2 + O_2$$

D Zersetzungsreaktion
F décomposition
P reakcja rozkładu, reakcja analizy
R реакция разложения

DECOMPOSITION REACTION. *See* DECOMPOSITION.

899
DECOMPOSITION VOLTAGE, U_d.
Value of overvoltage extrapolated to $I = 0$ corresponding to characteristic bend of curve for current of electrolysis I and electrolysis voltage U; especially well observed in case of gaseous products of electrolysis.

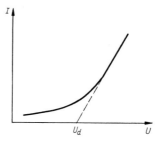

Remark: Term used mainly in technical electrochemistry; has no strict physical interpretation.

D Zersetzungsspannung
F tension de décomposition
P napięcie rozkładowe
R напряжение разложения электролита

900
DECONTAMINATION.
Removal of radioactive matter from skin, textiles, instruments and wrappings, i.e. from places where radioactivity threatens health and safety.

D Dekontamination, Entseuchung
F décontamination
P dekontaminacja, odkażanie
R дезактивация

901
DECONTAMINATION FACTOR.
Ratio of amount of radioactive impurities present in substance before purification to that present after purification.

D Dekontaminationsfaktor, Entseuchungsgrad
F facteur de décontamination
P współczynnik dekontaminacji
R коэффициент очистки

DEFECT CONDUCTOR. *See* ELECTRON-DEFECT SEMICONDUCTOR.

902
DEFORMATION OF ATOMIC NUCLEI.
Deviation of surface of atomic nuclei from sphericity. Deformed atomic nuclei are those with total number of protons and/or neutrons differing widely from a magic numbers.

D Atomkerndeformation
F déformation des noyaux atomiques
P deformacja jąder atomowych
R деформация атомных ядер

DEFORMATION POLARIZATION. *See* INDUCED POLARIZATION.

DEGENERACY. *See* DEGENERATION OF ENERGY LEVEL.

DEGENERACY TEMPERATURE. *See* FERMI TEMPERATURE.

903
DEGENERATE ELECTRON GAS.
Group of electrons in metal or
semiconductor which may be treated as
group of particles obeying laws of
quantum mechanics, travelling freely in
field of constant potential or in periodic
field (in range of whole crystal).
Distribution of energy in degenerate
electron gas is described by Fermi-Dirac
law.
D entartetes Elektronengas
F gaz électronique dégénéré
P gaz elektronowy zdegenerowany
R вырожденный электронный газ

904
DEGENERATE STATE.
Energetic state of quantum mechanical
system described simultaneously by more
than one linearly independent wave
function.
D entarteter Zustand
F état dégénéré
P stan zdegenerowany, stan zwyrodniały
R вырожденное состояние

905
DEGENERATE VIBRATIONS.
Two or more normal modes of the same
frequency.
D entartete Schwingungen
F vibrations dégénérées
P drgania zdegenerowane
R вырожденные колебания

906
DEGENERATION OF ENERGY LEVEL
(degeneracy).
For given energy level there are two or
more linearly independent quantum
states; in this case the energy value is not
sufficient for unique description of
quantum state of system on given energy
level.
D Entartung (des Energieniveaus)
F dégénérescence (du niveau énergétique)
P degeneracja (poziomu energetycznego),
zwyrodnienie (poziomu energetycznego)
R вырождение (уровня энергии)

907
DEGRADATION.
Decomposition of organic compound by
various methods resulting in formation of
compounds of simpler structure, e.g.

$$(CH_3)_3\overset{\oplus}{N}CH_2CH_2\overset{\ominus}{O}H \xrightarrow{\Delta} H_2O + N(CH_3)_3 + CH_2=CH_2$$

D Abbau
F dégradation
P degradacja
R деградация

908
DEGRADATION OF POLYMER.
Degradation of macromolecules caused
by physical, chemical or biological factors,
involving structural change of polymer
and lowering of polymerization degree.
Degradation generally decreases quality of
polymer.
D Abbau (des Polymers)
F dégradation (du polymère)
P degradacja (polimeru)
R деструкция (полимера), разложение
(полимера)

DEGREE ABSOLUTE. *See* KELVIN.

DEGREE KELVIN. *See* KELVIN.

DEGREE OF ADVANCEMENT OF
REACTION. *See* EXTENT OF REACTION.

909
DEGREE OF ASSOCIATION.
Ratio of number of associated monomer
molecules to total number of monomer
molecules.
D Assoziationsgrad
F degré d'association
P stopień asocjacji
R степень ассоциации

910
DEGREE OF DEGENERACY.
Number of linearly independent wave
functions corresponding to stationary
states of system with the same energy.
D Entartungsgrad
F degré de dégénérescence
P stopień degeneracji, stopień
zwyrodnienia
R степень вырождения

911
DEGREE OF DEPOLARIZATION (*of
light*).
Ratio of scattered light intensity in
direction parallel to incident light beam
and scattered light intensity perpendicular
to incident beam.
D Depolarisationsgrad
F facteur de dépolarisation
P stopień depolaryzacji
R степень деполяризации

912
DEGREE OF DISPERSION (dispersity), *D*.
Ratio of total surface of dispersed phase
to volume occupied by it.
D Dispersitätsgrad, Dispersionsgrad
F degré de dispersion, dispersité
P stopień dyspersji
R степень дисперсности, дисперсность

913
DEGREE OF DISSOCIATION, α.

$$\alpha = \frac{i-1}{v-1} = \frac{\Lambda_c}{\Lambda_0}$$

where i = van't Hoff isotonic coefficient, ν = number of ions arising from one molecule of electrolyte, Λ_0 and Λ_c = limiting equivalent conductivity and equivalent conductivity at a given concentration c respectively; this value is not identical with true degree of dissociation.

D Dissoziationsgrad nach Arrhenius
F coefficient d'ionisation, degré de dissociation électrolytique, taux d'ionisation
P stopień dysocjacji (według Arrheniusa)
R степень элекртолитической диссоциации (по Аррениусу)

914
DEGREE OF FORMATION, α_n.
Ratio of equilibrium concentration of complex $[ML_n]$ to overall metal concentration c_M

$$\alpha_n = \frac{[ML_n]}{c_M}$$

D Bildungsgrad
F ...
P stopień tworzenia kompleksu
R степень образования

915
DEGREE OF FREEDOM (degree of freedom of motion).
Every component of total energy of molecule, proportional to square of position coordinate or momentum coordinate. E.g. degree of freedom is 3 for translational motion of monoatomic molecule.

D Freiheitsgrad (der Bewegung)
F degré de liberté (de mouvement)
P stopień swobody (kinetyczny)
R степень свободы (движения)

DEGREE OF FREEDOM OF MOTION.
See DEGREE OF FREEDOM.

916
DEGREE OF FREEDOM OF THERMODYNAMIC SYSTEM.
Independent variable which may be changed without disturbance of thermodynamic equilibrium.

D Freiheitsgrad von thermodynamischen System
F degré de liberté du système thermodynamique
P stopień swobody układu termodynamicznego
R степень свободы термодинамического состояния, термодинамическая степень свободы

917
DEGREE OF HYDROLYSIS.
Fraction of molecules or ions which undergo hydrolysis process.

D Hydrolysengrad
F degré d'hydrolyse
P stopień hydrolizy
R степень гидролиза

918
DEGREE OF IONIZATION.
Ratio (chiefly expressed as percentage) of the number (or concentration) of ionized molecules to their total number (or concentration) in solution investigated.

D ...
F degré d'ionisation
P stopień jonizacji
R степень ионизации

919
DEGREE OF POLYMERIZATION OF MACROMOLECULE OF POLYMER.
Number of monomeric units in polymeric macromolecule.

D Polymerisationsgrad
F degré de polymérisation (de la macromolécule)
P stopień polimeryzacji (makro)cząsteczki polimeru
R степень полимеризации макромолекулы (полимера)

920
DEGREE OF POLYMERIZATION OF POLYMER.
Mean value of degrees of polymerization of separate polymeric macromolecules. Type of determination of degree of polymerization should be stated, e.g. number-average degree of polymerization.

D mittlerer Polymerisationsgrad
F degré de polymérisation moyen
P stopień polimeryzacji polimeru
R степень полимеризации полимера

921
DE HAAS-VAN ALPHEN EFFECT.
Oscillatory changes of magnetic susceptibility of certain metals with increasing magnetic field strength at low temperatures.

D de Haas-van Alphen-Effekt
F effet de Haas-van Alphen
P efekt de Haasa i van Alphena
R эффект де Хааза-ван Алфена

922
DEHALOGENATION.
Elimination of one or more halogen atoms (F, Cl, Br or I) from molecule of chemical compound, e.g.

$$BrCH_2CH_2Br + Zn \rightarrow CH_2{=}CH_2 + ZnBr_2$$

D Dehalogenierung, Abspaltung des Halogens
F dé(s)halogénation
P dehalogenacja, odszczepienie fluorowca
R отщепление галогена

923
DEHYDRATION.
Elimination of water molecule from molecule of chemical compound, e.g.

$$CH_3CH_2OH \xrightarrow[\Delta]{H_2SO_4} CH_2=CH_2 + H_2O$$

$$CuSO_4 \cdot 5H_2O \xrightarrow[-5\,H_2O]{\Delta} CuSO_4$$

D Dehydratisierung, Dehydratation, Wasserabspaltung
F déshydratation
P dehydratacja, odwodnienie
R дегидратация

924
DEHYDROCYCLIZATION.
Elimination of hydrogen molecule from hydrocarbon molecule (open-chain, aromatic or hydroaromatic) with formation of compounds containing one or more cycles in molecule, e.g.

$$CH_3(CH_2)_5CH_3 \rightarrow \text{[benzene ring]}-CH_3 + 4H_2$$

$$\text{[benzene ring]} \begin{matrix} -CH=CH_2 \\ -CH=CH_2 \end{matrix} \rightarrow \text{[naphthalene]} + H_2$$

D Cyclodehydrierung
F dehydrocyclisation
P cyklodehydrogenacja, dehydrocyklizacja
R дегидроциклизация

925
DEHYDROGENATION.
Removal of hydrogen molecule from molecule of organic compound, e.g.

$$CH_3OH \xrightarrow[\Delta]{Cu} HCHO + H_2$$

D Dehydrierung, Dehydrogenation
F déshydrogénation
P odwodornienie, dehydrogenacja
R дегидрогенизация

926
DEHYDROHALOGENATION.
Elimination of molecule of hydrogen halide from molecule of organic compound with formation of unsaturated compound, e.g.

$$CH_3CH_2X \xrightarrow{\overset{\ominus}{OH}} CH_2=CH_2 + HX$$

D Dehydrohalogenierung, Halogenwasserstoffabspaltung
F déshydrohalogénation
P dehydrohalogenacja, odszczepienie chlorowcowodoru
R отщепление галогеноводорода

927
DEIONIZATION (demineralization).
Process of removing ions from solution.
D Entsalzung, Entionisierung, Entmineralisierung
F déminéralisation

P dejonizacja, demineralizacja
R деионизация

928
DELAYED FLUORESCENCE.
Fluorescence with relatively long life-time; starting upper singlet state from which emission of light originates is achieved either through triplet-triplet annihilation process (P-type delayed fluorescence) or optical electrons are thermally excited from lower triplet electronic state (E-type delayed fluorescence).

D verzögerte Fluoreszenz
F fluorescence retardée
P fluorescencja opóźniona
R замедленная флуоресценция

929
DELAYED NEUTRONS.
Fission neutrons, emitted with delay, by primary and secondary fission fragments of atomic nucleus (half-life of order of seconds). Delayed neutrons constitute about 1% of all neutrons emitted in fission reactions.

D verzögerte Neutronen
F neutrons retardés, neutrons différés
P neutrony opóźnione
R запаздывающие нейтроны

930
DELÉPINE PRIMARY AMINE SYNTHESIS.
Preparation of primary amines by addition of organic halide to hexamethylenetetramine, followed by acid hydrolysis of resulting quaternary compound with alcoholic solution of hydrogen chloride, e.g.

$$RCH_2X + (CH_2)_6N_4 \longrightarrow [RCH_2(CH_2)_6N_4]^{\oplus}X^{\ominus}$$

$$\xrightarrow[C_2H_5OH]{HCl,\ H_2O} RCH_2NH_2$$

D Delépine Amin-Synthese
F réaction de Delépine
P synteza (amin pierwszorzędowych) Delépine'a
R реакция Делепина

931
DELOCALIZATION ENERGY.
Difference between energy of real molecule with conjugated double bonds and energy of hypothetical molecule with localized double bonds.

D Delokalisationsenergie
F énergie de délocalisation
P energia delokalizacji
R энергия делокализации

932
DELOCALIZED ELECTRONS.
Electrons assumed to occupy delocalized molecular orbitals.

D delokalisierte Elektronen
F électrons délocalisés
P elektrony zdelokalizowane
R делокализованные электроны,
   нелокализованные электроны

933
DELOCALIZED MOLECULAR ORBITAL
(delocalized orbital).
Molecular orbital spread over whole
molecule or substantial part of it, e.g.
π-orbitals in conjugated double bond
molecules.
D delokalisiertes Orbital, delokalisierte
   Bindung
F orbitale moléculaire délocalisée
P orbital zdelokalizowany
R делокализованная молекулярная
   орбиталь

DELOCALIZED ORBITAL. *See*
DELOCALIZED MOLECULAR ORBITAL.

DELTA ELECTRONS. *See* DELTA RAYS.

934
DELTA RAYS (delta electrons).
Non-nuclear electron radiation generated
when ionizing particles or photons pass
through matter and having energy
sufficient to cause further ionization of
atoms of medium.
D δ-Strahlen, δ-Elektronen
F rayons δ
P promienie δ, elektrony δ
R δ-лучи, δ-электроны

935
DEMETHYLATION.
Elimination of methyl group —CH_3 from
molecule of organic compound, e.g.

CH_3
|
S   SH
| |
$(CH_2)_2$ $\xrightarrow{ATP}$ $(CH_2)_2$
| |
$CHNH_2$ $CHNH_2$
| |
COOH   COOH

D Entmethylierung, Demethylierung
F déméthylation
P demetylowanie, odmetylowanie
R деметилирование

936
DEMIANOV REARRANGEMENT.
Rearrangement taking place concurrently
with deamination of primary alicyclic
amines, involving augmentation or
diminishing of ring, e.g.

D Demjanov-Umlagerung, Demjanov
   Ringweitenänderung
F transposition de Demjanov
P przegrupowanie Demjanowa
R перегруппировка Демьянова

DEMINERALIZATION. *See*
DEIONIZATION.

937
DENATURATION OF PROTEINS.
Irreversible coagulation of proteins.
D Denaturierung von Eiweiß
F dénaturation des protéines
P denaturacja białek
R денатурация белков

DENSITOMETER. *See*
MICROPHOTOMETER.

938
DENSITY, ρ.
Mass of unit volume of described
substance (gas, liquid or solid) at defined
temperature and pressure.
D Dichte
F densité
P gęstość, masa właściwa
R плотность

DENSITY MATRIX. *See* STATISTICAL
OPERATOR.

939
DENSITY OF PROBABILITY.
In quantum mechanics defined as limit of
ratio of probability of finding system in
some volume element to its volume.
Density of probability is given by modulus
square of corresponding wave function.
D Wahrscheinlichkeitsdichte
F densité de probabilité
P gęstość prawdopodobieństwa
R плотность вероятности

940
DENSITY OF STATES.
Ratio of number of quantum mechanical
states of system with continuous or
quasi-continuous energy spectrum in
a very small energy interval ΔE to value
of this interval. Derivative of number
of quantum mechanical states of system
with respect to its energy.
D Dichte der Zustände
F densité des états
P gęstość stanów
R плотность состояний

DENSITY OPERATOR. *See*
STATISTICAL OPERATOR.

DEPARTING GROUP. *See* LEAVING
GROUP.

941
DEPOLARIZER.
Agent preventing polarization of electrode,
e.g. manganese dioxide MnO_2 in Leclanché
cell.
D Depolarisator
F dépolarisant
P depolaryzator
R деполяризатор

942
DEPOLYMERIZATION.
Reversion of polymerization;
decomposition of polymer with formation
of simpler components: monomers, dimers
etc.
D Depolymerisation
F dépolymérisation
P depolimeryzacja
R деполимеризация

943
DEPOSITION POTENTIAL, ε_w.
Electrical potential which should be
exceeded to observe effective course of
a given electrode process, $\varepsilon_w = \varepsilon + \eta$,
where ε = reversible electrode potential
and η = overvoltage of tested electrode
process.
D Abscheidungspotential
F potentiel de dépôt
P potencjał wydzielania
R потенциал выделения

944
DEPTH DOSE.
Dose of ionizing radiation at defined depth
(mostly at 10 cm); usually depth dose
is defined relatively to surface dose as
percentage.
D Tiefendosis
F dose en profondeur
P dawka głębok(ościow)a
R глубинная доза

945
DERIVATOGRAPHY.
Branch of thermal analysis that examines
substances by simultaneous
thermogravimetry and differential
thermal analysis.
D Derivatographie
F ...
P derywatografia
R дериватографический анализ

DESAMINATION. See DEAMINATION.

946
DESMOTROPISM.
Obsolete term denoting tautomerism in
which equilibrium rate is relatively low,
thus enabling isolation of tautomers, e.g.
ketonic and enolic forms of ethyl
acetoacetate were called desmotropes.

D Desmotropie
F desmotropie
P desmotropia
R десмотропия

947
DESORPTION.
Reverse process to sorption, which
involves liberation of adsorbate from
adsorbent surface or absorbate from bulk
of absorbent.
D Desorption
F désorption
P desorpcja
R десорбция

948
DESTRUCTIVE ACTIVATION
ANALYSIS.
Activation method of analysis in which
activity of the element to be determined
is measured after its separation from an
irradiated sample by chemical methods.
D chemische Aktivierungsanalyse
F analyse destructive par activation
P analiza aktywacyjna destrukcyjna
R радиохимический вариант
активационного анализа

DESTRUCTIVE HYDROGENATION. *See*
HYDROGENOLYSIS.

949
DESULFURIZATION.
Elimination of sulfur atom from molecule
of chemical compound, usually effected
with Raney nickel, e.g.
$R_2S + Ni + C_2H_5OH \rightarrow 2RH + NiS + CH_3CHO$
D Entschwefelung
F désulfuration
P desulfuracja
R десульфирование

950
DETECTION (*of chromatogram*).
Process leading to making visible or
detection of chromatographic bands or
spots of normally colourless substances
in column, on filter paper etc.
D Entwickeln
F révélation
P wywołanie (chromatogramu)
R индикация, проявление пятен,
обнаружение

951
DETECTION OF RADIOACTIVITY.
Detection and characteristics of nature,
energy and number of particles or photons
emitted by sample examined.
D Nachweis der Kernstrahlung
F détection d'un rayonnement (radioactif)
P detekcja promieniowania jądrowego
R обнаружение и измерение ядерного
излучения

952
DETERMINATION (estimation).
Quantitative estimation of content of a given component in studied sample.
D Bestimmung
F dosage
P oznaczanie
R определение

953
DETONATION.
Explosion accompanied by formation and propagation of detonation wave.
D Detonation
F détonation
P detonacja
R детонация

954
DETONATION WAVE.
Shock wave on front of which very rapid exothermic chemical reaction proceeds.
D Detonationswelle
F onde de détonation
P fala detonacyjna
R детонационная волна

955
DEUTERATION (deuterium exchange).
Replacement of hydrogen atom in molecule of chemical compound by deuterium atom.
D Deuterierung
F deutération
P deuterowanie
R дейтерирование

956
DEUTERIUM (heavy hydrogen), D.
Isotope of hydrogen. Atomic number 1. Mass number 2.
D Deuterium, schwerer Wasserstoff
F deutérium, hydrogène lourd
P deuter, wodór ciężki
R дейтерий, тяжёлый водород

DEUTERIUM EXCHANGE. See DEUTERATION.

957
DEUTERON (deuton), d.
Nucleus of deuterium, consisting of one proton and one neutron.
D Deut(er)on
F deut(ér)on
P deuteron
R дейтрон

DEUTON. See DEUTERON.

958
DEVELOPMENT OF CHROMATOGRAM.
Passing of appropriate mobile phase through column or layer of stationary phase to get separation as result of differences in migration rates of components being separated.
D Entwicklung
F développement (de chromatogramme)
P rozwijanie chromatogramu
R проявление (хроматограммы)

959
DEVIATION.
Function describing difference between elements of series (e.g. values in series of measurements) and assumed mean value of this series.
D Abweichung
F écart, déviation
P odchylenie
R отклонение

960
DEW-POINT.
Temperature (pressure) at which superheated vapour reaches value for saturated vapour pressure; isobaric cooling of vapour below dew-point temperature or isothermal compression of vapour above dew-point pressure causes appearance of first droplets of liquid, or dew.
D Taupunkt
F point de rosée
P punkt rosy
R точка росы

DEXTRO-LAEVO MODIFICATION. See RACEMIC MODIFICATION.

961
DEXTRO-ROTATION.
Specific effect in optically active substances; when plane-polarized light, passing through an optically active substance, has its plane of polarization rotated in clockwise direction (when viewed from opposite side to light source), such substance is said to be (+), i.e. dextro-rotatory (obsolete notation: d-).
D Rechtsdrehung
F dextrorotation
P prawoskrętność
R правое вращение

962
DIALYSIS.
Process of selective diffusion through semi-permeable membrane, in which penetration of dissolved substances of low molecular weight, but not colloidal particles or macromolecules, takes place.
D Dialyse
F dialyse
P dializa
R диализ

963
DIAMAGNETIC ANISOTROPY.
Anisotropy of magnetic susceptibility of diamagnetic crystal given by three fundamental susceptibilities χ_1, χ_2 and χ_3 measured along three mutually perpendicular fundamental magnetic axes.
D diamagnetische Anisotropie
F anisotropie diamagnétique
P anizotropia diamagnetyczna
R диамагнитная анизотропия

DIAMAGNETIC BODIES. *See* DIAMAGNETICS.

964
DIAMAGNETICS (diamagnetic substances, diamagnetic bodies).
Substances with negative magnetic susceptibility and magnetic permeability less than 1.
D Diamagnetika, diamagnetische Körper, diamagnetische Stoffe
F diamagnétiques, corps diamagnétiques, substances diamagnétiques
P diamagnetyki, ciała diamagnetyczne, substancje diamagnetyczne
R диамагнетики, диамагнитные тела, диамагнитные вещества

DIAMAGNETIC SCREENING OF NUCLEUS. *See* DIAMAGNETIC SHIELDING OF NUCLEUS.

965
DIAMAGNETIC SHIELDING OF NUCLEUS (diamagnetic screening of nucleus).
Participation of electrons in magnetic shielding of nucleus, leading to nuclear magnetic resonance shifts in direction of higher static field or lower frequencies of alternating magnetic field.
D diamagnetische Abschirmung des Kerns
F écran diamagnétique du noyau
P ekranowanie jądra diamagnetyczne
R диамагнитное экранирование ядра

DIAMAGNETIC SUBSTANCES. *See* DIAMAGNETICS.

966
DIAMAGNETIC SUSCEPTIBILITY.
Negative magnetic susceptibility induced by external magnetic field in all bodies and related to Larmor precession.
D diamagnetische Suszeptibilität
F susceptibilité diamagnétique
P podatność diamagnetyczna
R диамагнитная восприимчивость

967
DIAMAGNETISM.
Phenomenon of induction of magnetization of opposite direction to external magnetic field; present in all substances placed in magnetic field.
D Diamagnetismus
F diamagnétisme
P diamagnetyzm
R диамагнетизм

968
DIASTEREOISOMERISM.
Type of stereoisomerism caused by different spatial arrangement of atoms or groupings of atoms (i.e. by different configuration) in compounds that are not enantiomers. This different spatial arrangement does not result from restricted rotation about single bond. Stereoisomers having two or more chiral centres are said to be diastereoisomeric when they have identical configuration at some chiral centres.
D Diastereo(iso)merie
F diastéréo-isomérie
P diastereoizomeria
R диастереоизомерия

969
DIASTEREOISOMERS.
Stereoisomers which are not enantiomers. The following combinations are possible: both diastereoisomers are chiral (I); one is chiral, the other achiral (II); both are achiral (III and IV)

```
      COOH          COOH          COOH          COOH
   H—|—OH        H—|—OH        H—|—OH        H—|—OH
   H—|—OH       HO—|—H        HO—|—H        H—|—OH
      CH₃           CH₃          COOH          COOH
        I                          II

      COOH          COOH
   H—|—OH        H—|—OH
   H—|—OH       HO—|—H
   H—|—OH        H—|—OH
      COOH          COOH
        III

   H—C—COOH             H—C—COOH
      ‖                    ‖
   H—C—COOH           HOOC—C—H
        IV
```

D Diastereoisomere
F diastéréo-isomères
P diastereoizomery
R диастереоизомеры

DIATHERMAL WALL. *See* DIATHERMIC PARTITION.

970
DIATHERMIC PARTITION (diathermal wall, thermally conducting wall).
Partition which allows energy transfer in form of heat but does not allow for mass or energy transfer in form of work.
D diathermische Wand, thermisch leitende Wand
F paroi diathermane
P odgraniczenie diatermiczne, ścianka diatermiczna, osłona diatermiczna

R теплопроводящая оболочка,
диатермическая диафрагма

DIAZO REACTION. *See*
DIAZOTIZATION.

971
DIAZOTIZATION (diazo reaction).
Reaction of primary aromatic amines with
nitrous acid in presence of mineral acid,
resulting in formation of diazonium salt,
e.g.

$$C_6H_5NH_2 + NaNO_2 + 2HCl \xrightarrow{0° -5°} C_6H_5\overset{\oplus}{N}\equiv N\overset{\ominus}{Cl} +$$
$$+ NaCl + 2H_2O$$

D Diazotierung
F diazotation, diazo-réaction
P dwuazowanie
R диазотирование

972
DICHROISM.
Feature of plate cut of optically
anisotropic material that, viewed in the
directions of principal optical axes, has
different colours. Dichroism is caused by
differences of light absorption coefficients
in directions of principal optical axes.

D Dichroismus
F dichroïsme
P dichroizm, dwubarwność
R дихроизм

DIECKMANN INTRAMOLECULAR
CYCLIZATION. *See* DIECKMANN
REACTION.

973
DIECKMANN REACTION (Dieckmann
intramolecular cyclization).
Intramolecular condensation of
dicarboxylic acid esters containing
hydrogen atoms at δ- or ε-carbon atoms,
activated by carbonyl groups, catalyzed
by alcoholates and resulting in formation
of cyclic β-ketoesters, e.g.

D Dieckmann-Reaktion, Dieckmann
intramolekulare Esterkondensation
F réaction de Dieckmann
P reakcja Dieckmanna, cyklizacja
Dieckmanna
R реакция Дикмана

974
DIELECTRIC.
Substance undergoing electric polarization
and of very low electric conductivity
(insulator).

D Dielektrikum
F diélectrique
P dielektryk
R диэлектрик

975
DIELECTRIC ABSORPTION.
Absorption of alternating electric field
energy by dielectric; measured by
dielectric losses.

D dielektrische Absorption
F absorption diélectrique
P absorpcja dielektryczna
R диэлектрическая абсорбция

DIELECTRIC CONSTANT. *See*
RELATIVE PERMITTIVITY.

976
DIELECTRIC DISPERSION.
Decrease of dielectric permittivity with
increasing frequency of electric field.

D dielektrische Dispersion
F dispersion diélectrique
P dyspersja dielektryczna
R диэлектрическая дисперсия

977
DIELECTRIC LOSSES, ε'.
Imaginary component of complex
permittivity; quantity defined by
$\varepsilon' = \varepsilon \tan \delta$, where: ε = permittivity,
δ = loss factor.

D dielektrische Verluste
F pertes diélectriques
P straty dielektryczne, współczynnik strat
dielektrycznych
R диэлектрические потери

DIELECTRIC PERMITTIVITY. *See*
PERMITTIVITY.

978
DIELECTRIC POLARIZATION
(polarization).
Formation of macroscopic dipole moment
in dielectric due to external electric field.
Remark: Value of induced dipole moment is
usually refered to unit volume or to 1 mole.

D (dielektrische) Polarisation
F polarisation (diélectrique)
P polaryzacja dielektryczna
R (диэлектрическая) поляризация

979
DIELECTRIC POLARIZATION
(polarization), *P*.
Vector quantity equal to electric dipole
moment per unit volume of dielectric.

D (dielektrische) Polarisation
F polarisation (diélectrique)
P polaryzacja dielektryczna, wektor
polaryzacji dielektrycznej
R (диэлектрическая) поляризация

980
DIELECTRIC RELAXATION.
Delay of dielectric polarization with
respect to electric field changes.

D dielektrische Relaxation
F relaxation diélectrique
P relaksacja dielektryczna
R диэлектрическая релаксация

981
DIELECTRIC RELAXATION TIME.
Time interval after electric field is
switched off in which dielectric
polarization decreases to 1/e its original
value.

D dielektrische Relaxationszeit
F temps de relaxation diélectrique
P czas relaksacji dielektrycznej
R время диэлектрической релаксации

982
DIELECTRIC SATURATION (electric
saturation).
Limit of the change of permittivity
induced by external electric field.

D dielektrische Sättigung
F saturation diélectrique
P nasycenie dielektryczne
R диэлектрическое насыщение

983
DIELECTRIC SPECTROSCOPY.
Branch of spectroscopy dealing with
investigation of dielectric properties of
matter as function of frequency of applied
electromagnetic fields; it covers
measurements of absorption and dispersion
dielectric spectra.

D DK-Messung
F absorption dipolaire de Debye,
absorption diélectrique, dispersion
diélectrique
P spektroskopia dielektryczna
R диэлектрическая спектроскопия

DIELS-ALDER REACTION. *See* DIENE
SYNTHESIS.

DIENE ANALOGS. *See* DIENE
ANALOGUES.

984
DIENE ANALOGUES (diene analogs).
Dienes containing conjugated bonds
between heteroatoms or between carbon
atom and a heteroatom, e.g.
α, β-unsaturated aldehydes or ketones

D Dienanaloga
F analogues diéniques
P analogi dienowe
R диены содержащие гетероатомы

985
DIENE SYNTHESIS (Diels-Alder
reaction).
Synthesis of six-membered unsaturated
ring system, resulting from 1,4-addition
of conjugated diene to compound
containing double or triple bond
(dienophile), e.g.

D Diensythese, Diels-Alder-Reaktion
F synthèse diénique, réaction de
Diels-Alder
P synteza dienowa, reakcja Dielsa i Aldera
R диеновый синтез, реакция
Дильса-Альдера

986
DIENONE-PHENOL REARRANGEMENT.
Acid-catalyzed rearrangement of alicyclic
dienones to corresponding phenols,
involving migration of two hydrogen
atoms, or in certain cases, of one hydrogen
atom and an alkyl group, e.g.

D Dienon→Phenol-Umlagerung
F transposition diénones-phénols
P przegrupowanie dienono-fenolowe
R . . .

987
DIETERICI'S EQUATION OF STATE.
Empirical (thermal) equation of state of
real gas

$$p(V_m - b) = RT \exp\left(\frac{-a}{RTV_m}\right)$$

where p = pressure, V_m = molar gas
volume, T = thermodynamic temperature,
R = universal gas constant,
a, b = constants characteristic for the
given gas.

D Zustandsgleichung von Dieterici
F équation d'état de Dieterici
P równanie stanu Dietericiego
R уравнение состояния Дитеричи

DIFFERENTIAL ABSORPTIOMETRY.
See DIFFERENTIAL
SPECTROPHOTOMETRY.

988
DIFFERENTIAL CAPACITY OF DOUBLE LAYER, C.
Capacity of electrical double layer at electrode — solution interface for a given electrode potential

$$C = \frac{\mathrm{d}q}{\mathrm{d}\varepsilon}$$

where ε = electrode potential, q = charge of the electrode.
D differentielle Kapazität der Doppelschicht
F capacité différentielle de la couche double
P pojemność warstwy podwójnej różniczkowa
R дифференциальная ёмкость двойного (электрического) слоя

DIFFERENTIAL COLORIMETRY. *See* DIFFERENTIAL SPECTROPHOTOMETRY.

989
DIFFERENTIAL CROSS-SECTION.
Cross-section for nuclear process whereby an angle is specified (relative to direction of incidence) for emission of particles or photons per unit angle or per unit solid angle.
D differentieller Wirkungsquerschnitt
F section efficace différentielle
P przekrój czynny różniczkowy
R дифференциальное (эффективное) сечение

990
DIFFERENTIAL DETECTOR.
Device that measures instantaneous concentration of substance in gas or liquid flowing out of chromatographic column.
D Differentialdetektor
F détecteur différentiel
P detektor różnicowy
R дифференциальный детектор

991
DIFFERENTIAL HEAT OF ADSORPTION.
Heat effect associated with adsorption process and calculated for one mole of absorbate when surface state of adsorbent remains practically unaffected.
D differentielle Adsorptionswärme
F chaleur partielle d'adsorption
P ciepło adsorpcji cząstkowe (molowe)
R дифференциальная теплота адсорбции

DIFFERENTIAL HEAT OF DILUTION.
See PARTIAL HEAT OF DILUTION.

DIFFERENTIAL HEAT OF SOLUTION.
See PARTIAL HEAT OF SOLUTION.

992
DIFFERENTIAL METHOD (*of determining the order of reaction*).
Method of determination of reaction order n, consisting in determination of reaction rate with respect to defined concentrations of reacting substance. Logarithm of rate v is linear function of logarithm of concentration c and slope of line is equal to order of reaction n

$$n = \frac{\lg v_2 - \lg v_1}{\lg c_2 - \lg c_1}$$

D ...
F méthode différentielle (de Van't Hoff)
P metoda różnicowa van't Hoffa
R дифференциальный метод (определения порядка реакции)

993
DIFFERENTIAL POLAROGRAPHY.
Polarography based on recording and analysis of derivatives of current with respect to potential on potential of dropping mercury electrode.
D Derivativpolarographie
F polarographie dérivée
P polarografia różniczkowa
R производная полярография, дифференциальная полярография

994
DIFFERENTIAL SPECTROPHOTOMETRY (differential colorimetry, differential absorptiometry).
Spectrophotometric method of determination of concentration of investigated solution by measurement of its absorption with respect to standard solution of substance to be determined and of known and similar concentration.
D differentielle Spektralphotometrie
F colorimétrie différentielle
P spektrofotometria różnicowa
R дифференциальный спектрофотометрический анализ

995
DIFFERENTIAL THERMAL ANALYSIS, DTA.
Thermal analysis carried out under comparison of increases in temperatures of specimen and reference substance on heating at a constant rate.
D Differentialthermoanalyse, differentiale thermische Analyse
F analyse thermique différentielle
P analiza termiczna różnicowa
R дифференциальный термический анализ

996
DIFFERENTIATING SOLVENT.
Solvent in which equivalent conductivity
of electrolytes is strongly dependent on
nature of electrolyte; for salt — solvent
which does not form stable crystalline
solvates, for acids — protogenic solvent —
and protophilic solvent for bases.
D differenzierendes Lösungsmittel
F solvant différenciant, solvant capable
de différencier des acides ou des bases
P rozpuszczalnik różnicujący
R дифференцирующий растворитель

997
DIFFRACTION OF ATOMS AND
MOLECULES.
Diffraction of monochromatic waves of
matter (atoms, molecules) on certain
diffraction centres, e.g. crystal lattice,
followed by interference of scattered
waves.
D Atom- und Molekülbeugung
F diffraction des atomes et molécules
P dyfrakcja atomów i cząsteczek
R дифракция атомов и молекул

DIFFRACTION OF LIGHT. See LIGHT
DIFFRACTION.

998
DIFFUSE LAYER.
Part of electrical double layer at metal —
electrolyte solution interphase next to
compact layer on solution side, identical
with model of Gouy and Chapman (see
theories of double layer).
D diffuse Schicht, diffuse Doppelschicht
F couche diffuse
P warstwa rozmyta
R диффузный двойной слой

999
DIFFUSE REFLECTION.
Scattering of radiation caused by
reflection in all possible directions. Diffuse
reflection is shown by materials with
inhomogeneous internal structure and by
mat surfaces.
D diffuse Reflexion
F réflexion diffuse
P odbicie dyfuzyjne
R диффузионное отражение

1000
DIFFUSION.
Spontaneous equalization of differences in
chemical potential occurring in
multicomponent systems due to
spontaneous, random movement of
particles.
D Diffusion
F diffusion
P dyfuzja
R диффузия

1001
DIFFUSION COEFFICIENT (diffusivity).
Magnitude D defined by Fick's first law,
equal to quantity of diffusing component
through unit area in unit time, when
change of one unit of concentration occurs
at unit length.
D Diffusionskoeffizient
F coefficient de diffusion
P współczynnik dyfuzji, stała dyfuzji
R коэффициент диффузии

1002
DIFFUSION LAYER.
Layer of electrolyte with different
concentration of potential-determining
ions in comparison to bulk of solution;
it forms near electrode as result of
diffusion polarization.
D Diffusionsschicht
F couche de diffusion
P warstwa dyfuzyjna
R диффузионный слой

1003
DIFFUSION MODEL OF RADIOLYSIS
(radical diffusion model).
Model explaining chemical effects of
ionizing radiation absorption by diffusion
of intermediate products of radiolysis
in bulk of solution from spurs where they
are formed in high concentration.
D Diffusionsmodell der Radiolyse
F . . .
P model radiolizy dyfuzyjny
R диффузионная модель радиолиза

1004
DIFFUSION POLARIZATION.
Type of concentration polarization; it is
due to transport of reactants to or from
electrode surface, slow in comparison
to electrode reaction rate. Diffusion
polarization is expressed by diffusion
overvoltage.
D Diffusionspolarisation
F polarisation de concentration
P polaryzacja dyfuzyjna
R диффузионная поляризация

DIFFUSION POTENTIAL. See LIQUID
JUNCTION POTENTIAL.

1005
DIFFUSION REGION OF REACTION.
Region of reaction where its rate is
determined by diffusion of reagents.
D Diffusionsgebiet des Prozesses
F . . .
P obszar dyfuzyjny reakcji
R диффузионная область реакции

DIFFUSIVITY. See DIFFUSION
COEFFICIENT.

1006
DIGONAL HYBRID (sp hybrid).
One of two equivalent hybridized atomic orbitals formed from one s and one p orbital. Angle between two possible digonal hybrids is 180°. Digonal hybrids are used in description of electronic structure of acetylene.

D digonale Hybride, sp Hybride
F hybride digonal, hybride sp
P hybryd digonalny, hybryd liniowy, hybryd sp
R дигональная гибридная орбиталь

1007
DILATANCY.
Ability of colloidal systems to stiffen after causing agitation in system and their return to initial state when agitation ceases.

D Dilatanz
F dilatance
P dylatancja
R дилатанция

1008
DILUTE SOLUTION METHOD.
Method of dipole moment measurement of polar compound in non-polar solvent based on assumption of additivity of molar polarizations of components.

D Methode der verdünnten Lösungen
F méthode des solutions diluées
P metoda rozcieńczonych roztworów
R метод разбавленных растворов

1009
DILUTION METHOD.
Method of colorimetric analysis for determination of concentration of coloured solution by equalizing its absorption with that of standard after dilution of one of these solutions. Then, $c_1 l_1 = c_2 l_2$, where l_1, l_2 correspond to absorbing layer thickness and c_1, c_2 are concentrations of investigated and comparative solution respectively.

D ...
F méthode par dilution
P metoda rozcieńczania
R метод разбавления, метод изменения концентрации растворов

1010
DIMERIZATION.
Combination of two identical simple molecules of chemical compound (monomers) resulting in formation of more complex molecule (dimer), e.g.

$$2(CH_3)_2C{=}CH_2 \xrightarrow{H_2SO_4} (CH_3)_3CCH_2\overset{\overset{CH_3}{|}}{C}{=}CH_2$$

D Dimerisation
F dimérisation
P dimeryzacja
R димеризация

1011
1,3-DIPOLAR ADDITION.
Reaction of $a^{\oplus}-\overset{..}{b}-c^{\ominus}$ dipole molecule with alkene molecule, resulting in formation of cyclic system, e.g.

$$\left[\overset{\oplus}{b}\overset{a}{\diagup}_{\underset{c:}{\ominus}} \leftrightarrow :b\overset{\overset{a^{\oplus}}{\diagup}}{\underset{c:}{\diagdown}} \right] + \overset{d}{\underset{e}{\parallel}} \longrightarrow :b\overset{a-d}{\underset{c-e}{\diagup|}}$$

D 1,3-dipolare Addition
F addition dipolaire 1,3
P przyłączenie dipolarne 1,3, addycja dipolarna 1,3
R 1,3-биполярное присоединение

DIPOLAR MOLECULE. *See* POLAR MOLECULE.

DIPOLE. *See* ELECTRIC DIPOLE.

1012
DIPOLE-DIPOLE INTERACTION (dipole interaction).
Electrostatic intermolecular interaction between two polar molecules; two permanent parallel dipole moments μ_1 and μ_2 separated by distance r in medium of electric permittivity ε interact with energy

(a) $U_{(r)} = \pm \dfrac{2\mu_1\mu_2}{\varepsilon r^3}$ (b) $U_{(r)} = \pm \dfrac{\mu_1\mu_2}{\varepsilon r^3}$

(formulae in e.s.u. CGS system)
where $+$ sign corresponds to repulsive energy for the same direction of both dipoles and $-$ sign represents attractive energy for antiparallel dipoles.

D Dipol-Wechselwirkung
F interaction dipolaire
P oddziaływanie dipol-dipol, oddziaływanie dipolowe
R дипольное взаимодействие

DIPOLE INTERACTION. *See* DIPOLE-DIPOLE INTERACTION.

DIPOLE MOMENT. *See* ELECTRIC DIPOLE MOMENT.

DIRADICALS. *See* BIRADICALS.

DIRECT FIELD EFFECT. *See* FIELD EFFECT.

1013
DIRECT ISOTOPE DILUTION.
Isotope dilution in which substance of
natural isotopic composition is determined
by addition of the substance labelled with
isotopic tracer.
D einfache Isotopenverdünnung, direkte
  Isotopenverdünnung
F dilution isotopique simple
P rozcieńczenie izotopowe proste
R прямое изотопное разбавление

1014
DIRECT NUCLEAR REACTION.
Nuclear reaction involving direct
interaction of incident particle with one
or a small number of nucleons of target
nucleus.
D direkte Kernreaktion
F reaction nucléaire directe
P reakcja jądrowa bezpośredniego
  oddziaływania
R прямая ядерная реакция, ядерная
  реакция прямого взаимодействия

DIRECT PIEZOELECTRIC EFFECT. See
PIEZOELECTRIC EFFECT.

1015
DISCHARGE OF ION.
Neutralization of electrical charge of ion
by exchange of electrons; usually this
description is limited to exchange of
electrons between ion and electrode.
D Ionenentladung
F décharge d'ion
P rozładowanie jonu
R разряд иона

DISCONTINUOUS SPECTRUM. See
DISCRETE SPECTRUM.

1016
DISCONTINUOUS SYSTEM.
System composed of parts with
discontinuoities of thermodynamic
properties at interfaces.
D diskontinuierliches System
F système discontinu
P układ nieciągły
R прерывная система

1017
DISCRETE SPECTRUM (discontinuous
spectrum).
In spectroscopic terminology, term used
for line and band spectra.
D diskontinuierliches Spektrum
F spectre discontinu
P widmo dyskretne
R дискретный спектр

DISINTEGRATION SCHEME. See DECAY
SCHEME.

DISK CHROMATOGRAPHY. See
RADIAL CHROMATOGRAPHY.

DISLOCATIONS. See LINE DEFECTS.

DISMUTATION. See
DISPROPORTIONATION.

1018
DISPERGATION.
Pulverization of relatively large particles
of a given substance in order to obtain
disperse system.
D Dispergieren, Dispergierungsprozeß
F dispergation
P dyspergowanie
R диспергирование

1019
DISPERSED PHASE.
Component or components of disperse
system which form non-continuous phase.
D disperse Phase
F phase dispersée
P faza rozproszona, składnik rozproszony
R дисперсная фаза

1020
DISPERSE SYSTEM.
Colloidal system physically inhomogeneous
consisting of two phases: dispersion
medium and dispersed phase.
D disperses System, Dispersoid
F dispersoïde
P układ dyspersyjny, dyspersoid
R дисперсная система

1021
DISPERSION (scatter).
Variability of population elements
(e.g. results of measurements or
determinations) due to random factors.
D Streuung, Dispersion
F dispersion
P rozrzut, dyspersja statystyczna
R дисперсия, рассеяние, разброс

1022
DISPERSION CURVE (dispersion signal).
In magnetic resonance, curve representing
dependence of real part χ' of magnetic
susceptibility on strength of static
magnetic field at a given fixed frequency
of alternating field or on frequency of
alternating field at fixed static magnetic
field strength.
D Dispersionskurve, Dispersionssignal
F courbe de dispersion, signal de
  dispersion
P krzywa dyspersji, sygnał dyspersji
R кривая дисперсии, сигнал дисперсии

1023
DISPERSION EFFECT.
One of the components of intermolecular
forces. Dispersion effect leads to attractive
forces between molecules of electrically
neutral non-polar character and can be
explained in terms of electron density
fluctuations in interacting subsystems.

D Dispersionseffekt
F effet de dispersion
P efekt dyspersyjny
R дисперсионный эффект

1024
DISPERSION FORCES.
Attractive intermolecular forces due to fluctuating assymmetry of electron density distribution.
D Dispersionskräfte
F forces de dispersion
P siły dyspersyjne
R дисперсионное взаимодействие

1025
DISPERSION MEDIUM (continuous phase).
Component or components of disperse system forming continuous phase.
D Dispersionsmittel
F milieu dispersif
P faza rozpraszająca, ośrodek rozpraszający, ośrodek dyspersyjny
R дисперсионная среда

1026
DISPERSION METHODS.
In colloid chemistry, methods consisting of disintegrating substance, which is to form dispersed phase in presence of dispersion medium, until colloidal particles are formed.
D Dispersionsmethoden
F méthodes de dispersion
P metody dyspersyjne
R дисперсионные методы, методы диспергирования

1027
DISPERSION OF CONDUCTANCE (Debye-Falkenhagen's effect).
Increase of equivalent conductance of electrolyte as a result of disappearance of relaxation effect in case of very large frequencies (above 10^6 Hz) of alternating electric field.
D Leitfähigkeitsdispersion, Debye-Falkenhagen-Effekt
F dispersion de la conductibilité, effet de Debye et Falkenhagen
P dyspersja przewodnictwa, efekt Debye'a i Falkenhagena
R эффект Дебая-Фалькенгагена, дисперсия электропроводности

DISPERSION OF LIGHT. *See* LIGHT DISPERSION.

DISPERSION SIGNAL. *See* DISPERSION CURVE.

DISPERSITY. *See* DEGREE OF DISPERSION.

1028
DISPLACEMENT (displacement development).
Chromatographic method that consists in washing out of mixture of components from column with aid of agent whose affinity to stationary phase is greater than that of any of components being separated.
D Verdrängungsverfahren
F déplacement
P rugowanie, wypieranie
R вытеснение

DISPLACEMENT DEVELOPMENT. *See* DISPLACEMENT.

1029
DISPROPORTIONATION (dismutation).
Process of simultaneous oxidation and reduction of chemical compound, e.g.

$$2C_6H_5CHO \xrightarrow{\overset{\ominus}{OH}} C_6H_5COOH + C_6H_5CH_2OH$$

D Disproportionierung, Dismutation
F disproportionation, dismutation
P dysproporcjonowanie, dysmutacja
R диспропорционирование, дисмутация

1030
DISPROPORTIONATION OF RADICALS.
Stabilization of free radicals, involving hydrogen atom transfer from one radical to another; two stable molecules are thus formed, e.g. ethylene and ethane from two ethyl radicals

$$2 \cdot C_2H_5 \rightarrow CH_2=CH_2 + CH_3CH_3$$

D Disproportionierung der Radikale, Dismutation der Radikale
F dismutation de radicaux libres
P dysproporcjonowanie rodników, dysmutacja rodników
R диспропорционирование радикалов, дисмутация радикалов

1031
DISSIMILAR CHIRAL CENTRES (non-identical chiral centres, unequal dissymmetry centres).
Non-identical chiral centres with respect to chemical structure, e.g. any two chiral centres in molecule of a simple sugar.
D ungleichwertige Chiralitätszentren
F ...
P centra chiralności nierównocenne
R ...

1032
DISSIPATION.
Scattering of energy or thermodynamic potential due to irreversible processes.
D Energiedissipation
F dissipation d'énergie
P dyssypacja
R диссипация энергии

1033
DISSIPATION COEFFICIENT, tan δ.
Tangent of loss factor.
D dielektrischer Verlustfaktor
F coefficient de dissipation
P współczynnik rozproszenia,
  współczynnik dyssypacji
R коэффициент рассеивания

1034
DISSIPATIVE PROCESS.
Process of transfer of matter (e.g.
diffusion, thermodiffusion) or energy (heat
conduction, Dufour effect) within system
or between system and surroundings. Also
relaxation processes (e.g. chemical
reactions) undergone in nonequilibrium
conditions.
D dissipative Vorgänge, dissipative Effekte
F phénomènes dissipatifs
P procesy dyssypatywne, zjawiska
  dyssypatywne
R диссипационные процессы,
  диссипативные процессы

1035
DISSIPATIVE WORK.
Part of free energy changed into internal
energy during thermodynamic process.
D dissipierte Arbeit
F fonction de dissipation de l'énergie
  mécanique
P praca dyssypowana
R диссипативная работа

1036
DISSOCIATION.
Splitting of molecule into smaller
fragments.
D Dissoziation
F dissociation
P dysocjacja
R диссоциация

DISSOCIATION CONSTANT. *See*
INSTABILITY CONSTANT.

1037
DISSOCIATION FIELD EFFECT.
Increase of equivalent conductance of
electrolyte as a result of increase of
number of ions in solution in case of high
electric field (10^4—10^5 V/cm).
D Dissoziationsspannungseffekt
F effet dissociant du champ électrique
P efekt dysocjacyjny pola elektrycznego,
  dysocjacja pod wpływem pola
  elektrycznego
R диссоциационный эффект
  напряжённости поля

1038
DISTECTIC POINT.
Congruent (melting) point at maximum
of liquidus curve.

D distektischer Punkt
F point distectique
P punkt dystektyczny, dystektyk
R дистектическая точка

DISTORTION POLARIZATION. *See*
INDUCED POLARIZATION.

1039
DISTRIBUTION COEFFICIENT (partition
coefficient *in chromatography*), D_g.
Ratio of equilibrium concentrations of
substance in stationary and mobile phase
respectively.
D Verteilungskoeffizient
F coefficient de partage
P współczynnik podziału
R коэффициент распределения

DISTRIBUTION FUNCTION. *See*
PROBABILITY DISTRIBUTION
FUNCTION.

1040
DISTRIBUTION ISOTHERM (*in
chromatography*).
Isotherm characterizing dependence
between equilibrium concentrations of
given substance in stationary and mobile
phase respectively.
D Verteilungsisotherme
F isotherme de distribution
P izoterma podziału
R изотерма распределения

dl-COMPOUND. *See* RACEMIC
COMPOUND.

1041
dl MOLECULAR MIXTURE (molecular
mixture of opposite forms).
One of the racemic modification in solid
state, monophase system of solution of one
enantiomer in another. Melting point of dl
molecular mixture equals melting point of
enantiomer.
D . . .
F solution solide des composés d et l
P roztwór stały racemiczny, kryształy
  mieszane pseudoracemiczne
R . . .

1042
DOEBNER REACTION (Doebner
synthesis).
Preparation of 2-substituted derivatives of
cinchonic acid by condensation of
arylamine, aliphatic (or aromatic)
aldehyde and pyruvic acid, e.g.

D Doebner Cinchoninsäure-
-Ringkondensation
F synthèse de Doebner
P synteza (pochodnych kwasu
cynchoninowego) Doebnera
R синтез Дёбнера

DOEBNER SYNTHESIS. *See* DOEBNER
REACTION.

DOERNER-HOSKINS DISTRIBUTION
LAW. *See* LOGARITHMIC
DISTRIBUTION LAW.

1043
DOMAINS.
Regions of spontaneous magnetization
in magnetic materials (e.g. in
ferromagnets) or spontaneous polarization
in ferroelectrics.
D Bezirke, Bereiche, Domänen
F domaines
P domeny
R домены

1044
DOMAIN STRUCTURE.
Structure exhibited by ferromagnets and
ferroelectrics due to presence of large
number of zones (domains) of different
mutual orientation in absence of external
magnetic field (for ferromagnets) or
electric field (for ferroelectrics).
D Bezirksstruktur, Bereichsstruktur,
Domänenstruktur
F structure en domaines
P struktura domenowa
R доменная структура

DONNAN POTENTIAL. *See* MEMBRANE
POTENTIAL.

DONOR. *See* DONOR CENTRE.

1045
DONOR-ACCEPTOR COMPLEX
(charge-transfer complex, molecular
complex).
Combination of molecules which are
electron donors with molecules which
are acceptors of electrons. Linked
molecules are capable of independent
existence.
D Elektronen-Donator-Akzeptor-Komplex,
Ladungsübertragungskomplex,
Komplexmolekel
F complexe de transfert de charge
P kompleks donorowo-akceptorowy,
kompleks cząsteczkowy, związek
addycyjny cząsteczkowy, kompleks
z przeniesieniem ładunku, kompleks
charge-transfer
R донорно-акцепторный комплекс,
молекулярное соединение

DONOR ATOM. *See* COORDINATING
ATOM.

1046
DONOR CENTRE (donor, donor impurity).
Atom of valency higher than normal
constituent of crystal (e.g. admixture
of arsenic in germanium crystal), the
presence of which causes formation and
occupation by electron of new allowed
energy band, located just below
conductivity band (in range of forbidden
band)
D Donator, Donator-Zentrum
F donneur
P centrum donorowe, donor
R донор

DONOR IMPURITY. *See* DONOR
CENTRE.

1047
DONOR LEVEL (donor state *in band
model of solid state*).
Local, filled energy level situated in area
of forbidden band and connected with
presence of donor centre in crystal.
D Donatorniveau, Donatorenterm
F niveau donneur
P poziom donorowy, stan donorowy
R уровень донора, донорный уровень

DONOR STATE. *See* DONOR LEVEL.

1048
DOPPLER EFFECT.
Change of observed wave frequencies due
to relative movement of source and
observer.
D Doppler-Effekt
F effet Doppler
P efekt Dopplera
R эффект Допплера

1049
DORABIALSKA AND
ŚWIĘTOSŁAWSKI'S ADIABATIC
MICROCALORIMETER.
Type of adiabatic microcalorimeter
adapted for measurements of heats
exchanged in slow-rate processes.
D Świętosławski und Dorabialska
adiabatisches Mikrokalorimeter
F microcalorimètre adiabatique construit
par Świętosławski et Dorabialska
P mikrokalorymetr Świętosławskiego
i Dorabialskiej
R микрокалориметр Свентославского
и Дорабяльской

DORN EFFECT. *See* SEDIMENTATION
POTENTIAL.

1050
DOSE DISTRIBUTION.
Distribution of energy per unit mass in
object absorbing ionizing radiation. Dose
distribution cannot be homogeneous in
object of finite dimensions.
D ...
F distribution des doses
P rozkład dawki
R распределение радиационной дозы,
распределение поглощённой энергии

1051
DOSE EQUIVALENT.
Product of absorbed dose of ionizing
radiation of a particular kind and
respective coefficient of quality of
radiation and, if required, other
correction factors. Dose equivalent of
mixed radiation is equal to sum of
equivalents of all radiations involved.
D äquivalente Dosis
F équivalent de dose
P równoważnik dawki
R эквивалентная доза (излучения)

1052
DOSE METER.
Electronic device usually based on
principle of some sort of amplification,
used for measurement of very low and
low doses and dose rates of ionizing
radiation, mainly for purpose of
radiological protection.
D Dosimeter, Dosismesser
F dosimètre
P dawkomierz, dozymetr
R дозиметр

1053
DOSE RATE.
Absorbed dose of ionizing radiation with
respect to unit time. Dose rate is
expressed in Gy/s (in SI system).
D Dosisleistung
F débit de dose
P moc dawki, szybkość dawkowania
R мощность дозы

1054
DOSE RATE EFFECT (in radiation
chemistry).
Influence of rate of supply of ionizing
radiation energy on radiation yield or on
biological effects of absorbed radiation.
D Dosisleistungseffekt
F effet du taux de dose, influence de
l'intensité de dose
P efekt szybkości dawkowania, efekt mocy
dawki
R влияние мощности дозы

1055
DOSIMETER.
Substance, usually solid or liquid, used for
semiquantitative indication or precise

measurement of absorbed ionizing
radiation in medium (10^2—10^3 Gy) or for
high dose range (10^3—10^5 Gy), operating
on principle of chemical changes in
material of dosimeter.
D Dosimeter, Dosismesser
F dosimètre
P dawkomierz, dozymetr
R дозиметр

1056
DOSIMETRY OF IONIZING RADIATION
(radiation dosimetry).
Analytical and/or measuring technique of
predominantly physical (e.g. in
calorimetric dosimeter) or chemical (e.g.
in Fricke dosimeter) or of mixed
character (e.g. in ionization chamber),
for purpose of determination of dose
absorbed by essential part of dosimeter.
D Dosimetrie (der ionisierenden Strahlung)
F dosimétrie (du rayonnement ionisant),
dosimétrie (de radiation ionisante)
P dozymetria (promieniowania
jonizującego)
R дозиметрия (ионизирующего
излучения)

DOUBLE ADIABATIC CALORIMETER.
See TWIN CALORIMETERS.

1057
DOUBLE BEAM
SPECTROPHOTOMETER.
Spectrophotometer with splitting of
radiation beam into two beams; one
passing through reference material and
the other through investigated sample.
The beam splitting may either be effected
in time or, more frequently, in space.
D Zweistrahlspektrometer,
Spektralphotometer mit
Zweistrahlverfahren
F spectrophotomètre bifaisceau
P spektrofotometr dwuwiązkowy
R двухлучевой спектрофотометр

1058
DOUBLE BOND.
Chemical bond formed by two pairs of
valence electrons; one pair is present in
molecular bonding σ-orbital, and the other
in molecular bonding π-orbital, e.g. C=C
bond in ethylene.
D Doppelbindung
F liaison double
P wiązanie podwójne
R двойная связь

1059
DOUBLE COMPARATOR (twin
comparator, comparator).
Device used to project and to compare
standard and sample spectrum.
D Doppelprojektor, Doppelkomparator
F comparateur de spectres

P komparator, spektroprojektor podwójny
R двойной спектропроектор

1060
DOUBLE ESCAPE PEAK.
Peak in gamma radiation spectrum
generated as a result of formation of
electron-positron pair, annihilation of
positron formed and escape of both
annihilation photons from detector.
D zwei-Quanten-Escape-Peak,
  zwei-Quanten-Escape-Linie
F pic de deuxième échappement
P pik ucieczki podwójnej
R пик двойного вылета, пик вылета двух
  квантов

1061
DOUBLE REFRACTION.
Splitting of light beam into two
components in optically anisotropic
medium.
D Doppelbrechung
F réfraction double
P załamanie światła podwójne
R двойное преломление

1062
DOUBLE RESONANCE.
Magnetic resonance with two alternating
magnetic fields, one H_1 of frequency equal
to resonance frequency of spin system I
(nuclear or electronic) and another H_2
corresponding to resonance frequency of
spin system II (nuclear or electronic).
If two systems interact this may lead
to substantial changes in resonance signal
of spin system I and in turn to its
simplification.
D Doppelresonanz
F double résonance
P rezonans podwójny
R двойной резонанс

1063
DOUBLET.
In atomic spectroscopy set of spectral
lines arising from energetic transitions
between two atomic doublet terms.
D Dublett
F doublet
P dublet widmowy
R спектральный дублет

1064
DOUBLY MAGIC NUCLEUS
(proton-and-neutron magic nucleus).
Atomic nucleus in which both number of
protons and number of neutrons are equal
to magic numbers, viz. nucleus with
occupied proton and neutron shells.
Doubly magic nucleus has particularly
high binding energy. Known doubly magic
nuclei: $^4_2$He, $^{16}_8$O, $^{40}_{20}$Ca, $^{48}_{20}$Ca, $^{208}_{82}$Pb.
D doppelt-magischer Kern
F noyau bimagique

P jądro podwójnie magiczne
R дважды магическое ядро

1065
DROP ERROR.
Error occurring due to last drop of
titrating solution introducing more reagent
than necessary for equivalence point of
titration.
D Tropfenfehler
F erreur d'une goutte
P błąd kropli
R капельная ошибка

1066
DROPPING MERCURY ELECTRODE.
Electrode constructed from glass capillary
with internal diameter of order of 0.05 mm
from which mercury flows in small drops.
D Quecksilbertropfelektrode
F électrode à goutte de mercure
P elektroda rtęciowa kroplowa (kapiąca)
R капельный ртутный электрод

1067
DROP TIME.
In polarography, time between detaching
of two successive drops of mercury from
polarographic capillary. Usually in region
1—8 seconds.
D Tropfzeit
F temps de goutte
P czas trwania kropli
R время жизни капли

DRUDE-LORENTZ THEORY. *See*
ELECTRON THEORY OF METALS.

1068
DUFF-BILLS PHENOLIC ALDEHYDE
SYNTHESIS.
Preparation of o-hydroxybenzoic aldehyde
derivatives from phenols and
hexamethyltetramine in acidic medium,
e.g.

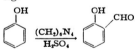

D Duff Phenol-Formylierung, Duff
  Aldehyd-Synthese
F réaction de Duff
P reakcja Duffa (i Billsa), formylowanie
  fenoli Duffa
R реакция Даффа

1069
DUROUR EFFECT.
Cross effect opposite to thermodiffusion,
i.e., heat flux accompanying isothermic
diffusion.
D Diffusionsthermoeffekt, Dufour-Effekt
F effet de Dufour
P efekt Dufoura
R эффект Дюфо

DUHEM-MARGULES EQUATION. *See*
DUHEM-MARGULES RELATION.

1070
DUHEM-MARGULES RELATION
(Duhem-Margules equation).
Equation which determines relation
between volatilities a_{pi} of components of
gas mixture at constant temperature T
and pressure p

$$\sum_i x_i (\mathrm{d} \ln a_{pi})_{T,p} = 0$$

where x_i = mole fraction of i-th
component and $\sum_i$ denotes summation over
all components.
D Duhem-Margulessche Beziehung
F relation de Duhem-Margules
P równanie Duhema i Margulesa
R уравнение Дюгема-Маргулеса

DUPLICATION METHOD. *See*
COLORIMETRIC TITRATION.

1071
DYNAMIC REACTIVITY INDICES.
Reactivity indices following from
consideration of electronic structure of
molecule in transition state of given
reaction. Dynamic reactivity indices
provide measure of activation energy,
e.g. localization energy.
D dynamische Indizes chemischer
Reaktionsfähigkeit
F indices dynamiques de réactivité
(chimique)
P indeksy reaktywności dynamiczne
R динамические индексы реакционной
способности

1072
DYSPROSIUM, Dy.
Atomic number 66.
D Dysprosium
F dysprosium
P dysproz
R диспрозий

1073
EBULLIOMETRY (ebullioscopy).
Methods for measurement of boiling
temperatures of mono- or multicomponent
systems. Ebulliometric methods are used
for determination of molar mass, degree
of purity of substance, for investigation
of liquid-vapour phase equilibrium, etc.
D Ebullioskopie
F ébulliométrie, ébullioscopie
P ebuliometria
R эбулиометрия, эбулиоскопия

1074
EBULLIOSCOPIC CONSTANT, E.
Quantity given by equation

$$E = \frac{RT^2}{\Delta H} M$$

where R = molar gas constant,
T = normal boiling temperature of solvent
(in kelvins), ΔH = molar evaporation heat
of solvent at boiling point, and
M = molar mass of solvent.
D ebullioskopische Konstante, molare
Siedepunktserhöhung
F constante ébullioscopique, constante
ébulliométrique
P stała ebulioskopowa, molalne
podwyższenie temperatury wrzenia
R эбулиоскопическая константа,
эбулиоскопическая постоянная

EBULLIOSCOPY. *See* EBULLIOMETRY.

1075
ECLIPSED CONFORMATION, ±sp.
Conformation in which the most bulky
substituents at two adjacent carbon (or
any other element, e.g. silicium or
nitrogen) atoms are sterically as near one
another as possible. In Newman's
projection formula, bond angles of less
remote and more remote substituents
equal 0°. This type of conformation is the
least stable

D ekliptische Konformation, verdeckte
Konformation
F conformation éclipsée, conformation
s-*cis*
P konformacja ekliptyczna, konformacja
± syn-periplanarna
R эклиптическая конформация,
полностью заслонённая конформация

1076
EDGE DISLOCATIONS.
Rows of atoms forming edges of lattice
planes running through parts of crystal
only.
D Stufenversetzungen
F . . .
P dyslokacje krawędziowe, dyslokacje
brzegowe
R краевые дислокации

1077
EFFECTIVE ATOMIC NUMBER.
Total number of electrons surrounding
central atom of metal in complex. This
number often is equal to atomic number
of nearest noble gas.

D effektive Atomnummer
F nombre atomique effectif
P liczba atomowa efektywna
R эффективный атомный номер

1078
EFFECTIVE COLLISION (*in chemical kinetics*).
Collision of substrate molecules bearing activation energy. Collision results in change of substrates into products if it takes place with appropriate geometrical orientation.

D erfolgreicher Stoß, erfolgreicher Zusammenstoß
F collision efficace, choc efficace
P zderzenie aktywne cząsteczek
R эффективное столкновение, активное столкновение (молекул)

1079
EFFECTIVE CROSS-SECTION FOR REACTIONS OF RADICALS.
Quantitative expression for probability of inelastic collisions between free radical and second reactant which is stable compound or (sometimes) another free radical.

D Reaktionswirkungsquerschnitt für Radikale
F section efficace de la réaction des radicaux
P przekrój czynny reakcji rodników
R эффективное сечение реакции радикалов

1080
EFFECTIVE HEIGHT OF THEORETICAL PLATE (plate height *in chromatography*).
Thickness of column segment in which mean concentration of substance is in equilibrium with its own effluent.

D Höhe des effektiven Bodens, effektive Bodenhöhe
F hauteur du plateau théorique
P wysokość półki teoretycznej
R высота эквивалентная теоретической тарелке

1081
EFFECTIVE NUMBER OF BOHR MAGNETONS, μ_{ef}.
Magnetic moment of atom, ion or molecule determined from experimentally measured magnetic susceptibility according to assumption of its temperature dependence given either by Curie law or by Curie-Weiss law

$$\mu_{ef} = 2.83 \sqrt{C_M} \ \mu_B = 2.83 \sqrt{\chi_M (T - \Theta)} \ \mu_B$$

where μ_B = Bohr magneton, χ_M = molar magnetic susceptibility at temperature T, C_M = molar Curie constant,
Θ = paramagnetic Curie temperature.

D effektives magnetisches Moment
F nombre effectif de magnétons de Bohr
P moment magnetyczny efektywny
R эффективный магнитный момент, эффективное число магнетонов Бора

1082
EFFECTIVE QUANTUM NUMBER.
Number characteristic for given spectroscopic term of many-electron atom, defined as $n_{ef} = \sqrt{R/T_n}$,
where R = Rydberg constant and T_n = term energy. Effective quantum number differs from principal quantum number of given type of electron shells by approximately the same value; it takes into account screening of nucleus by electrons.

D effektive Quantenzahl
F nombre quantique effectif
P liczba kwantowa efektywna
R эффективное квантовое число

EFFECT OF STERIC HINDRANCE. *See* HINDERING EFFECT.

1083
EFFECT OF THE STATE OF AGGREGATION (phase effect *in radiation chemistry*).
Differences in yield of products of radiolysis in the same system, but in different states of aggregation during irradiation.

D Zustandseffekt
F effet de phase
P efekt fazowy, efekt stanu skupienia (*radiacyjny*)
R влияние агрегатного состояния

EFFLUENT CONCENTRATION HISTORY. *See* BREAKTHROUGH CURVE.

EFFLUENT CONCENTRATION HISTORY. *See* ELUTION CURVE.

1084
EFFLUENT VOLUME.
Volume of solution collected from chromatographic column up to given moment during run.

D Volumen der Eluierlösung, Eluat-Volumen
F volume d'effluent, volume d'élution
P objętość wycieku
R объём элюата, объём фильтрата

1085
EIGENFUNCTION.
Solution of time-independent Schrödinger equation. It describes quantum-mechanical stationary states with definite energy values of system.

D Eigenfunktion
F fonction propre
P funkcja własna hamiltonianu
R собственная функция

1086
EIGENVALUE.
Number which satisfies equation
$O\Phi = c\Phi$, where O = operator,
c = eigenvalue, Φ = eigenfunction
corresponding to this eigenvalue c.

D Eigenwert
F valeur propre
P wartość własna operatora
R собственное значение

1087
EINSTEINIUM, Es.
Atomic number 99.

D Einsteinium
F einsteinium
P ajnsztajn
R эйнштейний

1088
EINSTEIN MASS-ENERGY
EQUIVALENCE LAW.
Each mass m corresponds to equivalent
amount of energy $E = mc^2$, where
c = velocity of light.

D Einstein-Gleichung,
  Masse-Energie-Beziehung
F équation d'Einstein, loi d'équivalence
  de la masse et de l'énergie
P prawo Einsteina, prawo równoważności
  masy i energii
R закон Эйнштейна, соотношение
  эквивалентности массы и энергии

1089
EINSTEIN-SMOLUCHOWSKI
EQUATION (Einstein-Smoluchowski
formula).

Formula for mean path projection $\overline{\Delta}_x$,
covered by particle in rectilinear
Brownian movement, on abscissa

$$\overline{\Delta}_x = \sqrt{\frac{RT}{N_A} \cdot \frac{1}{3\pi\eta r}} \cdot t$$

where R = gas constant,
T = thermodynamic temperature,
N_A = Avogadro constant, η = viscosity
coefficient of dispersion medium,
r = radius of colloidal particle, t = time.

D Einstein-Smoluchowskische Formel
F équation d'Einstein-Smoluchowski
P wzór Einsteina i Smoluchowskiego
R уравнение Эйнштейна-Смолуховского

EINSTEIN-SMOLUCHOWSKI FORMULA.
See EINSTEIN-SMOLUCHOWSKI
EQUATION.

1090
ELASTICITY OF SHAPE.
Property of material bodies of regaining
its original shape after removal of
external forces.

D Formelastizität
F élasticité de forme
P sprężystość postaci
R упругость формы

1091
ELASTIC SCATTERING.
Scattering which does not change kinetic
energy of scattered particles in system's
centre of mass.

D elastische Streuung
F diffusion élastique
P rozpraszanie sprężyste, rozpraszanie
  elastyczne
R упругое рассеяние

1092
ELASTIC WAVE.
Disturbance propagating without damping
in matter. It consists of transfer of
mechanical energy of particles which
vibrate around their positions of
equilibrium.

D elastische Welle
F onde élastique
P fala sprężysta
R упругая волна

ELBS ANTHRACENE SYNTHESIS. *See*
ELBS REACTION.

1093
ELBS PERSULFATE OXIDATION.
Oxidation of monohydroxyphenols
to dihydroxyphenols by potassium
persulfate in akaline medium, e.g.

D Elbs Persulfatoxydation von Phenolen
F réaction d'Elbs
P utlenianie Elbsa
R окисление (персульфатами) Эльбса

1094
ELBS REACTION (Elbs anthracene
synthesis).
Pyrolitic conversion of diaryl ketones
having methyl or methylenic substituent
in o-position with respect to carbonyl
group, involving dehydration and
cyclization to anthracene derivatives, e.g.

D Elbs-Reaktion, (Elbs)
  Anthracen-Ringschluß
F réaction d'Elbs
P synteza antracenu Elbsa
R реакция Эльбса

1095
ELECTRET.
Substance preserving macroscopic
field-induced dipole moment when electric
field is switched off.

D Elektret
F électret
P elektret
R электрет

1096
ELECTRICAL DOUBLE LAYER
(electrochemical double layer).
Layer which forms at interface of two
electrically conducting phases as result
of change of distribution of charges
in layers on both sides of phase boundary.
It has corresponding fall of internal
potential.

D elektr(ochem)ische Doppelschicht
F double couche électrochimique, couche
   double
P warstwa elektryczna podwójna, warstwa
   jonowa podwójna
R двойной электрический слой

1097
ELECTRIC CURRENT CARRIER (carrier).
Collective term for particles able to
transport electric charges. In general term
is applied only to mobile charge carriers,
i.e. free electrons and electron holes (see
electron conductivity, hole conductivity).

D Elektrizitätsträger, Ladungsträger
F porteur de charge, porteur d'électricité
P nośnik prądu
R носитель (электрического заряда)

1098
ELECTRIC DIPOLE (dipole).
System of charges with non-coinciding
centres of gravity of negative and positive
charge. In the simplest case, system of two
equal but opposite and displaced point
charges.

D (elektrischer) Dipol
F dipôle
P dipol elektryczny
R (электрический) диполь

1099
ELECTRIC DIPOLE MOMENT (dipole
moment), μ.
Vector directed from negative to positive
pole of dipole with absolute (scalar) value
equal to $|\mu| = |q| \cdot d$, where $|q|$ = absolute
value of one of pole charges; d = mutual
distance between poles.
Remark: Sometimes opposite definition of
dipole moment direction is utilized which is in
disagreement with definition of polarization
vector direction. In this case dipole moment
should be represented by $\rightarrow\!\!\!\!+$.

D (elektrisches) Dipolmoment
F moment (électrique) dipolaire

P moment dipolowy (elektryczny)
R (электрический) дипольный момент

ELECTRIC DOUBLE REFRACTION. *See*
KERR EFFECT.

ELECTRIC PERMITTIVITY. *See*
PERMITTIVITY.

ELECTRIC SATURATION. *See*
DIELECTRIC SATURATION.

ELECTROANALYSIS. *See*
ELECTROCHEMICAL ANALYSIS.

ELECTROCAPILLARITY. *See*
ELECTROCAPILLARITY EFFECT.

1100
ELECTROCAPILLARITY CURVE.
Curve which represents dependence of
surface tension of metal (e.g. mercury) at
metal-electrolyte interface on potential of
metal.

D Elektrokapillarkurve
F courbe électrocapillaire
P krzywa elektrokapilarna
R электрокапиллярная кривая

1101
ELECTROCAPILLARY EFFECT
(electrocapillarity).
Change of surface tension of metal
according to its potential.

D Elektrokapillareffekt, Elektrokapillarität
F effet électrocapillaire
P efekt elektrokapilarny,
   elektrokapilarność
R электрокапиллярный эффект

1102
ELECTROCHEMICAL ANALYSIS
(electroanalysis).
Methods of chemical analysis based on
phenomena occurring during electrode
reactions and/or flow of current through
electrolyte solutions.

D Elektroanalyse
F analyse électrochimique
P analiza elektrochemiczna, elektroanaliza
   chemiczna
R электроанализ

ELECTROCHEMICAL DOUBLE LAYER.
See ELECTRICAL DOUBLE LAYER.

1103
ELECTROCHEMICAL GENERATION OF
TITRANT.
Production of titrant by electrolytic
oxidation or reduction. Quantity of
produced titrant is calculated on basis
of charge which passes through circuit in
time of generation.

D elektrochemische Erzeugung des
   Reagenzes
F préparation de réactif par électrolyse
P generowanie elektrolityczne odczynnika
R электрогенерирование титранта

1104
ELECTROCHEMICAL POTENTIAL (of
a given sort of ions) $\tilde{\mu}_B$.
Intensive thermodynamic quantity

$$\tilde{\mu}_B = \mu_B + z_B F\varphi$$

where μ_B = chemical potential of ions B,
z_B = valency of ion and φ = inner
potential of phase.

D elektrochemisches Potential
F potentiel électrochimique
P potencjał elektrochemiczny
R электрохимический потенциал

1105
ELECTROCHEMICAL SERIES
(electromotive series).
Electrode reactions ordered according
to increasing values of corresponding
standard potentials.

D Spannungsreihe
F série des tensions, série des potentiels
P szereg napięciowy
R электрохимический ряд,
    ряд напряжений

1106
ELECTROCHEMISTRY.
Branch of physical chemistry dealing with
systems and processes in which
electrolytes participate.

D Elektrochemie
F électrochimie
P elektrochemia
R электрохимия

1107
ELECTROCHROMATOGRAPHY.
Method of separation consisting in
combination of chromatography and
electrophoresis.

D Elektrochromatographie
F électrochromatographie
P elektrochromatografia
R электрохроматография

1108
ELECTRODE.
Non-electrolytic conductor, usually
metallic, bordering with electrolyte or
other medium carrying electrons to or out
of that medium.

D Elektrode
F électrode
P elektroda
R электрод

1109
ELECTRODE (half-cell in
electrochemistry).
System of two or more conducting phases,
at least one of which is electrolytic
conductor. These phases border in such

a way as to enable flow of electric
charges or ions across interfacial surfaces.

D Elektrode, Halbzelle
F électrode, demi-cellule
P elektroda, półogniwo
R электрод, полуэлемент

1110
ELECTRODE (in spark and arc
spectroscopy).
One of electrodes of source of excitation.

D Elektrode
F électrode
P elektroda (spektralna)
R электрод

1111
ELECTRO-DECANTATION.
Modified electrodialysis in which
concentration of dispersed phase occurs
simultaneously with dialysis due to
electrophoresis and gravitational
sedimentation; clear solution lying above
concentrated dispersed phase is removed
by decantation.

D Elektrodekantation
F électrodécantation
P elektrodekantacja
R электродекантация

ELECTRODE GAP. See ANALYTICAL
GAP.

1112
ELECTRODE OF FIRST KIND.
Electrode (half-cell) of metallic
conductor|electrolyte solution type, e.g.
silver, copper, hydrogen electrodes.

D Elektrode erster Art
F électrode de première espèce, électrode
    du premier genre
P elektroda pierwszego rodzaju
R электрод первого рода

1113
ELECTRODE OF SECOND KIND.
Electrode (half-cell) composed of metal M
covered with thin layer of its poorly
soluble salt MA and dipped into solution
of well soluble salt of some metal M'A
with anion common with salt MA.
Solution of M'A is saturated with salt
MA, e.g. calomel electrode, silver chloride
electrode, oxide electrodes.

D Elektrode zweiter Art
F électrode de deuxième espèce, électrode
    du deuxième genre
P elektroda drugiego rodzaju
R электрод второго рода

1114
ELECTRODE OF THIRD KIND.
Electrode (half-cell) composed of metal A
covered with thin layer of its poorly

soluble salt MA, and layer of poorly soluble salt other metal M′A dipped into solution of well soluble salt of metal M′; solution is saturated with respect to salts MA and M′A; e.g. Zn, ZnC_2O_4, $CaC_2O_4|Ca^{2+}$.

D Elektrode dritter Art
F électrode de troisième espèce, électrode du troisième genre
P elektroda trzeciego rodzaju
R электрод третьего рода

1115
ELECTRODE REACTION (potential determining reaction).
Oxidation or reduction of substance occurring at electrode surface; reaction which takes place at electrode — electrolyte interface whose course determines potential of electrode.

D Elektrodenreaktion, potentialbestimmende (Brutto-)Reaktion
F réaction d'électrode
P reakcja elektrodowa, reakcja potencjałotwórcza
R электродная реакция, потенциалопределяющая реакция

1116
ELECTRODIALYSIS.
Dialysis combined with electrolysis through semi-permeable membrane.

D Elektrodialyse
F électrodialyse
P elektrodializa
R электродиализ

ELECTROENDOOSMOSIS. *See* ELECTROOSMOSIS.

1117
ELECTROGRAPHIC ANALYSIS.
Method of analysis of minerals and metals based on electrolytic transfer of ions from analysed sample onto surface of another material (for instance on papers saturated with electrolyte), where these ions are identified.

D ...
F analyse électrographique
P analiza elektrograficzna
R электрографический анализ

1118
ELECTROGRAVIMETRY (analysis by electrodeposition).
Weight determination of metallic elements quantitatively deposited from solution on cathode as metal or on anode as oxide.

D Elektrogravimetrie, Elektroanalyse
F électrogravimétrie, électro-analyse

P elektrograwimetria, analiza elektrograwimetryczna, elektroanaliza
R электровесовой анализ, электроанализ

1119
ELECTROKINETIC POTENTIAL (zeta potential), ζ.
Difference of electrical potentials between sliding surface and interior of liquid phase moving in respect to solid phase.

D elektrokinetisches Potential, ζ-Potential
F potentiel électrocinétique, potentiel ζ
P potencjał elektrokinetyczny, potencjał zeta, potencjał ζ
R электрокинетический потенциал, дзета-потенциал

1120
ELECTROLYSIS.
Process consisting in flow of electric current from external source across electrolytic conductor; this process is accompanied by electrochemical reactions at metallic conductor — electrolytic conductor phase boundary.

D Elektrolyse
F électrolyse
P elektroliza
R электролиз

1121
ELECTROLYTE.
Substance which conducts electrical current by means of free, stable ions (e.g. $NaCl_{(s)}$, $NaCl_{(l)}$) or conducting after formation of ions as a result of reaction between its molecules
(e.g. $2H_2O_{(l)} \rightleftarrows H_3O^+ + OH^-$)
or its molecules and solvent
(e.g. $HCl_{(g)} + H_2O_{(l)} \rightleftarrows H_3O^+ + Cl^-$).

D Elektrolyt
F électrolyte
P elektrolit
R электролит

ELECTROLYTIC BRIDGE. *See* SALT BRIDGE.

ELECTROLYTIC CONDUCTIVITY. *See* SPECIFIC CONDUCTANCE.

ELECTROLYTIC CONDUCTOR. *See* IONIC CONDUCTOR.

1122
ELECTROLYTIC DISSOCIATION.
Decomposition of substance into thermodynamically stable, free ions as result of interactions between dissolved substance and solvent.

D elektrolytische Dissoziation
F dissociation électrolytique
P dysocjacja elektrolityczna
R электролитическая диссоциация

1123
ELECTROLYTIC DISSOCIATION
CLASSICAL CONSTANT, K_c.
Equilibrium constant of electrolytic
dissociation reaction expressed by
concentrations c, e.g. for 1.1-electrolyte,

$$MA \rightleftarrows M^+ + A^-$$

$$K_c = \frac{c_{M^+} + c_{A^-}}{c_{MA}}$$

equation applies approximately to diluted
solutions of weak electrolytes (see
electrolytic dissociation constant).

D klassische Dissoziationskonstante des
Elektrolyts
F constante d'ionisation apparente
P stała dysocjacji elektrolitycznej
klasyczna
R константа электролитической
диссоциации (классическая)

1124
ELECTROLYTIC DISSOCIATION
CONSTANT, K_a.
Thermodynamic equilibrium constant
of electrolytic dissociation reaction; for
instance for 1.1-electrolyte $MA \rightleftarrows M^+ + A^-$

$$K_a = \frac{a_{M^+} a_{A^-}}{a_{MA}} = \frac{c_{M^+} c_{A^-}}{c_{MA}} \cdot \frac{f_{M^+} f_{A^-}}{f_{MA}} =$$

$$= K_c \frac{f_{M^+} f_{A^-}}{f_{MA}}$$

where K_c = electrolytic dissociation
classical constant; equation applies to all
electrolytic solutions.

D (thermodynamische)
Dissoziationskonstante des Elektrolyts
F constante de la dissociation
électrolytique, constante d'ionisation
P stała dysocjacji elektrolitycznej
(termodynamiczna)
R константа электролитической
диссоциации (термодинамическая)

1125
ELECTROLYTIC DISSOLUTION
PRESSURE.
Historical concept from Nernst theory
of electrode potentials; it describes
tendency of electrode material to send its
ions into solution.

D elektrolytischer Lösungsdruck
F ...
P prężność roztwórcza (elektrolityczna)
R электролитическая упругость
растворения

1126
ELECTROLYTIC OXIDATION (anodic
oxidation).
Oxidation taking place at anode in course
of electrolysis, e.g.

$$2Cl^{\ominus} \xrightarrow{-2e} Cl_2$$

$$R-CO_2^{\ominus} \xrightarrow{-e} R-CO_2 \cdot \begin{cases} CO_2 + R \cdot \rightarrow \text{ other products} \\ \xrightarrow{-e} CO_2 + R^{\oplus} \rightarrow \text{ other products} \end{cases}$$

D elektrolytische Oxydation
F oxydation anodique
P utlenianie elektrolityczne
R электролитическое окисление

1127
ELECTROLYTIC POLARIZATION.
Processes which occur during flow of
anodic or cathodic current exceeding
exchange current through electrode —
solution interphase; they result in
deviation of half-cell from equilibrium
due to one or several slow steps of
electrode processes.

D elektrolytische Polarisation
F polarisation électrolytique, polarisation
électrochimique
P polaryzacja elektrolityczna
R электродная поляризация,
электролитическая поляризация,
электрохимическая поляризация

1128
ELECTROLYTIC REDUCTION.
Reduction taking place at cathode in
electrolysis, e.g.

$$Na^{\oplus} \xrightarrow{+e} Na$$

$$C_6H_5N\overset{O}{\underset{O}{}} \xrightarrow[+2H^{\oplus}]{+2e} C_6H_5N\overset{O}{\underset{O}{}} OH \rightleftharpoons C_6H_5\ddot{N}\overset{OH}{\underset{OH}{}} \xrightarrow{+H^{\oplus}}$$

$$C_6H_5\ddot{N}\overset{\oplus OH_2}{\underset{OH}{}} \xrightarrow[-H_2O]{+H^{\oplus} + 2e} C_6H_5\ddot{N}\overset{H}{\underset{OH}{}}$$

D elektrolytische Reduktion
F réduction électrolytique
P redukcja elektrolityczna
R электролитическое восстановление

1129
ELECTROLYTIC THERMOCELL.
Galvanic cell constructed from two
identical electrodes, which however
remain at different temperatures.

D elektrolytische Thermozelle
F pile thermoélectrique, cellule
thermoélectrique
P termoogniwo galwaniczne
R электрохимическая термоцепь,
гальванический термоэлемент

1130
ELECTROMAGNETIC INTERACTION.
Interaction of particles which gives rise to
processes taking place within time range
of 10^{-18}—10^{-15} s. Real or virtual photons
participate in electromagnetic interaction.

D elektromagnetische Wechselwirkung
F interaction électromagnétique
P oddziaływanie elektromagnetyczne
R электромагнитное взаимодействие

1131
ELECTROMAGNETIC RADIATION.
Periodical changes of electromagnetic field
propagated through space with finite
velocity.
D elektromagnetische Strahlung
F rayonnement électromagnétique
P promieniowanie elektromagnetyczne
R электромагнитное излучение

1132
ELECTROMERIC EFFECT.
Shift of electron density in molecules with
multiple, and especially conjugated bonds,
induced by external electric field.
D Elektromerie-Effekt
F effet électromère
P efekt elektromeryczny
R электромерный эффект

1133
ELECTROMOTIVE FORCE, EMF.
Difference of potentials between
electrodes of thermodynamically
reversible cell.
D elektromotorische Kraft
F force électromotrice
P siła elektromotoryczna
R электродвижущая сила

ELECTROMOTIVE SERIES. *See*
ELECTROCHEMICAL SERIES.

1134
ELECTRON, e (e$^-$).
Stable elementary particle, a lepton, of
mass $m = 9.109\,558 \cdot 10^{-28}$ g =
= 0.511 0041 MeV, with negative
elementary charge $Q = 4.803\,250 \cdot 10^{-10}$ ES
units = $1.602\,1917 \cdot 10^{-19}$ C, baryon number
$B = 0$, electronic lepton number
$L_e = 1$ and spin quantum number $J = 1/2$.
D Elektron
F électron
P elektron, negaton
R электрон

1135
ELECTRON ACCEPTOR.
Atom accepting pair of electrons from
donor.
D Elektronenakzeptor
F accepteur d'électrons
P akceptor elektronów, elektronobiorca
R акцептор электронов

1136
ELECTRON AFFINITY.
Energy ensuing from electron capture by
atom or molecule resulting in formation
of negative ion.
D Elektronenaffinität
F affinité électronique
P powinowactwo elektronowe, energia
powinowactwa elektronowego
R электронное сродство, сродство
к электрону

1137
ELECTRON BEAM.
Fast electrons moving along straight,
almost parallel tracks. Electron beam is
produced in accelerating devices. Its power
is defined by intensity (in amperes) and
energy of electrons (in mega-electronvolts).
D Elektronenstrahl
F faisceau d'électrons
P wiązka elektronów
R электронный пучок

1138
ELECTRON CAPTURE.
Radioactive transition whereby nucleus
captures one of its orbital electrons (most
often from K shell, more seldom from L
or M shells) with emission of neutrino.
Excited nucleus returns to ground state
emitting characteristic X-rays and Auger
electrons.
D Elektroneneinfang
F capture électronique
P wychwyt elektronu orbitalnego
R электронный захват

1139
ELECTRON CLOUD.
Distribution of electrons around atomic
nucleus; it is usually accepted that they
occupy atomic orbitals.
D Elektronenwolke
F nuage électronique
P chmura elektronowa
R электронное облако

1140
ELECTRON CONDUCTIVITY
(n-conduction).
Transfer of electric charge through
crystal under external field. Electrons
occupying quantum states in range of
conduction band move to neighbouring
unoccupied quantum states in range of
this band in opposite direction to vector
of electric field.
D Elektronen-Leitung, n-Leitung
F conductivité électronique, n-conductivité
P przewodnictwo elektronowe
R электронная проводимость

1141
ELECTRON CONFIGURATION.
For atoms, way of distributing electrons among atomic electron shells; it is given by symbols of all occupied shells with superscript number equal to number of electrons in shell, e.g. electron configuration of ground electronic state of chlorine atom is $1s^2 2s^2 2p^6 3s^2 3p^5$.
D Elektronenkonfiguration
F configuration électronique
P konfiguracja elektronowa
R электронная конфигурация

1142
ELECTRON CORRELATION (correlation of electrons).
Difference between real electron motion and the one following from self-consistent field method description.
D Korrelation der Elektronen
F corrélation des électrons
P korelacja elektronów
R электронная корреляция

1143
ELECTRON-DEFECT SEMICONDUCTOR (p-type semiconductor, defect conductor).
Semiconductor in which electron holes formed in ground band are main carriers of electric current.
D Defekt(halb)leiter, p-Halbleiter
F semiconducteur du type p, p-(semi)conducteur
P półprzewodnik niedomiarowy, półprzewodnik typu p, półprzewodnik dziurowy
R дырочный полупроводник, полупроводник p-типа

1144
ELECTRON DEFICIENCY.
Deficiency of electrons characteristic of some elements, e.g. boron and aluminium; lack of sufficient number of valency electrons for formation of adequate number of atomic bonds between all atoms constituting a given molecule.
D Elektronenmangel
F déficit d'électrons
P deficyt elektronowy, niedobór elektronowy
R электронный дефицит

1145
ELECTRON DENSITY (electronic density).
In atom or molecule density of electron cloud. Defined by modulus square of atomic or molecular orbital respectively.
D Elektronendichte
F densité électronique
P gęstość elektronowa
R электронная плотность

ELECTRON DENSITY. *See* FREE ELECTRON DENSITY.

1146
ELECTRON DIFFRACTION.
Scattering of monochromatic stream of electrons on atoms, followed by interference of matter waves.
D Elektronenbeugung, Elektronendiffraktion
F diffraction des électrons, diffraction électronique
P dyfrakcja elektronów
R дифракция электронов

1147
ELECTRON DONOR.
Atom which gives pair of electrons to acceptor.
D Elektronendonator
F donneur d'électrons
P donor elektronów, elektronodawca
R донор электронов

ELECTRON DOUBLET. *See* ELECTRON PAIR.

1148
ELECTRONEGATIVITY.
Qualitative measure of electron attraction by atom or group of atoms.
D Elektronegativität
F électronégativité
P elektroujemność pierwiastków, elektronegatywność pierwiastków
R электроотрицательность

1149
ELECTRONEGATIVITY SCALE.
Ordering of atoms of elements according to their electronegativities.
D Elektronegativitätsskala
F échelle d'électronégativité
P skala elektroujemności pierwiastków
R шкала электроотрицательности

1150
ELECTRON-EXCESS SEMICONDUCTOR (n-type semiconductor).
Semiconductor in which electrons transferred to conductance band are main carriers of electric current.
D Überschußleiter, n-Halbleiter
F semiconducteur du type n, n-(semi)conducteur
P półprzewodnik nadmiarowy, półprzewodnik typu n, półprzewodnik elektronowy
R электронный полупроводник, полупроводник n-типа

1151
ELECTRON EXCHANGERS.
Solids able to participate in reversible oxidation-reduction reactions.

D Elektronenaustauscher
F résines échangeuses d'électrons
P wymieniacze elektronów
R электронообменные смолы

1152
ELECTRON GAS.
Group of electrons in metal or semiconductor which may be treated as group of particles obeying laws of classical mechanics, travelling freely in field of constant potential or in periodic field (in range of whole crystal). Distribution of energy in electron gas is described by Maxwell-Boltzmann law.

D Elektronengas
F gaz d'électrons
P gaz elektronowy
R электронный газ

1153
ELECTRONIC ANTINEUTRINO, $\bar{\nu}_e$.
Antiparticle corresponding to electronic neutrino of lepton electronic number $L_e = -1$ and positive (right-handed) helicity, stable.

D e-Antineutrino
F antineutrino ($\bar{\nu}_e$)
P antyneutrino elektronowe
R электронное антинейтрино

1154
ELECTRONIC BAND.
In molecular electronic spectra band corresponding to transition between rotation-vibration-electronic states.

D Elektronenband
F bande électronique
P pasmo elektronowe
R электронная полоса

ELECTRONIC CONDUCTOR. *See* METALLIC CONDUCTOR.

1155
ELECTRONIC DEFECTS (ionic defects).
Imperfections of crystal lattice consisting in occupation of certain interstitial positions by ions of abnormal charge (e.g. Fe^{3+} in FeO crystal), neutral atoms, electrons or localized holes respectively in anionic or cationic vacancies. The number of defects of one sign must be compensated by stoichiometric number of defects of opposite sign.

D elektronische Fehlordnung
F *(centres de perturbation à défaut ou excès d'électrons)*
P defekty elektryczne
R электрические дефекты

ELECTRONIC DENSITY. *See* ELECTRON DENSITY.

ELECTRONIC ENERGY LEVELS OF A MOLECULE. *See* MOLECULAR ELECTRONIC ENERGY LEVELS.

1156
ELECTRONIC FORMULA.
Structural formula with distribution of valency electrons marked by dots and lone electron pairs represented by bars, e.g. acetylene H:C:::C:H, ammonia H:N̄:H
Ḧ

D Elektronenformel
F formule électronique
P wzór elektronowy
R электронная формула

ELECTRONIC INTERNAL PARTITION FUNCTION. *See* ELECTRONIC PARTITION FUNCTION.

1157
ELECTRONIC ISOMERISM.
Existence of different electronic configurations (states of different multiplicity) in molecules of a given chemical compound; e.g. Chichibabin's hydrocarbon exists in singlet or triplet forms

$(C_6H_5)_2C$⟨⟩=⟨⟩$=C(C_6H_5)_2$ ⇌

⇌ $(C_6H_5)_2\dot{C}$⟨⟩—⟨⟩$—\dot{C}(C_6H_5)_2$

D Elektronenisomerie
F électromérie, tautomérie des électrons
P izomeria elektronowa
R электронная изомерия

1158
ELECTRONIC LEPTON NUMBER, L_e.
Number attributed to elementary particles; for electron and electronic neutrino, $L_e = 1$, for their antiparticles, $L_e = -1$. Other particles and antiparticles have $L_e = 0$. Electronic lepton number is conserved in all interactions.

D Elektronenleptonenzahl
F nombre leptonique électronique
P liczba leptonowa elektronowa
R электронное лептонное число

1159
ELECTRONIC NEUTRINO, ν_e.
Elementary particle, a lepton, with rest mass probably zero, electrically neutral, stable, of baryon number $B = 0$, quantum spin number $J = 1/2$, negative (left-handed) helicity and electronic lepton number $L_e = 1$.

D Elektron-Neutrino, e-Neutrino
F neutrino ν_e
P neutrino elektronowe
R электронное нейтрино

1160
ELECTRONIC PARTITION FUNCTION
(electronic internal partition function),
$e(T)$.
Function

$$e(T) = \sum_{\varepsilon_e} g(\varepsilon_e) \exp\left[-\frac{\varepsilon_e - \varepsilon_{e0}}{kT} \right]$$

where $g(\varepsilon_e)$ = degeneracy of electronic
level equal to its statistical weight,
ε_{e0} = energy of electronic ground state,
k = Boltzmann constant,
T = thermodynamic temperature, and
summation takes place overall electronic
states of molecule. Electronic partition
function determines influence of
molecular electronic structure of gas
molecules on its thermodynamic
properties. For $T \ll \dfrac{\varepsilon_{e1} - \varepsilon_{e0}}{k}$ electronic
partition function simplifies to
$e(T) \approx g(\varepsilon_0)$.
D Elektronenzustandssumme,
   Verteilungsfunktion der Elektronenhülle
F fonction de partition électronique
P funkcja rozdziału elektronowa
R электронная статистическая сумма,
   электронная сумма состояний

1161
ELECTRONIC POLARIZABILITY, α_E.
Dielectric polarization (per molecule)
arising in unit local field due to
deformation of electron density.
D elektronische Polarisierbarkeit
F polarisabilité électronique
P polaryzowalność elektronowa
R электронная поляризуемость

1162
ELECTRONIC POLARIZATION.
Polarization due to deformation of electron
density distribution.
D Elektron(verschiebungs)polarisation
F polarisation électronique
P polaryzacja elektronowa
R электронная поляризация

1163
ELECTRONIC POLARIZATION
(MOLAR), P_E.
Measure of electronic polarization defined
by

$$P_E = \frac{1}{3\varepsilon_0} N_A \alpha_E \text{ (in Si units)}$$

where N_A = Avogadro constant,
α_E = electronic polarizability,
ε_0 = permittivity of free space.
D Elektron(verschiebungs)polarisation
F polarisation électronique

P polaryzacja elektronowa (molowa)
R электронная поляризация

1164
ELECTRONIC SHELL.
Set of atomic orbitals with the same
principal quantum number.
D Elektronenschale
F couche électronique
P powłoka elektronowa atomu, warstwa
   elektronowa atomu
R электронная оболочка

1165
ELECTRONIC SPECIFIC HEAT.
Contribution of free or semi-free electrons
to total specific heat of metal. It is
experimentally determined from
low-temperature behaviour of specific
heat.
D Elektronenwärme
F chaleur spécifique électronique
P ciepło właściwe elektronowe
R электронная теплоёмкость

1166
ELECTRONIC SPECTROSCOPY.
Branch of spectroscopy dealing with study
of electronic spectra of atoms and
molecules.
D Elektronenspektroskopie
F spectroscopie électronique
P spektroskopia elektronowa
R электронная спектроскопия

1167
ELECTRONIC SPECTRUM.
Spectrum in visible and ultraviolet region
arising due to transitions between
different electronic energy levels of
a given system.
D Elektronenspektrum
F spectre électronique
P widmo elektronowe
R электронный спектр

1168
ELECTRONIC SUBSHELL.
Set of atomic orbitals with the same
principal and orbital quantum numbers.
D Elektronenunterschale
F sous-couche électronique
P podpowłoka elektronowa atomu,
   podwarstwa elektronowa atomu
R электронная оболочка

1169
ELECTRONIC THEORY OF CATALYSIS
Theory of heterogeneous catalysis which
ascribes principal role in catalytic process
to chemisorption of molecules of
substrates on surface of catalyst.
D Elektronentheorie der Katalyse
F théorie électronique de l'activite
   catalytique

P teoria elektronowo-chemiczna
R электронная теория катализа

1170
π-π^* ELECTRONIC TRANSITION.
According to one-electron excitation
model, transition involving electron jump
from bonding to antibonding π-orbital.
D π-π^* Übergang
F transition π-π^*
P przejście elektronowe typu π-π^*
R π-π^* переход

1171
σ-π^* ELECTRONIC TRANSITION.
According to one-electron excitation
model, transition involving electron jump
from bonding σ-orbital to antibonding
π-orbital.
D σ-π^* Übergang
F transition σ-π^*
P przejście elekronowe typu σ-π^*
R σ-π^* переход

1172
n-π^* ELECTRONIC TRANSITION.
According to one-electron excitation
model, transition corresponding to electron
jump from non-bonding lone pair orbital,
n, to antibonding π-orbital.
D n-π^* Übergang
F transition n-π^*
P przejście elektronowe typu n-π^*
R n-π^* переход

1173
n-σ^* ELECTRONIC TRANSITION.
According to one-electron excitation
model, transition involving electron jump
from non-bonding lone pair orbital, n,
to antibonding σ-orbital.
D n-σ^* Übergang
F transition n-σ^*
P przejście elektronowe typu n-σ^*
R n-σ^* переход

1174
ELECTRONIC-VIBRATION-ROTATION
SPECTRUM (rotation-vibration-electronic
spectrum).
Molecular spectrum corresponding to
transitions between electronic energy
levels with simultaneous change of
vibrational and rotational energy of
molecule.
D Elektronenspektrum
F spectre de bandes électroniques
P widmo rotacyjno-oscylacyjno-
-elektronowe
R электронно-колебательно-
-вращательный спектр

1175
ELECTRON MICROPROBE X-RAY
ANALYZER (electron probe
microanalyzer).
Device which enables qualitative and
quantitative analysis of chemical
composition of micro-regions (of diameter
of a few µm) to be performed on sample
surface from X-ray spectrum excited by
precisely focused electron beam.
D Elektronensonde-
-Röntgenmikroanalysator
F microanalyseur à sonde électronique
P mikrosonda elektronowa
R рентгеновский микроанализатор

1176
ELECTRON OCTET.
Eight electrons filling s- and p-subshells
of valence shell of atom.
D Elektronenoktett
F octet électronique
P oktet elektronowy
R электронный октет

1177
ELECTRON PAIR (electron doublet).
Two electrons with opposite orientation
of their spins; according to Lewis theory
they form single covalent bond, e.g. in H_2
molecule.
D Elektronenpaar, Zweierschale
F doublet électronique, paire d'électrons
P doublet elektronowy, para elektronów
R дублет электронов, электронная пара,
пара электронов

1178
ELECTRON PARAMAGNETIC
RESONANCE (electron spin resonance),
EPR (ESR).
Magnetic resonance exhibited by
paramagnetic substances and by
diamagnetic samples with paramagnetic
centres.
D paramagnetische Elektronenresonanz,
Elektronenspin-Resonanz
F résonance paramagnétique électronique
P rezonans (para)magnetyczny
elektronowy, rezonans spinowy
elektronowy
R электронный парамагнитный резонанс,
электронный спиновый резонанс

ELECTRON PROBE MICROANALYZER.
See ELECTRON MICROPROBE X-RAY
ANALYZER.

ELECTRON SPIN RESONANCE. *See*
ELECTRON PARAMAGNETIC
RESONANCE.

1179
ELECTRON THEORY OF METALS (free electron theory of metals, Drude-Lorentz theory).
Theory based on following assumptions: (a) crystal of metal consists of positive metal ions which are in equilibrium with electrons moving freely in lattice space (electron gas); (b) electrons obey laws of classical mechanics and kinetic theory of gases as concerns their movements in lattice space of crystal.
D Elektronentheorie der Metalle
F théorie électronique des métaux, théorie d'électrons libres dans les métaux
P teoria elektronowa metali, teoria elektronów swobodnych, teoria Drudego i Lorentza
R электронная теория металлов

1180
ELECTRONVOLT, eV.
Unit of energy. 1 eV is equal to the change of energy of an electron which passes through an electrical potential difference of one volt. In SI system
1 eV $= 1.60206 \cdot 10^{-19}$ J.
D Elektronenvolt
F électron-volt
P elektronowolt
R электронвольт

1181
ELECTROOSMOSIS (electroendoosmosis).
Motion of liquid with respect stationary porous (capillary) solid phase or through capillary under influence of applied electric field.
D Elektro(end)osmose
F électro-osmose
P elektroosmoza, (elektro)endoosmoza
R электро(энд)осмос

ELECTROPHILE. See ELECTROPHILIC REAGENT.

1182
ELECTROPHILIC REAGENT (electrophile).
Reagent which acts as acceptor of electron pair belonging to carbon atom (or to atom of another element). Positive ions of molecules, containing electron deficient atoms (Lewis acids) and molecules with highly polarized bonds, are electrophiles, e.g.

$H_3O^{\oplus}$, $R_3C^{\oplus}$, $AlCl_3$,

D elektrophiles Agens
F réactif électrophile
P czynnik elektrofilowy, odczynnik elektrofilowy
R электрофильный реагент

1183
ELECTROPHORESIS.
Motion of charged particles of dispersed phase under influence of electric field.
D Elektrophorese
F électrophorèse
P elektroforeza
R электрофорез

1184
ELECTROPHORETIC EFFECT.
Hindering of ionic motion in electric field as a result of movement in opposite direction of counter ions (ionic cloud) with their solvation sheaths.
D elektrophoretischer Effekt
F effet électrophorétique
P efekt elektroforetyczny
R (ионно-)электрофоретический эффект

ELECTROPHORETIC POTENTIAL. See SEDIMENTATION POTENTIAL.

1185
ELECTROSELECTIVITY.
Preference for ions of higher charge shown by ion exchanger in dilute solutions.
D Elektroselektivität
F électrosélectivité
P elektroselektywność
R электроселективность

1186
ELECTROSTATIC BOND.
Ionic bond, hydrogen bond and electrostatic interaction of multipoles.
D elektrostatische Bindung
F liaison ionique
P wiązanie elektrostatyczne
R электростатическая связь

ELECTROSTATIC FORCES. See COULOMB FORCES.

1187
ELECTROVALENCE.
Number of electrons given or accepted by atom during formation of ionic bond.
D Elektrovalenz, elektrochemische Wertigkeit
F électrovalence
P elektrowartościowość, wartościowość jonowa, wartościowość elektrochemiczna
R электровалентность

ELECTROVALENT BOND. See IONIC BOND.

1188
ELECTROVISCOUS EFFECT.
Relation between viscosity of sol and charge of particles of dispersed phase.
D elektroviskoser Effekt
F effet électrovisqueux

P efekt elektrowiskozowy
R электровискозный эффект

ELEMENT. *See* CHEMICAL ELEMENT.

1189
ELEMENT 104.
Atomic number 104.
D Element 104
F élément 104
P pierwiastek 104
R элемент 104

1190
ELEMENT 105.
Atomic number 105.
D Element 105
F élément 105
P pierwiastek 105
R элемент 105

ELEMENTAL ANALYSIS. *See*
ELEMENTARY ANALYSIS.

1191
ELEMENTARY ANALYSIS (elemental
analysis).
Determination of the amounts of various
elements constituting a given organic
substance.
D Elementaranalyse
F analyse élémentaire, dosage des
éléments
P analiza elementarna
R элемент(ар)ный анализ

1192
ELEMENTARY ENTITY (elementary
unit).
Elementary unit of matter: atom,
molecule, ion, free radical, elementary
particle.
D Teilchen
F particule
P cząstka (chemiczna)
R частица

1193
ELEMENTARY HEAT (elementary
quantity of heat), $\delta Q(Q_{el})$.
Infinitely small amount of heat exchanged
between system and its surroundings in
elementary thermodynamic process.
D ...
F ...
P ciepło elementarne
R ...

1194
ELEMENTARY PARTICLES.
Particles with structure which cannot be
explained on the basis of present status of
scientific knowledge. Elementary particles
may be considered as the smallest entities
of matter.

D Elementarteilchen
F particules élémentaires
P cząstki elementarne
R элементарные частицы

1195
ELEMENTARY PROCESSES OF
RADIOLYSIS.
Electronic excitations, especially ionization
occurring as result of interaction of
ionizing quanta or particles with
secondary electrons.
D Primärreaktionen der Radiolyse,
Elementarprozesse der Radiolyse,
strahlenchemische Primärreaktionen
F processus fondamentaux de radiolyse
P procesy radiolizy pierwotne, procesy
radiolizy elementarne
R элементарные процессы радиолиза,
первичные радиационно-химические
процессы

ELEMENTARY QUANTITY OF HEAT.
See ELEMENTARY HEAT.

ELEMENTARY QUANTITY OF WORK.
See ELEMENTARY WORK.

ELEMENTARY QUANTUM. *See* PLANCK
CONSTANT.

1196
ELEMENTARY REACTION.
Reaction running without intermediary
stages and in one direction only, from
substrates to products.
D Elementarreaktion
F réaction élémentaire
P reakcja elementarna, reakcja prosta
R элементарная реакция

1197
ELEMENTARY WORK (elementary
quantity of work), $\delta W(W_{el})$.
Infinitely small amount of work
exchanged between system and its
surroundings in elementary
thermodynamic process.
D elementare Arbeit
F travail élémentaire
P praca elementarna
R элементарная работа

ELEMENT OF MANIFOLD. *See*
ELEMENT OF SET.

ELEMENT OF POPULATION. *See*
ELEMENT OF SET.

1198
ELEMENT OF SET (element of manifold,
element of population).
Each member of a given set.
D ...
F élément d'une population
P element zbiorowości
R элемент совокупности

1199
ELIMINATION.
Reaction involving removal of atoms, groups of atoms or ions from molecules (or ions) with formation of unsaturated or cyclic compounds, e.g.

D Eliminierung
F élimination
P eliminacja
R отщепление

1200
ELUATE.
In paper- and column chromatography solution containing substances washed out of filter paper or column after separation.
D Eluat
F éluat
P eluat
R элюат

1201
ELUENT.
Gas or liquid used for selective washing out of substances retained on chromatographic column.
D Elutionsmittel, Eluens
F éluant
P eluent
R проявитель, элюент

1202
ELUTION.
Chromatographic method that consists in washing out of mixture from column (with simultaneous separation of components) by suitable mobile phase (eluent).
D Elution
F élution
P elucja, wymywanie
R элюирование, вымывание, проявительный анализ

1203
ELUTION CURVE (effluent concentration history).
Diagram showing dependence of substance concentration in eluate upon effluent volume or elution time.
D Elutionskurve
F courbe d'élution
P krzywa elucji
R кривая элюирования, кривая проявления

1204
EMANATING POWER.
Ratio of magnitude of emission of radioactive emanation released by solid

material to total emission produced in material in given time interval.
D Emaniervermögen
F pouvoir émanateur
P zdolność emanacyjna
R эманирующая способность

1205
EMANATION METHOD.
Method used for studying structure, surface, etc. of solid materials, and changes occurring in these solids.
It consists of the introduction into the material studied of a radionuclide (e.g. ^{226}Ra) which decays to give gaseous emanations, the rate of evolution of which is measured.
D Emanationsmethode
F méthode d'émanation
P metoda emanacyjna
R эманационный метод

1206
EMANATIONS.
Collective name for three radioactive gases (isotopes of the element radon) occurring in natural radioactive series of: radon — $^{22}_{86}$Rn, actinon — $^{219}_{86}$An and thoron — $^{220}_{86}$Th.
D Emanationen
F émanations
P emanacje
R эманации

1207
EMDE DEGRADATION.
Transformation of quaternary ammonium salt into tertiary amine due to reductive fission of carbon-nitrogen bond by sodium amalgam, e.g.

D Emde-Abbau (quartärer Ammoniumsalze)
F dégradation d'Emde
P degradacja Emdego
R расщепление по Эмде

1208
E₁ MECHANISM (mechanism of unimolecular elimination).
Mechanism involves formation of carbonium ion as first reaction step; in second step elimination of proton attached to β-carbon atom takes place with formation of unsaturated compound, e.g.

$$H-\overset{|}{\underset{|}{C}}-\overset{|}{\underset{|}{C}}\overset{\frown}{-X} \xrightarrow[-X^{\ominus}]{\text{slow}} H-\overset{|}{\underset{|}{C}}-\overset{|}{\underset{|}{C}}{}^{\oplus}$$

$$H-\overset{\frown}{\underset{|}{C}}-\overset{|}{\underset{|}{C}}{}^{\oplus} + OH^{\ominus} \xrightarrow{\text{fast}} \hspace{0.5em} \overset{}{\diagup}C=C\overset{}{\diagdown} + H_2O$$

D E_1-Mechanismus, monomolekularer Mechanismus der Eliminierung
F mécanisme de l'élimination E_1
P mechanizm E_1, mechanizm eliminacji jednocząsteczkowej
R механизм мономолекулярного отщепления E_1

1209
E_2 MECHANISM (mechanism of bimolecular elimination).
Mechanism involves attack of base on β-hydrogen atom with simultaneous elimination of group X as anion and formation of double bond. Rate of formation of active complex determines overall reaction rate

$$H\ddot{\underset{..}{O}}{:}^{\ominus}+ H-\overset{|\beta}{\underset{|}{C}}-\overset{|\alpha}{\underset{|}{C}}-X \rightarrow$$

$$\rightarrow \left[H\ddot{\underset{..}{O}}{:}\cdots H\cdots\overset{|}{\underset{|}{C}}\overset{}{\underset{}{\cdots}}\overset{|}{\underset{|}{C}}\cdots X \right] \rightarrow H_2O + \overset{}{\diagup}C=C\overset{}{\diagdown} + X^{\ominus}$$

activated complex

D E_2-Mechanismus, bimolekularer Mechanismus der Eliminierung
F mécanisme de l'élimination E_2
P mechanizm E_2, mechanizm eliminacji dwucząsteczkowej
R механизм E_2-отщепления, механизм бимолекулярного отщепления

1210
EMISSION SPECTRAL ANALYSIS (emission spectroanalysis).
Spectroanalysis based on investigation of emission spectra.
D Emissionsspektralanalyse
F analyse par spectroscopie d'émission
P analiza spektralna emisyjna
R эмиссионный спектральный анализ

EMISSION SPECTROANALYSIS. *See* EMISSION SPECTRAL ANALYSIS.

1211
EMISSION SPECTROSCOPY.
Branch of spectroscopy dealing with study of spectra emitted by excited systems.
D Emissionsspektroskopie
F spectroscopie d'émission
P spektroskopia emisyjna
R эмиссионная спектроскопия

1212
EMISSION SPECTRUM.
Spectrum of radiation emitted by a given source.
D Emissionsspektrum
F spectre d'émission
P widmo emisyjne
R спектр испускания

EMISSION WORK. *See* WORK FUNCTION.

1213
EMITTER.
Substance emitting a specified type of radiation.
D Strahler
F émetteur
P emiter
R излучатель

1214
EMPIRICAL DISTRIBUTION.
Probability distribution obtained empirically from sample investigation.

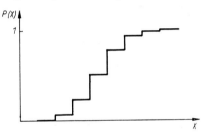

D empirische Verteilung
F distribution réelle
P rozkład empiryczny
R эмпирическое распределение

1215
EMPIRICAL FORMULA.
Chemical formula obtained from results of quantitative analysis; it gives type of atoms and their relative ratios, e.g. CH for acetylene.
D empirische Formel, Analyseformel
F formule empirique
P wzór empiryczny
R эмпирическая формула

EMPIRICAL TEMPERATURE. *See* TEMPERATURE.

1216
EMULSION.
Disperse system in which both dispersed phase and dispersion medium are liquid and dimensions of dispersed particles are larger than those characteristic for colloids.
D Emulsion
F émulsion
P emulsja
R эмульсия

1217
EMULSION CALIBRATION CURVE
(characteristic photographic curve,
characteristic curve).
Curve representing following dependence:
$S = f(\lg H)$, where S = blackening of
photographic emulsion, H = radiation
energy per unit surface.

D Schwärzungskurve
F courbe caractéristique (d'une émulsion),
courbe de noircissement (d'une émulsion)
P krzywa zaczernienia, krzywa
kalibrowania, krzywa charakterystyczna
(emulsji fotograficznej)
R характеристическая кривая эмульсии

1218
EMULSION POLYMERIZATION.
Polymerization taking place in emulsion
at considerable rate, even at low
temperature, thus enabling to obtain
a polymer of high molecular mass and
relatively small polymolecularity.

D Emulsionpolymerisation
F polymérisation en émulsion
P polimeryzacja emulsyjna
R эмульсионная полимеризация

1219
EMULSOID.
Historical term used to define lyophilic
sol.

D Emulsoid, Emulsionskolloid
F émulsoïde
P emulsoid
R эмульсоид

1220
ENANTIOMERIC CONFIGURATIONAL
UNIT.
Both of two substituted configurational
units, whose configuration of chiral and
prochiral atoms can be obtained as mirror
reflection of one from the other in plane
of main chain

$$-CH_2-\underset{\underset{CH_3}{|}}{\overset{\overset{H}{|}}{C}}- \quad \text{and} \quad -CH_2-\underset{\underset{H}{|}}{\overset{\overset{CH_3}{|}}{C}}-$$

D ...
F ...
P jednostka konfiguracyjna
enancjomeryczna
R энантиоморфная структурная единица

1221
ENANTIOMERS (antimers, optical
antipodes, enantiomorphs).
Two stereoisomers whose molecules are
mirror images but are non-superimposable
due to difference in configuration.

Enantiomers rotate plane of polarized
light by the same angle but in opposite
directions; they transmit at different
rates and differently absorb the right and
left circularly polarized light, also react
at different rates with optically active
compounds. Other physical and chemical
properties are identical.

D Enantiomere, Antimere, optische
Antipoden, Spiegelbildisomere
F antimères, antipodes optiques, isomères
optiques
P enancjomery, antymery, antypody
optyczne, izomery zwierciadlane
R энантиомеры, оптические антиподы,
зеркальные изомеры

ENANTIOMORPHS. See ENANTIOMERS.

1222
ENANTIOTROPIC PHASES.
Solid phases characterized by temperature
of transition of one into another lower
than melting temperatures of both phases.

D enantiotrope Phasen
F phases énantiotropes
P fazy enancjotropowe
R энантиотропные фазы

1223
ENANTIOTROPY.
Property possessed by substance existing
in two crystal forms, one stable below,
and other stable above a certain
temperature called transition point.

D Enantiotropie
F énantiotropie
P enancjotropia
R энантиотропия

1224
ENDO-EXO ISOMERISM.
Isomerism of norbornane
(bicyclo[2.2.1]heptane) derivatives and
their analogues; derivatives with
substituent in six-membered ring (at
positions: 2, 3, 5 or 6), located on side of
bridge, are commonly designated as exo-;
derivatives with substituents on opposite
side — as endo-, e.g.

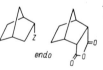

endo   exo

D endo-exo-Isomerie
F ...
P izomeria endo-egzo
R эндо-экзо изомерия

1225
ENDOTHERMAL REACTION
(endothermic reaction).
Process (chemical reaction or physical
process) accompanied by increase of
internal energy or enthalpy of system due
to heat absorption.
D endotherme Reaktion
F réaction endothermique
P przemiana endotermiczna
R эндотермическая реакция

ENDOTHERMIC REACTION. *See*
ENDOTHERMAL REACTION.

1226
END-POINT OF TITRATION.
Point corresponding to end of titration;
it may be detected e.g. on basis of change
of colour of indicator or jump of potential.
D Endpunkt der Titration
F fin de titrage
P punkt końcowy (miareczkowania)
R точка конца титрования, конечная
точка титрования

ENERGY BAND. *See* ALLOWED
ENERGY BAND.

1227
ENERGY CONSERVATION LAW.
In isolated system sum of all energies is
constant, provided no nuclear reactions
take place.
D Energie-Erhaltungssatz, Prinzip von der
Erhaltung der Energie
F loi de conservation d'énergie
P zasada zachowania energii, prawo
zachowania energii
R закон сохранения энергии

1228
ENERGY CONSERVATION LAW IN
CONTINUOUS SYSTEMS.
Equation

$$\frac{\partial}{\partial t}(\rho e) = -\mathrm{div}(J_Q + v \cdot P + \rho e v)$$

where t = time, v = velocity of centre of
mass, ρ = density, P = pressure tensor,
e = energy density, and J_Q = heat flux.
D Energie-Erhaltungssatz in
kontinuierlichen Systemen, Prinzip der
Erhaltung der Energie in
kontinuierlichen Systemen
F loi de conservation de l'énergie dans les
systèmes continus
P zasada zachowania energii w układach
ciągłych
R закон сохранения энергии
в непрерывных системах

1229
ENERGY EQUIPARTITION PRINCIPLE
(principle of equipartition of energy).
At sufficiently high thermodynamic
temperature T each additive energy term,
which is quadratic either in generalized
coordinate or in generalized momentum of
particle corresponds to average energy
$1/2\ kT$, where k = Boltzmann constant.
D Äquipartitionstheorem,
Gleichverteilungssatz
F principe de l'équipartition de l'énergie
P zasada ekwipartycji energii
R теорема о равномерном распределении

1230
ENERGY EQUIVALENT OF
CALORIMETER.
Amount of heat necessary for rise of
temperature of calorimeter for one degree,
under defined conditions.
D Wasserwert des Kalorimeters
F capacité calorifique du système
calorimétrique, équivalent en eau
P wartość cieplna kalorymetru
R энергетический эквивалент, тепловое
значение калориметра

1231
ENERGY FLUENCE.
Integral, over time, of intensity of
ionizing radiation.
D Energiefluenz
F fluence énergetique
P fluencja energii
R . . .

ENERGY GAP. *See* FORBIDDEN BAND.

1232
ENERGY OF NUCLEAR REACTION.
Difference between sum of kinetic
energies of nuclides (particles) after
reaction, and corresponding sum of
energies before reaction. This difference,
considering the equivalence of mass and
energy, is equal to the mass defect in the
reaction.
D Kernreaktionsenergie
F énergie de réaction nucléaire
P energia reakcji jądrowej
R энергия ядерной реакции

1233
ENERGY OF RADIOACTIVITY.
Difference between energies of mother
nuclide and transformation products
calculated from mass defect in a nuclear
reaction.
D radioaktive Zerfallsenergie
F énergie de désintégration radioactive
P energia przemiany promieniotwórczej
R энергия радиоактивного распада

1234
ENERGY SURFACE (ergodic surface).
In statistical mechanics a $2f-1$ dimensional
subspace of phase space, containing all
phase points which satisfy equation
$H(q_1, ..., q_f, p_1, ..., p_f) = E,$
where H = Hamilton function, E = energy,
q = coordinates, p = momenta, f = number
of degrees of freedom of system.

D Energiefläche
F hypersurface d'énergie
P powierzchnia ekwienergetyczna
R гиперповерхность постоянной энергии

1235
ENOLATE ANION.
Mesomeric anion formed as a result of
proton removal from the
electron-deficient atom bound with
carbonyl group. One of the resonance
structures has enol form, e.g.

$$CH_3-\underset{\underset{H}{|}}{C}=O \xrightarrow[-H^{\oplus}]{base} :\overset{\ominus}{C}H_2-\underset{\underset{H}{|}}{C}=O \leftrightarrow CH_2=\underset{\underset{H}{|}}{C}-\overset{\ominus}{O}$$

D ...
F ...
P anion enolanowy
R ...

1236
ENOL FORM.
Tautomer of constitution of
α, β-unsaturated alcohol, existing in
equilibrium with keto form. In most cases
equilibrium is shifted towards keto form,
so that enols cannot be isolated. Enolic
esters and ethers form stable derivatives,
e.g.

$$CH_3-\underset{\underset{O}{\|}}{C}-CH_3 \rightleftharpoons CH_3-\underset{\underset{OH}{|}}{C}=CH_2 \xrightarrow{CH_2=C=O}$$

$$CH_3-\underset{\underset{O-C-CH_3}{\underset{\|}{|}}}{C}=CH_2$$
$$\underset{O}{}$$

D Enol-Form
F forme énolique
P odmiana enolowa
R енольная форма

1237
ENOLIZATION.
Transformation of carbonyl compound
into unsaturated alcohol (enol), e.g.

$$RCOCH_2R' \rightleftharpoons RC=CHR'$$
$$\underset{OH}{|}$$

D Enoli(si)erung
F énolisation
P enolizacja
R енолизация

1238
ENRICHED MATERIAL.
In radiochemistry, material in which
content of one of the isotopes is increased
over its natural abundance.

D angereichertes Material
F matière enrichie
P materiał wzbogacony
R обогащённый материал

1239
ENRICHMENT FACTOR.
In radiochemistry, quotient of abundance
ratios of particular isotopes after and
before enrichment.

D Anreicherungsfaktor
F facteur d'enrichissement
P współczynnik wzbogacenia
R коэффициент обогащения

ENSEMBLE AVERAGE. *See*
STATISTICAL AVERAGE.

ENTHALPIMETRIC TITRATION. *See*
THERMOMETRIC TITRATION.

1240
ENTHALPY (enthalpy function), H.
Extensive thermodynamic function
$H = U+pV$, where U = internal energy,
p = pressure, and V = volume.

D Enthalpie
F enthalpie
P entalpia
R энтальпия

ENTHALPY FUNCTION. *See*
ENTHALPY.

1241
ENTHALPY OF ACTIVATION, $\Delta H^{\ddagger}$.
Intensive quantity equal to difference of
molar standard enthalpy of activated
complex and enthalpies of substrates. For
reactions in liquids enthalpy of activation
is connected with activation energy E_a of
collision theory: $\Delta H^{\ddagger} = E_a - RT$,
where R = molar gas constant and
T = thermodynamic temperature.

D Aktivierungsenthalpie
F enthalpie d'activation
P entalpia aktywacji
R энтальпия активации

1242
ENTROPY, S.
Extensive thermodynamic function whose
differential is defined by

$$dS = \frac{\delta Q}{T}$$

where δQ = elementary heat reversibly
exchanged with surroundings,
T = thermodynamic temperature.

D Entropie
F entropie
P entropia
R энтропия

1243
ENTROPY DENSITY, S_V.
Entropy per unit volume $S_V = S/V$,
where S = entropy and V = volume of
macroscopic element of system,
respectively.
D Entropiedichte
F densité d'entropie
P gęstość entropii
R плотность энтропии

1244
ENTROPY FACTOR.
In absolute reaction-rate theory, the value
$\exp(\Delta S^{\neq}/R)$, where $S^{\neq}$ = entropy of
activation, R = universal gas constant.
Entropy factor is counterpart to the steric
factor P in the collision theory.
D ...
F facteur de l'entropie
P współczynnik entropii
R энтропийный множитель

1245
ENTROPY FLOW (entropy flux), J_s.
Thermodynamic vector quantity, whose
component J_{s_n} perpendicular to any
surface at a given point is equal to
entropy transferred through unit element
of surface in unit time.
D Entropiestrom, Entropiefluß
F flux d'entropie, courant d'entropie
P przepływ entropii
R поток энтропии

ENTROPY FLUX. See ENTROPY FLOW.

1246
ENTROPY OF ACTIVATION (activation
entropy), $\Delta S^{\neq}$.
Intensive quantity equal to difference of
molar standard entropies of activated
complex and substrates, defined by

$$\Delta S^{\neq} = \frac{\Delta H^{\neq} - \Delta G^{\neq}}{T}$$

where $\Delta H^{\neq}$ = molar standard activation
enthalpy, $\Delta G^{\neq}$ = molar standard
thermodynamic potential of activation,
T = thermodynamic temperature.
Activation entropy can be correlated with
steric factor P of collision theory, i.e. for
bi-molecular reaction

$$\exp\left(\frac{\Delta S^{\neq}}{R}\right) \equiv P$$

where R = molar gas constant

D Aktivierungsentropie
F entropie d'activation
P entropia aktywacji
R энтропия активации

1247
ENTROPY OF MIXING, $S_{T,p}^M$.
Intensive quantity equal to entropy
increase S during isothermal-isobaric
formation of one mole of mixture from its
components.
D Mischungsentropie
F entropie de mélange
P entropia mieszania
R энтропия смешения

ENTROPY SOURCE STRENGTH. See
LOCAL ENTROPY PRODUCTION.

1248
ENVELOPE CONFORMATION.
Conformation in ring in which at least
four cycle-forming atoms are coplanar.
Such conformation is observed in
carbocyclic five-membered compounds (I)
and in heterocyclic six-membered boron
compounds having plane sterical
structure (II)

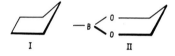

D Briefumschlag-Konformation
F conformation d'enveloppe
P konformacja kopertowa, konformacja
   zgiętego arkusza
R конформация конверта

1249
ENZYME.
Type of biochemical catalyst composed of
protein fragment (apoenzyme) and active
prosthetic group (coenzyme).
D Enzym, Ferment
F enzyme, ferment
P enzym, ferment
R фермент, энзим

1250
ENZYME CATALYSIS.
Phenomenon of acceleration of reaction
by enzyme.
D fermentative Katalyse
F catalyse enzymatique
P kataliza enzymatyczna, kataliza
   fermentacyjna
R ферментативный катализ

1251
EÖTVÖS RULE.
Temperature coefficient K of molar surface energy has approximately constant value, which for liquids not undergoing association is $K = 2.12$

$$K = -\frac{d(V_m^{2/3}\sigma)}{dT}$$

where V_m = molar volume of liquid, σ = surface tension, T = thermodynamic temperature.

D Regel von Eötvös
F relation d'Eötvös
P reguła Eötvösa
R правило Этвеша

1252
EPIMERIZATION.
Process changing configuration of atoms or groups of atoms at one of the chiral centres (at least two chiral centres should be present in molecule); e.g. epimerization at chiral carbon atom adjacent to aldehydic or ketonic group of a monose

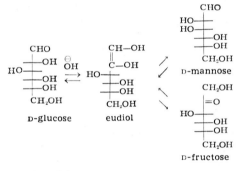

D-glucose   eudiol

D-mannose

D-fructose

D Epimerisierung
F épimérisation
P epimeryzacja
R эпимеризация

1253
EPIMERS.
Stereoisomeric compounds containing two or more chiral centres differing in configuration at one of them. Epimers are diastereoisomers, e.g. D-glucose and D-mannose.

D Epimere
F épimères
P epimery
R эпимеры

1254
EPITHERMAL ATOMS.
Atoms having kinetic energy slightly above their thermal agitation energy.

D epithermische Atome
F atomes épithermiques
P atomy epitermiczne
R надтепловые атомы

1255
EPITHERMAL NEUTRONS.
Neutrons of kinetic energy between 0.1 eV and 100 eV.

D epithermische Neutronen
F neutrons épithermiques
P neutrony epitermiczne
R эпитепловые нейтроны

1256
EPOXIDATION.
Oxidation of compound containing carbon-carbon double bond, resulting in formation of appropriate 1,2-epoxy (oxirane) derivative, e.g.

$$RCH=CHR \xrightarrow{[O]} R-\overset{\displaystyle O}{\overset{\diagup\ \diagdown}{CH-CH}}-R$$

D Epoxydation
F époxydation
P epoksydowanie
R эпоксидирование

EQUALIZING SOLVENT. *See* LEVELLING SOLVENT.

1257
EQUATION OF REFERENCE KINETIC CURVES.
Kinetic equation in which concentrations of substrates are expressed by fractions of initial concentrations.

D reduktive Reaktionsgeschwindigkeitsgleichung
F équation réduite de vitesse
P równanie kinetyczne zredukowane (Zawidzkiego)
R уравнение обобщённых (кинетических) кривых

1258
EQUATORIAL BOND.
One of the six bonds which extend outwards and away from three-fold axis of symmetry in chair form of cyclohexane. Equatorial bonds are declined 19.5° alternately upwards and downwards with respect to "mean" plane of ring, i.e. plane bisecting all C—C ring-forming bonds.

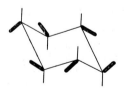

D äquatoriale Bindung
F liaison équatoriale
P wiązanie ekwatorialne, wiązanie e
R экваториальная связь

1259
EQUILIBRIUM ASYMMETRIC
TRANSFORMATION.
Change of configuration at one chiral
carbon atom in compound containing two
or more chiral carbon atoms, proceeding
till new thermodynamic equilibrium is
established.
D ...
F ...
P przekształcenie asymetryczne
   równowagowe
R равновесное асимметрическое
   превращение

1260
EQUILIBRIUM CONSTANT (approximate
equilibrium constant), K.
Ratio of product of equilibrium
concentrations (or partial pressures) of
products and product of equilibrium
concentrations (or partial pressures) of
substrates. Concentration (or partial
pressure) of each reactant is raised to
power equal to corresponding
stoichiometric coefficient. Equilibrium
constant can be expressed in terms of:
mole fractions (K_x), molar concentrations
(K_c), molalities (K_m) or partial pressures
(K_p); it depends on temperature, pressure
and composition of system.
D stöchiometrische
   Gleichgewichtskonstante
F constante d'équilibre (approximative)
P stała równowagi (chemicznej)
   przybliżona
R константа (химического) равновесия

1261
EQUILIBRIUM FLUCTUATIONS
(thermodynamic fluctuations).
Fluctuations in system in thermodynamic
equilibrium.
D Schwankungen in der Nähe des
   Gleichgewichtes
F fluctuations en équilibre
P fluktuacje równowagowe, fluktuacje
   termodynamiczne
R термодинамические флуктуации

EQUILIBRIUM QUOTIENT. *See*
SELECTIVITY COEFFICIENT.

1262
EQUIVALENCE POINT OF TITRATION
(theoretical end-point of titration).
Point for which amount of added titrant is
strictly equivalent to quantity of analysed
substance in solution.
D Äquivalenzpunkt der Titration,
   stöchiometrischer Punkt, theoretischer
   Endpunkt

F point équivalent
P punkt równoważnikowy
   (miareczkowania)
R точка эквивалентности

1263
EQUIVALENT CONDUCTIVITY, Λ_c.
Conductivity of 1 gram equivalent of
electrolyte in solution with a given
concentration

$$\Lambda_c = \frac{1000\varkappa}{c}$$

where $\varkappa$ = specific conductance of
solution with concentration c gram
equivalents/dm^3.
D Äquivalentleitfähigkeit
F conductivité équivalente, conductibilité
   équivalente
P przewodnictwo równoważnikowe
R эквивалентная электропроводность

EQUIVALENT CONDUCTIVITY AT
INFINITE DILUTION. *See* LIMITING
EQUIVALENT CONDUCTIVITY.

1264
ERBIUM, Er.
Atomic number 68.
D Erbium
F erbium
P erb
R эрбий

ERGODIC SURFACE. *See* ENERGY
SURFACE.

1265
ERGODIC SYSTEMS.
Systems with time average of any integral
of motion which is symmetric function of
dynamic variables of all particles,
depending solely on energy of system.
D Ergodensysteme
F systèmes ergodiques
P układy ergodyczne
R эргодические системы

1266
ERGODIC THEOREM.
In isolated system time averages are equal
to average values in microcanonical
ensemble.
Remark: Several conditions for validity of
ergodic theorem have been given. However,
they cannot be checked in the case of real
macroscopic systems.

D Ergodentheorem
F théorème ergodique
P twierdzenie ergodyczne
R эргодическая теорема

1267
ERLENMEYER-PLÖCHL AZLACTONE
SYNTHESIS.
Preparation of unsaturated azlactones by
condensation of aldehydes or ketones with
α-acylglycine, in presence of acetic
anhydride and sodium acetate, e.g.

$$RCHO + CH_2COOH \xrightarrow[CH_3COONa]{(CH_3CO)_2O}$$
$$\qquad\quad | $$
$$\qquad\quad NHCOR'$$

$$RCH{=}C{-}{-}C{=}O$$
$$\qquad | \qquad |$$
$$\qquad N \qquad O$$
$$\qquad\quad \diagdown C \diagup$$
$$\qquad\qquad |$$
$$\qquad\qquad R'$$

D Erlenmeyer-Plöchl
  Azlacton-Kondensation
F synthèse d'Erlenmeyer
P synteza azlaktonów Erlenmeyera
  i Plöchla
R синтез азлактонов
  Эрленмейера-Плёхля

1268
ERYTHRO FORM.
Diastereoisomer having two adjacent
non-equivalent chiral centres of
configuration shown below:

$$CH_3$$
$$Br{-}|{-}H$$
$$HO{-}|{-}H$$
$$CH_3$$

i.e. two identical or similar atoms (or
groups of atoms) are situated on the same
side of Fischer's projectional formula.

D erythro-Form
F forme érythro
P odmiana erytro
R эритро-форма

1269
ESCAPE PEAK.
Peak in gamma radiation spectrum caused
by escape of one or two annihilation
photons or X-rays from detector after
photoelectron emission.

D Escape-Peak
F pic de fuite, pic d'échappement
P pik ucieczki
R пик утечки, пик ухода, пик вылета

1270
ESCHWEILER-CLARKE REACTION.
Methylation of primary and secondary
amines with formaldehyde and formic
acid (in excess) at 100°C, e.g.

$$RNH_2 + 2HCHO + 2HCOOH \longrightarrow$$
$$RN(CH_3)_2 + 2CO_2 + 2H_2O$$

D Eschweiler-Clarke Amin-Methylierung
F réaction d'Eschweiler et Clarke
P reakcja Eschweilera i Clarke'a
R реакция Эшвейлера-Кларка

ESTER CONDENSATION. *See* CLAISEN
CONDENSATION.

1271
ESTERIFICATION.
Process of obtaining esters from acids
and alcohols (or phenols), e.g.

$$CH_3COOH + HOC_2H_5 \underset{}{\overset{H^{\oplus}}{\rightleftharpoons}} CH_3COOC_2H_5 + H_2O$$

D Veresterung
F estérification
P estryfikacja
R этерификация

ESTIMATE. *See* ESTIMATOR.

1272
ESTIMATION (point estimation).
Method of estimation of unknown
parameter Θ of population based on
assumption that Θ is equal to value
of estimator Z of this parameter obtained
from a given n-element random series.
D ...
F estimation (ponctuelle)
P estymacja (punktowa)
R точечная оценка

ESTIMATION. *See* DETERMINATION.

1273
ESTIMATOR (estimate), Z.
Any statistic utilized for estimation of
unknown parameter Θ of population.
D ...
F estimateur statistique
P estymator
R оценка

1274
ETHERIFICATION.
Process of obtaining ethers R—O—R',
usually by substituting hydrogen atom
of appropriate hydroxyl group with alkyl
or aryl group, e.g.

$$R{-}OH + CH_3OSO_2OCH_3 \xrightarrow{base} R{-}O{-}CH_3 +$$
$$+ CH_3OSO_3H$$

D Ätherifizierung
F formation d'éther-oxyde
P eteryfikacja
R эфирообразование

1275
ETHYLENE LINKAGE (olefinic link).
Carbon-carbon double bond.

D Äthylenbindung
F liaison éthylenique
P wiązanie etylenowe
R этиленовая связь

1276
ETHYNYLATION.
Introduction of ethynyl group —C≡CH
(or —C≡CR) into molecule of organic
compound, e.g.

$$-C\equiv CH + \!\!\!\! \diagdown\!\!C=O \xrightarrow{\text{KOH}} -C\equiv C-C\!\!\!\diagup \atop \underset{OH}{|}$$

D Äthinylierung
F éthynylation
P etynylowanie
R этинилирование

1277
EUROPIUM, Eu.
Atomic number 63.
D Europium
F europium
P europ
R европий

1278
EUTECTIC.
Liquid mixture coexisting with two or more solid phases at a temperature the change of which causes decay of one of phases.
D Eutektikum
F eutectique, mélange eutectique
P eutektyk (ciekły), mieszanina eutektyczna
R эвтектика, эвтектическая смесь

1279
EUTECTIC POINT.
Point on phase diagram which describes composition of solution coexisting with at least two solid phases.
D eutektischer Punkt
F point eutectique
P punkt eutektyczny
R эвтектическая точка

1280
EUTECTIC TEMPERATURE.
Temperature of equilibrium coexistence of liquid phase with two or more solid phases.
D eutektische Temperatur
F température eutectique
P temperatura eutektyczna
R эвтектическая температура

1281
EUTECTOID.
Solid eutectic formed as result of decay of solid solution during cooling.
D Eutektoid
F eutectoïde
P eutektoid
R эвтектоид

1282
EVEN-EVEN NUCLEUS.
Atomic nucleus containing even number of protons and even number of neutrons.

D doppelt-gerader Atomkern
F noyau pair-pair
P jądro parzysto-parzyste
R чётно-чётное ядро

1283
EVEN-ODD NUCLEUS.
Atomic nucleus containing even number of protons and odd number of neutrons.
D gerade-ungerader Atomkern
F noyau pair-impair
P jądro parzysto-nieparzyste
R чётно-нечётное ядро

EVEN SERIES. *See* MAIN GROUP.

1284
EXCESS FUNCTIONS (excess thermodynamic quantities), $Z_{T,\,p}^{E}$.
Mixing functions for a given mixture decreased by sum of mixing functions for the same mixture but calculated under assumption that this mixture is ideal.
D Zusatzfunktionen
F grandeurs d'excès
P funkcje nadmiarowe
R избыточные термодинамические функции

EXCESS THERMODYNAMIC QUANTITIES. *See* EXCESS FUNCTIONS.

1285
EXCHANGE ADSORPTION.
Exchange of one substance adsorbed on adsorbent surface by another.
D Austauschadsorption
F adsorption d'échange
P adsorpcja wymienna
R обменная адсорбция

EXCHANGE CAPACITY OF HIGHLY DISSOCIATED CATIONIC/ANIONIC GROUPS. *See* ION EXCHANGE CAPACITY OF STRONGLY ACIDIC/BASIC GROUPS.

1286
EXCHANGE CURRENT DENSITY, j_0.
Density of current which flows across electrode — electrolyte phase boundary when electrode reaction is at equilibrium:

$$j_0 = j_+ = |j_-|$$

where j_+ and j_- = densities of anodic and cathodic currents respectively corresponding to rates of electrode reactions at equilibrium.
D Austauschstromdichte
F densité du courant d'échange
P gęstość prądu wymiany
R плотность тока обмена

EXCHANGED ENTROPY. *See* SUPPLIED ENTROPY.

1287
EXCHANGE ENERGY.
Part of electrostatic electron interaction energy due to indistinguishability of electrons.
D Austauschungsenergie
F énergie d'échange
P energia wymiany
R обменная энергия

1288
EXCHANGE FRACTION.
In radiochemistry, ratio of concentration of molecules or atoms labelled at a given time to their concentration corresponding to their distribution at equilibrium.
D Austauschgrad
F fraction d'échange
P stopień wymiany izotopowej
R степень обмена

1289
EXCHANGE INTERACTION.
Quantum mechanical interaction of electrostatic nature which follows from theory of covalent bond and leads to several types of magnetic ordering in crystals.
D Austauschwechselwirkung
F interaction d'échange
P oddziaływanie wymienne
R обменное взаимодействие

EXCHANGE ISOPLANE. *See*
BREAKTHROUGH CURVE.

1290
EXCHANGE RATE.
Quantity given by reciprocal of the mean life time of atom in a given molecule $(1/\tau)$ in chemical exchange process.
D Austauschfrequenz
F fréquence d'échange
P szybkość wymiany chemicznej, częstość wymiany chemicznej
R скорость обмена

1291
EXCHANGE REACTION (replacement reaction).
Process of transformation of two or more substrate molecules, involving mutual exchange of their atoms or ions, e.g.

$$Na\,\overline{OH} + H\,\overline{Cl} = NaCl + H_2O$$

D Austauschreaktion
F réaction d'échange
P reakcja wymiany
R реакция обмена

1292
EXCHANGE REACTION (*nuclear*).
Nuclear reaction consisting in exchange of one or several nucleons between interacting atomic nuclei.
D Austauschreaktion
F réaction d'échange
P reakcja wymiany
R реакция обмена

1293
EXCHANGE SCATTERING.
Scattering in which particle colliding with system (atomic nucleus, atom) knocks out identical particle, but is absorbed itself by system. This effect cannot be distinguished from simple scattering and both processes are always observed together.
D Austauschstreuung
F diffusion d'échange
P rozpraszanie wymienne
R обменное рассеяние

1294
EXCIMERS.
Molecular associates, of two or several molecules, existing only in electronic excited states. Excimers play a role in fluorescence and different kinds of radiation-induced luminescence.
D Excimere
F excimères
P ekscymery
R ...

1295
EXCITATION.
Process of absorption of energy by atomic nuclei, atoms or molecules which is accompanied by their transition from ground state to states of higher energy (which are called excited states).
D Anregung
F excitation
P wzbudzenie
R возбуждение

1296
EXCITATION ENERGY.
Energy necessary to transfer molecule, atom or atomic nucleus from ground to excited state. In molecule, excitation energy is in general sum of electronic, oscillation and rotation excitations.
D Anregungsenergie
F énergie d'excitation
P energia wzbudzenia
R энергия возбуждения

1297
EXCITATION FUNCTION.
Relation between cross-section of a given nuclear reaction and energy of bombarding particles.
D Anregungsfunktion
F fonction d'excitation
P funkcja wzbudzenia
R функция возбуждения

1298
EXCITATION SOURCE (light source).
Energy source used to evaporate and excite investigated sample thus leading to emission of electromagnetic radiation.
D Strahlungsquelle, Lichtquelle
F source d'excitation
P źródło wzbudzenia, wzbudzalnik
R источник возбуждения, источник света

1299
EXCITED MOLECULE.
Molecule in which energy levels are higher than in ground state.
D angeregtes Molekül
F molécule excitée
P cząsteczka wzbudzona
R возбуждённая молекула

1300
EXCITED NUCLEUS.
Atomic nucleus in energetic state higher than the energy of ground state, unstable. Excited nucleus decays with emission of γ-quanta or other particles.
D angeregter Kern
F noyau excité
P jądro wzbudzone
R возбуждённое ядро

1301
EXCITED STATE.
Quantum mechanical state of system with total energy higher than in ground state.
D angeregter Zustand
F état éxcité
P stan wzbudzony
R возбуждённое состояние

1302
EXCITON.
Electronic state of excitation in crystal. Exciton is formed as electron-hole pair, e.g. in effect of absorption of ionizing energy radiation, it may move through crystal, thus transporting energy but not charge.
D Exciton
F exciton
P ekscyton
R экситон

EXCLUSION PRINCIPLE. See PAULI EXCLUSION PRINCIPLE.

1303
EXHAUSTIVE METHYLATION (*of amines*).
Methylation of amines (or ammonia) resulting in formation of quaternary ammonium salts, e.g.

D erschöpfende Methylierung
F méthylation totale
P metylowanie wyczerpujące
R исчерпывающее метилирование

1304
EXOTHERMAL REACTION (exothermic reaction).
Process (chemical reaction or physical process) accompanied by decrease of internal energy or enthalpy of system due to heat emission.
D exotherme Reaktion
F réaction exothermique
P przemiana egzotermiczna
R экзотермическая реакция

EXOTHERMIC REACTION. See EXOTHERMAL REACTION.

EXPECTATION. See EXPECTED VALUE.

EXPECTATION VALUE. See STATISTICAL AVERAGE.

1305
EXPECTED VALUE (expectation, mean of random variable), E.
Mean value of random variable of a given probability distribution. For discrete variable

$$E(X) = \sum_{x_i} [x_i \cdot P(X = x_i)]$$

and for continuous variable

$$E(X) = \int_{-\infty}^{+\infty} x f(x) dx$$

where $P(X = x_i)$ = probability that variable X has value x_i, $f(x)$ = probability density function of random variable X.
D Erwartungswert
F espérance mathématique, moyenne d'une variable aléatoire
P wartość przeciętna zmiennej losowej, wartość oczekiwana
R математическое ожидание случайной величины

1306
EXPLOSION.
Exothermal reaction proceeding at high rate.
D Explosion
F explosion
P wybuch
R взрыв

1307
EXPLOSION LIMITS.
Range of one of parameters of chemical reactions (pressure, composition or temperature) above which (lower explosion limit) or below which (upper explosion limit) reaction proceeds explosively at defined value of remaining parameters.
D Explosionsgrenzen
F limites d'explosion
P granice wybuchu, granice wybuchowości
R пределы взрыва

1308
EXPOSURE (exposure dose), X.
Amount of ionizing radiation (gamma or X rays) energy defined by ability to ionize air, $X = \Delta Q/\Delta m$, where ΔQ = sum of electric charges of all ions of the same sign, formed in unit of air of mass Δm, in conditions where all electrons (negatons and positrons) released by photons are totally stopped in that volume of air.
D Ionendosis
F exposition, dose d'exposition
P ekspozycja, dawka ekspozycyjna
R экспозиционная доза

EXPOSURE DOSE. See EXPOSURE.

EXTENSIVE PARAMETERS. See EXTENSIVE QUANTITIES.

EXTENSIVE PROPERTIES. See EXTENSIVE QUANTITIES.

1309
EXTENSIVE QUANTITIES (extensive properties, extensive variables, extensive parameters).
Thermodynamic parameters proportional to system's mass.
D extensive Größen, extensive Eigenschaften, extensive Zustandsvariablen, extensive Parameter, Quantitätsgrößen
F grandeurs extensives, variables extensives
P wielkości ekstensywne, parametry ekstensywne, zmienne ekstensywne
R экстенсивные величины, экстенсивные признаки, экстенсивные параметры, факторы экстенсивности, ёмкости

EXTENSIVE VARIABLES. See EXTENSIVE QUANTITIES.

1310
EXTENT OF REACTION (degree of advancement of reaction, reaction variable), ξ.
Parameter characterizing advance of chemical reaction

$$\xi = \frac{\Delta n_B}{\nu_B}$$

where Δn_B = number of moles of reactant B used or produced in chemical reaction, ν_B = stoichiometric coefficient of this reactant (positive for products, negative for substrates): value of extent of reaction is the same for all reactants and is proportional to number of gram-equivalents used in reaction.
D Reaktionslaufzahl
F degré d'avancement de réaction, avancement de réaction
P liczba postępu reakcji (chemicznej)
R степень полноты реакции, степень развития реакции, степень продвижения реакции, химическая переменная, число пробегов реакции, степень превращения

1311
EXTERNAL INDICATOR.
Indicator which is not added to titrated solution; reaction between external indicator and titrated solution is carried out externally by taking drop of analyzed solution.
D äußerer Indikator, Außen-Indikator
F indicateur externe
P wskaźnik zewnętrzny
R внешний индикатор

EXTERNAL PHOTOELECTRIC EFFECT.
See PHOTOELECTRIC EFFECT.

EXTERNAL REFERENCE. See EXTERNAL STANDARD.

1312
EXTERNAL RETURN.
Restitution of substrate by internally associated ionic pair, by recombination of $X^{\ominus}$ ions originating from different molecules of dissociated compounds. Intramolecular processes, such as inversion, racemization or rearrangements, often accompany external return

$$A\!-\!X \rightleftarrows A^{\oplus}X^{\ominus} \rightleftarrows A^{\oplus} \,\|\, X^{\ominus} \xrightarrow{\ X'^{\ominus}\ }$$
$$\rightleftarrows A^{\oplus} \,\|\, X'^{\ominus} \rightleftarrows A^{\oplus}X'^{\ominus} \rightleftarrows A\!-\!X' \quad (X\!=\!X')$$

D äußere Rückkehr
F retour externe
P powrót zewnętrzny
R . . .

1313
EXTERNAL STANDARD (in spectroanalysis).
Component absent in investigated sample but present on counter electrode. Its spectral line(s) is (are) used as reference for measurement of spectral line intensities of determined elements.

D äußerer Standard
F étalon externe
P wzorzec zewnętrzny
R внешний стандарт

1314
EXTERNAL STANDARD (external reference).
In NMR spectroscopy, reference substance external to investigated sample, used to determine chemical shifts.
D äußerer Standard
F étalon extérieur, référence extérieure
P wzorzec zewnętrzny
R внешний эталон

1315
EXTRACTION (liquid-liquid extraction).
Separation method that makes use of differences in partition of various substances between two immiscible liquids.
D (Flüssig-Flüssig) Extraktion
F extraction
P ekstrakcja (w układzie ciecz-ciecz)
R экстракция, экстрагирование

1316
EXTRACTION CHROMATOGRAPHY (reversed phase partition chromatography).
Chromatography in which organic extractant adsorbed on solid support serves as stationary phase.
D Extraktionschromatographie, Verteilungschromatographie mit umgekehrten Phasen
F chromatographie en phase inversée
P chromatografia ekstrakcyjna, chromatografia podziałowa z odwróconymi fazami
R распределительная хроматография с обращённой фазой

1317
EXTRINSIC SEMICONDUCTOR.
Semiconductor whose electrical properties are dependent on impurities added to the semiconductor crystal, in contrast to intrinsic semiconductor, whose properties are characteristic of ideal pure crystal.
D Störstellenhalbleiter
F semiconducteur extrinsèque
P półprzewodnik niesamoistny
R несобственный полупроводник

1318
E-Z SYSTEM.
Conventional system for specification of *cis-trans* configuration of stereoisomers of ethylene or other unsaturated derivatives. This convention (similarly to R-S nomenclature) is based on sequence rule. When atoms or groups of atoms of high atomic numbers are attached to double-bond carbon atoms from the same side, Z — configuration is ascribed to

system. In the reverse case, configuration is said to be (E)-, e.g.

$$(Cl > H) \quad \overset{Cl}{\underset{H}{\diagdown}} C = C \overset{Cl}{\underset{Br}{\diagup}} \quad (Br > Cl)$$

E-1-bromo-1,2-dichloroethen

$$\overset{CH_3}{\underset{\overset{\|}{N}}{\diagdown}} \overset{C_2H_5}{\underset{OH}{\diagup}} \quad (C_2H_5 > CH_3)$$
$$\quad (OH > zero)$$

Z-butanone oxime

Remark: Z stands for abreviation German zusammen, E stands for abreviation German entgegen.

D ...
F ...
P konwencja E-Z, system E-Z
R E-Z-система

1319
FABRY-PEROT INTERFEROMETER.
Instrument used for multi-beam interference of radiation. Main parts are two glass or quartz plates exactly parallel and neatly polished; each one slightly wedged (the angle between wedge planes is about 30′).
D Interferenzspektroskop von Fabry-Perot
F interféromètre de Fabry-Perot
P interferometr Fabry'ego i Perota
R интерферометр Фабри-Перо

1320
FALSE CHEMICAL EQUILIBRIUM (unstable chemical equilibrium).
Thermodynamic state of reacting system with reaction rates equal to zero due to kinetic hindrances but with non-zero affinity.
D labiles chemisches Gleichgewicht, instabiles chemisches Gleichgewicht
F équilibre chimique instable
P równowaga chemiczna pozorna, równowaga chemiczna nietrwała
R неустойчивое химическое равновесие

FAMILY. *See* GROUP.

1321
FAMILY OF ELEMENTS.
Set of elements with similar properties.
D Gruppe von Elementen
F famille des éléments
P rodzina pierwiastków
R семейство элементов

1322
FAMILY OF ELEMENTS.
Group of elements of radioactive series.
D Gruppe von Elementen
F famille des éléments
P rodzina pierwiastków
R семейство элементов

1323
FARADAIC RECTIFICATION METHOD.
Method of studying kinetics of electrode
processes based on non-linear and
asymmetric characteristics of electrodes.
When sinusoidal voltage of significant
frequency is applied to electrode system,
constant current component called
Faradaic rectification current is observed.
D Faradaysche Gleichrichtung
F ...
P metoda faradajowskiego prostowania
R метод фарадеевского выпрямления

FARADAY. *See* FARADAY'S CONSTANT.

1324
FARADAY'S CONSTANT (faraday), *F*.
Electrical charge of one gram-equivalent
of ions and also of one mole of electrons;
$F = 96\,494$ C.
D Faraday-Konstante, Faraday
F faraday
P stała Faradaya, faraday
R фарадей, число Фарадея, константа
Фарадея

1325
FARADAY'S LAWS.
(1) Amount of a given substance
undergoing reaction at electrodes are
proportional to quantity of electricity
passing through electric circuit.
(2) Amounts of different substances
undergoing reactions at electrodes, in
case of equal quantities of electricity
passing through circuit, are proportional
to their chemical equivalents.
D Faradaysche Gesetze
F lois de Faraday
P prawa elektrolizy Faradaya
R законы Фарадея

1326
FAR INFRARED.
Electromagnetic radiation range of
wavelength between ca. 25 μm to ca.
800 μm.
D langwelliges Infrarot
F infrarouge lointain
P podczerwień daleka
R дальняя инфракрасная область

1327
FAR ULTRAVIOLET, FUV.
Electromagnetic radiation range of
wavelength between ca. 170 nm to ca.
200 nm.
D fernes Ultraviolett, fernes UV
F ultraviolet lointain
P nadfiolet daleki
R дальняя ультрафиолетовая область

1328
FAST NEUTRONS.
Neutrons having kinetic energies greater
than 1 MeV.
D schnelle Neutronen
F neutrons rapides
P neutrony prędkie
R быстрые нейтроны

1329
FAVORSKII-BABAYAN ACETYLENIC
ALCOHOL SYNTHESIS.
Preparation of acetylenic alcohols by
condensation of ketones with acetylenic
hydrocarbons in presence of anhydrous
potassium hydroxide, e.g.

$$\begin{array}{c}R\\ \diagdown\\ \diagup \\ R'\end{array}C=O + HC\equiv CR'' \xrightarrow{KOH} \begin{array}{c}R\\ \diagdown\\ \diagup\\ R'\end{array}\overset{OH}{\underset{|}{C}}-C\equiv C-R''$$

D Favorskii-Babayan-Carboxo-
-Äthinierung
F synthèse de Favorsky-Babayan
P synteza alkoholi acetylenowych
Faworskiego i Babayana
R реакция Фаворского

1330
FAVORSKII REARRANGEMENT.
Rearrangement of α-haloketones into
carboxylic acids, effected by bases, e.g.

$$ClCH_2\overset{O}{\overset{\|}{C}}R + NaOH \rightarrow RCH_2C\overset{O}{\diagup}\underset{OH}{\diagdown} + NaCl$$

$$\text{[cyclohexanone with }=O, -Cl]\ \overset{+NaOH}{\underset{-NaCl}{\longrightarrow}}\ \text{[cyclopentane]}-COOH$$

D Favorskii (α-Halogenketon→
→Carbonsäure)-Umlagerung
F transposition de Favorsky
P przegrupowanie Faworskiego
R перегруппировка Фаворского

1331
F-CENTRES.
Colour centres formed as result of
trapping of electrons by anionic vacancies
in ionic crystal.
D F-Zentren
F centres F
P centra F
R F-центры

1332
F-CRITERION (*F*-test).
Test of agreement which allows decision
to be made wheather at a given level of
probability variances calculated from two
series are significantly (not randomly)
different.
D *F*-Test
F test *F*
P test *F*, test Snedecora
R *F*-критерий

1333
F-DISTRIBUTION.
Probability distribution of continuous random variable with density function given by

$$f(F) = \frac{\Gamma[(k_1+k_2)/2]}{\Gamma(k_1/2)\Gamma(k_2/2)} \left(\frac{k_1}{k_2}\right)^{k_1/2} \times$$

$$\times \frac{F^{(k_1-2)/2}}{[1+(k_1/k_2)F]^{(k_1+k_2)/2}}$$

where k_1 and k_2 = number of degrees of freedom.

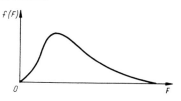

D (Fischersche) F-Verteilung
F loi de Fisher-Snedecor, distribution de Fisher-Snedecor
P rozkład F Snedecora
R F-распределение, распределение Фишера

1334
FEMTOMETRE, fm.
Unit of length, equal to 10^{-15} metre; used particularly in measuring nuclear distances.

D Femtometer
F femtomètre
P femtometr
R фемтометр

1335
FENTON REAGENT (Fenton system).
Aqueous solution containing hydrogen peroxide and ferrous ions. System generates free hydroxyl radicals ($\cdot$OH) without radiation of high energy.

D Fenton-Reagens
F réactif de Fenton
P odczynnik Fentona
R реактив Фентона

FENTON SYSTEM. *See* FENTON REAGENT.

1336
FERMAT PRINCIPLE.
Light beam in medium moves between points A and B in such way that its optical path, i.e. $\int_A^B 1/u \cdot ds$ is minimum, where u = phase frequency of light wave, ds = path element.

D Fermatsches Prinzip
F principe de Fermat
P zasada Fermata
R принцип Фэрма

1337
FERMI-DIRAC DISTRIBUTION.
Equilibrium distribution of particles among single-particle levels given by

$$\langle N_\varepsilon \rangle = \frac{g_\varepsilon}{\dfrac{1}{\lambda}\exp\dfrac{\varepsilon}{kT}+1}$$

where k = Boltzmann constant, T = thermodynamic temperature, λ = absolute activity and g_ε = degeneracy of energy level ε, $\langle N_\varepsilon \rangle$ = average occupation number for this level. Fermi-Dirac distribution can be used for systems of weakly interacting fermions.

D Fermi-Dirac-Verteilungsgesetz
F répartition de Fermi-Dirac
P rozkład Fermiego i Diraca
R распределение Ферми

1338
FERMI-DIRAC STATISTICS.
Statistics applicable to systems of quantum particles whose wave function is antisymmetric with respect to particle permutations (*cf.* Fermi-Dirac distribution).

D Fermi-Dirac-Statistik
F statistique de Fermi-Dirac
P statystyka Fermiego i Diraca
R статистика Ферми-Дирака

FERMI ENERGY. *See* FERMI LEVEL.

1339
FERMI-GAS MODEL.
Model of atomic nucleus which describes nucleus as consisting of mutually non-interacting nucleons being in constant movement (similar assumptions are made in consideration of gas). Fermi-gas model is applied in a simplified analysis of heavier atomic nuclei.

D ...
F modèle du gaz de Fermi
P model gazu Fermiego
R модель газа Ферми

1340
FERMI LEVEL (Fermi energy).
At 0 K highest occupied energy level in metals; at higher temperatures inflection point on energy dependence of density of states.

D Ferminiveau, Fermi-Energie
F niveau de Fermi, énergie de Fermi
P poziom Fermiego, energia Fermiego
R Ферми-уровень, энергия Ферми

1341
FERMIONS.
Elementary particles or complex systems of spin quantum number equal to one half unit, e.g. leptons, baryons, nuclei with odd number of nucleons. Fermions are described by Fermi-Dirac statistics.
D Fermionen
F fermions
P fermiony
R фермионы

1342
FERMI RESONANCE.
In vibrational molecular spectra abnormal increase of intensity of two absorption bands of the same symmetry of vibration and sufficiently close frequencies.
D Fermi-Resonanz
F résonance de Fermi
P rezonans Fermiego
R резонанс Ферми

1343
FERMI SURFACE.
Constant energy surface corresponding to Fermi level in system of fermions in k-space, where k is wave vector.
D Fermi-Fläche
F surface de Fermi
P powierzchnia Fermiego
R Ферми-поверхность

1344
FERMI TEMPERATURE (degeneracy temperature), T_F.
Parameter expressed in units of temperature

$$T_F = \frac{\varepsilon_F}{k}$$

where ε_F = Fermi level, k = Boltzmann constant. At temperatures low in comparison with Fermi temperature Fermi gas is degenerate.
D Fermi-Temperatur
F température de Fermi
P temperatura Fermiego, temperatura degeneracji
R температура Ферми, температура вырождения

1345
FERMIUM, Fm.
Atomic number 100.
D Fermium
F fermium
P ferm
R фермий

FERRIMAGNETICS. *See* FERRIMAGNETS.

FERRIMAGNETIC SUBSTANCES. *See* FERRIMAGNETS.

1346
FERRIMAGNETISM.
Physical phenomena in and properties of ferrimagnets.
D Ferrimagnetismus
F ferrimagnétisme
P ferrimagnetyzm
R ферримагнетизм

1347
FERRIMAGNETS (ferrimagnetic substances, ferrimagnetics).
Highly magnetic solids with properties similar to ferromagnets but of more complicated temperature dependence of saturation magnetization below Curie temperature and of magnetic susceptibility above Curie point (e.g. magnetite, pyrrhotite).
D Ferrimagnetika, ferrimagnetische Stoffe
F ferrimagnétiques
P ferrimagnetyki
R ферримагнетики

1348
FERRITES.
Oxides of complex structure containing Fe^{+3} and other metal ions, of crystal structure of spinel, garnet, magnetoplumbite etc. Numerous ferrites exhibit important magnetic properties and rather low electric conductivity; utilized i.a. in high frequency technology.
D Ferrite
F ferrites
P ferryty
R ферриты

1349
FERROELECTRIC.
Crystalline substance exhibiting spontaneous electric polarization, characterized by high permittivity and occurrence of polarization hysteresis; electric analogue of ferromagnets.
D Ferroelektrikum
F ferroélectrique
P ferroelektryk
R сегнетоэлектрик, ферроэлектрик

FERROMAGNETICS. *See* FERROMAGNETS.

FERROMAGNETIC SUBSTANCES. *See* FERROMAGNETS.

1350
FERROMAGNETISM.
Physical phenomena in and properties of ferromagnets.
D Ferromagnetismus
F ferromagnétisme
P ferromagnetyzm
R ферромагнетизм

1351
FERROMAGNETS (ferromagnetic substances, ferromagnetics).
Crystalline substances with magnetization in absence of external magnetic field which appears due to long-range ordering of elementary magnetic moments.
D Ferromagnetika, ferromagnetische Stoffe
F ferromagnétiques
P ferromagnetyki
R ферромагнетики, ферромагнитные вещества

1352
FERROUS METALS (first triad).
Three chemical elements grouped horizontally in eighth, additional, group of periodic system in period IV: iron, cobalt, nickel.
D Eisengruppe, Metalle der Eisengruppe
F groupe du fer, triade du fer
P żelazowce, triada żelaza
R триада железа, элементы триады железа, семейство железа

1353
FEYNMAN DIAGRAMS.
Diagrammatic representation of elements of matrix describing interaction processes, e.g. of elementary particles.
D Feynman-Graphen, Feynman-Diagramme
F diagrammes de Feynman
P diagramy Feynmana
R диаграммы Фейнмана, графики Фейнмана

1354
FICK'S FIRST LAW.
Transport by diffusion J_D, defined as number of moles, particles or units of mass of defined constituent flowing across unit area in unit time, is proportional to concentration gradient, grad c.

$$J_D = -D \operatorname{grad} c$$

where D = diffusion coefficient; in one-dimensional problems, when concentration gradient does not depend on time, Fick's first low is reduced to

$$m = -D \frac{dc}{dx} At$$

where m = amount of constituent flowing across area A perpendicular to direction of diffusion in time t (x coordinate).
D erstes Ficksches Diffusionsgesetz
F première loi de Fick
P prawo Ficka pierwsze
R первый закон диффузии Фика

1355
FICK'S SECOND LAW.
Relation resulting from Fick's first law and defining partial derivative of concentration c with respect to time t at defined unit volume as

$$\left(\frac{\partial c}{\partial t}\right)_{xyz} = -\operatorname{div}(D \operatorname{grad} c)$$

where D = diffusion coefficient, grad c = concentration gradient. In one-dimensional problems, when diffusion coefficient does not depend on concentration, Fick's second law is reduced to form:

$$\left(\frac{\partial c}{\partial t}\right)_x = -D \left(\frac{\partial^2 c}{\partial x^2}\right)_t$$

D zweites Ficksches Diffusionsgesetz
F deuxième loi de Fick
P prawo Ficka drugie
R второй закон Фика

1356
FIELD EFFECT (direct field effect).
Effect exerted by dipole of bond linking substituent with rest of molecule on distribution of electron density of molecule. Unlike inductive effect, this effect is not transferable by bond system but acts sterically in direct way.
D Effekt des Raumes
F effet de champ
P efekt pola
R . . .

1357
FIELD STRENGTH EFFECT (Wien's effect).
Increase of equivalent electrolytic conductance in high electric field (10^4—10^5 V/cm) as a result of progressive disappearance of ionic atmosphere.
D Feldstärkeeffekt, Wien-Effekt
F effet Wien
P efekt Wiena, efekt natężenia pola elektrycznego
R эффект Вина

1358
FILM DOSIMETER (photoemulsion dosimeter).
Photographic film, the relative darkness of which, measured after development, gives information about dose absorbed by photoactive layer.
D Filmdosimeter
F dosimètre de film, dosimètre photographique
P dawkomierz filmowy, dawkomierz fotograficzny, dawkomierz fotometryczny
R плёночный (фото)дозиметр

FILTERING. See FILTRATION.

1359
FILTRATE.
Solution separated from precipitate by filtration.
D Filtrat
F filtrat
P przesącz
R фильтрат

1360
FILTRATION (filtering).
Separation of precipitate from liquid by passing mixture through porous material, e.g. filters or crucibles with porous bottoms.
D Filtrieren, Filtration
F filtration
P sączenie
R фильтрование

1361
FINE STRUCTURE (*in atomic spectroscopy*).
Structure of spectral lines or atomic terms arising due to interaction between spin and angular momentum of electrons.
D Feinstruktur
F structure fine
P struktura subtelna
R тонкая структура

1362
FINGERPRINT REGION.
Spectral range in infrared from 650 to 1480 cm^{-1} with a number of bands characteristic for a molecule as a whole and non-characteristic for separate functional groups.
D ...
F empreinte digitale
P przedział daktyloskopowy, zakres odcisku palca
R область отпечатка пальцев

FIRST COORDINATION SPHERE. *See* COORDINATION SPHERE.

1363
FIRST HEAT OF SOLUTION.
Enthalpy change in process of dissolving one mole of a given substance in infinitely large bulk of solvent.
D erste Lösungswärme
F chaleur primaire de dissolution
P ciepło rozpuszczania pierwsze
R первая теплота растворения

1364
FIRST LAW OF THERMODYNAMICS.
In any process increase of internal energy ΔU of closed system is equal to energy supplied in this process in form of work W and heat Q

$$\Delta U = W + Q$$

D erster Hauptsatz der Thermodynamik
F premier principe de la thermodynamique
P zasada termodynamiki pierwsza
R первое начало термодинамики

1365
FIRST-ORDER ASYMMETRIC TRANSFORMATION.
Spontaneous epimerization taking place in solution of optically unstable diastereoisomers. Consequently, new equilibrium between epimers is established, e.g. if compounds A and B are able to form salt A·B, then first-order asymmetric transformation can be expressed by equation:

$$2(\pm)A + 2(-)B \rightarrow \underset{50\%}{(+)A\cdot(-)B} + \underset{50\%}{(-)A\cdot(-)B} \rightleftarrows$$

$$\underset{(50+x)\%}{(+)A\cdot(-)B} + \underset{(50-x)\%}{(-)A\cdot(-)B}$$

D ...
F transformation asymétrique de première espèce
P przekształcenie asymetryczne pierwszego rodzaju
R асимметрическое превращение первого рода

1366
FIRST-ORDER PHASE TRANSITION.
Change of physical state of matter accompanied by sudden change of first-order derivatives of Gibbs free energy with respect to pressure and temperature, i.e. volume and entropy.
D Phasenumwandlung erster Ordnung
F transformation du premier ordre
P przemiana (fazowa) pierwszego rzędu
R фазовый переход первого рода

1367
FIRST-ORDER REACTION.
Reaction with kinetic equation characterized by sum of powers in which concentrations of reactants occur being equal to one.
D Reaktion 1. Ordnung, Reaktion erster Ordnung
F réaction du premier ordre
P reakcja pierwszego rzędu
R реакция первого порядка

FIRST TRIAD. *See* FERROUS METALS.

1368
FISCHER-HEPP REARRANGEMENT.
Rearrangement of N-nitroso-N-alkyl or N-aryl derivatives of aniline into C-nitroso derivatives in acid medium, e.g.

$$CH_3-N=NO \qquad CH_3-N-H$$
$$\xrightarrow{H^{\oplus}}$$
$$N=O$$

D Fischer-Hepp (Nitrosamin)-Umlagerung
F transposition de Fischer et Hepp
P przegrupowanie Fischera i Heppa
R ...

1369
FISCHER INDOLE SYNTHESIS.
Preparation of indole and its derivatives by heating aldehyde or ketone phenylhydrazone in presence of catalysts (e.g. zinc chloride, polyphosphoric acid), e.g.

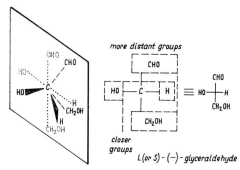

D Fischer Indol-Synthese
F synthèse de Fischer
P synteza (pochodnych) indolu Fischera
R синтез индола Фишера

1370
FISCHER PROJECTION FORMULA.
Projection formula of chiral molecule in which chiral atom is placed in plane of projection, so that its bonds with adjacent atoms of carbon chain are beyond projection plane, but at equal angles to it. The more oxidized part of chain is at the top of formula; remaining two atoms (or groups) linked with chiral carbon atom project towards viewer (at equal angles to plane).

D Fischersche Projektionsformel
F formule en projection de E. Fischer
P wzór (rzutowy) Fischera
R проекционная формула Фишера

FISCHER-TROPSCH PROCESS. *See* FISCHER-TROPSCH SYNTHESIS.

1371
FISCHER-TROPSCH SYNTHESIS (Fischer-Tropsch process).
Technological process of preparation of hydrocarbons, alcohols, acids etc., depending on conditions applied, based on reaction of carbon monoxide with hydrogen at 200—300°C, under pressure, in presence of metallic catalysts (Fe, Co)

$$n\mathrm{CO} + 2n\mathrm{H_2} \rightarrow -(\mathrm{CH_2})_n- + n\mathrm{H_2O}$$

D Fischer-Tropsch-Verfahren, Fischer-Tropsch Kohlenoxid-Druckhydrierung
F procéde de Fischer et Tropsch
P synteza Fischera i Tropscha
R синтез Фишера-Тропша

1372
FISSILE MATERIAL (fissionable material).
Material containing nuclides with nuclei undergoing fission with emission of neutrons in reaction with neutrons.

D Spaltmaterial, Spaltstoff
F matière fissile
P materiał rozszczepialny
R делящееся вещество, расщепляющийся материал

FISSION. *See* NUCLEAR FISSION.

FISSIONABLE MATERIAL. *See* FISSILE MATERIAL.

FISSION BOMB. *See* ATOMIC BOMB.

1373
FISSION CHAIN REACTION.
Chain of nuclear reactions: in neutron fission chain reaction, neutron hitting fissionable atom causes fission resulting in number of neutrons which in turn cause other fissions with emission of further neutrons; number of fissions may diminish with time, remain constant or grow rapidly.

D Kernspaltungskettenreaktion, Spaltungskettenreaktion
F réaction en chaine de fission
P reakcja rozszczepienia łańcuchowa
R цепная реакция деления

1374
FISSION FRAGMENTS.
Nuclei formed in nuclear fission process with kinetic energy obtained during this process.

D Kernbruchstücke, Spaltprodukte
F fragments de fission
P fragmenty rozszczepienia
R осколки деления

1375
FISSION NEUTRONS.
Neutrons emitted in atomic nuclear fission reactions, immediately or with delay. Spectrum of fission neutrons is continuous with maximum at 1 MeV.

D Spaltneutronen
F neutrons de fission
P neutrony rozszczepieniowe
R нейтроны деления

1376
FISSION PARAMETER (fissility parameter), x.
Ratio of square of atomic number Z divided by mass number A of nucleus to critical value of this quotient

$$x = \frac{Z^2}{A} \bigg/ \left(\frac{Z^2}{A}\right)_{cr}$$

where $\left(\dfrac{Z^2}{A}\right)_{cr} \approx 50$. In liquid-drop model, fission parameter gives expression for the stability of a nucleus of spherical shape, thus describing the ability of a nucleus to undergo fission.
D Parameter der Spaltbarkeit
F paramètre fissible
P parametr rozszczepialności
R параметр делимости

1377
FISSION PRODUCTS.
Fission fragments from atomic nucleus and products of their radioactive decay.
D Spaltprodukte
F produits de fission
P produkty rozszczepienia
R продукты деления

1378
FISSION YIELD.
Fraction of fissions leading to formation of product of specified mass number.
D Spaltausbeute
F rendement de fission
P wydajność produktu rozszczepienia
R выход осколков деления

1379
FITTIG REACTION.
Preparation of polycyclic aromatic compounds by treating aryl halogenides with sodium, e.g.

$2ArX + 2Na \longrightarrow Ar - Ar + 2NaX$

D Fittig-Synthese, Fittig-Reaktion
F réaction de Fittig et Tollens
P reakcja Fittiga
R реакция Фиттига

FIXATION PAIR. *See* CONTROL LINE PAIR.

1380
FLAME PHOTOMETER.
Instrument used to measure intensities of radiation in flame-excited spectra of samples.
D Flammenphotometer
F photomètre de flamme
P fotometr płomieniowy
R пламенный фотометр

1381
FLAME PHOTOMETRY.
Branch of emission spectral analysis using flame as excitation source.
D flammenphotometrische Analyse, Flammenphotometrie
F analyse par photométrie de flamme, photométrie de flamme
P analiza fotometryczna płomieniowa, fotometria płomieniowa
R анализ по фотометрии пламени, фотометрия пламени

1382
FLAME SPECTROPHOTOMETER.
Flame photometer which measures intensity of radiation of a given wavelength after its separation by monochromator.
D Flammenspektrophotometer, Flammenspektralphotometer
F spectrophotomètre à flamme
P spektrofotometr płomieniowy
R пламенный спектрофотометр

FLAME SPECTROPHOTOMETRY. *See* ANALYSIS BY FLAME SPECTROPHOTOMETRY.

1383
FLASH PHOTOLYSIS.
Photolysis generated by very short but high-intensity flash of light.
D Blitz(licht)photolyse, Lichtblitzphotolyse
F photolyse éclair
P fotoliza błyskowa
R импульсный фотолиз

1384
FLASH SPECTROSCOPY.
Branch of molecular spectroscopy which uses technique of light pulses to obtain high population of molecular excited states for observation of absorption spectrum.
D Lichtblitzspektroskopie, Blitzspektroskopie, Flash-Technik
F spectroscopie d'éclair
P spektroskopia błyskowa
R импульсная спектроскопия

1385
FLEXIBILITY OF MACROMOLECULES.
Capacity of changing conformation characteristic of macromolecules due to intramolecular thermal movements of certain fragments or action of external forces.
D Biegsamkeit von Makromolekülen
F flexibilité des macromolécules
P giętkość makrocząsteczek
R гибкость макромолекул

1386
FLOCCULATION.
Gathering of particles of dispersed phase into flocculent aggregations.
D Flockung
F floculation
P flokulacja
R флокуляция

FLOCCULATION VALUE. *See*
COAGULATION CONCENTRATION.

1387
FLOTATION.
Method for separation of crushed solid substances making use of difference in wettability of the particular substances in (solid — liquid — gas) three-phase system.
D Flotation
F flottation
P flotacja
R флотация

1388
FLOW METHOD OF RATE MEASUREMENTS.
Measurement of chemical reaction rates in flow, i.e. in system in which flow of reactants is enforced.
D Strömungsmethode, dynamische Methode
F méthode à courant, méthode dynamique
P metoda potokowa pomiarów kinetycznych
R проточный метод исследования кинетики реакции

1389
FLUCTUATIONS.
Differences between observed values of quantity A' and its expected value $\bar{A}$ in a given state.
D statistische Schwankungen
F fluctuations (d'une grandeur)
P fluktuacje statystyczne
R статистические флуктуации

1390
FLUENCE (particle fluence).
Integral, over time, of density of beam of particles or ionizing quanta.
D ...
F fluence
P fluencja (cząstek)
R ...

1391
FLUID.
Substance which at given conditions in macroscopic dimension has no defined shape and fills vessel in which it is contained or melts away in space. Fluids include: gases, liquids and also solid substances in delicate disintegration and suspension in gas or liquid.

D Fluid(um)
F fluide
P płyn
R текучая среда

1392
FLUIDITY.
Reciprocal of viscosity coefficient $\varphi = 1/\eta$.
D Fluidität
F fluidité
P płynność
R текучесть

1393
FLUORESCENCE.
Type of the photoluminescence depending on emission of light during transition of atoms or molecules from electronically excited state to other energetically lower electronic state of like multiplicity.
D Fluoreszenz
F fluorescence
P fluorescencja
R флуоресценция

1394
FLUORESCENCE ANALYSIS
(fluorometric analysis, fluorimetry, fluorescimetry, fluorophotometry).
Emission spectral analysis based on investigation of molecular fluorescence spectrum.
D fluorimetrische Analyse, Fluoreszenzanalyse, Fluorimetrie, Fluorometrie
F analyse par fluorescence, analyse fluorimétrique, fluorimétrie, spectrométrie de fluorescence
P analiza fluorymetryczna, analiza fluorescencyjna, fluorymetria
R флуоресцентный метод анализа, флуориметрия, флуорометрия

1395
FLUORESCENCE EXCITATION SPECTRUM.
Dependence of fluorescence intensity at some wavelength recorded as function of exciting radiation wavelength.
D ...
F spectre d'activation de fluorescence
P widmo wzbudzenia fluorescencji
R ...

1396
FLUORESCENCE SPECTRUM.
Spectrum of electromagnetic radiation resulting from re-emission of previously absorbed excitation radiation by a given sample.
D Fluoreszenzspektrum
F spectre de fluorescence
P widmo fluorescencyjne, widmo fluorescencji
R спектр флуоресценции

1397
FLUORESCENT X-RAY
SPECTROMETER.
Spectrometer for study of X-ray
fluorescence.

D Röntgenfluoreszenzspektrometer
F spectromètre à fluorescence de rayons X
P spektrometr rentgenofluorescencyjny
R рентгеновский флюоресцентный
спектрометр

FLUORESCIMETRY. *See*
FLUORESCENCE ANALYSIS.

1398
FLUORIMETER (fluorometer).
Instrument for measurement of intensity
of fluorescence radiation.

D Fluorimeter
F fluorimètre
P fluorymetr
R флуориметр, флуорометр

FLUORIMETRY. *See* FLUORESCENCE
ANALYSIS.

1399
FLUORINATION.
Introduction of fluorine atom into
organic compound molecule by
substituting hydrogen atom. Usually, this
reaction is effected indirectly in presence
of catalysts, e.g.

$$RH + 2CoF_3 \rightarrow RF + HF + 2CoF_2$$

or by halogen atom exchange (Swart
reaction), e.g.

$$C_6H_5CCl_3 + SbF_3 \xrightarrow{SbCl_5} C_6H_5CF_3 + SbCl_3$$

D Fluorieren, Fluorierung
F fluoration
P fluorowanie
R фторирование

1400
FLUORINE, F.
Atomic number 9.

D Fluor
F fluor
P fluor
R фтор

FLUOROMETER. *See* FLUORIMETER.

FLUOROMETRIC ANALYSIS. *See*
FLUORESCENCE ANALYSIS.

FLUOROPHOTOMETRY. *See*
FLUORESCENCE ANALYSIS.

1401
FLUX (fusing agent).
Substance added to other compound in
order to lower its melting point or to
transform it into soluble compound.

D Flußmittel
F fondant, flux
P topnik
R плавень, флюс

1402
FLUX CONVERTER (neutron converter).
Small quantity of fissionable material
introduced into thermal column of reactor
to produce local fast neutron flux.

D Flußumwandler, Flußkonverter
F convertisseur de neutrons
P konwerter strumienia (neutronów)
R преобразователь нейтронного потока

FLUX DENSITY. *See* RADIANT FLUX
DENSITY.

FLUXES. *See* THERMODYNAMIC
FLUXES.

1403
FLUX OF RADIATION.
Radiation power or energy (or number of
particles or photons) per unit time passing
through a surface.

D Strahlungsfluß, Fluß
F flux de rayonnement
P strumień (promieniowania)
R поток (излучения)

1404
FOAM.
Disperse system in which dispersion
medium is liquid or solid (solid foam) and
dispersed phase is gas.

D Schaum
F écume, mousse
P piana
R пена

1405
FORBIDDEN BAND (forbidden energy
band, band gap, energy gap *in band theory
of solids*).
Range of energy states which cannot be
occupied by electrons in ideally periodic
crystal lattice.

D verbotenes Band, verbotene Zone
F bande interdite
P pasmo wzbronione
R запрещённая зона, запретная зона,
энергетическая щель

FORBIDDEN ENERGY BAND. *See*
FORBIDDEN BAND.

1406
FORBIDDEN TRANSITION.
Transition between two energy levels of
system whose probability, according to
selection rules, is equal to zero.

D verbotener Übergang
F transition interdite
P przejście wzbronione
R запрещённый переход

1407
FORKED CHAIN (branched chain).
System of mutually linked atoms with
branching chain in a definite place, e.g.

$$
\begin{array}{cccc}
| & | & | & | \\
-C-C-C-C- \\
| & | & | & | \\
 & -C- & & \\
 & | & &
\end{array}
\qquad
\begin{array}{c}
| \\
-C- \\
| \quad | \quad | \\
-C-C-C- \\
| \quad | \\
-C- \\
|
\end{array}
$$

D verzweigte Kette
F chaîne ramifiée
P łańcuch rozgałęziony
R разветвленная цепь

1408
FORMATION CURVE (*in coordination
chemistry*).
Curve relating ligand number $\bar{n}$ and
equilibrium concentration of free ligand
[L], usually expressed in logarithmic
form: $\bar{n} = f(p[L])$, where p[L] is negative
logarithm of ligand concentration.

D Bildungskurve
F fonction de formation
P krzywa tworzenia
R кривая образования

1409
FORMYLATION.
Substitution of hydrogen atom in organic
compound molecule by formyl group
—CHO, e.g. Gattermann-Koch reaction.

D Formylierung
F formylation
P formylowanie
R формилирование

1410
FRACTIONAL FREE VOLUME (fractional
void volume).
Ratio of free volume to total volume of
chromatographic column.

D Zwischenraumkoeffizient der Säule
F *rapport du volume libre au volume
total d'une colonne*)
P objętość wolna ułamkowa
R доля свободного объёма (колонки)

FRACTIONAL VOID VOLUME. *See*
FRACTIONAL FREE VOLUME.

1411
FRACTION COLLECTOR.
Device for automatic collection of
fractions of effluent from column.

D Fraktionssammler
F collecteur de fractions
P kolektor frakcji
R коллектор фракций, сборник фракций

1412
FRAGMENTATION REACTION.
Elimination resulting in formation of
organic moieties, originating from fission
of carbon chain, e.g.

$$
\begin{array}{c}
\diagdown \quad | \quad \overset{\frown}{|} \quad | \quad \overset{\frown}{} \\
N-C-C-C-X \\
\diagup \quad \underset{\alpha|}{} \quad \underset{\beta|}{} \quad \underset{\gamma|}{}
\end{array}
\longrightarrow
\begin{array}{c}
\diagdown \overset{\oplus}{N}=C\diagdown \\
\diagup
\end{array}
+
\begin{array}{c}
\diagdown C=C\diagdown \\
\diagup
\end{array}
+ X^{\ominus}
$$

where X = halogen atom or tosyl group.
D Fragmentierung
F fragmentation
P reakcja fragmentacji
R реакция фрагментации

1413
FRANCIUM, Fr.
Atomic number 87.
D Francium, Frankium
F francium
P frans
R франций

1414
FRANCK-CONDON PRINCIPLE.
Time required for electronic transition in
molecule is so short (about 10^{-15} s) —
compared to period of vibration of
molecule (about 10^{-13} s) — that, during act
of absorption and excitation to excited
electronic state, nuclei do not appreciably
alter their relative positions (i.e.
internuclear distance remains practically
constant) or their kinetic energies.
D Franck-Condon-Prinzip
F principe de Franck-Condon
P zasada Francka i Condona
R принцип Франка-Кондона

1415
FREE ELECTRON DENSITY (electron
density).
Number of electrons in unit volume (e.g.
in 1 cm³) which may participate in charge
transfer.
D Konzentration der Elektronen,
Elektronendichte, Dichte der
Leitungselektronen
F densité d'électrons (libres), densité
électronique
P stężenie elektronów, gęstość
elektronowa, koncentracja nośników
ujemnych
R концентрация электронов, электронная
плотность

FREE ELECTRON THEORY OF
METALS. *See* ELECTRON THEORY OF
METALS.

FREE ENERGY. *See* HELMHOLTZ FREE
ENERGY.

FREE ENTHALPY. *See* GIBBS FREE
ENERGY.

FREE INTERNAL ROTATION. *See* FREE
ROTATION.

1416
FREE RADICAL (radical).
Atom or molecule with unpaired electron and electrically inert, e.g. $\cdot H$, $\cdot OH$, $\cdot CH_3$, or having electric charge (ion radical) and being sufficiently stable to be detected and investigated.
D (freies) Radikal
F radical (libre)
P rodnik (wolny)
R (свободный) радикал

1417
FREE-RADICAL POLYMERIZATION.
Polymerization in which free radicals or macroradicals take part in all stages of process.
D Radikalpolymerisation
F polymérisation radicalaire, polymérisation radicalique
P polimeryzacja wolnorodnikowa
R радикальная полимеризация

1418
FREE-RADICAL REACTION (radical reaction).
Process in which substrates, intermediates or end-products are free radicals.
D Radikalreaktion
F réaction radicalaire
P reakcja wolnorodnikowa
R (свободно)-радикальная реакция

1419
FREE ROTATION (free internal rotation).
Rotation of two parts of molecule about single bond, due to small difference in potential energy (at defined temperature) of conformers. Rotational barrier does not surpass 2.5 kJ/mol.
D freie Rotation, freie Drehbarkeit
F rotation libre
P rotacja swobodna
R свободное вращение

1420
FREE SURFACE.
Surface of liquid separating liquid and gaseous phases.
D freie Oberfläche, freie Flüssigkeitsoberfläche
F surface libre
P powierzchnia swobodna
R свободная поверхность

1421
FREE SURFACE ENERGY.
Work needed to increase liquid surface by unit area.
D Oberflächenenergie
F énergie superficielle
P energia powierzchniowa
R поверхностная энергия

1422
FREE VOLUME (interstitial volume, void volume).
Volume occupied by mobile phase in chromatographic column (total column volume minus volume of stationary phase).
D Zwischenkornvolumen
F espace vide, volume interstitiel d'une colonne
P objętość wolna, objętość swobodna, objętość międzyziarnowa
R свободный объём, промежуточный объём (колонки), мёртвый объём (колонки)

1423
FREEZING.
Transition from liquid to solid state of substance.
D Erstarrung
F solidification, cristallisation
P krzepnięcie
R отвердевание

FREEZING POINT. *See* MELTING POINT.

1424
FRENKEL DEFECTS.
Imperfections of crystal lattice consisting in localization of certain number of atoms or ions in interstitial positions. A part of nodes may remain unoccupied, *see* vacancy (atomic, ionic).
D Frenkelsche Fehlordnung
F défauts de Frenkel
P defekty Frenkla
R дефекты по Френкелю

1425
FREQUENCY.
In periodic (oscillation) movement, number of returns of system to initial position or state in unit time. Frequency is equal to reciprocal of oscillation period.
D Frequenz
F fréquence
P często(tliwo)ść
R частота

FREQUENCY FACTOR. *See* PRE-EXPONENTIAL FACTOR.

FREQUENCY FUNCTION. *See* PROBABILITY DENSITY FUNCTION.

1426
FREUDENBERG'S RULE OF SHIFT.
Chiral molecules of similar structure, undergoing the same chemical conversions, may display considerable changes in optical rotation in the same directions.
D Freudenbergscher Verschiebungssatz
F règle de Freudenberg

P reguła przesunięć Freudenberga
R правило сдвига Фрейденберга

FREUNDLICH ADSORPTION
ISOTHERM. *See* FREUNDLICH
ISOTHERM.

1427
FREUNDLICH EQUATION.
Empirical equation of the adsorption
isotherm:

$$x = kp^n \quad \text{or} \quad x = kc^n$$

where x = amount of gas adsorbed by
unit mass of adsorbent, p = equilibrium
pressure, c = equilibrium concentration of
adsorbate, k, n = constants characteristic
of a given system and temperature.
D Freundlichsche Gleichung
F formule de Freundlich
P równanie Freundlicha
R уравнение Фрейндлиха

1428
FREUNDLICH ISOTHERM (Freundlich
adsorption isotherm).
Adsorption isotherm corresponding to
Freundlich equation.
D Freundlichsche Adsorptionsisotherme,
Isotherme von Freundlich
F isotherme d'adsorption de Freundlich
P izoterma (adsorpcji) Freundlicha
R изотерма (адсорбции) Фрейндлиха

1429
FRICKE DOSIMETER.
Solution of ferrous salt in diluted sulfuric
acid; Fricke dosimeter is based on
oxidation of Fe^{2+} to Fe^{3+}; this reaction
proceeds proportionally to amount of
absorbed energy.
D Fricke-Dosimeter
F dosimètre de Fricke
P dawkomierz Frickego
R дозиметр Фрикке

1430
FRIEDEL-CRAFTS REACTION.
Alkylation or acylation of aromatic and
aliphatic compounds with alkyl and acyl
halides or anhydrides in presence of Lewis
acids as catalysts (usually anhydrous
$AlCl_3$, $FeCl_3$, $ZnCl_2$, BF_3, etc.), e.g.

$$ArH + RX \xrightarrow{AlCl_3} ArR + HX$$

$$ArH + (RCO)_2O \xrightarrow{AlCl_3} ArCOR + RCOOH$$

$$CH_3OCCH_2CH_2CCl + H_2C{=}CH_2 \xrightarrow[-HCl]{AlCl_3}$$

$$CH_2{=}CHCCH_2CH_2COCH_3$$

D Friedel-Crafts-Reaktion
F réaction de Friedel et Crafts
P reakcja Friedela i Craftsa
R реакция Фриделя-Крафтса

1431
FRIEDLÄNDER SYNTHESIS.
Condensation of *o*-aminobenzaldehyde
with compounds containing —CH_2—CO—
group, resulting in formation of quinoline
derivatives, e.g.

D Friedländer-Chinolin-Ringschluß
F synthèse de Friedlaender
P synteza (pochodnych chinoliny)
Friedländera
R синтез Фридлендера

1432
FRIES REACTION (Fries rearrangement).
Rearrangement of phenolic esters into
o- and *p*-hydroxyketones in presence of
anhydrous aluminium chloride, e.g.

D Fries-Reaktion, Fries Phenylester →
Acylphenol-Umlagrung
F réaction de Fries, transposition selon
Fries
P przegrupowanie Friesa
R реакция Фриса, перегруппировка
Фриса

FRIES REARRANGEMENT. *See* FRIES
REACTION.

1433
FRONTAL ANALYSIS (frontal
chromatography).
Chromatographic method in which
mixture to be separated is continuously
fed to column. This enables isolation of
only one component (that most weakly
retained by column) in pure form.
D Frontalanalyse, Frontaltechnik
F analyse frontale
P analiza czołowa, metoda czołowa
R фронтальный анализ, фронтальная
хроматография

FRONTAL CHROMATOGRAPHY. *See*
FRONTAL ANALYSIS.

1434
F STATISTIC.
Statistic given by equation

$$F = \frac{\chi_1^2/f_1}{\chi_2^2/f_2}$$

where variables χ_1^2 and χ_2^2 have
χ^2-distribution with f_1 and f_2 degrees of
freedom, respectively; F statistic, with
F-distribution is used for investigation
of degree of homogeneity of variances of
two independent measurement series.
D...
F variable aléatoire F
P statystyka F
R F-статистика, F-параметр

F-TEST. See F-CRITERION.

1435
FUEL CELL.
Cell with total reaction consisting in
oxidation of easily available fuels (e.g.
coal, hydrocarbons, water gas).
D Brennstoffelement
F pile à combustibles, pile à combustion
P ogniwo paliwowe
R топливный элемент

FUGACITY. See VOLATILITY.

FULL WIDTH AT HALF HEIGHT. See
HALF-VALUE WIDTH.

FULL WIDTH AT HALF MAXIMUM.
See HALF-VALUE WIDTH.

1436
FULLY STAGGERED CONFORMATION
(anti conformation, anti-periplanar
conformation), $\pm$ap.
Conformation in which the most bulky
substituents at two adjacent carbon (or
any other element, e.g. silicium or
nitrogen) atoms are sterically as far away
one another as possible. In Newman's
projection formula, bond angle of the most
bulky substituents equal 180°. Fully
staggered conformation is the most stable
when no hydrogen bond is present
between substituents

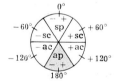

D gestaffelte Konformation,
anti-(periplanare) Konformation
F conformation opposée, conformation
s-trans
P konformacja anty, konformacja
$\pm$ anty-periplanarna
R трансоидная конформация,
заторможенная конформация,
анти-конформация

1437
FUNCTIONAL GROUP.
Grouping of atoms present in a compound,
conferring specific chemical properties on
its molecule. Organic compounds are often
systematized according to their functional
groups.
D funktionelle Gruppe
F groupe(ment) fonctionnel
P grupa funkcyjna
R функциональная группа

1438
FUNCTIONAL GROUP ANALYSIS
(analysis via functional groups).
Methods of analysis of organic compounds
consisting in identification and
determination of compounds according to
their functional groups.
D funktionelle Analyse, Bestimmung nach
den funktionellen Gruppen
F analyse fonctionelle
P analiza według grup funkcyjnych
R функциональный анализ

1439
FUNCTIONAL GROUP ISOMERISM.
Type of structural isomerism caused by
presence of different functional groups in
isomeric compounds, e.g. in ethanol
CH_3CH_2OH and dimethyl ether CH_3OCH_3.
D Metamerie
F métamérie
P metameria, izomeria funkcyjna, izomeria
grup funkcyjnych
R метамерия

1440
FUNCTIONS OF MIXING, $Z_{T, p}^M$.
Molar values of state quantity Z of
mixture decreased by sum of pure
components' Z functions for one mole of
mixture.
D Mischungsfunktionen
F grandeurs de mélange
P funkcje mieszania
R функции смешения

FUNCTIONS OF STATE. See STATE
QUANTITIES.

1441
FUNDAMENTAL BAND.
In IR spectroscopy band corresponding to
transition from state with vibrational

quantum number 0 to state with vibrational quantum number 1.

D Grundschwingungsband
F bande fondamentale
P pasmo podstawowe, pasmo fundamentalne
R основная полоса

FUNDAMENTAL CHAIN. *See* MAIN CHAIN.

FUNDAMENTAL CONSTANTS. *See* UNIVERSAL CONSTANTS.

FUNDAMENTAL EQUATION. *See* GIBBS RELATION.

FUSING AGENT. *See* FLUX.

FUSION. *See* MELTING.

GABRIEL PREPARATION OF PRIMARY AMINES. *See* GABRIEL SYNTHESIS.

1442
GABRIEL SYNTHESIS (Gabriel preparation of primary amines).
Preparation of primary aliphatic amines by hydrolysis of N-alkyl phthalimide derivatives, formed from alkyl halides and phthalimide potassium salts, e.g.

D Gabriel-Synthese, Gabriel Phthalimid-Spaltung
F synthèse de Gabriel
P synteza (amin pierwszorzędowych) Gabriela
R синтез Габриэля

1443
GADOLINIUM, Gd.
Atomic number 64.

D Gadolinium
F gadolinium
P gadolin
R гадолиний

1444
GALLIUM, Ga.
Atomic number 31.

D Gallium
F gallium
P gal
R галлий

1445
GALVANIC CELL.
System consisting of two half-cells of such construction that after their connection with metallic conductor, flow of electrons across this circuit is observed; then electroreduction occurs on one electrode and electrooxidation on the second one; for practical reasons one classifies galvanic cells as chemical or concentration cells, or as primary non-regenerated or secondary cells (*see* accumulator).

D galvanische Zelle, galvanische Kette
F pile, élément, cellule galvanique
P ogniwo galwaniczne
R гальваническая цепь, гальванический элемент

1446
GALVANI ELECTRIC POTENTIAL DIFFERENCE (Galvani potential).
Difference of inner potentials of two phases.

D Galvanispannung
F tension Galvani
P napięcie Galvaniego, potencjał Galvaniego, potencjał międzyfazowy
R гальвани-потенциал

GALVANI POTENTIAL. *See* GALVANI ELECTRIC POTENTIAL DIFFERENCE.

1447
GALVANOSTATIC METHODS.
Methods of study of kinetics of electrode reactions consisting in recording and analysis of potential changes of indicator electrode during electrolysis carried out with current of constant intensity.

D galvanostatische Einschaltmethoden
F méthodes galvanostatiques
P metody galwanostatyczne
R гальваностатические методы

1448
GAMMA CONSTANT, Γ.
Ratio of exposure dose rate of electromagnetic radiation of high energy emitted by a point source at distance of 1 m from source to its activity.

D Dosiskonstante
F constante de dose
P stała jonizacyjna, stała gamma
R гамма-постоянная

1449
GAMMA PHOTON ACTIVATION (γ-ray activation).
Activation by gamma quanta.

D Aktivierung mit γ-Quanten
F activation photonucléaire
P aktywacja fotojądrowa, fotoaktywacja
R фотоактивация

1450
GAMMA QUANTUM (γ-quantum).
High energy photon.
D Gammaquant
F quantum gamma
P kwant γ
R γ-квант

1451
GAMMA RADIATION (gamma rays).
Short-wave electromagnetic radiation of
wavelength approximately from 0.5 to
40 pm. It originates from quantum
transitions between nuclear energy levels,
or as a result of annihilation of a particle
and its antiparticle.
D γ-Strahlung, γ-Strahlen,
	Gammaquanten
F rayonnement γ, rayons γ, rayonnement
	gamma
P promie(niowa)nie γ, kwant γ
R γ-лучи, γ-кванты

1452
GAMMA-RAY ABSORPTION ANALYSIS.
Method of analysis based on measurement
of attenuation of a beam of γ-rays in layer
of material tested. It is used for measuring
thickness, density and for analysing
composition of mixtures (particularly
binary mixtures).
D Durchstrahlungsverfahren mit
	Quantenstrahlung
F analyse par absorption des rayons
	gamma
P analiza absórpcyjna gamma
R методы анализа по поглощению
	γ-излучения

GAMMA-RAY ACTIVATION. *See*
GAMMA PHOTON ACTIVATION.

GAMMA-RAY ACTIVATION ANALYSIS.
See PHOTOACTIVATION ANALYSIS.

1453
GAMMA-RAY CASCADE.
Process by which two (or more) γ-rays are
emitted successively from one radioactive
nucleus.
D γ-Kaskade
F photons gamma en cascade
P kaskada kwantów γ
R каскад гамма-квантов, каскадное
	гамма-излучение

GAMMA RAYS. *See* GAMMA
RADIATION.

1454
GAMMA-RAY SPECTROMETER.
Instrument designed for measuring energy
spectrum of gamma radiation.

D Gammastrahl-Spektrometer,
	Gamma-Spektrometer
F spectromètre (à rayons) gamma
P spektrometr (promieniowania) gamma
R гамма-спектрометр

GAP-CONDUCTIVITY. *See* HOLE
CONDUCTIVITY.

1455
GAS.
Substance with properties characteristic
for gaseous state.
D Gas
F gaz
P gaz
R газ

1456
GAS ANALYSIS.
Qualitative and quantitative determination
of composition of mixture of gases.
D Gasanalyse
F analyse de gaz
P analiza gazowa
R газовый анализ

1457
GAS CHROMATOGRAPHY.
Chromatography employing gas as mobile
phase.
D Gas-Chromatographie
F chromatographie des gaz,
	chromatographie en phase gazeuse
P chromatografia gazowa
R газовая хроматография

1458
GAS CONSTANT (molar gas constant,
universal gas constant), R.
Constant value in ideal gas equation

$$R = \frac{pV_m}{T} = 8.31433 \ \mathrm{JK^{-1}mol^{-1}}$$

where p = pressure, V_m = molar volume
of ideal gas, T = thermodynamic
temperature. Gas constant describes work
done by 1 mol of ideal gas when it
increases its volume due to temperature
increase of 1 K at constant pressure.
D (molare) Gaskonstante
F constante des gaz
P stała gazowa (molowa)
R (универсальная) газовая постоянная

1459
GAS DEGENERACY.
Violation of ideal gas equation $pV_m = RT$
due to inadequacy of Maxwell-Boltzmann
distribution for description of equilibrium
partition of gas particles among energy
levels when thermal de Broglie
wavelength is greater than or comparable

with cubic root of average volume per gas particle.

D Gasentartung
F dégénérescence quantique (de gaz)
P degeneracja gazu doskonałego kwantowa, zwyrodnienie gazu doskonałego kwantowe
R вырождение (газа)

1460
GAS ELECTRODE.
Electrode (half-cell) constructed usually from noble metal dipped into solution saturated with appropriate gas and containing ions resulting from oxidation or reduction of this gas, e.g. hydrogen, oxygen, or halogen electrodes.

D Gaselektrode
F électrode à gaz
P elektroda gazowa
R газовый электрод

1461
GASEOUS PHASE.
Phase composed of substance in gaseous state.

D Gasphase
F phase gazeuse
P faza gazowa
R газовая фаза

1462
GASEOUS STATE.
State of aggregation characterized by lack of resilience (elasticity) of shape, high compressibility coefficient and lack of arrangement of molecules.

D gasförmiger Zustand, Gaszustand
F état de gaz, état gazeux
P stan (skupienia) gazowy
R газообразное состояние

1463
GAS MULTIPLICATION.
Multiplication of charge value collected at electrodes of gas-filled counter tube, as a result of secondary ionization of gas.

D Gasverstärkung
F multiplication due au gaz
P wzmocnienie gazowe
R газовое усиление

1464
GASOMETRIC ANALYSIS.
Determination of components of sample based on measurement of volume of gas evolved in a given reaction.

D gasvolumetrische Methode
F gazométrie
P analiza gazomiernicza, analiza gazometryczna
R газообъёмный метод, газоволюметрический метод

1465
GATTERMANN ALDEHYDE SYNTHESIS.
Preparation of aromatic aldehydes from phenols (or phenyl ethers or some arenes), hydrogen cyanide and hydrogen chloride in presence of metal (zinc or aluminium) chlorides, e.g.

D Gattermann-Reaktion, Gattermann-Arylformylierung, Gattermann-Aldehyd-Synthese
F synthèse de Gattermann
P synteza aldehydów Gattermanna
R синтез альдегидов Гаттермана

1466
GATTERMANN DIAZO-REACTION.
Substitution of diazo group by halide, cyano group or other substituents in presence of metallic copper as catalyst, e.g.

$$ArN_2^{\oplus} Cl^{\ominus} \xrightarrow{Cu} ArCl + N_2$$

$$ArN_2^{\oplus} + CN^{\ominus} \xrightarrow{Cu} ArCN + N_2$$

D Gattermann-Synthese, Gattermann Diazonium-Austausch
F synthèse de Gattermann
P reakcja Gattermanna
R реакция Гаттермана

1467
GATTERMANN-KOCH ALDEHYDE SYNTHESIS.
Preparation of aromatic aldehydes by action of carbon monoxide — hydrogen chloride mixture on aromatic hydrocarbons, in presence of Lewis acids as catalysts (usually anhydrous aluminium chloride, e.g.

D Gattermann-Koch Aryl-Formylierung, Gattermann-Koch Aldehyd-Synthese
F réaction de Gattermann-Koch
P reakcja Gattermanna i Kocha
R реакция Гаттермана-Коха

1468
GAUCHE CONFORMATION (skew conformation), ± sc.
Conformation characteristic for staggered position of the most bulky substituents located at two adjacent carbon (or any other element, e.g. silicium or nitrogen) atoms. In Newman's projection formula, bond angles between less remote and more remote substituents equal 60°

D windschiefe Konformation, syn-Konformation
F conformation gauche, conformation oblique, conformation étoilée
P konformacja skośna, konformacja ± syn-klinalna
R скошенная конформация, гош конформация

GAUSSIAN DISTRIBUTION. See NORMAL DISTRIBUTION.

GAY-LUSSAC'S LAW. See CHARLES'S LAW.

GAY-LUSSAC'S LAW. See LAW OF COMBINING VOLUMES.

GEIGER-COUNTER. See GEIGER-MÜLLER COUNTER.

1469
GEIGER-MÜLLER COUNTER
(Geiger-counter, G-M counter).
Gas-filled counter operating at a gas amplification such that charge collected at electrodes during each pulse is constant and independent of charge liberated by initial ionizing event.

D Geiger-Müller-Zähler, Geiger-Müller Zählrohr, Auslösezählrohr
F compteur de Geiger(-Müller)
P licznik Geigera (i Müllera), licznik G-M
R счётчик Гейгера-Мюллера

1470
GEIGER-NUTTAL RULE.
Relation between range R of α-particles and disintegration constant λ of α-radioactive nuclide, expressed by equation

$$\ln R = A \ln \lambda + B$$

where A, B = constants, different for various radioactive series.
D Geiger-Nuttal-Regel
F relation de Geiger-Nuttal, loi de Geiger-Nuttal
P prawo Geigera i Nuttala
R закон Гейгера-Неттола

1471
GEL.
Disperse system in which a crosslinked, porous, spatial structure, filled with dispersion medium, is formed (see coagel, lyogel).
D Gel
F gel
P żel
R гель

1472
GELATION (gelling).
Stiffening of colloidal system into uniform mass without separating dispersed phase and dispersion medium.
D Gelatinierung
F gélatinisation, gélification
P żelatynowanie, galaretowacenie
R желатинирование, застуднвание

Ge(Li) DETECTOR. See LITHIUM-DRIFTED GERMANIUM DETECTOR.

GELLING. See GELATION.

1473
GENERAL ACID-BASE CATALYSIS.
Acid-base catalysis due to all acids or bases present in solution.
D allgemeine Säure-Base-Katalyse
F catalyse acido-basique généralisée
P kataliza kwasowo-zasadowa ogólna
R общий кислотно-основной катализ

1474
GENERAL CHEMISTRY.
Branch of chemistry dealing with the most general and fundamental chemical laws.
D allgemeine Chemie
F chimie générale
P chemia ogólna
R общая химия

GENERALIZED FLUXES. See THERMODYNAMIC FLUXES.

1475
GENERALIZED SPECTROPHOTOMETRY.
Method of analysis of two-component system with such a choice of conditions of analysis (wavelength, pH, temperature, equilibrium, etc.) that calculation of unknown concentrations from appropriate

equations is possible for results of two to four spectrophotometric measurements.

D ...
F spectrophotométrie généralisée
P spektrofotometria uogólniona
R ...

1476
GENERIC n-TH ORDER PHASE DISTRIBUTION (reduced n-molecule distribution function), ρ_n (1, 2, ..., n).
Partition function for n particles, where $0 \leqslant n \leqslant N$ and N is maximum number of particles in given system; determines probability that one particle is in state 1, one in state 2, ..., and one in state n, independently of way of ascribing particles to states and also independently of number of particles in other states. Particle state is defined by particle dynamic variables; if $n = N$, generic phase distribution contains maximum possible information about system.

D n-molekulare Verteilungsfunktion, reduzierte Verteilungsfunktion der Ordnung n
F fonction de distribution en position et impulsion de n corpuscules, fonction de distribution n-ple
P funkcja rozdziału n-molekularna, funkcja podziału n-molekularna, funkcja rozdziału n-cząstkowa
R n-частичная функция распределения

1477
GEOCHEMISTRY (geological chemistry).
Science dealing with occurrence and circulation of chemical elements in nature as well as with explanation of their role in geological processes.

D Geochemie
F géochimie
P geochemia
R геохимия

GEOLOGICAL CHEMISTRY. *See* GEOCHEMISTRY.

GEOMETRICAL ISOMERISM. *See* CIS-TRANS ISOMERISM.

GEOMETRIC ISOMERISM. *See* CIS-TRANS ISOMERISM.

1478
GEOMETRIC MEAN (*in series*), g.
Statistic given by equation

$$g = (x_1\, x_2\, ...\, x_n)^{1/n}$$

where $x_1, x_2, ..., x_n$ = values of series elements (e.g. results of measurements) and n = number of elements in series.

D geometrisches Mittel
F moyenne géométrique

P średnia geometryczna
R среднее геометрическое

1479
GEOMETRY FACTOR.
Ratio of solid angle at which detector "sees" radiation source, to solid angle of 4π steradians.

D Geometriefaktor, geometrische Ausbeute
F facteur de géométrie
P wydajność liczenia geometryczna
R геометрический коэффициент счётной установки

1480
GEOMETRY OF ABSORPTION OF IONIZING RADIATION.
Spatial distribution of radiation sources or of path and absorbing medium of particle beam.

D Geometrie von Strahlungsabsorption
F géométrie de l'absorption des rayonnements
P geometria absorpcji promieniowania
R геометрия поглощения излучения

1481
GERMANIUM, Ge.
Atomic number 32.

D Germanium
F germanium
P german
R германий

GIBBS ADSORPTION EQUATION. *See* GIBBS ADSORPTION THEOREM.

1482
GIBBS ADSORPTION THEOREM (Gibbs adsorption equation).
Quantitative relation describing isothermal adsorption at surface of liquid phase; for liquid two-component systems this relation is:

$$\Gamma = -\frac{c}{RT}\left(\frac{\partial \sigma}{\partial c}\right)_{q,\,T}$$

where Γ = surface concentration of adsorbate adsorbed at surface of liquid phase, in moles per unit area of surface, c = concentration of adsorbate in bulk liquid phase, R = gas constant, σ = surface tension, q = surface area of liquid phase, T = thermodynamic temperature.

D Gibbssches Adsorptionsgesetz
F théorème de Gibbs, équation de Gibbs
P reguła stężeń Gibbsa, równanie adsorpcji Gibbsa
R уравнение адсорбции Гиббса, адсорбционное уравнение Гиббса

1483
GIBBS CONCENTRATION TRIANGLE.
Equilateral triangle with each side equal
to unit length which makes possible
presentation of three-component mixture
composition

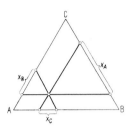

where x_A, x_B, x_C = molar fractions of
particular components of solution.
D Gibbs-Konzentrationsdreieck
F triangle de concentration de Gibbs
P trójkąt stężeń Gibbsa
R треугольник концентрации Гиббса

GIBBS-DUHEM IDENTITY. *See*
GIBBS-DUHEM RELATION.

1484
GIBBS-DUHEM RELATION
(Gibbs-Duhem identity).
Equation expressing relation between
intensive state parameters of a given
phase α

$$S^{\alpha}dT^{\alpha}-V^{\alpha}dp^{\alpha}+\sum_i n_i^{\alpha}\,d\mu_i^{\alpha}=0$$

where S = entropy, V = volume,
n_i = number of moles of i-th component,
μ_i = its chemical potential,
T = thermodynamic temperature,
p = pressure, α = indicator of a given
phase.
D Gibbs-Duhemsche Beziehung,
   Gibbs-Duhemsche Gleichung
F relation de Gibbs-Duhem
P równanie Gibbsa i Duhema
R уравнение Гиббса-Дюгема

GIBBS ENSEMBLE. *See* STATISTICAL
ENSEMBLE.

1485
GIBBS FREE ENERGY (free enthalpy,
Gibbs free energy function,
thermodynamic potential), G.
Extensive thermodynamic function
$G = H-TS$, where H = enthalpy,
T = thermodynamic temperature, and
S = entropy.
D freie Energie (nach Gibbs), Gibbssches
   (thermodynamisches) Potential
F enthalpie libre (à pression constante),
   potentiel de Gibbs, fonction de Gibbs

P potencjał termodynamiczny
   (izotermiczno-izobaryczny), entalpia
   swobodna, energia Gibbsa, potencjał
   termodynamiczny izobaryczny, potencjał
   termodynamiczny pod stałym ciśnieniem
R изобарно-изотермический потенциал,
   изобарно-изотермный потенциал,
   изобарный потенциал, свободная
   энтальпия

GIBBS FREE ENERGY FUNCTION. *See*
GIBBS FREE ENERGY.

GIBBS-HELMHOLTZ EQUATIONS. *See*
GIBBS-HELMHOLTZ RELATIONS.

1486
GIBBS-HELMHOLTZ RELATIONS
(Gibbs-Helmholtz equations).
Thermodynamic identities expressing
relations between Helmholtz free energy F
and internal energy U and between Gibbs
free energy G and enthalpy H

$$\left(\frac{\partial \dfrac{F}{T}}{\partial \dfrac{1}{T}}\right)_V = U \qquad \left(\frac{\partial \dfrac{G}{T}}{\partial \dfrac{1}{T}}\right)_p = H$$

where T = thermodynamic temperature
and V and p = volume and pressure
respectively.
D Gibbs-Helmholtzsche Gleichungen
F relations de Gibbs-Helmholtz, équations
   de Gibbs-Helmholtz
P relacje Gibbsa i Helmholtza, wzory
   Gibbsa i Helmholtza, równania Gibbsa
   i Helmholtza
R соотношения Гиббса-Гельмгольца,
   уравнения Гиббса-Гельмгольца

1487
GIBBS PHASE RULE.
Number of thermodynamic degrees of
freedom s of system containing α
independent components and β phases is
given by:

$$s = 2+\alpha-\beta$$

D Gibbssche Phasenregel, Gibbssches
   Phasengesetz
F règle des phases
P reguła faz Gibbsa
R правило фаз Гиббса

1488
GIBBS RELATION (fundamental
equation).
Equation

$$T^{\alpha}dS^{\alpha} = dU^{\alpha}+p^{\alpha}dV^{\alpha}-\sum_i \mu_i^{\alpha}\,dn_i^{\alpha}$$

where S = entropy, U = internal energy,
V = volume, n_i = number of moles of i-th
component of mixture, μ_i = its chemical
potential, and α numbers phases.

D Gibbssche Hauptgleichung,
  Fundamentalgleichung
F équation de Gibbs, équation
  fondamentale
P równanie Gibbsa
R уравнение Гиббса

1489
GLASS ELECTRODE.
Electrode (half-cell) constructed from
thin-wall glass bulb in which is placed
solution of known and constant pH and
also electrode of second kind to ensure
electrical contact.

D Glaselektrode
F électrode en verre
P elektroda szklana
R стеклянный электрод

GLASSY STATE. See VITREOUS STATE.

GLIDE PLANE. See
GLIDE-REFLECTION PLANE.

1490
GLIDE-REFLECTION PLANE (glide
plane).
Symmetry element of space group,
combined from two symmetry operations,
reflection in mirror plane and translation
in the given direction in space.

D Gleitspiegelebene
F plan de symétrie avec glissement
P płaszczyzna zwierciadlano-translacyjna,
  płaszczyzna poślizgu
R плоскость скольжения, плоскость
  скользящего отражения

G-M COUNTER. See GEIGER-MÜLLER
COUNTER.

GOLAY COLUMN. See CAPILLARY
COLUMN.

1491
GOLD, Au.
Atomic number 79.

D Gold
F or
P złoto
R золото

1492
GOLD NUMBER.
Smallest concentration of protective
colloid, expressed in mg per 10 cm^3 of
solution, which prevents colour change
of standard gold sol from poppy red to
violet after addition of 1 cm^3 of 10%
solution of NaCl or 3.7% solution of KCl.

D Goldzahl
F nombre d'or
P liczba złota
R золотое число

1493
GOMBERG-BACHMANN-HEY
SYNTHESIS OF DIARYL COMPOUNDS.
Preparation of diaryl compounds by action
of benzene (or its homologues) on
diazonium salt solutions in presence
of sodium hydroxide, e.g.

D Gomberg-Bachmann-Hey
  Diaryl-Synthese
F réaction de Gomberg et Bachmann
P reakcja Gomberga, Bachmanna i Heya
R реакция Гомберга-Бахмана-Хея

GO-NO-GO DOSIMETER. See
THRESHOLD DOSIMETER.

1494
GOUY-CHAPMAN'S THEORY OF
DOUBLE LAYER.
Theory referring to Helmholtz model (see
Helmholtz's theory of double layer), in
which it is assumed that neutralising
charge on electrolyte side is not located
strictly in interface, but is distributed
statistically in solution, analogously to
that in Debye-Hückel's ionic cloud.

D Gouy-Chapmansche Theorie der
  Doppelschicht
F théorie de la couche double de Gouy et
  Chapman
P teoria budowy warstwy podwójnej
  Gouya i Chapmana
R теория двойного (электрического) слоя
  Гуи-Чапмана

1495
GOUY MAGNETIC SUSCEPTIBILITY
BALANCE.
Instrument for the measurement of
magnetic susceptibility of cylindrical
samples of length of about 10—20 cm;
main parts are magnet and balance used
to measure forces acting on sample in
inhomogeneous magnetic field.

D magnetische Waage nach Gouy
F balance magnétique de Gouy
P waga magnetyczna Gouya
R магнитные весы Гуи

1496
GRADIENT ELUTION.
Elution in which composition
(concentration, pH, etc.) of eluent is
changed continuously.

D Gradienten-Elutions-Technik
F élution par gradient de pouvoir éluant
P elucja gradientowa
R градиентное элюирование

1497
GRAEBE-ULLMANN SYNTHESIS OF CARBAZOLES.
Preparation of carbazole by thermal decomposition of benzotriazole, obtained by diazotization of o-aminodiphenylamine

D Graebe-Ullmann Carbazol-Synthese
F synthèse de Graebe-Ullmann
P synteza (karbazolu) Graebego i Ullmanna
R синтез Гребе-Ульмана

1498
GRAFT POLYMER.
Polymer in whose macromolecules one or more species of block is attached as side chains, these side chains being different from the main chain, e.g.

```
··AAAAAAAA··
    |    |
    B    B
    B    B
    B    B
```

D Pfropf(co)polymer, Graft-Polymer, Graft-Copolymer
F (co)polymère greffé
P (ko)polimer szczepiony
R привитый сополимер

1499
GRAFT POLYMERIZATION.
Process in which one or more monomers and a polymer are converted together into graft polymer.
Remark: Term graft copolymerization has been reserved for case of graft polymerization with two or more participant monomers.

D Pfropfpolymerisation, Graftpolymerisation
F polymérisation greffée
P polimeryzacja szczepiona
R привитая сополимеризация

1500
GRAHAM'S DIALYZER.
Device for separation of colloids from dissolved substances of low molecular weight by means of dialysis.

D Dialysator von Graham
F dialyseur de Graham
P dializator Grahama
R диализатор Грэма

1501
GRAM-ATOM (gram-atomic weight).
Mass in grams of element equal to its relative atomic mass.

Remark: In view of introduction of mole as unit of amount of substance, molar mass should be used in preference of gram-atom.

D Grammatom, Atomgramm
F atome-gramme
P gramoatom
R грамм-атом

GRAM-ATOMIC WEIGHT. *See* GRAM-ATOM.

1502
GRAM EQUIVALENT.
Mass in grams of element equal to its relative atomic mass divided by its valency number.

D Grammäquivalent
F équivalent-gramme, valence-gramme
P gramorównoważnik, wal, val
R грамм-эквивалент

1503
GRAM METHOD (macro method, classical method).
Method of analysis where mass of investigated sample is $x \cdot 10^{-1}$—$x \cdot 10^0$ g and in the case of gases volume is $x \cdot 10^1$—$x \cdot 10^2$ cm^3.

D Makro-Methode
F méthode grammique, macrométhode, méthode classique
P metoda decygramowa, makrometoda
R макрометод

GRAM MOLE. *See* MOLE.

GRAM-MOLECULAR WEIGHT. *See* MOLE.

GRAM MOLECULE. *See* MOLE.

1504
GRAND CANONICAL ENSEMBLE (grand canonical statistical function).
Statistical ensemble with following probability distribution function

$$P(Q_N, N) = \exp\left[\frac{\Omega - E(Q_N, N) + \mu N}{kT}\right]$$

where $E(Q_N, N)$ = energy of system of N identical particles in state Q_N,
k = Boltzmann constant,
T = thermodynamic temperature,
μ = chemical potential of particle,
and Ω = grand thermodynamic potential. In this form formula refers to systems composed of identical particles. For systems composed of different particles this formula must be suitably generalized. Grand canonical ensemble is utilized for description of thermodynamic equilibrium in open systems and in systems undergoing phase transitions or chemical reactions.

D großkanonische Gesamtheit, große kanonische Gesamtheit
F ensemble grand canonique (de Gibbs)
P zespół grandkanoniczny, rozkład grandkanoniczny, wielki zespół kanoniczny
R большой канонический ансамбль (Гиббса), большое каноническое множество Гиббса, большое каноническое распределение Гиббса

GRAND CANONICAL STATISTICAL FUNCTION. *See* GRAND CANONICAL ENSEMBLE.

1505
GRAND PARTITION FUNCTION, Ξ.
Function defined by

$$\Xi(T, V, \mu) = \sum_N \sum_{Q_N} \exp\left[- \frac{E(N, Q_N) - \mu N}{kT} \right]$$

where T = thermodynamic temperature, V = volume of system, μ = molecular chemical potential, k = Boltzmann constant and $E(N, Q_N)$ = energy of system of N identical particles in state Q_N. Grand partition function determines all equilibrium thermodynamic properties of system.
D große Zustandssumme, große Verteilungsfunktion
F grande fonction de partition
P funkcja rozdziału duża, funkcja podziału duża, suma statystyczna wielka
R большая статистическая сумма, большая сумма состояний

GRAND POTENTIAL. *See* GRAND THERMODYNAMICAL POTENTIAL.

1506
GRAND THERMODYNAMICAL POTENTIAL (grand potential), Ω.
Ω function = $-kT \ln \Xi$, where
k = Boltzmann constant,
T = thermodynamic temperature and
Ξ = grand partition function. Grand thermodynamical potential determines all equilibrium properties of system. In thermodynamic limit $\Omega = -pV$, where p = pressure and V = volume of system.
D großes (thermodynamisches) Potential
F grand potentiel (thermodynamique)
P potencjał termodynamiczny duży, potencjał termodynamiczny wielki
R большой термодинамический потенциал

1507
GRAVIMETRIC ANALYSIS.
Determination of components of sample on basis of weight of precipitate obtained during analysis.
D Gewichtsanalyse, Gravimetrie
F gravimétrie, analyse pondérale

P analiza wagowa, analiza grawimetryczna, grawimetria
R весовой анализ

1508
GRAY, Gy.
Unit of absorbed dose of ionizing radiation corresponding to 1 J of absorbed energy per 1 kg of medium.
D Gray
F gray
P grej
R грэй

1509
GREY BODY.
Body which absorbs uniformly the same part of incident radiant energy falling upon it, no matter what the wavelength and spectral range of radiation is.
D grauer Körper
F corps gris
P ciało szare
R серое тело

1510
GRIGNARD REACTION (Grignard synthesis).
Preparation of organomagnesium compound from alkyl (or aryl) halide and magnesium in anhydrous diethyl ether (Grignard reagent); the latter is treated with compound containing multiple bond between atoms differing in electronegativity (C=O, C=S, C≡N, N=O), e.g.

$$RX + Mg \xrightarrow{\text{ether}} RMgX$$

$$\begin{array}{c} R' \\ \diagdown \\ \diagup \\ R'' \end{array} C{=}O + RMgX \longrightarrow R''{-}\overset{\displaystyle R'}{\underset{\displaystyle R}{C}}{-}OMgX \xrightarrow{H_2O}$$

$$R''{-}\overset{\displaystyle R'}{\underset{\displaystyle R}{C}}{-}OH + Mg(OH)X$$

D Grignard (Organomagnesium)-Addition
F réaction de Grignard
P reakcja Grignarda
R реакция Гриньяра

GRIGNARD SYNTHESIS. *See* GRIGNARD REACTION.

1511
GRIMM'S LAW.
Law defining conditions of isomorphism of two chemical substances. Two substances are isomorphous if: (a) types of chemical formula are identical; (b) types of unit cell of space lattices are the same; (c) dimensinos of unit cells are approximately equal in both compounds.
D Grimmsches Gesetz
F loi de Grimm
P prawo Grimma
R правило Гримма

1512
GROSS ERROR.
Error of value so enormously different from values of other errors that it can be treated as not belonging to investigated population and due to temporary source, e.g. improper measurement, use of erroneously working instrument or incorrect calculation.
D grober Fehler
F erreur
P błąd gruby, pomyłka
R грубая ошибка, промах

1513
GROSS SAMPLE (combined sample).
Sample composed of increments.
D Sammelprobe
F échantillon global
P próbka ogólna
R общая проба, генеральная проба

1514
GROTTHUS-DRAPER LAW.
Only that part of incident radiation may be photochemically active which is absorbed by reacting system.
D Grotthuss-Drapersches Gesetz
F loi de Grotthuss-Draper
P prawo Grotthussa i Drapera
R закон Гроттуса-Дрейпера

1515
GROUND STATE.
Quantum mechanical state of system with lowest possible energy.
D Grundzustand
F état fondamental
P stan podstawowy
R основное состояние

1516
GROUP (group of elements, periodic group, family).
Vertical column of elements in periodic system, characterized, with some exceptions, by the same maximum oxidation number. The group is subdivided into main and sub-groups.
D Familie, Gruppe
F groupe
P grupa
R группа

1517
GROUP DIPOLE MOMENT (group moment).
Contribution to total molecular dipole moment by given polar group.
D Gruppenmoment
F moment d'un groupement
P moment (dipolowy) grupy
R дипольный момент группы

GROUP MOMENT. *See* GROUP DIPOLE MOMENT.

GROUP OF ELEMENTS. *See* GROUP.

1518
GROUP REAGENT.
In systematic qualitative analysis, reagent which precipitates from solution a given group of cations or anions.
D Gruppenreagens
F réactif de groupe
P odczynnik grupowy
R групповой реагент

1519
GROUP VIBRATIONS.
Molecular vibrations with negligibly small vibration amplitudes of all atoms except for those of a given group.
D Gruppenschwingungen
F vibrations de groupements fonctionnels
P drgania grup atomów, drgania charakterystyczne, drgania grup funkcyjnych
R колебания групп атомов

GROWTH FACTOR. *See* SATURATION FACTOR.

1520
GULDBERG'S RULE.
Normal (thermodynamic) boiling temperature T_w of liquids not undergoing association amounts to about 2/3 of their critical temperature T_c.
D Regel von Guldberg
F règle de Guldberg
P reguła Guldberga
R правило Гульдберга

1521
GYROMAGNETIC EFFECTS (magnetomechanical effects).
Phenomena of magneto-mechanical character with enable gyromagnetic rations to be determined and depend on: (1) magnetization during rotation of ferromagnet (Barnett effect); (2) rotation following magnetization of ferromagnet change (Einstein-de Haas effect).
D gyromagnetische Effekte
F effets gyromagnétiques
P zjawiska giromagnetyczne
R гиромагнитные эффекты

1522
GYROMAGNETIC RATIO.
Ratio of system's magnetic moment and its mechanical momentum; for angular momenta gyromagnetic ratio is $\gamma_1 = e/2m_0c$, for spin momenta $\gamma_s = e/m_0c$, where e = electronic charge, m = electron mass, c = speed of light in vacuum.

D gyromagnetisches Verhältnis
F rapport gyromagnétique
P (współ)czynnik giromagnetyczny,
współczynnik żyromagnetyczny
R гиромагнитное отношение

1523
HADRONS.
Elementary particles (mesons and baryons)
which take part in strong interactions.
D Hadronen
F hadrons
P hadrony
R адроны

1524
HAFNIUM, Hf.
Atomic number 72.
D Hafnium
F hafnium, celtium
P hafn
R гафний

1525
HAHN'S PRECIPITATION RULE.
A microcomponent (e.g. radioactive
substance) contained in solution may be
incorporated in crystalline precipitate
to be separated when it is found in
structure of crystal lattice of
a macrocomponent, i.e. when it is
isomorphic or isodimorphic with
macrocomponent.
D Hahnsche Fällungsregel
F loi de Hahn
P prawo współstrącania Hahna
R закон соосаждения Хана

HALF-CELL. See ELECTRODE.

1526
HALF-INTENSITY WIDTH (half-width),
$\Delta\nu_{1/2}$.
Width of spectral range where intensity
of spectral line is equal to one half of
maximum intensity.
D Halbwertsbreite (der Spektrallinie)
F (de)mi-largeur
P szerokość linii widmowej połówkowa
R полуширина линии

1527
HALF LIFE (half-life period).
Time elapsing between start of reaction
to point where half of starting mass of
substrate is changed.
D Halbwertszeit
F temps de demi-réaction
P okres półtrwania reakcji, półokres
reakcji, okres połowicznej przemiany
R полупериод, период полураспада,
время полураспада

1528
HALF-LIFE (radioactive half-life).
Time in which activity of radioactive
substance decreases to half its initial
value, i.e. time in which half of the initial
number of nuclei undergoes radioactive
decay.
D Halbwertszeit
F période radioactive
P okres półtrwania, półokres trwania,
okres połowicznego zaniku, okres
półrozpadu
R период полураспада

HALF-LIFE PERIOD. See HALF-LIFE.

1529
HALF-THICKNESS (half-value layer).
Thickness of attenuating layer that
reduces intensity of radiation passing
through it to one half of its initial value.
D Halbwertsdicke
F couche de demi-atténuation
P warstwa półchłonna, warstwa
połówkowa, warstwa połowicznego
osłabienia
R слой половинного ослабления

HALF-VALUE LAYER. See
HALF-THICKNESS.

1530
HALF-VALUE WIDTH (full width at half
height, full width at half maximum).
Width of energy peak at half of its height.
D Halbwertsbreite
F largeur à (de)mi-hauteur
P szerokość połówkowa (piku)
R полуширина пика, ширина пика на
половине максимума

1531
HALF-WAVE POTENTIAL.
Potential corresponding to one-half
of amplitude of limiting current of
polarographic wave.
D Halbstufenpotential
F potentiel de demi-onde, potentiel de
demi-vague
P potencjał półfali
R потенциал полуволны

HALF-WIDTH. See HALF-INTENSITY
WIDTH.

1532
HALIDE COMPLEX (halo-complex).
Complex in which ions of halogens are
ligands (F, Cl, Br, I).
D Halogenokomplex
F complexe halogéné
P halogenokompleks
R галогенокомплекс

1533
HALL EFFECT.
Effect occurring in strip of conducting material carrying electric current I when constant magnetic field H is applied at right angles to face of strip. Potential difference (Hall potential V) is created across conductor in direction orthogonal to directions of both magnetic field and current flow. Sign of Hall potential depends on the kind of mobile charge carriers, i.e. free electrons or electron holes.

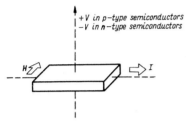

+V *in p–type semiconductors*
−V *in n–type semiconductors*

D Hall-Effekt
F effet Hall
P efekt Halla
R эффект Холла

HALO-COMPLEX. *See* HALIDE COMPLEX.

1534
HALOFORM REACTION.
Reaction of organic compounds containing —$COCH_3$ group, or corresponding alcohols with hypochlorites, hypobromites or hypoiodites of alkali metals, involving formation of $CHCl_3$, $CHBr_3$ or CHI_3, respectively; e.g.

$$RCOCH_3 \xrightarrow{\text{NaOCl}} RCOONa + CHCl_3$$

D Haloformreaktion
F réaction de Lieben
P reakcja haloformowa
R галоформная реакция

1535
HALOGENATION.
Substitution of hydrogen atom in molecule of organic compound by halogen atom (Cl, Br, I), e.g.

$$RH + X_2 \rightarrow RX + HX$$

D Halogenierung
F halogénation
P chlorowcowanie
R галогенирование

HALOGEN GROUP. *See* HALOGENS.

1536
HALOGENS (halogen group).
Elements in seventh main group of periodic system with structure of outer electronic shells of atoms ns^2np^5: fluorine, chlorine, bromine, iodine, astatine.

D Halogene, Fluorgruppe
F halogènes
P fluorowce, halogeny
R галогены, галоиды, элементы группы фтора

1537
HAMILTONIAN.
Quantum mechanical operator of total energy of system given as sum of kinetic and potential energy operators. Eigenfunctions of hamiltonian describe stationary states of system and its eigenvalues correspond to all possible total system energies.

D Hamilton-Operator
F opérateur d'Hamilton
P hamiltonian, operator energii
R гамильтониан, оператор Гамильтона

HAMMETT ACIDITY FUNCTION. *See* ACIDITY FUNCTION H.

1538
HAMMETT EQUATION.
Equation describing influence exerted by substituent on rate or equilibrium constants of reaction:

$$\lg k - \lg k_0 = \rho\sigma$$

where k, k_0 = reaction rate or equilibrium constants of substituted and non-substituted compounds respectively, σ = constant characteristic for substituent X, independent of type of reaction but dependent on type and location of substituent X with respect to reaction centre Y, ρ = reaction constant, dependent on reaction centre Y and reaction conditions (ρ expresses reaction sensitivity to substituent X). In case of benzene derivatives, Hammett equation is applicable to *meta-* and *para-*compounds.

D Hammett-Gleichung
F équation d'Hammett
P równanie Hammeta
R уравнение Гаммета

1539
HANGING MERCURY DROP ELECTRODE.
Electrode composed of hanging mercury drop usually with diameter within limits of 0.4—1.5 mm.

D hängende Hg-Tropfen-Elektrode
F électrode à goutte de mercure pendante
P elektroda rtęciowa (kroplowa) wisząca
R ртутный висящий капельный электрод

1540
HANTZSCH PYRIDINE SYNTHESIS.
Preparation of pyridine derivatives by condensation of two moles of β-ketoester with aldehyde and ammonia, followed by nitric acid oxidation of resulting condensation product, e.g.

ROOC—CH$_2$—CO—H$_3$C + R'—CHO + CH$_2$—CO—CH$_3$—COOR $\xrightarrow{\text{NH}_3}$

ROOC—R'—H—COOR / H$_3$C—N—CH$_3$ $\xrightarrow{\text{HNO}_3}$ ROOC—R'—COOR / H$_3$C—N—CH$_3$

D Hantzsch Pyridin-Synthese
F synthèse de Hantzsch
P synteza (pochodnych pirydyny) Hantzscha
R синтез пиридинов Ганчи

HARD CORE POTENTIAL. *See* HARD SPHERE POTENTIAL.

1541
HARD RADIATION.
Ionizing radiation of high energy and penetrating power, e.g. β- and γ-radiation of energy exceeding 1 MeV.

D harte Strahlung
F rayonnement dur
P promieniowanie twarde
R жёсткое излучение

HARD SPHERE CORE MODEL. *See* HARD SPHERE POTENTIAL.

1542
HARD SPHERE POTENTIAL (hard core potential, hard sphere core model).
Model intermolecular interaction potential

$$u(r) = \begin{cases} \infty & \text{for } r < a \\ 0 & \text{for } r > a \end{cases}$$

where u = potential energy of pair of molecules (particles), r = intermolecular (interparticle) distance and a = the closest approach distance.

D starres Kugel-Potential, starres Kugel-Modell
F potentiel de sphères rigides
P potencjał sztywnej kuli, model sztywnych kul
R потенциал взаимодействия упругих шаров

1543
HARMONIC OSCILLATOR.

Oscillator in which acting force directed towards equilibrium position is proportional to displacement from equilibrium position.

D harmonischer Oszillator
F oscillateur harmonique
P oscylator harmoniczny
R гармонический осциллятор

1544
HARTREE-FOCK METHOD
(self-consistent field method, SCF method).
Approximate quantum chemistry method based on one-electron approximation. Coulomb electron-electron interaction is replaced by effective, self-consistent interaction potential following from variational minimization of total average energy of many-electron system.

D Hartreesche Methode des Selfconsistentfield
F méthode du champ moléculaire self-consistent
P metoda pola samouzgodnionego, metoda Hartree i Focka
R метод самосогласованного поля

1545
HEAT.
Exchange of internal energy or enthalpy defined by appropriate equations which balance these quantities. In open systems concept of heat is not explicit.
Remark: Heat is sometimes, traditionally, identified with internal energy or enthalpy (i.e. kinetic energy of random movement of particles and potential energy of their mutual interactions).

D Wärme
F chaleur
P ciepło
R теплота

1546
HEAT BALANCE.
Comparison of total amount of energy supplied in form of heat to system in a given process with total sum of energy consumed in all particular conversions.

D Wärmebilanz
F bilan thermique
P bilans cieplny
R тепловой баланс

1547
HEAT CAPACITY.
Product of specific heat capacity and mass of system.

D Wärmekapazität
F capacité calorifique
P pojemność cieplna
R теплоёмкость

1548
HEAT CAPACITY AT CONSTANT PRESSURE, C_p.
Extensive thermodynamic function

$$C_p = \left(\frac{\partial H}{\partial T} \right)_p$$

where H = enthalpy, T = thermodynamic temperature, and p = pressure.

D Wärmekapazität bei konstantem Druck
F capacité calorifique à pression constante
P pojemność cieplna pod stałym ciśnieniem
R теплоёмкость при постоянном давлении

1549
HEAT CAPACITY AT CONSTANT VOLUME, C_V.
Extensive thermodynamic function

$$C_V = \left(\frac{\partial U}{\partial T} \right)_V$$

where U = internal energy,
T = thermodynamic temperature, and
V = volume.

D Wärmekapazität bei konstantem Volumen
F capacité calorifique à volume constant
P pojemność cieplna w stałej objętości
R теплоёмкость при постоянном объёме

1550
HEAT CONDUCTION.
Energy transport in form of heat due to temperature gradient.

D Wärmeleitung
F conduction thermique
P przewodzenie ciepła
R теплопроводность

1551
HEAT ENGINE (thermodynamic engine).
System in which cyclic process takes place such that in each cycle some portion of energy is transferred to surroundings by way of work and some portion is accepted in form of heat.

D Wärmekraftmaschine
F machine thermique
P maszyna cieplna, silnik cieplny
R тепловая машина

1552
HEAT OF ADSORPTION.
Heat effect accompanying process of adsorption.

D Adsorptionswärme
F chaleur d'adsorption
P ciepło adsorpcji
R теплота адсорбции

1553
HEAT OF CHANGE OF PHASE (latent heat).
Enthalpy change of system associated with isothermal and isobaric transition of a given amount of substance from one of its phases to another in equilibrium conditions.

D Umwandlungswärme, latente Wärme
F chaleur de changement de phase, chaleur latente
P ciepło przemiany fazowej, ciepło utajone
R теплота фазового перехода, скрытая теплота (процесса)

1554
HEAT OF COMBUSTION.
Heat effect accompanying complete combustion of defined amount of a given substance in oxygen to carbon dioxide, water (liquid) and corresponding oxidation products of other elements constituting substance under consideration. Numerical values of heats of combustion are usually given for $t = 25°C$ and $p = 1.013\,25 \cdot 10^{-1}$ MPa.

D Verbrennungswärme
F chaleur de combustion
P ciepło spalania
R теплота сгорания

1555
HEAT OF CONDENSATION.
Change of enthalpy associated with isothermal and isobaric transition of a given amount of substance from gaseous state either to liquid state (heat of liquefaction) or to solid state (heat of condensation to solid phase).

D Kondensationswärme
F chaleur de condensation
P ciepło kondensacji
R теплота конденсации

1556
HEAT OF CRYSTALLIZATION.
Change of enthalpy associated with isothermal and isobaric transition of a given amount of substance from liquid or gaseous phase to crystalline state.

D Kristallisationswärme
F chaleur de cristallisation
P ciepło krystalizacji
R теплота кристаллизации

1557
HEAT OF DILUTION (integral heat of dilution).
Enthalpy change of system associated with change in concentration of one mole of a given solute within defined interval of concentrations c_1 and c_2, caused by diluting solution under consideration with appropriate amount of solvent.

D (integrale) Verdünnungswärme
F chaleur (intégrale) de dilution
P ciepło rozcieńczania (całkowite)
R (интегральная) теплота разведения

1558
HEAT OF DISSOCIATION.
Change of enthalpy in process of breaking bonds in molecule, e.g. heat of atomization, heat of electrolytic dissociation. In photochemistry chemical heat of dissociation D_{ch} is obtained by subtracting vibrational zero-point energy E_0 of molecule from energy corresponding to dissociative vibrational level D, viz.:

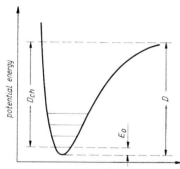

D Dissoziationswärme
F chaleur de dissociation
P ciepło dysocjacji
R теплота диссоциации

1559
HEAT OF EVAPORATION (heat of vaporization).
Increase of enthalpy of system during isothermal and isobaric transition (at saturated vapour pressure) of a given amount of substance from liquid state to saturated vapour state.

D Verdampfungswärme
F chaleur de vaporisation
P ciepło parowania
R теплота испарения, теплота парообразования

1560
HEAT OF FORMATION.
Change in enthalpy or internal energy of system in process of formation of one mole of chemical compound from corresponding constituent elements at $t = 25°C$ and $p = 1.013\,25 \cdot 10^{-1}$ MPa, and of their aggregate states that correspond to these conditions. Standard heats of formation of elements in their most stable allotropic form are taken arbitrarily equal to zero.

D Bildungswärme
F chaleur de formation
P ciepło tworzenia
R теплота образования

1561
HEAT OF FUSION.
Enthalpy increase of system during isothermal and isobaric transition of defined amount of substance from solid phase to liquid under conditions of phase equilibrium.

D Schmelzwärme
F chaleur de fusion
P ciepło topnienia
R теплота плавления

1562
HEAT OF GELATION.
Amount of heat released during conversion of sol into gel.

D Gelatinierungswärme
F chaleur de la gélification
P ciepło żelowania
R теплота желатинирования, теплота гелеобразования, теплота застуднения

1563
HEAT OF HYDRATION.
Heat of solvation with water as solvent.

D Hydratationswärme
F chaleur d'hydratation, enthalpie d'hydratation
P ciepło hydratacji
R теплота гидратации

1564
HEAT OF MIXING (molar), Q^M.
Change of enthalpy of system associated with formation of one mole of mixture from its liquid or gaseous components.

D Mischungswärme
F chaleur de mélange
P ciepło mieszania
R теплота смешения

1565
HEAT OF NEUTRALIZATION.
Enthalpy change of system in reaction of neutralization of one gram equivalent of acid by base, in dilute solution.

D Neutralisationswärme
F chaleur de neutralisation
P ciepło zobojętniania
R теплота нейтрализации

1566
HEAT OF RADIOACTIVITY.
Heat released continuously and spontaneously by any radioactive substance; it depends on half-life of a nuclide, kind of radiation and conditions of absorption. For radium, in equilibrium with its first four daughters, heat of radioactivity is about 175 J/kg s.

D radioaktive Strahlungswärme
F débit de chaleur du rayonnement
P ciepło promieniowania
R теплота радиоактивного излучения

1567
HEAT OF REACTION.
Heat evolved or absorbed during any
chemical reaction carried out either
isochorically (reaction energy) or
isobarically (reaction enthalpy) in which
the only possible work which may be done
by (or on) system is external work of
changing its volume, and all reactants and
products are at the same temperature
(usually, $t = 25°C$, $p = 1.013\ 25 \cdot 10^{-1}$ MPa).
Numerical value of heat of reaction
always corresponds to reaction of
substances in amounts as given by
stoichiometric equation.
D Reaktionswärme
F chaleur de réaction
P ciepło reakcji
R теплота реакции, тепловой эффект
реакции

1568
HEAT OF SOLUTION (integral heat of
solution).
Enthalpy change of system in process of
dissolving n_2 moles of a given substance
in n_1 moles of solvent, per 1 mole of solute
under consideration.
D (integrale) Lösungswärme
F chaleur (intégrale) de dissolution
P ciepło rozpuszczania (całkowite)
R (интегральная) теплота растворения

1569
HEAT OF SOLVATION.
Enthalpy change of system for one mole
of a given solute due to mutual
solute-solvent intermolecular interactions
in investigated solution.
D Solvatationswärme
F chaleur de solvatation
P ciepło solwatacji
R теплота сольватации

1570
HEAT OF SUBLIMATION.
Enthalpy increase of system during
isothermal and isobaric (at saturated
vapour pressure) transition of defined
amount of substance from solid phase into
saturated vapour state.
D Sublimationswärme
F chaleur de sublimation
P ciepło sublimacji
R теплота сублимации

HEAT OF VAPORIZATION. *See* HEAT
OF EVAPORATION.

1571
HEAT OF WETTING.
Heat effect associated with wetting of
solid body by liquid, calculated for unit
mass of wetted body.
D Benetzungswärme
F chaleur de mouillage
P ciepło zwilżania
R теплота смачивания

HEAVY HYDROGEN. *See* DEUTERIUM.

1572
HEAVY OXYGEN.
Isotope of oxygen. Atomic number 8.
Mass number 18.
D schwerer Sauerstoff
F oxygène lourd
P tlen ciężki
R тяжёлый кислород

1573
HEAVY PLATINUM METALS (third
triad).
Three chemical elements grouped
horizontally in eighth sub-group of
periodic system (in period VI): osmium,
iridium, platinum.
D schwere Platinmetalle
F métaux lourds de la mine de platine,
triade du platine
P platynowce ciężkie, osmowce, triada
osmu
R тяжёлые платиновые металлы

1574
HEAVY WATER, D_2O.
Water in the molecules of which hydrogen
atoms are replaced by deuterium atoms.
D schweres Wasser
F eau lourde
P woda ciężka
R тяжёлая вода

1575
HEDVALL EFFECT.
Increase of catalytic activity of solids
in vicinity of their phase transition
temperatures; also for ferromagnets in
vicinity of their Curie temperatures.
D Hedvall-Effekt
F effet Hedvall
P efekt Hedvalla
R эффект Гедвалла

HEISENBERG RELATION. *See*
UNCERTAINTY PRINCIPLE.

1576
HELICITY.
Quantum number determined by
projection of spin of particle onto
direction of its momentum.
D Helizität
F hélicité
P skrętność, spiralność
R спиральность, спиральное квантовое
число

1577
HELIUM, He.
Atomic number 2.
D Helium
F hélium
P hel
R гелий

HELIUM GROUP. *See* INERT GASES.

1578
HELL-VOLHARD-ZELINSKY
REACTION.
Preparation of α-chloro- or α-bromoacids
by treating carboxylic acids with chlorine
or bromine in presence of minute amount
of red phosphorus or appropriate
phosphorus trihalide, e.g.

$$RCH_2COOH \xrightarrow[Br_2]{P} RCH_2COBr \xrightarrow{Br_2}$$

$$RCHBrCOBr \xrightarrow{H_2O} RCHBrCOOH$$

D Hell-Volhard-Zelinsky α-Halogenierung
F réaction de Hell-Volhard-Zelinsky
P chlorowcowanie metodą Hella, Volharda
i Zielińskiego
R реакция Гелля-Фольгарда-Зелинского

HELMHOLTZ DOUBLE LAYER. *See*
COMPACT DOUBLE LAYER.

1579
HELMHOLTZ FREE ENERGY (free
energy, Helmholtz function, available
energy, Helmholtz potential, available
work at constant temperature, work
function), F.
Extensive thermodynamic function
$F = U - TS$, where U = internal energy,
T = thermodynamic temperature, and
S = entropy.
D freie Energie (nach Helmholtz)
F énergie libre (á temperature constante),
fonction de Helmholtz
P energia swobodna (Helmholtza),
potencjał termodynamiczny
izotermiczno-izochoryczny, potencjał
termodynamiczny izochoryczny,
potencjał termodynamiczny w stałej
objętości
R изохорно-изотермический потенциал,
изохорный потенциал, свободная
энергия

HELMHOLTZ FUNCTION. *See*
HELMHOLTZ FREE ENERGY.

HELMHOLTZ POTENTIAL. *See*
HELMHOLTZ FREE ENERGY.

1580
HELMHOLTZ'S THEORY OF DOUBLE
LAYER.
Theory which describes model of
electrical double layer at electrode —
electrolyte interface; on electrode surface

there exists excess of ions or electrons
which is accompanied by excess of ions of
opposite charge on solution side; this
system is analogous to plate condenser.
D Helmholtzsche Theorie der
Doppelschicht
F théorie de la couche double de
Helmholtz
P teoria budowy warstwy podwójnej
Helmholtza
R теория двойного (электрического) слоя
Гельмгольца

1581
HENDERSON'S EQUATION.
Formula which describes diffusion
potential, ε_d derived from assumption that
in boundary layer, concentrations of all
ions change linearly in direction of
diffusion

$$\varepsilon_d = -\frac{RT}{F} \frac{\sum \frac{\lambda_i}{z_i}(m_i^{II} - m_i^I)}{\sum \lambda_i(m_i^{II} - m_i^I)} \ln \frac{\sum \lambda_i m_i^{II}}{\sum \lambda_i m_i^I}$$

where m_i^I and m_i^{II} = concentrations of
ion i in neighbouring solutions I and II
respectively, λ_i = ionic conductivity.
D Hendersonsche Gleichung
F équation de Henderson
P równanie Hendersona
R формула Гендерсона

1582
HENDERSON'S METHOD FOR
CALCULATION OF DIFFUSION
POTENTIAL.
Method based on assumption of linear
change in direction of diffusion of
concentration of each type of ions in
boundary layer between two electrolytic
solutions.
D Hendersonsche Berechnungsmethode des
Diffusionspotentials
F méthode de l'évaluation du potentiel de
jonction d'Henderson
P metoda Hendersona obliczania
potencjału dyfuzyjnego
R метод Гендерсона расчёта
диффузионного потенциала

1583
HENRY'S LAW.
Mass m of gas dissolved at constant
temperature in defined volume of liquid
is proportional to partial pressure p of
that gas over solution; $m = kp$, where
k = proportionality factor, which depends
on gas and temperature.
D Henrysches Gesetz
F loi de Henry
P prawo Henry'ego
R закон Генри

1584
HERMAN-MAUGUIN SYMBOLS.
International symbols for representation of point and spatial symmetry of crystal lattice.
D Herman-Mauguin Symbole
F symboles d'Herman-Mauguin
P symbole Hermana i Mauguina, symbole międzynarodowe
R символы Германа-Могена, (международные) кристаллографические символы

1585
HESS'S LAW (law of constant heat summation).
Total heat effect in chemical reaction and other processes carried out at constant volume or at constant pressure, if no useful work is done, is independent of path of process but depends only on starting and final states of system under consideration.
D Hesssches Gesetz
F loi de Hess
P prawo Hessa
R закон Гесса

1586
HETEROATOM.
Atom of some element other than carbon such as, for instance, oxygen, present in cyclic or open-chain organic compound.
D Heteroatom
F hétéroatome
P heteroatom
R гетероатом

1587
HETEROAZEOTROPE (heteroazeotropic mixture).
Mixture composed of two or more liquid phases, distilling without change of mean composition.
D Heteroazeotrop, heteroazeotrope Mischung
F hétéroazéotrope
P heteroazeotrop, mieszanina heteroazeotropowa
R гетероазеотроп, гетероазеотропная смесь

HETEROAZEOTROPIC MIXTURE. *See* HETEROAZEOTROPE.

1588
HETEROAZEOTROPIC POINT.
Point on phase diagram describing composition and boiling temperature, or composition and vapour pressure of heteroazeotrope.
D heteroazeotroper Punkt
F point hétéroazéotropique
P punkt heteroazeotropowy
R гетероазеотропная точка

1589
HETEROAZEOTROPIC SYSTEM.
Azeotropic system of constituents forming heteroazeotrope.
D heteroazeotropes System
F système hétéroazéotropique
P układ heteroazeotropowy
R гетероазеотропная система

1590
HETEROCYCLIC RING.
Ring containing atoms of different elements, e.g. furane ring

CH—CH
‖ ‖
CH CH
\ /
O

D heterocyclischer Ring
F noyau hétérocyclique
P pierścień heterocykliczny
R гетероцикл

HETERODISPERSE SYSTEM. *See* POLYDISPERSE SYSTEM.

1591
HETERODYNE-BEAT METHOD.
Method of measurement of permittivity by recording condenser capacity with and without investigated substance.
D Überlagerungsmethode, Schwebungsmethode
F méthode des battements
P metoda dudnień
R метод биений

1592
HETEROGENEOUS CATALYSIS.
Catalysis in systems in which catalyst forms separate phase, most often solid.
D heterogene Katalyse
F catalyse hétérogène
P kataliza wielofazowa, kataliza heterogeniczna
R гетерогенный катализ

1593
HETEROGENEOUS ISOTOPIC EXCHANGE.
Isotopic exchange occurring in a polyphase system.
D heterogener Isotopenaustausch
F échange isotopique hétérogène
P wymiana izotopowa niejednorodna
R гетерогенный изотопный обмен

1594
HETEROGENEOUS SYSTEM (multiphase system).
Thermodynamic system composed of finite (but greater than one) number of homogeneous phases.
D heterogenes System, Mehrphasensystem
F système hétérogène

P układ heterogeniczny, układ
   wielofazowy
R гетерогенная система, многофазная
   система

HETEROLYSIS. *See* HETEROLYTIC
DISSOCIATION.

1595
HETEROLYTIC DISSOCIATION
(heterolysis, heterolytic fission).
Fission of covalent bond in molecule,
resulting in formation of two ions bearing
different charges.
D heterolytische Dissoziation, Heterolyse
F dissociation hétérolytique, hétérolyse
P rozpad heterolityczny, heteroliza
R гетеролитический распад, гетеролиз

HETEROLYTIC FISSION. *See*
HETEROLYTIC DISSOCIATION.

1596
HETEROMETRIC TITRATION
(turbidimetric titration).
Titration method with end point
determined from point of inflection on
curve representing dependence of opacity
changes on volume of added reagent.
D heterometrische Titration,
   turbidimetrische Titration
F . . .
P miareczkowanie turbidymetryczne,
   miareczkowanie heterometryczne
R турбидиметрическое титрование

1597
HETERONUCLEAR DIATOMIC
MOLECULE.
Molecule formed of two different atoms,
e.g. HCl.
D heteronukleares zweiatomiges Molekül
F molécule diatomique hétéronucléaire
P cząsteczka dwuatomowa heterojądrowa
R гетеронуклеарная двухатомная
   молекула

HETEROPOLAR BOND. *See* IONIC
BOND.

1598
HETEROPOLYACID.
Polyacid containing more than one
acid-forming element, e.g.
phospho-12-molybdic acid $H_3(PMo_{12}O_{40})$.
D Heteropolysäure
F hétéropolyacide
P heteropolikwas
R гетерополикислота

1599
HETEROPOLYMERIZATION.
Copolymerization of monomers of various
structures, when one monomer is not able
to undergo autopolymerization or
polymerizes with difficulty.

D Heteropolymerisation
F hétéropolymérisation
P heteropolimeryzacja
R гетерополимеризация

1600
HETEROZEOTROPIC SYSTEM.
Zeotropic system of limited mutual
miscibility of liquid constituents.
D heterozeotropes System
F système hétérozéotropique
P układ heterozeotropowy
R гетерозеотропная система

1601
HIGH-ELASTIC STATE (rubber-like
state *of polymers*).
State in which elastic deformations
involve dislocation of particular segments
only of macromolecule chain. High-elastic
state is achieved by heating polymer in
vitreous state to temperatures higher than
so called vitrification temperature, T_v.
Relaxation of strains in this state is
appreciably faster than in vitreous state.
D hochelastischer Zustand,
   kautschukelastischer Zustand
F état d'élasticité élevée
P stan (wysoko)elastyczny, stan
   kauczukopodobny
R высокоэластическое состояние

1602
HIGHER ORDER PHASE TRANSITION.
Transition taking place in some range of
temperature and characterized by
anomalous process of change of some
physico-chemical properties, e.g. molal
heat capacity, magnetic susceptibility.
D . . .
F . . .
P przemiana (fazowa) wyższego rzędu,
   przemiana fazowa ciągła
R . . .

1603
HIGHER ORDER REACTIONS.
Reaction with kinetic equation
characterized by sum of exponents in
which concentrations of reactants occur
being higher than three.
D Reaktion höherer Ordnung
F . . .
P reakcja wyższego rzędu
R реакция высшего порядка

HIGH FREQUENCY TITRATION. *See*
OSCILLOMETRIC TITRATION.

HIGH MOLECULAR WEIGHT
COMPOUND. *See* MACROMOLECULAR
COMPOUND.

HIGH-SPIN COMPLEX. *See* SPIN-FREE
COMPLEX.

HINDERED ROTATION. *See*
RESTRICTED ROTATION.

1604
HINDERED ROTATOR (restricted rotator).
Rigid rotator which during its motion around centre of mass has to overcome some potential barrier.
D abgehärteter Rotator
F rotateur restricte, rotateur empêché
P rotator zahamowany
R заторможённый ротатор

1605
HINDERING EFFECT (effect of steric hindrance).
Specific effect exerted by steric system of atoms in molecule, opposing conjugational or conformational changes, decreasing reaction rate and finally forcing reaction to proceed in specific direction.
D sterisch hindernde Wirkung
F . . .
P efekt przeszkody przestrzennej
R . . .

1606
HITTORF'S METHOD FOR DETERMINATION OF TRANSPORT NUMBERS.
Method based on determination of changes of anolite and catholite resulting from different mobilities and hence different transport numbers of anion and cation.
D Hittorfsche Bestimmungsmethode der Überführungszahlen
F méthode de la détermination des nombres de transport de Hittorf
P metoda Hittorfa oznaczania liczb przenoszenia
R метод Гитторфа определения чисел переноса

HOESCH SYNTHESIS. *See* HOUBEN-HOESCH SYNTHESIS.

1607
HOFMANN DEGRADATION.
Exhausting methylation of amine and thermal degradation of quaternary ammonium hydroxide to tertiary amine and alkene, e.g.

$$RCH_2CH_2NH_2 \xrightarrow{CH_3I} RCH_2CH_2\overset{\oplus}{N}(CH_3)_3\overset{\ominus}{I} \xrightarrow{Ag_2O}{H_2O}$$

$$RCH_2CH_2\overset{\oplus}{N}(CH_3)_3\overset{\ominus}{OH} \xrightarrow{100-200°C} RCH=CH_2 + N(CH_3)_3 + H_2O$$

D Hofmann-Abbau (quartärer Ammoniumhydroxid)
F . . .
P degradacja Hofmanna
R расщепление по Гофману

1608
HOFMANN-MARTIUS REARRANGEMENT OF ALKYLOANILINES.

Rearrangement of N-alkylaniline derivatives into C-alkyl derivatives, effected by heating aniline hydrogen chloride or aniline hydrogen bromide to 300°C

$$2C_6H_5\overset{\oplus}{N}H_2RX^{\ominus} \xrightarrow{250-300°C}$$

$$o\text{-}R\text{-}C_6H_4\overset{\oplus}{N}H_3X^{\ominus} + p\text{-}R\text{-}C_6H_4\overset{\oplus}{N}H_3X^{\ominus}$$

D Hofmann-Martius (N-Alkylanilin→ →C-Alkylanilin)-Umlagerung
F transposition de Hofmann et Martius
P przegrupowanie Hofmanna i Martiusa
R . . .

1609
HOFMANN REACTION.
Rearrangement of unsubstituted amides into primary amines having one carbon atom less, effected by bromine (or chlorine) in aqueous sodium hydroxide, e.g.

$$RCONH_2 \xrightarrow{Br_2, NaOH} RCONHBr \xrightarrow{-H^{\oplus}}$$

$$R\text{-}\overset{O}{\underset{||}{C}}\text{-}\overset{\ominus}{N}\text{-}Br \xrightarrow{-Br^{\ominus}} R\text{-}N=C=O \xrightarrow{H_2O}$$

$$R\text{-}NH_2 + CO_2$$

D Hofmann Carbonsäureamid→ Amin-Abbau
F dégradation d'Hofmann
P degradacja amidów Hofmanna
R перегруппировка Гофмана

1610
HOFMANN RULE.
Decomposition of quaternary ammonium base gives a predominance of the least substituted alkene (in elimination reactions), e.g.

$$\left[CH_3CH_2\text{-}\overset{CH_3}{\underset{CH_3}{N}}\text{-}CH_2CH_2CH_3 \right]^{\oplus} OH^{\ominus} \begin{array}{l} \nearrow CH_2=CH_2 \text{ main product} \\ \searrow CH_3CH=CH_2 \text{ by-product} \end{array}$$

D Hofmann-Regel
F règle d'Hofmann
P reguła Hofmanna
R правило Гофмана

1611
HOFMEISTER SERIES (lyotropic series).
Definite order of arrangement of ions according to their capability of bonding dispersion medium as function of their ionic radius, i.e. of their hydration. Univalent cations form Li^+, Na^+, K^+, Rb^+, Cs^+ series; divalent cations form Mg^{2+}, Ca^{2+}, Sr^{2+}, Ba^{2+} series; univalent anions form Cl^-, Br^-, NO_3^-, I^-, CNS^- series; Hofmeister series may be observed in coagulation, adsorption, and other processes.

D Hofmeistersche Reihen, lyotrope Reihen
F séries d'ions d'Hofmeister, séries
  lyotropiques
P szeregi Hofmeistera, szeregi liotropowe
R лиотропные ряды

1612
HOLD-BACK CARRIER.
Carrier added to solution in order to
retain a microcomponent in solution, e.g.
to prevent coprecipitation or sorption of
radioactive substances on to precipitates
to be separated.

D Rückhaltträger
F entraîneur de rétention, anti-entraîneur,
  entraîneur en retour
P nośnik zwrotny, nośnik zatrzymujący
R удерживающий носитель,
  антиноситель, обратный носитель

1613
HOLE (positive hole).
Empty quantum state in valence band.
It may move inside crystal under
influence of applied electric field.

D Defektelektron
F trou, lacune, vacance d'électron
P dziura elektronowa, dziura dodatnia,
  luka elektronowa
R дырка

1614
HOLE CONDUCTIVITY
(gap-conductivity).
Transfer of electric charge through
crystal under external electric field.
Electrons of not fully occupied valence
band move to unoccupied quantum levels
(positive holes) in range of this band, in
direction opposite to vector of electric
field, which is equivalent to movement
of positive charges in direction of vector
of electric field.

D Defektelektronen-Leitung
F conductivité de lacune
P przewodnictwo dziurowe
R дырочная проводимость

1615
HOLLOW-CATHODE LAMP
(hollow-cathode tube).
Lamp fed by direct current and filled with
rarified gas with cathode in shape of
hollow cylinder. Inside of cylinder emits
radiation characteristic for cathode
material.

D Hohlkathodenlampe
F lampe à cathode creuse
P lampa z katodą wnękową
R лампа с полым катодом

HOLLOW-CATHODE TUBE. *See*
HOLLOW-CATHODE LAMP.

1616
HOLMIUM, Ho.
Atomic number 67.
D Holmium
F holmium
P holm
R гольмий

1617
HOMOAZEOTROPE.
Homogeneous mixture of liquids, which
distils without change of composition.

D Homoazeotrop
F homoazéotrope
P homoazeotrop, mieszanina
  homoazeotropowa
R гомоазеотроп

1618
HOMOAZEOTROPIC SYSTEM.
Azeotropic system of constituents forming
homoazeotrope.

D homoazeotropes System
F système homoazéotropique
P układ homoazeotropowy
R гомоазеотропная система

1619
HOMOCYCLIC RING.
Ring composed of atoms of the same
element.

D homocyclischer Ring
F noyau homocyclique, noyau isocyclique
P pierścień izocykliczny, pierścień
  homocykliczny
R гомоциклическое кольцо,
  изоциклическое кольцо

1620
HOMOGENEOUS CATALYSIS.
Catalysis in systems where catalyst and
reactants constitute one phase.

D homogene Katalyse
F catalyse homogène
P kataliza jednofazowa, kataliza
  homogeniczna, kataliza jednorodna
R гомогенный катализ

1621
HOMOGENEOUS DISTRIBUTION
COEFFICIENT.
Magnitude D in Berthelot-Nernst
distribution law indicating how many
times ratio of content of a microcomponent
to a macrocomponent in a crystalline
precipitate is greater ($D > 1$) or smaller
($D < 1$) than analogous ratio in solution
containing precipitate.

D Kristallisationskoeffizient
F coefficient de recristallisation
P współczynnik krystalizacji
R коэффициент кристаллизации

HOMOGENEOUS DISTRIBUTION LAW.
See BERTHELOT-NERNST
DISTRIBUTION LAW.

1622
HOMOGENEOUS ISOTOPIC EXCHANGE.
Isotopic exchange occurring in
a monophase system.
D homogener Isotopenaustausch
F échange isotopique homogène
P wymiana izotopowa jednorodna
R гомогенный изотопный обмен

1623
HOMOGENEOUS PRECIPITATION
(precipitation from homogeneous solution).
Slow precipitation of solid by action of
suitable reagent which results from
chemical homogeneous reaction in whole
solution.
D Fällung aus homogener Lösung
F précipitation en milieu homogène
P strącanie homogeniczne, strącanie
  w roztworach homogenicznych
R гомогенное осаждение

1624
HOMOGENEOUS SYSTEM (one phase
system).
Thermodynamic system composed solely
of one homogeneous phase.
D homogenes System, Einphasensystem
F système homogène, système monophasé,
  système uniforme
P układ homogeniczny, układ
  jednofazowy
R гомогенная система, однофазная
  система

1625
HOMOGENIZATION OF EMULSIONS.
Crushing process of particles of disperse
system in emulsions.
D Homogenisation der Emulsionen,
  Homogenisierung der Emulsionen
F homogénéisation des émulsions
P homogenizacja emulsji
R гомогенизация эмульсий

1626
HOMOLOGOUS LINES (in emission
spectral analysis).
Lines whose intensity ratio is independent
on excitation conditions.
D homologe Linien
F raies homologues
P linie homologiczne
R гомологические линии

1627
HOMOLOGOUS SERIES.
Series of compounds with similar chemical
structure; each member of series differs
from the next by one (or more) definite
grouping of atoms added to molecule; e.g.
—CH₂— group in alkane carbon chain.
D homologe Reihe
F série homologue

P szereg homologiczny
R гомологический ряд

1628
HOMOLOGUES.
Compounds which are members of the
same homologous series.
D Homologe(n)
F homologues
P homologi
R гомологи

1629
HOMOLOGY.
Systematic change of physical properties
and similarity of chemical properties
displayed by members of homologous
series.
D Homologie
F homologie
P homologia
R гомология

HOMOLYSIS. See HOMOLYTIC
DISSOCIATION.

1630
HOMOLYTIC DISSOCIATION (homolysis,
homolytic fission).
Fission of covalent bond in molecule,
resulting in formation of two atoms or
free radicals.
D homolytische Dissoziation, Homolyse
F dissociation homolytique, homolyse
P rozpad homolityczny, homoliza
R гомолитический распад, гомолиз

HOMOLYTIC FISSION. See HOMOLYTIC
DISSOCIATION.

1631
HOMONUCLEAR DIATOMIC
MOLECULE.
Molecule formed of two identical atoms,
e.g. N₂.
D homonukleares zweiatomiges Molekül
F molécule diatomique homonucléaire
P cząsteczka dwuatomowa homojądrowa
R гомонуклеарная двухатомная молекула

HOMOPOLAR BOND. See COVALENT
BOND.

1632
HOMOPOLYMER.
Product of polymerization of single species
of monomer.
D Homopolymer, Homopolymerisat
F homopolymère
P homopolimer
R гомополимер

1633
HOMOPOLYMERIZATION.
Polymerization of single species of
monomer.

D Homopolymerisation
F homopolymérisation
P homopolimeryzacja
R гомополимеризация

1634
HOT-ATOM CHEMISTRY.
Branch of chemistry concerned with study of reactions occurring in systems containing hot atoms.
D Chemie der heißen Atome
F chimie des atomes chauds
P chemia atomów gorących
R химия горячих атомов

1635
HOT ATOMS (recoil atoms).
Atoms formed in nuclear reactions, and of energy corresponding to a temperature of the order of 10^5 K, and taking part in numerous specific chemical reactions.
D heiße Atome, Rückstoßatome
F atomes chauds, atomes de recul
P atomy gorące, atomy odrzutu
R горячие атомы, атомы отдачи

1636
HOT CELL.
Space or set of rooms, shielded biologically, used for handling by remote control of highly radioactive materials for purpose of chemical and mechanical processing, investigation, reloading etc.
D heiße Zelle
F cellule à haut activité, enceinte étanche
P komora gorąca
R горячая камера

1637
HOT LABORATORY.
Laboratory designed and equipped for handling of highly radioactive materials.
D heißes Laboratorium
F laboratoire chaud, laboratoire de haute activité
P laboratorium gorące
R горячая лаборатория

1638
HOT PLASMA.
Plasma of temperature of order of several milions kelvins.
D heißes Plasma
F plasma chaud
P plazma gorąca
R высокотемпературная плазма

1639
HOT RADICAL.
Free radical of substantial kinetic energy, higher than its share of thermal movement for the particular medium and temperature.

D heißes Radikal
F radical chaud
P rodnik gorący
R горячий радикал

1640
HOT REACTIONS.
Chemical reactions in which hot atoms of energy considerably in excess of 0.025 eV, characterized by their independence of surrounding temperature, state of aggregation and presence of radical scavengers take part.
D heiße Reaktionen
F reactions chaudes
P reakcje gorące
R горячие реакции

1641
HOUBEN-FISCHER NITRILE SYNTHESIS.
Preparation of aromatic cyanides by condensation of hydrocarbons with trichloroacetonitrile, in presence of hydrogen chloride and aluminium chloride, followed by alkaline hydrolysis of resulting ketoimine hydrochloride, e.g.

D Houben-Fischer Nitril-Synthese
F réaction d'Houben et Fischer
P synteza Houbena i Fischera
R реакция Губена-Фишера

1642
HOUBEN-HOESCH SYNTHESIS (Hoesch synthesis).
Preparation of aromatic hydroxyketones by condensation of phenols (or their ethers) with nitriles in presence of hydrogen chloride and zinc chloride (or aluminium chloride), e.g.

D Hoesch-(Houben Phenolketon-)Synthese
F synthèse d'Houben-Hoesch
P reakcja (Houbena i) Hoescha
R синтез Гёша

1643
H-THEOREM.
If at any time value of some function H relevant for system or statistical ensemble ascribed to it, is greater than minimum value of H, then it is very likely that value of H will increase with time. Minimum value of H is observed at equilibrium conditions. For systems where H-theorem applies, it gives a simple, although incomplete method of describing their evolution towards equilibrium.
D H-Theorem
F théorème H
P twierdzenie H
R H-теорема

1644
HÜCKEL'S EQUATION.
Semiempirical formula which describes mean activity coefficient $f_{\pm}$ of strong electrolyte:

$$\lg f_{\pm} = - \frac{A|z_+z_-|\sqrt{J}}{1+Ba\sqrt{J}} + CJ$$

where a and C = empirical constants; for other symbols, *see* Debye-Hückel's limiting law.
D Hückelsche Gleichung
F formule de Hückel
P równanie Hückla
R уравнение Гюккеля

1645
HÜCKEL'S RULE ($4n+2$ rule).
Aromatic character is displayed only by organic compounds containing ($4n+2$) π-electrons (where n is an integer).
D Hückel-Regel, aromatische Stabilitäts-Regel
F règle de Hückel
P reguła Hückla
R правило Гюккеля

1646
HUDSON ISOROTATION RULES.
Rules correlating rotation of the optically active compounds with their structure; these rules apply to anomers of simple sugars, glycosides and γ-lactones of aldonic acids.
D Isorotations-Regel von Hudson
F règle d'Hudson
P reguły izorotacji Hudsona
R правила Хадсона

1647
HUGONIOT EQUATION.
Equation $\Delta U_m = -\left(p + \frac{1}{2}\Delta p\right)\Delta V_m$ which determines dependence of molar internal energy change ΔU_m on pressure Δp and molar volume ΔV_m changes in shock wave.
D Hugoniot-Gleichung, Hugoniot-Relation
F équation de Hugoniot
P równanie Hugoniota, równanie adiabaty uderzeniowej
R уравнение ударной адиабаты, адиабата Гюгоньо

1648
HUME-ROTHERY'S PHASES.
Phases which occur in metallic alloys where besides phases of structures characteristic of pure constituents, also phase of regular space-centred structure, phase of complex regular lattice containing 52 atoms, and phase of hexagonal lattice and closest packing are to be seen.
D Hume-Rothery-Phasen
F phases de Hume-Rothery
P fazy Hume-Rothery'ego, związki Hume-Rothery'ego, związki elektronowe, fazy pośrednie
R фазы Юм-Розери, электронные соединения

1649
HUND'S RULES.
Rules which enable the lowest energy term of atom or ion to be determined for Russell-Saunders coupling, provided electronic configuration is known.
D Hundsche Regeln
F règles de Hund
P reguły Hunda
R правила Гунда

1650
HUNSDIECKER-BORODIN REACTION.
Decomposition of silver salts of carboxylic acids on treatment with halogens in anhydrous medium, resulting in formation of alkyl halides and carbon dioxide, e.g.

$$RCOOAg + X_2 \longrightarrow RX + CO_2 + AgX$$

D Hunsdiecker-Reaktion, (Hunsdiecker-Borodin) Silbersalz-Decarboxylierung
F . . .
P reakcja Hunsdieckera (i Borodina)
R реакция Хунсдикера-Бородина

1651
HYBRIDIZATION.
Mixing of atomic valence orbitals with different orbital quantum number l. Orbitals constructed in this way have more pronounced directional properties than original ones. Angles between them are different also from original ones.
D Hybridisierung, Bastardisierung
F hybridation des orbitales
P hybrydyzacja orbitali
R гибридизация орбитали

1652
HYBRIDIZED ORBITAL.
Atomic orbital constructed by hybridization of atomic orbitals with different orbital quantum numbers. It has more pronounced directional properties than component orbitals; e.g. sp^3 hybridized orbitals constructed of $2s$, $2p_x$, $2p_y$ and $2p_z$ atomic orbitals of carbon atom are used to explain geometry of methane.
D Hybridorbital, Bastardorbital
F orbitale hybride
P orbital zhybrydyzowany, hybryd
R гибридная орбиталь

1653
HYDRATED ELECTRON, $e_{aq.}^-$.
Electron attached to specific group of water molecules in liquid state.
D hydratisiertes Elektron
F électron hydraté
P elektron uwodniony
R гидратированный электрон

HYDRATE ISOMERISM. *See* HYDRATION ISOMERISM.

1654
HYDRATION.
Addition of water molecule (or hydrogen and oxygen in atomic ratio 2:1) to molecule of chemical compound.
D Hydratation, Hydratisierung
F hydratation
P reakcja hydratacji, hydratacja
R гидратация

1655
HYDRATION.
Solvation in aqueous medium.
D Hydratation
F hydratation
P hydratacja, uwodnienie
R гидратация

1656
HYDRATION ISOMERISM (hydrate isomerism).
Type of isomerism based on different arrangements of water molecules in first coordination sphere and beyond it, e.g. $[Cr(H_2O)_6]Cl_3$, $[Cr(H_2O)_5Cl]Cl_2 \cdot H_2O$, $[Cr(H_2O)_4Cl_2]Cl \cdot 2H_2O$.
D Hydratisomerie
F isomérie d'hydratation
P izomeria hydratacyjna
R гидратная изомерия

1657
HYDRATION NUMBER OF ION.
Solvation number of ion in aqueous solution.
D Hydratationszahl des Ions
F nombre d'hydratation d'ion
P liczba hydratacji jonu
R число гидратации иона

1658
HYDRATION OF ION.
Solvation of ion by water molecules.
D Hydratation des Ions
F hydratation d'ion
P hydratacja jonu
R гидратация иона

1659
HYDROAROMATIC RING.
Hydrogenated aromatic nucleus, e.g. cyclohexene, tetralin

D hydroaromatischer Ring
F cycle hydroaromatique
P pierścień hydroaromatyczny
R гидроароматическое кольцо

1660
HYDROBORATION.
Addition of diborane $(BH_3)_2$ to unsaturated organic compound with formation of trialkylborane, e.g.

$$6RCH=CH_2 + (BH_3)_2 \rightarrow 2(RCH_2CH_2)_3B$$

D Hydroborierung
F ...
P hydroborowanie, borowodorowanie
R гидроборирование

HYDRODYNAMICAL MODEL. *See* LIQUID-DROP MODEL.

1661
HYDRODYNAMIC STAGE.
In statistical mechanics, last stage of time evolution of macroscopic system in which dissipative processes occur leading from local equilibrium to thermodynamic equilibrium of the whole system.
D hydrodynamisches Stadium
F étape hydrodynamique
P stadium hydrodynamiczne
R гидродинамическая стадия

1662
HYDROFORMYLATION (oxo process).
Addition of carbon monoxide and hydrogen to unsaturated compound with formation of aldehyde, e.g.

$$>C=C< + CO + H_2 \xrightarrow[\substack{3\ MPa \\ Catalyst}]{200°C} >CH-\overset{|}{\underset{|}{C}}-C\overset{O}{\underset{H}{\diagdown}}$$

D Hydroformylierung, Oxosynthese
F réaction oxo
P hydroformylowanie, reakcja okso
R гидроформилирование, оксосинтез

1663
HYDROGEL.
Gel in which water is dispersion medium.
D Hydrogel
F hydrogel
P hydrożel
R гидрогель

1664
HYDROGEN, H.
Atomic number 1.
D Wasserstoff
F hydrogène
P wodór
R водород

1665
HYDROGENATION.
Process of catalytic addition of hydrogen molecule to organic compound molecule containing multiple bonds, e.g.

D Hydrogenation, Hydrierung
F hydrogénation
P uwodornianie, hydrogenizacja
R гидрогенизация

1666
HYDROGEN BOMB.
Explosion system based on synthesis reaction of light nuclei of deuterium and tritium, occurring at temperature of the order of 10^6 K.
D Wasserstoffbombe
F bombe à hydrogène
P bomba wodorowa
R водородная бомба

1667
HYDROGEN BOND.
Weak chemical bond formed between hydrogen and strongly electronegative atoms, such as nitrogen, oxygen or fluorine.
D Wasserstoffbindung, Wasserstoffbrücke, Wasserstoffbrückenbindung
F liaison hydrogène, pont hydrogène, liaison protonique
P wiązanie wodorowe, mostek wodorowy
R водородная связь

1668
HYDROGEN-BONDED CRYSTALS.
Crystals built of molecules kept together by hydrogen bonds. They are characterized by tendency to polymerize and by increased energy of crystal lattice.
D Kristalle mit Wasserstoffbrücken
F cristaux avec liaison hydrogène
P kryształy z wiązaniem wodorowym
R кристаллы с водородной связью

1669
HYDROGEN ELECTRODE.
Electrode (half-cell) constructed from platinum or palladium covered by platinum black, dipped into solution containing hydrogen ions and saturated with hydrogen, $Pt,H_2|H_3O^+$; electrode reaction is

$$1/2\,H_2 + H_2O \leftrightarrows H_3O^+ + e$$

and electrode potential

$$\varepsilon = \varepsilon^\circ + \frac{RT}{F}\ln\frac{a_{H_3O^+}}{a_{p_{H_2}}}$$

where ε° = standard electrode potential and $a_{p_{H_2}}$ = volatility of hydrogen.
D Wasserstoffelektrode
F électrode à hydrogène, électrode d'hydrogène
P elektroda wodorowa
R водородный электрод

HYDROGEN ION EXPONENT. *See* pH.

1670
HYDROGEN-LIKE ION.
Ion composed of nucleus and one electron, e.g. He^+, Li^{2+}, C^{5+}.
D wasserstoffähnliches Ion
F ion hydrogénoïde
P jon wodoropodobny
R водородоподобный ион

1671
HYDROGEN-LIKE ORBITAL.
Atomic orbital of hydrogen-like ion.
D wasserstoffähnliches Orbital
F orbitale hydrogénoïde
P orbital wodoropodobny
R водородоподобная волновая функция

1672
HYDROGENOLYSIS (destructive hydrogenation).
Catalytic hydrogen reduction with partial destruction of substrate and formation of molecules of lower molecular weight (than original ones) or with ring opening, e.g.

D Hydrogenolysis, Hydrogenolyse, destruktive Hydrierung
F hydrogénolyse
P hydrogenoliza
R гидрогенолиз

1673
HYDROLYSIS.
Decomposition of chemical compound by interaction with water, e.g.

$$RCOOR' + HOH \xrightarrow{\overset{\ominus}{OH}} RCOO^\ominus + R'OH$$

D Hydrolyse
F hydrolyse
P hydroliza
R гидролиз

1674
HYDROLYSIS (*of electrolytes*).
Reaction with water of following types of electrolytes:
salt of weak acid HA and strong base

$A^- + H_2O \rightleftarrows HA + OH^-$

salt of strong acid and weak base B

$BH^+ + H_2O \rightleftarrows B + H_3O^+$

salt of weak acid HA and weak base B

$$A^- + BH^+ \xrightarrow{\quad H_2O \quad} AH + B$$

D Hydrolyse
F hydrolyse
P hydroliza
R гидролиз

1675
HYDROLYSIS CONSTANT, K_h.
Equilibrium constant of hydrolysis reaction (*see* hydrolysis); for salt of weak acid (HA) and strong base:

$$K_h = \frac{a_{HA}\, a_{OH^-}}{a_{A^-}} = \frac{L_{H_2O}}{K_{HA}}$$

for salt of strong acid and weak base (B):

$$K_h = \frac{a_B\, a_{H_3O^+}}{a_{BH^+}} = \frac{L_{H_2O}}{K_B}$$

and for salt of weak acid (HA) and weak base (B):

$$K_h = \frac{a_{HA}\, a_B}{a_{A^-}\, a_{BH^+}} = \frac{L_{H_2O}}{K_{HA} K_B}$$

where K_{HA} and K_B = dissociation constants of weak acid and base respectively, L_{H_2O} = ionic product of water.

D Hydrolysenkonstante
F constante d'hydrolyse
P stała hydrolizy (elektrolitów)
R константа гидролиза

1676
HYDROLYTIC ADSORPTION.
Exchange adsorption with participation of hydrogen or hydroxylic ions.

D hydrolytische Adsorption
F adsorption hydrolytique
P adsorpcja hydrolityczna
R гидролитическая адсорбция

1677
HYDROSOL.
Sol in which water is dispersion medium.

D Hydrosol
F hydrosol
P hydrozol
R гидрозоль

HYDROSTATIC EQUILIBRIUM. *See*
MECHANICAL EQUILIBRIUM.

1678
HYDROXO-COMPLEX.
Metal complex in which hydroxyl anions are ligands.

D Hydroxokomplex
F complexe hydroxo
P hydroksokompleks
R гидроксокомплекс

1679
HYDROXYLATION
Reaction of alkenes with oxidizing agents ($KMnO_4$, H_2O_2 etc.) resulting in formation of 1,2-diols, e.g.

D Hydroxylierung
F hydroxylation
P hydroksylowanie, przyłączenie grup wodorotlenowych
R гидроксилирование

1680
HYDROXYMETHYLATION.
Introduction of hydroxymethyl group —CH_2OH into molecule of organic compound, e.g.

D Hydroxymethylierung
F hydroxyméthylation
P hydroksymetylowanie
R ...

1681
HYPERCHARGE, Y.
Additive dimensionless quantum number defined for hadrons and photons as sum of baryon number and strangeness $Y = B + S$. Hypercharge always has integral values.

D Hyperladung
F hypercharge
P hiperładunek
R гиперзаряд

13*

1682
HYPERCONJUGATION (no-bond resonance).
Conjugation of σ-electrons of C—H bonds in alkyl groups with adjacent π-electrons of double bonds, e.g. in toluene

D Hyperkonjugation, Baker-Nathan-Effekt
F hyperconjugaison
P hiperkoniugacja, rezonans bezwiązaniowy, efekt Bakera i Nathana
R гиперконьюгация, сверхсопряжение

1683
HYPERFINE INTERACTION.
Interaction between electron magnetic moments and nuclear magnetic moments in atom, ion or molecule which leads to hyperfine structure of spectral lines in atomic spectra and electron spin resonance spectra.
D Hyperfeinstrukturwechselwirkung
F interaction hyperfine
P oddziaływanie nadsubtelne
R сверхтонкое взаимодействие

1684
HYPERFINE SPLITTING.
Splitting of energy levels due to interaction between total angular momentum of electrons and nuclear angular momentum; in electron spin resonance spectra this splitting is caused by interaction of electron spin and nuclear spin.
D Hyperfeinstruktur-Aufspaltung
F écart hyperfin
P rozszczepienie nadsubtelne
R сверхтонкое расщепление

1685
HYPERFINE STRUCTURE (*in atomic spectroscopy*).
Splitting of atomic terms or spectral lines due to interaction between electronic and nuclear angular momentum.
D Hyperfeinstruktur
F structure hyperfine
P struktura nadsubtelna
R сверхтонкая структура

1686
HYPERFINE STRUCTURE (*in electron spin resonance*).
Splitting of resonance lines due to interaction between electron spin and nuclear spin.
D Hyperfeinstruktur
F structure hyperfine

P struktura nadsubtelna
R сверхтонкая структура

1687
HYPERONS.
Elementary particles, hadrons, having mass greater than that of neutron and less than that of deuteron (*see also* lambda, xi, sigma and omega hyperon).
D Hyperons
F hypérons
P hiperony
R гипероны

1688
HYPOCHROMIC EFFECT.
Decrease of absorption coefficient due to structural changes of absorbing substance.
D hypochromer Effekt
F effet hypochrome
P efekt hypochromowy
R гипохромный эффект

1689
HYPSOCHROME (hypsochromic group).
Atom or group of atoms which, when substituted into molecule of organic compound, shift its absorption spectrum towards shorter wavelengths.
D hypsochrome Gruppe
F hypsochrome
P hipsochrom, grupa hipsochromowa
R гипсохромная группа

HYPSOCHROMIC GROUP. *See* HYPSOCHROME.

1690
HYPSOCHROMIC SHIFT.
Shift of light absorption towards higher frequencies.
D hypsochrome Absorptionsverschiebung
F effet hypsochrome
P przesunięcie hypsochromowe, efekt hypsochromowy, przesunięcie niebieskie
R гипсохромное смещение

1691
HYSTERESIS.
Lack of coincidence of procession of change in opposite directions, due to dependence of state of system on its history.
D Hysteresis, Hysterese
F hystérésis, hystérèse
P histereza
R гистерезис

1692
IDEAL DILUTED SOLUTION.
Solution in which absolute activity λ_B of every dissolved constituent B is proportional to x_B/x_A where x_B = mole fraction of dissolved substance, x_A = mole fraction of solvent.

D ideal verdünnte Lösung
F solution diluée idéale, solution infinitement diluée
P roztwór rozcieńczony idealny, roztwór nieskończenie rozcieńczony
R идеальный разбавленный раствор, бесконечно разбавленный раствор

1693
IDEAL GAS (classical gas, perfect gas).
Gas which satisfies equation of state $pV_m = RT$, where p = pressure, V_m = molar volume, T = thermodynamic temperature, and R = molar gas constant. In statistical mechanics this means that there is no interaction between particles of such a gas.
D ideales Gas
F gas parfait
P gaz doskonały, gaz idealny
R идеальный газ, совершенный газ

1694
IDEAL GAS EQUATION OF STATE (perfect gas equation of state).
Thermal equation of state resulting from empirical laws of ideal gases

$$pV_m = RT$$

where p = pressure, V_m = molar gas volume, T = thermodynamic temperature, R = universal gas constant.
D Zustandsgleichung der idealen Gase
F équation d'état de gaz parfait
P równanie stanu gazu doskonałego, równanie Clapeyrona
R уравнение состояния идеального газа, уравнение (Менделеева-)Клапейрона

1695
IDEAL MIXTURE (ideal solution).
Mixture whose mixing thermodynamic potential G^M satisfies equation

$$G^M = RT \sum_B x_B \ln x_B$$

where R = molar gas constant, T = thermodynamic temperature, x_B = mole fraction of component B; mixing enthalpy and volume of mixing of ideal mixture are equal to zero.
D ideale Mischung
F mélange idéal
P mieszanina doskonała, roztwór doskonały, mieszanina idealna, roztwór idealny
R идеальная смесь, идеальный раствор, идеальная фаза

IDEAL SOLUTION. *See* IDEAL MIXTURE.

IDENTICAL CHIRAL CENTRES. *See* SIMILAR CHIRAL CENTRES.

1696
IDENTIFICATION.
In qualitative analysis determination of identity of investigated substance.
D Identifizierung
F identification
P identyfikacja
R идентификация

1697
IDENTIFICATION REACTION.
Chemical reaction which enables detection of a given chemical substance.
D Identitätsreaktion, Nachweisreaktion
F réaction d'identification
P reakcja charakterystyczna
R характерная реакция

1698
IGNITION.
Process of starting substance burning caused by external energy impulse.
D Entzündung, Zündung
F inflammation
P zapłon
R воспламенение, зажигание

IGNITION LIMITS. *See* LIMITS OF INFLAMMABILITY.

IGNITION POINT. *See* IGNITION TEMPERATURE.

1699
IGNITION TEMPERATURE (ignition point).
Lowest temperature necessary to start combustion of substance in precisely defined and usually normalized conditions.
D Entzündungstemperatur, Zündpunkt
F température d'inflammation, point d'inflammation
P temperatura zapłonu
R температура воспламенения, температура зажигания

1700
ILKOVIČ EQUATION.
Equation:
$i_l = 607\, n\, m^{2/3}\, t^{1/6}\, D^{1/2}\, C^\circ$ which relates polarographic limiting current i_l to concentration of reactant C°, its diffusion coefficient D; n = number of electrons transferred in electrode reaction, m = efficiency of capillary and t = drop time.
D Ilkovič Gleichung
F équation d'Ilkovič
P równanie Ilkoviča
R уравнение Ильковича

IMPERFECT GAS. *See* REAL GAS.

1701
IMPURITIES (chemical imperfections).
In solid state chemistry, imperfections in crystal lattice consisting in occupation of some nodes or interstitial positions by atoms or ions of other elements.
D chemische Fehlordnung, chemische Baufehler
F imperfections chimiques
P defekty chemiczne, domieszki
R химические дефекты, примеси

1702
INCLUSION COMPOUNDS.
Molecular compounds in which molecules of first ("guest") component are enclosed in intermolecular free spaces present in crystalline structure or crystal lattice of second ("host") component.
D Einschlußverbindungen
F ...
P związki włączeniowe
R соединения включения

1703
INCOHERENT SCATTERING.
Scattering on more than one scattering centre in which incident wave (particle) interacts with each centre separately. During incoherent scattering interference effects do not occur because intensity (but not amplitude) of scattered wave is sum of intensities of waves scattered on particular centres.
D inkohärente Streuung
F diffusion incohérente
P rozpraszanie niespójne, rozpraszanie niekoherentne
R некогерентное рассеяние

1704
INCONGRUENT MELTING POINT.
Point on phase diagram in which solid phase disintegrates partially into liquid phase and in part into another solid phase; new phases differ in chemical composition.
D *(Schmelzpunkt einer inkongruent schmelzenden Verbindung)*
F point de fusion incongruent
P punkt inkongruentny
R инконгруэнтная точка

1705
INCREMENT.
Portion of investigated material taken once from one part of material.
D Stichprobe
F prélèvement élémentaire, échantillon brut
P próbka pierwotna
R первичная проба, частичная проба

1706
INDEPENDENT COMPONENT OF PHASE SYSTEM.
Component which has to be defined qualitatively and quantitatively. Such definition is necessary and sufficient for description of chemical composition of every phase in system.
D unabhängige Komponente von Phasensystem
F composant indépendant du système de phases
P składnik (układu fazowego) niezależny
R независимый компонент фазовой системы

1707
INDEPENDENT PARTICLE MODEL (*in theory of atomic nucleus*).
Model of atomic nucleus based on postulate that each nucleon moves independently in a field determined by average positions of all other nucleons.
D Einteilchenmodell
F modèle à particules indépendantes
P model cząstek niezależnych
R модель независимых частиц

INDETERMINANCY PRINCIPLE. *See* UNCERTAINTY PRINCIPLE.

1708
INDICATOR (visual indicator).
In volumetric analysis, substance used to detect end-point of titration by observing change of colour, appearance or disappearance of fluorescence, turbidity, etc.
D Indikator
F indicateur
P wskaźnik
R индикатор

1709
INDICATOR ELECTRODE.
Electrode (half-cell) with potential dependent on the kind and concentration of electroactive substance.
D Indikatorelektrode
F électrode indicatrice
P elektroda wskaźnikowa
R индикаторный электрод

1710
INDICATOR ERROR (*of acid-base indicator*).
Error due to change of concentration of protons or hydroxyl ions by their reaction with indicator.
D Indikatorfehler
F erreur d'indicateur
P błąd wskaźnika
R индикаторная ошибка

1711
INDICATOR PAPER (test-paper).
Strip of filter-paper saturated with suitable reagent, e.g. pH indicator.

D Reagenzpapier, Indikatorpapier
F papier réactif
P papierek wskaźnikowy
R реактивная бумага

INDIFFERENT SOLVENT. *See* APROTIC SOLVENT.

1712
INDIUM, In.
Atomic number 49.
D Indium
F indium
P ind
R индий

1713
INDOPHENINE REACTION.
Coloured (blue) reaction of isatine with thiophene in presence of sulfuric acid, applied for identification of thiophene.
D Indopheninreaktion
F réaction d'indophénine
P reakcja indofeninowa
R индофениновая реакция

1714
INDUCED ACTIVITY.
Radioactivity induced in irradiated material by activation.
D induzierte Aktivität
F activité induite
P aktywność wzbudzona
R индуцированная активность, наведённая активность

1715
INDUCED DIPOLE.
Dipole formed by charge separation due to external electric field and vanishing when field is absent.
D induzierter Dipol
F dipôle induit
P dipol indukowany
R индуцированный диполь

1716
INDUCED DIPOLE MOMENT.
Dipole moment induced by electric field and determined by sum of electronic and atomic polarizations.
D induziertes Dipolmoment
F moment dipolaire induit
P moment dipolowy indukowany
R индуцированный дипольный момент

1717
INDUCED POLARIZATION (MOLAR) (deformation polarization, distortion polarization).
Sum of molar atomic and electronic polarizations, approximately equal to molar refraction.

D induzierte Polarisation, Deformationspolarisation
F polarisation induite
P polaryzacja indukcyjna (molowa), polaryzacja deformacyjna (molowa)
R индуцированная поляризация

1718
INDUCED RADIOACTIVITY (artificial radioactivity).
Radioactivity induced by irradiating substances with corpuscular radiation or with γ-rays.
D induzierte Radioaktivität, künstliche Radioaktivität
F radioactivité artificielle, radioactivité induite
P promieniotwórczość wzbudzona, promieniotwórczość sztuczna
R искусственная радиоактивность, наведённая радиоактивность

1719
INDUCED REACTION (secondary reaction).
In system of coupled reactions, path of secondary reaction which depends on simultaneous run of primary, initiating reaction.
D Sekundärreaktion, induzierte Reaktion
F réaction secondaire
P reakcja indukowana
R вторичная реакция, индуцированная реакция

INDUCED REACTIONS. *See* COUPLED REACTIONS.

1720
INDUCTION EFFECT.
One of the components of intermolecular forces. Induction effect leads to attractive intermolecular forces between pair of molecules one of which is charged or has permanent higher multipole moment inducing multipole moment in other molecule.
D Induktionseffekt
F effet induit
P efekt indukcyjny
R индукционный эффект

1721
INDUCTION FACTOR.
In coupled reactions, ratio of acceptor concentration loss to loss of inductor concentration.
D Induktionsfaktor
F . . .
P współczynnik indukcji chemicznej, czynnik indukcji
R фактор индукции

1722
INDUCTION PERIOD.
Time elapsing between point at which
system is ready to react and actual start
of reaction.
D Induktionsperiode
F période d'induction
P okres indukcyjny reakcji, okres indukcji
R период индукции

INDUCTION POLARIZABILITY. *See*
POLARIZABILITY.

1723
INDUCTIVE EFFECT (*in organic
chemistry*).
Polarization of bonds in organic molecule
by dipole existing in bond linking
electronegative substituent of
electronegativity other than that of
hydrogen with rest of molecule (it is
assumed that inductive effect of C—H
bond is zero). Inductive effect is more
pronounced in molecules containing more
polarizable double bonds

$$CH_3 \rightarrow CH_2 \rightarrow Cl \qquad CH_2 \overset{\frown}{=} CH \rightarrow CH_2 \rightarrow Cl$$

Arrows indicate direction of electron shift.
D Induktionseffekt
F effet induit, effet inducteur
P efekt indukcyjny, efekt $\pm$ I
R индукционный эффект

1724
INDUCTOR (*in coupled reaction*).
Substance which causes or accelerates
induced (secondary) reaction in system
of coupled reactions.
D Induktor (der induzierten Reaktion)
F ...
P induktor (reakcji sprzężonej)
R индуктор (сопряжённой реакции)

1725
INELASTIC SCATTERING.
Scattering of particles in which the kinetic
energies of scattered particles are changed
in system's centre of mass.
D unelastische Streuung
F diffusion inélastique
P rozpraszanie niesprężyste, rozpraszanie
nieelastyczne
R неупругое рассеяние

1726
INERT COMPLEX.
Complex characterized by small exchange
rate of ligands.
D inerter Komplex
F complexe inerte
P kompleks bierny
R инертный комплекс

1727
INERT GASES (noble gases, rare gases,
helium group).
Elements of eighth main group of closed
(octetic, except for helium) structure of
outer electronic shells of atoms ns^2np^6:
helium, neon, argon, krypton, xenon,
radon.
D Edelgase
F gaz inertes, gaz rares, gaz nobles
P helowce, gazy szlachetne
R инертные газы, благородные газы

1728
INFINITELY THICK LAYER.
Target of such thickness that there is
complete, or nearly complete, absorption
of incident particles or photons.
D unendlich dicke Schicht
F couche épaisse
P warstwa nieskończenie gruba
R слой насыщения

1729
INFINITELY THIN LAYER.
Target in which practically no reduction
of intensity of radiation due to absorption
or self-absorption is observed.
D unendlich dünne Schicht
F couche infiniment mince
P warstwa nieskończenie cienka
R бесконечно тонкий слой, слой „нулевой
толщины"

1730
INFINITESIMAL PROCESS (infinitesimal
transition).
Thermodynamic process with final state
only infinitesimally different from initial
one.
D infinitesimale Zustandsänderung
F transformation infinitésimale,
transformation élémentaire
P przemiana elementarna, przemiana
nieskończenie mała, proces elementarny
R элементарное изменение состояния

INFINITESIMAL TRANSITION. *See*
INFINITESIMAL PROCESS.

INFINITE VOLUME LIMIT. *See*
THERMODYNAMIC LIMIT.

1731
INFRARED (infrared region), IR.
Electromagnetic radiation range of
wavelength between ca. 0.75 μm to
ca. 800 μm.
D Infrarot, Ultrarot
F infrarouge, région infrarouge
P podczerwień
R инфракрасная область

INFRARED REGION. *See* INFRARED.

1732
INFRARED SPECTRUM (IR spectrum).
Spectrum in infrared wavelength region
with bands corresponding to molecular
vibration energy changes.

D Infrarot-Spektrum
F spectre infra-rouge
P widmo w podczerwieni, widmo IR
R инфракрасный спектр

1733
INHIBITOR (negative catalyst).
Catalyst which causes retardation of
chemical reaction rate.

D Inhibitor, Hemmstoff, Verzögerer,
negativer Katalysator
F inhibiteur, anticatalyseur, catalyseur
négatif
P inhibitor, katalizator ujemny
R ингибитор

1734
INHIBITOR (*of chain reaction*).
Substance reacting with intermediate or
propagator of chain, producing
non-reactive products responsible for
termination of reaction chain.

D Inhibitor der Kettenreaktion
F inhibiteur d'une réaction en chaîne
P inhibitor reakcji łańcuchowej
R замедлитель цепной реакции,
ингибитор цепной реакции

1735
INITIAL STAGE.
In statistical mechanics, earliest stage of
time evolution of macroscopic system.
No simple rules for behaviour of partition
functions during initial stage can be given.

D Anfangsstadium
F étape initiale
P stadium początkowe
R начальная стадия

INNER COORDINATION SPHERE. *See*
COORDINATION SPHERE.

1736
INNER ELECTRIC POTENTIAL, φ.
Electrical potential between point in phase
and point at infinity $\varphi = \psi + \chi$, where
ψ = outer potential and χ = surface
potential of phase.

D inneres elektrisches Potential
F potentiel électrique intérieur
P potencjał wewnętrzny fazy
R внутренний потенциал

1737
INNER ELECTRONS (inner-shell
electrons).
Electrons of atomic shells with principal
quantum number n less than that for
valence shell.

D innere Elektronen
F électrons internes
P elektrony wewnętrzne, elektrony
wewnętrznych powłok
R внутренние электроны, электроны
внутренних оболочек

1738
INNER-ORBITAL COMPLEX.
Complex in which d orbitals of smaller
principal quantum number than for s
and p orbitals take part in hybridization
of central ion orbitals; in octahedral
complex $(n-1)d^2nsp^3$.

D innerer orbitaler Komplex
F complexe à orbitales internes
P kompleks wewnętrznoorbitalowy
R комплекс внутренних орбиталей,
внутреннеорбитальный комплекс

INNER QUANTUM NUMBER. *See*
MAGNETIC SPIN QUANTUM NUMBER.

INNER-SHELL ELECTRONS. *See* INNER
ELECTRONS.

1739
INNER TRANSITION ELEMENTS.
Transition elements the atoms of which
have partially-filled d sub-shell and also
a farther partially unfilled f sub-shell.
Two series of these elements are
distinguished: lanthanoids and actinoids.

D Übergangselemente
F éléments de transition interne
P pierwiastki wewnętrznoprzejściowe,
metale wewnętrznoprzejściowe
R . . .

1740
INORGANIC CHEMISTRY.
Branch of chemistry dealing with studies
on properties and transformations of all
chemical elements and their compounds,
excluding only majority of carbon
compounds.

D anorganische Chemie
F chimie minérale, chimie inorganique
P chemia nieorganiczna
R неорганическая химия

1741
INORGANIC ION EXCHANGER.
Ion exchanger which consists of inorganic
skeleton bearing excess electric charge,
and mobile counter ions.

D anorganischer Ionenaustauscher
F échangeur minéral
P jonit nieorganiczny
R неорганический ионит

INSENSITIVE TIME. *See* DEAD TIME.

1742
INSERTION REACTION.
Reaction characteristic for carbenes and
nitrenes, involving insertion of
appropriate group between the carbon and
hydrogen atoms linked together, e.g.

D Einschiebung-Reaktion
F réaction d'insertion
P reakcja insercji
R реакция внедрения

1743
INSTABILITY CONSTANT (dissociation
constant *of coordination compound*).
Reciprocal of stability constant.

D Dissoziationskonstante (des Komplexes)
F constante d'instabilité, constante de
dissociation
P stała nietrwałości
R константа нестойкости

1744
INSTRUMENTAL ANALYSIS.
Analysis carried out with use of
instruments, based on analytical
application of measured physical or
physico-chemical quantities dependent on
content (concentration) of determined
component in investigated sample.

D Instrumentalanalyse,
Instrumentenanalyse
F analyse instrumentale
P analiza instrumentalna
R инструментальный анализ

1745
INSTRUMENTAL ERROR.
Measurement (determination) error due to
improper construction or working of
instrument.

D Fehlanzeige
F ...
P błąd przyrządu, błąd instrumentalny
R инструментальная ошибка

1746
INSULATOR.
Non-conducting material (gas, liquid or
solid) of high specific resistance, generally
accepted as $> 10^{10}$ Ωcm (dielectric).
According to band theory of solids, crystal
with broad forbidden band.

D Isolator
F isolant

P izolator, nieprzewodnik
R изолятор

1747
INTEGRAL CAPACITY OF DOUBLE
LAYER, K.
Capacity of electrical double layer at
electrode — solution interphase, defined
by equation:

$$K = \frac{q}{\varepsilon - \varepsilon_0}$$

where ε = electrode potential, ε_0 = zero
charge potential of electrode, q = charge
of electrode given by

$$q = \int_{\varepsilon_0}^{\varepsilon} C \, d\varepsilon$$

and C = differential capacity of double
layer.

D integrale Kapazität der Doppelschicht
F capacité intégrale de la couche double
P pojemność warstwy podwójnej całkowa
R интегральная ёмкость двойного слоя

1748
INTEGRAL DETECTOR.
Device that measures continuously total
amount of substance in gas or liquid
coming out of chromatographic column.

D Integraldetektor
F détecteur intégral
P detektor całkowy
R интегральный детектор

INTEGRAL DISTRIBUTION FUNCTION.
See PROBABILITY DISTRIBUTION
FUNCTION.

1749
INTEGRAL DOSE.
Amount of energy of any kind of ionizing
radiation imparted to sample or organism
and absorbed by it.

D Integraldosis
F dose intégrale
P dawka całkowita
R интегральная доза

1750
INTEGRAL HEAT OF ADSORPTION.
Total amount of heat which is liberated
(or absorbed) in adsorption process which
involves defined masses of adsorbate and
pure adsorbent engaged in process.

D integrale Adsorptionswärme
F chaleur totale d'adsorption
P ciepło adsorpcji całkowite
R (интегральная) теплота адсорбции

INTEGRAL HEAT OF DILUTION. *See*
HEAT OF DILUTION.

INTEGRAL HEAT OF SOLUTION. *See*
HEAT OF SOLUTION.

1751
INTEGRAL HEAT OF SWELLING.
Amount of heat released during total
swelling process of unit mass of dry gel.
D integrale Quellungswärme
F ...
P ciepło pęcznienia całkowite
R интегральная теплота набухания

INTEGRAL PROBABILITY
DISTRIBUTION FUNCTION. *See*
PROBABILITY DISTRIBUTION
FUNCTION.

INTENSITIES. *See* INTENSIVE
QUANTITIES.

INTENSITY OF MAGNETIZATION. *See*
MAGNETIZATION.

INTENSIVE PARAMETERS. *See*
INTENSIVE QUANTITIES.

INTENSIVE PROPERTIES. *See*
INTENSIVE QUANTITIES.

1752
INTENSIVE QUANTITIES (intensive
properties, intensive variables, intensive
parameters, intensities).
Thermodynamic quantities independent of
system's mass.
D intensive Größen, intensive
Eigenschaften, intensive
Zustandsvariablen, intensive Parameter,
Qualitätsgrößen
F variables intensives, variables de
tension, facteurs d'intensité
P wielkości intensywne, parametry
intensywne, zmienne intensywne
R интенсивные величины, интенсивные
признаки, интенсивные параметры,
факторы интенсивности

INTENSIVE VARIABLES. *See*
INTENSIVE QUANTITIES.

1753
INTERCOMBINATION TRANSITION.
Radiative electronic transition between
states of different multiplicity, e.g.
between triplet and singlet state.
D Interkombinationsübergang
F transition intercombinaison
P przejście interkombinacyjne
R интеркомбинационный переход

INTERCONVERSION OF CHAIR FORMS.
See CONVERSION OF CHAIR FORMS.

1754
INTERFACE (boundary between phases).
Surface which forms boundary between
two different phases.
D Phasengrenzfläche, Begrenzungsfläche
F interface de phases
P granica faz
R граница фаз, поверхность раздела фаз

1755
INTERFACIAL TENSION, σ_{AB}.
Tension at interface between two
immiscible liquids, $\sigma_{AB} = \sigma_A - \sigma_B$, where
σ_A and σ_B denote surface tensions (against
air or vapour) of two liquids A and B
respectively.
D Grenzflächenspannung
F tension interfaciale
P napięcie międzyfazowe
R межфазное натяжение

1756
INTERFERENCE FILTER.
Optical filter whose action is based on
principle of light interference in thin
plates or membranes.
D Interferenzfilter,
Interferenz-Verlauffilter
F filtre interférentiel
P filtr interferencyjny
R интерференционный светофильтр

1757
INTERFERING NUCLEAR REACTION.
In activation analysis, nuclear reaction
of constituent of a sample, other than that
to be determined, giving the same product
as that from the element to be determined
(result is too high), or reaction which
leads to decrease of product to be
determined (result is too low).
D störende Kernreaktion
F réaction nucléaire d'interférence
P reakcja jądrowa przeszkadzająca
R конкурирующая ядерная
реакция

1758
INTERFERING SECOND ORDER
REACTION.
In activation analysis, nuclear reaction in
which from the matrix atoms the same
radioisotope is formed in two successive
activations as that formed from
constituent to be determined.
D (störende) Reaktion zweiter Ordnung
F réaction secondaire (d'interférence)
P reakcja jądrowa przeszkadzająca
drugorzędowa
R вторичная (конкурирующая) реакция

1759
INTERFEROMETRIC ANALYSIS
(interferometry).
Method of analysis based on measurement
of difference of refractive indices for
investigated sample and standard sample
by using interferometer. Used mostly in
gas analysis, sometimes also for liquids.
D interferometrische Analyse,
Interferometrie
F analyse interférométrique,
interférométrie
P analiza interferometryczna
R интерферометрический анализ

INTERFEROMETRY. See
INTERFEROMETRIC ANALYSIS.

INTER-LABORATORY
REPRODUCIBILITY. See
REPRODUCIBILITY.

1760
INTERMEDIATE NEUTRONS.
Neutrons having kinetic energies in range
between 100 eV and 1 MeV.
D mittelschnelle Neutronen
F neutrons intermédiaires
P neutrony pośrednie
R промежуточные нейтроны

1761
INTERMETALLIC PHASES.
Solid phases in alloys of metals with space
lattices different from those of each
constituent.
D intermetallische Phasen
F phases intermétalliques
P fazy międzymetaliczne, fazy
intermetaliczne
R междуметаллические фазы,
интерметаллические фазы

1762
INTERMOLECULAR ENERGY
TRANSFER.
Transfer of different kinds of energy
(i.e. electronic vibrational, rotational and
translation energy) from one molecule
to another in different media.
D (zwischenmolekulare)
Energieübertragung
F transfert d'énergie (entre systèmes
moléculaires)
P przekazywanie energii
(międzycząsteczkowe), przenoszenie
energii
R (междумолекулярный) перенос энергии

1763
INTERMOLECULAR FORCES.
Interaction forces between molecules at
small distances (of order of 10^{-7} cm);
attractive (van der Waals) or repulsive
forces.

D zwischenmolekulare Kräfte
F forces intermoléculaires
P siły międzycząsteczkowe
R междумолекулярные силы

1764
INTERMOLECULAR HYDROGEN BOND.
Hydrogen bond in which hydrogen atom is
linked with two atoms of two different
molecules, e.g. in formic acid dimer

D intermolekulare
Wasserstoff(brücken)bindung,
intermolekulare Wasserstoffbrücke
F liaison hydrogène intermoléculaire
P wiązanie wodorowe międzycząsteczkowe
R межмолекулярная водородная связь

1765
INTERMOLECULAR POTENTIAL, $u(r)$.
Potential energy of pair of molecules as
function of their relative positions.
D zwischenmolekulares Potential,
Paarpotential
F potentiel d'interaction de deux
molécules
P potencjał oddziaływania
międzycząsteczkowego
R межмолекулярный потенциал,
потенциал взаимодействия молекул

1766
INTERNAL ADSORPTION.
Adsorption from solution of microamounts
of foreign ions on surface of a precipitate,
connected with incorporation of ions into
interior of crystals due to growth of
crystals.
D innere Adsorption
F adsorption interne
P adsorpcja wewnętrzna
R внутренняя адсорбция

1767
INTERNAL CONVERSION.
Transition of excited atomic nucleus to
lower energy state; the energy difference
between initial and final states being given
to an orbital electron (not to γ-photon).
D innere Konversion
F conversion interne
P konwersja wewnętrzna
R внутренняя конверсия

1768
INTERNAL CONVERSION (in
photochemistry).
Isoenergetic radiationless transition
between two electronic states of like
multiplicity.
D innere Umwandlung
F conversion interne

P konwersja wewnętrzna
R внутренняя конверсия

1769
INTERNAL CONVERSION COEFFICIENT, α.
Ratio of number of internal conversion electrons N_e to number of gamma quanta N_γ emitted in given transition, i.e. ratio of probability of electron emission λ_e to probability of photon emission λ_γ

$$\alpha = \frac{N_e}{N_\gamma} = \frac{\lambda_e}{\lambda_\gamma}$$

D Koeffizient der inneren Konversion
F coefficient de conversion interne
P współczynnik konwersji wewnętrznej
R коэффициент внутренней конверсии

1770
INTERNAL ELECTROLYSIS.
Electrolysis occurring spontaneously in cell constructed of two electrodes with appropriate potentials without applying external voltage to electrodes.

D innere Elektrolyse
F électrolyse interne
P elektroliza wewnętrzna
R внутренний электролиз

1771
INTERNAL ENERGY (internal energy function), U.
Extensive thermodynamic function equal to difference between total energy E of system and its macroscopic kinetic energy E_k and potential energy E_p

$$U = E - E_k - E_p$$

D innere Energie
F énergie interne
P energia wewnętrzna
R внутренняя энергия

INTERNAL ENERGY FUNCTION. *See* INTERNAL ENERGY.

INTERNAL FIELD. *See* LOCAL ELECTRIC FIELD.

INTERNAL FRICTION. *See* VISCOSITY.

1772
INTERNAL INDICATOR.
Indicator added to titrated solution.

D innerer Indikator, Innen-Indikator
F indicateur interne
P wskaźnik wewnętrzny
R внутренний индикатор

1773
INTERNAL PARTITION FUNCTION (*of a molecule*), $j(T)$.
Partition function for intramolecular degrees of freedom

$$j(T) = \sum_{\varepsilon} g_{\varepsilon} \exp\left[-\frac{\varepsilon}{kT}\right]$$

where g_{ε} = degeneracy of the energy level ε, k = Boltzmann constant, and T = thermodynamic temperature. Sum is taken over all internal (i.e. non-translational) degrees of freedom. Internal partition function determines influence of intramolecular degrees of freedom on thermodynamic properties of corresponding gas.

D innere Zustandssumme
F fonction de partition interne (de la molécule)
P funkcja rozdziału wewnętrzna (cząsteczki)
R внутренняя статистическая сумма

1774
INTERNAL PRESSURE (cohesion pressure).
Quantity defined by $\left(\dfrac{\partial U}{\partial V}\right)_T$, where U = internal energy of phase, V = its volume and T = phase temperature.

D innerer Druck, Binnendruck, Kohäsionsdruck
F pression interne, pression intérieure
P ciśnienie wewnętrzne
R внутреннее давление

INTERNAL REFERENCE. *See* INTERNAL STANDARD.

1775
INTERNAL RESISTANCE OF GALVANIC CELL.
Electrical resistance in internal circuit of galvanic cell; it depends on dimensions and separation of electrodes, electrical conductivity of electrolyte, resistance at electrode — electrolyte interphase and, if present, on nature of diaphragm.

D Innenwiderstand des Elements
F résistance interne de cellule
P opór wewnętrzny ogniwa
R внутреннее сопротивление гальванического элемента

1776
INTERNAL RETURN.
Recombination of internally associated ionic pair, leading to restitution of substrate. No ions coming from other molecules present in system take part in internal return. Intramolecular processes, such as inversion, racemization or rearrangements, often accompany internal return

$$A-X \leftrightarrows A^{\oplus}X^{\ominus} \leftrightarrows A^{\oplus} \| X^{\ominus}$$

D innere Rückkehr
F retour interne
P powrót wewnętrzny
R ...

1777
INTERNAL SOURCE OF RADIATION.
Radioactive substance distributed
uniformly in irradiated preparation.
D innere Strahlungsquelle
F source interne (du rayonnement)
P źródło (promieniowania) wewnętrzne
R внутренний источник (излучения)

1778
INTERNAL STANDARD
(*in spectroanalysis*).
Component added or already present in
the same concentration in standard and
investigated samples whose spectral
line(s) is (are) used as reference lines for
measurement of spectral line intensities
of determined elements.
D innerer Standard
F étalon interne, élément de référence
P wzorzec wewnętrzny
R внутренний стандарт

1779
INTERNAL STANDARD (internal
reference).
In NMR spectroscopy, reference substance
dissolved in investigated solution; used for
determination of chemical shifts.
D innerer Standard
F étalon intérieur, référence intérieure
P wzorzec wewnętrzny
R внутренний эталон

1780
INTERNAL-STANDARD METHOD.
Analytical method based on addition to
analysed sample of known quantity of
some substance (internal standard), for
which analytical curve in the given
conditions of determination is identical
with that for determined component.
Using linear relationship between quantity
measured for standard and analysed
substance and knowing concentration of
standard one may calculate amount of
analysed substance.
D Methode des inneren Standards
F méthode de l'étalonnage interne
P metoda wzorca wewnętrznego
R метод внутреннего стандарта

1781
INTERNAL STRAIN (I-strain).
Changes in steric interactions (angular,
conformational and transannular strains)
taking place in reactions of alicyclic
compounds, mainly due to change in
valence state of carbon atom forming
reaction centre. These strains may
stimulate or hinder reaction.
D innere Spannung, I-Spannung
F tension interne, influence I
P napięcie I
R ...

INTERNATIONAL STEAM TABLE
CALORIE. *See* IT CALORIE.

INTERNUCLEAR DISTANCE. *See* BOND
LENGTH.

INTERSTITIAL POSITION. *See*
INTERSTITIAL SITE.

1782
INTERSTITIAL SITE (interstitial
position).
Position in-between periodic nodes in
crystal lattice, occupied by ion or atom.
D Gitterzwischenraum,
 Zwischengitterraum,
 Gitterzwischenplatz, Zwischengitterplatz
F position interstitielle, interdistance, site
 interstitiel
P międzywęźle
R междоузлие

1783
INTERSTITIAL SOLID SOLUTION.
Solid solution in which atoms or ions of
dissolved substance occupy interstitial
positions in lattice in-between atoms or
ions of matrix substance.
D Einlagerungsmischkristalle,
 Überschußmischkristalle
F solution solide d'insertion
P roztwór stały międzywęzłowy, roztwór
 stały śródwęzłowy, roztwór stały
 addycyjny
R твёрдый раствор внедрения

INTERSTITIAL VOLUME. *See* FREE
VOLUME.

1784
INTERSYSTEM CROSSING (*in
photochemistry*).
Isoenergetic radiationless transition
between two electronic states of different
multiplicity.
D strahlungslose Interkombination
F transitions d'intercombinaisons
P konwersja interkombinacyjna
R интеркомбинационная конверсия

INTRA-LABORATORY
REPRODUCIBILITY. *See*
REPEATABILITY.

1785
INTRAMOLECULAR HYDROGEN BOND.
Hydrogen bond in which hydrogen atom is
bound with two different atoms of the
same molecule, e.g. in 2-nitrophenol.

D intramolekulare Wasserstoffbrücke
F liaison hydrogène intramoléculaire,
 liaison hydrogène interne

P wiązanie wodorowe
  wewnątrzcząsteczkowe
R внутримолекулярная водородная связь

1786
INTRAMOLECULAR REACTION.
Proces in which formation of product
results from reaction of functional groups
present in the same molecule, e.g.

D intramolekulare Reaktion
F réaction intramoléculaire
P reakcja wewnątrzcząsteczkowa
R внутримолекулярная реакция

1787
INTRINSIC EFFICIENCY OF DETECTOR.
Ratio of number of particles or photons
which give rise to counts, to number of
particles or photons reaching sensitive
part of detector.
D Ansprechwahrscheinlichkeit des
  Detektors
F efficacité du détecteur, rendement du
  détecteur
P wydajność detektora (wewnętrzna)
R эффективность счётчика, полная
  эффективность регистрации

1788
INTRINSIC SEMICONDUCTOR.
Semiconductor in which electrons
transferred to conduction band as well as
positive holes left in the same number in
ground band are carriers of electric
current. Intrinsic conductivity occurs in
very pure crystals, free from defects.
It may also occur in crystals with defects
at temperatures high enough to neglect
the number of carriers formed by
ionization of impurities, and in crystals
containing equivalent numbers of donor
and acceptor centres.
D Eigenhalbleiter
F semiconducteur intrinsèque
P półprzewodnik samoistny
R собственный полупроводник

1789
INTRINSIC VISCOSITY (limiting
Staudinger function), [η].
Limiting value of reduced viscosity
obtained by extrapolation to zero
concentration

$$[\eta] = \lim_{c \to 0} \eta_{red} = \lim_{c \to 0} \frac{\eta_{sp}}{c}$$

where η_{red} = reduced viscosity,
η_{sp} = specific viscosity, c = concentration.

D Grenzviskosität, Grundviskosität
F viscosité intrinsèque
P lepkość graniczna, liczba lepkościowa
  graniczna
R характеристическая вязкость,
  предельное число вязкости

1790
INVARIANT PHASE SYSTEM (invariant
system).
System in which number of constituents
and phases is determined explicitly by
temperature, pressure and concentration;
change of each of these parameters causes
change of number of constituents or
phases.
D invariantes Phasensystem, invariantes
  System
F système invariant (de phases)
P układ (fazowy) niezmienny, układ
  inwariantny
R инвариантная (фазовая) система,
  безвариантная (термодинамическая)
  система

INVARIANT SYSTEM. See INVARIANT
PHASE SYSTEM.

1791
INVERSION (of sugar).
Change of plane of rotation of
plane-polarized light passing through
aqueous solution of saccharose, caused by
hydrolysis of latter and formation of equal
amounts of D-(+)-glucose and
D-(−)-fructose.
Remark: Historical term, obsolescent.

D Inversion
F inversion
P inwersja cukru
R инверсия сахаров

1792
INVERSION AXIS (inversion symmetry
axis).
Coupling of n-fold rotation axis
(revolution by $360/n°$ angle) with reflection
in symmetry centre placed on axis.
D Inversions(dreh)achse
F axe de symétrie d'inversion
P oś (symetrii) inwersyjna
R инверсионная ось (симметрии)

1793
INVERSION LINE.
In spectroscopy, band corresponding to
transition between inversion levels of
lowest vibrational state of molecule; lies
in microwave region.
D Inversionslinie
F raie d'inversion
P pasmo inwersyjne
R инверсионная линия

1794
INVERSION OF CONFIGURATION.
Change in configuration of chiral centre of molecule caused by chemical reaction.
D Inversion der Konfiguration
F inversion de configuration
P inwersja konfiguracji
R обращение конфигурации

INVERSION POINT. *See* INVERSION TEMPERATURE.

1795
INVERSION SPECTRUM (*of ammonia*).
Spectrum of ammonia in microwave region corresponding to transition between two energy levels arising from splitting of zeroth vibrational level by inversion motion of molecule; inversion spectrum arises due to existence of two equivalent forms of molecule separated by energy barrier.
D Inversionsspektrum
F spectre d'inversion
P widmo inwersyjne
R инверсионный спектр

INVERSION SYMMETRY AXIS. *See* INVERSION AXIS.

1796
INVERSION TEMPERATURE (inversion point), T_i.
Temperature at which Joule-Thomson coefficient changes sign.
D Inversionstemperatur, Inversionspunkt
F température d'inversion
P temperatura inwersji
R температура инверсии, точка инверсии

IODIN. *See* IODINE.

1797
IODINATION.
Substitution of hydrogen atom in molecule of organic compound by iodine atom.
D Jodierung
F iod(ur)ation
P jodowanie
R иодирование

1798
IODINE (iodin *US*), I.
Atomic number 53.
D Jod
F iode
P jod
R иод

IODINE-131. *See* RADIOIODINE.

1799
ION.
Atom or group of atoms with electrical charge equal to elementary charge or its multiple.
D Ion
F ion
P jon
R ион

1800
ION ASSOCIATION COMPLEX (ion pair, outer sphere complex).
Species formed through association (mainly due to electrostatic interaction) of two oppositely charged ions, of which at least one is complex ion, e.g. $[Co(NH_3)_6]^{2+}[SO_4]^{2-}$
D Ionenassoziationskomplex, Ionenpaar
F ...
P kompleks jonowo-asocjacyjny, para jonowa, kompleks zewnątrzsferowy
R сверхкомплексное соединение

1801
ION CLUSTER (ionic cluster).
Hypothetical conglomerate of positive ion with attached neutral molecules in gases, recently experimentally confirmed to be solvation of ions in mass-spectrometry at higher than usual pressures (1.3 kPa and higher).
D Ionencluster
F essaim ionique
P rój jonowy
R ионные асоциации

ION EXCHANGE CAPACITY. *See* TOTAL ION EXCHANGE CAPACITY.

1802
ION EXCHANGE CAPACITY OF STRONGLY ACIDIC/BASIC GROUPS (exchange capacity of highly dissociated cationic/anionic groups).
Number of equivalents of ionogenic groups, able to exchange ions from neutral salt solutions per unit mass or volume of exchanger.
D Salzspaltkapazität
F capacité d'échange des groupes de fortes acidités/basicités
P zdolność wymienna grup silnie kwasowych (zasadowych), zdolność do rozszczepiania soli obojętnych
R обменная ёмкость по сильнодиссоциированным группам

1803
ION EXCHANGE CHROMATOGRAPHY.
Chromatography that uses solid ion exchanger as stationary phase.
D Ionenaustauschchromatographie
F chromatographie d'échange d'ions
P chromatografia jonitowa, chromatografia jonowymienna
R ионообменная хроматография

1804
ION EXCHANGE COLUMN.
Cylindrical tube filled with ion exchanger
and equipped with inlet and outlet for
solution being separated, purified etc.
D Ionenaustauschersäule
F colonne d'échangeur d'ions
P kolumna jonitowa
R ионообменная колонка

1805
ION-EXCHANGE ISOTHERM (*in
chromatography*).
Isotherm characterizing dependence of
equilibrium concentrations of counter-ion
in ion exchanger and external solution
respectively (in given conditions).
D Austauschisotherme
F isotherme d'échange
P izoterma wymiany
R изотерма обмена

1806
ION EXCHANGER (*solid*).
Insoluble solid containing ions that can be
exchanged for other ions from solution.
D Ionenaustauscher
F échangeur d'ions
P jonit, wymieniacz jonów
R ионит, ионообменник

1807
ION EXCHANGE RESIN (organic ion
exchanger).
Ion exchanger whose skeleton is made of
tri-dimensional net of hydrocarbon chains
to which ionogenic groups are attached;
ion exchange resin skeleton can be of
condensation or polymerization type.
D Kunstharz-Ionenaustauscher,
Ionenaustauscherharz
F échangeur organique, résine échangeuse
P jonit organiczny, żywica jonowymienna
R органический ионит, ионообменная
смола

1808
ION-EXCHANGER MEMBRANE.
Ion exchange material of arbitrary
geometrical form (usually sheet or foil)
which can be used as boundary between
two solutions.
D Ionenaustauscher-Membrane,
F membrane échangeuse d'ions
P membrana jonitowa, membrana
jonowymienna
R ионообменная мембрана, ионитовая
диафрагма

1809
ION EXCLUSION.
Method for separation of strong
electrolytes from weak electrolytes and

non-electrolytes on ion exchangers. Strong
electrolytes are excluded from exchanger's
phase (Donnan effect) and are eluted from
column ahead of weak electrolytes and
non-ionic compounds.
D Elektrolytvorlaufverfahren
F exclusion d'ions, occlusion d'ions
P ekskluzja jonów, wykluczanie jonów
R исключение ионов

1810
IONIC ATMOSPHERE (ionic cloud).
In Debye-Hückel's theory, layer of
electrolytic solution in immediate
neighbourhood of ion with a given charge
(central ion), in which mean with time
density of ions with opposite charge is
larger than of ions with identical charge.
D Ionenatmosphäre, Ionenwolke
F atmosphère ionique
P atmosfera jonowa, chmura jonowa
R ионная атмосфера, ионное облако

1811
IONIC BOND (electrovalent bond,
heteropolar bond, polar bond).
Chemical bond formed between atoms of
distinctly different electronegativity.
Valence electron of less negative atom
passes to more negative one and ions thus
formed are held together by electrostatic
forces, e.g. in molecules of NaCl or KCl.
D Ionenbindung, ionogene Bindung,
(hetero)polare Bindung, Ionenbeziehung
F liaison ionique, liaison électrovalente,
liaison d'électrovalence, liaison
hétéropolaire
P wiązanie jonowe, wiązanie
heteropolarne, wiązanie
elektrowalencyjne, wiązanie biegunowe,
wiązanie polarne
R ионная связь, гетерополярная связь,
электровалентная связь

IONIC CLOUD. *See* IONIC
ATMOSPHERE.

IONIC CLUSTER. *See* ION CLUSTER.

IONIC COMPLEX. *See* NORMAL
COMPLEX.

1812
IONIC CONDUCTANCE (λ_+ or λ_-).
$\lambda_+ = Fu_+$ or $\lambda_- = Fu_-$
where u_+ and u_- = mobilities of cation
and anion respectively; also conductance
of 1 gram equivalent of a given sort of ion
in solution of these ions with
concentration c.
D Ionen(äquivalent)leitfähigkeit
F conductivité ionique équivalente
P przewodnictwo jonowe równoważnikowe
R ионная электропроводность

1813
IONIC CONDUCTOR (electrolytic conductor).
Conductor conducting electrical current by ions.
D Ionenleiter, elektrolytischer Leiter
F conducteur ionique, conducteur électrolytique
P przewodnik jonowy, przewodnik elektrolityczny
R ионный проводник, электролитический проводник

1814
IONIC CRYSTALS.
Crystals composed of positive and negative simple or complex ions; they are characterized by low electric conductivity at low temperature and high ionic conductance at high temperatures.
D Ionenkristalle
F cristaux ioniques, cristaux d'ions
P kryształy jonowe
R ионные кристаллы

IONIC DEFECTS. *See* ELECTRONIC DEFECTS.

1815
IONIC MOBILITY (absolute ionic mobility), u.
Ratio of velocity of ion in cm/s to electrical field in V/cm in which ion is travelling.
D Ionenbeweglichkeit, Wanderungsgeschwindigkeit des Ions
F mobilité d'ion
P ruchliwość jonu
R подвижность иона, абсолютная скорость движения иона

1816
IONIC POLYMERIZATION.
Polymerization catalyzed by acids, bases or complexes in which ions or macroions take part in all stages of process.
D ionische Polymerisation
F polymérisation ionique
P polimeryzacja jonowa
R ионная полимеризация

1817
IONIC PRODUCT, $K(L)$.
Product of activities of cations and anions resulting from dissociation of weak or weakly soluble electrolytes

$$K = a_+^{\nu_+} \cdot a_-^{\nu_-}$$

where a_+ and a_- = concentration activities of cation and anion, respectively, ν_+ and ν_- = quantities of cations and anions resulting from one molecule of electrolyte. Value of ionic product depends on temperature.

D Ionenprodukt
F produit ionique
P iloczyn jonowy
R ионное произведение

1818
IONIC PRODUCT OF SOLVENT.
Equilibrium constant for autoprotolysis reaction of solvent if activity of undissociated part is equal to one.
D Ionenprodukt des Lösungsmittels
F produit ionique du solvant
P iloczyn jonowy rozpuszczalnika
R ионное произведение растворителя

1819
IONIC PRODUCT OF WATER.
Equilibrium constant for water autoprotolysis reaction, written as

$$L_{H_2O} = a_{H_3O^+} \cdot a_{OH^-}$$

since $a_{H_2O} = 1$ due to negligable self-dissociation of water.
D Ionenprodukt des Wassers
F produit ionique de l'eau
P iloczyn jonowy wody
R ионное произведение воды

1820
IONIC RADIUS.
Radius of ion in crystal or ionic molecule. Sum of ionic radii of two neighbouring ions determines ionic bond length.
D Ionenradius
F rayon ionique
P promień jonowy
R ионный радиус

1821
IONIC REACTION.
Reaction in which substrates, intermediates or products are ions.
D Ionenreaktion
F réaction ionique
P reakcja jonowa
R ионная реакция

1822
IONIC REFRACTIVITY.
Part of experimental value of molar refraction for ion in ionic compound computed according to additive scheme.
D Ionenrefraktion
F réfraction ionique
P refrakcja jonowa
R ионная рефракция

1823
IONIC SIEVES.
Ion exchangers (most often inorganic) of such pore dimensions which enable exchange of smaller ions while mechanically excluding larger ones; in case of big differences in counter ion dimensions also organic ion exchangers can act as ionic sieves.

D Ionensiebe
F tamis ioniques
P sita jonowe
R ионные сита

1824
IONIC STRENGTH, J or J^*.
Parameter which describes electrostatic interactions in solution of strong electrolyte:

$$J \equiv \frac{1}{2} \sum c_i z_i^2 \quad \text{or}$$

$$J^* \equiv \frac{1}{2} \sum m_i z_i^2$$

where c = molar concentration, m = molality of ions in solution, z = ion charge; sum is taken over all ions in solution, e.g. for 1 m $CaCl_2$ $J^* = 3$, for 1 m $K_4[Fe(CN)_6]$ $J^* = 10$ etc.

D Ionenstärke
F force ionique
P moc jonowa, siła jonowa
R ионная сила

1825
IONIC VALENCY.
Number of positive or negative charges of given ion.

D Ionen-Wertigkeit, Ionenladungszahl
F valence d'ion, valence ionique
P wartościowość jonu
R валентность иона

1826
IONIC YIELD, M/N.
Ratio of number of molecules which undergo change in irradiated gas to number of the ion pairs formed by ionizing radiation in the same volume.

D Ionenausbeute
F rendement ionique
P wydajność jonowa
R ионный выход

1827
IONIUM, Io.
Isotope of thorium. Atomic number 90. Mass number 230.

D Ionium
F ionium
P jon
R ионий

1828
IONIZATION.
Process in which atom or molecule losses one or more electrons and forms positive ion; also dissociation process of neutral molecule to ions.

D Ionisation, Ionisierung
F ionisation
P jonizacja
R ионизация

1829
IONIZATION CHAMBER.
Hollow chamber with electrodes, filled with suitable gas, enabling measurement of intensity of ionizing radiation from intensity of current due to ions formed.

D Ionisationskammer
F chambre d'ionisation
P komora jonizacyjna
R ионизационная камера

1830
IONIZATION DENSITY.
Number of ion pairs in unit volume of irradiated medium, also the number of ions per unit length of track of the ionizing particle.

D Ionisationsdichte
F densité d'ionisation, densité ionique
P gęstość jonizacji
R плотность ионизации

1831
IONIZATION DETECTOR.
Device that records composition of gas coming out of chromatographic column by ionization of gas molecules and electrometric measurement of ion current.

D Ionisationsdetektor
F détecteur par ionisation
P detektor jonizacyjny
R ионизационный детектор

1832
IONIZATION ENERGY.
Smallest amount of energy necessary for single ionization of atom or molecule.

D Ionisierungsenergie
F énergie d'ionisation
P energia jonizacji
R энергия ионизации

1833
IONIZATION ISOMERISM.
Type of isomerism of complex compounds concerned with different distribution of anions in first coordination sphere and beyond it, e.g. $[Co(SO_4)(NH_3)_5]Br$, and $[CoBr(NH_3)_5]SO_4$.

D Ionisationsisomerie
F isomérie d'ionisation
P izomeria jonizacyjna
R ионизационная изомерия

1834
IONIZATION POTENTIAL.
Energy in electronvolts necessary to detach the weakest bound electron from atom or molecule.

D Ionisierungsspannung, Ionisationspotential
F potentiel d'ionisation, tension d'ionisation
P potencjał jonizacji
R потенциал ионизации

1835
IONIZING PARTICLE.
Elementary particle, fast-moving atom or ion with energy sufficient to produce ionization of medium.
D ionisierende Strahlenpartikel
F particule ionisante
P cząstka jonizująca
R ионизирующая частица

1836
IONIZING RADIATION.
Electromagnetic or corpuscular radiation with energy of a photon or particle high enough to remove an electron from atom or molecule.
D ionisierende Strahlung
F rayonnement ionisant
P promieniowanie jonizujące
R ионизирующее излучение

1837
ION LINE (spark line *in emission spectral analysis*).
Spectral line in emission spectrum of excited ions.
D Ionenlinie, Funkenlinie
F raie d'ion, raie d'étincelle
P linia jonowa
R ионная линия, искровая линия

1838
ION-MOLECULE REACTION.
Reaction of ion-radical with stable molecule.
D Ion-Molekül-Reaktion
F ...
P reakcja jonocząsteczkowa
R ион-молекулярная реакция

1839
IONOPHORESIS.
Movement of ions in liquid under influence of electric field gradient.
D Ionophorese
F ionophorèse
P jonoforeza
R ионофорез

1840
ION PAIR (*in radiology*).
Parent positive ion and electron formed as result of ionization.
D Ionenpaar
F paire d'ions
P para jonowa
R ионная пара

ION PAIR. *See* ION ASSOCIATION COMPLEX.

1841
ION RADICAL (radical ion).
Free radical possessing electric charge, e.g. O^-, O_2^-, CO_2^+, CH_4^+, R_2C-O^-.

D Ionradikal, Radikalion
F ion radical
P jonorodnik
R ион-радикал

1842
ION RETARDATION.
Method for separation of strong electrolytes from weak electrolytes and non-electrolytes on amphoteric ion exchangers, which makes use of differences in sorption strength of individual components; strong electrolytes are more strongly retarded when passing column with amphoteric exchanger than weak electrolytes and non-ionic compounds.
D Ionenverzögerungs-Verfahren
F ion retardation
P hamowanie jonów
R способ отстающего электролита

1843
ION-SELECTIVE ELECTRODE.
Electrode (half-cell) in which potential, dependent on activity of a certain ion of electrolyte, arises at the interphase of electrode material and electrolyte.
D ionensensitive Elektrode
F ...
P elektroda jonoselektywna
R ионоселективный электрод

1844
ION SPECTRUM.
Spectrum of singly or multiply ionized atom.
D Ionenspektrum
F spectre ionique
P widmo jonowe
R ионный спектр

1845
IRIDIUM, Ir.
Atomic number 77.
D Iridium
F iridium
P iryd
R иридий

1846
IRON, Fe.
Atomic number 26.
D Eisen
F fer
P żelazo
R железо

IRON-59. *See* RADIOIRON.

1847
IRON GROUP.
Three elements in eighth sub-group of periodic system with structure of outer electronic shells of atoms: iron — $3d^64s^2$, ruthenium — $4d^75s^1$, osmium — $5d^66s^2$.

D...
F...
P żelazowce
R подгруппа железа, металлы подгруппы железа, семейство железа

1848
IRRADIATION.
Treatment with ionizing radiation.

D Bestrahlung
F irradiation
P napromienianie
R облучение

1849
IRRADIATION CHAMBER.
Space in irradiation device in which objects are placed for purpose of ionizing radiation treatement.

D Bestrahlungskammer, Bestrahlungsraum
F chambre d'irradiation
P komora radiacyjna
R камера для облучения

1850
IRRADIATION LOOP.
Closed-circuit system of pipes passing through or close to core of nuclear reactor. Irradiation loop is used for: (1) direct irradiation of through-flowing liquids or gases, (2) (n, γ) activation of circulating molten metal which releases energy as pure gamma radiation (i.e. without neutrons) in irradiation chamber through which loop passes.

D Bestrahlungsschleife
F boucle d'irradiation
P pętla radiacyjna
R радиационный контур

1851
IRREGULAR POLYMER.
Polymer is irregular when its macromolecules cannot be split into constitutional units of one species and in one sequential arrangement.

D...
F polymère irrégulier
P polimer nieregularny
R нерегулярный полимер

IRREVERSIBLE CHANGE. See
IRREVERSIBLE PROCESS.

1852
IRREVERSIBLE COAGULATION.
Coagulation in which product (precipitate or gel) cannot be reversibly transformed into sol state, e.g. denaturation of proteins, coagulation of lyophobic colloids by action of electrolytes.

D irreversible Koagulation
F coagulation irréversible
P koagulacja nieodwracalna
R необратимая коагуляция

1853
IRREVERSIBLE COLLOID.
Colloidal system from which, after removal of dispersion medium by careful evaporating and its renewed addition, initial system is not formed.

D irreversibles Kolloid
F colloïde irréversible
P koloid nieodwracalny
R необратимый коллоид

1854
IRREVERSIBLE PROCESS (irreversible change, natural process).
Thermodynamic process accompanied by production of entropy either in system or at boundary between system and surroundings; if irreversible process takes place, then according to second law of thermodynamics, simultaneous return of both system and surroundings to initial state is impossible.

D irreversibler Prozeß, irreversibler Vorgang, nichtumkehrbarer Vorgang
F transformation irréversible
P przemiana nieodwracalna, proces nieodwracalny
R необратимый процесс

1855
IRREVERSIBLE REACTION.
Reaction that proceeds until one of reactants has been exhausted.

D irreversible Reaktion, unumkehrbare Reaktion, nicht umkehrbare Reaktion
F réaction irréversible
P reakcja nieodwracalna
R необратимая реакция

IRREVERSIBLE THERMODYNAMICS.
See THERMODYNAMICS OF
IRREVERSIBLE PROCESSES.

IR SPECTRUM. See INFRARED
SPECTRUM.

1856
IRVING-WILLIAMS SERIES
(Irving-Williams stability series).
Divalent cations of metals of second half of first transition series arranged by stability of their complexes: $Mn^{2+} < Fe^{2+} < Co^{2+} < Ni^{2+} < Cu^{2+} < Zn^{2+}$

D Irving-Williams-Reihe
F série d'Irving-Williams, ordre des stabilités d'Irving-Williams
P szereg Irvinga i Williamsa
R ряд Ирвинга-Уильямса

IRVING-WILLIAMS STABILITY
SERIES. See IRVING-WILLIAMS
SERIES.

ISENTROPIC CHANGE. See
ISENTROPIC PROCESS.

1857
ISENTROPIC PROCESS (isentropic change).
Thermodynamic process with constant entropy of system.
D isentrope Zustandsänderung
F transformation isentropique
P przemiana izoentropowa, proces izoentropowy
R изоэнтропийный процесс, изоэнтропический процесс

1858
ISING MODEL.
Abstract model of crystal composed of space-fixed atoms whose spins that is also magnetic moments have only two possible opposite directions with respect to some selected direction. It is assumed that only nearest-neighbours interact. Ising model is used in statistical mechanics to describe higher-order phase transitions, e.g. in ferromagnetics and antiferromagnetics.
D Ising-Modell
F modèle d'Ising
P model Isinga
R модель Изинга

1859
ISOABSORPTIVE POINT.
Point on wavelength scale with equal absorption coefficients for two or more substances.
D ...
F ...
P punkt izoabsorpcyjny
R ...

1860
ISOBAR.
Set of states of system with the same pressure value.
D Isobare
F isobare
P izobara
R изобара

1861
ISOBARIC PROCESS.
Thermodynamic process at constant pressure.
D isobarer Prozeß, isobare Zustandsänderung
F transformation isobare, transformation à pression constante
P przemiana izobaryczna, proces izobaryczny
R изобарный процесс

ISOBARIC SPIN. *See* ISOSPIN.

1862
ISOBARS.
Nuclides having identical mass numbers A, but different atomic numbers Z, viz.

having identical number of nucleons but different number of protons.
D Isobare
F izobares
P izobary
R изобары

1863
ISOCHORE.
Set of states with identical volume of system.
D Isochore
F isochore
P izochora
R изохора

1864
ISOCHORIC PROCESS.
Thermodynamic process with constant volume of system.
D isochorer Prozeß
F transformation isochore, transformation à volume constante
P przemiana izochoryczna, proces izochoryczny
R изохорный процесс

ISOCYANIDE REACTION. *See* CARBYLAMINE REACTION.

1865
ISODIMORPHISM.
Ability to form substitutional solid solutions (in limited range of compositions) by two compounds which occur in different crystallographic systems (e.g. $NaClO_3$ of regular lattice, with $AgClO_3$ of tetragonal lattice).
D Isodimorphie
F isodimorphisme
P izodimorfizm
R изодиморфизм

ISODISPERSE SYSTEM. *See* MONODISPERSE SYSTEM.

1866
ISODOSE.
Curve or surface connecting points of equal dose rate in irradiated system.
D Isodosis
F isodose
P izodoza
R изодоза

1867
ISOELECTRIC POINT.
State of colloidal system characterized by zero value of electrokinetic potential of colloidal particles. The point is reached by discharging colloidal particles through addition of suitable quantity of electrolyte or by changing pH value of solution through addition of acid or base (e.g. in the case of proteins).

D isoelektrischer Punkt
F point isoélectrique
P punkt izoelektryczny
R изоэлектрическая точка

1868
ISOELECTRONIC MOLECULES.
Molecules with the same number of valence electrons, e.g. N_2 and CO.

D iso-elektronische Moleküle
F molécules isoélectroniques
P cząsteczki izoelektronowe
R изоэлектронные молекулы

1869
ISOELECTRONIC PRINCIPLE.
Molecules with the same number of electrons have similar molecular orbitals, and consequently, also similar physicochemical properties, e.g. CO and N_2 molecules.

D isoelektronisches Prinzip
F principe isoélectronique
P zasada izoelektronowości
R принцип изоэлектронных орбиталей

1870
ISOLATED DOUBLE BONDS.
Double bonds isolated one from another by two or more single bonds.

D isolierte Doppelbindungen
F liaisons doubles isolées
P wiązania podwójne izolowane
R изолированные двойные связи

1871
ISOLATED SYSTEM.
System which cannot exchange either energy or mass with surroundings.

D abgeschlossenes System, isoliertes System
F système isolé
P układ izolowany
R изолированная система, замкнутая система

1872
ISOLATION METHOD OF OSTWALD.
Method of determination of partial order of reaction n with respect to one substrate, consisting in elimination of changes in concentration of remaining substrates by introducing them to system in high stoichiometric excess. During run of reaction, its rate, defined as decay of concentration of deficient reactant, should be proportional to its concentration to power n.

D ...
F méthode d'isolement (d'Ostwald)
P metoda izolacyjna Ostwalda
R ...

ISOMERIC NUCLEUS. See
METASTABLE NUCLEUS.

1873
ISOMERIC TRANSITION.
Radioactive transition from one nuclear isomer to another of lower energy, with emission of gamma radiation or conversion electrons.

D Isomerenübergang
F transition isomérique
P przemiana izomeryczna, przejście izomeryczne
R изомерный переход

1874
ISOMERISM.
Different arrangement of atoms in molecules having the same elemental composition.

D Isomerie
F isomérie
P izomeria
R изомерия

1875
ISOMERIZATION.
Transformation of a chemical compound into another differing in structure or configuration but of identical molecular formula, e.g.

$$CH_3CH_2CH_2CH_3 \xrightarrow[50-100°C]{AlCl_3} \begin{array}{c} CH_3 \\ CH_3 \end{array} CHCH_3$$

D Isomerisation
F isomérisation
P izomeryzacja
R изомеризация

1876
ISOMERS.
Two or more substances which have the same elementary composition, but differ in structures, and hence in properties.

D Isomere
F isomères
P izomery
R изомеры

1877
ISOMORPHISM.
Similarity of crystalline forms of two chemical compounds which enables them to form substitutional solid solutions and alternate crystal growth of both compounds on each other's crystals. Necessary conditions of isomorphism are: identical type of unit cell of both compounds, similar dimensions of unit cells and the same type of chemical formula (e.g. KNO_3 and $CaCO_3$).

D Isomorphismus, Isomorphie
F isomorphisme, isomorphie
P izomorfizm
R изоморфизм

ISOMULTIPLETS. See CHARGE
MULTIPLETS.

1878
ISOPOLYACID.
Polyacid containing only one acid-forming
element, e.g. 7-molybdic acid $H_6[Mo_7O_{24}]$.
D Isopolysäure
F isopolyacide
P izopolikwas
R изополикислота

1879
ISOSBESTIC POINT.
Point on wavelength scale of equal
absorption coefficients of two or more
compounds existing in chemical
equilibrium.
D isosbestischer Punkt
F ...
P punkt izozbestyczny
R изобестическая точка

1880
ISOSPIN (isotopic spin, isobaric spin), I.
Quantum number attributed to elementary
particles and atomic nuclei according to
formula $n = 2I+1$, where n denotes
number of states in isospin multiplet.
D Isospin, Isotopenspin, Isobarenspin
F isospin, spin isotopique, spin isobarique
P izospin, spin izotopowy, spin izobaryczny
R изоспин, изотопический спин,
изобарический спин

1881
ISOSPIN CONSERVATION LAW.
In strong interactions, vectorial sums of
isospins of initial and final particles are
equal.
D Erhaltungssatz des Isospin
F loi de conservation d'isospin
P prawo zachowania izospinu
R закон сохранения изоспина

1882
ISOSTRUCTURAL COMPOUNDS.
Two or more chemical compounds
crystallizing in the same type of structure
in spite of differences in chemical
composition.
D isostrukturelle Verbindungen, isotype
Verbindungen
F composés isostructuraux
P związki izostrukturalne, związki
izotypowe
R изоструктурные соединения

1883
ISOTACTIC POLYMER.
Regular polymer composed of
macromolecules having regular repetition
of identical configurational base units at
least at one main-chain chiral or prochiral
atom in each configurational base unit.
In isotactic polymer molecule, stereobase
unit is identical with configurational base
unit

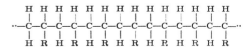

D isotaktisches Polymer
F polymère isotactique
P polimer izotaktyczny
R изотактический полимер

1884
ISOTHERM (isothermal).
Set of states of system of identical
temperature.
D Isotherme
F isotherme
P izoterma
R изотерма

ISOTHERMAL. *See* ISOTHERM.

1885
ISOTHERMAL CALORIMETER.
Type of calorimeter in which there is in
practice no temperature change when
chemical reaction or physical process is
carried out inside it. Amount of heat
evolved or absorbed is determined by
change in volumetric ratio of two phases
in equilibrium with each other of
appropriate calorimetric substance filling
jacket.
D isothermes Kalorimeter
F calorimètre isotherme
P kalorymetr izotermiczny
R изотермический калориметр

ISOTHERMAL CHANGE. *See*
ISOTHERMAL PROCESS.

ISOTHERMAL COMPRESSIBILITY. *See*
COEFFICIENT OF COMPRESSIBILITY.

1886
ISOTHERMAL PROCESS (isothermal
change, isothermic transition).
Thermodynamic process at constant
temperature of system.
D isotherme Zustandsänderung, isothermer
Vorgang
F transformation isotherme
P przemiana izotermiczna, proces
izotermiczny
R изотермический процесс

ISOTHERMIC TRANSITION. *See*
ISOTHERMAL PROCESS.

1887
ISOTONES.
Nuclides having the same number of
neutrons $A-Z$ but different number of
protons Z in their nuclei.
D Isotonen
F isotones
P izotony
R изотоны

1888
ISOTOPE DILUTION (isotopic dilution analysis).
Method for determination of substance by tracing change in its isotopic composition caused by addition of known amount of this substance of different isotopic composition.

D Isotopenverdünnung, Isotopen-Verdünnungsanalyse, Methode der Isotopenverdünnung
F dilution isotopique, analyse par dilution isotopique, méthode de dilution isotopique
P rozcieńczenie izotopowe, analiza metodą rozcieńczenia izotopowego, metoda rozcieńczenia izotopowego
R изотопное разбавление, анализ методом изотопного разбавления, метод изотопного разведения

1889
ISOTOPE EFFECT (*in chemical reactions*).
Change of reaction rate caused by introduction into molecule of isotope of atom whose removal is kinetically the slowest process. Isotope is located at the same place original atom occupied. Deuterium and tritium are applied as substituents for hydrogen due to the largest difference in their isotopic mass.

D kinetischer Isotopeneffekt
F effet isotopique (*dans les réactions chimiques*)
P efekt izotopowy kinetyczny
R изотопический эффект (*в химических реакциях*)

1890
ISOTOPE EFFECTS.
Changes in properties of substance owing to replacement of an element by its isotope. Effects are most distinct when hydrogen is replaced by deuterium or tritium (the largest relative difference in mass between an element's isotopes).

D Isotopeneffekte
F effets isotopiques
P efekty izotopowe
R изотопные эффекты

1891
ISOTOPES.
Nuclides of the same atomic number Z but different mass number A, i.e. of the same number of protons but different number of neutrons.

D Isotope
F isotopes
P izotopy
R изотопы

1892
ISOTOPIC ANALYSIS.
Determination of particular isotopes in a given element or its compounds.

D Isotopenanalyse
F analyse isotopique
P analiza izotopowa
R анализ изотопов, изотопный анализ

1893
ISOTOPIC CARRIER.
Carrier for radioactive substance chemically identical with substance, e.g. inactive SO_4^{2-} ions for $^{35}SO_4^{2-}$ ions.

D isotop(isch)er Träger
F entraîneur isotopique
P nośnik izotopowy
R изотопный носитель

1894
ISOTOPIC CROSS-SECTION.
Cross-section for specified isotope of a given element for a definite nuclear reaction.

D isotoper Wirkungsquerschnitt
F section efficace isotopique
P przekrój czynny izotopowy
R изотопное (эффективное) сечение

ISOTOPIC DILUTION ANALYSIS. *See* ISOTOPE DILUTION.

1895
ISOTOPIC ENRICHMENT.
Increase in content of specified isotope in mixture of isotopes.

D Isotopenanreicherung
F enrichissement isotopique
P wzbogacanie izotopowe
R изотопное обогащение

1896
ISOTOPIC EXCHANGE.
Process whereby distribution of isotopes between different phases, different chemical compounds or inside molecules changes.

D Isotopenaustausch
F échange isotopique
P wymiana izotopowa
R изотопный обмен

1897
ISOTOPIC NEUTRON SOURCE.
Source emitting neutrons (usually besides radiation of other types) as a result of a nuclear reaction occurring in material of source.

D radioaktive Neutronenquelle
F source isotopique de neutrons
P źródło neutronów izotopowe
R ампульный источник нейтронов

ISOTOPIC SPIN. *See* ISOSPIN.

1898
ISOTOPIC TRACER.
Isotope which is introduced into a system in order to examine it.
D Leitisotop, isotoper Indikator
F indicateur isotopique, traceur isotopique
P wskaźnik izotopowy
R изотопный индикатор

1899
ISOTOPIC TRACER METHOD.
General experimental method of wide use based on introduction of isotopic tracer (most frequently a radioactive one) into system, followed by examination of its behaviour during different transformations in system.
D Leitisotopenmethode, Indikatormethode
F méthode des indicateurs istotopiques
P metoda wskaźników izotopowych
R метод изотопных индикаторов

ISOTROPISM. *See* ISOTROPY.

1900
ISOTROPY (isotropism).
Independence of physical property of direction (most often crystallographic). All scalar physical properties are isotropic. Also in regular crystal all properties which may be described by vectors are isotropic.
D Isotropie
F isotropie
P izotropia
R изотропия

I-STRAIN. *See* INTERNAL STRAIN.

1901
IT CALORIE (international steam table calorie), cal (cal_{IT}).
Obsolete energy unit; in SI system
1 cal = 4.1868 J.
D internationale Kalorie, MKS-Kalorie
F calorie internationale, calorie I.T.
P kaloria (międzynarodowa), kaloria termotechniczna
R теплотехническая (международная) калория

1902
IVANOV REACTION.
Reaction of organomagnesium compound (obtained from phenylacetic acid salt — Ivanov reagent) with carbonyl compounds, resulting in formation of β-substituted α-phenyl-β-hydroxypropionic acids, e.g.

$C_6H_5CH_2COONa \xrightarrow{(CH_3)_2CHMgX}$

$C_6H_5CHCOOMgX \xrightarrow{\underset{RCR'}{\overset{O}{\|}}} R-\underset{\underset{C_6H_5}{|}}{\overset{\overset{OH}{|}}{C}}-CH-COOH$
$\quad\quad |$
$\quad\; MgX$ $\quad\quad\quad\quad R'$

D Iwanow-Reaktion, Ivanoff Hydroxycarbonsäure-Synthese
F ...
P reakcja Iwanowa
R реакция Иванова

JACOBSEN REACTION. *See* JACOBSEN REARRANGEMENT.

1903
JACOBSEN REARRANGEMENT
(Jacobsen reaction).
Migration of alkyl group or halogen atom in polyalkyl- or halogenopolyalkylbenzenesulfonic acids, effected by concentrated sulfuric acid, e.g.

D Jacobsen-Reaktion, Jacobsen Alkylwanderung
F transposition de Jacobsen
P przegrupowanie Jacobsena
R реакция Якобсена

1904
JAHN-TELLER EFFECT.
Inseparability of nuclear and electronic motions in degenerate electronic states of non-linear polyatomic molecules.
D Jahn-Teller-Effekt, Jahn-Teller-Kopplung
F effet de Jahn-Teller
P efekt Jahna i Tellera
R эффект Яна-Теллера

1905
JAHN-TELLER EFFECT (*in coordination chemistry*).
Decrease in symmetry of complex (e.g. from octahedral to tetragonal) caused by internal asymmetry due to incomplete filling with electrons of nonbinding d orbitals of central ion.
D Jahn-Teller-Effekt, Jahn-Teller-Verzerrung
F effet Jahn-Teller
P efekt Jahna i Tellera
R эффект Яна-Теллера

1906
JAHN-TELLER THEOREM.
The most favourable energetic state of non-linear paramagnetic molecule is the one having lowest non-degenerate energy level of paramagnetic atom. This leads to spontaneous deformation of molecules (exception — *see* Kramers' theorem).
D Jahn-Teller-Theorem
F théorème de Jahn-Teller
P teoremat Jahna i Tellera
R теорема Яна-Теллера

1907
JAPP-KLINGEMANN REACTION.
Preparation of arylhydrazones by coupling in alkaline medium of aryldiazonium chlorides with β-dicarbonyl compounds containing active hydrogen atoms, e.g.

$$CH_3COCHCOOC_2H_5 \rightleftharpoons CH_3C{=}CCOOC_2H_5 +$$
$$\overset{|}{CH_3} \qquad\qquad \overset{|}{HO}\ \overset{|}{CH_3}$$

$$+ \overset{\oplus}{ArN}{\equiv}NCl^{\ominus} \xrightarrow{\overset{\ominus}{OH}} CH_3COC{-}\overset{\overset{COOC_2H_5}{|}}{N}{=}NAr \xrightarrow{H_2O}$$
$$\overset{|}{CH_3}$$

$$CH_3COOH + CH_3C{=}N{-}NHAr$$
$$\overset{\overset{COOC_2H_5}{|}}{}$$

D Japp-Klingemann
    Arylhydrazon-Darstellung
F réaction de Japp-Klingemann
P reakcja Jappa i Klingemanna
R ...

1908
j-j COUPLING.
Interaction between spin and angular momentum of electron giving total angular momentum determined by quantum number j; each total angular momentum interacts with total angular momenta of other electrons in atom or ion and contributes to total angular momentum of whole system.

D jj-Kopplung
F couplage jj
P sprzężenie jj
R связь jj

1909
JOULE'S LAW.
Internal energy of ideal gas depends only on its temperature.

D zweites Gay-Lussacsches Gesetz
F loi de Joule
P prawo Joule'a
R закон Джоуля

1910
JOULE-THOMSON COEFFICIENT, μ.
Coefficient characterizing quantitatively Joule-Thomson effect, viz.:

$$\mu = \left(\frac{\partial T}{\partial p}\right)_{H_m}$$

where T = thermodynamic temperature, p = pressure, H_m = molal enthalpy of gas. Numerical value of μ is measure of departure of real gases from ideal gas state.

D Joule-Thomsonscher Koeffizient
F coefficient de Joule-Thomson
P współczynnik Joule'a i Thomsona
R коэффициент Джоуля-Томсона

1911
JOULE-THOMSON EFFECT.
Temperature change in adiabatic process of free expansion of real gas without any external work being done.

D Joule-Thomson Effekt
F effet Joule-Thomson
P efekt Joule'a i Thomsona
R эффект Джоуля-Томсона

KAON. See K-MESON.

KATHAROMETER. See THERMAL CONDUCTIVITY CELL.

1912
KAYSER, K.
Wavelength unit, 1 K = 1 cm^{-1}.

D Kayser
F kayser
P kajzer
R кайзер

1913
K BAND.
Intensive absorption band characteristic of conjugated systems corresponding to allowed π-π^* electronic transition.

D K-Band
F bande K
P pasmo K
R полоса K

1914
K CAPTURE (K electron capture).
Electron capture from K shell by nucleus of atom.

D K-Einfang
F capture K
P wychwyt K, przemiana K
R K-захват

1915
KÉKULÉ STRUCTURES.
Formulae of benzene and its derivatives, proposed by Kékulé, describing benzene as a cyclohexatriene in which double bonds are in constant oscillation between adjacent positions (which were to explain lack of expected isomers of the two and higher substituted derivatives), e.g.

D Kékulé-Formeln
F formules de Kékulé, formes de Kékulé
P struktury Kékulégo, wzory Kékulégo
R структуры Кекуле

K ELECTRON CAPTURE. See K CAPTURE.

1916
KELVIN (degree Kelvin, degree absolute),
K.
Unit of thermodynamic temperature equal
to 1/273.16 of thermodynamic temperature
of triple phase point of water.
Remark: Term kelvin and symbol K are used
both for denoting temperature value
(previously marked °K), and for denoting
temperature interval (previously marked deg).

D (Grad) Kelvin
F kelvin, degré (absolu) Kelvin
P kelwin, stopień Kelvina
R градус Кельвина, кельвин

1917
KELVIN EQUATION.
Equation describing dependence of
saturated vapour pressure (at constant
temperature) on curvature of surface of
a liquid; the equation enables maximum
radius of capillary filled with liquid of
given vapour pressure to be determined
for capillary condensation, viz.:

$$\ln \frac{p}{p_0} = -\frac{2\sigma V}{rRT}\cos\Theta$$

where V = molar volume of liquid,
σ = surface tension of adsorbate,
Θ = wetting angle of adsorbent, r = radius
of capillary, T = thermodynamic
temperature, R = gas constant,
p_0 = saturated vapour pressure of
adsorbate for flat liquid surface.

D Thomsonsche Gleichung
F équation de Kelvin
P równanie Kelvina
R уравнение (Томсона-)Кельвина

1918
KERMA.
Energy supplied to defined medium by
gamma or neutron radiation as kinetic
energy of secondary charged particles.
Remark: Abbreviation from kinetic energy
released in material.

D Kerma
F kerma
P kerma
R керма

1919
KERMA RATE.
Kerma per unit time.

D Kermarate
F débit de kerma
P moc kermy
R ...

1920
KERR CONSTANT, B.
Measure of Kerr effect; proportionality
factor in equation

$$n_A - n_N = B\lambda E^2$$

which expresses dependence between
difference of refractive indices for normal
n_N and abnormal n_A ray and wavelength λ
and electric field strength E.

D Kerr-Konstante
F constante de Kerr
P stała Kerra
R константа Керра

1921
KERR EFFECT (electric double
refraction).
Electric field-induced anisotropy of
refractive index of liquids.

D Kerr-Effekt, elektrische Doppelbrechung
F effet Kerr, biréfringence électrique
P efekt Kerra, dwójłomność Kerra
R явление Керра, электрическое двойное
преломление

1922
KETO-ENOL TAUTOMERISM.
Type of prototropic tautomerism
connected with reversible rearrangement
of α,β-unsaturated alcohols (enols) into
ketones and, in wider sense, of enols into
carbonyls.

$$\mathrm{C=C} \begin{array}{c} \\ \mathrm{OH} \end{array} \rightleftarrows \begin{array}{c} \mathrm{C-C-} \\ | \quad || \\ \mathrm{H} \quad \mathrm{O} \end{array}$$

D Keto-Enol-Tautomerie
F tautomérie céto-énolique
P tautomeria keto-enolowa
R кето-енольная таутомерия

1923
KETO FORM
Tautomeric form having carbonyl group
in molecule and at least one hydrogen

atom at α-carbon $\mathrm{H-C-C} \overset{\displaystyle O}{\diagdown}$. Keto form

exists in tautomeric equilibrium with enol
form

$$\mathrm{CH_3-C-H} \underset{}{\longleftrightarrow} \mathrm{CH_2=C-H}$$
$$\quad\quad || \qquad\qquad\qquad |$$
$$\quad\quad \mathrm{O} \qquad\qquad\qquad \mathrm{OH}$$

D Keto-Form
F forme cétonique
P odmiana ketonowa, odmiana
karbonylowa
R кетоформа

1924
KETONIC HYDROLYSIS.
Decomposition of β-ketoacid esters by
water, leading to formation of
corresponding ketones, alcohols and
carbon dioxide; process involves
hydrolysis of esters with subsequent
decarboxylation of acid formed, e.g.

$$\overset{\displaystyle O}{\underset{||}{}} \qquad\qquad\qquad \overset{\displaystyle O}{\underset{||}{}}$$
$$\mathrm{CH_3CCH_2COOC_2H_5} \xrightarrow[\mathrm{H_2O}]{\mathrm{H}^{\oplus}} \mathrm{CH_3CCH_3 + CO_2 + C_2H_5OH}$$

D Ketonspaltung
F dédoublement cétonique
P rozpad ketonowy
R кетонное расщепление

1925
KILIANI-FISCHER METHOD FOR EXTENDING AN ALDOSE CHAIN.
Method for extending aldose chain by formation of epimeric cyanohydrins, followed by hydrolysis to onic acids and consecutive reduction

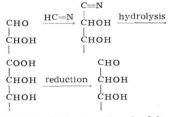

D Kiliani-Fischer Cyanhydrin-Synthese
F synthèse de Kiliani-Fischer
P reakcja (wydłużenia łańcucha aldoz) Kilianiego i Fischera
R удлинение цепи альдоз Килиани-Фишера

1926
KINEMATIC VISCOSITY, ν.
Ratio of dynamic viscosity to density of liquid, $\nu = \eta/\rho$ where η = dynamic viscosity, ρ = density. Unit of kinematic viscosity in SI system is m^2/s.

D kinematische Viskosität
F viscosité cinématique
P lepkość kinematyczna, współczynnik lepkości kinematycznej
R кинематическая вязкость, кинематический коэффициент вязкости

KINETICALLY UNSTABLE COMPLEX.
See LABILE COMPLEX.

1927
KINETIC CURRENT.
Current resulting from electrode process of substance which forms at electrode surface from other electrodically inactive specia.

D kinetischer Strom
F courant cinétique
P prąd kinetyczny
R кинетический ток

1928
KINETIC EQUATION (rate equation).
Equation describing relation between reaction-rate constant and concentrations of substrates.

D Reaktionsgeschwindigkeitsgleichung
F équation de vitesse

P równanie kinetyczne, równanie szybkości reakcji
R кинетическое уравнение (реакции), уравнение скорости реакции

1929
KINETIC PLOT.
Diagrammatic presentation of progress of reaction as function of time.

D ...
F courbe de vitesse
P krzywa kinetyczna
R ...

1930
KINETIC REGION OF REACTION.
Region of reaction where diffusion does not determine rate of progress.

D kinetisches Gebiet des Prozesses
F ...
P obszar kinetyczny reakcji
R кинетическая область реакции

1931
KINETIC SPECTROPHOTOMETRY.
Absorption or emission spectrophotometry of high time resolution, which enables investigation of short-living species (radicals and radical ions) to be carried out.

D kinetische Spektroskopie
F spectroscopie cinétique
P spektrofotometria kinetyczna
R кинетическая спектрофотометрия

1932
KINETIC STAGE.
In statistical mechanics, intermediate stage of time evolution of macroscopic system. During this stage bi- and polymolecular partition functions depend only on unimolecular partition function but the whole system does not yet reach local equilibrium.

D kinetisches Stadium
F étape cinétique
P stadium kinetyczne
R кинетическая стадия

1933
KINETIC THEORY OF GASES.
Branch of statistical mechanics which deals with properties of rarified gases by application of unimolecular partition function; e.g. in order to describe macroscopic properties of ideal gas by this theory one uses a molecular model, according to which gas is treated as ensemble of either dimensionless points or hard spheres moving randomly and mutually non-interacting except for elastic collisions. From kinetic theory of gases one can deduce equation of state of ideal gas, Maxwell-Boltzmann distribution law, expression for mean free path, etc.

D kinetische Gastheorie
F théorie cinétique des gaz
P teoria kinetyczna gazów
R кинетическая теория газов

1934
KIRCHHOFF'S EQUATION.
Heat effect of chemical reaction depends
on temperature and is determined by
change in relevant sums of heat capacities
of products and reactants of reaction in its
final and initial states.

D Kirchhoffsches Gesetz
F formule de Kirchhoff, loi de Kirchhoff
P prawo Kirchhoffa
R уравнение Кирхгофа

1935
KIRCHHOFF'S RADIATION LAW (for
thermal radiation).
At a given thermodynamic temperature T
and for a given wavelength λ ratio
between emissive and absorptive power
$E(\lambda, T)/A(\lambda, T)$ is the same for all
thermally excited bodies and is equal to
emissive power of perfectly black body
$E_{bb}(\lambda, T)$.

D Kirchhoffsches Gesetz
F loi de Kirchhoff
P prawo Kirchhoffa
R закон излучения Кирхгофа

1936
K-MESON (kaon).
Elementary particle, a hadron, of baryon
number $B = 0$, lepton number $L = 0$,
quantum spin number $J = 0$, strangeness
$S = 1$, isospin $I = 1/2$, parity $P = -1$,
unstable. K-meson occurs in two states of
charge (isospin doublet): of positive
elementary electric charge (K⁺-meson;
mass $m = 493.84$ MeV, third component
of isospin $I_3 = 1/2$, mean lifetime
$1.237 \cdot 10^{-8}$ s); and electrically neutral
(K⁰-meson; $m = 497.79$ MeV, $I_3 = -1/2$).
Neutral K⁰-mesons are mixture of K_s^0
particles with mean lifetime $0.862 \cdot 10^{-10}$ s
and of K_L^0 particles with mean lifetime
$5.172 \cdot 10^{-8}$ s. K⁺-meson and K⁰-meson
have corresponding antiparticles, K⁻
and $\overline{K}^0$.

D K-Meson
F méson K, kaon
P mezon K, kaon
R к-мезон, ка-мезон

1937
KNIGHT SHIFT.
Shift of resonance frequency or resonance
magnetic field strength in nuclear
magnetic resonance of metals, measured
with respect to position of resonance
signal for identical nuclei but in
non-metallic substance. Knight shift arises

due to interaction between nuclei and
electrons of conduction band.

D Knight-Shift, Knight-Verschiebung
F déplacement de Knight
P przesunięcie Knighta
R найтовский сдвиг

1938
KNOCK-OUT REACTION.
Nuclear reaction of direct interaction in
which incident nucleon interacts with
nucleon in target nucleus, knocking it out
but itself becoming bound in nucleus.

D Knock-out-Reaktion
F réaction knock-out
P reakcja knock-out, reakcja wybicia
R реакция выбивания

KNOEVENAGEL CONDENSATION. See
KNOEVENAGEL REACTION.

1939
KNOEVENAGEL REACTION
(Knoevenagel condensation).
Condensation of aliphatic or aromatic
aldehyde with compounds having active
hydrogen (e.g. malonic and cyanoacetic
acids and their esters) in presence of
catalyst such as piperidine

$$\text{\textbackslash}C=O + CH_2 \overset{CN}{\underset{COOR}{\diagdown}} \xrightarrow{\text{piperidine}} \text{\textbackslash}C=C \overset{CN}{\underset{COOR}{\diagdown}} + H_2O$$

D Knoevenagel-Reaktion, Knoevenagel
  Aldol-Kondensation, Knoevenagel
  Crotonisierung
F condensation de Knoevenagel
P kondensacja Knoevenagla
R реакция Кневенагеля

1940
KNORR PYRROLE SYNTHESIS.
Preparation of pyrrole and its derivatives
by condensation of α-aminoketones with
compounds containing —CO—CH₂— group,
e.g.

$$R'-C=O \\ | \\ R-C-NH_2 + \overset{CH_2-R''}{\underset{CO-R'''}{|}} \xrightarrow{-2H_2O} \overset{R'}{\underset{R}{}} \text{pyrrole} \overset{R''}{\underset{R'''}{}} \\ | \\ H$$

D Knorr Pyrrol-Synthese
F réaction de von Knorr
P synteza (pochodnych) pirolu Knorra
R синтез пирролов Кнорра

1941
KOENIGS-KNORR ACETYLGLYCOSIDE
SYNTHESIS.
Preparation of β-acetylglycosides as result
of reaction of α-acetylhalogenoses with
alcohols (or phenols), in presence of silver
carbonate, e.g.

$$2\left(\begin{array}{c} \text{------O} \\ \text{------} \\ \text{CH}_3\text{CO--O} \end{array}\Big\rangle X\right) + 2\text{CH}_3\text{OH} + \text{Ag}_2\text{CO}_3 \longrightarrow$$

$$2\left(\begin{array}{c} \text{------O} \quad \text{OCH}_3 \\ \text{------} \\ \text{CH}_3\text{CO--O} \end{array}\right) \left\{\begin{array}{l} + 2\text{AgX} \\ + \text{H}_2\text{O} \\ + \text{CO}_2 \end{array}\right.$$

D Koenigs-Knorr β-Glykosidierung
F synthèse de Koenigs-Knorr
P synteza glikozydów Koenigsa i Knorra
R синтез глюкозидов Кёнигс-Кнорра

KOHLRAUSCH'S ADDITIVE LAW. *See*
KOHLRAUSCH'S LAW OF
INDEPENDENT IONIC MOBILITIES.

1942
KOHLRAUSCH'S LAW OF
INDEPENDENT IONIC MOBILITIES
(Kohlrausch's additive law).
Limiting equivalent conductivity Λ_0 of
electrolyte is sum of limiting ionic
equivalent conductivities λ_{0+}, λ_{0-}

$$\Lambda_0 = \lambda_{0+} + \lambda_{0-}$$

D Kohlrauschsches Gesetz der
  unabhängigen Ionenwanderung,
  Kohlrauschsches Additivitätsgesetz
F loi d'additivité de Kohlrausch
P prawo niezależnego ruchu jonów
  Kohlrauscha
R закон Кольрауша независимого
  движения ионов

KOLBE ELECTROCHEMICAL
REACTION. *See* **KOLBE SYNTHESIS.**

1943
KOLBE-SCHMITT REACTION.
Preparation of phenoloacids (e.g. salicylic
acid) from sodium phenolate and carbon
dioxide at 125°C, under pressure of
0.4—0.7 MPa, e.g.

D Kolbe-Schmitt (Salicylsäure)-Synthese,
  Kolbe-Schmitt Phenolat-Carboxylierung
F réaction de Kolbe-Schmitt, carbonation
  des phénols selon Kolbe
P synteza Kolbego i Schmitta
R реакция Кольбе-Шмитта

1944
KOLBE SYNTHESIS (Kolbe
electrochemical reaction).
Method of preparation of hydrocarbons by
electrolysis of aqueous solutions of
carboxylic acid salts, e.g.

$$2\text{CH}_3\text{COONa} + 2\text{H}_2\text{O} \xrightarrow{\text{electrolysis}} \text{CH}_3\text{—CH}_3 + \\ + 2\text{CO}_2 + 2\text{NaOH} + \text{H}_2$$

D Kolbe Fettsäure-Elektrolyse
F réaction de Kolbe
P synteza (elektrolityczna węglowodorów)
  Kolbego
R электрохимический синтез Кольбе

1945
KONOWALOFF'S LAWS.
Laws describing liquid-vapour equilibrium
in systems of unlimited miscibility in
liquid phase: (1) Gas phase is enriched in
that component which, when added to
liquid, causes increase of vapour pressure
above it (i.e. in component which lowers
boiling point of liquid); (2) Maximum on
curve of vapour pressure corresponds to
minimum on the curve of boiling points
and vice versa; these are azeotropic
points.

D Konowalowsche Gesetze,
  Konowalowsche Regeln
F lois de Konovalov
P prawa Konowałowa
R законы Коновалова

1946
KOPP'S RULE.
Molal heat capacity of solid compound is
approximately equal to sum of atomic
heat capacities of its constituent elements.

D Kopp-Neumannsche Regel
F règle de Neumann-Kopp
P reguła Koppa i Neumanna
R правило Нейманна-Коппа

1947
KOSTANECKI SYNTHESIS OF
CHROMONES AND COUMARINS.
Preparation of chromones and coumarins
by treating o-acylphenols with fatty acid
anhydrides and fatty acid sodium salts, in
presence of triethylamine, e.g.

D Kostanecki-Robinson Chromon-Synthese
F réaction de Kostanecki-Robinson
P synteza chromonów i kumaryn
  Kostaneckiego
R ...

1948
KOSTANECKI SYNTHESIS OF FLAVONES.
Preparation of flavones or benzylidenecoumaranones by action of alkalis on dibromides of o-hydroxy- and o-acetoxychalcones. Flavones or coumaranones are formed depending on relative reactivity of both bromine atoms.

D ...
F ...
P synteza flawonów Kostaneckiego
R ...

1949
KRAMERS' THEOREM.
In atom with odd number of electrons electric field of any symmetry leaves at least two-fold spin degeneracy which can be destroyed by magnetic field only.
D Kramersches Theorem
F théorème de Kramers
P teoremat Kramersa
R теорема Крамерса

1950
KRYPTON, Kr.
Atomic number 36.
D Krypton
F krypton
P krypton
R криптон

1951
KUCHEROV SYNTHESIS OF CARBONYL COMPOUNDS.
Direct addition of water to acetylenic hydrocarbons taking place in dilute sulfuric acid medium, in presence of mercuric salts as catalysts. Unstable adduct isomerizes to form carbonyl compound, e.g.

$$HC \equiv CH + H_2O \xrightarrow{Hg}{H_2SO_4}$$

$$[CHOH = CH_2] \rightarrow CH_3 - CHO$$

D Kutscheroff Acetylen-Hydratisierung
F réaction de Kutscheroff, hydratation de l'acétylène selon Kutscheroff
P synteza Kuczerowa
R реакция Кучерова

1952
LABELLED ATOMS (tagged atoms).
Atoms which differ in characteristic nuclear properties (mass or radioactivity) to atoms of the element's principal isotopes.
D markierte Atome
F atomes marqués
P atomy znaczone
R меченые атомы

1953
LABELLED COMPOUND.
Chemical compound with some molecules containing isotopic tracer (radioactive or stable).
D markierte Verbindung
F composé marqué
P związek znaczony
R меченое соединение

1954
LABELLING OF COMPOUNDS.
Preparation of compounds containing labelled atoms in molecules using chemical synthesis, isotope exchange, biological synthesis, radiation induced chemical reactions, hot atom reactions, etc.
D Markierung von Verbindungen
F marquage de composés
P znakowanie związków
R получение меченых соединений

1955
LABILE COMPLEX (kinetically unstable complex).
Complex in which ligands may easily be replaced by other ligands. Labile complex is characterized by high exchange rate of ligands.
D labiler Komplex
F complexe labile
P kompleks labilny
R лабильный комплекс

1956
LABORATORY SAMPLE (average sample).
Sample obtained from gross sample by its decrease in way corresponding to nature of a given material.
D Durchschnittsprobe
F échantillon pour laboratoire, échantillon moyen
P próbka laboratoryjna średnia
R средняя лабораторная проба

1957
LACTIC FERMENTATION.
Type of oxidative fermentation of hexoses, e.g. of glucose, caused by *Bacillus Delbrückii*, resulting in decomposition of carbohydrate to lactic acid.

D Milchsäuregärung
F fermentation lactique
P fermentacja mlekowa
R молочнокислое брожение

1958
LADDER POLYMER.
Polymer containing double main chains
due to formation of condensed rings, e.g.
polyacrolein

D Leiterpolymer
F ...
P polimer drabinkowy
R лестничный полимер

1959
LAEVO-ROTATION.
Specific effect in optically active
substances; when plane-polarized light
passing through an optically active
substance has its plane of polarization
rotated in anti-clockwise direction (when
viewed from opposite side to light source),
such subsance is said to be (−), i.e.
laevo-rotatory (obsolete notation: $l-$).

D Linksdrehung
F lévorotation
P lewoskrętność
R левое вращение

1960
LAMBDA HYPERON (Λ-hyperon).
Elementary particle, a hadron, the lightest
of hyperons, of mass $m = 1\,115.59$ MeV,
electric charge $Q = 0$, baryon number
$B = 1$, lepton number $L = 0$, spin quantum
number $J = 1/2$, strangeness $S = -1$,
isospin $I = 0$, third component of isospin
$I_3 = 0$, unstable, of mean lifetime
$2.517 \cdot 10^{-10}$ s (has antiparticle $\bar{\Lambda}$).

D Λ-Hyperon
F hypéron Λ, hypéron lambda
P hiperon Λ
R Λ-гиперон, лямбда-гиперон

1961
LAMBERT-BEER LAW (Bouguer-Beer
law).
For monochromatic light passing through
solution of absorbing substance in
non-absorbing solvent absorbance A is
proportional to concentration c and path
length b, $A = acb$, where a = absorptivity.

D (Bouguer-)Lambert-Beersches Gesetz
F loi de Lambert-Beer, loi de
   Bouguer-Beer
P prawo Beera i Waltera, prawo Lamberta
   i Beera
R закон Бугера-Ламберта-Бера

LAMBERT'S LAW. See BOUGUER'S
LAW.

1962
LANDÉ FACTOR (spectroscopic splitting
factor g_J).
Dimensionless quantity in formula for
magnetic moment of paramagnetic atom;
depends on quantum numbers J, L and S.

D Landé-Faktor, Landéscher
   Aufspaltungsfaktor, spektroskopischer
   Aufspaltungsfaktor g_J
F facteur de Landé, facteur de
   décomposition spectrale
P czynnik Landégo, czynnik rozszczepienia
   spektroskopowego g_J
R фактор Ланде, фактор
   спектроскопического расщепления

1963
LANGEVIN'S THEORY.
Classical theory of dia- and
paramagnetism.

D Langevin-Theorie, Langevinsche
   Theorie
F théorie de Langevin
P teoria Langevina
R теория Ланжевена

LANGMUIR ADSORPTION ISOTHERM.
See LANGMUIR ISOTHERM.

1964
LANGMUIR EQUATION.
Equation of adsorption isotherm derived
on assumption that only monolayer
adsorption occurs:

$$x = x_m \frac{Kp}{1+Kp}$$

where x = amount of gas adsorbed by unit
mass of adsorbent, x_m = amount of gas
necessary for formation of monomolecular
layer at surface of unit mass of adsorbent,
p = equilibrium pressure, K = adsorption
coefficient.

D Langmuirsche Gleichung
F formule de Langmuir
P równanie Langmuira
R уравнение Ленгмюра

1965
LANGMUIR ISOTHERM (Langmuir
adsorption isotherm). Adsorption isotherm
corresponding to Langmuir equation.

D Langmuirsche Adsorptionsisotherme,
   Langmuirsche Isotherme
F isotherme d'adsorption de Langmuir
P izoterma (adsorpcji) Langmuira
R изотерма (адсорбции) Ленгмюра

LANTHANIDE ELEMENTS. See
LANTHANOIDS.

LANTHANIDES. See LANTHANOIDS.

LANTHANIDE SERIES. See
LANTHANOIDS.

1966
LANTHANOIDS (lanthanides, lanthanide series, lanthanide elements, lanthanons).
Series of elements in sixth group of periodic system: lanthanum, cerium, praseodymium, neodymium, promethium, samarium, europium, gadolinium, terbium, dysprosium, holmium, erbium, thulium, ytterbium, lutetium. Lanthanoids are transition elements and, except for lanthanum and lutetium, form first series of inner transition elements (filling of 4f electronic sub-shell).

D Lanthan(o)ide, Elemente der Lanthanreihe
F lanthanides, série du lanthane
P lantanowce
R лантан(о)иды, семейство лантанидов

LANTHANONS. *See* LANTHANOIDS.

1967
LANTHANUM, La.
Atomic number 57.

D Lanthan
F lanthane
P lantan
R лантан

1968
LARMOR FREQUENCY.
Frequency of nuclear or electronic magnetic moment precession about direction of external magnetic field (*cf.* Larmor precession).

D Larmor-Frequenz
F fréquence de Larmor
P częstość Larmora, częstość larmorowska
R ларморовская частота

1969
LARMOR PRECESSION.
Precession of vector of spin, orbital or total angular momentum in external magnetic field arising when axis of momentum vector does not coincide with magnetic field direction.

D Larmor-Präzession
F précession de Larmor
P precesja Larmora, precesja larmorowska
R ларморовская прецессия

1970
LAST HEAT OF SOLUTION.
Enthalpy change of system in process of dissolving one mole of a given substance in a large bulk of almost saturated solution which finally becomes saturated.

D letzte Lösungswärme
F chaleur dernière de dissolution
P ciepło rozpuszczania ostatnie
R последняя теплота растворения

LATENT HEAT. *See* HEAT OF CHANGE OF PHASE.

1971
LATENT HEAT OF SURFACE, q_s.
Enthalpy increase during increase of surface area by 1 cm². Relation between latent heat of surface and surface tension is: $q_s = -T \cdot d\sigma/dT$ where σ = surface tension, T = thermodynamic temperature.

D latente Oberflächenwärme
F chaleur latente superficielle
P ciepło tworzenia się powierzchni
R тепловой эффект образования единицы поверхности

1972
LATENT IMAGE (latent photographic image). Invisible image formed by minute particles of silver separated out during exposure of photographic emulsion.

D latentes photographisches Bild
F image latente
P obraz utajony
R скрытое фотографическое изображение

LATENT PHOTOGRAPHIC IMAGE. *See* LATENT IMAGE.

1973
LATIMER'S EQUATION.
Empirical formula which describes standard partial molar entropy $S_B^{\ominus}$ of monoatomic ion in aqueous solution:

$$S_B^{\ominus} = \frac{3}{2} R \ln A_r + 37 - 270 \frac{|z_B|}{r_B^2}$$

where r_B = effective ionic radius solution: larger by 0.1 nm and 0.2 nm for anions and cations respectively than their crystallographic radius, z_B = valency of ion.

D Latimersche Gleichung
F relation de Latimer
P równanie Latimera
R уравнение Латимера

1974
LATTICE DEFECTS (crystal defects, lattice imperfections).
All deviations from ideal periodicity of crystal lattice.

D Fehlordnung, Gitterfehler, Kristallbaufehler
F défauts dans les cristaux
P defekty sieciowe, defekty sieci krystalicznej
R дефекты кристаллической решётки

1975
LATTICE ENERGY (*of crystal*).
Work which must be done to displace elements of crystal from positions which they occupy in structure to infinity.

D Gitterenergie (der Kristalle)
F énergie de réseau

P energia sieci (krystalicznej)
R энергия (кристаллической) решётки

LATTICE IMPERFECTIONS. *See* LATTICE DEFECTS.

1976
LATTICE PLANE (net plane of lattice). Plane incorporating at least three nodes which are not placed on the same straight line and which is parallel to one of possible faces of crystal.

D Gitterebene, Netzebene
F plan réticulaire
P płaszczyzna sieciowa
R плоскость решётки

LAUE PATTERN. *See* LAUE PHOTOGRAPH.

1977
LAUE PHOTOGRAPH (Laue pattern). Population of dark spots in photographic layer, of characteristic distribution, often symmetrical, obtained by irradiating fixed flat crystal with narrow beam of X rays of continuous spectrum.

D Laue-Aufnahme, Laue-Photogramm, Laue-Diagramm
F Laue photographie, diagramme de Laue
P lauegram, diagram Lauego, rentgenogram Lauego
R лауэграмма

1978
LAVES PHASES.
Solid phases formed from atoms of constant ratio of diameters, e.g. Cu_2Mg, Au_2Bi, KBi_2.

D Laves-Phasen
F phases de Laves
P fazy Lavesa
R фазы Лавеса

1979
LAW OF COMBINING VOLUMES (Gay-Lussac's law).
In any chemical reaction, volumes of all gaseous substrates and products measured at the same constant pressure and temperature may be expressed in ratios of small integers.

D Gesetz der konstanten Volumenverhältnisse
F loi de Gay-Lussac
P prawo stosunków objętościowych
R закон объёмных отношений

1980
LAW OF CONSTANT ANGLES.
Angles between characteristic faces of particular kind of crystal at constant temperature remain independent of size of crystal, and of defects connected with process of growing.

D Gesetz der Konstanz der Kantenwinkel
F loi de constance des angles

P prawo stałości kątów
R закон постоянства углов

1981
LAW OF CONSTANT COMPOSITION (law of constant proportions, law of definite proportions, Proust's law). Proportion of masses of constituent elements or components in any definite chemical compound is fixed and constant.

D Gesetz der konstanten Proportionen
F loi des proportions définies, loi des nombres proportionnels, loi de Proust
P prawo stosunków stałych, prawo stałości składu, pierwsze prawo Daltona
R закон постоянства состава, закон постоянных отношений

LAW OF CONSTANT HEAT SUMMATION. *See* HESS'S LAW.

1982
LAW OF CONSTANT PROPERTIES.
Properties of a pure substance are independent of its origin or previous treatment (strictly observed only for gases and liquids because in solids certain properties may vary depending on their treatment).

D ...
F ...
P prawo stałych własności
R закон постоянства свойств

LAW OF CONSTANT PROPORTIONS. *See* LAW OF CONSTANT COMPOSITION.

LAW OF DEFINITE PROPORTIONS. *See* LAW OF CONSTANT COMPOSITION.

LAW OF DULONG AND PETIT. *See* RULE OF DULONG AND PETIT.

LAW OF EQUIVALENT PROPORTIONS. *See* LAW OF RECIPROCAL PROPORTIONS.

1983
LAW OF MASS ACTION.
In state of chemical equilibrium reactant activities satisfy equation

$$\prod_C (a_C)^{\nu_C} / \prod_B (a_B)^{\nu_B} = K_a$$

where a_B, a_C = activities,
ν_B, ν_C = stoichiometric coefficients of substrates B and products C, respectively.
K_a = thermodynamic chemical equilibrium constant.

D Massenwirkungsgesetz
F loi d'action de masse
P prawo działania mas, prawo równowagi chemicznej, prawo Guldberga i Waagego
R закон действия масс, закон действующих масс

1984
LAW OF MULTIPLE PROPORTIONS.
When two elements combine to form more than one compound, the several masses of one element that combine with a fixed mass of second one are in ratio of small whole numbers.
D Gesetz der multiplen Proportionen
F loi des proportions multiples, loi de Dalton
P prawo stosunków wielokrotnych, drugie prawo Daltona
R закон кратных отношений

1985
LAW OF RADIOACTIVE DECAY (decay law).
Number of nuclei which undergo radioactive decay in unit time $\dfrac{dN}{dt}$ is proportional to number N of unchanged nuclei present.

$$-\frac{dN}{dt} = \lambda N$$

where λ is disintegration constant.
A statistical law.
D Zerfallsgesetz
F loi de décroissance
P prawo rozpadu promieniotwórczego
R закон радиоактивного распада

1986
LAW OF RATIONAL INDICES.
If face P of crystal, chosen as unitary, intercepts on three chosen crystallographic axis sections a, b, c respectively, then every other face R of this crystal intercepts on these axes sections a', b', c' respectively, related according to equation $a/a' : b/b' : c/c' = h : k : l$; at the same time ratio of face numbers $h : k : l$ (called indices of face R) is always rational, i.e. may be expressed as integers.
D Gesetz der Rationalität der Indizes
F loi des indices rationnels
P prawo wymiernych wskaźników
R правило рациональных индексов, закон рациональных параметров

1987
LAW OF RECIPROCAL PROPORTIONS (law of equivalent proportions).
Elements combine with one another in definite amounts by weight corresponding to their equivalent weights.
D ...
F loi des proportions équivalentes
P prawo udziałów, prawo stosunków równoważnikowych
R закон паёв

LAW OF THE RECTILINEAR DIAMETER. *See* CAILLETET AND MATHIAS LAW.

1988
LAWRENCIUM, Lr.
Atomic number 103.
D Lawrencium, Lawrentium
F lawrencium
P lorens
R лоуренсий

LCAO APPROXIMATION. *See* LCAO METHOD.

1989
LCAO METHOD (LCAO approximation).
Method for calculation of approximate one-electron wave functions in form of linear combination of atomic orbitals.
D LCAO-Methode
F méthode CLOA
P metoda kombinacji liniowej orbitali atomowych, metoda LCAO
R метод МО-ЛКАО

1990
LEAD, Pb,
Atomic number 82.
D Blei
F plomb
P ołów
R свинец

1991
LEAVING GROUP (departing group).
Group displaced by entering group (or atom) in substitution reaction at carbon atom

$$-\overset{|}{\underset{|}{C}}-Cl + OH^{\ominus} \rightarrow -\overset{|}{\underset{|}{C}}-OH + Cl^{\ominus} \quad \text{(leaving group)}$$

D abgespaltener Substituent
F groupe à déplacer, groupe à substituer
P grupa opuszczająca
R уходящая группа, отщепляющаяся группа

1992
LE BEL-VAN'T HOFF THEORY.
Theory constituing foundation of stereochemistry and explaining optical activity as consequence of steric structure of organic compounds. Theory assumes that four atoms (or groups of atoms) on carbon atom are distributed round it at positions corresponding to corners of tetrahedron, carbon atom being located in centre of that tetrahedron. When all substituents a, b, c, d are different, molecule $Cabcd$ is not superimposable on its mirror image, displays optical activity and forms two isomers.
D van't Hoff und Le Bel Theorie
F principe Le Bel et Van't Hoff
P teoria van't Hoffa i Le Bela
R теория Вант-Гоффа и Ле Беля

1993
LE CHÂTELIER-BRAUN'S PRINCIPLE
(principle of mobile equilibrium).
Shift of chemical reaction equilibrium
state proceeds in such direction that
influence of externally imposed change of
a given intensive quantity (e.g.
temperature, pressure, concentration of
component) is diminished.

D Le Châtelier-Braunsches Prinzip,
Prinzip des kleinsten Zwanges
F principe de Le Châtelier (-Braun), règle
de Le Châtelier, loi de déplacement
d'équilibre
P zasada Le Châteliera i Brauna, reguła
przekory, zasada przekory Le Châteliera
i Brauna
R принцип Ле Шателье(-Брауна),
принцип подвижного равновесия,
закон модерации

1994
LECLANCHÉ CELL.
Cell of $C|MnO_2, NH_4Cl\ (20^0/o)|Zn$ type.

D Leclanché-Element
F pile Leclanché
P ogniwo Leclanchégo
R элемент Лекланше

1995
LENNARD-JONES POTENTIAL.
Semiempirical potential for intermolecular
interactions

$$u(r) = \frac{A}{r^n} - \frac{B}{r^m} \qquad (m < n)$$

where r = intermolecular distance;
A, B, m, n are constants. If van der Waals
forces are present it is usually assumed
that $m = 6$ and $n = 12$.

D Lennard-Jones-Potential
F potentiel de Lennard-Jones
P potencjał Lennarda-Jonesa
R потенциал Леннарда-Джонса

1996
LEPTON NUMBER, L.
Number attributed to elementary
particles; for neutrino, electron and
μ^--lepton, $L = 1$, for their antiparticles,
$L = -1$; other particles and antiparticles
have $L = 0$. Lepton number is conserved
in all interactions.

D Leptonenzahl
F nombre leptonique
P liczba leptonowa
R лептонное число, лептонный заряд

1997
LEPTON NUMBER CONSERVATION
LAW.
Sums of lepton numbers of initial and
final particles in given process are equal.
Lepton number conservation law is valid
for all interactions.

D Erhaltungssatz der Leptonenzahl
F loi de conservation du nombre
leptonique
P prawo zachowania liczby leptonowej
R закон сохранения лептонного числа

1998
LEPTONS.
Group of lightest elementary particles
taking part in electromagnetic and weak
interactions, of baryon number $B = 0$,
lepton number $L = 1$, quantum spin
number $J = 1/2$. To group of leptons
belong: electronic neutrino, ν_e; muonic
neutrino, ν_μ; electron, e^-; and
mu-lepton, μ^-. Antiparticles are
respectively: $\bar{\nu}_e, \bar{\nu}_\mu, e^+, \mu^+$.

D Leptonen
F leptons
P leptony
R лептоны

1999
LETHAL DOSE.
Dose of ionizing radiation which, when
absorbed by given population of living
organisms, causes death of certain part of
this population, after fixed period of time
(*c.f.* median lethal dose).

D Lethaldosis
F dose léthale, dose mortelle
P dawka śmiertelna
R летальная доза

2000
LEUCKART-WALLACH REACTION.
Conversion of aldehydes and ketones into
amines by heating them with formamide
or ammonium formate, e.g.

$$RR'CO + 2HCOONH_4 \rightarrow RR'CHNHCHO + NH_3 +$$
$$+ CO_2 + 2H_2O$$

$$RR'CHNHCHO \xrightarrow{H_2O} RR'CHNH_2 + HCOOH$$

D Leuckart(-Wallach)-Reaktion,
Leuckart Amin-Alkylierung, Leuckart
reduktive Carbonyl-Aminierung
F réaction de Leuckart
P reakcja Leuckarta (i Wallacha)
R реакция Лейкарта

2001
LEVELLING SOLVENT (equalizing
solvent).
Solvent in which different electrolytes
have large equivalent conductivities of the
same order; for salts — solvent forming
crystalline solvates, for acids —
protophilic solvent, for bases — protogenic
solvent.

D nivellierendes Lösungsmittel
F solvant-niveleur, solvant capable de
niveler des acides ou des bases
P rozpuszczalnik niwelujący,
rozpuszczalnik wyrównujący
R нивелирующий растворитель

2002
LEWIS-RANDALL'S IONIC STRENGTH
LAW.
In diluted solutions of strong electrolytes
of the same ionic strength $J < 0.02$, mean
activity coefficient of a given electrolyte
has identical value independent of type of
solution.
D Ionenstärke-Gesetz von Lewis und
Randall
F règle de la force ionique de Lewis et
Randall
P reguła mocy jonowej Lewisa i Randalla
R закон ионной силы Льюиса и Рендалла

2003
LEWIS-SARGENT'S EQUATION.
Formula which describes liquid junction
potential ε_d of bordering solutions of 1.1
electrolytes of equal concentration with
common anion or cation

$$\varepsilon_d = \frac{RT}{F} \ln \frac{\Lambda^{(1)}}{\Lambda^{(2)}}$$

where $\Lambda^{(1)}$ and $\Lambda^{(2)}$ are equivalent
conductivities of solution (1) and (2)
respectively.
D Lewis-Sargentsche Formel,
Lewis-Sargentsche Gleichung
F équation de Lewis et Sargent
P wzór Lewisa i Sargenta
R уравнение Льюиса и Сарджента

2004
LEWIS' THEORY OF ACIDS AND
BASES.
Theory according to which acids are
substances which may accept free electron
pair from atom or group of atoms,
forming coordinate bond; substances
which have such an electron pair form
group of bases.
D Lewissche Säuren-Basen-Theorie
F théorie des acides et des bases de Lewis,
théorie électronique des acides et des
bases de Lewis
P teoria kwasów i zasad Lewisa
R теория кислот и оснований Льюиса

2005
LIESEGANG RINGS.
Concentric bands of precipitate formed in
gel due to rhythmic precipitation of
sparingly soluble sols.
D Liesegangsche Ringe
F anneaux de Liesegang
P pierścienie Lieseganga
R кольца Лизеганга

LIGANCY. *See* COORDINATION
NUMBER.

2006
LIGAND.
Ion or molecule directly attached to
central atom in complex.
D Ligand, koordinierte Gruppe
F ligand
P ligand, addend, podstawnik, grupa
skoordynowana
R лиганд, адденд

2007
LIGAND FIELD.
Electric field defined by symmetry of
complex formed by ligands in complex
and causing splitting of energy levels
(terms) of central ion.
D Ligandenfeld
F champ des coordinats
P pole (krystaliczne) ligandów
R поле лигандов

LIGAND FIELD STABILIZATION
ENERGY. *See* CRYSTAL FIELD
STABILIZATION ENERGY.

2008
LIGAND FIELD THEORY.
Theory of electronic structure of complex
compounds which is modification of
crystal field theory and which takes into
account covalence of coordinate bond (on
basis of principal concepts of molecular
orbital theory).
D Ligandenfeldtheorie
F théorie du champ des ligandes
P teoria pola ligandów
R теория поля лигандов

2009
LIGAND NUMBER (average ligand
number), $\bar{n}$.
Average number of ligands of given type
linked with central atom

$$\bar{n} = \frac{c_L - [L]}{c_M}$$

where c_L is total ligand concentration,
[L] is equilibrium ligand concentration,
c_M is total concentration of metal in
solution.
D durchschnittliche Ligandenzahl
F nombre moyen de coordination
P liczba ligandowa, funkcja tworzenia
R функция образования

LIGATING ATOM. *See* COORDINATING
ATOM.

2010
LIGHT DIFFRACTION (diffraction of
light).
Deviation of linear beam of light in region
of geometric shadow of object arising
from wave character of light.

D Lichtbeugung, Diffraktion des Lichtes
F diffraction
P dyfrakcja światła, ugięcie światła
R дифракция света

2011
LIGHT DISPERSION (dispersion of light).
Splitting of white light beam in medium
into separate colours arising from
dependence of refractive index of a given
medium on wavelength.
D Dispersion des Lichtes
F dispersion de la lumière
P dyspersja światła
R дисперсия света

2012
LIGHT FILTER (optical filter).
Optical system transparent only for some
wavelengths of radiation.
D Filter, Lichtfilter
F filtre optique
P filtr (świetlny), filtr optyczny
R светофильтр

LIGHT INTENSITY. *See* LUMINOUS
INTENSITY.

2013
LIGHT PLATINUM METALS (second
triad).
Three chemical elements grouped
horizontally in eighth sub-group of
periodic system (in period V): ruthenium,
rhodium, palladium.
D leichte Platinmetalle
F métaux légers de la mine de platine,
triade du palladium
P platynowce lekkie, rutenowce, triada
rutenu
R лёгкие платиновые металлы

LIGHT QUANTUM. *See* PHOTON.

LIGHT SOURCE. *See* EXCITATION
SOURCE.

LIMITED COMPATIBILITY. *See*
PARTIAL MISCIBILITY.

LIMITING CONCENTRATION. *See*
LIMITING DILUTION.

2014
LIMITING CURRENT.
In electroanalysis, current which for given
conditions is controlled by transport rate
of reactant to electrode.
D Grenzstrom
F courant limité
P prąd graniczny
R предельный ток

2015
LIMITING DIFFUSION CURRENT.
In electroanalysis, maximum value of
current, independent over some range of

electrolysis voltage; it arises if in system
there exists mainly diffusion polarization.
D Diffusionsgrenzstrom
F courant de diffusion limite
P prąd graniczny dyfuzyjny
R предельный диффузионный ток

2016
LIMITING DILUTION (limiting
concentration).
Dilution of solution of analysed substance
for which one half of performed tests still
give positive result.
D Verdünnungsgrenze, Grenzkonzentration
F limite de dilution
P rozcieńczenie graniczne, stężenie
graniczne
R предельное разбавление

2017
LIMITING EQUIVALENT
CONDUCTIVITY (equivalent conductivity
at infinite dilution), Λ_0.
Limiting value to which the equivalent
conductance approaches when electrolyte
concentration tends to zero.
D Grenzleitfähigkeit,
Äquivalentleitfähigkeit bei
verschwindender Konzentration
F conductibilité équivalente limite
P przewodnictwo równoważnikowe
graniczne
R эквивалентная электропроводность
при бесконечном разведении,
эквивалентная электропроводность
при нулевой концентрации

LIMITING STAUDINGER FUNCTION.
See INTRINSIC VISCOSITY.

2018
LIMIT OF DETECTION.
Lowest concentration of analysed
component of sample which can be
determined by a given method.
D Nachweisgrenze, Erfassungsgrenze
F sensibilité (relative), limite de sensibilité
P granica wykrywalności, wykrywalność
R относительная чувствительность,
порог чувствительности

2019
LIMITS OF AUTOIGNITION.
Limiting values of combustible constituent
concentration, or pressure above which
(upper limit) or below which (lower limit)
autoignition of gas mixture occurs, at the
given temperature.
D ...
F ...
P granice samozapłonu
R пределы самовоспламенения

2020
LIMITS OF INFLAMMABILITY (ignition limits).
Boundary concentrations of inflammable constituent in gas mixture or (seldom) range of pressures of this mixture, at which inflammation may occur.
D Zündgrenzen
F limites d'inflammabilité
P granice zapalności
R пределы воспламенения

LINEAR ADSORPTION ISOTHERM. *See* LINEAR ISOTHERM.

2021
LINEAR ATTENUATION COEFFICIENT, μ.
Decrease in intensity of beam of ionizing radiation J (expressed as fraction of intensity of primary beam) after passing through absorber of unit thickness x

$$\frac{\mathrm{d}J}{J} = -\mu\,\mathrm{d}x$$

D linearer Schwächungskoeffizient
F coefficient d'atténuation linéique
P współczynnik osłabienia liniowy
R линейный коэффициент ослабления

LINEAR CHAIN. *See* NORMAL CHAIN.

LINEAR DEFECTS. *See* LINE DEFECTS.

2022
LINEAR DISPERSION (*of spectrograph*).
Derivative $\mathrm{d}x/\mathrm{d}\lambda$, where $\mathrm{d}x$ = distance, along surface of detection, of two closely spaced spectral lines of wavelength difference $\mathrm{d}\lambda$; usually expressed in milimetres per nanometre.
D lineare Dispersion
F dispersion linéaire
P dyspersja liniowa
R линейная дисперсия

2023
LINEAR ELECTRON ACCELERATOR.
Accelerator producing fast electrons in microwave field.
D Linearbeschleuniger von Elektronen
F accélérateur linéaire d'électrons
P akcelerator liniowy elektronów
R линейный ускоритель электронов

2024
LINEAR ENERGY TRANSFER, LET.
Amount of energy transferred to defined medium by ionizing particle or quantum per unit length of track of particle or path of quantum propagation.
D lineare Energieübertragung
F transfert linéique d'énergie
P przeniesienie energii liniowe
R линейная передача энергии

2025
LINEAR ISOTHERM (linear adsorption isotherm).
BET isotherm in the case of adsorption on ideal adsorbent (flat, non-porous surface).
D lineare Isotherme
F isotherme linéaire
P izoterma (adsorpcji) liniowa
R линейная изотерма

2026
LINEAR MOLECULE.
Molecule, whose atoms lie on straight line, e.g. $O{=}C{=}O$.
D lineares Molekül
F molécule linéaire
P cząsteczka liniowa
R линейная молекула

LINEAR PHENOMENOLOGICAL LAWS. *See* PHENOMENOLOGICAL EQUATIONS.

2027
LINEAR POLYMER.
Polymer composed of macromolecules forming long unbranched chains.
D lineares Polymer
F polymère linéaire
P polimer liniowy
R линейный полимер

2028
LINEAR REGRESSION.
Regression of form of linear function $y = ax + b$, where a, b = regression coefficients, most frequently determined by least squares method.
D lineare Regression
F régression linéaire
P regresja liniowa
R линейная регрессия

2029
LINEAR STOPPING POWER.
Average energy lost by charged particle of given energy in traversing certain path length in matter divided by that path length.
D lineares Bremsvermögen
F pouvoir d'arrêt linéique
P zdolność hamowania liniowa
R линейная тормозная способность

2030
LINE DEFECTS (linear defects, dislocations).
Irregularities in formation of crystal, occurring in structure along certain lines.
D Liniendefekte, Versetzungen
F défauts linéaires
P defekty liniowe, dyslokacje
R линейные дефекты, дислокации

2031
LINE INTERFERENCE (*in emission spectral analysis*).
Partial or complete overlap of two or more spectral lines of close wavelength.

D Linienkoinzidenz
F coïncidence de raies, interférence de raies
P koincydencja linii
R наложение линий, перекрытие линий

2032
LINE PAIR (*in spectral emission analysis*).
Pair consisting of analytical line and standard line whose intensities are compared for analytical purposes.

D Linienpaar
F couple de raies
P para linii analityczna
R аналитическая пара линий

2033
LINE WIDTH (breadth of spectral line).
Quantity which defines degree of monochromacity of a given line in spectrum; usually given as half-width of spectral line.

D Linienbreite
F largeur d'une raie
P szerokość linii widmowej
R ширина спектральной линии

2034
LINKAGE ISOMERISM (structural isomerism, salt isomerism).
Type of isomerism of complex compounds which may occur when unidentate ligand contains two different atoms capable of coordinating, e.g. SCN^- ion may be coordinated either through sulfur atom M—SCN (thiocyanates) or through the nitrogen atom M—NCS (isothiocyanates).

D Bindungsisomerie, Salzisomerie
F isomérie de liaison, isomérie structurale
P izomeria wiązania, izomeria strukturalna, izomeria solna
R структурная изомерия, солевая изомерия

2035
LIOUVILLE'S EQUATION.
Conservation equation for generic n-th order phase distribution

$$\frac{\partial \rho}{\partial t} + \sum_k \left(\frac{\partial H}{\partial p_k} \frac{\partial \rho}{\partial q_k} - \frac{\partial H}{\partial q_k} \frac{\partial \rho}{\partial p_k} \right) = 0$$

where ρ = generic n-th order phase distribution in phase space of system, H = Hamilton function of system, t = time, p = momenta, and q = coordinates. Summation extends over all degrees of freedom of system.
In classical statistical mechanics Liouville's equation determines time evolution of isolated system.

D Liouville-Gleichung
F équation de Liouville
P równanie Liouville'a
R уравнение Лиувилля

2036
LIOUVILLE'S THEOREM.
Conservation principle for phase volume — any given volume in phase space does not change in time, although its shape usually changes.

D Liouvillescher Satz, Prinzip von der Erhaltung der Phasenausdehnung
F théorème de Liouville
P twierdzenie Liouville'a, zasada zachowania objętości fazowej
R теорема Лиувилля, теорема о сохранении фазового объёма

LIOUVILLE-VON NEUMANN EQUATION. *See* VON NEUMANN'S EQUATION.

2037
LIPPMANN'S EQUATION.
Formula which expresses dependence between tangent to electrocapillary curve and charge density on both sides of metal η^M — solution η^s interface:

$$\left(\frac{\partial \sigma}{\partial \varepsilon} \right)_{p, T, \eta} = -\eta^M = \eta^s$$

where σ = surface tension and ε = potential of electrode.

D Lippmann-Gleichung
F équation de Lippmann
P równanie Lippmanna
R уравнение Липпмана

2038
LIQUEFACTION TEMPERATURE.
Temperature at which substance is converted from gaseous into liquid state.

D Verflüssigungspunkt
F température de condensation, température de liquéfaction
P temperatura skraplania
R температура конденсации

2039
LIQUID.
Substance with properties of liquid state.

D Flüssigkeit
F liquide
P ciecz
R жидкость

2040
LIQUID CHROMATOGRAPHY.
Chromatography employing liquid as mobile phase.

D Flüssigkeit-Chromatographie
F chromatographie en phase liquide
P chromatografia cieczowa
R жидкостная хроматография

2041
LIQUID COUNTER.
Counter of the Geiger-Müller type used to measure ionizing radiation emitted by liquids.
D Flüssigkeitszählrohr
F compteur à liquide, compteur à jupe
P licznik cieczowy
R жидкостный счётчик

2042
LIQUID CRYSTALS (mesomorphic phases).
Substances which form two separate phases in liquid state: nematic phase — of high linear order and smectic phase — of high planar order.
D flüssige Kristalle, kristalline Flüssigkeiten, mesomorphe Phasen
F cristaux liquides, phases mésomorphes
P kryształy ciekłe, ciecze krystaliczne, fazy mezomorficzne
R жидкие кристаллы, мезоморфные фазы

2043
LIQUID-DROP MODEL (hydrodynamical model).
Model of atomic nucleus in which nucleus is imagined as behaving like drop of liquid. Liquid-drop model takes into account strong coupling between nucleons. It explains approximate constancy of binding energy per nucleon, enables approximate calculation of mass of nuclei to be made and explains fission; it is valid for medium and heavy nuclei.
D Tröpfchen-Modell
F modèle de la goutte liquide
P model kroplowy
R гидродинамическая модель, капельная модель

2044
LIQUID ION EXCHANGER.
Solution of compound containing ionogenic groups in water immiscible organic solvent. E.g. solution of long chain aliphatic, sulfonic and phosphoric acids are used as cation exchangers, and solution of long chain aliphatic amine salts as anion exchangers, respectively.
D flüssiger Ionenaustauscher
F échangeur d'ions liquide
P jonit ciekły
R жидкий ионит

2045
LIQUID JUNCTION POTENTIAL (diffusion potential).
Difference of electrical potential at junction of two electrolytes resulting from different mobilities of diffusing ions.
D Diffusionspotential
F potentiel de jonction

P potencjał dyfuzyjny
R диффузионный потенциал

LIQUID-LIQUID EXTRACTION. See EXTRACTION.

2046
LIQUID PHASE.
Phase composed of substance in liquid state.
D flüssige Phase
F phase liquide
P faza ciekła
R жидкая фаза

2047
LIQUID STATE.
State of aggregation characterized by lack of resilience (elasticity) of shape, low compressibility coefficient and lack of crystalline structure.
D flüssiger Zustand, flüssiger Aggregatzustand
F état liquide
P stan (skupienia) ciekły
R жидкое состояние

2048
LIQUIDUS (liquidus curve).
In temperature-concentration diagram, line connecting temperatures at which fusion is just completed for various compositions.
D Liquiduskurve
F liquidus
P likwidus
R ликвидус, линия ликвидуса

LIQUIDUS CURVE. See LIQUIDUS.

2049
LITHIUM, Li.
Atomic number 3.
D Lithium
F lithium
P lit
R литий

2050
LITHIUM-DRIFTED GERMANIUM DETECTOR (Ge(Li) detector).
Spectrometric semiconductor detector of γ radiation consisting of germanium (type p) in which an active layer is produced by drifting lithium atoms.
D lithiumgedrifteter Germaniumdetektor
F détecteur semi-conducteur Ge(Li), détecteur au germanium compensé au lithium
P detektor germanowo-litowy
R литий-германиевый детектор, (полупроводниковый) Ge(Li)-детектор

2051
LOCAL ELECTRIC FIELD (internal field).
Mean electric field acting on molecule in dielectric. Local field is sum of cavity and reaction fields.

D lokales (elektrisches) Feld, inneres Feld
F champ local, champ interne
P pole lokalne, pole wewnętrzne (elektryczne)
R локальное электрическое поле, внутреннее поле

2052
LOCAL ENTROPY PRODUCTION
(entropy source strength), σ.
Rate of increase of locally produced entropy due to dissipative processes; according to second law of thermodynamics local entropy production is always non-negative.

D lokale Entropieerzeugung
F source d'entropie, création locale d'entropie
P źródło entropii, kreacja entropii lokalna
R источник энтропии, локальное возникновение энтропии

2053
LOCAL EQUILIBRIUM.
State of volume element in which its measured thermodynamic parameters correspond to values which could be obtained after isolation of this element from surroundings and when its thermodynamic equilibrium is reached.

D lokales Gleichgewicht
F équilibre local
P równowaga termodynamiczna lokalna
R локальное равновесие

2054
LOCALIZATION ENERGY.
Dynamic reactivity index which gives some measure of reactivity of given position in molecule.

D Lokalisierungsenergie
F énergie de localisation
P energia lokalizacji
R энергия локализации

2055
LOCALIZED ADSORPTION.
Adsorption localized at definite sites of surface preventing motion of molecules along surface.

D lokalisierte Adsorption
F adsorption localisée
P adsorpcja zlokalizowana
R локализованная адсорбция

2056
LOCALIZED ELECTRONS.
Electrons assumed to occupy localized molecular orbitals.

D lokalisierte Elektronen
F électrons localisés
P elektrony zlokalizowane
R локализованные электроны

2057
LOCALIZED MOLECULAR ORBITAL
(localized orbital).

Molecular orbital limited to some part of molecule, e.g. σ orbital of C—H bond in methane.

D lokalisiertes Orbital, lokalisierte Bindung
F orbitale moléculaire localisée
P orbital zlokalizowany
R локализованная молекулярная орбиталь

LOCALIZED ORBITAL. See LOCALIZED MOLECULAR ORBITAL.

2058
LOCAL POTENTIAL, $\Psi(t)$.
Functional

$$\Psi(t) = \int_V d^3x\, L(T, T_o; \mu_1, \mu_{1o}, \ldots, \mu_n, \mu_{no}; \boldsymbol{v}, \boldsymbol{v}_o)$$

where V = volume of system, $T(\boldsymbol{x}, t)$ = temperature as function of position and time, $\mu_i = \mu_i(\boldsymbol{x}, t)$ = chemical potential of i-th component as function of position and time, $\boldsymbol{v} = \boldsymbol{v}(\boldsymbol{x}, t)$ = local velocity of centre of mass as function of position and time; quantities supplied with subscript "o" refer to stationary state and quantities without this subscript change over all comparative states satisfying the same boundary conditions. Local potential reaches its minimum value at $T \equiv T_o$, $\mu_i \equiv \mu_{io}$, $\boldsymbol{v} \equiv \boldsymbol{v}_o$; in statistical mechanical interpretation local potential is measure of information gain which would have to accompany the system transition from a given state to comparative state.

D lokales Potential
F potentiel local
P potencjał lokalny
R локальный потенциал

2059
LOGARITHMIC DISTRIBUTION LAW
(Doerner-Hoskins distribution law).
Law of inhomogeneous distribution of a microcomponent between crystals and solution for two substances able to form mixed crystals; it is obeyed during process of slow cocrystalization and is expressed by equation

$$\ln \frac{a}{a-x} = \lambda \ln \frac{b}{b-y}$$

where x, y = quantities of micro- and macrocomponent which passed into crystalline phase; a, b = initial quantities of micro- and macrocomponent in system; λ = distribution constant.

D logarithmisches Verteilungsgesetz, Doerner-Hoskins-Verteilungsgesetz
F loi de Doerner-Hoskins
P logarytmiczne prawo podziału, prawo Doernera i Hoskinsa
R закон Дёрнера-Хоскинса

2060
LOGARITHMIC-NORMAL DISTRIBUTION.
Probability distribution of positive random variable, whose logarithm has normal distribution.

D logarithmische Normalverteilung
F loi lognormale
P rozkład logarytmiczno-normalny
R логарифмически-нормальное распределение

2061
LONG DISTANCE EFFECT.
Influence exerted by remote substituent in steroidal systems on reaction rate of reacting centre, e.g.

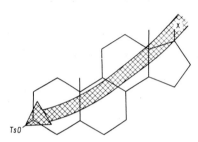

Ts = tosyl ($-SO_2C_6H_4CH_3$)

D Bartonscher Effekt
F effet de Barton
P efekt Bartona, efekt transmisji konformacyjnej, efekt dalekiego zasięgu
R эффект Бартона

LONGITUDINAL RELAXATION. *See* **SPIN-LATTICE RELAXATION.**

2062
LONG-LIVED RADIOISOTOPE.
Term which conventionally defines radioactive nuclides of half-lives longer than 100 hours.

D langlebiges Isotop
F radioisotope de période longue
P izotop długożyciowy
R долгоживущий изотоп

2063
LONG-RANGE ORDER.
Arrangement of atoms or molecules in crystal within 10^4—10^5 unit cells. Concept used in defining superstructures.

D Fernordnung
F ordre à longue échéance
P uporządkowanie dalekiego zasięgu
R дальний порядок

2064
LORENTZ-LORENZ EQUATION.
Equation relating molar refraction R_m of chemical compound to its electronic polarizability α

$$R_m = \frac{n^2-1}{n^2+2} \frac{M}{\rho} = \frac{4}{3} \pi N_A \alpha$$

where n = refractive index, M = relative molecular mass, ρ = density, N_A = Avogadro constant.

D Formel von Lorentz-Lorenz
F formule de Lorentz-Lorenz
P równanie Lorentza i Lorenza
R формула Лоренц-Лоренца

LORENTZ RELATION. *See* **WIEDEMANN-FRANZ LAW.**

2065
LORENZ FIELD.
Local field of strength E_L acting on molecule in isotropic dielectric

$$E_L = E + 4/3\pi P$$

where: E = external electric field vector, P = dielectric polarization.

D Lorenz-Feld
F champ de Lorenz
P pole Lorenza
R поле Лоренца

LOSS ANGLE. *See* **LOSS FACTOR.**

2066
LOSSEN REARRANGEMENT OF HYDROXAMIC ACIDS.
Conversion of hydroxamic acids (or their acyl derivatives) into isocyanates, caused by heating or thionyl chloride, e.g.

$$RC \overset{O}{\underset{N}{\big|}} H \longrightarrow RN{=}C{=}O + H_2O$$

$$\underset{OH}{}$$

D Lossen-Reaktion, (Lossen) Hydroxamsäure→Isocyanat-Abbau
F transposition de Lossen
P przegrupowanie Lossena
R перегруппировка Лоссена

2067
LOSS FACTOR (loss angle), δ.
Phase shift angle between current vector and its capacity component (angle of lead $\pi/2$ rad) in dielectric placed in alternating electric field.

D Verlustwinkel
F angle de perte
P kąt strat
R угол диэлектрических потерь

2068
LOW ENERGY REACTIONS.
Term in nuclear chemistry denoting ordinary chemical reactions dependent on temperature, state of aggregation and other factors, in contradistinction to hot reactions.

D thermische Reaktionen
F réactions thermiques

P reakcje termiczne
R тепловые реакции

2069
LOW-SPIN COMPLEX.
Complex in which some or all of unpaired electrons of free metal ions are paired.

D magnetischer anomaler Komplex
F complexe à spins appariés, complexe à spins faibles
P kompleks niskospinowy, kompleks o spinach sparowanych
R нискоспиновый комплекс

2070
L-S COUPLING (Russel-Saunders coupling).
Interaction of total spin and total angular momenta of all electrons giving total angular momentum of atom
$p_J = \hbar \sqrt{J(J+1)}$, where J is total momentum quantum number.

D Russell-Saunders-Kopplung
F couplage de Russell-Saunders
P sprzężenie LS, sprzężenie Russella i Saundersa
R Рессел-Саундеровская связь, связь LS

2071
LUGGIN CAPILLARY.
Capillary extension of reference electrode or salt bridge in measurements of polarization (overvoltage); tip of capillary is placed very near and perpendicularly to investigated electrode in order to decrease pseudo-resistance (overvoltage) polarization.

D Luggin-Kapillare
F siphon de Haber et Luggin, capillaire de Luggin
P kapilara Ługgina
R капилляр Луггина

2072
LUMINANCE, L.
Measure of intensity of visible part of radiation spectrum radiated by surface element in direction of observations. Luminance is given as ratio of light intensity in direction of observation and area of projection of radiating element onto plane perpendicular to this direction; SI unit is 1 cd/m².

D Leuchtdichte
F luminance
P luminancja, jaskrawość, jasność, blask
R яркость

2073
LUMINESCENCE.
Emission of light independently of thermal radiation of emitting body which follows no earlier than 10^{-10} s after absorption of energy by atoms or molecules of body.

D Lumineszenz
F luminescence
P luminescencja
R люминесценция

2074
LUMINESCENCE ANALYSIS (luminescent analysis).
Emission spectroanalysis by investigation of luminescence spectrum.

D Lumineszenzanalyse
F analyse par luminescence
P analiza luminescencyjna
R люминесцентный анализ

2075
LUMINESCENCE DOSIMETER.
Liquid or solid matter, luminescence of which in ionizing radiation field is proportional to dose rate of radiation being absorbed by dosimeter.

D Lumineszenzdosimeter
F dosimètre luminescent, luminodosimètre
P dawkomierz luminescencyjny
R люминесцентный дозиметр

LUMINESCENCE OF CRYSTALLIZATION. See CRYSTALLOLUMINESCENCE.

LUMINESCENT ANALYSIS. See LUMINESCENCE ANALYSIS.

2076
LUMINOPHORES.
Substances displaying luminescence after their excitation, e.g. solid solutions of some fluorescent dyes or activated crystal phosphors.

D Luminophore, Leuchtstoffe
F luminophores
P luminofory
R люминофоры

2077
LUMINOUS INTENSITY (light intensity, radiation intensity).
Flux of light radiated into unit solid angle. It is fundamental quantity in visual photometry; SI unit is 1 candle (cd).

D Lichtstärke
F intensité lumineuse
P natężenie światła, światłość
R сила света

2078
LUTETIUM, Lu.
Atomic number 71.

D Lutetium
F lutétium
P lutet
R лютеций

2079
LYATE ION.
Product of autoprotolysis; base formed from molecule of solvent after proton removal, coupled with equivalent amount of simultaneously formed acid, i.e. lyate ion; e.g. lyate ions of water and formic acid

$$2H_2O \rightleftarrows \boxed{OH^\ominus} + H_3O^\oplus$$

$$2HCOOH \rightleftarrows \boxed{HCOO^\ominus} + H-\underset{OH}{\overset{OH}{C^\oplus}}$$

D Lyat-Ion
F lyate
P jon liatowy
R лиат-ион

2080
LYOGEL.
Gel rich in liquid produced in gelation process when bonds between molecules of sol, making structure rigid, are formed.
D Lyogel, Gallerte
F lyogel, gelée
P liożel, galareta
R лиогель, студень

2081
LYONIUM ION.
Product of autoprotolysis; acid formed from solvent molecule and proton, coupled with equivalent amount of simultaneously formed base, i.e. lyonium ion; e.g. lyonium ions of water and formic acid

$$2H_2O \rightleftarrows OH^\ominus + \boxed{H_3O^\oplus}$$

$$2HCOOH \rightleftarrows HCOO^\ominus + \boxed{H-\underset{OH}{\overset{OH}{C^\oplus}}}$$

D Lyonium-Ion
F lyonium
P jon lionowy
R лионий-ион

2082
LYOPHILIC COLLOID.
Colloid in which particles of dispersed phase are solvated by dispersion medium, e.g. sol of gelatin in water, sol of caoutchouc in benzene.
D lyophiles Kolloid
F colloïde lyophile
P koloid liofilowy
R лиофильный коллоид

2083
LYOPHOBIC COLLOID.
Colloid in which particles of dispersed phase are not solvated by dispersion medium, e.g. sols of noble metals in water.

D lyophobes Kolloid
F colloïde lyophobe
P koloid liofobowy
R лиофобный коллоид

LYOTROPIC SERIES. *See* HOFMEISTER SERIES.

2084
MACRO-ANALYSIS.
Analysis of sample with mass exceeding 0.1 g or, in the case of gases, volume larger than 10 cm³.
D Makroanalyse
F macroanalyse
P makroanaliza, analiza w skali makro
R макроанализ

2085
MACRO-COMPONENT.
Component of system which can be detected or determined by analytical methods.
D Makrokomponente
F macrocomposant
P makroskładnik
R макрокомпонент

MACRO METHOD. *See* GRAM METHOD.

MACROMOLECULAR COLLOID. *See* MOLECULAR COLLOID.

2086
MACROMOLECULAR COMPOUND (high molecular weight compound).
Chemical compound composed of macromolecules.
D makromolekulare Verbindung
F composé macromoléculaire
P związek wielkocząsteczkowy
R высокомолекулярное соединение

2087
MACROMOLECULE.
Molecule composed of great number of atoms, the relative molecular mass of such a molecule exceeding several thousands.
D Makromolekül, Makromolekel
F macromolécule
P makrocząsteczka, makrodrobina, makromolekuła
R макромолекула

2088
MACROPOROUS ION EXCHANGER.
Ion exchanger having pores that are big in comparison with atomic dimensions.
D makro-retikularer Ionenaustauscher
F échangeur macroporeux
P jonit makroporowaty
R крупнорешётчатый ионит

2089
MACROSCOPIC CROSS-SECTION.
Product of microscopic cross-section σ and
number of particles N contained in 1 cm³.
D makroskopischer Wirkungsquerschnitt
F section efficace macroscopique
P przekrój czynny makroskopowy
R макроскопическое (эффективное)
   сечение

MACROSCOPIC EQUILIBRIUM STATE.
See THERMODYNAMIC EQUILIBRIUM.

MACROSCOPIC PARAMETER. *See*
STATE VARIABLE.

2090
MACROSCOPIC STATE (thermodynamic
state).
State of macroscopic system characterized
by such macroscopic properties as volume
of system, its density, internal energy,
electric current strength, etc.
D Makrozustand
F état macroscopique
P stan makro(skopowy), stan
   termodynamiczny
R макроскопическое состояние

MACROSCOPIC SYSTEM. *See*
THERMODYNAMIC SYSTEM.

2091
MACROSCOPIC THERMODYNAMICS
(phenomenological thermodynamics).
Theory of macroscopic systems based on
experimentally justified
phenomenological laws such as laws of
thermodynamics, phenomenological
equations, etc.
D phänomenologische Thermodynamik
F thermodynamique phénoménologique,
   thermodynamique macroscopique
P termodynamika fenomenologiczna
R феноменологическая термодинамика

MACROSCOPIC VARIABLES. *See*
STATE QUANTITIES.

2092
MADELUNG CONSTANT.
Coefficient in expression for potential
energy in ionic crystal, depending on mode
of distribution of positive and negative
ions in crystal lattice.
D Madelungsche Konstante
F constante de Madelung
P stała Madelunga
R постоянная Маделунга

2093
MAGIC NUCLEUS.
Atomic nucleus containing a magic
number of protons or neutrons, viz.

nucleus with completed proton or neutron
shells. Magic nucleus has high binding
energy.
D magischer Kern
F noyau magique
P jądro magiczne
R магическое ядро

2094
MAGIC NUMBERS.
Numbers of protons or neutrons in atomic
nuclei of high binding energy, equal to 2,
8, 20, 28, 50, 82, 126. Nuclei with magic
number of protons or neutrons have
completed proton or neutron shells.
D magische Zahlen, magische
   Nukleonenzahlen
F nombres magiques
P liczby magiczne
R магические числа

2095
MAGNESIUM, Mg.
Atomic number 12.
D Magnesium
F magnésium
P magnez
R магний

2096
MAGNETIC ANISOTROPY.
Anisotropy of magnetic properties of
crystal.
D magnetische Anisotropie
F anisotropie magnétique
P anizotropia magnetyczna
R магнитная анизотропия

2097
MAGNETIC CRITERION.
Criterion which enables division of
complexes into ionic or outer-orbital and
covalent or inner-orbital, basing on
comparison of number of unpaired
electrons in free and central metal ions.
D magnetisches Kriterium
F ...
P kryterium magnetyczne
R магнитный критерий

2098
MAGNETIC DIPOLE.
System of magnetic mass with distinct
negative and positive magnetic centres of
gravity. In the simplest case system of two
displaced point magnetic monopoles,
positive and negative, with equal magnetic
mass.
D magnetischer Dipol
F dipôle magnétique
P dipol magnetyczny
R магнитный диполь

2099
MAGNETIC FIELD.
Vector field described by magnetic
induction vector, produced by each
permanent magnet or conductor in current
within environment.
Remark: Term magnetic field frequently
denotes magnetic field strength.

D Magnetfeld, magnetisches Feld
F champ magnétique
P pole magnetyczne
R магнитное поле

MAGNETIC FIELD INTENSITY. *See*
MAGNETIC FIELD STRENGTH.

2100
MAGNETIC FIELD STRENGTH
(magnetic field intensity), **H**.
Vector quantity together with magnetic
induction **B** describing magnetic field and
defining value and direction of force
acting on magnetic pole at a given point
in field.

D magnetische Feldstärke
F intensité du champ magnétique
P natężenie pola magnetycznego
R напряжённость магнитного поля

2101
MAGNETIC FLUX, **Φ**.
Flux of induction vector **B** through
surface S

$$\Phi = \int_S B \cos\Theta \, dS$$

where $B \cos\Theta$ = perpendicular component
of vector **B** with respect to surface S.
D magnetischer Fluß, Induktionsfluß
F flux magnétique, flux d'induction
     magnétique
P strumień indukcji magnetycznej
R магнитный поток

2102
MAGNETIC HYSTERESIS.
Hysteresis of magnetization changes in
ferro- or ferrimagnets with respect to
magnetic field strength.
D magnetische Hysterese
F hystérèse magnétique, hystérésis
     magnétique
P histereza magnetyczna
R магнитный гистерезис

2103
MAGNETIC INDUCTION, **B**.
Quantity characterizing effective magnetic
field in any substance

$$B = H + 4\pi J$$

if **B**, **H** and **J** are parallel

$$B = \mu H$$

where μ = magnetic permeability,
H = magnetic field strength,
J = magnetization.
D magnetische Induktion
F induction magnétique
P indukcja magnetyczna
R магнитная индукция

2104
MAGNETIC MOMENT.
Quantity which provides measure of
magnetic field strength due to electric
current or magnetic body.
D magnetisches Moment
F moment magnétique
P moment magnetyczny
R магнитный момент

2105
MAGNETIC MOMENT OF ATOM (atomic
magnetic moment).
Vector quantity responsible for interaction
between atoms and external magnetic
field; for atoms with low atomic number
magnetic moment of atom is given by
$\sqrt{J(J+1)}\, g(L, S, J)\mu_B$ (Russel-Saunders
coupling), where J = total angular
momentum quantum number of atom,
$g(L, S, J)$ = Lande factor and μ_B = Bohr
magneton; for atoms with high atomic
number magnetic moment of atom is
equal to sum of magnetic moments of
individual electrons (j-j coupling).
D magnetisches Atommoment
F moment magnétique atomique
P moment magnetyczny atomu
R магнитный момент атома

2106
MAGNETIC PERMEABILITY, μ.
Tensor quantity defined by $B = \mu H$,
where **B** = magnetic induction,
H = external magnetic field strength.
D magnetische Permeabilität
F perméabilité magnétique
P przenikliwość magnetyczna,
     przenikalność magnetyczna
R магнитная проницаемость

2107
MAGNETIC QUANTUM NUMBER,
m (m_l).
Number which defines allowed values of
projection of electron angular momentum
onto some chosen direction, e.g. direction
of external magnetic field. For given value
of orbital quantum number l allowed
values of m are $0, \pm 1, \mp 2, ..., \pm l$.
D magnetische Quantenzahl
F nombre quantique magnétique
P liczba kwantowa magnetyczna
R магнитное квантовое число

2108
MAGNETIC RESONANCE (paramagnetic resonance).
Absorption of alternating electromagnetic field energy of characteristic frequency ν which satisfies resonance condition $h\nu = \Delta E = g\mu_B H$, where $\Delta E =$ spacing of adjacent Zeeman levels, $g = g$-factor, $\mu_B =$ Bohr or nuclear magneton, $H =$ magnetic field strength.

D magnetische Resonanz
F résonance magnétique
P rezonans (para)magnetyczny
R магнитный резонанс

2109
MAGNETIC SPIN QUANTUM NUMBER (inner quantum number), m_s.
Number which defines allowed values of projection of particle spin angular momentum onto given direction, e.g. direction of external magnetic field; for electrons m_s assumes only two numerical values $+1/2$ and $-1/2$.

D magnetische Spinquantenzahl
F nombre quantique de spin
P liczba kwantowa magnetyczna spinowa
R магнитное спиновое квантовое число

2110
MAGNETIC STRUCTURE.
Spatially ordered distribution of atomic magnetic moments in crystals of ferro-, ferri- or antiferromagnets.

D magnetische Struktur
F structure magnétique
P struktura magnetyczna
R магнитная структура

2111
MAGNETIC SUSCEPTIBILITY, $\varkappa$.
Tensor quantity characterizing ability of a given substance to change magnetization J upon change of magnetic field strength H, $J = \varkappa H$, where $\varkappa =$ magnetic susceptibility of 1 cm³ of substance. Magnetic susceptibility is also defined sometimes for 1 g (χ_g) or 1 mole (χ_M) of substance.

D magnetische Suszeptibilität
F susceptibilité magnétique
P podatność magnetyczna
R магнитная восприимчивость

2112
MAGNETIZATION (intensity of magnetization).
Magnetic moment of 1 cm³ (J), 1 g (σ) or 1 mole (M) of substance.

D Magnetisierung, Magnetisierungsintensität
F aimantation, intensité d'aimantation
P namagnesowanie, magnetyzacja, natężenie magnetyzacji
R намагниченность

2113
MAGNETOCALORIC EFFECT.
Decrease or increase of para- or ferromagnet temperature resulting from adiabatic change of magnetic field strength.

D magnetokalorischer Effekt
F effet magnétocalorique
P efekt magnetokaloryczny
R магнитокалорический эффект

2114
MAGNETOCHEMISTRY.
Branch of chemistry utilizing interaction of magnetic field with solids, liquids and gases for study of their physical and chemical properties.

D Magnetochemie
F magnétochimie
P magnetochemia
R магнетохимия

MAGNETOMECHANICAL EFFECTS. *See* GYROMAGNETIC EFFECTS.

2115
MAGNETOSTRICTION.
Change of body dimensions in magnetic field.

D Magnetostriktion
F magnétostriction
P magnetostrykcja
R магнетострикция

2116
MAGNONS.
Quasi-particles ascribed to spin waves introduced in quantum theory of magnetically ordered crystals; obey Bose-Einstein statistics.

D Magnone
F magnons
P magnony
R магноны

2117
MAIN CHAIN (backbone chain, trunk chain, parent chain, fundamental chain).
The longest system of chain-like linked atoms in branched chain compounds, e.g.

$$CH_3-CH_2-CH_2-\underset{\underset{CH_3}{|}}{CH}-\overset{\overset{CH_2-CH_3}{|}}{CH}-CH_2-CH_2-CH_3$$

D Hauptkette, Stammkette
F chaîne unique, tronc
P łańcuch główny
R главная цепь, основная цепь

2118
MAIN GROUP (even series, A family).
Part of group of periodic system
comprising elements, the atoms of which
have exclusively closed inner electronic
sub-shells and partly filled outer
electronic shells, of structure from ns^1 to
ns^2np^5.
D Hauptgruppe
F famille principale, groupe principal
P grupa główna
R главная группа

2119
MANGANESE, Mn.
Atomic number 25.
D Mangan(ium)
F manganèse
P mangan
R марганец

2120
MANGANESE GROUP.
Elements in seventh sub-group of periodic
system with structure of outer electronic
sub-shells of atoms: manganese — $3d^54s^2$,
technetium — $4d^65s^1$, rhenium — $5d^56s^2$.
D Mangangruppe
F métaux de la famille du manganèse
P manganowce
R подгруппа марганца, элементы
подгруппы марганца

MANNICH REACTION. *See*
AMINOMETHYLATION.

2121
MANUAL SPECTROPHOTOMETER.
Spectrophotometer, most frequently single
beam, with spectrum recorded point by
point at some chosen wavelengths.
D punktweise messendes
Spektralphotometer
F spectrophotomètre non enregistreur
P spektrofotometr punktowy
R нерегистрирующий спектрофотометр

MARIOTTE'S LAW. *See* BOYLE'S LAW.

2122
MARKOWNIKOFF'S RULE.
In reaction of hydrogen halide with
asymmetrical alkene, hydrogen atom
always adds to that carbon which already
carries more hydrogen atoms and halide
atom to the carbon with the lesser
number of hydrogen atoms, e.g.

$CH_3-CH=CH_2 + HX \longrightarrow CH_3-CHX-CH_3$

D Markownikoff-Regel
F règle de Markownikoff
P reguła Markownikowa
R правило Марковникова

2123
MASS ATTENUATION COEFFICIENT.
Quotient of linear attenuation coefficient
and density of absorbing medium.
D Massenschwächungskoeffizient
F coefficient d'atténuation massique
P współczynnik osłabienia masowy
R массовый коэффициент ослабления

2124
MASS AVERAGE RELATIVE
MOLECULAR MASS (weight average
molecular weight).
Average molecular mass calculated from
measurement of effects which depend on
mass of macromolecules contained in unit
volume (methods: diffusion of light, rate
of diffusion and, under certain conditions,
viscosity method).
D Gewichtsmittelwert, Gewichtsmittel
F poids moléculaire moyen au poids
P masa cząsteczkowa wagowo średnia
R средний весовой молекулярный вес

2125
MASS CONSERVATION LAW.
In isolated system sum of masses is
constant, provided no nuclear reactions
take place.
D Massenerhaltungssatz, Prinzip von der
Erhaltung der Masse
F loi de conservation de la masse
P zasada zachowania masy, prawo
zachowania masy
R закон сохранения массы

2126
MASS CONSERVATION LAW IN
CONTINUOUS SYSTEMS (conservation
equation for mass).
Equation

$$\frac{\partial \rho}{\partial t} = -\text{div}(\rho v)$$

where t = time, ρ = density, and
v = velocity of centre of mass.
D Massenerhaltungssatz in homogenen
Systemen
F loi de conservation de la masse, équation
de continuité
P zasada zachowania masy w układach
ciągłych
R закон сохранения массы
в непрерывных системах

2127
MASS DEFECT.
Difference between sum of masses of
constituent particles (nucleons) of nucleus
and atomic mass of this nucleus.
D Massendefekt
F défaut de masse

P defekt masy, niedobór masy, deficyt masy
R дефект массы

2128
MASS FRACTION, w_B.
Ratio of mass of constituent B to general mass of mixture

$$w_B = \frac{m_B}{\sum_i m_i}$$

where m_B = mass of constituent B,
$\sum_i m_i$ = sum of masses of all constituents
of mixture.
D Massenbruch
F fraction en masse
P ułamek masowy, ułamek wagowy, stężenie wagowe
R весовая доля

2129
MASSIEU FUNCTION, J.
Extensive thermodynamic function $J = -F/T$, where F = Helmholtz free energy and T = thermodynamic temperature.
D Massieusche Funktion
F fonction de Massieu
P funkcja Massieu
R функция Массье

2130
MASS NUMBER, A.
Integer closest to relative atomic mass of given atom.
D Massenzahl
F nombre de masse
P liczba masowa
R массовое число

2131
MASS SPECTRAL ANALYSIS (mass spectrometric analysis).
Qualitative and quantitative determination of sample's chemical composition from its mass spectrum.
D Massenspektralanalyse, massenspektrometrische Analyse
F analyse par spectroscopie de masse, analyse par spectrométrie de masse
P analiza spektralna masowa, analiza spektrometryczna masowa
R масс-спектральный анализ, масс-спектрометрический анализ

2132
MASS SPECTROGRAPH.
Device for separation of charged particles according to their mass to charge ratio and recording mass spectrum on photographic plate.

D Massenspektrograph
F spectrographe de masse
P spektrograf mas(owy)
R масс-спектрограф

2133
MASS SPECTROMETER.
Device for separation of charged particles (ions) according to their mass to charge ratio. It enables qualitative and quantitative analysis as well as study of structure of molecules to be conducted.
D Massenspektrometer
F spectromètre de masse
P spektrometr mas(owy)
R масс-спектрометр

MASS SPECTROMETRIC ANALYSIS.
See MASS SPECTRAL ANALYSIS.

2134
MASS SPECTROMETRY.
Analytical and structural research method based on measurement of mass to charge ratio of ions produced from a given sample.
D Massenspektrometrie
F spectrométrie de masse
P spektrometria mas(owa)
R масс-спектрометрия

2135
MASS SPECTRUM, MS.
Distribution of charged particles (ions) according to their mass to charge ratio, obtained in spectrograph or mass spectrometer.
D Massenspektrum
F spectre de masse
P widmo masowe
R масс-спектр

2136
MASS STOPPING POWER.
Average energy lost by charged particle in traversing certain path length in matter divided by that path length and density of medium.
D Massenbremsvermögen
F pouvoir d'arrêt massique
P zdolność hamowania masowa
R массовая тормозная способность

2137
MATRIX.
In chemical analysis, main component (components) of material in which trace components are to be determined.
D Matrix
F matrice
P matryca
R основа

MATTER. *See* SUBSTANCE.

2138
MATTOX-KENDALL METHOD (*for double bond introduction into steroids*). Preparation of α, β-unsaturated ketones by elimination of hydrogen bromide from α-bromoketones (which are converted first into 2,4-dinitrophenylhydrazones to increase mobility of α-bromine atom), e.g.

This method is mainly used in chemistry of steroids.

D Mattox-Kendall HBr-Abspaltung aus α-Bromketonen
F réaction de Mattox
P reakcja Mattoxa i Kendalla
R реакция Маттокса

2139
MAXIMUM COORDINATION NUMBER. Maximum number of coordinate bonds that can be formed by central atom (*cf.* characteristic coordination number).

D maximale Koordinationszahl
F nombre maximum de coordination
P liczba koordynacyjna maksymalna
R максимальное координационное число

MAXIMUM MECHANICAL WORK. *See* MAXIMUM WORK.

2140
MAXIMUM PERMISSIBLE DOSE. Maximum dose of defined ionizing radiation absorbed in defined time by the whole body or particular organs, not resulting — according to present stage of knowledge — in permanent recognisable damage to human body.

D höchstzulässige Dosis
F dose maximale admissible
P dawka dopuszczalna największa
R предельно допустимая доза

MAXIMUM USEFUL WORK. *See* NET MAXIMUM WORK.

2141
MAXIMUM WORK (maximum mechanical work). Maximum amount of work which can be performed by closed system on surroundings on isothermal transition from state I to state II; equal to loss of Helmholtz free energy of system $\Delta F = F_\mathrm{I} - F_\mathrm{II}$.

D maximale Arbeit
F travail maximum
P praca (zewnętrzna) maksymalna
R максимальная работа

2142
MAXWELL-BOLTZMANN DISTRIBUTION (Boltzmann distribution function). Equilibrium form of unimolecular partition function

$$\rho_1(1) = \lambda \exp\left[-\frac{\varepsilon(1)}{kT}\right]$$

where k = Boltzmann constant, T = thermodynamic temperature, λ = absolute activity and ε = energy of molecule, ρ_1 = unimolecular partition function and (1) stands for all molecule dynamic variables which characterize its state. Maxwell-Boltzmann distribution is valid for systems of weakly interacting particles and for rarified gases provided quantum effects are negligibly small.

D Maxwell-Boltzmannsches Verteilungsgesetz
F répartition de Maxwell-Boltzmann, répartition de Boltzmann
P rozkład Maxwella i Boltzmanna
R распределение Больцмана

2143
MAXWELLIAN VELOCITY DISTRIBUTION (Maxwell's distribution law). Equilibrium distribution of molecule velocities

$$\langle N(v_x, v_y, v_z)\rangle = \langle N\rangle \left(\frac{m}{2\pi kT}\right)^{3/2} \exp\left[-\frac{mv^2}{2kT}\right] dv_x\, dv_y\, dv_z$$

where $\langle N(v_x, v_y, v_z)\rangle$ = average number of molecules with velocity (v_x, v_y, v_z), $\langle N\rangle$ = average number of molecules in the system, m = molecule mass, k = Boltzmann constant, and T = thermodynamic temperature.

D Maxwellsche Geschwindigkeitsverteilung, Maxwellsche Verteilung
F distribution maxwellienne
P rozkład Maxwella
R распределение Максвелла, максвелловское распределение (молекул по скоростям)

MAXWELL'S DISTRIBUTION LAW. *See* MAXWELLIAN VELOCITY DISTRIBUTION.

MAXWELL'S EQUATIONS. *See* MAXWELL'S THERMODYNAMIC RELATIONS.

2144
MAXWELL'S THERMODYNAMIC
RELATIONS (Maxwell's equations).
Thermodynamic identities expressing
dependence of phase entropy S on its
volume V and pressure p in terms of
easily measurable thermodynamic
quantities

$$\left(\frac{\partial S}{\partial V}\right)_T = \left(\frac{\partial p}{\partial T}\right)_V, \quad \left(\frac{\partial S}{\partial p}\right)_T = -\left(\frac{\partial V}{\partial T}\right)_p$$

D Maxwellsche Relationen, Maxwellsche
   Beziehungen
F relations de Maxwell
P relacje Maxwella
R соотношения Максвелла

2145
MAYER'S RELATION.
Formula which for ideal gas determines
relation between molar heat capacity at
constant pressure C_p and molar heat
capacity at constant volume C_V,
$C_p - C_V = R$, where $R = $ molar gas
constant.

D Mayersche Gleichung
F formule de Mayer, relation de Mayer
P wzór Mayera
R формула Майера

2146
McFADYEN-STEVENS REDUCTION.
Reduction of aromatic carboxylic acids to
corresponding aldehydes, according to
following sequence of reactions:

$ArCOOC_2H_5 \xrightarrow{NH_2NH_2} ArCONHNH_2 \xrightarrow{C_6H_5SO_2Cl}$

$ArCONHNHSO_2C_6H_5 \xrightarrow{Na_2CO_3} ArCHO + N_2 +$
$+ C_2H_5SO_2Na$

D McFadyen-Stevens-Synthese,
   McFadyen-Stevens
   Carbonsäure-Reduktion
F synthèse de Mac Fadyen-Stevens
P redukcja McFadyena i Stevensa
R восстановление по Мак
   Фадиену-Стивенсу

2147
MEAN ACTIVITY COEFFICIENT OF
IONS, $f_{\pm}$.
Quantity defined by formula

$$f_{\pm} = \sqrt[\nu]{f_+^{\nu^+} f_-^{\nu^-}}$$

where f_+ and $f_- = $ activity coefficients of
ions, ν^+ and $\nu^- = $ numbers of cations and
anions respectively which are formed from
one molecule of electrolyte, $\nu = \nu^+ + \nu^-$.

D mittlerer Ionenaktivitätskoeffizient
F coefficient moyen d'activité ionique
P współczynnik aktywności jonowej średni
R средний коэффициент активности
   ионов

2148
MEAN FREE FLIGHT TIME (mean time
between collisions).

Time interval between two subsequent
collisions for typical gas particle.
D mittlere Freibewegungszeit
F (temps moyen qui sépare deux
   collisions)
P czas ruchu swobodnego średni
R (среднее) время свободного пробега

2149
MEAN FREE PATH (between
collisions), λ.
Average path covered by molecule
between two collisions

$$\lambda = \frac{\bar{c}}{\bar{z}} = \frac{1}{\sqrt{2}\,\pi d^2 n}$$

where $\bar{c} = $ average speed, $\bar{z} = $ mean
number of collisions of one molecule
during unit time, $d = $ sum of radii of
colliding molecules, $n = $ number of
molecules in unit volume.
D mittlere freie Weglänge
F libre parcours moyen
P droga swobodna średnia
R средний (свободный) пробег, средняя
   длина свободного пробега

2150
MEAN IONIC CONCENTRATION, $c_{\pm}$.

$$c_{\pm} = \sqrt[\nu]{c_+^{\nu^+} c_-^{\nu^-}}$$

where c_+ and $c_- = $ concentrations of
cations and anions respectively,
ν^+ and $\nu^- = $ number of cations and anions
respectively which arise from one
molecule of electrolyte, $\nu = \nu^+ + \nu^-$.
D mittlere Ionenkonzentration
F concentration ionique moyenne
P stężenie jonowe średnie
R средняя концентрация ионов

2151
MEAN IONIC DIAMETER.
According to Debye-Hückel's theory of
strong electrolytes, distance of closest
approach of arbitrary ion of ionic cloud to
central ion; this parameter is determined
by using Debye-Hückel's equation, thus its
physical meaning is only approximately as
suggested by the name (see Debye-
Hückel's equation).
D mittlerer Ionendurchmesser
F sphère d'activité, rayon ionique
P średnica jonu efektywna, średnica jonu
   średnia
R средний ионный диаметр

2152
MEAN LIFE (average life).
Average time in which transition occurs
from one physical state to another.
D mittlere Lebensdauer
F vie moyenne
P czas życia średni
R среднее время жизни

2153
MEAN LIFE OF RADIONUCLIDE.
Reciprocal of disintegration constant λ.
D mittlere Lebensdauer
F vie moyenne
P czas życia średni (radionuklidu)
R среднее время жизни

MEAN LIFE-TIME OF LUMINESCENCE.
See OBSERVED LIFE-TIME OF
LUMINESCENCE.

2154
MEAN MOLAL HEAT CAPACITY, $\overline{C}$.
Proportion between amount of heat Q
received isochorically or isobarically by
one mole of substance over temperature
range from T_1 to T_2

$$\overline{C} = \frac{Q}{T_2 - T_1}$$

D mittlere Molwärme
F chaleur spécifique molaire moyenne
P ciepło molowe średnie
R средняя теплоёмкость

MEAN OF RANDOM VARIABLE. *See*
EXPECTED VALUE.

2155
MEAN SQUARE DEVIATION (variance,
mean square fluctuation, second moment
of fluctuating variable).
Mean square of distribution of actual
values of quantity A (e.g. energies of
system) with respect to its statistical
average $\langle A \rangle$: $\langle (A - \langle A \rangle)^2 \rangle$.
D mittleres Schwankungsquadrat
F fluctuation quadratique moyenne
P fluktuacja kwadratowa średnia,
odchylenie kwadratowe średnie
R средняя квадратичная флуктуация,
среднеквадратичная флуктуация,
средний квадрат флуктуации

MEAN SQUARE FLUCTUATION. *See*
MEAN SQUARE DEVIATION.

MEAN TIME BETWEEN COLLISIONS.
See MEAN FREE FLIGHT TIME.

2156
MEAN TIME OF COLLISION.
Time interval spent by typical gas particle
in region of interaction with target
particle.
D Stoßdauer, Zeitdauer eines Stoßes
F durée moyenne d'une collision
P czas trwania zderzenia średni
R (средняя) длительность столкновения,
(среднее) время столкновения

2157
MECHANICAL EQUILIBRIUM
(hydrostatic equilibrium).

State of system with local velocity v of
centre of mass independent of time,
$$\frac{dv}{dt} = 0$$

D mechanisches Gleichgewicht
F équilibre mécanique
P równowaga mechaniczna
R механическое равновесие

2158
MECHANISM IN RADIATION
CHEMICAL PROCESSES.
Sequence of reactions or of several
parallel sequences of competing reactions
in which reactants are primary and
secondary products of radiolysis; all such
reactions finally end with chemical
compounds stable in actual conditions.
D Mechanismus von strahleninduzierten
Reaktionen
F mécanisme des réactions radiochimiques
P mechanizm chemicznych reakcji
radiacyjnych
R механизм радиационно-химических
реакций

MECHANISM OF BIMOLECULAR
ELECTROPHILIC SUBSTITUTION. *See*
S_E2 MECHANISM.

MECHANISM OF BIMOLECULAR
ELIMINATION. *See* E_2 MECHANISM.

MECHANISM OF BIMOLECULAR
NUCLEOPHILIC SUBSTITUTION. *See*
S_N2 MECHANISM.

MECHANISM OF BIMOLECULAR
NUCLEOPHILIC SUBSTITUTION WITH
ALLYLIC REARRANGEMENT. *See* S_N2'
MECHANISM.

2159
MECHANISM OF ELECTROPHILIC
ADDITION TO CARBON-CARBON
MULTIPLE BONDS.
The first step in this reaction involves
addition of electrophile to carbon-carbon
multiple bond with formation of cation.
In the second step, *trans*-addition of
nucleophile to cation takes place.
Formation of π-complex precedes
addition, e.g.

π-complex   bromonium cation

addition product

D elektrophiler Mechanismus der Addition
an Kohlenstoff-Kohlenstoff-
-Mehrfachbindungen
F mécanisme de l'addition électrophile
aux liaisons multiples
P mechanizm przyłączenia elektrofilowego
do wiązań wielokrotnych C=C i C≡C
R механизм электрофильного
присоединения к кратным
углерод-углеродным связям

2160
MECHANISM OF ELECTROPHILIC
AROMATIC SUBSTITUTION.
The following reaction steps are
postulated: first, formation of π-complex
due to interaction of ring π-electrons with
electrophile $X^{\oplus}$; secondly, transformation
of π-complex into carbonium ion
(σ-complex) in which $X^{\oplus}$ is bound to one
atom of ring (positive charge of carbonium
ion is shared over whole system); thirdly,
proton elimination from carbonium ion
(proton is consumed by nucleophile $Y^{\ominus}$
present in system) with returning to
energetically poor aromatic system.
σ-complex formation or proton elimination
may be the rate determining step

D Mechanismus der elektrophilen
aromatischen Substitution
F mécanisme de la substitution
électrophile aromatique
P mechanizm elektrofilowego
podstawienia aromatycznego
R механизм электрофильного
ароматического замещения

2161
MECHANISM OF FREE-RADICAL
ADDITION TO UNSATURATED
HYDROCARBONS.
This mechanism, typical for addition in
the presence of free radical initiators
(e.g. light, peroxides), involves the
following steps: decomposition of added
molecule into free radicals (atoms),
addition of free radical (atom) to π-bond
and formation of new free radical with
unpaired electron on carbon atom,
reaction of carbon free radical with added
molecule with formation of new addition
product and new free radical. The
free-radical addition is a typical chain
reaction, e.g.

D Mechanismus der radikalischen Addition
an ungesättigte Systeme
F mécanisme de l'addition radicalaire aux
oléfines
P mechanizm przyłączenia
wolnorodnikowego do węglowodorów
nienasyconych
R механизм радикального присоединения
к кратным углеродным связям

2162
MECHANISM OF FREE-RADICAL
POLYMERIZATION.
Polymerization in which free radicals
devoid of electric charge play dominant
role.
D Radikalmechanismus der Polymerisation
F mécanisme de la polymérisation par les
radicaux
P mechanizm polimeryzacji rodnikowy
R полимеризация по свободно-
-радикальному механизму

MECHANISM OF FREE-RADICAL
SUBSTITUTION. *See* S_R MECHANISM.

MECHANISM OF INTERNAL
NUCLEOPHILIC SUBSTITUTION. *See*
S_{Ni} MECHANISM.

2163
MECHANISM OF NUCLEOPHILIC
ADDITION TO CARBONYL GROUP.
Addition to carbonyl group is initiated by
nucleophile attack on carbonyl carbon
atom, which is electron-deficient due to
polarization of carbonyl bond. In the
second reaction step, cation adds to ion
with negatively charged oxygen atom, e.g.

D Mechanismus der nucleophilen Addition
an Karbonylgruppe
F mécanisme de l'addition nucléophile au
carbonyle
P mechanizm przyłączenia nukleofilowego
do grupy karbonylowej
R механизм присоединения
нуклеофильных реагентов
к карбонильной группе

2164
MECHANISM OF NUCLEOPHILIC ADDITION TO ETHYLENIC BOND.
Addition of nucleophiles to ethylenic bonds conjugated with $\diagdown C=O$, $-C\equiv N$ and similar groups. The first reaction step involves addition of anion to terminal carbon atom of the conjugated system. In the second step, addition of cation to atom takes place at position 2- or 4-, e.g.

D Mechanismus der nucleophilen Addition an C=C Bindung
F mécanisme de l'addition nucléophile à la liaison éthylénique
P mechanizm przyłączenia nukleofilowego do wiązania etylenowego
R механизм нуклеофильного присоединения к этиленовой связи

2165
MECHANISM OF NUCLEOPHILIC AROMATIC SUBSTITUTION.
This reaction may proceed in two different ways: (1) In first, rate-determining step, nucleophile attacks carbon atom linked with leaving group, forming intermediate anion with delocalized charge (Meisenheimer complex). In second step, aromatic system is restored due to elimination of leaving group as anion. (2) Complex mechanism involving elimination with subsequent addition. In first, base-catalized stage, elimination takes place with formation of intermediate dehydrobenzene. The latter reacts readily with nucleophile forming addition product. Nucleophilic aromatic substitution may proceed parallelly by both routes

D Mechanismus der nucleophilen aromatischen Substitution

F mécanisme de la substitution nucléophile aromatique
P mechanizm nukleofilowego podstawienia aromatycznego
R механизм нуклеофильного замещения в бензольном кольце

MECHANISM OF UNIMOLECULAR ELECTROPHILIC SUBSTITUTION. *See* S$_E$1 MECHANISM.

MECHANISM OF UNIMOLECULAR ELIMINATION. *See* E$_1$ MECHANISM.

MECHANISM OF UNIMOLECULAR NUCLEOPHILIC SUBSTITUTION. *See* S$_N$1 MECHANISM.

2166
MECHANOCHEMISTRY.
Branch of chemistry dealing with chemical reactions effected by mechanical treatment, in particular reactions of macromolecular compounds.

D Mechanochemie
F mécanochimie
P mechanochemia
R механохимия

2167
MEDIAN, m.
Medium value in series; e.g. in series of measurement (determination) results ordered according to their increasing values x_1, x_2, ..., x_n for n odd, $m = x_{(n+1)/2}$ and for n even, $m = (x_{n/2}+x_{(n/2)+1})/2$

D Median
F médiane
P mediana z próby
R медиана

2168
MEDIAN LETHAL DOSE, LD$_{50}$.
Dose of ionizing radiation absorbed (by the whole body, human or animal) which after uninterrupted, single irradiation, causes death in 50% of cases after defined period of time. Also dose after which defined population of living organisms (e.g. of bacteria) suffers 50% death rate after defined period of time.

D mittlere Lethaldosis
F dose léthale moyenne, dose léthale 50%
P dawka śmiertelna średnia, dawka śmiertelna medianowa
R доза половинной выживаемости, среднелетальная доза

2169
MEERWEIN CONDENSATION.
Reaction of aromatic diazonium chlorides or bromides with α, β-unsaturated carbonyl compounds, esters and nitriles, in presence of copper salts, resulting in

formation of aromatic derivatives at
α-position, e.g.

$$[Ar—\overset{\oplus}{N}\equiv N]\overset{\ominus}{X} + CH_2=CH—C\equiv N \xrightarrow{Cu^{2\oplus}}$$

$$Ar—CH_2—\underset{\underset{X}{|}}{CH}—C\equiv N + N_2$$

D Meerwein-Arylierung
F condensation de Meerwein
P arylowanie Meerweina
R конденсация Меервейна

2170
MEERWEIN-PONNDORF-VERLEY
REDUCTION.
Reduction of carbonyl group to hydroxyl
group (in case of a ketones with
isopropanol and aluminium isopropylate),
e.g.

$$CH_2=CHCH_2COCH_3 + (CH_3)_2CHOH \xrightarrow{((CH_3)_2CHO)_3Al}$$

$$CH_2=CHCH_2CHOHCH_3 + CH_3COCH_3$$

This reaction is reversible (see also
Oppenauer oxidation).
D Meerwein-Ponndorf-Verley
  (Carbonyl)-Reduktion
F réduction selon
  Meerwein-Ponndorf-Verley
P redukcja Meerweina, Ponndorfa
  i Verleya
R восстановление по Меервейну-
  -Понндорфу-Верлею

2171
MELTING (fusion).
Transition from solid to liquid state.
D Schmelzen
F fusion
P topnienie
R плавление

2172
MELTING POINT (freezing point).
Temperature at which liquid and solid
phases are in equilibrium.
D Schmelztemperatur,
  Erstarrungstemperatur
F température de fusion, température de
  solidification
P temperatura topnienia, temperatura
  krzepnięcia
R температура плавления, температура
  замерзания

2173
MEMBRANE EQUILIBRIUM.
Thermodynamic state of system composed
of two phases partitioned by
semipermeable membrane and with
identical value of chemical potential of
any constituent transferred through phase
boundary in both phases.

D Membrangleichgewicht
F équilibre membraneux
P równowaga membranowa, równowaga
  Donnana
R мембранное равновесие, доннаново
  равновесие

2174
MEMBRANE POTENTIAL (Donnan
potential), $\Delta\varphi$.
Difference of inner potentials of solutions
(I) and (II) separated by semipermeable
diaphragm; if diaphragm is permeable for
B ions, then

$$\Delta\varphi = \frac{RT}{zF}\ln\frac{a_B^I}{a_B^{II}}$$

D Membranpotential, Donnan-Potential
F potentiel de membranes
P potencjał przeponowy, potencjał
  Donnana, potencjał membranowy
R мембранный потенциал, потенциал
  Доннана

2175
MENDELEVIUM, Md.
Atomic number 101.
D Mendelevium
F mendélévium
P mendelew
R менделевий

2176
MENISCUS.
Shape exhibited by free surface of liquid
in vicinity of wall of vessel as a result of
surface tension.
D Meniskus
F ménisque
P menisk
R мениск

MER. See CONSTITUTIONAL UNIT.

2177
MERCURY, Hg.
Atomic number 80.
D Quecksilber
F mercure
P rtęć
R ртуть

2178
MERCURY CATHODE.
Mercury electrode used in reduction of
ions or molecules, as dropping mercury
electrode or with large surface for
electrolytic separation of certain metals or
preparation of certain substances.
D Quecksilberkathode
F cathode de mercure
P katoda rtęciowa
R ртутный катод

2179
MESO FORM.
Diastereoisomer containing symmetrically located pairs of equivalent chirality centres of opposite configurations, so that one half of molecule is mirror image of other half. Molecule is therefore achiral (identical with its mirror image) and optically inactive, e.g. *meso*-tartaric acid

D meso-Form
F forme méso
P odmiana mezo
R мезо-форма

2180
MESO-IONIC COMPOUNDS.
5- or 6-membered heterocyclic compounds whose structure cannot be described by simple covalent arrangement of bonds. Such compounds possess electronic sextet covering all atoms of ring and atom linked to it; e.g. sydnon or nitron

sydnon   nitron
D mesoionische Verbindungen
F composés mésoioniques
P związki mezojonowe
R мезоионные соединения

2181
MESOMERIC EFFECT.
Change of electronic density in organic molecules in which substituents containing atoms with free electron pairs (or multiple bonds) are located close to double bonds (especially conjugated double bond systems). Mesomeric effect results from conjugations of π-electrons with free electron pairs of substituent, e.g.

+M effect (electron shift toward the ring)   −M effect (electron shift toward the substituent)

D Mesomerieeffekt
F effet mésomère
P efekt mezomeryczny, efekt $\pm$ M
R мезомерный эффект

2182
MESOMERISM.
Static model describing configuration and electronic structure of molecule by using resonance structures expressed by classical formulae. The real structure of molecule lies somewhere in between resonance structures. Mesomerism is qualitatively identical with the comparatively more modern resonance method, which is based on quantum mechanical assumptions.
D Mesomerie
F mésomérie
P mezomeria, rezonans
R мезомерия

MESOMORPHIC PHASES. *See* LIQUID CRYSTALS.

2183
MESOMORPHIC STATE.
State of considerable order of molecules in liquid and of resulting anisotropy, e.g. optical properties. It occurs in melts of certain organic compounds close to solidification point.
D mesomorpher Zustand, kristallin-flüssiger Zustand, liquokristalliner Zustand
F état mésomorphe
P stan mezomorficzny, stan kryształów ciekłych
R мезоморфное состояние, жидкокристаллическое состояние

2184
MESON PHOTOPRODUCTION.
Formation of mesons as a result of interaction between electromagnetic radiation with hadrons.
D Meson-Photoerzeugung
F photoproduction de mésons
P fotoprodukcja mezonów
R фоторождение мезонов

2185
MESON RESONANCES.
Resonances with baryon number $B = 0$.
D Mesonenresonanzen
F résonances des mésons
P rezonanse mezonowe
R мезонные резонансы

2186
MESONS.
Group of strongly interacting particles (hadrons) of baryon number $B = 0$, lepton number $L = 0$, and integral value of quantum spin number; e.g. π-mesons, K-mesons.
D Mesonen
F mésons
P mezony
R мезоны

2187
MESOTHORIUM I, MsTh I.
Isotope of rad. Atomic number 88. Mass number 228.

D Mesothorium I
F mésothorium I
P mezotor I
R мезоторий I

2188
MESOTHORIUM II, MsTh II.
Isotope of actinium. Atomic number 89. Mass number 228.

D Mesothorium II
F mésothorium II
P mezotor II
R мезоторий II

METAL AMMINE. *See* AMMINE COMPLEX.

2189
METAL CARBONYL.
Compound formed between metal of zero oxidation number with carbon monoxide; such a compound possesses M—CO bond. Metal carbonyl may be mononuclear $M(CO)_x$ or polynuclear $M_x(CO)_y$.

D Metallcarbonyl
F métal carbonyle
P karbonylek metalu
R карбонил металла

2190
METAL ELECTRODE.
Electrode (half-cell) constructed from metal dipped into solution containing its ions, $M|M^{z+}$; electrode reaction is $M \rightleftarrows M^{z+} + ze$ and potential of electrode,

$$\varepsilon = \varepsilon^0 + \frac{RT}{zF} \ln a_{M^{z+}}$$

where $\varepsilon^0 = $ standard electrode potential.

D Metallelektrode
F électrode de métal, électrode métallique
P elektroda metalowa
R металлический электрод

2191
METALLIC BOND.
Bond occurring in metals due to interaction of delocalized valence electrons with positively-charged atomic cores.

D metallische Bindung, Metallbindung
F liaison métallique
P wiązanie metaliczne
R металлическая связь

2192
METALLIC CONDUCTOR (electronic conductor).
Conductor conducting electrical current by electrons.

D Elektronenleiter, metallischer Leiter
F conducteur métallique
P przewodnik metaliczny, przewodnik elektronowy
R металлический проводник, электронный проводник

2193
METALLIC STATE. State of solid substance in which valence electrons of atoms have freedom to move inside metal and may be accelerated by applied electric field, causing flow of current.

D metallischer Zustand
F état métallique
P stan metaliczny
R металлическое состояние

2194
METALLOIDS (semimetals).
Nonmetallic elements of moderate electronegativity; with properties characteristic for metals, e.g. electrical conductivity; their oxides often have amphoteric properties.
Remark: Term was used to denote non-metals.

D Halbmetalle
F métalloïdes
P metaloidy
R металлоиды

2195
METALS.
Electropositive elements, i.e. exhibiting tendency to lose electrons and to transform into positive ions. Metals occur as simple substances which form metallic phases, have good thermal and electrical conductivities, their vapours are often monoatomic.

D Metalle
F métaux
P metale
R металлы

2196
METAMAGNETIC COMPOUNDS.
Antiferromagnetic materials which become ferromagnetic in low external magnetic field strength.

D metamagnetische Stoffe
F métamagnétiques
P metamagnetyki
R метамагнетики

2197
METAMAGNETISM.
Physical phenomena in and properties of metamagnetic compounds.

D Metamagnetismus
F métamagnétisme
P metamagnetyzm
R метамагнетизм

2198
METASTABLE ELECTRONIC STATE
(metastable state).
Excited electronic state of atom or
molecule with radiative dipole transition
to lower electronic states forbidden.
D metastabiler Zustand
F état métastable
P stan metastabilny, stan metatrwały
R метастабильное состояние

2199
METASTABLE NUCLEUS (isomeric
nucleus).
Atomic nucleus in excited nuclear state of
long lifetime in comparison to normal
excited state of this nucleus.
D metastabiler Kern, isomerer Kern
F noyau métastable
P jądro metastabilne, jądro metatrwałe,
jądro izomeryczne
R метастабильное ядро

2200
METASTABLE PHASE.
Phase in state of metastable equilibrium,
e.g. supercooled liquid, superheated liquid,
supersaturated vapour.
D metastabile Phase
F phase métastable
P faza metastabilna
R метастабильная фаза

2201
METASTABLE STATE.
Stable state of system not in
thermodynamic equilibrium. Finite
disturbance causes transition of system
from metastable state to state of
thermodynamic equilibrium.
D metastabiler Zustand
F état métastable
P stan metastabilny, stan metatrwały
R метастабильное состояние

METASTABLE STATE. *See*
METASTABLE ELECTRONIC STATE.

2202
METHODIC ERROR.
Systematic error due to physico-chemical
properties of system and measurement
procedure (analytical determination)
applied affected by them.
D methodischer Fehler
F erreur propre à la méthode
P błąd metody(czny)
R ошибка метода

2203
METHOD OF CELLS OF VARIABLE
PATH LENGTH.
Colorimetric analysis method by
equalization of colour of investigated and
standard solution observed in direction
perpendicular to cell base and obtained by
changing path length in one of cells, for
instance in Dubosque colorimeter.
Unknown concentration is calculated from
$c_1l_1 = c_2l_2$, where l_1, l_2 denote path lengths
and c_1, c_2 are concentrations of
investigated and standard solution
respectively.
D Methode mit dem Eintauschkolorimeter
F méthode de variation d'épaisseur de la
solution
P metoda zmiany grubości warstwy
R метод изменения толщины слоя

2204
METHOD OF COMPARATIVE
MEASUREMENTS.
Physico-chemical method which depends
on direct comparison of a given numerical
value characterizing defined substance
investigated with corresponding value for
other known substance taken as standard.
Both results should be obtained from
measurements carried out under identical
conditions.
D Vergleichungsmethode
F méthode des mesures comparatives
P metoda pomiarów porównawczych
R сравнительный метод измерений

2205
METHOD OF DIFFUSION FLAMES.
Flow method of measuring kinetics of fast
reactions of high activation energy,
applied in particular to reactions of alkali
metals with organic halides and
consisting in measurement of diffusion
path of metal towards excess of halides.
D ...
F ...
P metoda płomieni dyfuzyjnych
R ...

2206
METHOD OF HIGHLY DILUTE
FLAMES.
Flow method of measuring kinetics of fast
reactions, consisting in measurement of
partial pressures of gases (vapours) of
substances reacting in countercurrent and
of luminescence accompanying reaction.
D Methode der hochverdünnten Flammen
F ...
P metoda płomieni rozcieńczonych
R метод разрежённого пламени

2207
METHOD OF INTEGRATION (*of
determining the order of reaction*).
Method of determining order n of reaction
(in constant volume of system) consisting
in looking by trial and error for such

a value of n which, when inserted into general rate equation:

$$k = \frac{1}{(n-1)(t_2-t_1)}\left(\frac{1}{c_2^{n-1}} - \frac{1}{c_1^{n-1}}\right)$$

would give constant value (within limits of experimental confidence) for absolute rate constant k for all experimentally determined concentrations c and corresponding times of reaction t.

D Methode der Integration
F méthode d'intégration
P metoda całkowa
R интегральный метод (*определения порядка реакции*)

METHOD OF INTERNAL ADDITION. *See* METHOD OF STANDARD ADDITION.

2208
METHOD OF LEAST SQUARES.
Method for determination of parameters a_i of function $\hat{y} = f(x; a_1, a_2, ..., a_k)$ of independent variable x, in such a way that following condition is satisfied:

$$\sum_{i=1}^{n} [y_i - f(x_i; a_1, a_2, ..., a_k)]^2 = \min$$

where $\{y_i\}$ = set of measurement (determination) results and x_i = fixed values.

D Methode der kleinsten Quadrate
F méthode des moindres carrés
P metoda najmniejszej sumy kwadratów, metoda najmniejszych kwadratów
R метод наименьших квадратов

2209
METHOD OF STANDARD ADDITION (method of internal addition).
Method of determination used in various methods of instrumental analysis within limits of concentration of analysed substance where analytical curve is linear; measured quantity is recorded twice, in original solution and after addition of known quantity of analysed substance.

D Methode der bekannten Zusätze
F méthode à étalon interne
P metoda dodawania wzorca, metoda dodatków
R метод добавок

2210
METHYLATION.
Introduction of methyl group $—CH_3$ into molecule of chemical compound.

D Methylierung
F méthylation
P metylowanie
R метилирование

2211
MEYER-SCHUSTER REARRANGEMENT.
Rearrangement of tertiary alcohols with alkynyl group at α-position into α, β-unsaturated ketones in presence of acid catalysts (formic acid, sulfuric acid, acetyl chloride, thionyl chloride), e.g.

D Meyer-Schuster (Äthinylcarbinol→ Keton)-Umlagerung
F transposition de Meyer-Schuster
P przegrupowanie Meyera i Schustera
R перегруппировка Мейера-Шустера

2212
MICELLE. Colloidal particle with double electric layer and solvation sphere.
Remark: This term was previously ambiguous.

D Mizelle
F micelle
P micela
R мицелла

2213
MICHAEL REACTION.
Addition of compounds possessing active hydrogen atom to α, β-unsaturated ketones, aldehydes, esters or nitriles catalyzed by bases (piperidine), e.g.

$$CH_3COCH_2COOC_2H_5 + CH_2{=}CH—CN \xrightarrow{\text{base}}$$

$$CH_3COCHCOOC_2H_5$$
$$|$$
$$CH_2CH_2CN$$

D Michael-Reaktion, Michael nucleophile Methylen-Addition
F condensation de Michael
P addycja Michaela
R реакция Михаэля

2214
MICHALSKI REACTION.
Preparation of oxophosphoranesulfenyl chlorides by treating trialkyl thionophosphoranes with chlorination agents (Cl_2, SO_2Cl_2), e.g.

$$(RO)_3P{=}S + Cl_2 \longrightarrow$$

D Michalski-Reaktion
F réaction de Michalski
P reakcja Michalskiego
R реакция Михальского

2215
MICHALSKI REARRANGEMENT.
Preparation of thionopyrophosphoranes
(and their analogues) resulting from
condensation of oxophosphoranosulfenyl
chlorides with di- or trialkyl phosphites,
also alkylphosphines. Symmetric
thiopyrophosphorane, formed as
intermediate, undergoes this reaction

$$(RO)_2P-SCl + RO-P\overset{R}{\underset{}{-}}OR \longrightarrow$$

(with $\|$ O below first P)

$$\left[(RO)_2P-S-P-OR\right] \xrightarrow{-RCl} \left[(RO)_2P-S-P-OR\right]$$

$$\longrightarrow (RO)_2P-O-P-OR$$

D Michalski-Umlagerung
F transposition de Michalski
P przegrupowanie Michalskiego
R перегруппировка Михальского

2216
MICRO-ANALYSIS.
Analysis of sample with mass of
0.001—0.01 g or volume of 0.1—1 cm³ for
gas.
D Mikroanalyse
F microanalyse
P mikroanaliza, analiza w skali mikro
R микроанализ

2217
MICROANALYTICAL REAGENT.
Reagent of special purity prepared for
microanalytical determinations (mainly in
elemental analysis).
D mikroanalytisches Reagens
F ...
P odczynnik do mikroanalizy, odczynnik
mikroanalityczny
R реагент для микроанализа

2218
MICROCALORIMETER.
Calorimeter adapted for measuring small
heat effects.
D Mikrokalorimeter
F microcalorimètre
P mikrokalorymetr
R микрокалориметр

2219
MICROCALORIMETRY.
Branch of calorimetry dealing with very
precise measurements of small heat
effects which usually occur in slow-rate
processes, e.g. heats of some radioactive
transformations.
D Mikrokalorimetrie
F microcalorimétrie
P mikrokalorymetria
R микрокалориметрия

MICROCANONICAL ASSEMBLY. *See*
MICROCANONICAL ENSEMBLE.

2220
MICROCANONICAL ENSEMBLE
(microcanonical assembly).
Statistical ensemble given by following
probability distribution function
$$P(Q) = const \cdot \delta(E(Q) - E_0)$$
where δ = Dirac delta-function,
$E(Q)$ = energy of system in state Q,
E_0 = energy eigenvalue of system;
Microcanonical ensemble is utilized for
statistical description of thermodynamic
equilibrium in isolated systems.
D mikrokanonische Gesamtheit,
mikrokanonische Verteilung
F ensemble microcanonique de Gibbs
P zespół mikrokanoniczny, rozkład
mikrokanoniczny
R микроканонический ансамбль,
микроканоническое распределение

2221
MICROCHEMISTRY.
Branch of experimental chemistry
concerned with practical techniques with
substances of quantities from 10^{-3} to
10^{-2} g.
D Mikrochemie
F microchimie
P mikrochemia
R микрохимия

2222
MICROCOMPONENT.
Component present in system in great
dilution which may be detected or
determined only by special methods.
D Mikrokomponente
F microcomposant
P mikroskładnik
R микрокомпонент

2223
MICROCRYSTALLINE REACTION.
Chemical reaction resulting in formation
of crystalline precipitate; one may detect
various chemical species by microscopic
inspection of shape of crystals.
D mikrochemische Nachweisreaktion
F (*réaction de cristallisation sous le
microscope*)
P reakcja mikrokrystaliczna
R микрокристаллоскопическая реакция

MICRODENSITOMETER. *See*
MICROPHOTOMETER.

2224
MICROGRAM METHOD (ultramicro
method).

Method of analysis where mass of investigated sample in $x \cdot 10^{-6}$—$x \cdot 10^{-4}$ g, or volume $x \cdot 10^{-5}$—$x \cdot 10^{-3}$ cm^3 in the case of gases.

D Ultramikromethode
F méthode microgrammique, ultramicrométhode
P metoda mikrogramowa, ultramikrometoda
R ультрамикрометод

2225
MICRO-HETEROGENEOUS CATALYSIS.
Multi-phase catalysis due to presence of catalyst in colloidal dispersion.

D ...
F catalyse microhétérogène
P kataliza mikrowielofazowa, kataliza mikroheterogeniczna
R микрогетерогенный катализ

MICRO METHOD. *See* MILLIGRAM METHOD.

2226
MICROPHOTOMETER
(microdensitometer, densitometer).
Device used to measure degree of darkening of spectral lines recorded on photographic plate.

D Spektrenphotometer, (Spektrallinien-)Photometer
F microphotomètre, densitomètre
P mikrofotometr
R микрофотометр

2227
MICROSCOPIC CROSS-SECTION.
Cross-section with respect to a single particle.

D mikroskopischer Wirkungsquerschnitt
F section efficace microscopique
P przekrój czynny mikroskopowy
R микроскопическое (эффективное) сечение

2228
MICROSCOPIC STATE (microstate).
State of system in which number of dynamic quantities such as momenta, particle coordinates, etc., of a given value is maximum; according to classical statistical mechanics all dynamic variables have precisely defined values at microscopic state; in quantum case not all of them are determined precisely.

D Mikrozustand
F état microscopique
P stan mikro(skopowy)
R микроскопическое состояние

2229
MICROSPECTROPHOTOMETRY.
Method of measurement of absorption by objects with very small surface

(ca. 2—20 μm^2) or by very thin objects (e.g. fibres) using microspectrophotometer, i.e. combination of spectrometer and microscope.

D Mikrospektroskopie
F microspectrophotométrie
P mikrospektrofotometria
R ...

MICROSTATE. *See* MICROSCOPIC STATE.

2230
MICROWAVE SPECTROSCOPY.
Branch of spectroscopy dealing with study of absorption between rotational (in gases) or vibrational (in solid state) energy levels of dipolar molecules in range of electromagnetic spectrum between 300 kHz and 300 GHz.

D Mikrowellen-Spektroskopie
F spectroscopie de microondes
P spektroskopia mikrofalowa
R микроволновая спектроскопия

2231
MICROWAVE SPECTRUM.
Spectrum registered in range of frequencies from about $3 \cdot 10^9$ Hz to about $3 \cdot 10^{12}$ Hz, associated with transitions between molecular rotational energy levels.

D Mikrowellenspektrum
F spectre de microondes
P widmo mikrofalowe
R микроволновой спектр

2232
MIDDLE INFRARED.
Electromagnetic radiation range of wavelength between ca. 3 μm to ca. 25 μm.

D Infrarot, Ultrarot
F infrarouge, région infra-rouge
P podczerwień (średnia)
R инфракрасная область

2233
MIGRATION CURRENT.
Current resulting from electrode process of ions which migrate to electrode surface.

D Migrationsstrom
F courant de migration
P prąd migracyjny
R миграционный ток

2234
MIGRATION OF IONS.
Motion of ions under influence of electrical field.

D Überführung der Ionen, Ionenwanderung im Feld
F migration des ions
P migracja jonów
R миграция ионов

2235
MILLER INDICES.
Indices of planes, h, k, l, three numbers describing position of any plane of lattice in chosen system of crystallographic axes a, b, c; e.g. plane which intercepts axis a at $x = a$ and parallel to b, c has indices (100), a plane parallel to a, intercepting axis b at $y = b/2$ and axis c at $z = c/3$ has indices (023).

D Millersche Indizes, kristallographische Indizes
F indices inverses
P wskaźniki Millera, wskaźniki płaszczyzny
R миллеровские индексы

2236
MILLIGRAM METHOD (micro method).
Method of analysis where mass of investigated sample is $x \cdot 10^{-3}$ g, or volume $x \cdot 10^{-2} - x \cdot 10^{-1}$ cm^3 in the case of gases.

D Mikromethode, mikroanalytische Methode
F méthode milligrammique, microméthode
P metoda miligramowa, mikrometoda
R микрометод

2237
MINERALIZATION.
Complete decomposition and oxidation of analysed organic substance in order to determine its mineral components.

D Mineralisation
F minéralisation
P mineralizacja
R минерализация

2238
MIRROR NUCLEI.
Pairs of isobars in which number of neutrons in one nucleus is equal to number of protons in second one (and vice-versa, number of protons in first nucleus equal to number of neutrons in second nucleus).

D Spiegelkerne
F noyaux miroirs
P jądra zwierciadlane
R зеркальные ядра, зеркальные изотопы

2239
MIRROR REFLECTION SYMMETRY (reflection plane symmetry).
Symmetry of system in which every point has as its counterpart its reflection in the same mirror plane.

D Spiegelungssymmetrie
F symétrie par rapport à un plan
P symetria zwierciadlana
R симметрия зеркальной плоскости

2240
MITSCHERLICH LAW OF ISOMORPHISM.

Substances which have similar crystalline structure and similar chemical properties, usually have similar chemical formulae.

D Mitscherlichsches Gesetz der Isomorphie
F loi (d'isomorphie) de Mitscherlich
P prawo izomorfizmu Mitscherlicha
R правило изоморфизма Митчерлиха

2241
MIXED BED.
Mixture of cation and anion exchanger in one column. Mixed bed of cation exchanger in hydrogen form and anion exchanger in hydroxyl form is used for deionization of aqueous solutions.

D Mischbett
F lit mélangé
P złoże mieszane
R смешанный слой

2242
MIXED CATALYST (compound catalyst).
Mixture of catalysts exhibiting different catalytic activity (usually higher) than the one exhibited by each constituent of mixture individually.

D Mischkatalysator
F catalyseur mélangé
P katalizator mieszany
R смешанный катализатор

2243
MIXED COMPLEX (mixed ligand complex).
Complex containing central atom and more than one kind of ligands, e.g. MX_iY_j.

D gemischter Komplex, Mischkomplex
F complexe polysubstitué
P kompleks mieszany, kompleks wielopodstawny, kompleks różnoligandowy
R смешанный комплекс

MIXED CRYSTALS. *See* SOLID SOLUTION.

2244
MIXED ELEMENT.
Chemical element which consists inherently of mixture of its natural isotopes.

D Mischelement
F mélange d'isotopes
P pierwiastek mieszany
R смешаный элемент

MIXED LIGAND COMPLEX. *See* MIXED COMPLEX.

MIXED POLYELECTRODE POTENTIAL. *See* MIXED POTENTIAL.

2245
MIXED POTENTIAL (mixed polyelectrode potential).

Potential of multiple electrode, resultant of potentials of electrode reactions occurring in multiple electrode.

D Mischpotential
F tension mixte (d'une polyélectrode)
P potencjał mieszany
R стационарный потенциал, смешанный потенциал (полиэлектрода)

2246
MIXED RADIATION.
Ionizing radiation composed of different particles, or particles and photons.

D Mischstrahlung
F rayonnement mixte
P promieniowanie mieszane
R смешанное излучение

2247
MOBILE PHASE (*in chromatography*).
Gas or liquid flowing through chromatographic column or layer of stationary phase.

D mobile Phase, bewegliche Phase
F phase mobile
P faza ruchoma
R подвижная фаза

2248
MODELS OF NUCLEUS (nuclear models).
Set of assumptions on basis of which properties of atomic nuclei are explained qualitatively and quantitatively.

D Modelle des Atomkerns
F modèles du noyau, modèles nucléaires
P modele jądra atomowego
R ядерные модели

MODERATION. *See* SLOWING DOWN.

2249
MODERATOR.
Medium able to slow down fast neutrons, applied in thermal nuclear reactors, e.g. graphite, water, heavy water.

D Moderator, Bremssubstanz
F modérateur, ralentisseur
P moderator (reaktorowy), spowalniacz
R замедлитель

2250
MODIFIER OF A CHAIN REACTION.
Substance which easily reacts with radicals and facilitates transfer of reaction chain.

D Modifikator der Kettenreaktion
F modificateur d'une réaction en chaîne
P modyfikator reakcji łańcuchowej
R модификатор цепной реакции

MOLAL CONCENTRATION. *See* MOLALITY.

2251
MOLAL HEAT CAPACITY (true molal heat capacity), C.

Intensive quantity equal to ratio of elementary heat δQ received isochorically or isobarically by one mole of substance to corresponding temperature change δT caused by this process, viz.:

$$C = \frac{\delta Q}{\delta T}$$

D (wahre) Molwärme
F chaleur spécifique molaire, chaleur spécifique vraie
P ciepło molowe (rzeczywiste)
R (истинная) мольная теплоёмкость

2252
MOLAL HEAT CAPACITY AT CONSTANT PRESSURE, C_p.
Intensive quantity defined as:

$$C_p = \left(\frac{\partial H_m}{\partial T} \right)_p$$

where H_m = molal enthalpy,
T = thermodynamic temperature,
p = pressure.

D Molwärme bei konstantem Druck
F chaleur spécifique molaire à pression constante
P ciepło molowe pod stałym ciśnieniem
R мольная теплоёмкость при постоянном давлении, изобарная мольная теплоёмкость

2253
MOLAL HEAT CAPACITY AT CONSTANT VOLUME, C_V.
Intensive quantity defined as follows:

$$C_V = \left(\frac{\partial U_m}{\partial T} \right)_V$$

where U_m = molal internal energy,
T = thermodynamic temperature,
V = volume.

D Molwärme bei konstantem Volumen
F chaleur spécifique molaire à volume constant
P ciepło molowe w stałej objętości
R мольная теплоёмкость при постоянном объёме, изохорная мольная теплоёмкость

2254
MOLALITY (molal concentration), m_B.
Concentration m_B of a solution expressed in moles of solute B per 1 kg of solvent.

D Molalität
F molalité
P molalność (roztworu), stężenie molalne
R моляльность, моляльная концентрация

MOLAR ABSORBANCY INDEX. *See* MOLAR ABSORPTIVITY.

2255
MOLAR ABSORPTIVITY (molar
absorbancy index), ε.
Absorbance of solution of unit molar
concentration when path length equals
1 cm. Its numerical value is usually
calculated from the Lambert-Beer law
applied to measurement results for fairly
diluted solutions.
D dekadischer molarer
  Absorptionskoeffizient, dekadischer
  molarer Extinktionskoeffizient
F coefficient d'extinction moléculaire
P współczynnik absorpcji molowy
R молярный коэффициент поглощения,
  молярный коэффициент погашения,
  десятичный показатель погашения

2256
MOLAR CONCENTRATION (molarity),
c_B ([B]).
Concentration c_B of a solution expressed
in moles of solute B per 1 dm^3 of solution.
D molare Konzentration, Molarität
F concentration molaire, molarité
P stężenie molarne, molarność (roztworu),
  stężenie molowe, molowość (roztworu)
R молярная концентрация, молярность

2257
MOLAR CONDUCTIVITY, Λ'_c.
Conductivity of 1 mole of electrolyte in
solution with a given concentration c.

$$\Lambda'_c = \frac{1000\varkappa}{c}$$

where $\varkappa$ = specific conductance of solution
with concentration c mol/dm^3.
D molare Leitfähigkeit
F conductivité molaire
P przewodnictwo molowe
R молекулярная электропроводность

2258
MOLAR ENTHALPY, H_m.
Enthalpy of one mole of system.
D molare Enthalpie
F enthalpie molaire
P entalpia molowa
R мольная энтальпия, молярная
  энтальпия

2259
MOLAR ENTROPY (proper entropy), S_m.
Entropy of one mole of system.
D molare Entropie
F entropie spécifique molaire
P entropia molowa
R мольная энтропия

MOLAR GAS CONSTANT. *See* GAS
CONSTANT.

MOLARITY. *See* MOLAR
CONCENTRATION.

2260
MOLAR MASS, M.
Mass, in grams, of one mole of given type
of particles, i.e. atoms, molecules, ions,
free radicals, elementary particles, or
atomic groups. Numerically molar mass
is equal to corresponding relative atomic
or molecular mass — however, in contrast
to former molar mass has dimension.
D molare Masse
F masse molaire
P masa molowa
R мольная масса

2261
MOLAR POLARIZATION (molecular
polarization), P_m.
Measure of dipole moment induced by
electric field of unit strength in 1 mole of
substance; scalar quantity defined by

$$P_m = \frac{1}{3\varepsilon_0} N_A[\alpha_I + \alpha_O) \qquad \text{(in SI units)}$$

where N_A = Avogadro constant,
α_I = induced polarizability,
α_O = orientation polarizability,
ε_0 = permittivity of free space.
D Mol(ekul)arpolarisation
F polarisation molaire
P polaryzacja molowa (całkowita
  dielektryczna)
R мол(екул)ярная поляризация

2262
MOLAR QUANTITIES (proper
quantities), Z_m.
Intensive quantities defined by equation

$$Z_m = \frac{Z}{\sum_i n_i}$$

where Z = extensive function of state of
a given phase and $\sum_i n_i$ denotes sum of

number of moles of all phase
constituents.
D molare Zustandsfunktionen, mittlere
  molare Zustandsgrößen
F grandeurs molaires, propriétés moyennes
P wielkości molowe (termodynamiczne),
  wielkości molowe średnie
R мольные величины, (средние)
  молярные величины

2263
MOLAR REFRACTION (molecular
refraction), R_m.
Constant characteristic for a given
substance and defined by

$$R_m = \frac{n^2-1}{n^2+2} V_m$$

where n = refractive index, V_m = molar
volume.

D Molrefraktion
F réfraction moléculaire, réfractivité moléculaire
P refrakcja molowa
R молекулярная рефракция

2264
MOLAR ROTATION.
Quantity equal to product of specific rotation and 1/100 of molar mass of a given compound.
D molare Drehung
F rotation moléculaire
P skręcalność molowa
R молекулярное вращение

2265
MOLAR VOLUME, V_m.
Volume of one mole of substance (either single- or multi-component).
D Molvolumen, molares Volumen
F volume mol(écul)aire
P objętość molowa
R мол(екул)ярный объём, мольный объём

2266
MOLE.
Unit of amount of substance; amount of substance containing number of particles equal to number of atoms in 0.012 kg of pure ^{12}C nuclide. Type of particles should be given, i.e., atoms, molecules, ions, free radicals, elementary particles, atomic groups in molecule, etc.
D Mol
F mole
P mol
R моль

2267
MOLE (gram mole, gram molecule, gram-molecular weight).
Number of grams of substance equal numerically to its relative molecular mass.
Remark: In view of introduction of mole as unit of amount of substance, term molar mass should be used in preference to mole = gram molecule.
D Mol, Gramm-Molekül
F mole, molécule-gramme
P mol, gramocząsteczka
R моль, грамм-молекула, грамм-моль

2268
MOLECULAR BEAM METHOD.
Method of dipole moment determination utilizing deviation by inhomogeneous electric field of beam of molecules emitted by heated container in high vacuum.
D Molekularstrahlverfahren, Molekularstrahlmethode
F méthode des rayons moléculaires
P metoda wiązki molekularnej, metoda wiązek molekularnych
R метод молекулярного пучка, метод молекулярных пучков

MOLECULAR BOND. *See* VAN DER WAALS BOND.

2269
MOLECULAR CHAOS.
Simplifying assumption used in kinetic theory of gases implying complete lack of correlation between particle states.
D zwischenmolekulare Unordnung
F chaos moléculaire
P chaos molekularny
R молекулярный хаос

2270
MOLECULAR COLLOID (macromolecular colloid).
Colloidal system, physically homogeneous, in which macromolecules form one component.
D Molekülkolloid, makromolekulares Kolloid
F colloïde macromoléculaire
P koloid cząsteczkowy
R макромолекулярный коллоид

MOLECULAR COMPLEX. *See* DONOR-ACCEPTOR COMPLEX.

MOLECULAR COMPOUNDS. *See* ADDITION COMPOUNDS.

2271
MOLECULAR CRYSTALS
(van der Waals's crystals).
Crystals whose molecules are bound by van der Waals forces; they are characterized by low melting points and high compressibility.
D Molekularkristalle, Molekülkristalle
F cristaux moléculaires
P kryształy cząsteczkowe, kryształy molekularne
R молекулярные кристаллы, кристаллы Ван-дер-Ваальса

2272
MOLECULAR DIAGRAM.
Scheme of π-bonds in conjugated systems. Atoms are given their effective charges, and bonds their bond order values, and numbers at arrows leaving carbon atoms correspond to free valencies, e.g. molecular diagram of aniline

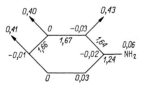

D Molekulardiagramm
F diagramme moléculaire
P diagram molekularny
R молекулярная диаграмма

2273
MOLECULAR ELECTRONIC ENERGY
LEVELS (electronic energy levels of
a molecule).
Discrete values of molecular electronic
energy determined by quantum
mechanical conditions.

D Elektronenniveaus des Moleküls
F niveaux (énergétiques) électroniques de
la molécule
P poziomy energetyczne cząsteczki
elektronowe
R электронные уровни молекулы

2274
MOLECULAR ENERGY LEVELS.
Discrete values of molecular energy
determined by quantum mechanical
conditions.

D Energieniveaus des Moleküls
F niveaux énergétiques de la molécule
P poziomy energetyczne cząsteczki
R энергетические уровни молекулы

2275
MOLECULAR FIELD.
Hypothetical internal magnetic field,
whose strength is proportional to
magnetization of substance (*cf.* molecular
field theory).

D Molekularfeld
F champ moléculaire
P pole molekularne
R молекулярное поле

2276
MOLECULAR FIELD THEORY.
Phenomenological theory of ferro-, ferri-
and antiferromagnetic crystals in which
exchange interaction leading to magnetic
ordering is replaced by interaction of
atomic magnetic moments with molecular
field.

D Molekularfeldtheorie
F théorie du champ moléculaire
P teoria pola molekularnego
R теория молекулярного поля

2277
MOLECULAR FORMULA.
Chemical formula which determines type
and number of atoms in given molecule,
e.g. C_2H_2 for acetylene.

D Summenformel, Bruttoformel
F formule brute
P wzór sumaryczny
R молекулярная формула

2278
MOLECULARITY OF REACTION.
The smallest number of elementary
entities necessary to create elementary
entity (entities) of product (or products)
of elementary reaction.

D Reaktionsmolekularität
F molécularité de la réaction
P cząsteczkowość reakcji
R молекулярность реакции

MOLECULAR MIXTURE OF OPPOSITE
FORMS. *See* dl MOLECULAR MIXTURE.

2279
MOLECULAR ORBITAL.
One-electron molecular wave function
extended over whole molecule.

D Molekularorbital
F orbitale moléculaire
P orbital molekularny, orbital
cząsteczkowy, orbital drobinowy
R молекулярная орбиталь

2280
MOLECULAR ORBITAL METHOD
(MO method).
Quantum chemistry method based on
one-electron approximation. In molecular
orbital method each electron is ascribed
to some molecular orbital and these are
used to construct total molecular wave
function.

D Molekülorbitalmethode, MO-Methode
F méthode des orbitales moléculaires
P metoda orbitali molekularnych, metoda
MO
R метод молекулярных орбиталей

MOLECULAR POLARIZATION. *See*
MOLAR POLARIZATION.

2281
MOLECULAR PRODUCTS OF
RADIOLYSIS.
Stable structure products of radiolysis,
e.g. H_2 and H_2O_2 in case of water.

D Molekularprodukte der Radiolyse
F produits moléculaires de radiolyse
P produkty radiolizy cząsteczkowe
R молекулярные продукты радиолиза

MOLECULAR REFRACTION. *See*
MOLAR REFRACTION.

2282
MOLECULAR ROTATIONAL ENERGY
LEVELS (rotational energy levels of
a molecule).
Discrete values of molecular rotational
energy determined by quantum
mechanical conditions.

D Rotationsniveaus des Moleküls
F niveaux (énergétiques) de rotation de la
molécule
P poziomy energetyczne cząsteczki
rotacyjne
R вращательные уровни молекулы

2283
MOLECULAR SIEVES.
Sorbents of regular crystalline structure
and defined pore dimensions. Molecular

sieves can sorb small molecules but exclude molecules which are greater than their pore size.

D Molekülsiebe
F tamis moléculaires
P sita molekularne, sita cząsteczkowe
R молекулярные сита

MOLECULAR SOLUTION. *See* TRUE SOLUTION.

2284
MOLECULAR SPECTROANALYSIS.
Spectroanalysis based on investigation of molecular spectra.

D Molekülspektralanalyse
F spectranalyse moléculaire
P analiza spektralna cząsteczkowa
R молекулярный спектральный анализ

2285
MOLECULAR SPECTROSCOPY.
Branch of spectroscopy dealing with investigation of molecular emission and absorption spectra in spectral range from 500 nm to 2000 nm.

D Molekülspektroskopie
F spectroscopie moléculaire
P spektroskopia cząsteczkowa, spektroskopia molekularna
R молекулярная спектроскопия

2286
MOLECULAR SPECTRUM.
General term for rotation, vibration-rotation and electronic-vibration-rotation spectra of a given molecule.

D Molekülspektrum
F spectre moléculaire
P widmo cząsteczkowe
R молекулярный спектр

2287
MOLECULAR VIBRATIONAL ENERGY LEVELS (vibrational energy levels of a molecule).
Discrete values of molecular vibrational energy determined by quantum mechanical conditions.

D Schwingungsniveaus des Moleküls
F niveaux (énergétiques) de vibration de la molécule
P poziomy energetyczne cząsteczki oscylacyjne
R колебательные уровни молекулы

MOLECULAR WEIGHT. *See* RELATIVE MOLECULAR MASS.

2288
MOLECULE.
Electrically neutral system of chemically bound atoms.

D Molekül, Molekel
F molécule
P cząsteczka, drobina, molekuła
R молекула

2289
MOLE FRACTION, x_B.
Ratio of number of moles of constituent B to sum of number of moles of all constituents of mixture

$$x_B = \frac{n_B}{\sum\limits_i n_i}$$

where n_B = number of moles of constituent B, $\sum\limits_i n_i$ = number of moles of all constituents of mixture.

D Mollenbruch
F fraction molaire
P ułamek molowy
R мольная доля

2290
MOLYBDENUM, Mo.
Atomic number 42.

D Molybdän
F molybdène
P molibden
R молибден

2291
MOMENTUM CONSERVATION LAW.
In isolated mechanical system without friction momentum of centre of mass is constant.

D Impulssatz
F loi de conservation de l'impulsion
P zasada zachowania pędu
R закон сохранения количества движения

MO METHOD. *See* MOLECULAR ORBITAL METHOD.

2292
MONITOR (*of ionizing radiation*).
Device for continuous checking and possible recording of intensity of ionizing radiation of a given kind.

D Monitor
F moniteur
P monitor
R монитор

2293
MONOCHROMATIC RADIATION.
Beam of electromagnetic radiation with identical energy quanta, i.e. of the same frequency or wavelength.

D monochromatische Strahlung
F radiation monochromatique
P promieniowanie monochromatyczne
R монохроматическое излучение

2294
MONOCHROMATOR.
Device used for selection of
monochromatic radiation from
polychromatic beam of electromagnetic
radiation.
D Monochromator
F monochromateur
P monochromator
R монохроматор

2295
MONODISPERSE SYSTEM (isodisperse
system).
Disperse system in which particles of
dispersed phase have the same sizes.
D monodisperses System, isodisperses
System, homodisperses System
F système monodispersé, système
isopegmatique, système isodispersé,
système homéodispersé
P układ monodyspersyjny, układ
izodyspersyjny
R монодисперсная система

MONOFUNCTIONAL LIGAND. *See*
UNIDENTATE LIGAND.

MONOLAYER. *See* MONOMOLECULAR
LAYER.

2296
MONOLAYER ADSORPTION
(unimolecular adsorption).
Adsorption on surface of solid or liquid
bodies (adsorbents) with formation of
unimolecular layer of adsorbate.
D monomolekulare Schicht-Adsorption
F adsorption d'une couche
monomoléculaire
P adsorpcja jednowarstwowa, adsorpcja
jednocząsteczkowa, adsorpcja
monomolekularna
R мономолекулярная адсорбция

2297
MONOMER.
Substance consisting of molecules each of
which can provide one or more
constitutional units.
D Monomer
F monomère
P monomer
R мономер

2298
MONOMERIC UNIT.
Unit comprising atom or group of atoms
contributed by monomer molecule to
growing polymer chain.
D Monomereinheit, monomere Einheit
F ...
P jednostka monomeryczna
R мономерная единица, мономерное звено

MONOMOLECULAR FILM. *See*
MONOMOLECULAR LAYER.

2299
MONOMOLECULAR LAYER (monolayer,
monomolecular film).
Layer consisting of densely-packed
molecules whose thickness is equal to size
of molecules when it is viewed from
vertical direction to surface of layer.
D monomolekulare Schicht,
monomolekularer Film
F couche monomoléculaire
P warstwa monomolekularna, warstwa
jednocząsteczkowa
R мономолекулярный слой

2300
MONOMOLECULAR REACTION
(unimolecular reaction).
Elementary unit reaction of molecularity
equal to one.
D monomolekulare Reaktion,
unimolekulare Reaktion
F réaction monomoléculaire, réaction
unimoléculaire
P reakcja jednocząsteczkowa, reakcja
monomolekularna
R мономолекулярная реакция,
одномолекулярная реакция

2301
MONONUCLEAR COMPLEX.
Complex containing one central atom.
D einkerniger Komplex
F complexe mononucléaire
P kompleks jednordzeniowy
R одноядерный комплекс

2302
MONOTROPIC PHASES.
Polymorphic phases among which one may
be melted before appearence of second
phase.
D monotrope Phasen
F phases monotropes
P fazy monotropowe
R монотропные фазы

2303
MONOTROPY.
Property of substance by which it occurs
only in one form (commonly only one
solid form).
D Monotropie
F monotropie
P monotropia
R монотропия

2304
MORSE CURVE (Morse function).
Approximate potential energy curve $U(r)$
for diatomic molecule given by

$$U(r) = D\{1-\exp[-a(r-r_0)]\}^2$$

where D = energy of dissociative
molecular oscillation energy level,
a = constant, r_0 = internuclear distance
corresponding to curve minimum.

D Morse-Kurve, Morse-Funktion
F courbe de Morse, fonction de Morse
P krzywa Morse'a, funkcja Morse'a
R кривая Морзе, функция Морзе

MORSE FUNCTION. *See* MORSE CURVE.

2305
MOSAIC CRYSTALS.
Crystals composed of perfect crystals in blocks 100—1000 nm in size, mutually inclined at very low angles.
D Mosaikkristalle
F cristaux mosaïques
P kryształy mozaikowe
R мозаичные кристаллы

2306
MOSAIC STRUCTURE.
Structure consisting in random deviations (misalignment) of crystal lines from average statistical direction by angles up to 1—2°.
D Mosaikstruktur
F structure mosaïque
P struktura mozaikowa
R мозаичная структура

2307
MOSELEY LAW.
X-ray frequency of given line of atomic spectral series emitted by atom is quadratic function of corresponding atomic number.
D Moseleysches Gesetz
F loi de Moseley
P prawo Moseleya
R закон Мозелея, закон Мозли

2308
MÖSSBAUER EFFECT.
Recoilless emission or absorption of gamma radiation originating from certain radionuclides (e.g. ^{191}Ir, ^{57}Fe) incorporated in crystal lattice which takes over the recoil momentum; it is characterized by exceptionally small width of spectrum lines (10^{-10}—10^{-5} eV) and is used for studying structure of solids.
D Mössbauer-Effekt
F effet Mössbauer
P efekt Mössbauera
R эффект Мёссбауэра

2309
MÖSSBAUER SPECTROSCOPY.
Branch of nuclear spectroscopy based on effect of recoilless absorption and emission of γ-rays by atomic nuclei (Mössbauer effect).
D Mössbauer-Spektroskopie
F spectroscopie Mössbauer
P spektroskopia Mössbauera

R спектроскопия Мёссбауэра, мёссбауэровская спектроскопия

2310
MOST PROBABLE VELOCITY (*of gas molecules*), c_{max}.
Velocity exhibited by majority of molecules in the given population, corresponding to maximum on Maxwell-Boltzmann curve equal to

$$c_{max} = \sqrt{\frac{2RT}{M}}$$

where R = universal gas constant,
T = thermodynamic temperature,
M = molar mass.
D wahrscheinlichste Geschwindigkeit
F vitesse la plus probable
P prędkość najbardziej prawdopodobna
R наиболее вероятная скорость

2311
MOVING BOUNDARY METHOD.
Method of determination of transference numbers of ions based on shift of boundary surface between two electrolyte solutions in contact under influence of electrical field perpendicular to it.
D Methode der wandernden Grenzfläche
F méthode du déplacement des surfaces limites
P metoda poruszającej się granicy
R метод перемещающейся границы

2312
MOVING BOUNDARY METHOD.
In colloid chemistry, method for determining rate of electrophoresis by measuring shift of boundary between colloidal solution and contact liquid. Observations can be either visual or refractometric.
D Grenzbewegungsmethode, Methode der wandernden Grenzfläche
F méthode des surfaces limites mobiles
P metoda poruszającej się granicy, metoda ruchomej granicy
R метод подвижной границы

2313
MU⁻ LEPTON (μ⁻-lepton, muon).
Elementary particle of mass $m = 105.659\,9$ MeV, negative elementary electric charge, baryon number $B = 0$, lepton muonic number $L_\mu = 1$, quantum spin number $J = 1/2$, unstable, of mean lifetime $2.198\,3 \cdot 10^{-6}$ s; μ⁻-lepton has antiparticle, μ⁺.
D μ⁻-Lepton, Myon, Muon
F lepton μ⁻, muon
P lepton μ⁻, mion μ⁻
R лептон μ⁻, мюон, мю-мезон

2314
MULLIKEN'S ELECTRONEGATIVITY SCALE.
Ordering of atoms according to increasing electronegativity defined as average of atom's ionization potential and its electron affinity.
D Mullikensche Elektronegativitätsskala der Elemente
F échelle d'électronégativité de Mulliken
P skala elektroujemności Mullikena
R шкала электроотрицательности Малликена

2315
MULTICENTRE REACTION.
Reaction proceeding through cyclic transition state, whose formation from substrates and transformation into products is effected by simultaneous electron displacement (with no intervention of ions or radicals), e.g.

D synchroner Mehrzentrenprozeß
F transfert électronique circulaire à plusieurs centres
P reakcja wielocentrowa
R ...

MULTICOMPONENT IDEAL GAS. See PERFECT GASEOUS MIXTURE.

2316
MULTICOMPONENT SYSTEM.
System composed of many components which may coexist in one or several phases.
D Mehrstoffsystem
F système multicomposant
P układ wieloskładnikowy
R многокомпонентная система

2317
MULTIDENTATE LIGAND (polyfunctional ligand).
Ligand which occupies more than one coordination position around central atom.
D mehrzähniger Ligand, mehrwertiger Ligand
F coordinant multidenté
P ligand wielodonorowy, ligand wielofunkcyjny, ligand wielokleszczowy
R полидентатный адденд

2318
MULTILAYER ADSORPTION (multimolecular adsorption).
Adsorption which results in formation of multimolecular layers on surface of adsorbent.

D Mehrschichtenadsorption
F adsorption à plusieurs couches
P adsorpcja wielowarstwowa, adsorpcja wielocząsteczkowa
R полимолекулярная адсорбция

MULTIMOLECULAR ADSORPTION. See MULTILAYER ADSORPTION.

MULTIPHASE SYSTEM. See HETEROGENEOUS SYSTEM.

2319
MULTIPLE BONDING.
Chemical bond formed by more than one pair of electrons, e.g. double or triple bond.
D Mehrfachbindung
F liaison multiple
P wiązanie wielokrotne
R кратная связь

MULTIPLE ELECTRODE. See POLYELECTRODE.

2320
MULTIPLE PRODUCTION.
Process occurring during collision of high energy particles, leading to formation of more than two particles.
D Vielfacherzeugung
F production plurale, production multiple
P produkcja wielorodna, proces wielociałowy
R множественное рождение

MULTIPLE RANGE INDICATOR. See UNIVERSAL INDICATOR.

2321
MULTIPLET.
Group of energetic electronic states in atom of close energies, which belong to the same atomic term.
D Multiplett
F multiplet
P multiplet
R мультиплет

2322
MULTIPLET (in atomic spectroscopy).
Group of spectral lines arising from transitions between energy levels of two atomic multiplet terms.
D Multiplett
F multiplet
P multiplet widmowy
R спектральный мультиплет

MULTIPLET. See BALANDIN MULTIPLET.

2323
MULTIPLICITY (term multiplicity).
Number of components of multiplet electronic term; equal to $2S+1$ where S

denotes total spin quantum number; it is given as left superscript of term symbol.

D Multiplizität
F multiplicité
P multipletowość (termu)
R мультиплетность (терма)

2324
MULTIPLICITY.
Number of particles formed as a result of high energy particle interaction.

D Multiplizität
F multiplicité
P krotność produkcji wielorodnej
R множественность рождения

MUON. *See* MU⁻ LEPTON.

2325
MUONIC ANTINEUTRINO, $\bar{\nu}_\mu$.
Antiparticle corresponding to muonic neutrino of muon number $L_\mu = -1$ and positive (right-handed) helicity, stable.

D μ-Antineutrino
F antineutrino ($\bar{\nu}_\mu$), antineutretto
P antyneutrino mionowe
R мюонное антинейтрино

2326
MUONIC NEUTRINO, ν_μ.
Elementary particle, a lepton, with rest mass probably zero, electrically neutral, of baryon number $B = 0$, quantum spin number $J = 1/2$, muon number $L_\mu = 1$ and negative (left-handed) helicity, stable.

D My-Neutrino, μ-Neutrino
F neutrino ν_μ, neutretto
P neutrino mionowe
R мюонное нейтрино

2327
MUON NUMBER, L_μ.
Number attributed to elementary particles; for μ⁻-lepton, and muonic neutrino, $L_\mu = 1$, for their antiparticles, $L_\mu = -1$. Other particles and antiparticles have $L_\mu = 0$. Muon number is conserved in all interactions.

D Myonenzahl
F nombre muonique
P liczba leptonowa mionowa
R мюонное лептонное число

2328
MUREXIDE REACTION.
Reaction applied for detection of uric acid, involving its transformation into coloured murexide (reddish-purple or violet) under action of nitric acid and ammonia.

D Murexidreaktion
F réaction de murexide
P reakcja mureksydowa
R мурексидная реакция

2329
MUTAROTATION.
Slow change of optical rotation power of solution due to changes of equilibrium between diastereoisomers of solute caused by solvent.

D Mutarotation
F mutarotation
P mutarotacja
R мутаротация

NAMETKIN CHANGE. *See* NAMETKIN REARRANGEMENT.

2330
NAMETKIN REARRANGEMENT (Nametkin change).
Rearrangement of retropinacoline type of transient carbocation, formed from halide alcohol or alkene in acidic medium.
In case of bicyclic terpenes, this reaction does not change ring system, e.g.

camphenilol   santen

D Nametkin (Retropinakolin)-Umlagerung
F transposition de Nametkine
P przegrupowanie Namiotkina
R перегруппировка Намёткина, камфеновая перегруппировка второго рода

2331
NANOGRAM METHOD (submicro method, ultraultramicro method).
Method of analysis where mass of investigated sample is $x\cdot10^{-9}$—$x\cdot10^{-7}$ g, or volume $x\cdot10^{-8}$—$x\cdot10^{-6}$ cm³ in the case of gases.

D Nanogrammethode
F méthode nanogrammique, submicrométhode, ultraultramicrométhode
P metoda nanogramowa, ultraultramikrometoda, submikrometoda
R . . .

NATURAL BREADTH OF SPECTRAL LINE. *See* NATURAL LINE WIDTH.

2332
NATURAL LINE WIDTH (natural breadth of spectral line).
Spectral line width as determined by Heisenberg uncertainty principle.

D natürliche Linienbreite
F largeur propre de raie spéctrale
P szerokość linii widmowej naturalna
R естественная ширина спектральной линии

NATURAL PROCESS. *See* IRREVERSIBLE PROCESS.

2333
NATURAL RADIATION.
Ionizing radiation arising from extraterrestrial sources and from naturally occurring radionuclides contained in earth crust, air and living organisms.
D natürliche Strahlung
F rayonnement naturel
P promieniowanie naturalne
R естественное излучение, природное излучение

2334
NATURAL RADIOACTIVITY.
Radioactivity of naturally occurring nuclides.
D natürliche Radioaktivität
F radioactivité naturelle
P promieniotwórczość naturalna
R природная радиоактивность, естественная радиоактивность

NATURAL RADIOCARBON. *See* RADIOCARBON.

2335
NATURAL RADIONUCLIDE.
Radioactive nuclide occurring naturally.
D natürliches Radionuklid
F radionucléide naturel
P nuklid promieniotwórczy naturalny
R естественный радиоактивный изотоп, природный радиоактивный изотоп

N-CONDUCTION. *See* ELECTRON CONDUCTIVITY.

2336
NEAR INFRARED (overtone region), NIR.
Electromagnetic radiation range of wavelength between 0.75 μm to ca. 3 μm.
D nahes Infrarot, kurzwelliges IR
F infrarouge proche
P podczerwień bliska, zakres nadtonów
R ближняя инфракрасная область

2337
NEAR ULTRAVIOLET (ultraviolet).
Electromagnetic radiation range of wavelength between ca. 200 nm to ca. 400 nm.
D Ultraviolett
F ultraviolet, région ultraviolette
P nadfiolet (bliski)
R ближняя ультрафиолетовая область

NÉEL POINT. *See* NÉEL TEMPERATURE.

2338
NÉEL TEMPERATURE (Néel point, antiferromagnetic Curie point).
Temperature of maximum magnetic susceptibility of antiferromagnet related to loss of its magnetic ordering.

D Néel-Temperatur, Néel-Punkt, antiferromagnetische Curie-Temperatur
F température de Néel, point de Néel
P temperatura Néela, punkt Néela, temperatura Curie antyferromagnetyczna
R температура Нееля, точка Нееля

2339
NEF REACTION.
Conversion of sodium (or potassium) salts of primary or secondary nitroparaffins to aldehydes or ketones under action of strong acids

$$2RCH_2-N{\nearrow O \atop \searrow O} \xrightleftharpoons{NaOH} 2RCH=N{\nearrow O \atop \searrow ONa} \xrightarrow{H^{\oplus}}$$

$$2RCHO + N_2O + H_2O$$

D Nef Acinitroalkan-Spaltung
F synthèse de Nef
P synteza (związków karbonylowych) Nefa
R реакция Нефа

2340
NEGATIVE AZEOTROPE.
Azeotrope of composition at which there is minimum vapour pressure and correspondingly, maximum boiling temperature.
D negatives Azeotrop
F azéotrope négatif
P azeotrop ujemny
R отрицательная азеотропная смесь

2341
NEGATIVE CATALYSIS.
Phenomenon of retardation of reaction by inhibitor.
D negative Katalyse
F catalyse négative
P kataliza ujemna
R отрицательный катализ

NEGATIVE CATALYST. *See* INHIBITOR.

2342
NEGATIVE VALENCY.
Number of electrons moved to atom from atom of positive valency during formation of chemical bond.
D negative Wertigkeit
F valence négative
P wartościowość ujemna
R отрицательная валентность

2343
NEIGHBOURING-GROUP PARTICIPATION.
Adjacent groups to reaction centre display inductive, mesomeric and steric effects and sometimes have influence on reaction rate by accelerating or slowing it down (*see* anchimeric assistance effect) or may cause rearrangements in molecules; e.g.

$$(CH_3)_2\overset{..}{N} \quad CH-Cl \xrightarrow{-Cl^{\ominus}} \quad (CH_3)_2\overset{\oplus}{N} \quad CH$$
$$\overset{|}{CH_2}\,\overset{|}{C_2H_5} \qquad\qquad \overset{|}{CH_2}\,\overset{|}{C_2H_5}$$

$$\xrightarrow{\ominus OH} \quad (CH_3)_2N-CH-CH_2OH$$
$$\overset{|}{C_2H_5}$$

D Nachbargruppenbeteiligung
F intervention des groupes voisins
P udział grup sąsiadujących
R влияние соседних групп

2344
NENCKI REACTION.
Acylation of phenols with carboxylic acids (reaction is similar to that of Friedel-Crafts), catalyzed with $ZnCl_2$ or $FeCl_3$, e.g.

D Nencki C-Acylierung
F réaction de Nencki
P reakcja Nenckiego
R реакция Ненцкого

2345
NENITZESCU ACYLATION REACTION.
Preparation of acylcycloalkanes from cycloalkenes by treating the latter with acid chlorides in presence of cyclohexane (acting as hydrogen atom donor) and catalytic amount of anhydrous aluminium chloride, e.g.

$$(2C_6H_{12} \rightarrow C_{12}H_{22} + 2H)$$

Reaction also takes place with cycloalkanes often resulting in contraction of ring system, e.g.

D Nenitzescu hydrierende Acylierung
F ...
P reakcja (acylowania) Nenitescu
R реакция Неницеску

2346
NEODYMIUM, Nd.
Atomic number 60.

D Neodym
F néodyme
P neodym
R неодим

2347
NEON, Ne.
Atomic number 10.

D Neon
F néon
P neon
R неон

2348
NEPHELAUXETIC SERIES OF LIGANDS.
Series of ligands arranged by increasing share of covalent bond in bond between central atom and ligand: $F^- < H_2O < (NH_2)_2CO < NH_3 < NH_2CH_2CH_2NH_2 < C_2O_4^{2-} < NCS^- \sim Cl^- \sim CN^- < Br^- < (C_2H_5O)_2PS_2^- \sim S^{2-} \sim I^- < (C_2H_5O)_2PSe_2^-$.

D nephelauxetische Reihe der Liganden, nephelauxetische Serie der Liganden
F ...
P szereg nefeloauksetyczny ligandów
R ...

2349
NEPHELOMETER.
Instrument for determination of suspension concentrations in solutions by measurement of intensity of radiation scattered at angle different from $180°$ with respect to incident ray.

D Nephelometer
F néphélomètre
P nefelometr
R нефелометр

2350
NEPHELOMETRY.
Method of analysis used to determine concentration of suspension or its particle dimensions by measurement of light scattered at a given angle.

D Nephelometrie, Tyndallometrie
F néphélométrie
P analiza nefelometryczna, nefelometria
R нефелометрия

2351
NEPTUNIUM, Np.
Atomic number 93.

D Neptunium
F neptunium
P neptun
R нептуний

2352
NERNST DISTRIBUTION LAW.
In the state of interphase equilibrium the ratio of the activity of a component in two different phases depends only on temperature and pressure.

D Nernstscher Verteilungssatz
F loi de distribution de Nernst
P prawo podziału (Nernsta), prawo równowagi fazowej
R закон распределения

NERNST-PLANCK THEOREM. *See* THIRD LAW OF THERMODYNAMICS.

2353
NET MAXIMUM WORK (net useful work, available net work, maximum useful work).
Maximum amount of work which can be performed by closed system on surroundings during isothermal-isobaric transition from state I to state II; net maximum work is equal to loss of Gibbs free energy of system $\Delta G = G_I - G_{II}$.
D maximale Nutzarbeit
F (travail maximum diminué du travail d'expansion)
P praca maksymalna użyteczna
R максимальная полезная работа

NET PLANE OF LATTICE. See LATTICE PLANE.

NET USEFUL WORK. See NET MAXIMUM WORK.

2354
NETWORK POLYMER (space polymer).
Polymer having three-dimentional structure, formed by cross-linking of main chains.
D vernetztes Polymer, Raumgitterpolymer
F polymère réticulé
P polimer usieciowany, polimer przestrzenny
R сшитый полимер, трёхмерный полимер, пространственный полимер, сетчатый полимер

2355
NEUTRAL CHELATE.
Chelate in which positive charge of central ion is compensated for by negative charge of ligand.
D inneres Komplexsalz
F complexe interne
P sól wewnątrzkompleksowa, chelat wewnętrzny, kompleks wewnętrzny
R внутрикомплексное соединение

2356
NEUTRAL FILTER.
Filter diminishing radiation intensity to the same extent independently of wavelength (it has this feature only in some wavelength region).
D Neutralfilter, Graufilter
F filtre neutre
P filtr neutralny, filtr szary
R нейтральный светофильтр

2357
NEUTRALIZATION.
Reaction of acid and base which produces neutral solution, i.e. solution for which activities of both acidic and basic forms of solvent are equal (see autoprotolysis).
D Neutralisation
F neutralisation

P zobojętnianie, neutralizacja
R нейтрализация

2358
NEUTRINO, ν.
General term for two elementary particles: electronic neutrino ν_e, and muonic neutrino ν_μ.
D Neutrino
F neutrino
P neutrino
R нейтрино

2359
NEUTRON, n.
Elementary particle, a hadron, one of isospin doublet state of nucleon, of mass $m = 939.552\ 7$ MeV, electrically neutral, of baryon number $B = 1$, lepton number $L = 0$, quantum spin number $J = 1/2$, strangeness $S = 0$, isospin $I = 1/2$, third component of isospin $I_3 = -1/2$. Neutron is constituent of atomic nuclei. A free neutron is unstable, its mean lifetime is $0.932 \cdot 10^3$ s.
D Neutron
F neutron
P neutron
R нейтрон

NEUTRON ABSORPTIOMETRY. See NEUTRON ABSORPTION ANALYSIS.

2360
NEUTRON ABSORPTION ANALYSIS (neutron absorptiometry).
Method of analysis based on measuring of attenuation of neutron flux in layer of material tested. It is applied for determination of elements of high neutron absorption cross-sections.
D Neutronenabsorptionsmethode, Neutronenabsorptionsverfahren
F analyse par absorption des neutrons
P analiza absorpcyjna neutronowa
R нейтронно-абсорбционный метод анализа

2361
NEUTRON ACTIVATION.
Activation by neutron irradiation, occurring most often in an (n, γ) nuclear reaction.
D Neutronenaktivierung
F activation neutronique
P aktywacja neutronowa
R нейтронная активация

2362
NEUTRON ACTIVATION ANALYSIS.
Activation method of analysis in which a flux of neutrons is applied for activation.
D Neutronenaktivierungsanalyse
F analyse par activation neutronique

P analiza aktywacyjna neutronowa
R нейтронный активационный анализ, нейтроноактивационный анализ

NEUTRON CONVERTER. *See* FLUX CONVERTER.

2363
NEUTRON COUNTER. Counter for measuring neutron flux. In case of slow neutrons, counter is usually filled with boron compounds and its work is based on detection of α-particles from the $^{10}B(n, α)^7Li$ reaction.

D Neutronenzählrohr, Neutronenzähler
F compteur des neutrons
P licznik neutronów
R счётчик нейтронов

2364
NEUTRON DETECTOR.
Detector registering charged particles liberated in the course of a nuclear reaction occurring with participation of a specified number of neutrons.

D Neutronendetektor
F détecteur de neutrons
P detektor neutronów
R детектор нейтронов

2365
NEUTRON DIFFRACTION.
Scattering of monochromatic stream of neutrons followed by interference of their matter waves.

D Neutronenbeugung
F diffraction des neutrons
P dyfrakcja neutronów
R дифракция нейтронов

2366
NEUTRON GENERATOR.
Accelerator producing neutrons in appropriate nuclear reactions, e.g. $^2H(d, n)^3He$, $^3H(d, n)^4He$.

D Neutronengenerator
F générateur de neutrons
P generator neutronów
R нейтронный генератор

2367
NEUTRON OUTPUT (neutron yield).
Number of neutrons emitted from source in unit time at solid angle of 4π steradians.

D Neutronenausbeute
F débit de neutrons
P wydatek neutronów
R выход нейтронов

2368
NEUTRON RADIATIVE CAPTURE.
Nuclear reaction of (n, γ) type, i.e. reaction consisting in absorption of

neutron, giving rise to excitation of nucleus, and eventually to return of nucleus to ground state due to emission of γ-quantum.

D Neutronenstrahlungseinfang
F capture radiative du neutron
P wychwyt radiacyjny neutronu
R радиационный захват нейтрона

2369
NEUTRON SPECTRUM.
Curve representing energy dependence of neutron beam density.

D Neutronenspektrum
F spectre de neutrons
P widmo neutronów
R спектр нейтронов

2370
NEUTRON THERMALIZATION.
Slowing down of neutrons to reduce their energy to value corresponding to thermal energy of atoms or molecules of medium.

D Neutronenthermalisierung
F thermalisation des neutrons
P termalizacja neutronów
R термализация нейтронов

NEUTRON YIELD. *See* NEUTRON OUTPUT.

2371
NEWMAN FORMULA.
Projection formula portraying molecule in a plane, in way convenient for conformational analysis, e.g. conformation of *anti meso*-2,3-dibromobutane

Formula is viewed from side along the C^2—C^3 bond. C^3 atom, more distant to observer, is represented as circle; rest of C^2—C^3 atomic bonds form symmetrical system of lines.

D Newmansche Formel
F projection de Newman
P wzór (rzutowy) Newmana
R формула Ньюмена

2372
NEWTONIAN VISCOSITY.
Dynamic viscosity independent of change in velocity gradient dv/dx (*see* viscosity coefficient).

D Newtonsche Viskosität
F viscosité newtonienne
P lepkość newtonowska, lepkość normalna
R ньютоновская вязкость

2373
NICKEL, Ni.
Atomic number 28.
D Nickel
F nickel
P nikiel
R никель

2374
NICKEL GROUP.
Three elements in eighth sub-group of periodic system with structure of outer electronic shells of atoms: nickel — $3d^8 4s^2$, palladium — $4d^{10}$, platinum — $5d^9 6s^1$.
D ...
F ...
P niklowce
R ...

2375
NIEMENTOWSKI QUINOLINE SYNTHESIS.
Preparation of quinoline and its derivatives by condensation of anthranilic acid with compounds containing —CH_2—CO— group, e.g.

D Niementowski Chinolin-Ringschluß
F réaction de Niementowski
P synteza (pochodnych) chinoliny Niementowskiego
R ...

2376
NIERENSTEIN REACTION.
Preparation of ω-chloroacetophenon (and its derivatives) by treating aroyl chlorides with diazomethane

$$ArCOCl + CH_2N_2 \rightarrow ArCOCH_2Cl + N_2$$

D ...
F réaction de Nierenstein
P reakcja Nierensteina (i Clibbensa)
R реакция Ниренштейна

2377
NILSSON MODEL.
Model of atomic nucleus using shell concept, describing deformed nuclei with ellipsoidal, axially symmetrical deformation of nuclear potential.
D Nilsson-Modell
F modèle de Nilsson
P model Nilssona
R модель Нильссона

2378
NINHYDRIN REACTION.
Coloured (blue, violet or red) reaction of most α-amino and imino acids, also of amines with ninhydrin in basic medium; applied for detection of proteins.
D Ninhydrinreaktion
F réaction de ninhydrine
P reakcja ninhydrynowa
R нингидринная реакция

2379
NIOBIUM, Nb.
Atomic number 41.
D Niob
F niobium
P niob
R ниобий

2380
NITRATION.
Substitution of hydrogen atom in molecule of organic compound by nitro group —NO_2, e.g.

$$C_6H_6 + HNO_3 \xrightarrow{H_2SO_4} C_6H_5NO_2 + H_2O$$

D Nitrierung
F nitration
P nitrowanie
R нитрование

2381
NITRO-ACI-NITRO TAUTOMERISM.
Type of prototropic tautomerism connected with reversible rearrangement of primary and secondary nitro compounds into *aci*-form

D Nitro-Acinitro-Tautomerie
F tautomérie dérivé nitré-forme aci, prototropie dérivé nitré-forme aci
P tautomeria nitro-acinitrowa
R нитро-изонитро таутомерия

2382
NITROGEN, N.
Atomic number 7.
D Stickstoff
F azote
P azot
R азот

NITROGEN FAMILY. *See* NITROGEN GROUP.

2383
NITROGEN GROUP (nitrogen family).
Elements of fifth main group in periodic system with structure of outer electronic

shells of atoms ns^2np^3: nitrogen, phosphorus, arsenic, antimony, bismuth.
D Stickstoffgruppe, Stickstoff-Phosphor-Gruppe
F famille de l'azote, azotides
P azotowce
R группа азота, элементы (под)группы азота

2384
NITROSATION.
Substitution of hydrogen atom in molecule of organic compound by nitroso group —NO, resulting in formation of nitroso compounds or their tautomers, oximes, e.g.

$$RCH_2NO_2 \xrightarrow[-H_2O]{HNO_2} RCHNO_2 | NO$$

$$\underset{||}{\overset{O}{RCCH_2COOR}} \xrightarrow[-H_2O]{HNO_2} \underset{||}{\overset{O}{RCCCOOR}} \rightleftarrows \underset{|}{\overset{OH}{RC=CCOOR}}$$
$$NOH \qquad NO$$

D Nitrosierung
F nitrosation
P nitrozowanie
R нитрозирование

2385
NITROSO-OXIMINO TAUTOMERISM.
Type of prototropic tautomerism connected with reversible rearrangement of primary and secondary nitroso compounds into oximes (isonitroso compounds)

$$-\underset{H}{\overset{|}{C}}-N=O \rightleftarrows \overset{}{C}=N-OH$$

D ...
F tautomérie dérivé nitrosé-oxime, prototropie dérivé nitrosé-oxime
P tautomeria nitrozowo-oksymowa
R нитрозо-изонитрозо таутомерия

NMR SPECTROMETER. *See* NUCLEAR MAGNETIC RESONANCE SPECTROMETER.

NMR SPECTRUM. *See* NUCLEAR MAGNETIC RESONANCE SPECTROSCOPY.

NMR SPECTRUM. *See* NUCLEAR MAGNETIC RESONANCE SPECTRUM.

2386
NOBELIUM, No.
Atomic number 102.
D Nobelium
F nobélium

P nobel
R нобелий

NOBLE GASES. *See* INERT GASES.

NO-BOND RESONANCE. *See* HYPERCONJUGATION.

2387
NO-MECHANISM REACTION.
Reaction possessing unidentified transition states, e.g. Claisen rearrangement.
D ...
F ...
P reakcja bezmechanizmowa
R ...

2388
NON-BONDING ELECTRONS.
Electrons assumed to occupy molecular orbitals which do not contribute to chemical bonds.
D nichtbindende Elektronen, n-Elektronen, nonbonding Elektronen
F électrons non liants
P elektrony niewiążące
R несвязывающие электроны

2389
NON-BONDING ORBITAL.
Molecular orbital built of identical atomic orbitals of the same orbital energy; its energy is equal to that of component atomic orbitals. Appears in Hückel theory of odd alternant hydrocarbons; also atomic orbital of lone electron pairs.
D nichtbindendes Molekülorbital
F orbitale non liante
P orbital niewiążący
R несвязывающая молекулярная орбиталь

2390
NON-CLASSICAL IONS.
Ions with delocalized σ-electrons, introduced to explain certain non-typical rearrangements and reactivity of alicyclic systems; e.g. non-classical norbornyl cation

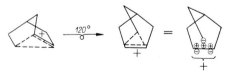

D nichtklassische Ionen
F ions non-classiques
P jony nieklasyczne
R неклассические ионы

2391
NON-DESTRUCTIVE ACTIVATION
ANALYSIS.
Activation method of analysis in which
selective measurement of activity of
element to be determined is carried out
by physical methods without chemical
separation of irradiated sample.
D zerstörungsfreie Aktivierungsanalyse
F analyse non destructive par activation,
analyse par activation instrumentale
P analiza aktywacyjna niedestrukcyjna,
analiza aktywacyjna instrumentalna
R инструментальный активационный
анализ

NON-EQUILIBRIUM
THERMODYNAMICS. *See*
THERMODYNAMICS OF
IRREVERSIBLE PROCESSES.

NON-IDEAL GAS. *See* REAL GAS.

NON-IDEAL SOLUTION. *See* REAL
SOLUTION.

NON-IDENTICAL CHIRAL CENTRES. *See*
DISSIMILAR CHIRAL CENTRES.

2392
NON-ISOTOPIC CARRIER.
Carrier for radioactive substance
chemically non-identical with substance
(usually isomorphic or isodimorphic),
e.g. Ba^{2+} ions for Ra^{2+} ions which
coprecipitate with precipitates of barium
salts.
D nichtisotop(isch)er Träger
F entraîneur nonisotopique
P nośnik nieizotopowy
R неизотопный носитель

2393
NON-METALS.
Elements not having properties of metals
and, except for noble gases,
electronegative.
D Nichtmetalle
F metalloïdes, éléments non-métalliques,
non-métaux
P niemetale
R неметаллы

NON-NEWTONIAN VISCOSITY. *See*
STRUCTURAL VISCOSITY.

2394
NON-OSMOTIC MEMBRANE
EQUILIBRIUM.
Thermodynamic state of system with
electrochemical potential of any kind of
ions identical in all phases.
D elektrochemisches Gleichgewicht
F équilibre électrochimique
P równowaga elektrochemiczna
R электрохимическое равновесие

NON-RADIATIVE TRANSITION. *See*
RADIATONLESS TRANSITION.

2395
NON-STATIONARY STATE.
State with at least one time-dependent
parameter.
D unstationärer Zustand
F état nonstationnaire
P stan niestacjonarny
R нестационарное состояние

2396
NON-STOICHIOMETRIC COMPOUNDS
(bertholides).
Compounds of composition depending in
some degree on mode of preparation.
These compounds do not fulfill law of
definite proportions.
D nichtstöchiometrische Verbindungen,
Berthollidverbindungen
F berthollides
P związki niestechiometryczne, bertolidy
R бертоллиды

2397
NORMAL BOILING POINT.
Boiling point at pressure of
$p = 1.013\,25 \cdot 10^{-1}$ MPa (formerly
$p = 1$ atm).
D normale Siedetemperatur, normaler
Siedepunkt
F température normale d'ébullition, point
normal d'ébullition
P temperatura wrzenia normalna
R нормальная температура кипения

2398
NORMAL CHAIN (linear chain, straight
chain).
System of mutually linked atoms forming
straight chain (with no branching), e.g.
$CH_3-CH_2-CH_2-CH_3$
D normale Kette, gerade(linige) Kette,
lineare Kette
F chaîne droite, chaîne linéaire
P łańcuch prosty, łańcuch normalny
R нормальная цепь, прямая цепь

2399
NORMAL COMPLEX (ionic complex).
Complex of ionic character which
dissociates in solution into components,
e.g. $[Cd(CN)_4]^{2-}$, $[Co(NH_3)_6]^{2+}$.
D Normalkomplex, Anlagerungskomplex
F complexe normal
P kompleks normalny
R нормальный комплекс, равновесный
комплекс

2400
NORMAL CONCENTRATION (normality
of solution).
Number of gram-equivalents of substance
dissolved in 1 dm^3 of solution.

D normale Konzentration, Normalität
F concentration normale, normalité de
solution
P stężenie równoważnikowe, normalność
R нормальность

NORMAL CONDITIONS. *See* STANDARD
CONDITIONS.

2401
NORMAL DISTRIBUTION (Gaussian
distribution),
Probability distribution of random
variable with probability density
function given by

$$f(X) = \frac{1}{\sigma \sqrt{2\pi}} \exp\left[-(X-\mu)^2/2\sigma^2\right]$$

where X denotes random variable,
μ = its mean value, and σ = its standard
deviation.

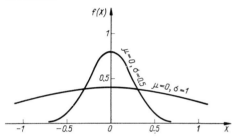

D Normalverteilung, Gaußsche Verteilung
F distribution normale, distribution de
Laplace-Gauss, répartition normale
P rozkład normalny, rozkład Laplace'a
i Gaussa
R нормальное распределение, гауссово
распределение

2402
NORMAL ELECTRODE POTENTIAL
(standard electrode potential).
Potential of electrode (half-cell) with
activity of components which take part
in electrode potential equal to unity.
Normal potentials are expressed usually
with respect to potential of normal
hydrogen electrode or saturated calomel
electrode.

D Standardpotential (der Elektrode),
Normalpotential (der Elektrode)
F tension standard (d'électrode), potentiel
normal (d'électrode)
P potencjał normalny elektrody
R стандартный (электродный) потенциал,
нормальный (электродный) потенциал

NORMAL ELECTROMOTIVE FORCE.
See STANDARD ELECTROMOTIVE
FORCE.

2403
NORMAL HYDROGEN ELECTRODE
(standard hydrogen electrode).
Hydrogen electrode taken as having
a zero potential at all temperatures at
one atmosphere pressure of hydrogen,
i.e. in SI system $p_{H_2} = 0.1325 \cdot 10^5$ Pa, in
solution of unit hydrogen activity,
$a_{H_3O^+} = 1$.

D Standard-Wasserstoffelektrode,
Normal-Wasserstoffelektrode
F électrode standard à hydrogène,
électrode normale à hydrogène
P elektroda wodorowa normalna,
elektroda wodorowa standardowa
R стандартный водородный электрод,
нормальный водородный электрод

NORMALITY. *See* NORMAL
CONCENTRATION.

2404
NORMALIZATION OF WAVE
FUNCTION.
Procedure to obtain from given wave
function Ψ its normalized counterpart
$N\Psi$, where N is determined from
normalization condition $\int |N\Psi|^2 d\tau = 1$;
$d\tau$ = integration volume element and
integration is over all spatial and spin
variables of system.

D Normierung der Wellenfunktion
F normalisation de la fonction d'onde
P normalizacja funkcji falowych
R нормировка волновой функции

NORMAL PHOTOELECTRIC EFFECT.
See PHOTOELECTRIC EFFECT.

2405
NORMAL VIBRATIONS.
Harmonic oscillations in system of
material points with all points moving
with the same frequency and constant,
time-independent phase. Normal
vibrations in molecules follow from
consideration of molecule as set of
material points (atoms).

D Normalschwingungen
F vibrations normales
P drgania normalne
R нормальные колебания

N-R TRANSITION. *See* RYDBERG
TRANSITION.

n-TYPE SEMICONDUCTOR. *See*
ELECTRON-EXCESS SEMICONDUCTOR.

NUCLEAR ACTIVITY. *See* ACTIVITY.

NUCLEAR ANGULAR MOMENTUM. *See*
NUCLEAR SPIN.

NUCLEAR ATOM. *See* CENTRAL ATOM.

2406
NUCLEAR BINDING ENERGY.
Energy that would be necessary to separate atomic nucleus into nucleons. Nuclear binding energy is equal to mass defect of that nucleus.
D Kernbindungsenergie
F énergie de liaison nucléaire
P energia wiązania jądra
R энергия связи ядра

2407
NUCLEAR CHEMISTRY.
Collective name for a number of particular branches of chemistry, the common feature of which is research and practical utilization of nuclear phenomena.
D Kernchemie
F chimie nucléaire
P chemia jądrowa
R ядерная химия

2408
NUCLEAR ENERGY.
Internal energy of atomic nucleus connected with interaction and movement of nucleons in nucleus.
Remark: Commonly, this term denotes also nuclear energy released during nuclear reactions.
D Kernenergie
F énergie nucléaire
P energia jądrowa
R ядерная энергия, энергия атомного ядра, внутриядерная энергия

2409
NUCLEAR ENERGY LEVEL.
Eigenvalue of atomic nucleus hamiltonian, viz. energy value defining state of nucleus. Nuclear energy level is often defined as difference between the energy of nucleus in any given state and in lowest (ground) state.
D Kernenergieniveau
F niveau énergétique du noyau
P poziom energetyczny jądra
R уровень энергии ядра

2410
NUCLEAR FISSION (fission *of atomic nucleus*).
Splitting of nucleus into several (generally two) fragments of which at least two have mass number greater than 20. Fission can be spontaneous or induced by absorption of particle or γ-ray by nucleus.
D Kernspaltung
F fission (nucléaire)
P rozszczepienie (jądra atomowego), reakcja rozszczepienia
R деление атомного ядра

2411
NUCLEAR FORCES.
Short-range forces occurring among nucleons, in particular binding nucleons into nuclei.
D Kernkräfte
F forces nucléaires
P siły jądrowe
R ядерные силы

2412
NUCLEAR FUEL.
Material containing fissionable nuclides in chemical form and geometrical shape ready for use in the particular type of nuclear reactor.
D Kernbrennstoff
F combustible nucléaire
P paliwo jądrowe
R ядерное топливо

2413
NUCLEAR FUEL CYCLE.
Sequence of operations and processes of preparation, exploitation and reprocessing of nuclear fuel.
D Brennstoffkreislauf, Brennstoffzyklus
F cycle de combustible
P cykl paliwowy
R топливный цикл

NUCLEAR FUSION. *See*
THERMONUCLEAR REACTION.

2414
NUCLEAR g FACTOR (nuclear Landé factor).
Factor in formula for nuclear magnetic moment of paramagnetic nucleus, corresponding to Landé factor.
D Kern-g-Faktor
F facteur de Landé nucléaire
P czynnik g jądra
R ядерный g-фактор

2415
NUCLEAR INDUCTION.
Old term used to describe magnetic nuclear resonance; now obsolescent.
D Kerninduktion
F induction nucléaire
P indukcja jądrowa
R ядерная индукция

NUCLEAR INTERACTION. *See* STRONG INTERACTION.

2416
NUCLEAR ISOMERISM.
Existence of metastable excited states of atomic nucleus.
D Kernisomerie
F isomérie nucléaire

P izomeria jądrowa
R изомерия атомных ядер

2417
NUCLEAR ISOMERS.
Atomic nuclei of identical mass number A
and atomic number Z in different
metastable energetic states.
D Kernisomere
F isomères nucléaires
P izomery jądrowe
R ядерные изомеры

NUCLEAR LANDÉ FACTOR. *See*
NUCLEAR g FACTOR.

2418
NUCLEAR MAGNETIC MOMENT, μ_I.
Vector quantity, defined by $\mu_I = g_I I \mu_N$,
where g_I = nuclear g factor, I = spin
quantum number, μ_N = nuclear magneton.
D magnetisches Kernmoment
F moment nucléaire
P moment magnetyczny jądra
R магнитный момент ядра

2419
NUCLEAR MAGNETIC RESONANCE,
NMR.
Magnetic resonance observed in
substances containing paramagnetic nuclei
with resonance frequencies different for
each nucleus. For static magnetic field
strength of 795 kA m^{-1} resonance
frequencies appear in range of a few to
a few tens of MHz and correspond either
to short waves or radiowaves.
D kernmagnetische Resonanz
F résonance magnétique nucléaire
P rezonans (para)magnetyczny jądrowy
R ядерный магнитный резонанс

2420
NUCLEAR MAGNETIC RESONANCE
SPECTROMETER (NMR spectrometer).
Instrument used for obtaining and
recording NMR spectra.
D kernmagnetisches Resonanz-
-Spektrometer
F spectromètre à résonance magnétique
nucléaire
P spektrometr NMR
R спектрометр ЯМР

2421
NUCLEAR MAGNETIC RESONANCE
SPECTROSCOPY (NMR spectroscopy).
Branch of radiospectroscopy which utilizes
nuclear paramagnetic resonance for study
of physico-chemical properties of
materials; e.g. study of structure of
molecules and crystals, study
of autodiffusion processes, hindered
rotation, etc.

D kernmagnetische Resonanzspektroskopie
F spectroscopie de résonance magnétique
nucléaire
P spektroskopia NMR
R спектроскопия ядерного магнитного
резонанса, ЯМР-спектроскопия

2422
NUCLEAR MAGNETIC RESONANCE
SPECTRUM (NMR spectrum).
Curve (signal) of absorption or of first
derivative of energy absorbed from
radio-frequency electromagnetic field,
usually registered at fixed resonance
frequency and linearly varying (increasing
or decreasing) strength of external static
magnetic field.
D kernmagnetisches Resonanzspektrum
F spectre de résonance magnétique
nucléaire
P widmo magnetycznego rezonansu
jądrowego, widmo NMR
R спектр ядерного магнитного резонанса

2423
NUCLEAR MAGNETON, μ_N (β_N).
Unit of nuclear magnetic moment;
$1\ \mu_N = \dfrac{1}{1837}\ \mu_B$, where μ_B denotes Bohr
magneton; $1\ \mu_N = 5.050 \cdot 10^{-27}$ A m^2.
D Kernmagneton
F magnéton nucléaire
P magneton jądrowy
R ядерный магнетон

NUCLEAR MODELS. *See* MODELS OF
NUCLEUS.

NUCLEAR PARENT. *See* PARENT
NUCLIDE.

2424
NUCLEAR POTENTIAL.
Effective potential energy of single
nucleon inside atomic nucleus caused by
strong nuclear interactions with
remaining nucleons.
D Kernpotential
F potentiel nucléaire
P potencjał jądrowy
R ядерный потенциал

2425
NUCLEAR PURITY.
Absence in material of elements of high
absorption cross-section for neutrons to
such a degree that it can be used in
reactor technique.
D Kernreinheit, Nuklearreinheit
F pureté nucléaire
P czystość jądrowa
R ядерная чистота

2426
NUCLEAR QUADRUPOLE MOMENT.
Electric quadrupole moment displayed by
nuclei with spin $I > 1/2$ due to
non-spherical distribution of electric
charge.
D Kernquadrupolmoment
F moment quadrupolaire nucléaire
P moment kwadrupolowy jądra
R квадрупольный момент ядра

2427
NUCLEAR QUADRUPOLE RESONANCE.
Phenomenon associated with quantum
transitions between energy levels of
electric nuclear quadrupole moments in
internal inhomogeneous electric field of
molecules and crystals.
D Kernquadrupolresonanz
F résonance quadrupolaire nucléaire
P rezonans kwadrupolowy jądrowy
R ядерный квадрупольный резонанс

2428
NUCLEAR RADIATION.
Corpuscular radiation and gamma
radiation emitted by naturally and
artificially radioactive substances.
D Kernstrahlung, radioaktive Strahlung
F rayonnement nucléaire, rayonnement
radioactif
P promieniowanie jądrowe
R ядерное излучение

2429
NUCLEAR RADIATION SPECTRUM.
Curve representing energy dependence of
number of particles or photons.
D Kernstrahlungsspektrum
F spectre du rayonnement (nucléaire)
P widmo promieniowania (jądrowego)
R энергетический спектр (излучения)

2430
NUCLEAR REACTION.
Transformation of atomic nucleus A by
action of elementary particle b (or other
nucleus) leading to emission of elementary
particle (or several particles) c, and to
formation of new nucleus (or nuclei) D.
The record is $A(b, c)D$.
D Kernreaktion
F réaction nucléaire
P reakcja jądrowa
R ядерная реакция

2431
NUCLEAR REACTION CHANNEL.
Nuclear reaction path from initial to final
state, i.e. determined by reaction by
reaction products and their quantum
numbers.

D Kernreaktionskanal
F canal de réaction nucléaire
P kanał reakcji jądrowej
R канал ядерной реакции

2432
NUCLEAR REACTION ENERGY
BALANCE.
Comparison of total energy liberated in
nuclear reaction in different forms to
energy of system before reaction (with
consideration of energies and rest masses
of system).
D Kernreaktionsenergiebilanz
F bilan énergétique de réaction nucléaire
P bilans energetyczny reakcji jądrowej
R энергетический баланс ядерной
реакции

2433
NUCLEAR REACTOR COOLANT.
Liquid or gas removing heat released in
core of nuclear reactor. Nuclear reactor
coolant may act at the same time as
moderator.
D Kernreaktorkühlmittel
F réfrigérant du réacteur nucléaire
P chłodziwo reaktorowe
R реакторный теплоноситель

2434
NUCLEAR SPECTROSCOPY.
Branch of nuclear physics dealing with
study of absorption and emission of
radiation by atomic nuclei.
D Kernspektroskopie
F spectroscopie nucléaire
P spektroskopia jądrowa
R ядерная спектроскопия

2435
NUCLEAR SPIN (nuclear angular
momentum), J.
Vector equal to geometric sum of total
angular momenta of nucleons in nucleus,
expressed in units
$\hbar = 1.054\ 591\ 9 \cdot 10^{-27}$ erg s. Nuclear spin
takes values which are multiples of 1/2 in
these units. Nuclear spin describes nucleus
and its state. For nuclei in ground state
and even number of nucleons, $|J|$ is an
integral number; for nuclei with an odd
number of nucleons, $|J| = 1/2,\ 3/2,\ 5/2 \ldots$
D Kernspin, Eigendrehimpuls
F spin nucléaire, moment angulaire du
noyau
P spin jądra atomowego, moment pędu
jądra
R спин ядра, момент количества
движения ядра

2436
NUCLEAR SYMMETRY ENERGY.
Difference between binding energy of
nucleus with equal number of protons and
neutrons and binding energy of nucleus
of the same mass number A but higher
number of neutrons N than number of
protons Z. Nuclear symmetry energy
is defined as second derivative of binding
energy per nucleon over parameter of
excess of neutrons $(N-Z)/A$ at the same
mass number.
D Kern-Symmetrieenergie
F énergie de symétrie nucléaire
P energia symetrii jądra atomowego
R энергия симметрии ядра

2437
NUCLEON, N.
Elementary particle: proton or neutron
(isospin doublet) of baryon number
$B = 1$, lepton number $L = 0$, quantum
spin number $J = 1/2$, strangeness $S = 0$,
isospin $I = 1/2$, third component of isospin
$I_3 = 1/2$ (for proton) and $I_3 = -1/2$ (for
neutron).
D Nukleon
F nucléon
P nukleon
R нуклон

NUCLEOPHILE. See NUCLEOPHILIC
REAGENT.

2438
NUCLEOPHILIC REAGENT
(nucleophile).
Reagent which acts as donor of its
electron pair to form new bond with
carbon atom of another molecule.
Negative ions and molecules containing
atoms with free electron pairs are
nucleophiles, e.g.

$Cl^\ominus$, $OH^\ominus$, $:CH_2-C\overset{O}{\underset{H}{}} \leftrightarrow CH_2=C\overset{\ddot{O}:^\ominus}{\underset{H}{}}$,

$H_2\ddot{O}$, $R\ddot{O}H$, $\dot{N}H_3$

D nukleophiles Agens
F réactif nucléophile
P (od)czynnik nukleofilowy
R нуклеофильный реагент

2439
NUCLIDE.
Species of atoms characterized by their
mass number, atomic number and nuclear
energy state; each isotope of an element
is an individual nuclide; atoms whose
nuclei decay in times too short to be
measured are not nuclides.
D Nuklid
F nucl(é)ide

P nuklid
R нуклид, изотоп

2440
NULL HYPOTHESIS, H_0.
Fundamental statistical hypothesis
verified by a given test.
D Nullhypothese
F hypothèse nulle
P hipoteza zerowa
R нулевая гипотеза, исходная гипотеза

NULL POTENTIAL. See ZERO CHARGE
POTENTIAL.

NUMBER AVERAGE MOLECULAR
WEIGHT. See NUMBER AVERAGE
RELATIVE MOLECULAR MASS.

2441
NUMBER AVERAGE RELATIVE
MOLECULAR MASS (number average
molecular weight).
Average molecular mass calculated from
measurement of effects that depend on
number of macromolecules contained in
unit volume, i.e. on molar concentration
(methods: osmotic, ebullioscopic,
cryoscopic, end groups, etc.)
D Zahlenmittel-Molekulargewicht,
   Zahlenmittelwert, Zahlenmittel
F poids moléculaire moyen au nombre
P masa cząsteczkowa liczbowo średnia
R средний числовой молекулярный вес

2442
NUMBER OF DEGREES OF FREEDOM.
Number of independent random variables
minus number of redundancies. When
computing variance of series of n
independent measurement results, number
of degrees of freedom is $n-1$ since one
of results is related to others by
$\bar{x} = (x_1+x_2+...+x_n)/n$.
D Anzahl der Freiheitsgrade
F nombre de degrés de liberté
P liczba stopni swobody
R число степеней свободы

2443
NUMBER OF INDEPENDENT
CHEMICAL REACTIONS.
Number of chemical reactions in system,
whose stoichiometric equations are
linearly independent of each other.
D Anzahl der unabhängigen (chemischen)
   Reaktionen
F nombre des réactions chimiques
   indépendantes
P liczba niezależnych reakcji chemicznych
R число независимых (химических)
   реакций

2444
NUMBER OF INDEPENDENT
COMPONENTS (number of independent
constituents).
Number of types of particles whose
equilibrium concentrations in system can
be changed independently, i.e. number of
types of particles less number of
independent chemical reactions and
number of additional bounds limiting
concentrations.
D Anzahl der (unabhängigen)
  Komponenten
F nombre de constituants (indépendants)
P liczba składników (niezależnych)
R число (независимых) компонентов

NUMBER OF INDEPENDENT
CONSTITUENTS. *See* NUMBER OF
INDEPENDENT COMPONENTS.

NUMBER OF INDEPENDENT
INTENSIVE VARIABLES. *See* NUMBER
OF THERMODYNAMIC DEGREES OF
FREEDOM.

2445
NUMBER OF THERMODYNAMIC
DEGREES OF FREEDOM (variance,
variability, number of independent
intensive variables).
Number of intensive state parameters
whose values can be changed
independently of each other (in some
range) in such a way that number of
phases in equilibrium is not changed.
D Anzahl der (thermodynamischen)
  Freiheitsgrade, Varianz
F variance d'un système, nombre de
  variables (intensives) indépendantes
P liczba (termodynamicznych) stopni
  swobody, zmienność układu
R число термодинамических степеней
  свободы, вариантность системы

2446
OBSERVED LIFE-TIME OF
LUMINESCENCE (mean life-time of
luminescence).
Relaxation time of luminescence given by
reciprocal of total sum of rate constants
of all processes which deactivate any
given electronically excited state.
D mittlere Lebensdauer
F durée de vie, durée d'existence
P czas życia luminescencji obserwowany,
  czas życia luminescencji średni
R время жизни

2447
OCCUPATION NUMBER.
Number of particles in given
quantum-mechanical state.

D Besetzungszahl
F nombre d'occupation
P liczba obsadzenia
R число заполнения

2448
OCTAHEDRAL COMPLEX.
Complex of coordination number 6 in
which donor atoms of ligands are
arranged at vertices of regular or
distorted octahedron. Hybridization of
d^2sp^3 or sp^3d^2 type occurs in this complex.
D oktaedrischer Komplex
F complexe octaédrique
P kompleks oktaedryczny
R октаэдрический комплекс

2449
OCTANT RULE.
Rule permitting forecast of sign and
approximate amplitude of Cotton effect in
cyclohexanone derivatives.
D Octant-Regel
F règle des octants
P reguła oktantów
R правило октантов

OCTET RULE. *See* ABEGG RULE.

2450
ODD-EVEN NUCLEUS.
Atomic nucleus containing odd number of
protons and even number of neutrons.
D ungerade-gerader Atomkern
F noyau impair-pair
P jądro nieparzysto-parzyste
R нечётно-чётное ядро

2451
ODD-ODD NUCLEUS.
Atomic nucleus containing odd number
of protons and odd number of neutrons.
D doppelt-ungerader Atomkern
F noyau impair-impair
P jądro nieparzysto-nieparzyste
R нечётно-нечётное ядро

ODD SERIES. *See* SUB-GROUP.

OHMIC POLARIZATION.
See RESISTANCE POLARIZATION.

2452
OHMIC PSEUDO-POLARIZATION
(pseudo-resistance polarization).
Ohmic drop of voltage in layer of
electrolyte between electrode and tip of
Luggin capillary during flow of polarizing
current through electrolyte.
D Pseudo-Widerstandspolarisation
F pseudo-polarisation de résistance, chute
  ohmique
P polaryzacja pseudooporowa
R псевдо-омическая поляризация

2453
OLEFINE COMPLEX.
Complex of heavy metal (particularly platinum) with hydrocarbons containing double bonds. In this complex coordination bond is formed with participation of π electrons.
D Olefinkomplex
F complexe alcène
P kompleks olefinowy
R координационное соединение олефинов

OLEFINIC LINK. *See* ETHYLENE LINKAGE.

2454
OLIGOMER.
Substance composed of molecules containing a few of one or more species of atoms or groups of atoms (called constitutional units or mers) repetitively linked to each other. Oligomer molecules are of such sizes that physical properties of oligomer are different from those of both the corresponding monomer and polymer.
D Oligomer
F oligomère
P oligomer
R олигомер

2455
OLIGOMERIZATION.
Process of converting monomer or mixture of monomers into oligomer.
D Oligomerisation
F oligomérisation
P oligomeryzacja
R олигомеризация

2456
OMEGA$^-$ HYPERON (Ω^--hyperon).
Elementary particle, a hadron, of mass $m = 1\,672.5$ MeV, negative elementary electric charge, baryon number $B = 1$, lepton number $L = 0$, quantum spin number $J = 3/2$, strangeness $S = -3$, isospin $I = 0$, third component of isospin $I_3 = 0$, unstable, of mean lifetime $1.3 \cdot 10^{-10}$ s.
D Ω^--Hyperon
F hypéron Ω^-, hypéron omega
P hiperon Ω^-
R Ω^--гиперон, омега-гиперон

2457
ONE-COMPONENT SYSTEM.
System composed of one constituent.
D Einkomponentensystem, Einstoffsystem
F système unaire
P układ jednoskładnikowy
R однокомпонентная система

2458
ONE-ELECTRON BOND.
Chemical bond formed by valence electron present in molecular bonding orbital, e.g. ionized molecule H_2^+ is supposed to have one-electron bond.
D Einelektronenbindung
F liaison monoélectronique, liaison par un électron
P wiązanie jednoelektronowe
R одноэлектронная связь

2459
ONE-ELECTRON EXCITATION.
Excitation of molecule by raising one of its electrons either from bonding or non-bonding orbital to higher-energetic antibonding orbital.
D Einelektronanregung
F excitation monoélectronique
P wzbudzenie jednoelektronowe
R одноэлектронное возбуждение

2460
ONE PARTICLE APPROXIMATION (single particle approximation, single particle model).
Approximation which replaces real interparticle interaction by some self-consistent effective field for each particle.
D Einteilchenmodell, Einkörpermodell
F modèle à particules indépendantes
P przybliżenie jednocząstkowe, model cząstek niezależnych
R метод самосогласованного поля

ONE PHASE SYSTEM. *See* HOMOGENEOUS SYSTEM.

ONSAGER PRINCIPLE. *See* ONSAGER RECIPROCAL RELATIONS.

2461
ONSAGER RECIPROCAL RELATIONS (Onsager reciprocity relations, Onsager principle).
Relations determining symmetry of matrix of phenomenological coefficients of linear phenomenological equations $L_{ik} = L_{ki}$; it follows from Onsager reciprocal relations that each cross effect (e.g. determined by coefficient L_{ik}) has reciprocal cross effect (determined by coefficient L_{ki}).
D Onsagersche Reziprozitätsbeziehungen
F relations de réciprocité d'Onsager
P relacje Onsagera
R соотношения взаимности Онзагера

ONSAGER RECIPROCITY RELATIONS. *See* ONSAGER RECIPROCAL RELATIONS.

2462
ONSAGER THEORY.
Theory of electric polarization in which Lorenz field is replaced by local (Onsager) field.
D Onsagersche Theorie
F théorie d'Onsager
P teoria Onsagera
R теория Онзагера

2463
OPEN SYSTEM.
System which can exchange energy and mass with surroundings.
D offenes System
F système ouvert
P układ otwarty
R открытая система

2464
OPERATIVE ERROR (personal error).
Error due to improper work of experimentalist.
D persönlicher Fehler
F erreur personnelle
P błąd osobisty
R субъективная ошибка

2465
OPENAUER OXIDATION.
Oxidation of primary and secondary alcohols to corresponding aldehydes or ketones, effected by carbonyl compounds in presence of aluminium butanolate, isopropanolate or phenolate, e.g.

$$RR'CHOH + CH_3COCH_3 \xrightarrow{[(CH_3)_2CHO]_3Al}$$
$$RCOR' + (CH_3)_2CHOH$$

anisaldehyde

This reaction is reversible (see also Meerwein-Ponndorf-Verley reduction).
D Oppenauer-Oxydation, Oppenauer Alkohol-Dehydrierung
F oxydation d'Oppenauer
P utlenianie Oppenauera
R окисление по Оппенауеру

2466
OPTICAL ACTIVITY.
Property of substance of rotation of plane of polarized light passing through it. Optical activity is caused by chirality of crystals or molecules. All chiral molecules exhibit optical activity in gaseous and liquid states, also in solutions.
D optische Aktivität
F activité optique
P czynność optyczna, aktywność optyczna
R оптическая активность

OPTICAL ANTIPODES. *See* ENANTIOMERS.

2467
OPTICAL AXIS.
Direction of optical isotropy in crystal. Beam of electromagnetic radiation polarized linearly, which runs in crystal in direction of optical axis has speed independent of orientation of electric vector *E*. Crystals may have one or two optical axes.
D optische Achse
F axe optique
P oś optyczna
R оптическая ось

OPTICAL DENSITY. *See* ABSORBANCE.

OPTICAL DENSITY. *See* BLACKENING.

2468
OPTICAL ELECTRONS.
Electrons occupying atomic valence shell orbitals. They determine character of atomic spectrum.
Remark: Term optical electrons commonly used for alkali metals.
D Leuchtelektronen, Valenzelektronen
F électrons optiques
P elektrony optyczne, elektrony świetlne
R валентные электроны

2469
OPTICAL EXALTATION.
Anomalous increase in molar refraction value of given compound with respect to theoretically calculated value on the basis of additivity.
D Exaltation der Molrefraktion
F exaltation
P egzaltacja optyczna
R оптическая экзальтация, экзальтация молекулярной рефракции

OPTICAL FILTER. *See* LIGHT FILTER.

2470
OPTICAL INTERFEROMETER.
Instrument for measurement of difference of refractive indices of investigated sample and standard from position of interference lines which appear due to difference of radiation velocities in these two media.
D (optisches) Interferometer
F interféromètre optique
P interferometr optyczny
R (оптический) интерферометр

2471
OPTICAL ISOMERISM.
Type of stereoisomerism caused by chirality of molecules, giving them optical activity. Optical isomers form two main groups: enantiomers (these are mutual

mirror images) and diastereoisomers (not mirror images). Diastereoisomers are sub-divided into epimers (including anomers) and meso-isomers, which are optically inactive due to internal compensation of rotation, caused by even number of chiral centres.

D optische Isomerie, Spiegelbildisomerie
F isomérie optique
P izomeria optyczna
R оптическая изомерия

2472
OPTICAL MODEL.
Model representing atomic nucleus as semitransparent optical medium. Optical model describes deflection of particles by the nucleus as process analogous to dispersion of light.

D optisches Modell, optisches Kernmodell
F modèle optique
P model optyczny
R оптическая модель

2473
OPTICAL PERMITTIVITY, ε_∞.
Electric permittivity at very high $(\nu > 10^{12}$ Hz) frequencies of alternating electric field.

D optische Dielektrizitätszahl
F permittivité optique
P przenikalność elektryczna optyczna
R оптическая диэлектрическая проницаемость

2474
OPTICAL PURITY.
Estimate of degree of contamination of enantiomer in terms of another, expressed in percent. When enantiomeric mixture contains 95% of (−)enantiomer and 5% of (+)enantiomer, the former is said to be 90% optically pure.

D optische Reinheit
F pureté optique
P czystość optyczna
R оптическая чистота

2475
OPTICAL ROTATORY DISPERSION, ORD.
Variation with wavelength of ability to rotate polarization plane of polarized light.

D optische Rotationsdispersion
F dispersion rotatoire
P dyspersja skręcalności optycznej, dyspersja rotacyjna
R дисперсия вращения

2476
OPTICAL SENSITIZATION (in photography).
Sensitization of photographic emulsion to light of longer wavelength than that which is absorbed by silver halide itself.

D Sensibilisierung
F sensibilisation
P sensybilizacja optyczna
R оптическая сенсибилизация, спектральная сенсибилизация

2477
OPTICAL SPECTRUM.
Most frequently denotes absorption or emission spectrum in visible or ultraviolet region; sometimes also used in the case of infrared spectra.
Remark: Because of its rather imprecise character this term is only rarely used.

D optisches Spektrum
F spectre optique
P widmo optyczne
R оптический спектр

2478
OPTICAL TRANSITION.
Radiative transition between two energy states of system accompanied by emission or absorption of light in visible or ultraviolet region.

D Elektronenübergang
F transition optique
P przejście optyczne
R оптический переход

2479
ORBITAL.
Wave function of electron in atom, molecule or solid state, computed by any one-electron approximation method.

D Orbital
F orbitale
P orbital
R орбиталь, орбитальная собственная функция

ORBITAL ANGULAR MOMENTUM. *See* ANGULAR MOMENTUM.

2480
ORBITAL ELECTRONS.
Electrons in extranuclear structure of atom.

D Hüllenelektronen
F électrons orbitaux, électrons planétaires
P elektrony orbitalne, elektrony atomowe
R орбитальные электроны

2481
ORBITAL MOMENT (of atom), $\boldsymbol{p}_L$.
Vector quantity, defined by

$$\boldsymbol{p}_L = \hbar \sqrt{L(L+1)}$$

where L = orbital quantum number. Orbital magnetic moment
$\boldsymbol{\mu}_L = \sqrt{L(L+1)}\,\mu_B$, corresponds to orbital moment, where μ_B = Bohr magneton.

D Bahndrehimpuls, orbitaler Drehimpuls
F moment orbital
P moment orbitalny atomu
R орбитальный момент

2482
ORBITAL QUANTUM NUMBER
(azimuthal quantum number), l.
Number which defines allowed quantized
values of orbital angular momentum of
system. For given value of principal
quantum number n, l assumes following
values 0, 1, ..., $n-1$.
D Bahndrehimpulsquantenzahl,
  Nebenquantenzahl, Azimutalquantenzahl
F nombre quantique azimutal, nombre
  quantique secondaire
P liczba kwantowa orbitalna, liczba
  kwantowa poboczna
R орбитальное квантовое число

2483
ORDERED SOLID SOLUTION.
Solid solution which exhibits long-range
order.
D geordnete Mischkristalle
F solution solide ordonnée
P roztwór stały uporządkowany
R упорядоченный твёрдый раствор

ORDER OF INCREASING TRANS
EFFECT. See TRANS EFFECT SERIES.

2484
ORDER OF INTERFERENCE.
Value of parameter $n = 1, 2, 3, ...$ in
formula which gives diffraction maxima
for diffraction grating
$n\lambda = a(\sin \beta - \sin \alpha)$
where λ = wavelength of incident light,
a = diffraction grating constant,
β = incidence angle, α = diffraction angle.
D Interferenz-Ordnungszahl,
  Beugungsordnung
F ordre d'interférence, ordre de réflexion
P rząd widma
R порядок дифракционного спектра

2485
ORDER OF REACTION (overall reaction
order).
Sum of exponents over concentrations of
reactants in kinetic equation.
D (kinetische) Reaktionsordnung,
  Gesamtordnung der Reaktion
F ordre de la réaction, ordre global de la
  réaction
P rząd reakcji
R (суммарный) порядок реакции

2486
ORDER WITH RESPECT TO
A PARTICULAR SUBSTANCE.
Order of reaction equal to exponent with
which concentration of this substance
appears in kinetic equation.
D (kinetische) Einzelordnung
F ordre partiel de la réaction
P rząd reakcji cząstkowy
R порядок реакции по данному веществу

2487
ORGANIC CHEMISTRY.
Branch of chemistry dealing with studies
on numerous carbon compounds,
excluding only a few discussed in
inorganic chemistry.
D organische Chemie
F chimie organique
P chemia organiczna
R органическая химия

ORGANIC ION EXCHANGER. See ION
EXCHANGE RESIN.

2488
ORGANIC SEMICONDUCTOR.
Solid organic compound in crystalline or
amorphous form, exhibiting properties of
semiconductor.
D organischer Halbleiter
F semiconducteur organique
P półprzewodnik organiczny
R органический полупроводник

2489
ORGANOGEL.
Gel in which organic liquid forms
dispersion medium.
D Organogel
F organogel
P organożel
R органогель

2490
ORGANOGENIC ELEMENT.
Fundamental element (C, H, O, N, S, P
etc.) playing the most important part in
structure of organic matter.
D Organogen
F ...
P pierwiastek organogeniczny
R органогенный элемент

2491
ORGANOLEPTIC ANALYSIS
(organoleptic test).
Investigation or identification of substance
through the senses, e.g. taste and smell.
D organoleptischer Test, Sinnenprüfung
F analyse organoleptique
P analiza organoleptyczna
R органолептический анализ

ORGANOLEPTIC TEST. See
ORGANOLEPTIC ANALYSIS.

2492
ORGANOSOL.
Sol in which organic liquid forms
dispersion medium.
D Organosol
F organosol
P organozol
R органозоль

2493
ORIENTATION EFFECT.
One of the components of intermolecular forces. Orientation effect leads to attractive forces between pair of molecules each one having permanent dipole or higher multipole moment.
D Orientierungseffekt
F effet d'orientation
P efekt orientacyjny
R ориентационный эффект

2494
ORIENTATION IN AROMATIC SUBSTITUTION.
Position taken by entering substituent depends on substituents already present in aromatic ring. Electron-donating substituents ($-NR_2$, $-\overline{O}H$, $-\overline{O}R$, alkyl) direct entering substituent to *ortho*- and *para*-positions; electron-withdrawing groups ($-\overset{\oplus}{N}R_3$, $-NO_2$, $-COOH$, $-SO_3H$) direct entering substituent to *meta*-position.
D Orientierung bei der aromatischen Substitution
F orientation de substitution aromatique
P orientacja w podstawieniu aromatycznym
R ориентация ароматического замещения

2495
ORIENTATION POLARIZABILITY, α_O.
Dielectric polarization (per molecule) arising in unit local field due to dipole orientation. According to Debye theory it is given by

$$\alpha_O = \frac{\mu^2}{3kT}$$

where: μ = permanent dipole moment, k = Boltzmann constant, T = thermodynamic temperature.
D ...
F polarisabilité d'orientation
P polaryzowalność orientacyjna
R ориентационная поляризуемость

2496
ORIENTATION POLARIZATION.
Dielectric polarization due to orientation of polar molecules or polar groups in molecules with some freedom of internal rotation.
D Orientierungspolarisation
F polarisation d'orientation
P polaryzacja orientacji, polaryzacja ustawienia
R ориентационная поляризация

2497
ORIENTATION POLARIZATION (MOLAR), P_O.
Measure of orientation polarization defined by

$$P_O = \frac{1}{3\varepsilon_0} N_A \alpha_O \quad \text{(in SI units)}$$

where N_A = Avogadro constant,
α_O = orientation polarizability,
ε_0 = electric permittivity of free space.
D Orientierungspolarisation
F polarisation d'orientation
P polaryzacja orientacji (molowa)
R ориентационная поляризация

2498
ORTHOGONALITY OF WAVE FUNCTIONS.
Two functions φ_1 and φ_2 are orthogonal if $\int \varphi_1^+ \varphi_2 d\tau = 0$, where + denotes complex conjugate, $d\tau$ = integration volume element and integration is in region where both functions are defined.
D Orthogonalität der Wellenfunktionen
F orthogonalité des fonctions d'onde
P ortogonalność funkcji falowych
R ортогональность волновых функций

2499
ORTHOKINETIC AGGREGATION (orthokinetic flocculation).
Coagulation that occurs due to collision of particles under influence of external force, e.g. gravitational force or centrifugal force in ultra-centrifuge.
D orthokinetische Koagulation
F coagulation orthocinétique
P koagulacja ortokinetyczna
R ортокинетическая коагуляция

ORTHOKINETIC FLOCCULATION. *See* ORTHOKINETIC AGGREGATION.

2500
ORTON REARRANGEMENT OF N-HALOGENOACYLAMIDES.
Conversion of N-chloroacylamides into mixture of *o*- and *p*-chloro-N-acylamides, effected by hydrochloric acid, e.g.

D Orton-Halogenwanderung, Orton-Umlagerung
F transposition d'Ortone
P przegrupowanie Ortona
R перегруппировка Ортона

2501
OSCILLATING CRYSTAL METHOD.
X-ray technique for determination of
crystal structure. X-rays are diffracted
from crystal oscillating through angle of
5° to 15°. Oscillating crystal method is
used in place of rotating crystal method
when elementary cell dimensions are
relatively large.
D Schwenkverfahren
F méthode d'oscillation, méthode
   mécanique d'oscillation du cristal
P metoda kryształu wahanego, metoda
   kryształu kołysanego, metoda kryształu
   oscylującego
R метод качания

2502
OSCILLATION PHOTOGRAPH.
Photograph of diffracted X-rays obtained
by oscillating crystal method.
D Schwenkaufnahme
F (photographie radioscopique obtenue par
   la méthode d'oscillation)
P dyfraktogram kryształu wahanego
R рентгенограмма качания

OSCILLATIONS. See VIBRATIONS.

OSCILLATION SPECTRUM. See
VIBRATION SPECTRUM.

2503
OSCILLATOR (in classical mechanics).
Mechanical system e.g. material point,
oscillating around equilibrium position.
D Oszillator
F oscillateur
P oscylator
R осциллятор

2504
OSCILLOGRAPHIC POLAROGRAPH.
Apparatus used for studies of dependence
of current on fast linear changes of
voltage applied to electrodes;
oscillographic polarography is also used
with relation $dE/dt = f(E)$,
where E = indicator electrode potential
and t = time.
D Kathodenstrahlpolarograph,
   oszillographischer Polarograph
F oscillopolarographe
P oscylopolarograf
R осциллографический полярограф

2505
OSCILLOMETRIC TITRATION (high
frequency titration).
Titration for which end point is
determined from changes of frequency
and intensity of current in special
oscillometric circuit.
D Hochfrequenztitration
F conductimétrie à haute fréquence

P miareczkowanie oscylometryczne,
   analiza w szybkozmiennym polu
   elektrycznym
R высокочастотное титрование

2506
OSCILLOPOLAROGRAPHY (cathode-ray
polarography).
Group of electrochemical methods using
an oscillograph to observe relationships:
$i-E$, $(dE/dt)-E$, or $E-t$, where
E = electrode potential, i = electrolysis
current, t = time of electrolysis.
D oszillographische Polarographie
F polarographie oscillographique
P oscylopolarografia
R осциллографическая полярография

2507
OSMIUM, Os.
Atomic number 76.
D Osmium
F osmium
P osm
R осмий

2508
OSMOMETER.
Apparatus for measurement of osmotic
pressure.
D Osmometer
F osmomètre
P osmometr
R осмометр

2509
OSMOSIS.
Diffusion of constituent from solution of
higher chemical potential to solution of
lower chemical potential, through
semi-permeable membrane.
D Osmose
F osmose
P osmoza
R осмос

2510
OSMOTIC COEFFICIENT, g.
Coefficient defined by formula

$$g = \frac{\Pi}{\Pi_{id}}$$

where Π = osmotic pressure in real
solution, Π_{id} = osmotic pressure in ideal
solution of the same concentration as real
solution at the same temperature.
D osmotischer Koeffizient
F coefficient d'osmose
P współczynnik osmotyczny
R осмотический коэффициент

2511
OSMOTIC PRESSURE.
Difference of pressures necessary for
thermodynamic equilibrium of two phases

with different concentrations of permeable component across their interface (semi-permeable membrane).

D osmotischer Druck
F pression osmotique
P ciśnienie osmotyczne
R осмотическое давление

2512
OSMOTIC SCALE.
Device for gravimetric determination of height of rise of liquid column in osmometer.

D osmotische Waage
F balance osmotique
P waga osmotyczna
R осмотические весы

2513
OSTWALD'S DILUTION LAW.
Formula which describes classical electrolytic dissociation constant K_c

$$K_c = \frac{\Lambda_c^2 c}{(\Lambda_0 - \Lambda_c)\Lambda_c}$$

where Λ_0 and Λ_c = equivalent conductivity of infinitely diluted electrolyte and electrolyte at concentration c respectively; Ostwald's dilution law applies approximately to diluted solutions of weak electrolytes.

D Ostwaldsches Verdünnungsgesetz
F loi de dilution d'Ostwald
P prawo rozcieńczeń Ostwalda
R закон разведения Оствальда

OUTER COORDINATION SPHERE. *See* SECOND COORDINATION SPHERE.

2514
OUTER ELECTRIC POTENTIAL OF PHASE, ψ.
Electric potential between point separated 10^{-4} cm from surface of a given phase and point at infinity.

D äußeres elektrisches Potential der Phase
F potentiel électrique extérieur d'une phase
P potencjał zewnętrzny fazy
R внешний потенциал (фазы)

2515
OUTER-ORBITAL COMPLEX.
Complex in which, in hybridization of orbitals of central ion, d orbitals have the same principal quantum number as s and p orbitals; in octahedral complex nsp^3d^2.

D outerer orbitaler Komplex
F complexe à orbitales externes
P kompleks zewnętrznoorbitalowy
R комплекс внешних орбиталей, внешнеорбитальный комплекс

OUTER SPHERE COMPLEX. *See* ION ASSOCIATION COMPLEX.

2516
OVERALL INSTABILITY CONSTANT (cumulative instability constant), K_n.
Equilibrium constant of dissociation reaction of complex

$$ML_n \rightleftharpoons M + nL$$

$$K_n = \frac{[M][L]^n}{[ML_n]}$$

where [M], [L] and $[ML_n]$ = equilibrium concentrations of metal, ligand and complex respectively. This constant is equal to product of consecutive instability constants.

D Bruttodissoziationskonstante (des Komplexes)
F constante d'instabilité globale
P stała nietrwałości całkowita, stała dysocjacji (kompleksu) całkowita
R общая константа нестойкости

OVERALL REACTION ORDER. *See* ORDER OF REACTION.

OVERALL STABILITY CONSTANT. *See* STABILITY CONSTANT.

2517
OVER-EXPOSURE REGION.
Curved part of blackening curve of photographic emulsion corresponding to region of excessive exposure.

D Überexpositionsgebiet
F région de surexposition
P obszar prześwietleń
R область передержек

2518
OVERLAP.
Two orbitals φ_1 and φ_2 have non-zero overlap if $\int \varphi_1^+ \varphi_2 d\tau \neq 0$,
where $+$ = complex conjugate and $d\tau$ = volume element of integration region where both functions are defined.

D Überlappung
F recouvrement des orbitales
P nakładanie się orbitali
R перекрывание орбиталей

2519
OVERPOTENTIAL (overvoltage), η.
Measure of electrolytic polarization defined by equation $\eta \equiv \varepsilon_i - \varepsilon$, where ε_i = potential of electrode (half-cell) across which there flows electrical current larger than exchange current, ε = potential of this electrode at equilibrium.

D Überspannung
F surtension
P nadnapięcie
R перенапряжение, сверхпотенциал

OVERTONE REGION. *See* NEAR INFRARED.

OVERVOLTAGE. *See* OVERPOTENTIAL.

2520
OXIDANT (oxidizing agent, oxidizer).
Substance that gains electrons from
another (oxidized) substance called
reducing agent; thus, oxidant is reduced
in electron transfer process.
D Oxydationsmittel, Oxydans
F oxydant, agent oxydant
P utleniacz, środek utleniający
R окислитель

2521
OXIDATION.
Reaction involving removal of electron
(or electrons) from molecule (atom or
ion); in narrower sense: process of
enrichment of compound by adding
oxygen to it, or its impoverishment by
removal of hydrogen.
D Oxydation
F oxydation
P utlenianie
R окисление

2522
OXIDATION NUMBER (oxidation state).
Number of electrons accepted or donated
by atoms during formation of molecule,
provided all bonds are ionic, e.g.
$\overset{+1+6-2}{K_2SO_4}$.
D Oxydationsstufe, Oxydationszahl,
 Oxydationswert, Ladungswert,
 elektrochemische Wertigkeit
F degré d'oxydation, état d'oxydation,
 nombre d'oxydation
P stopień utlenienia, liczba utlenienia
R степень окисления, окислительное
 число, электрохимическая валентность

2523
OXIDATION-REDUCTION BUFFER.
Solution which contains redox couple in
a such concentration that its redox
potential changes only slightly after
addition of small quantities of oxidant
(reductant).
D...
F solution-tampon redox
P bufor redoks
R...

2524
OXIDATION-REDUCTION CATALYSIS.
Catalysis by oxidizing or reducing
substances. In this type of catalysis,
reaction of catalyst with substrate is
connected with transfer of electron.
D Redoxkatalyse
F catalyse d'oxydo-réduction
P kataliza utleniająco-redukująca
R окислительно-восстановительный
 катализ

2525
OXIDATION-REDUCTION TITRATION
(redox titration).
Volumetric determination of oxidants or
reductants present in solution.
D Oxydations-Reduktions-Titration,
 Redoxtitration
F titrage par oxydo-réduction
P miareczkowanie redoks, miareczkowanie
 redoksymetryczne
R оксидиметрическое титрование

OXIDATION STATE. See OXIDATION
NUMBER.

2526
OXIDATION STATE OF CENTRAL ION.
Conventional charge of central ion in
complex compound calculated according
to definite rules.
D Oxydationszahl des Zentralions,
 Oxydationsstufe des Zentralions
F état d'oxydation du ion central
P stopień utlenienia jonu centralnego
R степень окисления центрального иона

2527
OXIDATIVE FERMENTATION.
Decomposition of hexoses to compounds
of simpler structure, occurring under
action of bacteria, in which initial
degradation of carbohydrate is similar to
alcoholic fermentation; final stage leads,
however, to products of higher oxidation
number.
D...
F fermentation oxydative
P fermentacja tlenowa
R...

OXIDIMETRIC TITRATION. See
OXIDIMETRY.

2528
OXIDIMETRY (oxidimetric titration).
Determination of reducer in solution by
oxidants.
D Oxydimetrie
F oxydimétrie
P oksydymetria, miareczkowanie
 oksydymetryczne
R оксидиметрия

OXIDIZER. See OXIDANT.
OXIDIZING AGENT. See OXIDANT.

2529
OXO COMPLEX.
Complex in which ligand is (formally)
O^{2-} anion.
D Oxokomplex
F complexe oxo
P oksokompleks
R оксокомплекс

OXO PROCESS. *See*
HYDROFORMYLATION.

2530
OXYGEN, O.
Atomic number 8.
D Sauerstoff
F oxygène
P tlen
R кислород

OXYGEN-BOMB CALORIMETER. *See*
BOMB CALORIMETER.

2531
OXYGEN ELECTRODE.
Electrode (half-cell) consisting of metallic
conductor immersed in solution of oxygen
and hydroxyl ions, $M,O_2|OH^-$; electrode
reaction is

$$4\,OH^- \rightleftarrows O_2 + 2\,H_2O + 4e$$

and its potential

$$\varepsilon = \varepsilon^0 + \frac{RT}{F}\ln\frac{a_{pO_2}^{1/4}}{a_{OH^-}}$$

where ε^0 = standard electrode potential
and a_{pO_2} = volatility of oxygen.
D Sauerstoffelektrode
F électrode à oxygène
P elektroda tlenowa
R кислородный электрод

OZONATION. *See* OZONIZATION.

2532
OZONIZATION (ozonation).
Addition of ozone to unsaturated organic
compound, resulting in formation of
ozonide, e.g.

$$R_2C{=}CR_2 \xrightarrow{O_2} R_2C\!\!\!\overset{O}{\underset{O\text{---}O}{\diagup\diagdown}}\!\!\!CR_2$$

D Ozonisation, Ozonisierung
F ozon(is)ation
P ozonowanie
R озонирование

2533
OZONOLYSIS.
Degradation (fission) of alkene molecule
with formation of carbonyl compounds,
effected by ozone and water or any other
reducing agent, e.g.

$$R_2C{=}CR_2 \xrightarrow{O_2} R_2C\!\!\!\overset{O}{\underset{O\text{---}O}{\diagup\diagdown}}\!\!\!CR_2 \xrightarrow[-H_2O_2]{H_2O} 2R_2C{=}O$$

D Ozonolyse, Ozonspaltung
F ozonolyse
P ozonoliza
R озонолиз

2534
PACKING (*in chromatography*).
Stationary phase placed in column.
D Füllmaterial
F matériel de remplissage
P wypełnienie kolumny
R насадка, набивка

PAIR CREATION. *See* PAIR
PRODUCTION.

2535
PAIRED ELECTRONS.
Two electrons with different spin
functions occupying simultaneously the
same atomic or molecular orbital.
D gepaarte Elektronen
F électrons appariés, électrons couplés
P elektrony sparowane
R спаренные электроны

PAIRING CORRELATIONS MODEL. *See*
SUPERCONDUCTIVITY MODEL.

2536
PAIRING ENERGY, *Π*.
In coordination chemistry, energy
necessary for greatest possible pairing of
d electrons of central atom.
D Paarungsenergie, Spinpaarungsenergie
F énergie de parité
P energia sparowania
R энергия спаривания

2537
PAIR PEAKS.
All the escape peaks generated through
annihilation of a positron produced during
formation of pairs.
D Paarbildungspeaks, Paarlinien
F pics de paire
P piki par
R пики от пар

2538
PAIR PRODUCTION (pair creation).
Formation of particle-antiparticle pair.
Process opposite to annihilation, which
may occur during collision of particles
and lead to formation of
baryon-antibaryon or lepton-antilepton
pair.
D Paarerzeugung
F production de paires
P tworzenie pary, kreacja pary
R рождение пары (частиц), образование
пары

2539
PALLADIUM, Pd.
Atomic number 46.
D Palladium
F palladium
P pallad
R палладий

2540
PANETH AND HEVESY METHOD.
Method of determination of critical potential of deposition or decomposition voltage during electrolysis of extremally diluted solutions based on direct measurement of rate of formation of substance at electrode.
D Hevesy-Paneth-Methode
F méthode de Paneth-Hevesy
P metoda Hevesy'ego i Panetha
R метод Хевеши-Панета

2541
PAPER CHROMATOGRAPHY.
Chromatography in which stationary phase consists of, or contains filter paper.
D Papierchromatographie
F chromatographie sur papier
P chromatografia bibułowa, chromatografia na bibule
R хроматография на бумаге, бумажная хроматография

2542
PARACHOR, P.
Property defined by formula:

$$P = MC^{1/4} = \left(\frac{M}{\rho_l - \rho_v} \right) \sigma^{1/4} = V_m \sigma^{1/4}$$

where M = molar mass, σ = surface tension, ρ_l = density of liquid, ρ_v = density of vapour, V_m = molar volume of liquid, C = characteristic constant independent of temperature for substances not undergoing association. Constant C for associating liquids increases with increase of temperature. Parachor is additive quantity, i.e. it is sum of atomic bond and ring parachors.
D Parachor
F parachor
P parachora
R парахор

PARACHOR EQUIVALENT. *See* ATOMIC PARACHOR.

2543
PARACYCLOPHANES.
Carbocyclic compounds containing two benzene rings linked at *para*-positions by saturated carbon chains, thus forming ring system of general formula (I), e.g. bi-*p*-xylylene (II)

D ...
F cyclophanes

P paracyklofany
R парациклофаны

2544
PARALLAX ERROR.
Error due to incorrect reading of liquid level because of improper positioning of eye with respect to liquid meniscus.
D Parallaxenfehler, Meniskusfehler
F erreur de parallaxe
P błąd paralaksy
R ошибка от параллакса

2545
PARALLEL BAND.
In vibration spectra band corresponding to vibration with transition dipole moment parallel to molecular symmetry axis.
D Parallel-Band
F bande du type parallèle
P pasmo równoległe, pasmo typu równoległego
R параллельная полоса

2546
PARALLEL REACTIONS (simultaneous reactions, competing reactions).
System of two or more elementary reactions proceeding simultaneously with one or more common substrates which are changed into different products, e.g.

D Parallelreaktionen
F réactions parallèles, réactions simultanées
P reakcje równoległe, reakcje współbieżne
R параллельные реакции

PARAMAGNETIC BODIES. *See* PARAMAGNETS.

2547
PARAMAGNETIC CURIE POINT (asymptotic Curie point), Θ.
Constant, in kelvins, entering Curie-Weiss law, positive for ferromagnets and, in general, negative for antiferromagnets, provided they satisfy Curie-Weiss law above Curie or Néel temperature, respectively.
D paramagnetische Curie-Temperatur
F point de Curie paramagnétique
P temperatura Curie paramagnetyczna, temperatura Curie asymptotyczna
R парамагнитная точка Кюри

PARAMAGNETIC RESONANCE. *See* MAGNETIC RESONANCE.

PARAMAGNETIC SCREENING OF NUCLEUS. *See* PARAMAGNETIC SHIELDING OF NUCLEUS.

2548
PARAMAGNETIC SHIELDING OF
NUCLEUS (paramagnetic screening of
nucleus).
Participation of electrons in magnetic
shielding of nucleus leading to chemical
shifts of nuclear magnetic resonance in
direction of lower static fields or higher
frequency alternating magnetic fields.
D paramagnetische Abschirmung des
   Kerns
F écran paramagnétique du noyau
P ekranowanie jądra paramagnetyczne
R парамагнитное экранирование ядра

2549
PARAMAGNETIC SUSCEPTIBILITY, χ.
Positive contribution to magnetic
susceptibility of paramagnets.
In alternating magnetic field
paramagnetic susceptibility becomes
complex quantity $\chi = \chi' + i\chi''$, where
χ' = real part, χ'' = imaginary part.
D paramagnetische Suszeptibilität
F susceptibilité paramagnétique
P podatność paramagnetyczna
R парамагнитная восприимчивость

2550
PARAMAGNETISM.
Physical phenomena in and properties of
paramagnets.
D Paramagnetismus
F paramagnétisme
P paramagnetyzm
R парамагнетизм

2551
PARAMAGNETS (paramagnetic bodies).
Solids, liquids or gases of positive value
of magnetic susceptibility, and magnetic
permeability slightly larger than 1. They
are divided into normal paramagnets,
whose magnetic susceptibility satisfies
Curie-Weiss law, and metallic
paramagnets, whose magnetic
susceptibilities slightly increase or
decrease with temperature.
D Paramagnetika, paramagnetische
   Körper
F paramagnétiques, corps
   paramagnétiques
P paramagnetyki, ciała paramagnetyczne,
   substancje paramagnetyczne
R парамагнетики, парамагнитные тела

2552
PARAMETER OF POPULATION.
Characteristic function describing
population.
D statistische Kennzahl
F paramètre (de population)

P parametr (populacji)
R параметр

2553
PARASITIC FERROMAGNETISM.
Weak ferromagnetism of antiferromagnets
which display small deviations from
colinearity of magnetic moments of
sublattices.
D (schwacher Ferromagnetismus in
   antiferromagnetischen Substanzen)
F ferromagnétisme parasite
P ferromagnetyzm pasożytniczy
R (слабый ферромагнетизм
   антиферромагнетиков)

PARENT CHAIN. See MAIN CHAIN.

2554
PARENT ION.
Positive ion formed by detachment of
electron in interaction of ionizing
radiation with molecule or atom.
D primäres Ion
F ...
P jon macierzysty
R ...

2555
PARENT NUCLIDE (nuclear parent).
Radioactive nuclide from which another
nuclide is formed by decay.
D Mutternuklid
F père nucléaire, nucléide père
P nuklid macierzysty
R материнский нуклид

2556
PARITY, P.
Quantum number ascribed to particles of
integral quantum spin number (photons
and mesons), which gives behaviour of
wave function for particle undergoing
coordinate inversion. If wave function
does not change, $P = 1$, but if sign is
changed, $P = -1$.
D Parität
F parité
P parzystość
R чётность

2557
PARITY CONSERVATION LAW.
In strong and electromagnetic
interactions, parities of system of initial
and final particles in given process are
equal. Parity conservation law is not
observed in weak interactions.
D Erhaltungssatz der Parität
F loi de conservation de la parité
P prawo zachowania parzystości
R закон сохранения чётности

2558
PARTIAL HEAT OF DILUTION
(differential heat of dilution).
Enthalpy change of system in process of adding one mole of solvent to such a large bulk of solution of a given concentration that it may be assumed in practice that its concentration remains unaffected.

D differentielle Verdünnungswärme
F chaleur molaire partielle de dilution
P ciepło rozcieńczania cząstkowe (molowe), ciepło rozcieńczania różniczkowe
R дифференциальная теплота разведения, парциальная теплота разведения

2559
PARTIAL HEAT OF SOLUTION
(differential heat of solution).
Enthalpy change of system in process of dissolving one mole of a given substance in such a large bulk of solution of a given concentration that it may be assumed in practice that its concentration remains unaffected.

D differentielle Lösungswärme
F chaleur molaire partielle de dissolution
P ciepło rozpuszczania cząstkowe (molowe), ciepło rozpuszczania różniczkowe
R дифференциальная теплота растворения, парциальная теплота растворения

2560
PARTIAL HEAT OF SWELLING.
Amount of heat released during the imbibition of unit mass of liquid by large quantity of dry or partially swollen gel.

D differentielle Quellungswärme
F ...
P ciepło pęcznienia cząstkowe
R дифференциальная теплота набухания

2561
PARTIALLY ECLIPSED CONFORMATION (partly eclipsed conformation, anti-clinal conformation), ± ac.
Conformation in which the most bulky substituents at two adjacent carbon (or any other element, e.g. silicium or nitrogen) atoms are eclipsed by smaller substituents. In Newman's projection formula, bond angles of less remote and more remote substituents equal 0°. This type of conformation is less stable than fully staggered conformation, but more stable than eclipsed

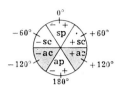

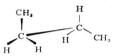

D teilweise verdeckte Konformation, anti-clinale Konformation
F conformation partiellement éclipsée
P konformacja częściowo ekliptyczna, konformacja ± anty-klinalna
R частично заслонённая конформация, анти-клинальная конформация

2562
PARTIAL MISCIBILITY (limited compatibility).
Ability of mixture to undergo mixing in limited proportion.

D begrenzte Mischbarkeit, teilweise Mischbarkeit
F miscibilité limitée
P mieszalność ograniczona
R ограниченная смешиваемость, ограниченная взаимная растворимость

PARTIAL MOLAL QUANTITIES. *See* PARTIAL MOLAR QUANTITIES.

2563
PARTIAL MOLAL VOLUME (*of component* B), V_B.
Intensive quantity, defined by

$$V_B = \left(\frac{\partial V}{\partial n_B} \right)_{T,\,p,\,n_C}$$

where n_B = number of moles of component B in a given phase, n_C = number of moles of other components in this phase, V = phase volume, T = thermodynamic temperature, and p = pressure.

D partielles Molvolumen.
F volume molaire partiel
P objętość cząstkowa molowa
R парциальный мольный объём

2564
PARTIAL MOLAR QUANTITIES (partial molal quantities *of component* B), Z_B.
Intensive quantities defined by

$$Z_B = \left(\frac{\partial Z}{\partial n_B} \right)_{T,\,p,\,n_C}$$

where Z = extensive function of state of a given phase, n_B = number of moles of component B in this phase, n_C = number of moles of all remaining components in this phase, T = thermodynamic temperature, p = pressure.

D partielle molare Größen, partielle molare Zustandsfunktionen
F quantités partielles molaires, grandeurs molaires partielles, propriétés molaires partielles
P wielkości cząstkowe molowe
R парциальные мольные величины

2565
PARTIAL PRESSURE, p_B.
Intensive thermodynamic function $p_B = x_B p$, where x_B = mole fraction of component B and p = pressure in a given phase.

D Partialdruck
F pression partielle
P ciśnienie cząstkowe, prężność cząstkowa, ciśnienie parcjalne
R парциальное давление

2566
PARTICLE ENERGY.
Kinetic energy of fast particle (e.g. in beam of ionizing radiation), expressed in electron volts and in SI system in joules.

D Teilchenenergie
F énergie de la particule
P energia cząstki
R энергия частицы

PARTICLE FLUENCE. *See* FLUENCE.

2567
PARTICLE NUMBER DENSITY (average density).
Average number of particles in element of unit volume V, $n = \dfrac{\langle N \rangle}{V}$

D Teilchendichte, Moleküldichte, molekulare Dichte
F densité des particules
P gęstość cząstek
R плотность молекул, средняя плотность числа частиц

2568
PARTICLE PRODUCTION.
Process occurring during collision of high energy particles and leading to formation of other particles.

D Erzeugung der Teilchen
F production des particules
P produkcja cząstek
R рождение частиц

PARTICLE RADIATION. *See* CORPUSCULAR RADIATION.

2569
PARTICLE SIZE DETERMINATION.
Determination of sizes of small particles and grains of substances as well as determination of quantitative ratios of individual fractions.

D Korngrößen-Bestimmung
F analyse granulométrique, granulométrie
P analiza mechaniczna
R механический анализ, гранулометрический анализ

2570
PARTICLE TRAJECTORY.
Path along which particle changes its position during movement.

D Teilchenbahn
F trajectoire
P tor cząstki
R траектория частицы

2571
PARTITION CHROMATOGRAPHY.
Chromatography employing stationary phase, which consists of thin layer of liquid retained on solid support.

D Verteilungschromatographie
F chromatographie de partage
P chromatografia podziałowa
R распределительная хроматография

PARTITION COEFFICIENT. *See* DISTRIBUTION COEFFICIENT.

2572
PARTITION FUNCTION (configurational partition function, configurational integral), $Q(T, V, N)$.
Function used to express thermodynamic quantities in terms of intermolecular interactions

$$Q(T, V, N) = \int \exp\left[-\frac{1}{kT} \sum_{i > j} \varepsilon(r_{ij}) \right] \prod_{i=1}^{N} d^3 \boldsymbol{x}_i$$

where k = Boltzmann constant, T = thermodynamic temperature of N-particle system in equilibrium in volume V, $d^3 x$ = volume element, r_{ij} = distance between i-th and j-th particle, and $\varepsilon(r)$ = intermolecular potential.

D Konfigurationsintegral, Verteilungsfunktion der potentiellen Energie
F fonction de partition potentielle
P całka konfiguracyjna, funkcja rozdziału konfiguracyjna
R конфигурационный интеграл, интеграл взаимодействия

PARTLY ECLIPSED CONFORMATION. *See* PARTIALLY ECLIPSED CONFORMATION.

2573
PASCHEN-BACK EFFECT.
Splitting of energy levels in very strong
magnetic fields with mutually
independent orientation of angular and
spin momenta with respect to magnetic
field.
D Paschen-Back Effekt
F effet Paschen-Back
P efekt Paschena i Backa
R явление Пашена-Бака

2574
PASSERINI REACTION.
Preparation of α-acyloxycarboxylic acid
N-arylamides from aryl isocyanates,
aldehydes or carboxylic acids, e.g.

$$
\begin{array}{l}
\text{R}\\
\quad\searrow\\
\qquad \text{C=O + Ar−N≡C + R''COOH} \longrightarrow\\
\text{R'}\nearrow
\end{array}
$$

$$
\begin{array}{c}
\text{R}\\
|\\
\text{R'−C−C−NHAr}\\
|\ \ ||\\
\text{O O}\\
|\\
\text{R''C=O}
\end{array}
$$

$$
\left[\xrightarrow{\text{H}_2\text{O}} \begin{array}{c}\text{R}\\|\\ \text{R'−C−C−NH−Ar + R''COOH}\\ |\ \ ||\\ \text{HO O}\end{array}\right]
$$

D Passerini (α-Hydroxy-*N*-
-Arylamid)-Synthese
F réaction de Passerini
P reakcja Passeriniego
R реакция Пассерини

2575
PASSIVITY OF METAL.
State of metal characterized by practically
complete inhibition of reaction of this
metal with corroding medium, e.g. of iron
and nickel with concentrated nitric acid,
aluminium with air and some metals
anodically dissolved in certain acids etc.
D Passivität des Metalls
F passivité du métal
P pasywność metalu
R пассивность металла

2576
PAULI EXCLUSION PRINCIPLE
(exclusion principle).
In system of identical fermions each
quantum mechanical state can be occupied
at most by single fermion.
D Ausschliessungsprinzip, Pauli-Prinzip,
Pauli-Verbot
F principe (d'exlusion) de Pauli
P zakaz Pauliego, zasada Pauliego
R принцип Паули, принцип исключения

2577
PAULING'S ELECTRONEGATIVITY
SCALE.
Ordering of atoms according to increasing
electronegativity computed from Pauling's
definition.
D Paulingsche Elektronegativitätsskala
der Elemente
F échelle d'électronégativité de Pauling
P skala elektroujemności Paulinga
R шкала электроотрицательности
Полинга

2578
P-BRANCH.
In rotation-oscillation band, set of
transitions corresponding to change
of rotational quantum number by −1.
D P-Zweig
F branche P
P gałąź P, gałąź ujemna
R отрицательная ветвь P

2579
PEAK.
Maximum on curve (e.g. elution curve).
D Peak, Pik
F pic
P pik
R пик

2580
PEAK RESOLUTION (resolution *in
chromatography*), R_s.
Quantity describing degree of separation
of two chromatographic peaks.
D Pikauflösung
F résolution de pics
P zdolność rozdzielcza
R степень разделения, разделение

2581
PECHMANN-DUISBERG COUMARIN
SYNTHESIS.
Preparation of coumarine and its
derivatives by condensation of phenols
with apple acid or β-ketoesters, in
presence of dehydrating agents (sulfuric
acid, aluminium chloride, phosphorus
pentoxide etc.), e.g.

$$
\begin{array}{c}
\text{OH} \qquad \text{COOC}_2\text{H}_5\\
\bigcirc \ + \ \underset{\text{COCH}_3}{|}\,\text{CH}_2 \longrightarrow
\end{array}
\begin{array}{c}
\text{OH}\\
\bigcirc\text{OH}\\
\text{C−CH}_2\text{−COOC}_2\text{H}_5\\
|\\
\text{CH}_3
\end{array}
$$

$$
\xrightarrow[-\text{C}_2\text{H}_5\text{OH}]{-\text{H}_2\text{O}}
\begin{array}{c}
\text{O}\ \ \text{O}\\
\bigcirc\\
\text{CH}_3
\end{array}
$$

D Pechmann-Synthese (von Cumarin),
Pechmann-Duisberg Cumarin-
-Ringschluß

F condensation de Pechmann
P synteza (pochodnych) kumaryny
  Pechmanna, kondensacja Pechmanna
R конденсация Пехмана

PEHAMETER. See pH METER.

PEHAMETRY. See pH METRY.

2582
PENETRATION COMPLEX.
Complex of generally covalent character
which does not decompose in solution, e.g.
$[Fe(CN)_6]^{4-}$, $[Co(NH_3)_6]^{3+}$.
D Durchdringungskomplex
F complexe de pénétration
P kompleks penetracyjny
R неравновесный комплекс, комплекс
  проникновения

2583
PEPTIZATION.
Conversion of gel or freshly precipitated
colloidal precipitate into sol, e.g. under
influence of washing with pure solvent or
with solution of appropriate electrolyte.
D Peptisation
F peptisation
P peptyzacja
R пептизация

2584
PERCENTAGE CONCENTRATION.
Number of units of mass, volume or
quantity of substance per 100 respective
units of mixture. Mass-, volume-, and
molar- percentage concentrations may be
distinguished, depending on units chosen.
D...
F pourcentage (de concentration)
P stężenie procentowe
R процентная концентрация

PERCENTAGE CONVERSION. See
CONVERSION.

2585
PERCENT ERROR.
Relative error in per cent.
D prozentualer Fehler
F erreur pour-cent
P błąd procentowy
R процентная ошибка

2586
PERFECT CRYSTAL.
Crystal with ideal periodic lattice.
D idealer Kristall, Idealkristall
F cristal parfait, cristal idéal
P kryształ doskonały
R идеальный кристалл, совершенный
  кристалл

PERFECT GAS. See IDEAL GAS.

2587
PERFECT GASEOUS MIXTURE
(multicomponent ideal gas).
Mixture in which chemical potential of
each component is equal to chemical
potential of this component in state of
pure ideal gas at the same temperature
and pressure as mixture.
D ideales Gasgemisch, Gemisch idealer
  Gase
F mélange idéal des gaz parfaits
P mieszanina gazowa doskonała
R идеальная газовая смесь, смесь
  идеальных газов

PERFECT GAS EQUATION OF STATE.
See IDEAL GAS EQUATION OF STATE.

2588
PERFECT GAS LAWS.
Empirical laws established for so-called
stable gases by Boyle and Mariotte,
Gay-Lussac, Charles, Avogadro, Dalton,
and Amagat; set of these laws describes
properties of perfect gases.
D ideale Gasgesetze
F lois des gaz parfaits
P prawa gazów doskonałych
R законы идеальных газов

2589
PERIKINETIC AGGREGATION
(perikinetic flocculation).
Coagulation that occurs due to collision of
particles caused only by Brownian
movement.
D perikinetische Koagulation
F coagulation péricinétique
P koagulacja perikinetyczna
R перикинетическая коагуляция

PERIKINETIC FLOCCULATION. See
PERIKINETIC AGGREGATION.

2590
PERIOD.
In periodic system, horizontal series of
elements showing regular changes in
properties; ordinal number of period
corresponds to particular value of main
quantum number; in conventional periodic
system period begins with an alkali metal
and ends with a noble gas.
D Periode
F période
P okres
R период

PERIODIC CLASSIFICATION. See
PERIODIC SYSTEM.

PERIODIC GROUP. See GROUP OF
ELEMENTS.

2591
PERIODIC LAW.
In its present form it states that physical and chemical properties of elements are not accidental but depend on structure of atom and regularly change with increasing atomic number.
D Gesetz der Periodizität
F loi périodique
P prawo okresowości
R периодический закон

2592
PERIODIC SYSTEM (periodic classification).
Arrangement of elements by increasing atomic number in successive horizontal sequences, lying under each other, called periods. By this arrangement, elements of like properties are simultaneously ordered into vertical sequences called groups.
D Periodensystem, natürliches System der Elemente
F classification périodique des éléments, système périodique
P układ okresowy pierwiastków
R периодическая система элементов

2593
PERITECTIC POINT.
Incongruent melting point identical with discontinuity on liquidus curve.
D peritektischer Punkt, Peritektikum
F point péritectique
P punkt perytektyczny, perytektyk
R перитектическая точка, перитектика

2594
PERITECTIC PROCESS.
Process of melting of polyconstituent phase in which both liquid phase and another solid phase are formed; new phases are of different chemical composition.
D peritektischer Vorgang
F processus péritectique
P proces perytektyczny
R перитектический процесс

2595
PERKIN REACTION.
Aldol-type condensation of aromatic aldehyde with carboxylic acid anhydride containing hydrogen atoms at α-carbon, in presence of base, resulting in formation of α, β-unsaturated acid, e.g.

$$C_6H_5CHO + (CH_3CO)_2O \xrightarrow{CH_3COOK}$$

$$C_6H_5CH{=}CHCOOH + CH_3COOH$$

D Perkin-Reaktion, Perkin Zimtsäure-Synthese, Perkin Aldol-Kondensation

F réaction de Perkin
P synteza Perkina
R реакция Перкина

2596
PERKOV REACTION.
Preparation of O,O-dialkyl-O-vinylphosphoranes by treating α-haloaldehydes, α-haloketones or α-haloesters with trialkyl phosphites, e.g.

$$(CH_3O)_3P + ClCH_2CHO \rightarrow (CH_3O)_3\overset{\oplus}{P}{-}CH{-}CH_2{-}Cl$$

$$\underset{O^{\ominus}}{|}$$

$$(CH_3O)_2P{-}O{-}CH{=}CH_2$$

$$\underset{O}{\|}$$

D Perkow (Trialkylphosphit→ →Vinylphosphat)-Umwandlung
F réaction de Perkov
P reakcja Perkowa
R реакция Перкова

2597
PERMANENT DIPOLE.
Electric dipole existing independently of presence of external electric field.
D permanenter Dipol
F dipôle permanent
P dipol trwały
R постоянный диполь

2598
PERMANENT DIPOLE MOMENT.
Dipole moment existing independently of presence of external electric field.
D permanentes Dipolmoment
F moment dipolaire permanent
P moment dipolowy trwały
R постоянный дипольный момент

2599
PERMITTIVITY (electric permittivity, dielectric permittivity), ε.
Ratio of dielectric shift and electric field strength in dielectric or ratio of capacity of condenser filled with given dielectric to that of empty one.
D (absolute) Dielektrizitätskonstante, Dielektrizitätszahl
F permittivité
P przenikalność elektryczna (bezwzględna), przenikalność dielektryczna
R диэлектрическая проницаемость

2600
PERMITTIVITY OF FREE SPACE
(permittivity of vacuum), ε_0.
In SI units $\varepsilon_0 = 8.855 \cdot 10^{-12}$ F/m (previously
in CGS electrostatic units $\varepsilon_0 = 1$).

D Dielektrizitätskonstante des Vakuums,
Dielektrizitätszahl des Vakuums
F permittivité du vide
P przenikalność elektryczna próżni
(bezwzględna)
R диэлектрическая проницаемость
вакуума

PERMITTIVITY OF VACUUM. *See*
PERMITTIVITY OF FREE SPACE.

2601
PERPENDICULAR BAND.
In vibration spectra band corresponding
to rotation-vibration transition with
transition dipole moment perpendicular to
molecular symmetry axis.

D senkrechtes Band
F bande du type perpendiculaire
P pasmo prostopadłe, pasmo typu
prostopadłego
R перпендикулярная полоса

2602
PERPETUUM MOBILE OF THE FIRST
KIND.
Heat engine which could perform cyclic
process and could do work during this
process without accepting any energy;
such an engine cannot exist according to
the first law of thermodynamics.

D Perpetuum mobile erster Art
F mouvement perpétuel de première
espèce
P perpetuum mobile pierwszego rodzaju
R вечный двигатель первого рода

2603
PERPETUUM MOBILE OF THE SECOND
KIND.
Heat engine which could perform cyclic
process with energy accepted in form of
heat from bath of constant (positive)
thermodynamic temperature and returned
quantitatively in form of work; such an
engine cannot exist according to the
second law of thermodynamics.

D Perpetuum mobile zweiter Art
F mouvement perpétuel de seconde espèce
P perpetuum mobile drugiego rodzaju
R вечный двигатель второго рода

PERSISTENT LINES. *See* RAIES
ULTIMES.

PERSONAL ERROR. *See* OPERATIVE
ERROR.

PERSONNEL DOSE MONITOR. *See*
PERSONNEL DOSIMETER.

2604
PERSONNEL DOSIMETER (personnel
dose monitor).
Dosimeter (e.g. film dosimeter) carried by
persons being in danger of ionizing
radiation.
D Personendosimeter
F dosimètre individuel
P dawkomierz osobisty, dawkomierz
indywidualny
R индивидуальный дозиметр

2605
PERSPECTIVE FORMULA.
Formula portraying three-dimentional
model of molecule as simplified
perspective drawing with all interatomic
bonds and distances, e.g. formula of chair
or boat forms of cyclic compounds, or
sawhorse formula of open-chain
compounds.
D perspektivische Strukturformel
F ...
P wzór perspektywiczny
R перспективная формула

PERTURBATION METHOD. *See*
PERTURBATION THEORY.

2606
PERTURBATION THEORY (perturbation
method).
Method used to compute small
corrections to properties of system
arising from small perturbation.
D Störungsrechnung
F méthode des perturbations, théorie des
perturbations
P metoda perturbacyjna, rachunek
zaburzeń
R метод возмущений

2607
PFITZINGER REACTION.
Synthesis of cinchonic acid derivatives
from isatin (or isatinic acid) and
compounds containing —CH_2—CO—
group, e.g.

D Pfitzinger Reaktion, Pfitzinger
Chinolin-Ringschluß
F réaction de Pfitzinger
P synteza (pochodnych chinoliny)
Pfitzingera
R реакция Пфитцингера

2608
pH (hydrogen ion exponent, power hydrogen).
Negative logarithm of hydrogen-ion activity, $pH \equiv -lg\ a_{H_3O^+}$, used for convenience, since individual-ion activities cannot be measured, as a scale based on standard buffer solutions and EMF's measured with electrodes reversible with respect to hydrogen ions

$$pH(x) = pH(s) + \frac{(E_x - E_s)F}{2.303\ RT}$$

pH(s) and pH(x) being the pH's of the standard buffer and the test solution and E_s and E_x their respective EMP's.
D pH, Wasserstoffionenexponent
F pH, cologarithme de l'activité des ions hydrogènes
P pH, wykładnik jonów wodorowych
R pH, водородный показатель

2609
PHASE.
Part of system separated from the rest by distinct boundaries and differing from the rest discontinuously as concerns thermodynamic properties.
D Phase
F phase
P faza
R фаза

2610
PHASE CELL.
Volume in phase space which represents microscopic states of system, indistinguishable by macroscopic methods.
D Phasenzelle
F cellule de l'espace des phases
P komórka fazowa
R ячейка фазового пространства

PHASE CHANGE. See PHASE TRANSITION.

2611
PHASE DIAGRAM (phase equilibrium diagram).
Plot showing zones of existence of particular phases and conditions of change of one phase into another in the given system, as function of qualitative and quantitative composition and of state variables.
D Phasendiagramm, Zustandsdiagramm, Phasenschaubild
F diagramme de phases
P wykres fazowy, diagram fazowy, wykres równowagi fazowej
R фазовая диаграмма, диаграмма состояния

PHASE EFFECT. See EFFECT OF THE STATE OF AGGREGATION.

2612
PHASE EQUILIBRIUM (transfer equilibrium).
Thermodynamic state of multiphase system with the same value of chemical potential of each component in all coexisting phases.
D Phasengleichgewicht, stoffliches Gleichgewicht
F équilibre entre phases, équilibre physique
P równowaga (między)fazowa
R фазовое равновесие

PHASE EQUILIBRIUM DIAGRAM. See PHASE DIAGRAM.

2613
PHASE EQUILIBRIUM LINE.
Curve in phase equilibrium diagram, separating areas of presence of particular phase.
D Phasengleichgewichtslinie
F ligne d'équilibre de phases
P krzywa równowagi fazowej
R кривая фазового равновесия

PHASE OF THE SYSTEM. See PHASE POINT.

PHASE ORBIT. See PHASE TRAJECTORY.

2614
PHASE POINT (phase of the system, representative point, specific phase).
In statistical mechanics point in phase space. Phase point represents microscopic state of system and is determined by values of all its coordinates and momenta.
D Phasen(raum)punkt, (spezielle) Phase
F point de l'espace des phases
P punkt fazowy, faza
R фазовая точка, точка фазового пространства

2615
PHASE SPACE (classical phase space).
In statistical mechanics $2f$-dimensional space; each point of this space represents some microscopic state of system with f degrees of freedom.
D Phasenraum
F espace des phases
P przestrzeń fazowa
R фазовое пространство

2616
PHASE TRAJECTORY (phase orbit).
Curve in phase space whose points are phase points representing states of system in its time evolution.
D Phasenbahn, Phasenkurve, Phasenlinie
F trajectoire dans l'espace des phases
P trajektoria fazowa
R фазовая траектория

2617
PHASE TRANSITION (phase change).
Thermodynamic process of vanishing
(gradually or suddenly) of one phase and
simultaneous appearance of another.
D Phasenumwandlung
F changement de phase
P przemiana fazowa, przejście fazowe
R фазовый переход, фазовое
   превращение

2618
PHENOMENOLOGICAL COEFFICIENTS,
L_{ik}.
Quantities $L_{ik} = \left(\dfrac{\partial J_i}{\partial X_k} \right)_0$, where
J_i = thermodynamic flux and
X_k = thermodynamic force; "0" subscript
denotes derivative calculated at
equilibrium.
D phänomenologische Koeffizienten
F coefficients phénoménologiques
P współczynniki fenomenologiczne
R феноменологические коэффициенты

2619
PHENOMENOLOGICAL EQUATIONS
(phenomenological relations, linear
phenomenological laws).
Equations expressing thermodynamic
fluxes J_i in terms of thermodynamic
forces X_i or vice versa; they give
dependence of thermodynamic forces on
fluxes in system; in linear
thermodynamics of irreversible processes
phenomenological equations have linear
form, most frequently expressed by

$$J_i = \sum_k L_{ik} X_k$$

where L_{ik} = phenomenological
coefficients.
D phänomenologische Gleichungen,
   phänomenologische Ansätze
F lois phénoménologiques
P równania fenomenologiczne
R феноменологические соотношения

PHENOMENOLOGICAL RELATIONS.
See PHENOMENOLOGICAL
EQUATIONS.

PHENOMENOLOGICAL
THERMODYNAMICS. *See*
MACROSCOPIC THERMODYNAMICS.

2620
PHENONIUM ION.
Ion formed in reactions of nucleophilic
substitution of compounds containing
aromatic ring in position adjacent
to substituted group. Such an ion is

formed due to interaction of π-electrons
(of ring) with electrophilic reaction centre.
Nucleophile attack in the second stage of
reaction may take place at 1- or
2-position

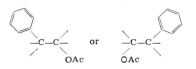

phenonium ion

where
Ts = tosyl $(-SO_2C_6H_4CH_3)$, Ac = acyl
$(-COR)$

D Phenonium-Ion
F ion phénonium
P jon fenoniowy
R феноний-ион

2621
pH METER (pehameter).
Apparatus used for measurements of pH
or potential with high input impedance
and large amplification of electrical
signal.
D pH-Messer, pH-Meter, Pehameter
F pH-mètre
P pehametr
R pH-метр

2622
pH METRY (pehametry).
Group of methods based on pH
measurements.
D pH-Metrie, Pehametrie
F pH-métrie
P pehametria
R pH-метрия

PHONOCHEMISTRY. *See*
SONOCHEMISTRY.

2623
PHONON.
Quantum of harmonic vibrations viz. the
imaginary boson, which describes various
types of quantized vibrations, e.g.
crystalline lattice vibrations or vibrations
on surface of atomic nucleus.
D Phonon
F phonon
P fonon
R фонон

2624
PHOSPHORESCENCE.
Type of photoluminescence which consists
in emission of light accompanying
spin-forbidden radiative transition
between two electronic states of different
multiplicity, e.g. from triplet to singlet
state.
D Phosphoreszenz
F phosphorescence
P fosforescencja
R фосфоресценция

2625
PHOSPHORUS, P.
Atomic number 15.
D Phosphor
F phosphore
P fosfor
R фосфор

PHOSPHORUS-32. See
RADIOPHOSPHORUS.

2626
PHOSPHORYLATION.

Introduction of phosphoryl group $-PO$

into chemical compound molecule,
usually by displacing hydrogen atom.
D Phosphorylierung
F phosphoryl(is)ation
P fosforylowanie, fosforylacja
R фосфорилирование

2627
PHOTOACTIVATION ANALYSIS (γ-ray
activation analysis, photonuclear
activation analysis).
Activation method of analysis in which
a beam of γ-quanta is used for activation.
D Aktivierungsanalyse mit γ-Quanten
F analyse par activation dans les photons
γ, analyse par activation photonucléaire,
analyse par activation aux rayons γ
P analiza fotoaktywacyjna
R фотоактивационный анализ,
гамма-активационный анализ

PHOTOCHEMICAL DISSOCIATION. See
PHOTOLYSIS.

2628
PHOTOCHEMICAL QUANTUM YIELD.
Ratio between number of molecules which
undergo a given photochemical reaction
and total number of photons previously
absorbed.
D (photochemische) Quantenausbeute
F rendement photochimique
P wydajność kwantowa reakcji
(fotochemicznej)
R квантовый выход реакции

2629
PHOTOCHEMICAL REACTION.
Reaction initiated by absorption of
photons with appropriate energy by at
least one component of reacting system
composed of atoms or molecules.
D photochemische Reaktion
F réaction photochimique
P reakcja fotochemiczna
R фотохимическая реакция

2630
PHOTOCHEMISTRY.
Branch of chemistry dealing with
reactions induced by action of light
(wavelengths ranging from ca. 100 nm to
ca. 1000 nm) in which electronically
excited atoms or molecules take part
inherently.
D Photochemie
F photochimie
P fotochemia
R фотохимия

2631
PHOTOCOLORIMETER (photoelectric
photometer).
Device used to perform colorimetric
analyses with monochromator of low
resolution and photoelectric detector.
D Photo(kolori)meter, Elektrophotometer
F colorimètre photoélectrique
P kolorymetr fotoelektryczny,
fotokolorymetr, absorpcjometr
R фотоколориметр

2632
PHOTOCONDUCTIVE EFFECT
(photoconductivity).
Increase semiconductor's conductivity
when exposed to light source.
D innerer Photoeffekt, Photoleitung
F effet photoélectrique interne,
photoconductivité
P efekt fotoelektryczny wewnętrzny,
fotoprzewodnictwo
R внутренний фотоэффект,
фотопроводимость

PHOTOCONDUCTIVITY. See
PHOTOCONDUCTIVE EFFECT.

2633
PHOTOCONDUCTOR.
Semiconductor exhibiting increase of
electrical conductivity under illumination.
D Photoleiter, lichtelektrischer Halbleiter
F photo-conducteur
P foto(pół)przewodnik
R фотопроводник, фотопроводящий
полупроводник

PHOTODISINTEGRATION. See
PHOTONUCLEAR REACTION.

PHOTODISSOCIATION. See
PHOTOLYSIS.

2634
PHOTOELECTRIC EFFECT (external photoelectric effect, normal photoelectric effect).
Emission of electrons from metal surface due to its bombardment by stream of photons.
D äußerer Photoeffekt
F effet photoélectrique externe, effet photoémissif normal
P efekt fotoelektryczny zewnętrzny
R внешний фотоэффект

PHOTOELECTRIC PEAK. *See* PHOTOPEAK.

PHOTOELECTRIC PHOTOMETER. *See* PHOTOCOLORIMETER.

2635
PHOTOELECTRON.
Orbital electron emitted by atom under the influence of electromagnetic radiation, i.e. through the external photoelectric effect.
D Photoelektron
F photoélectron
P fotoelektron
R фотоэлектрон

PHOTOEMULSION DOSIMETER. *See* FILM DOSIMETER.

2636
PHOTOFISSION.
Division of atomic nucleus into two heavy fragments induced by absorption of γ-quantum of sufficiently high energy.
D Photospaltung
F photofission
P fotorozszczepienie
R фотоделение ядра

2637
PHOTOFRACTION.
Ratio of number of counts in a photopeak to total number of counts in a gamma radiation spectrum.
D Photoanteil
F photofraction
P fotofrakcja
R фоточасть

2638
PHOTOGRAPHIC PHOTOMETRY.
Branch of photometry; radiation intensity is determined quantitatively from blackening of photographic materials.
D photographische Photometrie
F photométrie photographique
P fotometria fotograficzna
R фотографическая фотометрия

2639
PHOTOGRAPHIC PLATE.
Glass plate covered with photo-sensitive emulsion; frequently is especially sensitized to some wavelength regions of electromagnetic radiation.
D Photoplatte
F plaque photographique
P płyta fotograficzna
R фотографическая пластинка

2640
PHOTOIONIZATION.
Ionization initiated by absorption of photons of equal or greater energy than that of ionization energy.
D Photoionisation
F photo-ionisation
P fotojonizacja
R фотоионизация

2641
PHOTOISOMERIZATION.
Photochemical isomerization reaction of electronically excited molecules.
D Photoisomerisierung
F photoisomérisation
P fotoizomeryzacja
R фотохимическая изомеризация

2642
PHOTOLUMINESCENCE.
Luminescence induced by absorption of light.
D Photolumineszenz
F photoluminescence
P fotoluminescencja
R фотолюминесценция

2643
PHOTOLYSIS (photodissociation, photochemical dissociation).
Fission (dissociation) of molecule induced by photon absorption, e.g.

$$HI + h\nu \rightarrow H + I$$

D Photolyse, Photodissoziation
F photolyse, photodissociation
P fotoliza, dysocjacja fotochemiczna
R фотолиз, фотодиссоциация, фотохимическая диссоциация

2644
PHOTOMETER.
Instrument which enables ratio of intensities (or their functions) of two electromagnetic radiation beams to be measured through controlled intensity changes.
Remark: Term sometimes used for colorimeter.
D Photometer
F photomètre
P fotometr
R фотометр

PHOTOMETRIC TITRATION. *See* SPECTROPHOTOMETRIC TITRATION.

2645
PHOTOMETRY.
Group of measurement techniques for determination of radiation intensity in visible, ultraviolet and infrared region.

D Photometrie
F photométrie
P fotometria
R фотометрия

2646
PHOTON (light quantum).
Smallest indivisible portion of radiation energy E of electromagnetic field of given frequency v, $E = hv$, where h = Planck constant. Photons are usually considered as elementary particles with rest mass equal to zero and spin quantum number $J = 1$.

D Photon, Lichtquant
F photon, quantum de lumière
P foton, kwant promieniowania elektromagnetycznego
R фотон, световой квант

PHOTON ENERGY. *See* QUANTUM ENERGY.

PHOTONUCLEAR ACTIVATION ANALYSIS. *See* PHOTOACTIVATION ANALYSIS.

2647
PHOTONUCLEAR REACTION (photodisintegration).
Nuclear reaction caused by absorption of γ quantum by nucleus.

D Photokernreaktion
F réaction photonucléaire, photodésintégration
P reakcja fotojądrowa, fotorozpad jądrowy
R фотоядерная реакция

2648
PHOTOOXIDATION.
Photochemical reaction of oxidation which consists either in transfer of electron from excited molecule or addition of molecular oxygen to it or, alternatively, in its dehydrogenation.

D Photooxydation
F photooxydation
P fotoutlenianie
R фотоокисление

2649
PHOTOPEAK (photoelectric peak, total absorption peak).
Peak in a gamma radiation spectrum recorded as a result of total absorption in detector of γ-photons of a given energy.

D Photopeak, Photolinie, Fotopeak
F pic photoélectrique, photopic, pic d'énergie totale

P fotopik, pik fotoelektryczny
R фотопик, пик полного поглощения

2650
PHOTOPEAK EFFICIENCY.
Ratio of number of counts in photopeak to number of incident quanta on spectrometric detector.

D Photoansprechvermögen
F efficacité photoélectrique
P wydajność fotoelektryczna, wydajność fotopiku
R фотоэффективность

2651
PHOTOPOLAROGRAPHY.
Polarographic investigations of photochemical processes.

D Photopolarographie
F . . .
P fotopolarografia
R фотополярография

2652
PHOTOPOLYMERIZATION (photosensitized polymerization).
Photochemical reaction of polymerization initiated by free radicals or excited molecules very often in triplet state.

D Photopolymerisation
F photopolymérisation
P fotopolimeryzacja, polimeryzacja fotosensybilizowana
R фотосенсибилизированная полимеризация

2653
PHOTOREDUCTION.
Photochemical reduction reaction of electronically excited molecules, e.g. photoreduction of benzophenone to benzopinacone.

D Photoreduktion
F photoréduction
P fotoredukcja
R фотовосстановление

2654
PHOTOSENSITIZATION.
Effect due to presence of some radiation-absorbing atoms or molecules on evident changes in some other coexistent atoms or molecules which are incapable by themselves to absorb incident radiation directly.

D Photosensibilisierung
F sensibilisation photochimique
P sensybilizacja optyczna, fotosensybilizacja
R фотосенсибилизация, оптическая сенсибилизация

PHOTOSENSITIZED POLYMERIZATION. *See* PHOTOPOLYMERIZATION.

2655
PHOTOSENSITIZED REACTION.
Reaction initiated by radiative energy
absorbed by certain atoms or molecules
which, however, do not react themselves
but transfer their excessive excitation
energy to appropriate coexistent
substrates of reaction.
D photosensibilisierte Reaktion
F réaction sensibilisée
P reakcja (fotochemiczna) sensybilizowana
R фотосенсибилизированная реакция

2656
PHOTOSENSITIZER (sensitizer).
Molecule or the atom which is capable by
itself to absorb the radiative energy of
light and to transfer it to coexistent
substances which then undergo
photochemical reactions
(*cf.* photosensitized reaction).
D Sensibilisator
F sensibilisateur
P sensybilizator optyczny,
    fotosensybilizator
R фотосенсибилизатор, оптический
    сенсибилизатор

2657
PHOTOSYNTHESIS.
Multi-stage process of assimilation of
carbon dioxide by green plants and some
bacteria under influence of light whose
energy is absorbed in the first stage
of photosynthesis by chlorophyll in its
role as photosensitizer. Carbohydrates are
final products of photosynthesis.
D Photosynthese
F photosynthèse
P fotosynteza
R фотосинтез

2658
PHYSICAL ADSORPTION (physisorption,
van der Waals adsorption).
Increase in concentration of any adsorbed
substance (adsorbate) at interfacial
surface of adsorbent caused by attractive
intermolecular forces.
D physikalische Adsorption, van der
    Waalssche Adsorption
F adsorption physique, physisorption,
    adsorption de Van der Waals
P adsorpcja fizyczna, fizysorpcja,
    adsorpcja (siłami) van der Waalsa
R физическая адсорбция

2659
PHYSICAL CHEMISTRY.
Science dealing with studies on
interdependence between physical
properties and chemical structure of
substances or systems as well as on
physical phenomena accompanying
chemical transformations.
D physikalische Chemie
F chimie physique

P chemia fizyczna, fizykochemia
R физическая химия

2660
PHYSICAL PHOTOMETRY.
Branch of photometry; radiation intensity
is determined by measurement of changes
of some physical quantity, e.g.
photoelectric or thermoelectric current,
caused by radiation.
D physikalische Photometrie
F photométrie énergétique
P fotometria obiektywna
R объективная фотометрия

2661
PHYSICAL SCALE OF ATOMIC
WEIGHTS.
Previously used scale of atomic masses
with 1/16 of ^{16}O nuclide mass as unit
of mass.
D physikalische Massenskala
F échelle physique des masses atomiques,
    échelle physique des poids atomiques
P skala mas atomowych fizyczna
R физическая шкала атомных весов

2662
PHYSICAL STAGE OF RADIOLYSIS.
Primary stage of phenomena in sequence
of changes caused by absorption of
ionizing radiation, lasting 10^{-15}—10^{-13} s.
D ...
F ...
P stadium radiolizy fizyczne
R ...

PHYSICAL STATE. *See* STATE OF
AGGREGATION.

PHYSISORPTION. *See* PHYSICAL
ADSORPTION.

2663
PI BOND (π-bond).
Chemical bond formed by valence
electrons present in molecular π-bonding
orbital, antisymmetric with respect to
nodal plane passing through bond axis,
e.g. bond between carbon atoms in
ethylene.

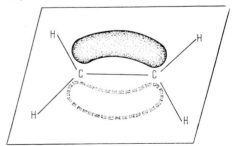

D π-Bindung
F liaison π
P wiązanie π
R π-связь

2664
PICK-UP REACTION.
Nuclear reaction in which incident
particle removes nucleon from surface
shell of target nucleus, and proceeds
together with picked up nucleon in
slightly changed direction.
D Pick-up-Reaktion
F rapt, enlèvement
P reakcja pick-up, reakcja porywania,
reakcja wychwytu
R реакция подхвата

2665
PICOGRAM METHOD (subultramicro
method).
Method of analysis where mass of
investigated sample is $x \cdot 10^{-12}$—$x \cdot 10^{-10}$ g,
or volume $x \cdot 10^{-9}$ cm³ in the case of gases.
D Submikromethode
F méthode picogrammique,
subultramicrométhode
P metoda pikogramowa,
subultramikrometoda
R . . .

2666
PI COMPLEX (π-complex).
Labile complex formed of π-electrons of
unsaturated molecule and cation, or
electron-deficient atom of polarized
molecule, e.g.

D π-Komplex
F complexe π
P kompleks π
R π-комплекс

2667
PI ELECTRONS (π-electrons).
Electrons assumed to occupy π molecular
orbitals.
D π-Elektronen
F électrons π
P elektrony π
R π-электроны

2668
PIEZOELECTRIC.
Solid exhibiting piezoelectric effect.
D Piezoelektrikum
F piézo-électrique
P piezoelektryk
R пьезоэлектрик

2669
PIEZOELECTRIC EFFECT (direct
piezoelectric effect, piezoelectricity).
Dielectric polarization of crystal due to its
mechanical deformation.

D piezoelektrischer Effekt,
Piezoelektrizität
F piézo-électricité
P efekt piezoelektryczny (prosty)
R пьезоэлектрический эффект,
пьезоэлектричество

PIEZOELECTRICITY. See CONVERSE
PIEZOELECTRIC EFFECT.

PIEZOELECTRICITY. See
PIEZOELECTRIC EFFECT.

2670
PI MESON (π-meson, pion).
Elementary particle, a hadron, of baryon
number $B = 0$, lepton number $L = 0$,
quantum spin number $J = 0$, strangeness
$S = 0$, isospin $I = 1$, parity $P = -1$,
unstable. π-meson occurs in three states
of charge (isospin triplet): with positive
elementary electric charge (π^+-meson;
mass $m = 139.576$ MeV, third component
of isospin $I_3 = 1$, mean lifetime
$2.602\,4 \cdot 10^{-8}$ s); electrically neutral
(π^0-meson; $m = 134.972$ MeV, $I_3 = 0$, mean
lifetime $0.84 \cdot 10^{-16}$ s); and of negative
elementary electric charge (π^--meson;
$I_3 = -1$, mass and mean lifetime as
π^+-meson).

D π-Meson, Pion
F méson π, pion
P mezon π, pion
R π-мезон, пи-мезон

PINACOLONE REARRANGEMENT. See
PINACOL REARRANGEMENT.

2671
PINACOL REARRANGEMENT
(pinacolone rearrangement).
Acid-catalyzed rearrangement of
α-glycols (most often di-ternary)
to ketones, involving elimination of
water molecule and anionic migration
(with rearrangement-1,2) of alkyl or aryl
group, or of hydride anion from diol
molecule, e.g.

D Pinakolinumlagerung
F transposition pinacolique
P przegrupowanie pinakolinowe
R пинаколиновая перегруппировка

PION. See PI MESON.

2672
PI ORBITAL (π-orbital).
Molecular orbital antisymmetric with
respect to nodal plane which contains
bond axis.

D π-Orbital
F orbitale π
P orbital molekularny (typu) π
R π-орбиталь

2673
PITZER STRAIN (torsional strain).
Strain resulting from total (or partial)
overwhelming of rotational barrier
hindering free rotation about single bond
in different conformational systems.
D Pitzer-Spannung
F tension de Pitzer
P naprężenie Pitzera, napięcie Pitzera
R торсионное напряжение, пицеровское
напряжение

2674
pK.
Negative logarithm of equilibrium
constant of reaction.
D pK
F pK
P pK
R pK

2675
PLANCK CONSTANT (elementary
quantum), h.
The smallest, indivisible amount of action
$h = (6.6254 \pm 0.0002)\ 10^{-27}$ erg s
D Plancksche Konstante, Plancksches
Wirkungsquantum, Elementarquantum
F constante de Planck
P stała Plancka, kwant działania
R постоянная Планка, квант действия

2676
PLANCK FUNCTION, Y.
Extensive thermodynamic function
$Y = -G/T$, where $G =$ isothermal-isobaric
thermodynamic potential (free enthalpy)
and $T =$ thermodynamic temperature.
D Plancksche Funktion
F fonction de Planck
P funkcja Plancka
R функция Планка

2677
PLANCK'S METHOD FOR
CALCULATION OF DIFFUSION
POTENTIAL.
Method based on assumption that
boundary layer between two electrolyte
solutions may be considered as porous
layer in contact with both electrolytes
with concentrations constant with time.
D Plancksche Berechnungsmethode des
Diffusionspotentials
F méthode du calcul du potentiel de
jonction de Planck
P metoda Plancka obliczania potencjału
dyfuzyjnego
R метод Планка расчёта диффузионного
потенциала

2678
PLANCK'S OSCILLATOR (classical
harmonic oscillator).
In old quantum theory, harmonic
oscillator of total energy $E_n = nh\nu$,
where $n =$ principal quantum number,
$h =$ Planck's constant, $\nu =$ frequency of
oscillations.
D Planckscher Oszillator
F oscillateur de Planck
P oscylator Plancka
R осциллятор Планка

2679
PLASMA.
Quasi-neutral highly ionized gas
(i.e. population of positive ions and
electrons) of such density, that every
charge is surrounded by charges of
opposite sign (i.e. screening takes place);
thus electric field induced by any charge
in plasma is shielded by neighbouring
charges of opposite sign.
D Plasma
F plasma
P plazma
R плазма

2680
PLASMA SPECTROSCOPY.
Branch of physics covering investigation
of components of radiation emitted by
ionized plasma in order to determine its
composition, concentration of charged
particles, temperature strength of electric
and magnetic field.
D Plasma-Spektroskopie
F spectroscopie de plasma
P spektroskopia plazmy
R спектроскопия плазмы

2681
PLASMA STATE.
State of matter at high temperatures in
which atoms undergo total or almost total
ionization and matter consists of electrons
and positive ions (atomic nuclei).
D Plasmazustand
F état plasma
P stan plazmy
R состояние плазмы

2682
PLASTIC STATE.
State of solid crystalline or amorphous
substance in which its properties resemble
liquid of high viscosity and low rigidity.
It occurs e.g. in metals beyond range of
elastic strain (deformation), or in
polymers above flow temperature.
D plastischer Zustand
F état plastique
P stan plastyczny
R пластическое состояние, вязкотекучее
состояние

2683
PLASTIC STATE (*of polymers*).
State in which macromolecules are capable of performing not only oscillations caused by free revolution of particular segments but may also perform translatory motion, "sliding" with respect to each other. Plastic state is achieved by heating polymer in elastic state to temperatures higher than so-called flow temperature T_f. Relaxation of strains in this state is fast and inner strains quickly disappear.
D plastischer Zustand
F état plastique
P stan lepkopłynny, stan plastyczny
R вязкотекучее состояние

2684
PLATEAU OF COUNTER.
Range of amplifying potential of counter in which counting rate undergoes relatively small changes.
D Plateau des Zählrohres
F plateau du compteur
P plateau licznika
R плато счётчика

PLATE HEIGHT. *See* EFFECTIVE HEIGHT OR THEORETICAL PLATE.

2685
PLATINUM, Pt.
Atomic number 78.
D Platin
F platine
P platyna
R платина

2686
PLATINUM ELECTRODE.
Electrode constructed from platinum wire, foil or gauze used as indicator electrode in potentiometric or voltammetric measurements and in electrogravimetry for electrodeposition of electrolyzed substance.
D Platinelektrode, Pt-Elektrode
F électrode de platine
P elektroda platynowa
R платиновый электрод

2687
PLATINUM METALS.
Elements forming two triads in eighth sub-group of periodic system (in periods V and VI): ruthenium, rhodium, palladium and osmium, iridium, platinum.
D Platinmetalle
F métaux de la mine de platine, platinoïdes
P platynowce
R платиновые металлы, элементы подгруппы платины

2688
PLUTONIUM, Pu.
Atomic number 94.
D Plutonium
F plutonium
P pluton
R плутоний

2689
PNEUMATIC SAMPLE TRANSFER SYSTEM (pneumatic tube system, pneumatic tube facility).
Pneumatic pipe system for fast transfer of samples, e.g. short-lived isotopes, from reactor to laboratory.
D pneumatisches Rohrpostsystem
F système pneumatique de transfert
P poczta pneumatyczna
R пневмопочта

PNEUMATIC TUBE FACILITY. *See* PNEUMATIC SAMPLE TRANSFER SYSTEM.

PNEUMATIC TUBE SYSTEM. *See* PNEUMATIC SAMPLE TRANSFER SYSTEM.

2690
POGGENDORFF'S COMPENSATION METHOD.
Method based on compensation of EMF of cell by oppositely directed known potential difference.
D Poggendorffsche Kompensationsmethode
F méthode d'opposition de la mesure de f.e.m.
P metoda kompensacyjna pomiaru SEM, metoda Poggendorffa
R компенсационный метод измерения Э.Д.С.

2691
pOH.
Negative logarithm of hydroxyl ion activity $pOH = -lg\ a_{OH^-}$
D pOH
F pOH
P pOH, wykładnik jonów wodorotlenowych
R pOH

2692
POINT DEFECTS (point imperfections).
Irregularities in structure of crystal lattice with dimensions of order of space occupied by one atom.
D Störstellen
F défauts ponctuels
P defekty punktowe
R точечные дефекты

POINT ESTIMATION. *See* ESTIMATION.

2693
POINT GROUP.
Set of symmetry operations (rotation around axis, reflection across plane or combination of planes), not including translation, which when carried out on periodic arrangement of points in space, brings that system of points to self-coincidence.
D Punktsymmetrie, Punktgruppe
F groupe ponctuel
P grupa punktowa
R точечная группа

POINT IMPERFECTIONS. See POINT DEFECTS.

2694
POISSON DISTRIBUTION.
Probability distribution of step-like random variable with probability function given by

$$P(X = k) = \lambda^k \exp(-\lambda)/k!$$

for $k = 0, 1, 2, \ldots$ $(\lambda > 0)$

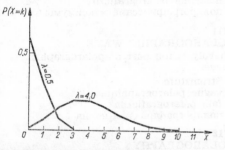

D Poissonverteilung
F loi de Poisson, distribution de Poisson, répartition de Poisson
P rozkład Poissona
R распределение Пуассона, пуассоновское распределение

2695
POISSON'S EQUATION.
Equation of adiabatic of perfect gas, whose heat capacity is temperature-independent, $pV^\varkappa = \text{const.}$, where $p = $ pressure, $V = $ volume, and $\varkappa = C_p/C_V$, $C_p = $ heat capacity at constant pressure, $C_V = $ heat capacity at constant volume.
D Poissonsches Gesetz
F équation de Laplace, loi de Poisson-Laplace
P równanie Poissona
R уравнение адиабаты Пуассона

POLAR BOND. See IONIC BOND.

2696
POLAR GROUP.
Arrangement of atoms in molecule, having permanent dipole moment, e.g. C—H, C—Cl, C=O, C—NH₂, C—C≡N.
D polare Gruppe
F groupement polaire
P grupa polarna
R полярная группа

2697
POLARIMETER.
Device used to measure angle of rotation of polarization plane of polarized light of a given frequency (usually sodium D line) by optically active substance.
D Polarimeter
F polarimètre
P polarymetr
R поляриметр

2698
POLARIMETRY.
Method of analysis used for identification (specific rotation measurement) or quantitative determination of optically active substances based on measurement of angle of rotation of polarization plane of polarized light.
D Polarimetrie
F polarimétrie
P analiza polarymetryczna, polarymetria
R поляриметрия

2699
POLARIZABILITY (induction polarizability), α_I.
Sum of atomic and electronic polarizabilities.
D Polarisierbarkeit
F polarisabilité
P polaryzowalność (indukcyjna), polaryzowalność deformacyjna
R поляризуемость

POLARIZATION. See DIELECTRIC POLARIZATION.

2700
POLARIZATION OF LIGHT (in general).
Ordering of light wave vibrations.
D Polarisation des Lichtes
F polarisation de la lumière
P polaryzacja światła
R поляризация света

2701
POLARIZATION OF LIGHT.
Limitation of direction of light wave electric vector E oscillations to some plane.
D Polarisation des Lichtes
F polarisation de la lumière
P polaryzacja światła liniowa
R линейная поляризация света

2702
POLARIZED BOND.
Chemical bond with non-zero dipole moment.
D polarisierte Bindung
F liaison polarisée
P wiązanie spolaryzowane
R поляризованная связь

2703
POLARIZED COVALENT BOND.
Intermediate bond between the covalent and ionic bonds, characterized by various probabilities of finding bonding electrons near linked atoms. This bond occurs always between different atoms and is often identified (when strongly polarized) with ionic bond.
D polarisierte Kovalenz, polare Atombindung
F liaison de covalence polarisée, liaison atomique dissymétrique
P wiązanie kowalencyjne spolaryzowane
R полярная ковалентная связь

2704
POLARIZED LIGHT.
Light without properties of cylindrical symmetry in direction of propagation.
D polarisiertes Licht
F lumière polarisée
P światło spolaryzowane
R поляризованный свет

2705
POLARIZING MICROSCOPE.
Microscope for investigation of optically anisotropic materials in linearly polarized light.
D Polarisationsmikroskop
F microscope polarisant
P mikroskop polaryzacyjny
R поляризационный микроскоп

2706
POLAR MOLECULE (dipolar molecule).
Molecule with permanent dipole moment.
D polares Molekül, Dipolmolekül
F molécule (di)polaire
P cząsteczka polarna, cząsteczka dipolowa
R полярная молекула, дипольная молекула

2707
POLAROGRAM.
Curve representing dependence of current on potential, recorded during polarographic experiment.
D polarographische Stromspannungskurve, Polarogramm
F courbe polarographique, polarogramme
P krzywa polarograficzna, polarogram
R полярограмма

2708
POLAROGRAPH.
Apparatus for automatic recording of current changes with slowly linearly changing voltage (0.1—1.0 V/min) applied to electrodes.
D Polarograph
F polarographe
P polarograf
R полярограф

2709
POLAROGRAPHIC ANALYSIS.
Analytical methods based on application of polarography.
D polarographische Analyse
F analyse polarographique
P analiza polarograficzna
R полярографический анализ

2710
POLAROGRAPHIC MAXIMA.
In polarography, characteristic increases of current above value predicted by Ilkovič equation.
D polarographische Maxima
F maxima polarographiques
P maksima polarograficzne
R полярографические максимумы

2711
POLAROGRAPHIC WAVE.
Sharply rising part of polarographic curve.
D Stromstufe
F vague polarographique
P fala polarograficzna
R полярографическая волна

2712
POLAROGRAPHY.
Voltammetric method based on investigation of current-potential relationship (polarogram) recorded during electrolysis carried out with linearly changing voltage applied to electrodes, one of which is dropping mercury electrode, the second being constant potential electrode with large surface.
D Polarographie
F polarographie
P polarografia
R полярография

2713
POLONIUM, Po.
Atomic number 84.
D Polonium
F polonium
P polon
R полоний

2714
POLYACID.
Product of condensation of certain number of simple molecules of oxy-acid,

having more than one mole of acid anhydride per one mole of water.
D Polysäure
F polyacide
P polikwas
R поликислота

2715
POLYADDITION.
Generally — addition polymerization. Particular meaning — polymerization by repeated addition process of monomers not containing carbon-carbon double bonds (e.g. reaction of epoxy monomers, isocyanates or lactams).
D Polyaddition
F polyaddition
P poliaddycja
R полиприсоединение, полимеризация

POLYCONDENSAT. *See* CONDENSATION POLYMER.

POLYCONDENSATION. *See* CONDENSATION POLYMERIZATION.

2716
POLYCRYSTALLINE AGGREGATE.
Solid substance composed of a large number of small, irregularly shaped and differently oriented crystals.
D (viel)kristallines Aggregat
F agrégat polycristallin
P ciało polikrystaliczne
R поликристаллическое тело, (поли)кристаллический агрегат

2717
POLYDISPERSE SYSTEM (heterodisperse system).
Disperse system in which particles of dispersed phase have different sizes.
D polydisperses System, heterodisperses System
F système polydispersé
P układ polidyspersyjny, układ heterodyspersyjny
R полидисперсная система

2718
POLYDISPERSITY.
Dispersion of mass of particles in suspensions and colloids.
D Polydispersität
F polydispersité
P polidyspersyjność
R полидисперсность

POLYDISPERSITY OF MOLECULAR WEIGHT. *See* POLYMOLECULARITY.

2719
POLYELECTRODE (multiple electrode).
Electrode (half-cell) at which simultaneously two or more electrode reactions determining potential occur.

D Mehrfachelektrode, Mischelektrode
F polyélectrode, électrode multiple
P elektroda złożona, elektroda mieszana
R смешанный электрод, полиэлектрод

2720
POLYELECTROLYTE.
Macromolecular substance which dissociates in water or other ionizing solvent with formation of ions with multiple charge — polyions (polyanions, polycations or polyzwitterions) and equivalent number of mobile ions with low opposite charge; polyelectrolytes may dissociate also to polycations and polyanions.
D Polyelektrolyt
F polyélectrolyte
P polielektrolit
R полиэлектролит

POLYFUNCTIONAL LIGAND. *See* MULTIDENTATE LIGAND.

2721
POLYMER.
Substance composed of molecules containing many of one or more species of atoms or groups of atoms (called constitutional units or mers) repetitively linked to each other.
D Polymer
F polymère
P polimer
R полимер

2722
POLYMERIZATION.
Process of converting monomer or mixture of monomers into polymer.
D Polymerisation
F polymérisation
P polimeryzacja
R полимеризация

2723
POLYMERIZATION ISOMERISM.
Occurrence complex compounds of the same empirical formula but different molecular masses, it may result from coordination polymerization, e.g. $[Co(NH_3)_3(NO_2)_3]$ and $[Co(NH_3)_6][Co(NO_2)_6]$; or nuclear polymerization, e.g.

$$\left[(H_3N)_3Co \underset{OH}{\overset{OH}{\underset{\diagdown}{\overbrace{}}}} Co(NH_3)_3 \right] X_3$$

and

$$\left[Co \left\{ \underset{OH}{\overset{OH}{\diagup}} Co(NH_3)_4 \right\}_3 \right] X_6$$

D Polymerisationsisomerie
F isomérie de polymérisation
P polimeria koordynacyjna, izomeria polimeryzacyjna
R координационная полимерия

2724
POLYMOLECULARITY (polydispersity of molecular weight).
Dispersion of molecular mass of polymer.
D Polymolekularität
F polymolecularité, polydispersité (moléculaire)
P polimolekularność, polidyspersyjność polimerów
R полимолекулярность, полидисперсность полимера

2725
POLYMORPHIC FORM.
One of the forms in which substance may occur in defined conditions.
D polymorphische Form
F forme polymorphe
P odmiana polimorficzna
R полиморфная форма

2726
POLYMORPHISM.
Occurrence of chemical compound in different crystallographic forms.
D Polymorphie, Vielgestaltigkeit
F polymorphisme
P polimorfizm, wielopostaciowość
R полиморфизм, многообразие

2727
POLYNUCLEAR COMPLEX (bridged complex).
Complex composed of two or more central atoms with ligands forming coordination spheres linked by atoms or bridge groups.
D mehrkerniger Komplex
F complexe polynucléaire
P kompleks wielordzeniowy, kompleks mostkowy
R многоядерный комплекс

2728
POLYTROPIC PROCESS.
Quasistatic process with constant heat capacity $C = \dfrac{\delta Q}{\delta T}$. In ideal gas of constant C_V polytropic process is described by equation

$$pV^n = \text{const.}$$

where δQ = elementary heat of quasistatic process, T = thermodynamic temperature and $n = (C_p - C)/(C_V - C)$; C_p = heat capacity at constant pressure and C_V = heat capacity at constant volume.
D polytrope Zustandsänderung
F transformation polytropique
P przemiana politropowa, proces politropowy
R политропный процесс, политропический процесс

2729
POMERANZ-FRITSCH SYNTHESIS OF ISOQUINOLINES.
Preparation of isoquinoline and its derivatives by condensation of aromatic aldehyde (or ketone) with aminoacetal, followed by cyclization of resulting benzylideneaminoacetal, caused by acids, e.g.

D Pomeranz-Fritsch Isochinolin--Ringschluß
F réaction de Pomeranz-Fritsch
P synteza (pochodnych) izochinoliny Pomeranza i Fritscha
R реакция Померанца-Фрича

2730
POPULATION.
Statistical ensemble, i.e. set of arbitrary elements non-identical from point of view of a given feature.
D Grundgesamtheit
F population
P populacja (generalna), zbiorowość generalna
R генеральная совокупность

2731
POSITION ISOMERISM.
Type of structural isomerism caused by different location of substituents on identical carbon skeleton (straight-chained or cyclic) of isomeric compounds, e.g. propan-1-ol and propan-2-ol or o-xylene and m-xylene.
D Stellungsisomerie
F isomérie de position
P izomeria położenia, izomeria podstawienia
R изомерия положения

2732
POSITIVE AZEOTROPE.
Azeotrope of composition at which there is maximum vapour pressure and correspondingly, minimum boiling temperature.
D positives Azeotrop
F azéotrope positif
P azeotrop dodatni
R положительная азеотропная смесь

2733
POSITIVE CATALYSIS.
Phenomenon of acceleration of reaction by catalyst.

D positive Katalyse
F catalyse positive
P kataliza dodatnia
R положительный катализ

POSITIVE HOLE. *See* HOLE.

2734
POSITIVE VALENCY.
Number of electrons in one atom moved
in direction of other atom of negative
valency during formation of chemical
bond.
D positive Wertigkeit
F valence positive
P wartościowość dodatnia
R положительная валентность

2735
POSITRON, e^+.
Antiparticle of electron, a lepton, of
positive elementary electric charge,
baryon number $B = 0$, electronic lepton
number $L_e = 1$, stable.
D Positron, positives Elektron
F posit(r)on, électron positif
P pozyton, antyelektron, elektron dodatni
R позитрон, положительный электрон

2736
POSITRONIUM.
Unstable system consisting of electron
and positron, analogous to hydrogen atom
(proton is replaced by positron); it is
formed, e.g. during the slowing down of
positrons in matter.
D Positronium
F positonium
P pozyt(onium)
R позитроний

POST-EFFECT. *See* AFTEREFFECT.

POST-IRRADIATION EFFECT. *See*
AFTEREFFECT.

POSTULATES OF QUANTUM
MECHANICS. *See* QUANTUM
MECHANICAL POSTULATES.

POSTURANIC ELEMENTS. *See*
TRANSURANIC ELEMENTS.

2737
POTASSIUM, K.
Atomic number 19.
D Kalium
F potassium
P potas
R калий

2738
POTENTIAL.
Ratio of potential energy of system in
potential field to some scalar quantity
characteristic for system and denoting its
interaction with field.

D Potential
F potentiel
P potencjał
R потенциал

2739
POTENTIAL BARRIER.
Spatial region with local maximum of
potential energy. Repulsive forces tend to
remove particle from this region.
D Potentialwall
F barrière de potentiel
P bariera potencjału
R потенциальный барьер

POTENTIAL BARRIER. *See* POTENTIAL
ENERGY BARRIER.

POTENTIAL BOX. *See* POTENTIAL
WELL.

POTENTIAL DETERMINING
REACTION. *See* ELECTRODE
REACTION.

2740
POTENTIAL ENERGY BARRIER
(potential barrier *in chemical kinetics*).
Point of maximum potential energy on
reaction path, i.e. of potential energy of
activated complex.
D Energiebarriere, Energieberg
F barrière d'énergie
P próg energetyczny (reakcji), próg
potencjalny (reakcji), bariera
energetyczna
R потенциальный барьер (реакции),
энергетический барьер

2741
POTENTIAL ENERGY CURVE, $U(r)$.
Curve representing functional dependence
of potential energy U of pair of particles
on their relative distance r, e.g. potential
energy curve for diatomic molecule.
D Potentialkurve, potentielle Energiekurve
F courbe d'énergie potentielle
P krzywa energii potencjalnej
R потенциальная кривая,
кривая потенциальной энергии

2742
POTENTIAL ENERGY SURFACE (*in
chemical kinetics*).
Population of all points of coordinates
r_1, r_2, $E(r_1,r_2)$ where r_1 = distance between
centres of atoms X and Y, r_2 = distance
between centres of atoms Y and Z,
$E(r_1,r_2)$ = potential energy depending on
r_1 and r_2, in reacting systems
$XY + Z \rightarrow X + YZ$.
D Potential(ober)fläche, Potentialgebirge
F surface d'énergie potentielle
P powierzchnia energii potencjalnej
R поверхность потенциальной энергии

2743
POTENTIAL FIELD.
Field derived as negative gradient of
some scalar function (potential, potential
energy).
D Potentialfeld
F champ potentiel
P pole potencjalne
R потенциальное поле

2744
POTENTIAL-TIME CURVE.
Dependence of potential of indicator
electrode on time of chronopotentiometric
analysis.
D Potential-Zeitkurve
F courbe potentiel-temps
P chronopotencjogram
R хронопотенциограмма

2745
POTENTIAL WELL (potential box).
Potential given by $U(x) = 0$ for $0 \leqslant x \leqslant a$,
$U(x) = U_0$ for either $x < 0$ or $x > a$

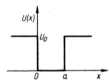

D Rechteckpotential
F puits de potentiel
P jama potencjału, studnia potencjału
R потенциальная яма

2746
POTENTIOMETER.
In electrochemistry, apparatus for
measurement or comparison of EMF of
cells.
D Potentiometer, Kompensationsgerät
F potentiomètre
P potencjometr
R потенциометр

2747
POTENTIOMETRIC TITRATION.
Titration in which end point is
determined from changes of potential of
indicator electrode.
D potentiometrische Titration
F titrage potentiométrique
P miareczkowanie potencjometryczne,
analiza potencjometryczna,
potencjometria
R потенциометрическое титрование

2748
POTENTIOMETRY.
Methods based on measurements of EMF
of suitably chosen cells (chemical analysis,
dissociation constants of weak electrolytes,
pH, solubility of salts etc.).
D Potentiometrie
F potentiométrie
P potencjometria
R потенциометрия

2749
POTENTIOSTAT.
Apparatus which controls automatically
potential of working electrode.
D Potentiostat
F potentiostat
P potencjostat
R потенциостат

2750
POTENTIOSTATIC METHODS.
Methods of study of kinetics of electrode
reactions consisting in recording and
analysis of current changes during
electrolysis carried out at constant
potential of electrode.
D potentiostatische Einschaltmethoden
F méthodes potentiostatiques
P metody potencjostatyczne
R потенциостатические методы

POWDER DIFFRACTION METHOD. *See*
DEBYE-SCHERRER METHOD.

2751
POWDER PHOTOGRAPH.
X ray photograph obtained by Debye,
Scherrer and Hull: diffraction image in
shape of concentric rings on strip of
photographic film, produced by
monochromatic X-ray beam passing
through powdered crystalline sample.
Powder is usually placed in thin tube
made of appropriate glass.
D Pulveraufnahme
F radiogramme de poudre
P rentgenogram proszkowy, dyfraktogram
proszkowy, diagram proszkowy,
debajogram
R рентгенограмма порошка, порошковая
рентгенограмма, дебаеграмма

2752
POWER FACTOR, $\sin \delta$.
Sine of loss factor.
D Leitungsfaktor
F coefficient de puissance
P współczynnik mocy
R коэффициент мощности

POWER HYDROGEN. *See* pH.

2753
PRASEODYMIUM, Pr.
Atomic number 59.
D Praseodym
F praséodyme

P prazeodym
R празеодим

2754
PRE-ARC PERIOD (*in emission spectral analysis*).
Time interval between arc initiation and spectrum recording for analytical purposes
D Vorbrennungszeit
F (pré)flambage
P czas przedpalenia
R время предварительного обжига

PRECIPITATION FROM HOMOGENEOUS SOLUTION. *See* **HOMOGENEOUS PRECIPITATION.**

2755
PRECIPITATION TITRATION.
Titration with solution of substance of known concentration which forms poorly soluble compound with analysed substance.
D Fällungstitration
F titrage par précipitation
P miareczkowanie strąceniowe
R преципитационное титрование, титрование по методу осаждения

PRECISION ABSORPTIOMETRY. *See* **PRECISION SPECTROPHOTOMETRY.**

PRECISION COLORIMETRY. *See* **PRECISION SPECTROPHOTOMETRY.**

2756
PRECISION OF METHOD.
Measure of agreement between experimental values obtained in whole investigation performed in specified conditions.
D Präzision der Methode
F fidélité de la méthode
P precyzja metody
R точность метода

2757
PRECISION SPECTROPHOTOMETRY
(precision absorptiometry, precision colorimetry).
Spectrophotometric method of determination of concentration based on measurements with respect to one or two standard solutions of known concentration, used to fix 100 per cent and/or 0 per cent point of transmittance scale.
D...
F...
P analiza spektrofotometryczna precyzyjna, spektrofotometria precyzyjna

R метод предельной точности, максимально точный метод

2758
PRECRITICAL STATE.
State of substance under conditions close to critical state, characterized by considerable fluctuations of physical properties.
D vorkritischer Zustand
F état subcritique
P stan przedkrytyczny
R докритическое состояние

2759
PRECURSOR.
Chemical compound which forms in first stage of chemical process and serving as substrate in its further stages; e.g. carotenes are precursors of vitamin A; phenylacetic acid is the precursor of benzylic group in biosynthesis of benzylpeniciline.
D Vorstufe (von) einer Substanz
F promoteur
P prekursor
R прекурсор

2760
PREDISSOCIATION.
Blurring of vibrational energetic levels of molecule due to radiationless transitions between two electronic states, the first of which is stationary excited state of molecule, whereas the second corresponds to energetic continuum for a given configuration of nuclei.
D Prädissoziation
F prédissociation
P predysocjacja
R предиссоциация

2761
PRE-EXPONENTIAL FACTOR
(frequency factor), A.
Coefficient occuring in Arrhenius equation, specific for given reaction.
Remark: Pre-exponential factor is interpreted in different ways in theory of absolute reaction rate (*see* entropy factor) and collision theory (*see* steric factor).

D Häufigkeitsfaktor
F facteur de fréquence
P współczynnik (przedwykładniczy w równaniu) Arrheniusa
R предэкспоненциальный множитель, фактор частоты

PREIONIZATION. *See* **AUTOIONIZATION.**

2762
PRELOG'S RULE.
Rule permitting to forecast of configuration of asymmetric synthesis product formed in excess from dicarbonyl compounds. According to rule, substituents at chiral centre ($L > M > S$) acquire conformation in which S is coplanar with two anti-parallelly oriented carbonyl groups. Nucleophile $Y^\ominus$ will preferentially attack carbonyl group from the less covered side, i.e. from M substituent

D Prelogsche Regel
F règle de Prelog
P reguła Preloga
R правило Прелога

2763
PREPOLYMER.
Polymer having degree of polymerization between that of monomer (or monomers) and the final product.
D Prepolymer
F prépolymère
P prepolimer
R преполимер

2764
PRE-SPARK PERIOD (in emission spectral analysis).
Time interval between spark initiation and spectrum recording for analytical purposes.
D Vorfunkenzeit
F préétincelage, flambage
P czas przediskrzenia
R время предварительного обыскривания

2765
PRESSURE COEFFICIENT (of gas), β.
Relative change of pressure p of gas, caused by change of thermodynamic temperature T by 1 K at constant volume V

$$\beta = \frac{1}{p} \left(\frac{\partial p}{\partial T} \right)_V \quad [\text{K}^{-1}]$$

Pressure coefficient of ideal gas is

$$\beta = \frac{1}{273.15} \text{ K}^{-1}$$

D (thermischer) Spannungskoeffizient
F (coefficient d'accroissement de pression sous volume constant)

P współczynnik rozprężliwości
R термический коэффициент давления, коэффициент возрастания давления

2766
PRÉVOST REACTION.
Method of conversion of alkenes into trans-glycol esters by action of silver salt of acid and by iodine, e.g.

D Prévost Jod-Silbersalz-Addition
F réaction de Prévost
P reakcja Prévosta, hydroksylowanie Prévosta
R реакция Прево

2767
PRILESCHAJEW OXIRANE SYNTHESIS.
Preparation of oxiranes by oxidation of alkenes with organic peroxides, e.g.

D Prileschajew Olefin-Epoxydation
F réaction de Prileschajew
P synteza Prileżajewa
R реакция Прилежаева

PRIMARY INTERFERENCE. See **PRIMARY INTERFERING REACTION.**

2768
PRIMARY INTERFERING REACTION (primary interference).
In activation analysis, nuclear interfering reaction caused by stream of bombarding primary particles.
D (störende) Primärreaktion
F réaction (d'interférence) du premier ordre, interférence du premier ordre
P reakcja jądrowa przeszkadzająca pierwotna
R конкурирующая ядерная реакция, побочная ядерная реакция

2769
PRIMARY IONIZATION.
Ionization produced directly by interaction of charged particle with atoms of matter through which it is passing.

D Primärionisation
F ionisation primaire
P jonizacja pierwotna
R первичная ионизация

2770
PRIMARY PHOTOCHEMICAL PROCESS.
Photochemical reaction which takes place
in electronically excited state immediately
after absorption of photon by a given
atom or molecule, e.g. photolysis,
photoisomerization.

D photochemische Primärreaktion
F processus primaire
P reakcja fotochemiczna pierwotna
R первичная фотохимическая реакция

2771
PRIMARY REACTION.
Initiating reaction in system of coupled
reactions which determines path of
secondary reaction i.e. of induced reaction.

D Primärreaktion
F réaction inductrice
P reakcja indukująca
R первичная реакция, самопроизвольная
реакция

2772
PRIMARY SALT EFFECT.
Influence of non-reacting electrolytes
present in solution on change in rate of
proceeding ionic reaction. Primary salt
effect is expressed by difference of
activity coefficients of substrates and
activated complex.

D primärer Salzeffekt
F effet de sel primaire
P efekt solny pierwotny
R первичный солевой эффект

2773
PRIMARY STANDARD (*in volumetric
analysis*).
Substance of high purity samples of which
are used to standardize solution or to
prepare solutions of known
concentrations.

D primäre Urtitersubstanz
F solution titrée étalon
P wzorzec pierwotny, substancja
podstawowa
R первичный эталон

2774
PRIMARY STANDARD SUBSTANCE (*in
thermochemistry*).
High-purity substance recommended by
IUPAC for calibration of calorimetric
systems, e.g. pure benzoic acid is
recommended as standard for
determination of combustion heats of
solids, whereas hydrogen is used for
volatile liquids and gases.

D primäre Standardsubstanz, primäre
Eichsubstanz

F étalon primaire
P wzorzec podstawowy termochemiczny
R основное стандартное вещество

2775
PRIMARY VALENCY.
According to Werner's theory, valency
characteristic for dissociating compounds
(with ionic bonds) and corresponding to
atom's oxidation number in given
compound.

D Hauptvalenz
F valence primaire
P wartościowość główna, wartościowość
pierwszorzędowa, wartościowość
pierwotna
R главная валентность

2776
PRIMITIVE CELL (primitive unit cell).
Unit cell in which identical points are on
corners of parallelepiped only.

D einfache primitive Elementarzelle
F forme (fondamentale) primitive
P komórka prymitywna
R примитивная ячейка, пустая ячейка

PRIMITIVE UNIT CELL. *See* PRIMITIVE
CELL.

2777
PRINCIPAL POLARIZABILITIES.
Molecular polarizability values along
three mutually perpendicular axes chosen
in such a way that the first polarizability
value is maximum and the third one
minimum for given molecule.
Polarizability is arithmetic mean of three
principal components.

D Hauptpolarisierbarkeiten
F polarisabilités principales
P polaryzowalności główne
R главные поляризуемости

2778
PRINCIPAL QUANTUM NUMBER, *n*.
Positive integer which determines energy
levels of atom or ion.

D Hauptquantenzahl
F nombre quantique principal
P liczba kwantowa główna
R главное квантовое число

2779
**PRINCIPLE OF CORRESPONDING
STATES.**
For real gases there is some identical
equation of state formulated in terms of
reduced temperature, reduced pressure
and reduced volume.

D Theorem der übereinstimmenden
Zustände
F loi d'états correspondantes
P prawo stanów odpowiadających (sobie)
R закон соответственных состояний

2780
PRINCIPLE OF DETAILED
BALANCING.
Fundamental principle of statistical
mechanics according to which in state of
thermodynamic equilibrium the same
(average) number of particles passes in
time unit from state I to state II and from
state II to state I. In particular each
elementary chemical reaction proceeds
with the same velocity in both directions.
D Prinzip des detaillierten Gleichgewichtes
F principe de la réversibilité
   microscopique
P zasada szczegółowej równowagi
R принцип детального равновесия,
   принцип микроскопической
   обратимости

PRINCIPLE OF EQUIPARTITION OF
ENERGY. *See* ENERGY
EQUIPARTITION PRINCIPLE.

2781
PRINCIPLE OF GEOMETRIC
AGREEMENT.
In heterogeneous catalysis, effectiveness
of catalyst is determined by arrangement
of adsorption centres of geometry related
to arrangement of atoms or ions in
reacting molecules participating in
catalytic reactions.
D ...
F ...
P zasada zgodności geometrycznej, zasada
   zgodności strukturalnej
R принцип геометрического соответствия,
   принцип структурного соответствия

PRINCIPLE OF INACCESSIBILITY. *See*
CARATHÉODORY'S PRINCIPLE.

2782
PRINCIPLE OF INCREASE OF
ENTROPY.
Spontaneous adiabatic transition is always
accompanied by system entropy increase.
D Prinzip von der Erzeugung der
   Entropie, Prinzip von der Vermehrung
   der Entropie
F principe de création de l'entropie
P zasada wzrostu entropii, prawo wzrostu
   entropii
R принцип возрастания энтропии,
   закон возрастания энтропии

2783
PRINCIPLE OF LEAST STRUCTURAL
CHANGES.
In chemical reactions formation of new
bond (at atom of reactive centre) is
accompanied by cleavage of bond, taking
place at the same atom. Rearrangement
reactions proceed against this principle.
D Prinzip der geringsten strukturellen
   Änderungen

F principe du moindre changement
   structurel
P zasada najmniejszych zmian
   strukturalnych
R принцип наименьших структурных
   изменений

PRINCIPLE OF MOBILE EQUILIBRIUM.
See LE CHÂTELIER-BRAUN'S
PRINCIPLE.

PRINCIPLE OF SUPERPOSITION. *See*
SUPERPOSITION PRINCIPLE.

2784
PRINS REACTION.
Preparation of 1,3-glycol and 1,3-dioxane
derivatives based on reaction of aliphatic
aldehydes (mainly formaldehyde) with
alkenes, in presence of acids, e.g.

$$2CH_3CH\!=\!CH_2 + 3HCHO + 2CH_3COOH \xrightarrow{H^{\oplus}}$$

$$\underset{\underset{CH_3COO}{|}}{CH_3CHCH_2CH_2} + \underset{\underset{OOCCH_3}{|}}{} $$

D Prins-Reaktion, Kriewitz-Prins
   Olefin-Formaldehyd-Addition
F réaction de Prins
P synteza 1,3-glikoli Prinsa, reakcja
   Prinsa
R реакция Принса

2785
PRIORITY RULE (sequence rule).
Rule on which R-S and E-Z systems and
names of conformation systems are based.
Atoms and groups of atoms attached to
chiral centre (R-S system) or to double
bond system (E-Z system), are arranged
in sequence of decreasing atomic number.
If two or more of these atoms have the
same atomic number, selection is made by
comparing atomic numbers of second
group of atoms attached to first atoms.
If ambiguity still persists, the third,
fourth, etc., sets (working outwards from
asymmetric carbon atom) are compared
until selection is made. Substituents are
given in order of decreasing priority: I,
Br, Cl, SR, SH, F, OH, NO_2, NH_2, CCl_3,
COCl, CHO, CH_2OH, C_6H_5, CH_3, D, H.
D Prioritätsfolgeregel, Rangfolgeregel
F ...
P reguła kolejności, reguła starszeństwa
R правило старшинства, правило
   последовательности

PRIOR PROBABILITY. *See*
PROBABILITY A PRIORI.

2786
PROBABILITY A PRIORI (prior
probability), P.
Ratio of number of elementary events
favourable for a given event and total

number of elements in set of elementary events, provided that all events have the same chance of occurrence.
D mathematische Wahrscheinlichkeit
F probabilité à priori, probabilité mathématique
P prawdopodobieństwo matematyczne, prawdopodobieństwo a priori
R априорная вероятность

PROBABILITY DENSITY. *See* PROBABILITY DENSITY FUNCTION.

2787
PROBABILITY DENSITY FUNCTION (probability density, frequency function). First order derivative of integral distribution function of random variable.
D Verteilungsdichte, Verteilungsdichtefunktion
F densité de probabilité, fonction densité (de la fréquence), fonction densité de répartition
P gęstość prawdopodobieństwa zmiennej losowej, funkcja rozkładu zmiennej losowej różniczkowa
R плотность распределения, плотность вероятности

2788
PROBABILITY DISTRIBUTION FUNCTION (integral probability distribution function, integral distribution function, distribution function).
Function $F(x)$ which determines probability P of event that random variable X will have value less than x: $F(x) = P(X < x)$.
D Verteilungsfunktion, Verteilungssummenfunktion
F fonction des probabilités totales, fonction de répartition
P dystrybuanta zmiennej losowej, funkcja rozkładu zmiennej losowej całkowa
R распределение полной вероятности

PROBABILITY FACTOR. *See* STERIC FACTOR.

PROBABILITY LEVEL. *See* CONFIDENCE LEVEL.

2789
PROCHIRALITY.
Formation of chirality centre in molecule, resulting from substitution of one of the identical ligands linked with given moiety of molecule (named prochirality centre) by another ligand.
D Prochiralität
F prochiralité

P prochiralność
R прохиральность

2790
PROJECTION FORMULA.
Formula portraying three-dimentional model of molecule as perpendicular projection on a plane.
D Projektionsformel
F formule projetée
P wzór rzutowy
R проекционная формула

2791
PROMETHIUM, Pm.
Atomic number 61.
D Promethium
F prométhium
P promet
R прометий

2792
PROMPT GAMMA (prompt gamma radiation).
In activation analysis, γ-quanta emitted by atoms from target material instantaneously after radiative neutron capture reaction, i.e. (n, γ) reaction.
D prompte γ-Strahlung
F rayonnement γ instantané
P promieniowanie γ natychmiastowe
R γ-лучи захвата

PROMPT GAMMA RADIATION. *See* PROMPT GAMMA.

2793
PROMPT NEUTRONS.
Fission neutrons emitted in time shorter than 10^{-13} s. Prompt neutrons constitute about $99^0/0$ of all neutrons emitted in fission reactions.
D prompte Neutronen
F neutrons prompts, neutrons immédiats
P neutrony natychmiastowe
R мгновенные нейтроны

PROPER ENTROPY. *See* MOLAR ENTROPY.

PROPER QUANTITIES. *See* MOLAR QUANTITIES.

2794
PROPORTIONAL COUNTER.
Gas-filled counter operating at such a gas amplification that charge collected at electrodes during one pulse is proportional to charge liberated by initial ionizing event.
D Proportionalzählrohr, Proportionalzähler
F compteur proportionnel
P licznik proporcjonalny
R пропорциональный счётчик

2795
PROTACTINIUM, Pa.
Atomic number 91.
D Protactinium, Protaktinium
F protactinium
P protaktyn
R протактиний

2796
PROTECTING GROUP (blocking group, covering group).
Grouping of atoms attached temporarily to functional group to protect the latter from entering into unwanted reaction.
D Schutzgruppe
F ...
P grupa zabezpieczająca, grupa osłaniająca, grupa blokująca
R защитная группа

2797
PROTECTION OF FUNCTIONAL GROUP.
Conversion of one of the several functional groups present in molecule of organic compound into another (protective) group, which can easily be removed to restore the primary functional group. This protective group is to be left unchanged in course of reactions carried out with other functional groups present in molecule.
D Schutz
F protection
P zabezpieczenie grupy funkcyjnej, osłanianie grupy funkcyjnej
R защита

2798
PROTECTIVE COLLOID.
Lyophilic colloid which, when added in small quantities to lyophobic colloid, increases its stability, e.g. gelatin, sodium protalbinate.
D Schutzkolloid
F colloïde protecteur
P koloid ochronny
R защитный коллоид

2799
PROTECTIVE EFFECT (in radiation chemistry).
Influence of minor additions of chemical compounds affecting yield of radiolysis of main constituent of system or diminution of biological consequences of irradiation.
D Schutzeffekt
F effet de protection
P efekt ochronny
R предохранительный эффект, защитное действие

2800
PROTEIN ERROR.
Indication error of acid-base indicators due to presence of proteins in solution.

D ...
F ...
P błąd białkowy, efekt białkowy
R белковая ошибка

2801
PROTEOLYSIS.
Catalytic (e.g. enzymatic) hydrolysis of protein resulting in formation of peptones, polypeptides or amino acids.
D Proteolyse
F protéolyse
P proteoliza
R протеолиз

2802
PROTIUM. Isotope of hydrogen.
Atomic number 1. Mass number 1.
D Protium, leichter Wasserstoff
F protium, hydrogene normal
P prot
R протий

2803
PROTOGENIC SOLVENT (proton donor solvent, acidic solvent).
Solvent which gives proton to dissolved substance, e.g. glacial acetic acid, liquid hydrogen fluoride, anhydrous sulfuric acid (see Brönsted's theory of acids and bases).
D Donorlösungsmittel, saures Lösungsmittel
F solvant acide, solvant proton-donneur
P rozpuszczalnik protonogenowy, rozpuszczalnik protonodonorowy, rozpuszczalnik kwaśny
R протогенный растворитель, кислый растворитель

2804
PROTOLYSIS (protolytic reaction).
Exchange of protons between two pairs of acids and conjugates bases; second acid-base pair is usually the solvent.

$$HA + B \rightleftarrows A + HB$$

HA and A as well as HB and B represent these conjugated pairs (see Brönsted's theory of acids and bases).
D Protolyse, protolytische Reaktion
F protolyse, réaction acido-basique, réaction prototropique
P protoliza, reakcja protolizy
R протолиз, реакция протолиза

PROTOLYTIC REACTION. See PROTOLYSIS.

2805
PROTON, p.
Elementary particle, a hadron, one of isospin doublet state of nucleon, stable, of mass $m = 938.259\ 2$ MeV, positive elementary electric charge, baryon number $B = 1$, lepton number $L = 0$,

quantum spin number $J = 1/2$, strangeness $S = 0$, isospin $I = 1/2$, third component of isospin $I_3 = 1/2$. It is assumed that parity of proton is $P = +1$. Proton is constitutent of all atomic nuclei.

D Proton
F proton
P proton
R протон

PROTON ACCEPTOR SOLVENT. *See* PROTOPHILIC SOLVENT.

PROTON-AND-NEUTRON MAGIC NUCLEUS. *See* DOUBLY MAGIC NUCLEUS.

2806
PROTONATED DOUBLE BOND.
One of the conceptions of boron-boron bond in B_2H_6 molecule, proposed by Pitzer: two bridge-forming protons are closely located to B—B bond axis

D Doppelbindung mit Protonen
F liaison double protonique
P wiązanie podwójne protonowane
R протонированная двойная связь

PROTON DONOR SOLVENT. *See* PROTOGENIC SOLVENT.

2807
PROTON-PROTON CHAIN.
Series of thermonuclear reactions leading to synthesis of helium from hydrogen

$^1H(p,\beta^+\nu)\ ^2H(p,\gamma)\ ^3He$

$^3He, 2p)\ ^4He$
$(\alpha,\gamma)\ ^7Be \begin{cases}(\beta^-\nu)\ ^7Li(p,\alpha)\ ^4He \\ (p,\gamma)\ ^8B(\beta^+\nu)\ ^8Be(\alpha)\ ^4He\end{cases}$

According to Salpeter, proton-proton chain is source of energy of stars.

D Wasserstoff-Zyklus
F cycle de l'hydrogène
P cykl protonowo-protonowy
R водородный цикл, протон-протонная цепочка

2808
PROTOPHILIC SOLVENT (proton acceptor solvent, basic solvent).
Solvent with affinity for proton, e.g. liquid ammonia, amines, water, ether, dioxane (*see* Brönsted's theory of acids and bases).

D Akzeptorlösungsmittel, basisches Lösungsmittel

F solvant basique, solvant proton-accepteur
P rozpuszczalnik protonofilowy, rozpuszczalnik protonoakceptorowy, rozpuszczalnik zasadowy
R протофильный растворитель, основной растворитель

2809
PROTOTROPIC TAUTOMERISM.
The most often encountered type of tautomerism in which equilibrium between tautomers differing in location of hydrogen atom and multiple bond in molecule is caused by proton migration (addition or removal), e.g.

$H—X—CH=Y \rightleftarrows X=CH—Y—H$

D Protonentautomerie, prototrope Tautomerie
F tautomérie prototropique
P tautomeria prototropowa
R прототропная таутомерия

2810
PROTOTROPISM.
Reversible migration of proton from one position in molecule to another

D Prototropie
F prototropie
P proto(no)tropia
R прототропия

PROUST'S LAW. *See* LAW OF CONSTANT COMPOSITION.

2811
PSCHORR SYNTHESIS OF PHENANTHRENE DERIVATIVES.
Preparation of phenanthrene derivatives from *o*-nitrobenzaldehyde and arylacetic acid, e.g.

D Pschorr Phenanthren-Ringschluß
F synthèse de Pschorr
P synteza Pschorra
R синтез Пшорра

2812
PSEUDOCRYSTALLINE STATE.
Intermediate state between gas and crystalline solid, specific for liquids, especially close to solidification temperature. Liquid in this state consists of microscopic zones exhibiting short-range order.
D Zustand der quasikristallinen Ordnung, quasikristalliner Zustand
F état pseudo-cristallin
P stan pseudokrystaliczny
R псевдокристаллическое состояние

2813
PSEUDOHALOGENS.
Compounds with molecules containing at least two electronegative atoms. Such compounds display considerable similarity to fluorine derivatives in their properties and chemical reactions, e.g. $(CN)_2$, $(SCN)_2$, IN_3, $(SCSN_3)_2$, ClO_2.
D Pseudo-Halogene
F ...
P pseudochlorowce
R псевдогалогены

PSEUDO-ORDER OF REACTION. See APPARENT ORDER OF REACTION.

PSEUDO-RESISTANCE POLARIZATION. See OHMIC PSEUDO-POLARIZATION.

2814
PSEUDO-UNIMOLECULAR REACTION.
First-order reaction of molecularity higher than one.
D pseudomonomolekulare Reaktion, quasiunimolekulare Reaktion
F réaction pseudo-monomoléculaire
P reakcja pseudojednocząsteczkowa
R псевдомономолекулярная реакция

p-TYPE SEMICONDUCTOR. See ELECTRON-DEFECT SEMICONDUCTOR.

2815
PULSED REACTOR.
Nuclear reactor reaching high power for periods of milliseconds.
D Impulsreaktor
F réacteur pulsatoire
P reaktor (jądrowy) impulsowy
R импульсный реактор, мигающий реактор

2816
PULSE IRRADIATION.
Irradiation with ionizing radiation in pulses of microsecond, nanosecond or as short as pico-second duration.
D Pulsbestrahlung
F irradiation pulsée
P napromienianie impulsowe
R импульсное облучение

2817
PULSE POLAROGRAPHY.
Variation of polarography where current or charge is measured during application to indicator electrode of short pulses potential of amplitude linearly increasing with time. In another modification short pulses of low-amplitude potential are superimposed on potential linearly increasing with time.
D Pulspolarographie
F polarographie á impulsions
P polarografia (im)pulsowa
R пульсполярография

2818
PULSE RADIOLYSIS.
Radiolysis with a large dose of ionizing radiation over period of micro- or nanoseconds.
D Pulsradiolyse
F radiolyse éclair
P radioliza impulsowa
R импульсный радиолиз

2819
PURE CHEMISTRY.
Branch of chemistry which covers research carried out for purely scientific purposes.
D reine Chemie
F chimie pure
P chemia czysta
R чистая химия

PURE ELEMENT. See SIMPLE ELEMENT.

2820
PURE REAGENT.
D reines Reagens
F ...
P odczynnik czysty
R чистый реактив

PURE SUBSTANCE. See SUBSTANCE.

2821
PUSH-PULL MECHANISM.
Trimolecular mechanism of nucleophilic substitution describing reaction as result of simultaneous pushing of carbon atom engaged in substitution by nucleophile Y, and pulling of substituent X by solvent (or electrophile), e.g.

$$Y: + \overset{|}{\underset{|}{-C}}-X + ROH \rightarrow \left[Y \cdots \overset{|}{\underset{\diagdown}{C}} \cdots X \cdots H - OR \right] \rightarrow$$
$$\text{activated complex}$$

$$\rightarrow Y - \overset{|}{\underset{|}{C}} - + X^{\ominus} \cdots HOR$$

D Stoß-Zug-Mechanismus
F mécanisme push-pull

P mechanizm push-pull, mechanizm
  nacisku i odciągania
R пуш-пульный механизм,
  ударно-тянущий механизм

2822
PYROELECTRIC.
Solid exhibiting pyroelectric effect.
D Pyroelektrikum
F pyro-électrique
P piroelektryk
R пироэлектрик

2823
PYROELECTRIC EFFECT.
(pyroelectricity).
Appearance of macroscopic dipole
moment due to temperature changes.
D pyroelektrischer Effekt, Pyroelektrizität
F pyro-électricité
P efekt piroelektryczny
R пироэлектрический эффект,
  пироэлектричество

PYROELECTRICITY. *See*
PYROELECTRIC EFFECT

2824
PYROLYSIS.
Decomposition of chemical compound by
heat.
D Pyrolyse
F pyrolyse, thermolyse
P piroliza
R пиролиз

2825
PYROSOL.
Colloidal system stable only at higher
temperatures, e.g. gold in fused borax.
D Pyrosol
F pyrosol
P pyrozol
R пирозоль

2826
Q-BRANCH.
In rotation-oscillation band, set of
transitions which do not involve any
change of rotational quantum number.
D Q-Zweig
F branche Q
P gałąź Q, gałąź zerowa
R нулевая ветвь Q

2827
QUADRUPOLE.
System of two mutually compensating
electric (electric quadrupole) or magnetic
(magnetic quadrupole) moments.
The following quadrupoles may be
distinguished according to relative
position of dipoles: (a) regular, (b) axial,
(c) general.

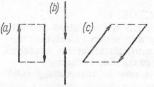

D Quadrupol
F quadrupôle
P kwadrupol
R квадруполь

2828
QUADRUPOLE MOMENT, Q.
Quantity which determines electric field
of quadrupole. Symmetric second-rank
tensor, traceless. In Cartesian coordinate
system quadrupole moment of any charge
distribution is given by:

$$Q_{xx} = \sum_i q_i x_i^2, \qquad Q_{xy} = Q_{yx} = \sum_i q_i x_i y_i$$

$$Q_{xz} = Q_{zx} = \sum_i q_i x_i z_i, \qquad Q_{yy} = \sum_i q_i y_i^2$$

$$Q_{yz} = Q_{zy} = \sum_i q_i y_i z_i, \qquad Q_{zz} = \sum_i q_i z_i^2$$

where $q_i = $ i-th charge value,
$x_i, y_i, z_i = $ its coordinates.
D Quadrupolmoment
F moment quadrupolaire
P moment kwadrupolowy
R квадрупольный момент

2829
QUADRUPOLE RELAXATION.
Relaxation in nuclear magnetic
resonance determined by interaction
between quadrupole nuclear moments and
surrounding inhomogeneous electric field.
D Quadrupolrelaxation
F relaxation quadrupolaire
P relaksacja kwadrupolowa
R квадрупольная релаксация

2830
QUALITATIVE ANALYSIS.
Detection or identification of components
of investigated sample.
D qualitative Analyse
F analyse qualitative
P analiza jakościowa
R качественный состав

2831
QUALITATIVE COMPOSITION (*of*
sample).
Chemical individuals which form sample.
D chemische Art der Bestandteile
F composition qualitative
P skład jakościowy
R качественный анализ

2832
QUALITY FACTOR, k.
Factor from which hypothetical radiation
dose producing any given effect may be
calculated on multiplication by absorbed
dose from the particular radiation.
Remark: Term is used in radiobiology only;
it is used seldom in radiation chemistry, since
it is not precise enough.

D Qualitätsfaktor
F facteur de qualité
P współczynnik jakości promieniowania
R коэффициент качества излучения

2833
QUANTITATIVE ANALYSIS.
Quantitative determination of components
of analysed sample.

D quantitative Analyse
F analyse quantitative
P analiza ilościowa
R количественный анализ

2834
QUANTITATIVE COMPOSITION (of
sample).
Proportions (for instance in per cent) of
chemical individuals forming sample.

D Mengenverhältnis
F composition quantitative
P skład ilościowy
R количественный состав

2835
QUANTIZATION.
Choice of physically acceptable solutions
of eigenvalue problem of given quantum
mechanical operator. Quantization leads to
conclusion that only some of
mathematically acceptable eigenvalues
are realized in nature.

D Quantelung
F quantification
P kwantowanie, kwantyzacja
R квантование

2836
QUANTUM CHEMISTRY.
Branch of theoretical chemistry which
deals with study of chemical systems
by using quantum-mechanical methods.

D Quantenchemie
F chimie quantique
P chemia kwantowa
R квантовая химия

2837
QUANTUM ENERGY (photon energy).
Photon energy usually given in
electronvolts; in SI units expressed in
joules.

D Quantenenergie
F énergie quantique
P energia kwantu promieniowania
R энергия кванта излучения

QUANTUM LENGTH. *See* THERMAL
DE BROGLIE WAVELENGTH.

QUANTUM-MECHANICAL LIOUVILLE
EQUATION. *See* VON NEUMANN'S
EQUATION.

2838
QUANTUM MECHANICAL OPERATORS.
In quantum mechanics, operators
replacing dynamic variables of classical
mechanics; constructed according to
quantum mechanics postulates.

D Operatoren der Quantenmechanik
F opérateurs de la mécanique quantique
P operatory kwantowo-mechaniczne
R квантово-механические операторы

2839
QUANTUM MECHANICAL
POSTULATES (postulates of quantum
mechanics).
Four postulates which establish following
assumptions:
(1) for each quantum mechanical state of
system there is wave function depending
on either generalized coordinates, or
generalized momenta of particles;
(2) superposition principle is valid;
(3) each dynamic variable is represented
by quantum mechanical operator with
given rule for calculation of average
values;
(4) time evolution of system is given by
solution of time-dependent Schrödinger
equation.

D quantenmechanische Postulate
F postulats de la mécanique ondulatoire
P postulaty mechaniki kwantowej
R постулаты квантовой механики

2840
QUANTUM MECHANICAL STATE.
State of system with definite wave
function.

D quantenmechanischer Zustand
F état quantique
P stan kwantowo-mechaniczny układu
R квантовое состояние

2841
QUANTUM MECHANICS (wave
mechanics).
Branch of theoretical physics dealing with
description of micro-particle systems.

D Quantenmechanik
F mécanique quantique, mécanique
   ondulatoire
P mechanika kwantowa, mechanika
   falowa
R квантовая механика

2842
QUANTUM THEORY.
Theory of physical phenomena exhibiting
energy discontinuity.

D Quantentheorie
F théorie des quanta
P teoria kwantów
R квантовая теория

2843
QUANTUM YIELD OF LUMINESCENCE.
Ratio between number of photons emitted
in luminescent process (*see* luminescence)
and total number of photons absorbed.

D Quantenausbeute der Lumineszenz
F rendement quantique de la luminescence
P wydajność kwantowa luminescencji
R квантовый выход люминесценции

2844
QUARKS.
Hypothetical particles of electric charge
equal to 1/3, −1/3 or −2/3 of elementary
charge. Quarks are considered to be
building blocks of strongly interacting
elementary particles (hadrons).

D Quarks
F quarks
P kwarki
R кварки

2845
QUASI-RACEMATES.
Equimolecular compounds formed of two
optically active compounds of similar
structure and opposite configurations,
existing in solid state only, e.g.
(+)-chlorosuccinic and (−)-bromosuccinic
acids form quasi-racemates in solid state.

D Quasiracemate
F composés quasi-racémiques
P związki quasi-racemiczne
R квазирацемические соединения

QUASISTATIC CHANGE. *See*
QUASISTATIC PROCESS.

2846
QUASISTATIC PROCESS (quasistatic
change).
Thermodynamic process with
thermodynamic equilibrium (at any
instant) inside system and without any
dissipative processes within this system.
According to second law of
thermodynamics each quasistatic process
is reversible and each reversible process
is quasistatic.

D quasistatischer Prozeß
F transformation quasi-statique
P przemiana quasi-statyczna, proces
   quasi-statyczny
R равновесный процесс,
   квазистатический процесс

2847
QUENCHING OF LUMINESCENCE.
Drop in quantum yield of luminescence of
substance A caused either by admixture
of substance B, due to intermolecular
transfer of energy from A to B or to their
chemical reaction, or by increase in
temperature (due to increased rates of
non-radiative transitions).

D Löschung der Lumineszenz
F extinction de luminescence
P wygaszanie luminescencji
R затухание люминесценции

2848
QUINHYDRONE ELECTRODE.
Electrode (half-cell) constructed from
platinum (or gold) dipped into
quinhydrone solution; electrode reaction is

$$C_6H_4(OH)_2 + 2 H_2O \rightleftarrows C_6H_4O_2 + 2 H_3O^+ + 2e$$

and potential of electrode

$$\varepsilon = \varepsilon^0 + \frac{RT}{2F} \ln \frac{a_{quinone} a^2_{H_3O^+}}{a_{quinhydrone}}$$

where ε^0 = standard electrode potential.

D Chinhydronelektrode
F électrode à quinhydrone
P elektroda chinhydronowa
R хингидронный электрод

2849
RABBIT.
Device used for transfer of short-lived
radioisotopes by means of pneumatic-tube
conveyor.

D Rohrpostbüchse, Rohrpostkapsel
F furet, cartouche
P kaseta poczty pneumatycznej
R контейнер пневмотранспортера

RACEMATE. *See* RACEMIC COMPOUND.

RACEMATE. *See* RACEMIC
MODIFICATION.

RACEMATION. *See* RACEMIZATION.

2850
RACEMIC COMPOUND (*dl*-compound,
racemate).
One form of racemic modification in solid
state — mixture consisting of two
opposite enantiomers, paired up in unit
cell of crystal, which thus contains equal
numbers of (+) and (−) molecules.
Melting point of racemic compound is
generally higher than those of
enantiomers; other physical properties
are also different.

D racemische Verbindung, (echtes)
   Racemat
F composé racémique, racémique (vrai)
P związek racemiczny, związek (±),
   racemat
R рацемическое соединение, рацемат

2851
RACEMIC MIXTURE (conglomerate).
One form of racemic modification in solid
state — gross mixture of enantiomer
crystals; racemic mixture forms in
crystallization from solution of racemic
modification, each enantiomer
crystallizes separately. Melting point of
racemic mixture is lower than those of
enantiomers and eutectic is formed.
Solubility of racemic mixture is higher
than those of enantiomers.

D (racemisches) Konglomerat
F conglomérat racémique
P mieszanina racemiczna, konglomerat
R рацемическая смесь, конгломерат

2852
RACEMIC MODIFICATION (dextro-laevo
modification, racemate).
Equimolecular, optically inactive mixture
of enantiomeric molecules, existing in
gaseous and liquid states, also in
solution. In solid state, racemic
modification exists in one of three kinds
of crystals (depending on its chemical
character and crystallization conditions —
temperature and solvent): racemic
mixture, racemic compound or *dl* racemic
mixture.

D racemische Modifikation, Racemat
F mélange racémique, mélange inactif
dédoublable, racémique
P odmiana racemiczna
R рацемическая форма, рацемат

2853
RACEMIZATION (racemation).
Transformation of one of the enantiomers
of chiral compound, or non-equimolar
mixture of two enantiomers into
equimolar mixture with loss of optical
activity.

D Racemisierung
F racémisation
P racemizacja
R рацемизация

2854
RAD.
Obsolete unit of absorbed dose of ionizing
radiation corresponding to 100 ergs of
absorbed energy per 1 g of medium.
In SI system 1 rad = 0.01 Gy (*see* gray).
Remark: Abbreviation from radiation absorbed
dose.

D Rad
F rad
P rad
R рад

2855
RADIAL CHROMATOGRAPHY (disk
chromatography, circular technique).

Technique of paper chromatography in
which sample is transferred onto centre
(or close to it) of filter paper disk and
chromatogram is developed in direction
from centre towards edges.

D Rundfilter-Papierchromatographie,
Rundfiltertechnik
F chromatographie circulaire
P chromatografia krążkowa, technika
krążkowa
R радиальная хроматография

2856
RADIAL DISTRIBUTION FUNCTION, g(**r**).
Probability density of finding particle at
point **r** and another particle at position
r = 0.

D radiale Verteilungsfunktion
F fonction radiale de distribution, fonction
de distribution radiale
P funkcja korelacji radialna
R радиальная функция распределения

2857
RADIANT FLUX DENSITY (flux density).
Product of density of particles or photons
(number of particles or photons per unit
volume) and their mean velocity, its
dimension being $m^{-2}s^{-1}$.

D Strahlungsflußdichte, Flußdichte
F densité de flux (de radiation)
P gęstość strumienia (promieniowania)
R плотность (лучистого) потока

2858
RADIATION CATALYSIS.
Catalysis caused by influence of ionizing
radiation or by modification of catalyst by
ionizing radiation.

D Strahlenkatalyse
F catalyse sous irradiation
P kataliza radiacyjna
R радиационный катализ

2859
RADIATION CHEMISTRY.
Branch of chemistry dealing with
chemical effects resulting from absorption
of ionizing radiation.

D Strahlenchemie
F chimie des radiations
P chemia radiacyjna
R радиационная химия

2860
RADIATION COUNTER (counter).
Device for measuring radiation in which
an electrical pulse is produced when
a charged particle or photon is passing
through it.

D Zähler
F compteur
P licznik (promieniowania)
R счётчик

2861
RADIATION DAMAGE.
Undesirable permanent changes in
a material caused by absorption of
ionizing radiation.
D Bestrahlungsschäden, Strahlenschäden
F défaut dû à l'irradiation, dégat par
rayonnement
P uszkodzenie radiacyjne
R радиационное повреждение

2862
RADIATION DEGRADATION OF
POLYMER.
Degradation of polymer caused by ionizing
radiation.
D strahlenchemischer Abbau des Polymers
F dégradation du polymère par radiation
P degradacja polimeru radiacyjna
R радиационная деградация полимера

2863
RADIATION DETECTOR.
Device in which interaction of radiation
with matter is transformed into an easily
measurable physical quantity.
D Strahlendetektor, Strahlungsdetektor
F détecteur de rayonnement
P detektor promieniowania
R детектор излучения

2864
RADIATION DOSE.
General description of extent of influence
of ionizing radiation on object of
consideration; an unprecise definition.
D Strahlendosis, Dosis
F dose (de rayonnement)
P dawka promieniowania, doza
promieniowania
R доза (излучения)

RADIATION DOSIMETRY. *See*
DOSIMETRY.

2865
RADIATION EQUILIBRIUM.
In radiation chemistry, state in which
concentration of at least one of the
components of irradiated system is
constant with time. In radiation
equilibrium, reaction rates of formation
and disappearance of a chemical
individual (or individuals) are equal.
D strahlenchemisches Gleichgewicht
F ...
P równowaga radiacyjna
R радиационное равновесие

2866
RADIATION FIELD.
Space in which ionizing radiation is
dissipated.
D Strahlenfeld
F champ de rayonnement, champ
d'irradiation
P pole radiacyjne
R поле излучения

2867
RADIATION INDUCED ACTIVATION.
Process of incidental induction of
radioactivity in material irradiated with
ionizing radiation which produces
chemical or biological effects.
D Aktivierung mit Strahlung
F activation par la radiation
P aktywacja radiacyjna
R радиационная активация

RADIATION INDUCED
LUMINESCENCE. *See*
RADIOLUMINESCENCE.

2868
RADIATION INDUCED MUTATION.
Abrupt change of properties of plant or
animal organisms, inherited by progeny
and caused by absorption of energy
of ionizing radiation in certain stage of
developement or reproduction of living
organism.
D strahleninduzierte Mutation
F mutation radio-induite
P mutacja radiacyjna
R мутация вызванная излучением

2869
RADIATION INDUCED REACTIONS.
Chemical reactions initiated by ionizing
radiation.
D strahlenchemische Reaktionen,
strahlungschemische Prozesse
F réactions chimiques des radiations
P reakcje radiacyj(nochemicz)ne
R радиационно-химические реакции

RADIATION INTENSITY. *See*
LUMINOUS INTENSITY.

2870
RADIATIONLESS TRANSITION
(non-radiative transition).
Transition between energy states of
system without emission of photon.
D strahlungsloser Übergang
F transition non radiative, transition sans
radiation
P przejście bezpromieniste, przejście
bezradiacyjne
R безызлучательный процесс,
безызлучательный переход

2871
RADIATION PROTECTION.
All provisions to protect personnel
against action of ionizing radiation and
personnel and laboratories against
radioactive contamination; and for
decontamination.
D Strahlenschutz
F protection contre les radiations,
radioprotection
P ochrona przed promieniowaniem,
ochrona radiologiczna
R защита от излучений

2872
RADIATION QUALITY.
Value or values characterizing ionizing
radiation in terms of its penetrability.
D Strahlenqualität, Qualität
F qualité (du rayonnement)
P jakość promieniowania
R качество (излучения)

RADIATION SENSITIZER. See
RADIOSENSITIZER.

2873
RADIATION SOURCE.
Device or substance capable of emitting
ionizing radiation. Sources most often
used are: isotopic sources, nuclear
reactors, accelerators and X-ray tubes.
D Strahlenquelle
F source de rayonnement
P źródło promieniowania (jonizującego)
R источник излучения

2874
RADIATION YIELD, G_x.
Number of indexed chemical changes (x)
due to irradiation, per 100 eV of absorbed
energy of ionizing radiation.
D (chemische) Ausbeute der
Strahlenwirkung
F rendement radiolytique, rendement
radiochimique
P wydajność radiacyjna
R радиационный выход,
радиационно-химический выход

2875
RADIATIVE LIFE-TIME.
Life-time of electronically excited state
given by reciprocal of rate constant of
luminescence originating from this
excited state.
D ...
F durée de vie moyenne
P czas życia radiacyjny
R излучательное время жизни

2876
RADIATIVE TRANSITION.
Transition between higher and lower
energy state of system accompanied by
photon emission.
D strahlender Übergang
F transition radiative
P przejście promieniste, przejście
radiacyjne
R переход с излучением

RADICAL. See FREE RADICAL.

RADICAL DIFFUSION MODEL. See
DIFFUSION MODEL OF RADIOLYSIS.

RADICAL ION. See ION RADICAL.

2877
RADICAL PRODUCTS OF
RADIOLYSIS.
Radiolysis products containing unpaired
electron, e.g. e^-_{aq}, H·, HO· in case of
water.
D Radikalprodukte der Radiolyse
F produits radicalaires de radiolyse
P produkty radiolizy rodnikowe
R радикальные продукты радиолиза

RADICAL REACTION. See
FREE-RADICAL REACTION.

2878
RADIOACTINIUM, RdAc.
Isotope of thorium. Atomic number 90.
Mass number 227.
D Radio-Actinium, Radioaktinium
F radio-actinium
P rad(i)oaktyn
R радиоактиний

RADIOACTIVATION. See ACTIVATION.

RADIOACTIVATION ANALYSIS. See
ACTIVATION ANALYSIS.

2879
RADIOACTIVE DATING.
Determination of age of minerals,
materials or objects from their content of
radioactive nuclides or daughter products.
D radioaktive Altersbestimmung
F détermination de l'âge radioactif
P radiochronologia, radiodatowanie
R радиологическое определение возраста

2880
RADIOACTIVE DECAY (radioactive
disintegration).
Spontaneous transformation of unstable
nucleus into another nucleus with
emission of corpuscular or electromagnetic
radiation.
D radioaktive Umwandlung, radioaktiver
Zerfall
F désintégration radioactive, décroissance
radioactive
P przemiana promieniotwórcza, rozpad
promieniotwórczy

R радиоактивное превращение,
радиоактивный распад

2881
RADIOACTIVE DECAY CONSTANT, λ.
Proportionality coefficient in equation
describing law of radioactive decay; it
expresses probability of transformation
of the particular nucleus of
a radioactive nuclide per unit time.

D Zerfallskonstante
F constante de désintégration, constante
radioactive
P stała przemiany promieniotwórczej
R постоянная распада

RADIOACTIVE DISINTEGRATION. *See*
RADIOACTIVE DECAY.

RADIOACTIVE ELEMENT. *See*
RADIOELEMENT.

2882
RADIOACTIVE EQUILIBRIUM.
Statistical equilibrium between activities
of genetically connected radioactive
substances. It occurs between members of
radioactive series.

D radioaktives Gleichgewicht
F équilibre radioactif
P równowaga promieniotwórcza
R радиоактивное равновесие

2883
RADIOACTIVE FALL-OUT.
Radioactive material deposited on the
earth's surface following the explosion of
nuclear weapons.

D radioaktiver Niederschlag
F retombée radioactive
P opad radioaktywny, opad
promieniotwórczy
R радиоактивные осадки

RADIOACTIVE FAMILIES. *See*
RADIOACTIVE SERIES.

RADIOACTIVE HALF-LIFE. *See*
HALF-LIFE.

2884
RADIOACTIVE KRYPTONATE.
Solid into which radioactive krypton ^{85}Kr
in the loosely bonded form has been
introduced.

D radioaktives Kryptonat
F kryptonate radioactif
P kryptonat promieniotwórczy
R радиоактивный криптонат

2885
RADIOACTIVE PURITY.
Amount of radioactive nuclides of desired
type, expressed as percentage of all
radioactive nuclides present in substance.

D Kernreinheit
F pureté radioactive
P czystość promieniotwórcza
R радиоактивная чистота

2886
RADIOACTIVE SERIES (radioactive
families).
Series of genetically connected radioactive
nuclides each of which is formed by
radioactive decay of its predecessor. Each
series starts with a long-lived radioactive
nuclide and ends with a stable nuclide.

D radioaktive Zerfallsreihen
F familles radioactives, séries radioactives
P szeregi promieniotwórcze, rodziny
promieniotwórcze
R радиоактивные ряды,
радиоактивные семейства

2887
RADIOACTIVE SOURCE.
Source of ionizing radiation produced by
radioactive nuclide.

D isotope Strahlenquelle
F source radioactive
P źródło izotopowe
R изотопный источник

RADIOACTIVE STANDARD. *See*
REFERENCE SOURCE.

2888
RADIOACTIVE SUBSTANCE.
Substance containing natural or artificial
radioactive nuclides.

D radioaktiver Stoff
F matière radioactive
P substancja promieniotwórcza
R радиоактивное вещество

2889
RADIOACTIVE TRACER.
Radioactive nuclide which is introduced
into a system in order to examine it.

D radioaktives Leitisotop, radioaktiver
Indikator
F indicateur radioactif, traceur radioactif
P wskaźnik promieniotwórczy
R радиоактивный индикатор

2890
RADIOACTIVE TRACER METHOD.
Method for tracing course of chemical
and physical processes by means of
compounds labelled with radioactive
tracer. Absence of isotopic effects is
essential for applicability of this method.

D radioaktive Indikatormethode
F méthode des indicateurs radioactifs
P metoda wskaźników
promieniotwórczych
R метод радиоактивных индикаторов

2891
RADIOACTIVE WASTES.
Materials or objects containing radioactive substances, further use of which is impossible or unprofitable.

D radioaktive Abfälle
F déchets radioactifs
P odpady promieniotwórcze
R радиоактивные отходы

2892
RADIOACTIVITY.
Spontaneous nuclear transformation of nuclide into another nuclide, accompanied by emission of nuclear radiation, either corpuscular or electromagnetic.

D Radioaktivität
F radioactivité
P promieniotwórczość, radioaktywność
R радиоактивность

RADIOAUTOGRAPHY. See
AUTORADIOGRAPHY.

2893
RADIOCALCIUM (calcium-45).
Isotope of calcium. Atomic number 20.
Mass number 45.

D Radiocalcium, Radiokalzium,
   Kalzium-45
F radiocalcium
P radiowapń, wapń-45
R кальций-45, радиокальций

2894
RADIOCARBON (natural radiocarbon).
Isotope of carbon. Atomic number 6. Mass number 14.

D Radiokohlenstoff, Radiokarbon
F radiocarbone
P radiowęgiel, węgiel-14
R радиоуглерод

2895
RADIOCHEMICAL ANALYSIS
(radiochemical analytical methods,
analytical radiochemistry).
Detection or determination of components of substance examined based on detecting and measuring radiation emitted by radioactive nuclides.

D radiochemische Analyse
F analyse radiochimique
P analiza radiochemiczna
R радиохимический анализ

RADIOCHEMICAL ANALYTICAL
METHODS. See RADIOCHEMICAL
ANALYSIS.

2896
RADIOCHEMICAL PURITY (of
a radioactive preparation).
Absence of impurities from other radioactive substances.

D radiochemische Reinheit
F pureté radiochimique
P czystość radiochemiczna
R радиохимическая чистота

2897
RADIOCHEMICAL YIELD.
Percentage recovery of an element after chemical separation.

D chemische Ausbeute
F rendement chimique
P wydajność (radio)chemiczna
R химический выход

2898
RADIOCHEMISTRY.
Chemistry and physical chemistry of radioactive substances.

D Radiochemie
F chimie radioactive
P radiochemia
R радиохимия

2899
RADIOCHROMATOGRAPHY.
Chromatography of substances labelled with radioactive tracers, whose radioactivity is used subsequently for qualitative and/or quantitative evaluation of chromatograms.

D Radiochromatographie
F radiochromatographie
P radiochromatografia
R радиохроматография, хроматография
   радиоактивных веществ

2900
RADIOCOBALT (cobalt-60).
Isotope of cobalt. Atomic number 27.
Mass number 60.

D Radiokobalt
F radiocobalt
P radiokobalt, kobalt-60
R кобальт-60, радиокобальт

2901
RADIOCOLLOIDS.
Colloids composed of particles of radioactive substance with colloidal dispersion, or pseudoradiocolloids which are products of adsorption of radioactive substance on foreign colloidal particles.

D Radiokolloide
F radiocolloïdes
P koloidy promieniotwórcze, radiokoloidy
R радиоколлоиды

2902
RADIOELEMENT (radioactive element).
Chemical element having no stable natural isotopes.

D Radioelement
F radioélément
P pierwiastek promieniotwórczy
R радиоактивный элемент

2903
RADIOGRAPH.
Result of radiography, shadow image of
examined object recorded most often on
photographic plate.
D Radiogramm
F radiogramme
P radiogram
R радиограмма

2904
RADIOGRAPHY.
Use of ionizing radiation for studying
internal structure of solids;
a photographic emulsion being most often
used as the detector of radiation.
D Radiographie
F radiographie
P radiografia
R радиография

2905
RADIOIODINE (iodine-131).
Isotope of iodine. Atomic number 53.
Mass number 131.
D Radiojod, Jod-131
F radioiode
P radiojod, jod-131
R иод-131, радиоиод

2906
RADIOIRON (iron-59).
Isotope of iron. Atomic number 26. Mass
number 59.
D Radioeisen, Eisen-59
F radiofer
P radiożelazo, żelazo-59
R железо-59, радиожелезо

2907
RADIOLUMINESCENCE (radiation
induced luminescence).
Any kind of luminescence caused by
absorption of ionizing radiation.
D Radio(photo)lumineszenz
F radioluminescence
P luminescencja radiacyjna
R радиолюминесценция

2908
RADIOLYSIS.
Chemical decomposition of materials by
ionizing radiation.
D Radiolyse
F radiolyse
P radioliza
R радиолиз

2909
RADIOMETRIC ANALYSIS.
(1) Analytical techniques in which use is
made of various nuclear phenomena and
interaction between ionizing radiation and
matter, and also methods of radiotracers
for determination of non-radioactive
elements. (2) Method for determination of
a radioactive component of a mixture
based on measurement of its radioactivity.
D radiometrische Analyse
F analyse radiométrique
P analiza radiometryczna
R радиометрический анализ

2910
RADIOMETRIC TITRATION.
Titration in which titrant or substance
titrated is labelled with radioactive tracer
and end point is determined by
radiometric measurement.
D radiometrische Titration
F titrage radiométrique
P miareczkowanie radiometryczne
R радиометрическое титрование

2911
RADIOMETRY.
Methods for measuring radiation and
determining activity of radioactive
preparations.
D Radiometrie
F radiométrie
P radiometria
R радиометрия

2912
RADIONUCLIDE.
Unstable nuclide undergoing spontaneous
radioactive transformation with emission
of particle or photon.
D Radionuklid
F nucléide radioactif, radionucléide
P nuklid promieniotwórczy, radionuklid
R радиоактивный нуклид,
    радиоактивный изотоп

2913
RADIOPHOSPHORUS (phosphorus-32).
Isotope of phosphorus. Atomic number 15.
Mass number 32.
D Radiophosphor, Phosphor-32
F radiophosphore
P radiofosfor, fosfor-32
R фосфор-32, радиофосфор

2914
RADIORESISTANCE.
Property of material responsible for
specially low yields of radiolysis.
D Strahlenbeständigkeit
F radiorésistance
P odporność radiacyjna
R радиорезистентность

2915
RADIOSENSITIVITY.
Qualitative character and quantitative
extent of changes caused by absorption
of ionized radiation in pure chemical
compounds, mixtures of such compounds,
synthetic materials and living organisms.
D Strahlenempfindlichkeit
F radiosensibilité
P wrażliwość na promieniowanie
R радиочувствительность

2916
RADIOSENSITIZER (radiation sensitizer).
Chemical compound added to irradiated
system for purpose of enhancing
a particular chemical effect.
D Sensibilisator
F sensibilisateur pour la réaction
radiochimique
P sensybilizator radiacyjny, uczulacz
reakcji radiacyjnej
R сенсибилизатор для радиационной
реакции

2917
RADIOSODIUM (sodium-24).
Isotope of sodium. Atomic number 11.
Mass number 24.
D Radionatrium, Natrium-24
F radiosodium
P radiosód, sód-24
R натрий-24, радионатрий

2918
RADIOSPECTROSCOPY.
Branch of spectroscopy comprising
nuclear magnetic, electron spin and
nuclear quadrupole resonance phenomena
and others in which one observes
stimulated quantum transitions between
energy levels produced according to given
conditions (e.g. in magnetic field) and
corresponding to absorption or emission of
electromagnetic radiation in
radio-frequency region.
D Hochfrequenzspektroskopie
F spectroscopie hertzienne
P radiospektroskopia
R радиоспектроскопия

2919
RADIOSTRONTIUM (strontium-90).
Isotope of strontium. Atomic number 38.
Mass number 90.
D Strontium-90, Sr-Isotop 90
F radiostrontium
P radiostront, stront-90
R стронций-90

2920
RADIOSULFUR (sulfur-35).
Isotope of sulfur. Atomic number 16. Mass
number 35.

D Radioschwefel, Schwefel-35
F radiosoufre
P radiosiarka, siarka-35
R сера-35, радиосера

2921
RADIOTHERMOLUMINESCENCE.
Thermoluminescence of solid substance
irradiated previously with ionizing
radiation.
D Radiothermolumineszenz
F radiothermoluminescence
P termoluminescencja radiacyjna
R радиотермолюминесценция

2922
RADIOTHORIUM, RdTh.
Isotope of thorium. Atomic number 90.
Mass number 228.
D Radiothorium
F radiothorium
P radiotor
R радиоторий

2923
RADIOTOXICITY.
Toxicity connected with radiation of
radioactive nuclide present in organism.
D Radiotoxizität
F radiotoxicité
P radiotoksyczność
R токсичность радиоактивных веществ

2924
RADIUM, Ra.
Atomic number 88.
D Radium
F radium
P rad
R радий

2925
RADIUM D, RaD.
Isotope of lead. Atomic number 82. Mass
number 210.
D Radium D
F radioplomb, radium D
P radioołów, rad D, RaD
R радий D

2926
RADIUM EMANATION, RaEm.
Isotope of radon. Atomic number 86. Mass
number 222.
D Radium-Emanation
F émanation du radium
P emanacja radowa
R радон-222, эманация радия

2927
RADIUM G (radium lead), RaG.
Isotope of lead. Atomic number 82. Mass
number 206.
D Uranblei, Radium G, Radiumblei
F urano-plomb, plomb de l'uranium,
radium G

P radoołów, rad G
R радий G

RADIUM LEAD. *See* RADIUM G.

2928

RADIUS OF IONIC ATMOSPHERE, $\frac{1}{\varkappa}$.

Concept from Debye-Hückel theory of strong electrolytes

$$\varkappa^2 = \frac{8\pi e_0^2 N_A}{1000 \varepsilon_r kT} J$$

Introduction of numerical values of constants gives

$$\frac{1}{\varkappa} = 1.988 \cdot 10^{-10} \sqrt{\frac{\varepsilon_r T}{J}} \text{ [cm]}$$

where e_0 = charge of electron, N_A = Avogadro's constant, ε_r = relative permittivity, k = Boltzmann constant and J = ionic strength.

D Radius der Ionenatmosphäre
F épaisseur de l'atmosphère ionique
P promień atmosfery jonowej
R радиус ионной атмосферы

2929
RADON, Rn.
Atomic number 86.

D Radon
F radon
P radon
R радон

2930
RADZISZEWSKI AMIDE
PREPARATION.
Synthesis of amides by treating nitriles with hydrogen peroxide in basic medium, e.g.

$$RCN + 2H_2O_2 \xrightarrow[40°]{NaOH} RCONH_2 + O_2 + H_2O$$

D ...
F ...
P synteza amidów Radziszewskiego
R ...

2931
RADZISZEWSKI SYNTHESIS OF
IMIDAZOLES.
Preparation of imidazole derivatives by heating mixture of 1,2-diketone, ammonia and aldehyde in ethanolic solution, e.g.

D ...
F ...

P synteza pochodnych imidazolu
Radziszewskiego
R ...

2932
RAIES ULTIMES (persistent lines *in emission spectral analysis*).
Analytical lines of a given element which disappear last when concentration of this element in analytical gap is reduced to zero.

D letzte Linien
F raies ultimes
P linie ostatnie
R последние линии, чувствительные линии

2933
RAMAN EFFECT (Raman scattering).
Inelastic scattering of electromagnetic radiation. In spectrum of scattered light original frequency is accompanied by lower and higher frequencies.

D Raman-Effekt
F effet Raman
P efekt Ramana, rozpraszanie ramanowskie
R комбинационное рассеяние света, эффект Рамана

2934
RAMAN FREQUENCY (Raman shift).
Wave number of rotational or vibrational transition recorded from Raman spectrum as difference of Rayleigh and Raman line positions.

D Raman-Frequenz
F fréquence de Raman
P częstość ramanowska, przesunięcie ramanowskie
R частота линии комбинационного рассеяния

2935
RAMAN LINE.
Spectral line of scattered electromagnetic radiation with frequency either decreased (Stokes line) or increased (anti-Stokes line) with respect to frequency of excitation line (Rayleigh line). Frequency of Raman line corresponds to rotational or oscillation transitions in target molecules.

D Raman-Linie
F raie (de) Raman
P linia ramanowska
R линия комбинационного рассеяния

RAMAN SCATTERING. *See* RAMAN
EFFECT.

RAMAN SHIFT. *See* RAMAN
FREQUENCY.

2936
RAMAN SPECTROMETRY.
Method of analysis utilizing Raman
spectra for identification and qualitative
and quantitative analysis, and for
determination of primary components in
multi-component mixtures.
D Raman-Spektrometrie
F spectrométrie Raman
P analiza ramanowska
R анализ по спектрам комбинационного
рассеяния

2937
RAMAN SPECTROSCOPY.
Branch of molecular spectroscopy dealing
with investigation of Raman spectra.
D Raman-Spektroskopie
F spectroscopie Raman
P spektroskopia ramanowska
R спектроскопия комбинационного
рассеяния

2938
RAMAN SPECTRUM.
Spectrum of inelastic radiation scattering
on molecules of a given sample; besides
original frequency of scattered radiation,
it also contains lower and higher
frequencies.
D Raman-Spektrum
F Raman spectre
P widmo Ramana
R спектр комбинационного рассеяния

RANDOM CHOICE. *See* RANDOM
SELECTION.

2939
RANDOM ERROR (accidental error).
Error due to set of random factors active
during measurement. Value and direction
of random error does not follow any
regularity.
D zufälliger Fehler, Zufallsfehler
F erreur accidentelle, erreur aléatoire
P błąd przypadkowy
R случайная ошибка

2940
RANDOM NUMBERS.
Values of random variable which depend
on presence or absence of some random
event of a given probability distribution.
D Zufallszahlen
F nombres aléatoires
P liczby losowe
R случайные числа

2941
RANDOM SELECTION (random choice).
Method of sampling with probability of
drawing into series being at time of
selection identical for all units drawn
belonging to a given population.

D zufällige Auswahl
F échantillonnage simple, échantillonnage
au hasard
P wybór losowy
R случайный выбор

2942
RANDOM SOLID SOLUTION.
Solid solution in which atoms or ions of
constituents are randomly distributed in
lattice.
D ...
F solution solide désordonné
P roztwór stały nieuporządkowany
R неупорядоченный твёрдый раствор

2943
RANDOM VARIABLE (variate).
Real function which assigns given
numerical value (e.g. result of
measurement, analytical determination) to
random event (e.g., measurement,
analytical determination).
D Zufallsveränderliche, zufällige Variable
F variable aléatoire
P zmienna losowa
R случайная переменная

2944
RANGE.
Statistic characterizing dispersion; equal
to difference between the largest and the
smallest value in series (e.g. series of
measurement results).
D Variationsbreite
F étendue d'une série statistique
P rozstęp
R ширина дисперсии

2945
RANGE OF NUCLEAR FORCES.
Radius of potential well of interaction
between nucleons, i.e. distance of order of
10^{-13} cm, outside which interaction
between two nucleons may be neglected
in comparison to other interactions.
D Reichweite der Kernkräfte
F rayon d'action des forces nucléaires
P zasięg sił jądrowych
R радиус (взаимо)действия ядерных сил

2946
RANGE OF PARTICLE.
Maximum distance to which a particle
(α, β or heavy ion) will penetrate a given
medium. Range is a function of energy
of particles of specified type.
D Reichweite des Teilchen
F parcours de particule
P zasięg cząstki
R пробег частицы

2947
RAOULT'S LAW.
Partial pressures at constant temperature
of vapours of constituents of perfect

solution are proportional to concentrations of these constituents in liquid phase, expressed in mole fractions

$$p_B = p_B^* x_B$$

where p_B = partial pressure of constituent B in gas over solution, p_B^* = vapour pressure of pure constituent B, x_B = mole fraction of constituent B in liquid phase.

D Raoultsches Gesetz
F loi de Raoult
P prawo Raoulta
R закон Рауля

2948
RAPID COAGULATION.
Coagulation of colloidal particles completely discharged when their electrokinetic potential ζ is equal to zero. Rate of coagulation is independent of both electrokinetic concentration and particular features of ions causing coagulation.

D schnelle Koagulation
F coagulation rapide
P koagulacja szybka
R быстрая коагуляция

RARE GASES. *See* INERT GASES.

RATE CONSTANT. *See*
REACTION-RATE CONSTANT.

2949
RATE-DETERMINING STEP (*in the reaction*).
Slowest step determinating overall rate of process.

D geschwindigkeitsbestimmendes
   Reaktionsstadium
F (*stade qui règle la vitesse de réaction*)
P stadium określające szybkość reakcji
R стадия определяющая скорость
   (реакции)

RATE EQUATION. *See* KINETIC
EQUATION.

RATES. *See* THERMODYNAMIC
FLUXES.

2950
RAYLEIGH LINE.
Scattered electromagnetic radiation spectral line of the same frequency as that of incident radiation.

D Rayleigh-Linie
F raie de Rayleigh
P linia rayleighowska
R рэлеевская линия

2951
RAYLEIGH SCATTERING.
Scattering of light on molecules without frequency change in scattered light.

D Rayleigh-Streuung
F diffusion de Rayleigh

P rozpraszanie klasyczne, rozpraszanie
   rayleighowskie
R рэлеевское рассеяние

2952
R BAND.
Band in electronic spectra of organic molecules containing heteroatoms; R band corresponds to n-π^* electronic transitions and is of much lower intensity than K band.

D R-Band
F bande R
P pasmo R
R полоса R

2953
R-BRANCH.
In rotation-oscillation band, set of transitions corresponding to change of rotational quantum number by +1.

D R-Zweig
F branche R
P gałąź R, gałąź dodatnia
R положительная ветвь R

REACTANT. *See* SUBSTRATE.

2954
REACTING SUBSTANCE.
Substance undergoing chemical reaction. This term covers both substrates and products of reaction.

D Reagent
F substance réagissante
P reagent, substancja reagująca
R реагирующее вещество, реагент

2955
REACTION (*of solution*).
Characteristic feature of electrolyte solution displayed by its value of hydrogen ion exponent (i.e. negative logarithm of hydrogen ion activity called pH or p_H in aqueous solutions; neutral reaction is achieved when pH = 7).

D Reaktion
F réaction
P odczyn roztworu
R реакция

2956
REACTION COORDINATE.
Coordinate (e.g. distance between atoms or time) denoting path of potential energy changes in system of particles of elementary reaction during transition from substrates over activated complex to products.

D Reaktionskoordinate,
   Reaktionsparameter
F ...
P współrzędna reakcji
R координата реакции

2957
REACTION CROSS-SECTION.
Measure of probability of definite nuclear reaction.

D Reaktionsquerschnitt, Wirkungsquerschnitt der Reaktion
F section efficace de la réaction
P przekrój czynny reakcji jądrowej
R сечение ядерного процесса, эффективное сечение реакции

REACTION EQUILIBRIUM. See CHEMICAL EQUILIBRIUM.

2958
REACTION FIELD.
Part of local electric field induced in cavity obtained by removing one molecule of dielectric.

D Reaktionsfeld
F champ de réaction
P pole reakcji, pole oddziaływań
R реактивное поле

REACTION ISOCHORE. See VAN'T HOFF'S ISOCHORE.

2959
REACTION ISOTHERM.
Relation between chemical affinity A and thermodynamic chemical equilibrium constant K_a and activities a_B of reagents before reaching chemical equilibrium

$$A = RT \ln \frac{K_a}{\prod_B a_B^{\nu_B}}$$

where R = molar gas constant,
T = thermodynamic temperature,
ν_B = stoichiometric coefficient of reactant
B, $\prod_B$ = product of all reactants.

D van't Hoffsche Reaktionsisotherme
F isotherme de réaction
P izoterma reakcji, izoterma van't Hoffa
R изотерма реакции, уравнение изотермы химической реакции

REACTION KINETICS. See CHEMICAL KINETICS.

2960
REACTION MECHANISM.
Description of subsequent changes which substrate undergoes to yield reaction product.

D Reaktionsmechanismus
F mécanisme de réaction
P mechanizm reakcji
R механизм реакции

2961
REACTION OF FRACTIONAL ORDER.
Reaction with kinetic equation characterized by fractional sum of exponents in which concentrations of reactants occur. Examples of reactions of fractional order are some chain reactions and catalyzed reactions.

D Reaktion gebrochener Ordnung
F réaction d'ordre fractionnaire, réaction d'ordre non entier
P reakcja ułamkowego rzędu
R реакция дробного порядка

2962
REACTION PATH.
Population of all subsequent values of potential energy in system of reacting molecules during its transition from stage of substrates to stage of products through transition state.

D Reaktionsweg
F voie de réaction, chemin de réaction, cours de réaction
P droga reakcji
R путь реакции, реакционный путь

2963
REACTION POLARIZATION (chemical polarization).
Type of concentration polarization consisting in slow achievement of equilibrium of chemical reactions preceding or following electrode process proper, in comparison to rate of electrode reaction. Reaction polarization is expressed by reaction overvoltage.

D Reaktionspolarisation
F polarisation de réaction
P polaryzacja reakcyjna
R реакционная поляризация, химическая поляризация

2964
REACTION PRODUCT.
Substance formed as result of chemical reaction. In stoichiometric equation corresponding symbol (formula) appears on its right.

D Reaktionsprodukt, Produkt, Endprodukt, Endstoff
F produit de réaction
P produkt reakcji
R (конечный) продукт реакции

2965
REACTION RATE (chemical reaction rate), v.
Diminution of mass of substrate referred to unit time and volume. In homogeneous system, v is equal to loss of substrate concentration in unit time and is

expressed as derivative of concentration c of substrate with respect to time t, $v = -dc/dt$.

D Reaktionsgeschwindigkeit
F vitesse de réaction
P szybkość reakcji (chemicznej)
R скорость (химической) реакции

2966
REACTION-RATE CONSTANT (rate constant), k.
Coefficient, constant at defined temperature and pressure, specific for given reaction, occurring in kinetic equation for reaction $v = kc^n$, where v = reaction rate, c = substrate concentration, n = order of reaction.

D Reaktionsgeschwindigkeitskonstante, Geschwindigkeitskonstante
F constante de vitesse
P stała szybkości reakcji
R константа скорости реакции

2967
REACTION RATE METHODS.
Methods of analysis based on measurement of reaction rates and interpretation of values obtained in terms of concentration.

D ...
F ...
P metody analizy kinetyczne
R кинетические методы анализа

REACTION VARIABLE. *See* EXTENT OF REACTION.

2968
REACTIVITY INDICES (chemical reactivity indices).
Numbers describing quantitatively reactivity of given position in molecule for different reaction types.

D Indizes chemischer Reaktionsfähigkeit
F indices de réactivité (chimique)
P indeksy reaktywności
R индексы реакционной способности

REACTOR CHANNEL FOR EXPERIMENTS. *See* BEAM HOLE.

2969
REAGENT.
Chemical substance used in laboratories.

D (chemisches) Reagens
F réactif (chimique)
P odczynnik
R реактив, реагент

2970
REAGENT'S PURITY.
Purity determined as content of impurities in reagent (in weight per cent).

D Reinheit (des Reagens)
F pureté (des réactifs)
P czystość odczynnika
R чистота (реагентов)

2971
REAL GAS (imperfect gas, non-ideal gas).
Gas which does not satisfy equation of state for ideal gas. Particles of real gas interact with each other.

D reales Gas
F gaz réel
P gaz rzeczywisty, gaz niedoskonały
R реальный газ, неидеальный газ

2972
REAL SOLUTION (non-ideal solution).
Solution with properties differing from those of ideal mixture.

D reale Lösung, reale kondensierte Mischphase, nichtideale Lösung
F solution réelle, solution non idéale
P roztwór rzeczywisty, roztwór niedoskonały
R реальный раствор, неидеальный раствор

2973
REARRANGEMENT.
Process proceeding against principle of least structural changes, involving modification of systems of arrangements of atoms (or groups of atoms) and covalent bonds in molecule, thus causing structural isomerization of the latter.

D Umlagerung
F transposition
P przegrupowanie
R перегруппировка

2974
REARRANGEMENT-1,2.
Rearrangement of organic molecule involving migration of atom or group of atoms from given atom to adjacent one, e.g.

$$X-A-B \rightarrow A-B-X$$

D 1,2-Umlagerung, 1,2-Verschiebung
F déplacement-1,2
P przegrupowanie 1,2
R перегруппировка-1,2

2975
RECIPROCAL LINEAR DISPERSION (*of spectrograph*).
Reciprocal value of linear dispersion usually expressed in nanometres per milimetre.

D reziproke (lineare) Dispersion
F dispersion linéaire réciproque, dispersion inverse
P dyspersja odwrotna
R обратная дисперсия

2976
RECIPROCITY LAW (Bunsen-Roscoe law).
Amount of products obtained in any photochemical reaction is directly proportional to product of intensity of incident light and time of its action.

D Bunsen-Roscoesches Gesetz
F loi de reciprocité
P prawo Bunsena i Roscoe
R закон Бунзена-Роско

RECOIL ATOMS. *See* HOT ATOMS.

RECOIL EFFECT. *See*
SZILARD-CHALMERS EFFECT.

2977
RECOMBINATION OF FREE RADICALS.
Reaction between two free radicals resulting in molecules or ions.

D Rekombination der Radikale
F récombinaison de radicaux
P rekombinacja rodników
R рекомбинация (свободных) радикалов

2978
RECORDING SPECTROPHOTOMETER.
Spectrophotometer, most frequently double beam, recording spectrophotometric curve in form of plot in a given coordinate system.

D registrierendes Spektralphotometer
F spectrophotomètre enregistreur
P spektrofotometr rejestrujący
R регистрирующий спектрофотометр

2979
REDOX ELECTRODE.
Electrode (half-cell) usually constructed from noble metal dipped into solution which contains ions or compounds of different oxidation states, e.g. $Pt|Fe^{3+}$, Fe^{2+} or $Pt|$quinone, hydroquinone; electrode reaction is

$$R \rightleftharpoons Ox + ze$$

where R and Ox = reduced and oxidized form respectively; electrode potential is

$$\varepsilon = \varepsilon^0 + \frac{RT}{zF} \ln \frac{a_{Ox}}{a_R}$$

where ε^0 = standard electrode potential.

D Redoxelektrode
F électrode à système oxydo-reducteur, électrode du type redox
P elektroda utleniająco-redukująca, elektroda redoks(owa)
R окислительно-восстановительный электрод, редокси-электрод

REDOX TITRATION. *See*
OXIDATION-REDUCTION TITRATION.

2980
REDUCED EQUATION OF STATE.
Thermal equation of state of real gas expressed by reduced parameters: reduced pressure p_r, reduced volume V_r and reduced thermodynamic temperature T_r, e.g. reduced van der Waals equation has the form

$$\left(p_r + \frac{3}{V_r} \right)(3V_r - 1) = 8T_r$$

D reduzierte Zustandsgleichung
F équation d'état réduite
P równanie stanu zredukowane
R приведённое уравнение состояния

2981
REDUCED MASS, μ.
Quantity used in description of relative motion of two particles of mass m_1 and m_2, with respect to their centre of mass

$$\mu = \frac{m_1 m_2}{m_1 + m_2}$$

D reduzierte Masse
F masse réduite
P masa zredukowana
R приведённая масса

REDUCED n-MOLECULE
DISTRIBUTION FUNCTION. *See*
GENERIC n-TH ORDER PHASE
DISTRIBUTION.

2982
REDUCED PARAMETERS (reduced variables).
Dimensionless quantities defined as ratios of parameters of state and corresponding critical parameters of the given substance $X_r = X/X_c$, where X_r = reduced parameter, X = parameter of state, X_c = corresponding critical parameter of discussed substance.

D reduzierte Größen
F variables réduites
P parametry zredukowane, wielkości zredukowane, zmienne zredukowane
R приведённые величины, приведённые переменные

2983
REDUCED PRESSURE, p_r.
Dimensionless quantity $p_r = \dfrac{p}{p_c}$, where p = pressure and p_c = critical pressure.

D reduzierter Druck
F pression réduite
P ciśnienie zredukowane
R приведённое давление

2984
REDUCED TEMPERATURE, T_r.
Dimensionless magnitude $T_r = T/T_c$ where T = thermodynamic temperature,

T_c = critical temperature of a given substance.

D reduzierte Temperatur
F température réduite
P temperatura zredukowana
R приведённая температура

REDUCED VARIABLES. *See* REDUCED PARAMETERS.

2985
REDUCED VISCOSITY (Staudinger function), η_{red}.
Ratio of specific viscosity to concentration of solution

$$\eta_{red} = \frac{\eta_{sp}}{c} = \frac{\eta - \eta_0}{\eta_0 c}$$

where η = viscosity coefficient of solution, η_0 = viscosity coefficient of solvent, η_{sp} = specific viscosity, c = concentration of solution.

D Viskositätszahl
F viscosité réduite
P lepkość zredukowana, liczba lepkościowa
R приведённая вязкость

2986
REDUCED VOLUME, V_r.
Dimensionless quantity $V_r = V_m/V_c$, where V_m = molar volume and V_c = critical volume.

D reduziertes Volumen
F volume réduit
P objętość zredukowana
R приведённый объём

2987
REDUCER (reducing agent, reductant).
Substance giving up its electrons to another (reduced) substance called oxidizing agent which gains electrons; thus, reducer is oxidized in this process of electron transfer.

D Reduktionsmittel
F réducteur, agent réducteur, désoxydant
P reduktor, środek redukujący
R восстановитель

REDUCING AGENT. *See* REDUCER.

REDUCTANT. *See* REDUCER.

2988
REDUCTION.
Reaction involving transfer of electron (or electrons) from one molecule (atom or ion) to another. In narrower meaning: the enrichment of molecule with hydrogen atoms, or impoverishment of molecule by removal of oxygen atoms.

D Reduktion
F réduction
P redukcja
R восстановление

REDUCTOMETRIC TITRATION. *See* REDUCTOMETRY.

2989
REDUCTOMETRY (reductometric titration).
Determination of oxidants in solution by means of titration with standardized solution of reductants.

D Reduktionsanalyse
F réductométrie
P reduktometria, miareczkowanie reduktometryczne
R редуктометрия

2990
REFERENCE (reference material).
In optical absorption technique, precisely defined object used as reference for absorption measurements of a given object in order to eliminate certain factors like reflection, scattering, etc.

D ...
F ...
P odnośnik
R относитель

2991
REFERENCE BAND.
Spectral band of auxiliary sample of known frequency, superimposed on investigated spectrum in order to obtain reference point on frequency scale.

D Eichband
F bande de référence
P pasmo kalibracyjne, pasmo odniesienia
R калибровочная полоса

2992
REFERENCE ELECTRODE.
Electrode with respect to which potentials of other electrodes are measured as EMF of cell constructed with that electrode. Usually it is electrode of second kind with stable and reproducible potential with respect to normal hydrogen electrode, whose potential is assumed equal to zero.

D Vergleichselektrode, Bezugselektrode
F électrode de référence
P elektroda porównawcza, elektroda odniesienia
R электрод сравнения

REFERENCE MATERIAL. *See* REFERENCE.

2993
REFERENCE SOLUTION.
Solution with respect to which relative measurements of studied solution are conducted.

D Vergleichslösung, Bezugslösung
F solution de référence
P roztwór porównawczy
R раствор сравнения

2994
REFERENCE SOURCE (radioactive
standard).
Radiation source of parameters (e.g.
activity, purity) determined with
sufficient accuracy to apply source as
standard.
D radioaktives Standardpräparat
F source de référence
P źródło (promieniowania) wzorcowe
R эталонный источник, стандартный
источник

REFLECTION PLANE SYMMETRY. *See*
MIRROR REFLECTION PLANE
SYMMETRY.

2995
REFLECTION SPECTROSCOPY.
Method of optical analysis used for
investigation of solids. Quality of a given
product (e.g. whiteness of sugar or paper)
or concentration of a given component
is evaluated from measurement of
diffusionally reflected radiation at some
wavelength or in some spectral range.
D Reflexionsspektroskopie
F spectroscopie par réflexion
P analiza refleksometryczna,
reflaksometria, reflektometria
R отражательная спектроскопия

2996
REFORMATSKY REACTION.
Method of preparation of β-hydroxyesters
of carboxylic acids by treating carbonyl
compounds with organo-zinc compounds,
e.g.

$$C_2H_5OOC—CH_2Br + Zn \rightarrow [C_2H_5OOC—CH_2—ZnBr]$$

$$O=C\begin{subarray}{l}R'\\ \\R''\end{subarray} \xrightarrow{\hspace{2cm}} C_2H_5OOC—CH_2—C\begin{subarray}{l}R'\\ | \\OH\end{subarray}R''$$

D Reformatsky
(β-Hydroxycarbonsäureester)-Synthese
F réaction de Réformatsky
P reakcja Reformatskiego
R реакция Реформатского

2997
REFRACTION OF LIGHT.
Change of propagation direction of light
beam at boundary between two phases in
which velocity of light is not the same.
D Lichtbrechung
F réfraction de lumière
P refrakcja, załamanie światła
R преломление света, светопреломление

2998
REFRACTIVE INDEX (relative refractive
index), n.
Ratio of velocities of light in two media;
numerically equal to ratio of sine of
incidence angle α at border between these
media to sine of angle of refraction β in
second medium, $n = \sin \alpha/\sin \beta$; it depends
on temperature, presure and wavelength.
Remark: Refractive index is most frequently
determined at 20°C for wavelength 589.3 nm.
D Brechungsverhältnis,
Brechungsexponent
F indice de réfraction
P współczynnik refrakcji (względny),
współczynnik załamania światła
(względny)
R (относительный) показатель
преломления, коэффициент рефракции

2999
REFRACTOMETER.
Instrument used for measurement of
refractive index in liquids; based on
principle of measurement of total internal
angle of reflection.
D Refraktometer,
Strahlenbrechungsmesser
F réfractomètre
P refraktometr
R рефрактометр

3000
REFRACTOMETRY.
Analytical methods based on the
measurement of refractive index. Used
for identification purposes, purity control
and for quantitative analysis of mostly
two-component mixtures of components
with substantially different refractive
indices.
D Refraktometrie
F réfractométrie
P analiza refraktometryczna,
refraktometria
R рефрактометрия

3001
REGIOSELECTIVE REACTION.
Formation of one of possible isomers in
predominant amount with respect to other
isomeric products, e.g.

$$CH_3—CH_2—\overset{\overset{CH_3}{|}}{\underset{\underset{Br}{|}}{C}}—CH_3 \xrightarrow[-HBr]{t-BuO^{\ominus}} CH_3—CH_2—\overset{\overset{CH_3}{|}}{C}{=}CH_2 + 72\%$$

$$+ CH_3—CH{=}\overset{\overset{CH_3}{|}}{C}—CH_3$$
$$23\%$$

Electrophilic, nucleophilic and
free-radical aromatic substitutions can be
o-, m- or p-regioselective.
D regioselektive Reaktion
F réaction regioselective
P reakcja regioselektywna
R региоселективная реакция

3002
REGIOSPECIFIC REACTION.
Formation of one of possible isomers, due to absence of rearrangements in system, e.g.

$(CH_3)_3C—CH=CH—CH_3 + IN_3 \longrightarrow$

$$(CH_3)_3C—\underset{\underset{I}{|}}{C}H—\underset{\underset{N_3}{|}}{C}H—CH_3$$

D regiospezifische Reaktion
F réaction regiospécifique
P reakcja regiospecyficzna
R региоспецифическая реакция

3003
REGRESSION.
Function which determines character and form of dependence between distribution of features (variables). It allows prediction of value of one feature (variable) provided second feature (variable) has a fixed value.

D Regression
F régression
P regresja, funkcja regresji
R перрессия

3004
REGULAR POLYMER.
Polymer is regular when its macromolecules can be split into constitutional units of one species and in one sequential arrangement only.

D ...
F polymère régulier
P polimer regularny
R регулярный полимер

3005
REGULAR SOLUTION.
Solution with enthalpy of mixing H^M (see functions of mixing) different from zero, i.e. having value other than in ideal solution, but with entropy of mixing S^M equal to entropy of mixing of ideal solution

$$S^M = -R \sum_i x_i \ln x_i$$

where R = universal gas constant, x_i = molar fraction of constituent of solution, $\sum_i$ = summation of all constituents of solution.

D reguläre Mischung
F solution régulière
P roztwór regularny, roztwór prawidłowy
R регулярный раствор

3006
REIMER-TIEMANN REACTION.
Method of preparation of phenolaldehydes from chloroform and phenol in alkaline medium

$$CHCl_3 \xrightarrow[-HCl]{HO^{\ominus}} :CCl_2$$

D Reimer-Tiemann-Reaktion, Reimer-Tiemann Phenol-C-Formylierung, Reimer-Tiemann Aldehyd-Synthese
F réaction de Reimer et Tiemann
P synteza Reimera i Tiemanna
R реакция Реймера-Тимана

3007
REISSERT REACTION.
Method of reduction of acyl chlorides to aldehydes involving reaction of former compounds with potassium cyanide and quinoline in anhydrous medium. "Reissert compound" thus obtained yields upon acid hydrolysis aldehyde and quinaldic acid, e.g.

(Reissert compound)

D Reissert-Grosheintz-Fischer-Reaktion
F réaction de Reissert
P reakcja Reisserta (, Grosheintza i Fischera), redukcja Reisserta
R реакция Рейссерта

3008
RELATIVE ATOMIC MASS (atomic weight), A_r.
Ratio of mean atomic mass of given element and 1/12 of mass of ^{12}C nuclide; mean atomic mass is calculated as average of natural abundance of nuclides; e.g. $A_r(Cl) = 35.453$.
Remark: Concept of relative atomic mass can be extended to any mixture of nuclides. However, natural mixture is assumed if not stated otherwise.

D (relative) Atommasse, relative atomare Masse, Atomgewicht
F masse atomique (relative), poids atomique, poids relatif d'atome
P masa atomowa (względna)
R атомная масса, атомный вес

3009
RELATIVE CONFIGURATION.
Determination of configuration around
chiral centre, related to conventionally
adopted model; for instance, in the case
of hydroxyaldehydes — to one of the
enantiomers of glyceraldehyde. It has
been arbitrarily adopted that
dextro-rotary glyceraldehyde enantiomer
possesses configuration shown in
projection formula, in which hydroxyl
group is located on right side of carbon
chain; this configuration is denoted as D;
configuration of laevo-rotary
glyceraldehyde enantiomer is denoted
as L.

D-(+)-glyceraldehyde   L-(−)-glyceraldehyde

D relative Konfiguration
F configuration relative
P konfiguracja względna
R относительная конфигурация

3010
RELATIVE ELECTRODE POTENTIAL.
EMF of cell constructed from normal
hydrogen electrode and particular given
electrode.
Remark: Term traditional; more appropriate:
EMF of reference cell.

D Bezugs-EMK
F potentiel relatif d'électrode, tension
relative d'électrode
P potencjał elektrody (względny)
R потенциал электрода (в водородной
шкале), относительный электродный
потенциал

3011
RELATIVE ERROR.
Error given as fraction of value assumed
as true.

D relativer Fehler
F erreur relative
P błąd względny
R относительная ошибка

RELATIVE FLUCTUATION. *See*
RELATIVE ROOT-MEAN-SQUARE
FLUCTUATION.

3012
RELATIVE FREQUENCY.
Ratio of number of elements
corresponding to a given numerical value
to total number of elements in sample.

D relative Häufigkeit
F fréquence relative

P częstość (względna)
R относительная частота

3013
RELATIVE MOLECULAR MASS
(molecular weight), M_r.
Ratio of mean mass of molecule, as given
by its formula and natural abundance of
nuclides, to 1/12 of mass of ^{12}C nuclide;
e.g. M_r (KCl) = 74.555.
Remark: Concept of relative molecular mass
can be extended to any mixture of nuclides.
However, natural mixture is assumed if not
stated otherwise.

D (relatives) Molekulargewicht,
Mol(ar)gewicht
F masse moléculaire, poids moléculaire
P masa cząsteczkowa (względna)
R молекулярный вес

3014
RELATIVE PERMITTIVITY (dielectric
constant), ε_r.
Ratio of permittivity and permittivity of
free space.

D Dielektrizitätszahl, relative
Dielektrizitätskonstante
F permittivité relative, constante
diélectrique
P przenikalność elektryczna względna,
stała dielektryczna
R относительная диэлектрическая
проницаемость, диэлектрическая
постоянная

3015
RELATIVE REACTION RATE.
Ratio of rate of one selected chemical
reaction from among several reactions
proceeding in identical conditions to rate
of reaction chosen from that group as
reference reaction. Generally,
reaction-rate constants of reactions of the
same order are compared, at the same
temperature.

D relative Reaktionsgeschwindigkeit
F vitesse relative de réaction
P szybkość reakcji względna
R . . .

3016
RELATIVE REACTIVITY.
Reactivity of certain atoms (or groups of
atoms) in chemical compound as compared
to reactivity of identical atoms (or groups
of atoms) in model compound (reactivities
are evaluated from rate constants).

D relative Aktivität
F réactivité relative
P reaktywność względna
R относительная реакционная
способность

RELATIVE REFRACTIVE INDEX. *See*
REFRACTIVE INDEX.

3017
RELATIVE ROOT-MEAN-SQUARE
FLUCTUATION (relative fluctuation).
Relative measure of distribution of actual
values of quantity A (e.g. energy of
system) with respect to its statistical
average $\langle A \rangle$:

$$\frac{\sqrt{\langle (A-\langle A \rangle)^2 \rangle}}{|\langle A \rangle|}$$

D relative mittlere Schwankung
F fluctuation relative
P fluktuacja względna
R относительная флуктуация

3018
RELATIVE STANDARD DEVIATION (*in
series*).
Standard deviation in series divided by
corresponding arithmetic mean value.
D relative Standardabweichung (*der
Stichprobe*)
F écart quadratique moyen relatif,
écart-type empirique relatif
P odchylenie standardowe względne
R относительное стандартное отклонение

3019
RELATIVE VISCOSITY (viscosity ratio),
η_{rel}.
Ratio of viscosity coefficient of solution to
viscosity coefficient of pure solvent
$\eta_{rel} = \eta/\eta_0$, where η = viscosity coefficient
of solution, η_0 = viscosity coefficient of
solvent.
D relative Viskosität, Viskosität-Verhältnis
F viscosité relative
P lepkość względna
R относительная вязкость

3020
RELAXATION.
Tendency of perturbed system to acquire
stationary state.
D Relaxation
F relaxation
P relaksacja
R релаксация

3021
RELAXATION EFFECT.
Electrostatic hindering of movement of
ion in electric field, which results from
assymetry of ionic atmosphere around
central ion due to movement of ion.
D Relaxationseffekt
F effet de relaxation
P efekt relaksacyjny, efekt asymetrii
R релаксационный эффект

3022
RELAXATION METHODS.
In chemical kinetics, methods of
investigation of very fast chemical
reactions. Relaxation methods consist in
disturbance of the equilibrium state in
the system, which may be sudden (pulse
methods, e.g. method of pressure jump) or
periodically changing (steady state
methods, e.g. ultrasonic absorption) and
investigation of chemical relaxation.
D Relaxationsmethoden
F méthodes de relaxation
P metody relaksacyjne
R релаксационные методы

3023
RELAXATION TIME, τ.
Time describing return of disturbed
system to steady state

$$\Delta f_t = \Delta f_{t=0}\, e^{-\frac{t}{\tau}}$$

where relaxation time is period in which
initial change $\Delta f_{t=0}$ of investigated
value f from value in steady state
diminishes e times

$$\Delta f_{t=\tau} = \frac{\Delta f_{t=0}}{e}$$

where t = time, e = base of natural
logarithms.
D Relaxationszeit
F temps de relaxation, période de
relaxation
P czas relaksacji
R время релаксации

3024
REM.
Obsolete unit of relative biological
efficiency equal to absorbed dose of any
ionizing radiation causing the same
biological effect as 1 rad of X-ray
radiation with specific ionization of
100 ions pairs in water layer of 1 nm
thickness.
Remark: Abbreviation from röntgen equivalent
man.

D Rem
F rem
P rem
R бэр

REMANENT MAGNETIZATION. *See*
RESIDUAL MAGNETIZATION.

3025
RENNER EFFECT.
Inseparability of nuclear and electronic
motions in degenerate electronic states of
linear polyatomic molecules.
D Renner-Effekt
F effet de Renner
P efekt Rennera
R эффект Реннера

3026
REP.
Obsolete unit of absorbed dose of ionizing radiation of any kind (in SI system substituted by unit gray); dose of 1 rep is equal to 93 erg/g, i.e. to dose absorbed by approximately 1 g of substance from exposure dose of 1 röntgen of hard gamma or X-ray radiation.
Remark: Abbreviation from röntgen equivalent physical.

D Rep
F rep
P rep
R фэр

3027
REPEATABILITY (intra-laboratory reproducibility).
Precision of method in the case of one man measuring in a given laboratory and obtaining subsequent results for identical product and by using the same instrument.
D Wiederholbarkeit
F répétabilité
P powtarzalność
R воспроизводимость

REPLACEMENT REACTION. *See* EXCHANGE REACTION.

3028
REPPE REACTIONS.
Syntheses applying acetylene under pressure and elevated temperatures for vinylation, ethynylation, carbonylation and cyclic polymerization.
D Reppe-Synthesen
F méthode de Reppe
P syntezy Reppego
R реакции Реппе

3029
REPRECIPITATION.
Repetitive precipitation of analysed component using the same or in succession different precipitating reagents to obtain better separation from impurities.
D Umfällung
F double précipitation
P strącanie wielokrotne
R переосаждение

REPRESENTATIVE ENSEMBLE. *See* STATISTICAL ENSEMBLE.

REPRESENTATIVE POINT. *See* PHASE POINT.

3030
REPRESENTATIVE SAMPLE (representative series).

Sample whose structure from point of view of a given feature does not differ significantly from that of whole population.
D repräsentative Stichprobe
F échantillon représentatif
P próba reprezentatywna, próba reprezentacyjna
R репрезентативная выборка, представительная выборка

REPRESENTATIVE SERIES. *See* REPRESENTATIVE SAMPLE.

3031
REPRODUCIBILITY (inter-laboratory reproducibility).
Precision of method which gives measure of agreement of results obtained by different persons in different laboratories or in the same laboratory but at different times provided individual results refer to the same product and method.
D Reproduzierbarkeit
F reproductibilité
P odtwarzalność
R воспроизводимость

3032
RESIDUAL CURRENT.
In electroanalysis, usually small electrolysis current at potentials not exceeding decomposition potential. Residual current is due to charging current and reduction or oxidation of impurities.
D Reststrom
F courant résiduel
P prąd szczątkowy, prąd resztkowy
R остаточный ток

3033
RESIDUAL MAGNETIZATION (remanent magnetization).
Magnetization of previously saturated ferro- or ferrimagnet which remains when external magnetic field is removed.
D remanente Magnetisierung
F aimantation rémanente
P namagnesowanie szczątkowe
R остаточная намагниченность

3034
RESISTANCE POLARIZATION (ohmic polarization).
Ohmic drop of voltage at electrode during polarizing current flow through it resulting from presence on electrode surface of poorly-conducting layers of salts, oxides, hydroxides etc. Resistance polarization is expressed by resistance overvoltage.
D Widerstandspolarisation
F polarisation de résistance, polarisation ohmique

P polaryzacja oporowa
R омическая поляризация

3035
RESOLUTION (resolving power).
Quantity which expresses ability of
resolution of two closely spaced spectral
lines by a given spectroscopic instrument.
D Auflösungsvermögen, Auflösung
F (pouvoir de) résolution
P zdolność rozdzielcza, rozdzielczość
R разрешающая способность

RESOLUTION. *See* PEAK RESOLUTION.

RESOLVING POWER. *See* RESOLUTION.

3036
RESONANCE CAPTURE.
Capture of neutron of resonance energy.
D Resonanzeinfang
F capture de résonance
P wychwyt rezonansowy
R резонансный захват

3037
RESONANCE EFFECT.
In coordination chemistry, increase in
stability constant of chelate complex
occurring with increasing number of
possible resonance structures.
D Resonanzeffekt
F effet de résonance
P efekt rezonansowy
R резонансный эффект

3038
RESONANCE ENERGY.
Difference between the experimentally
measured energy of real molecule having
conjugated double bond system and
calculated value of energetically poorest
resonance structure.
D Resonanzenergie, Mesomerieenergie
F énergie de résonance, énergie de
mésomérie
P energia mezomerii, energia rezonansu,
energia stabilizacji
R энергия резонанса, энергия
стабилизации

3039
RESONANCE METHOD.
Method of approximate description of
electronic structure of molecules, based
on quantum mechanics. Valence electrons
in such molecules are able to acquire
slightly different energy levels. These
levels can be described by various
non-existing resonance structures. The
real electronic structure of molecule lies
somewhere in-between resonance

structures and is called resonance hybrid,
e.g.

$$CH_2=CH-Cl \leftrightarrow \overset{\ominus}{C}H_2-CH=\overset{\oplus}{Cl}$$

D Resonanzmethode
F méthode de la résonance
P metoda rezonansu
R метод резонанса

3040
RESONANCE NEUTRON ACTIVATION.
Activation by resonance neutrons, the
energy of which causes the resonance
induction of the particular kind of atomic
nuclei.
D Resonanzneutronenaktivierung
F activation par résonance
P aktywacja rezonansowa, aktywacja
neutronami rezonansowymi
R активация резонансными нейтронами,
резонансная активация

3041
RESONANCE NEUTRON DETECTOR.
Neutron detector in which intensity of
neutron flux of different energies is
determined by measuring activity of
a metal foil (e.g. cobalt, manganese,
copper) activated by neutrons of specified
energy.
D Resonanzneutronendetektor
F détecteur de neutrons à résonance
P detektor neutronów rezonansowy
R резонансный детектор нейтронов

3042
RESONANCE PEAK.
Peak in curve relating cross-section of
a nuclear reaction to energy of
bombarding particle (excitation function);
it corresponds to resonance absorption.
D Resonanzspitze, Resonanzlinie
F pic de résonance
P pik rezonansowy
R резонансный максимум,
резонансный пик

3043
RESONANCES.
Systems of hadrons in resonance state
decaying in 10^{-23}—10^{-22} s, characterized
by discrete quantum numbers.
D Resonanzen
F résonances
P rezonanse
R резонансы

3044
RESONANCE STRUCTURES (canonical structures, contributing structures).
Structures of compounds containing conjugated bonds (π-π, p-π), differing in their electronic pair arrangements. No arrangement describes the actual structure of molecule, which lies somewhere in between resonance structures; e.g. the Kekulé formulae are energetically the poorest resonance structures for benzene.

D Grenzstrukturen, kanonische Strukturen
F structures limites, structures de résonance
P struktury graniczne, struktury rezonansowe, struktury kanoniczne
R предельные структуры, канонические структуры

RESTRICTED INTERNAL ROTATION. *See* RESTRICTED ROTATION.

3045
RESTRICTED ROTATION (restricted internal rotation, hindered rotation).
Lack of free rotation about single bond in molecule, resulting from considerable difference in potential energies at defined temperature of conformers (unstable one with high energy and stable with low energy). Rotational barrier of restricted rotation is ca. 75—85 kJ/mol.

D gehemmte Rotation
F rotation empêchée, rotation supprimée
P rotacja zahamowania
R заторможённое вращение

RESTRICTED ROTATOR. *See* HINDERED ROTATOR.

3046
RETENTION.
In radiochemistry, ratio of activity which cannot be separated from irradiated target to total primary activity of target; frequently expressed as a percentage.

D Retention
F rétention
P retencja, zatrzymanie
R удержание

3047
RETENTION OF CONFIGURATION.
Retention of configuration at chiral centre in chemical reaction, when cleavage of previous and formation of new bonds with chiral atom take place.

D Retention, Beibehalten der Konfiguration
F rétention de configuration
P retencja konfiguracji
R сохранение конфигурации

3048
RETENTION VOLUME (total retention volume), V_R.
Volume of gas or liquid needed for elution of given compound from chromatographic column (measured up to moment when its concentration in effluent reaches maximum).

D (unkorrigiertes) Retentionsvolumen
F volume de rétention
P objętość retencji
R удерживаемый объём

3049
RETROGRADE CONDENSATION.
Partial liquefaction of two component mixtures of gases during their isothermal expansion. Retrograde condensation occurs in neighbourhood of critical state in the case of two component systems in which liquid phase may exist in a certain range of temperatures above critical temperature due to condition $(\partial^2 H_m/\partial x_i^2) \neq 0$ (where H_m = molar enthalpy at critical state, x_i = mole fraction of constituent in phase).

D rückläufige Kondensation
F condensation rétrograde
P kondensacja wsteczna
R ретроградная конденсация

REVERSED PHASE PARTITION CHROMATOGRAPHY. *See* EXTRACTION CHROMATOGRAPHY.

3050
REVERSE ISOTOPE DILUTION.
Isotope dilution in which substance labelled with isotopic tracer is determined by addition of the substance with natural isotopic composition.

D umgekehrte Isotopenverdünnung
F dilution isotopique inverse
P rozcieńczenie izotopowe odwrotne
R обратное изотопное разбавление

REVERSIBLE ADIABATIC. *See* ADIABATIC.

3051
REVERSIBLE CELL.
Galvanic cell in which total reaction and all other processes are thermodynamically reversible.

D reversible Zelle, reversible Kette
F pile réversible
P ogniwo odwracalne
R обратимая цепь, обратимый элемент

REVERSIBLE CHANGE. *See* REVERSIBLE PROCESS.

3052
REVERSIBLE COAGULATION.
Coagulation in which product (usually gel) can be reconverted into sol state, e.g.

coagulation of lyophilic colloids by action of electrolytes.

D reversible Koagulation
F coagulation réversible
P koagulacja odwracalna
R обратимая коагуляция

3053
REVERSIBLE COLLOID.
Colloidal system in which, after removal of dispersion medium by careful evaporating and its renewed addition, initial system is again formed.

D reversibles Kolloid
F colloïde réversible
P koloid odwracalny
R обратимый коллоид

3054
REVERSIBLE CYCLE (reversible cyclic process).
Simultaneously cyclic and reversible thermodynamic process.

D reversibler Kreisprozeß
F cycle réversible
P obieg odwracalny
R обратимый цикл

REVERSIBLE CYCLIC PROCESS. *See* REVERSIBLE CYCLE.

3055
REVERSIBLE ELECTRODE.
Electrode (half-cell) in state of thermodynamic equilibrium.

D reversible Elektrode, umkehrbare Elektrode
F électrode réversible
P elektroda odwracalna
R обратимый электрод

3056
REVERSIBLE PROCESS (reversible change).
Thermodynamic process with entropy source vanishing in system as well as at boundary between system and surroundings.

D reversibler Prozeß, reversibler Vorgang, umkehrbarer Vorgang, umkehrbarer Prozeß
F transformation réversible
P przemiana odwracalna, proces odwracalny
R обратимый процесс

3057
REVERSIBLE REACTION (balanced reaction).
Reaction composed of two elementary reactions which proceeds simultaneously in opposite directions in such a way that substrates of one reaction are products of other reaction and vice versa $A \rightleftarrows B$, so that state of equilibrium is finally reached.

D reversible Reaktion, umkehrbare Reaktion, Gleichgewichtsreaktion
F réaction réversible, réaction d'équilibre
P reakcja odwracalna
R обратимая реакция

3058
RHENIUM, Re.
Atomic number 75.

D Rhenium
F rhénium
P ren
R рений

3059
RHEOLOGY.
Branch of science concerned with studies of processes involving flow of matter, or more generally, of time-dependent deformation processes. Rheological studies concern mainly mechanical properties (viscosity, plasticity and disappearing of stresses).

D Rheologie
F rhéologie
P reologia
R реология

3060
RHEOPEXY.
Reverse process to thixotropy, isothermal reversible conversion of some sols into gels under influence of very slow flowing or mixing.

D Rheopexie
F rhéopexie
P reopeksja
R реопексия

3061
RHODIUM, Rh.
Atomic number 45.

D Rhodium
F rhodium
P rod
R родий

3062
RIGID ROTATOR.
Two material points at constant distance from each other rotating around their centre of mass.

D Rotator
F rotateur rigide
P rotator sztywny
R жёсткий ротатор

3063
RILEY OXIDATION.
Selenium dioxide oxidation of active methyl or methylene group to carbonyl group, e.g.

$$CH_3CH_2CHO + SeO_2 \longrightarrow CH_3COCHO + H_2O + Se$$

D Riley Selendioxyd-Oxydation
F méthode de Riley
P utlenianie Rileya
R ...

3064
RING-CHAIN TAUTOMERISM.
Type of prototropic tautomerism caused by reversible rearrangement of the ring compounds into the open-chain compounds, e.g.

D Oxo-Cyclo-Tautomerie, Ring-Ketten-Tautomerie
F tautomérie noyau-chaîne
P tautomeria pierścieniowo-łańcuchowa, tautomeria okso-cykliczna
R кольчато-цепная таутомерия

RING CLOSURE. *See* CYCLIZATION.

RING FORMATION. *See* CYCLIZATION.

3065
RING OPENING.
Reaction involving fission of ring with formation of open-chain compound, e.g.

D Ringöffnung
F décyclisation, ouverture d'un cycle
P decyklizacja, rozerwanie pierścienia
R дециклизация

3066
RITTER AMIDE PREPARATION.
Preparation of N-substituted carboxylic acid amides involving addition of nitriles to alkenes in presence of concentrated sulfuric acid, followed by hydrolysis, e.g.

D Ritter Nitril-Olefin-Addition
F réaction de Ritter
P reakcja Rittera
R реакция Риттера

3067
RÖNTGEN, R.
Obsolete unit of exposure dose. In SI system 1 R = 0.258 mC/kg.

D Röntgen
F röntgen
P rentgen
R рентген

ROOT MEAN SQUARE DEVIATION. *See* STANDARD DEVIATION.

3068
ROOT MEAN SQUARE VELOCITY (*of gaseous molecules*).
Velocity, square of which is equal to mean of squares of velocities of all molecules in particular population, equal to

$$\sqrt{\overline{c^2}} = \sqrt{\frac{3RT}{M}}$$

where R = universal gas constant, T = thermodynamic temperature, M = molar mass.

D Wurzel aus dem mittleren Geschwindigkeitsquadrat
F vitesse quadratique moyenne
P prędkość kwadratowa średnia
R среднеквадратичная скорость

3069
ROSENMUND REACTION (Rosenmund reduction).
Catalytic reduction of carboxylic acid chlorides to aldehydes, effected by hydrogen in presence of palladium on barium sulfate, containing small amount of catalyst's poison (thiourea) to stop the reduction process at aldehyde stage, e.g.

$$RCOCl + H_2 \xrightarrow[SC(NH_2)_2]{Pd/BaSO_4} RCHO + HCl$$

D Rosenmund-Reaktion, Rosenmund-Saytzeff Säurechlorid-Reduktion
F réduction de Rosenmund
P redukcja Rosenmunda
R реакция Розенмунда

ROSENMUND REDUCTION. *See* ROSENMUND REACTION.

3070
ROTATED RING DISC ELECTRODE.
Rotating disc electrode, with disc surrounded by layer of insulator and ring electrode around disc; in electrochemical experiments both electrodes are placed exactly on the same surface and are polarized independently from two circuits.
D rotierende Scheibenelektrode mit Ring
F ...
P elektroda dyskowa z pierścieniem
R вращающийся дисковый электрод с кольцом

3071
ROTATING CRYSTAL METHOD.
X-ray technique for determination of crystal structure. X-rays are diffracted by crystal rotating $360°$.

D Drehkristall-Verfahren
F méthode du crystal rotatif, méthode de rotation, méthode mécanique de rotation du cristal
P metoda kryształu obracanego
R метод вращения кристалла

3072
ROTATING-CRYSTAL PHOTOGRAPH.
X-ray image obtained by rotating crystal method; population of spots placed on parallel lines (strata) formed on cylindrical photographic film. Film surrounds crystal which rotates slowly around chosen crystallographic axis. Wavelength of applied radiation (monochromatic beam) and angle of inclination of beam against rotation axis are defined.

D Schichtliniendiagramm, Schichtlinienaufnahme, Drehkristallaufnahme
F diagramme du cristal tournant
P rentgenogram warstwicowy, dyfraktogram warstwicowy, dyfraktogram obrotowy, diagram warstwicowy
R рентгенограмма вращения

3073
ROTATING PLATINUM ELECTRODE.
Electrode constructed from platinum wire rotated in examined solution at constant rate; this electrode is used for estimation of current-potential relationship at positive potential of indicator electrode and also in amperometric titrations.

D rotierende Platinelektrode
F électrode de platine tournante
P elektroda platynowa wirująca
R вращающийся игольчатый электрод

3074
ROTATING SECTOR METHOD.
Method of experimental investigation of photolysis or radiolysis: light or beam of ionizing radiation is supplied to system in intermittent way due to revolution of disc with cut-off window (sector).

D Drehsektormethode
F méthode à secteur rotatif
P metoda wirującego sektora
R метод вращающегося сектора

3075
ROTATIONAL BARRIER.
Energy necessary to overcome restriction of rotation about single or double bond. In the case of single bond, rotational barrier is ca. 8—40 kJ/mol in conformers and 80—150 kJ/mol in atropisomers;

rotational barrier of ethylenic bond (change of configuration from cis- to trans-) is approximately 160—250 kJ/mol.

D Rotationsbarriere
F barrière de rotation
R барьер вращения
P bariera rotacji, bariera obrotu

3076
ROTATIONAL CHARACTERISTIC TEMPERATURE (of a molecule), Θ_r.
Parameter expressed in units of temperature, $\Theta_r = \dfrac{\hbar}{2kI}$ where $\hbar$ = Planck constant divided by 2π, k = Boltzmann constant, I = momentum of inertia of molecule. Rotational characteristic temperature enters expression for molecular rotational partition function.

D charakteristische Temperatur der Rotation
F température caractéristique de rotation
R характеристическая вращательная температура
P temperatura rotacji (cząsteczki) charakterystyczna

3077
ROTATIONAL ENERGY.
Kinetic energy of rotational movement of molecule.

D Rotationsenergie
F énergie de rotation
R вращательная энергия
P energia rotacyjna

ROTATIONAL ENERGY LEVELS OF A MOLECULE. See MOLECULAR ROTATIONAL ENERGY LEVELS.

3078
ROTATIONAL PARTITION FUNCTION, $r(T)$.
Function defined by

$$r(T) = \sum_{J=0}^{\infty} (2J+1) \exp\left[-J(J+1)\frac{\Theta_r}{T}\right]$$

where T = thermodynamic temperature, Θ_r = characteristic rotational temperature and J = rotational quantum number. Rotational partition function determines influence of rotation of nuclei in diatomic molecule on thermodynamic properties of corresponding gas. For $T \gg \Theta_r$ rotational partition function simplifies to $r(T) = \dfrac{T}{\Theta_r}$.

D Rotationszustandssumme, Verteilungsfunktion der Rotation
F fonction de partition de rotation, somme des états de rotation
P funkcja rozdziału rotacyjna
R вращательная статистическая сумма, вращательная сумма состояний

3079
ROTATIONAL QUANTUM NUMBER, J.
In quantum mechanics, number which determines energy of rigid rotator; J assumes non-negative integral values.

D Rotationsquantenzahl
F nombre quantique de rotation
P liczba kwantowa rotacji
R вращательное квантовое число

ROTATIONAL SPECTRUM. *See* ROTATION SPECTRUM.

ROTATION OF PLANE OF POLARIZATION OF LIGHT. *See* ROTATORY POLARIZATION.

3080
ROTATION SPECTRUM (rotational spectrum).
Spectrum corresponding to transitions between rotation energy levels of the same vibrational-electronic state; it refers either to far infrared or to microwave region.

D Rotationsspektrum
F spectre de rotation
P widmo rotacyjne
R вращательный спектр

3081
ROTATION-VIBRATION BAND.
In vibration spectra band corresponding to simultaneous transition between rotational and vibrational molecular energy levels; if difference of vibrational quantum numbers is constant then rotational quantum numbers change.

D Rotation-Schwingungsband
F bande de rotation-vibration
P pasmo rotacyjno-oscylacyjne
R колебательно-вращательная полоса

ROTATION-VIBRATION-ELECTRONIC SPECTRUM. *See* ELECTRONIC-VIBRATION-ROTATION SPECTRUM.

3082
ROTATORY POLARIZATION (rotation of plane of polarization of light).
Rotation of plane of oscillations of electric vector E of linearly polarized electromagnetic wave travelling through optically active medium.

D optische Drehung, Drehung der Polarisationsebene des Lichtes
F polarisation rotatoire, rotation du plan de polarisation
P skręcenie płaszczyzny polaryzacji światła
R вращение плоскости поляризации света

3083
R-S NOMENCLATURE (R-S system).
Cahn-Ingold-Prelog unambiguous system for specification of absolute configuration.

If molecule contains chirality centre, four atoms attached to that centre are arranged in sequence of decreasing atomic numbers, e.g. $a > b > c > d$.
Three-dimensional model of molecule is viewed from side remote to group of the lowest priority and sequence (decreasing priority) of other three groups is recorded as clockwise or anti-clockwise. When it is clockwise, symbol R (**rectus**) is used to denote configuration. When it is anti-clockwise, symbol S (**sinister**) is employed.

R configuration   S configuration

D Cahn-Ingold-Prelogsche Konvention
F ...
P konwencja R-S, system Cahna, Ingolda i Preloga
R система Кана-Ингольда-Прелога, R-S система

R-S SYSTEM. *See* R-S NOMENCLATURE.

RUBBER-LIKE STATE. *See* HIGH-ELASTIC STATE.

3084
RUBIDIUM, Rb.
Atomic number 37.

D Rubidium
F rubidium
P rubid
R рубидий

3085
RUBIN NUMBER (Congo rubin number).
Amount of protective colloid, expressed in mg per 100 g of solution of Congo red at final concentration of 0.01%, which prevents colour change during 10 min on addition of 160 mmoles of KCl.

D Rubinzahl
F ...
P liczba rubinowa
R рубиновое число

3086
RUFF-FENTON DEGRADATION OF SUGARS.
Shortening of carbon chain in aldoses, involving oxidation of calcium salt of aldonic acid by hydrogen peroxide in presence of ferric acetate

D Ruff-Fentonscher Abbau, Ruff-Fenton
Zucker-Abbau
F dégradation de Ruff-Fenton
P degradacja Ruffa i Fentona
R расщепление по методу Руффа-
-Фентона

4n+2 RULE. *See* HÜCKEL RULE.

3087
RULE OF DULONG AND PETIT (law of
Dulong and Petit).
Rule applied for approximate calculations
of molal heat capacities of solid
substances. According to this rule all solid
elements have nearly the same atomic
heat capacity — ca. 26.8 J gramatom^{-1}K^{-1}.
D Dulong-Petitsche Regel
F loi de Dulong et Petit
P reguła Dulonga i Petita
R правило Дюлонга и Пти

3088
RULES FOR RESONANCE.
Rules determining choice of appropriate
resonance structures in resonance method.
D Regeln der Rezonanzmethode
F règles de résonance
P reguły rezonansu
R правила резонанса

RUSSELL-SAUNDERS COUPLING. *See*
L-S COUPLING.

3089
RUTHENIUM, Ru.
Atomic number 44.
D Ruthen(ium)
F ruthénium
P ruten
R рутений

3090
RUTHERFORD, Rd.
Obsolete unit of activity, quantity of
radioactive material which undergoes 10^6
disintegrations per second; in SI system
1 Rd = 1 MBq, *see* becquerel.
D Rutherford
F rutherford
P rezerford, rutherford
R резерфорд

3091
R$_f$ VALUE.
In paper and thin-layer chromatography,
ratio of distance migrated by given
substance to that migrated by solvent
(both measured from place of application
of sample). R$_f$ value characterizes
substance in given conditions of obtaining
chromatogram.

D R$_f$-Wert
F coefficient R$_f$
P współczynnik R$_f$
R величина R$_f$

3092
R$_M$ VALUE.
In paper and thin-layer chromatography,
quantity linearly dependent upon free
energy of transfer of substance from one
phase to another (R$_M$ = lg 1/R$_f$−1).
R$_M$ value is additive quantity for separate
component parts of chemical compound in
given chromatographic system.
D R$_M$-Wert
F symbole R$_M$
P współczynnik R$_M$
R величина R$_M$

3093
RYDBERG, Ry.
Energy unit used in atomic and molecular
spectroscopy equal to product of Rydberg
constant R, Planck's constant h, and
velocity of light in vacuum c
1 Ry = Rhc = (2.17972 ± 0.00017)·10^{-18} J
D Rydberg
F rydberg
P rydberg
R ридберг

3094
RYDBERG CONSTANT, R.
Constant with dimension of wavenumber,
used in atomic and molecular
spectroscopy

$$R = \frac{\mu_0^2 m_e e^4 c^3}{8h^3} =$$

$$= (1.097373\,1 \pm 0.000\,000\,3)10^{-7} \text{ m}^{-1}$$

where μ_0 = magnetic permittivity in
vacuum, m_e = electronic rest mass,
e = electronic charge, c = velocity of light
in vacuum, h = Planck's constant.
D Rydberg-Konstante
F constante de Rydberg
P stała Rydberga
R постоянная Ридберга

3095
RYDBERG TRANSITION
(N-R transition).
High-energy electronic transition from
molecular ground state to excited state
whose wave function is approximately
described by atomic orbital of
one-electron atom.
D Rydberg-Übergang
F transition Rydberg
P przejście rydbergowskie, przejście
N → R
R переход Ридберга

3096
SACHSE-MOHR THEORY (theory of strainless rings).
Theory maintaining that six-membered rings of cyclohexane type and more-membered ones can be strainless, on assumption that cycles are not planar but buckled.
D Sachse-Mohr-Theorie
F théorie de Sachse-Mohr
P teoria Sachsego i Mohra
R теория Саксе-Мора

3097
SALT BRIDGE (electrolytic bridge).
Usually u-shaped glass tubing filled with solution of electrolyte (mostly saturated KCl); its role to connect electrolytically two solutions and to decrease diffusion potential.
D Salzbrücke
F pont électrolytique
P klucz elektrolityczny, mostek elektrolityczny
R электролитический ключ

3098
SALT ERROR.
Indication error of acid-base indicators due to high concentration of ions in solution.
D Salzfehler
F erreur de sels
P błąd solny
R солевая ошибка

3099
SALTING OUT.
Coagulation of lyophilic colloids on addition of electrolytes at considerable concentrations.
D Aussalzen
F relargage
P wysalanie
R высаливание

SALT ISOMERISM. See LINKAGE ISOMERISM.

3100
SAMARIUM, Sm.
Atomic number 62.
D Samarium
F samarium
P samar
R самарий

3101
SAMPLE (series).
Part or subset of population, which is directly investigated from point of view of a given feature in order to reach conclusions about character of this feature in whole population; e.g. series of measurements (determinations) of the same quantity by using the same method.
D Stichprobe, Meßreihe
F échantillon
P prób(k)a, zbiorowość próbna, seria
R выборка

SAMPLE CELL. See ABSORPTION CELL.

3102
SAMPLE PATH LENGTH.
In spectrophotometric analysis, distance between internal surfaces of absorption cell windows.
D Schichtdicke
F épaisseur de solution absorbante
P grubość warstwy absorbującej
R толщина слоя

3103
SAMPLING ERROR.
Estimation error for a given feature of population due to non-representative sampling. Sampling error can be either random (depending on amount of sample and distribution of investigated feature in population) or systematic (depending on method of sampling).
D ...
F erreur aléatoire d'échantillonnage
P błąd pobrania próbki
R ошибка выборочного обследования

3104
SANDMEYER DIAZO REACTION.
Preparation of haloaromatic (or cyanoaromatic) derivative from diazonium salt by catalytic action of cuprous halide or cyanide, e.g.

$$C_6H_5NH_2 \xrightarrow{\text{NaNO}_2 / \text{HCl}} C_6H_5N_2Cl \xrightarrow{\text{CuCN + KCN} / 50°} C_6H_5CN + N_2$$

D Sandmeyer-Reaktion, Sandmeyer Diazonium-Austausch
F réaction de Sandmeyer
P reakcja Sandmeyera
R реакция Зандмейера

3105
SANDWICH COMPOUND.
Complex compound in which metal atom is between two planar (or nearly planar) organic ring molecules. Bond in such compound is result of coordination of π electrons of organic molecule by central atom, e.g. in ferrocene.

D Sandwich-Verbindung,
  Sandwich-Struktur
F molécule sandwich
P związek sandwiczowy
R сандвичевое соединение

3106
SAPONIFICATION OF FATS.
Process of basic hydrolysis of fats
(triesters of glycerol and fatty acids),
resulting in formation of soaps.
D Verseifung von Fetten
F saponification des corps gras
P zmydlanie tłuszczów
R омыление жиров

3107
SATURATED SOLUTION.
Solution remaining in equilibrium with
another phase (solid, liquid or gaseous)
which contains dissolved constituent.
D gesättigte Lösung
F solution saturée
P roztwór nasycony
R насыщенный раствор

3108
SATURATED VAPOUR.
Vapour in equilibrium with condensed
phase.
D gesättigter Dampf
F vapeur saturée, vapeur saturante
P para nasycona
R насыщенный пар

3109
SATURATION ACTIVITY.
Radioactivity induced in a given material
under specified conditions of irradiation,
reached after an infinitely long time of
activation.
D Sättigungsaktivität
F activité à saturation
P aktywność nasycenia
R активность насыщения

3110
SATURATION FACTOR (growth factor).
Given by the equation $1-e^{-\lambda t}$, where λ is
disintegration constant of radioactive
isotope formed during activation, and t
is time of activation. Saturation factor
defines ratio of activity induced in time t,
to activity induced under the same
conditions after an infinite time
of irradiation (i.e. until saturation activity
is reached).
D Zeitfaktor, Wachstumsfaktor
F facteur de saturation
P współczynnik nasycenia
R фактор насыщения

3111
SATURATION MAGNETIZATION.
Maximum magnetization asymptotically
approached by magnetization of ferro- or
ferrimagnet during increase of external
magnetic field strength.
D Sättigungsmagnetisierung
F aimantation à saturation
P namagnesowanie nasycenia
R намагниченность насыщения

3112
SAWHORSE FORMULA.
Perspective formula portraying
conformations of open-chain compounds,
e.g. *anti-meso*-2,3-dibromobutane
conformation

D ...
F ...
P wzór kozłowy
R ...

3113
SAYTZEFF RULE.
Hydrogen halide elimination from alkyl
halides results in formation of alkene with
the highest number of alkyl substituents
at double bond (when formation of two
isomeric alkenes is possible)

$$\underset{CH_3CH_2CHCH_3}{\overset{Br}{|}} \xrightarrow{OH^{\ominus}} \begin{array}{l} CH_3CH=CHCH_3 \quad \text{main product} \\ CH_3CH_2CH=CH_2 \quad \text{byproduct} \end{array}$$

D Saizew-Regel, Abspaltungsregel von
  Saytzeff
F règle de Saytzeff
P reguła Zajcewa
R правило Зайцева

SCALE OF ATOMIC MASSES. *See*
ATOMIC MASS SCALE.

SCALE OF ATOMIC WEIGHTS. *See*
ATOMIC MASS SCALE.

3114
SCANDIUM, Sc.
Atomic number 21.
D Scandium, Skandium
F scandium
P skand
R скандий

3115
SCANDIUM GROUP.
Elements in third sub-group of periodic
system with structure of outer electronic
shells of atoms $(n-1)d^1ns^2$: scandium,
yttrium, lanthanum, actinium.
D Scandiumgruppe
F sous-groupe du scandium
P skandowce
R подгруппа скандия, элементы
  подгруппы скандия

SCATTER. *See* DISPERSION.

3116
SCATTERING OF PARTICLES.
Interaction of particles (also of atomic nuclei) with other particles or systems (atomic nuclei, atoms etc.) leading in general to change of direction of their movement and to change of their internal state.

D Streuung der Teilchen
F diffusion des particules
P rozpraszanie cząstek
R рассеяние частиц

3117
SCATTERING OF RADIATION.
In quantum electrodynamics general concept describing all possible forms of interaction between radiation and matter. In spectroscopy radiation scattering usually denotes effects arising when radiation propagates through optically inhomogeneous media; if frequency of radiation does not change upon scattering, scattering is said to be elastic; in the case of inelastic scattering either frequency change (Raman effect) or complete disappearance of scattered radiation (absorption) is observed.

D Streuung der Strahlung
F diffusion du rayonnement
P rozpraszanie promieniowania
R рассеяние излучения

3118
SCAVENGER.
Chemical compound reacting at high rate constant with specified free radicals.

D Radikalfänger
F intercepteur (de radicaux libres)
P zmiatacz rodnikowy, akceptor wolnych rodników
R акцептор

3119
SCAVENGER.
In radiochemistry, substance used to remove unwanted radioactive impurities by their co-precipitation or adsorption. The scavenger is usually a precipitate with a well developed surface.

D Scavenger
F scavenger, entraîneur
P zmiatacz
R изоморфный носитель

3120
SCAVENGING.
In radiochemistry, removal of carrier-free radioactive impurities from solutions by co-precipitation or adsorption on precipitates with well developed surfaces.

D Scavenging, Scavengen, Spülung
F scavenging, entraînement, balayage
P zmiatanie
R (изоморфное или адсорбционное соосаждение)

3121
SCAVENGING OF FREE RADICALS.
Remowal of free radicals from solution with reagents (scavengers) reacting at high rate constants with specific radicals or groups of radicals.

D ...
F entraînement
P zmiatanie wolnych rodników
R ...

SCF METHOD. *See* HARTREE-FOCK METHOD.

3122
SCHIEMANN REACTION.
Preparation of fluoro-aromatic derivatives by pyrolytic decomposition of diazonium fluoroborates, e.g.

$$ArNH_2 + HNO_2 + [BF_4]^{\ominus} \longrightarrow [ArN_2]^{\oplus}[BF_4]^{\ominus} \xrightarrow{\Delta}$$
$$ArF + N_2 + BF_3$$

D Schiemann-Reaktion, Schiemann-Kernfluorierung
F réaction de Schiemann
P reakcja Schiemanna
R реакция Шимана

3123
SCHMIDT HYDRAZOIC ACID REACTION.
Conversion of carbonyl compounds by hydrazoic acid in sulfuric (or polyphosphoric) acid medium; amines are formed from carboxylic acids, nitriles from aldehydes and acid amides from ketones, e.g.

$$RCOOH + HN_3 \xrightarrow{H_2SO_4} RNH_2 + CO_2 + N_2$$

In presence of great excess of HN_3, substituted tetrazoles from aldehydes and ketones are formed.

D Schmidt-Reaktion, Schmidt Carbonyl-Abbau
F réaction de Schmidt
P reakcja Schmidta
R реакция Шмидта

3124
SCHMIDT'S RULE.
Double bond strengthens adjacent single bond, simultaneously weakening next single bond; e.g. free-radical fission of alkenes takes place most readily in allylic position.

D Schmidt-Regel
F règle de Schmidt
P reguła Schmidta
R правило Шмидта

3125
SCHOENFLIES' SYMBOLS.
Symbols for representation of symmetry elements of point group.

D Schoenfliessche Symbole
F symboles de Schoenflies
P symbole Schoenfliesa
R символы Шенфлиса

3126
SCHOTTEN-BAUMANN ACYLATION REACTION.
Acylation of hydroxylic and amine groups with acid chlorides in presence of dilute alkaline solutions, e.g.

$C_6H_5OH + C_6H_5COCl + NaOH \rightarrow C_6H_5COOC_6H_5 + NaCl + H_2O$

$C_4H_9NH_2 + CH_3COCl + NaOH \rightarrow CH_3CONHC_4H_9 + NaCl + H_2O$

D Schotten-Baumann Acylierung
F réaction de Schotten-Baumann
P reakcja Schottena i Baumanna
R реакция Шоттена-Баумана

3127
SCHOTTKY DEFECTS.
Defects of crystal lattice consisting in transfer of certain number of atoms or ions from nodes inside crystal to its surface. Therefore some nodes in lattice remain unoccupied, see vacancy (atomic, ionic).

D Schottkysche Fehlordnung
F défauts de Schottky
P defekty Schottky'ego
R дефекты по Шотки

SCHRÖDINGER EQUATION WITH TIME. See TIME-DEPENDENT SCHRÖDINGER EQUATION.

3128
SCHULZE-HARDY RULE.
Coagulation power of electrolyte increases considerably with increasing valency of ions of sign opposite to that of colloidal particles (counter-ions).

D Hardy-Schulze-Flockungsregel
F loi de Hardy-Schulze
P reguła Hardy'ego i Schulzego
R правило (значности) Шульце-Гарди

3129
SCINTILLATION COUNTER.
Counter in which light flashes produced in a scintillator by ionizing radiation are transformed into electrical pulses by means of a photomultiplier.

D Szintillationszähler
F compteur à scintillation
P licznik scyntylacyjny
R сцинтилляционный счётчик

3130
SCINTILLATION SPECTROMETER.
Instrument with scintillation counter designed for measuring spectrum of nuclear radiation.

D Szintillationsspektrometer
F spectromètre à scintillation
P spektrometr scyntylacyjny
R сцинтилляционный спектрометр

3131
SCINTILLATOR.
Phosphor which emits a light flash under the influence of absorbed particle or photon.

D Szintillator
F scintillateur
P scyntylator
R сцинтиллятор

3132
S-CIS-S-TRANS CONFORMATION.
Conformation in conjugated double-bond systems, e.g. butadiene

s-cis conformation is less stable.
Remark: s stands for abreviation single

D s-*cis*-s-*trans* Konformation, cisoide und transoide Form
F conformation s-*cis*-s-*trans*
P konformacja s-*cis* i s-*trans*
R s-*цис*-s-*транс* конформация, цисоидная и трансоидная конформация

SCREEN ANALYSIS. See SIEVE ANALYSIS.

SCREENING CONSTANT. See SHIELDING CONSTANT.

SCREENING OF NUCLEUS. See SHIELDING OF NUCLEUS.

3133
SCREW AXIS.
Coupling of n-fold rotation axis with translation in direction parallel to axis by fraction of period indicated by axis symbol; e.g. 3_2 denotes three-fold screw axis; this action consists in combination of revolution by angle of 120° with shift by 2/3 of period in direction parallel to axis.

D Schraubenachse
F axe hélicoïdal, axe spiral
P oś (symetrii) śrubowa
R винтовая ось

3134
SCREW DISLOCATIONS.
Rows of atoms around which vertical lattice planes change into screw surfaces.

D Schraubenversetzungen
F ...
P dyslokacje śrubowe
R винтовые дислокации

3135
SEALED SOURCE.
Radioactive substance applied in sealed capsule.
D geschlossenes radioaktives Präparat
F source scellée
P źródło (promieniowania) zamknięte
R закрытый источник

SECONDARY CELL. *See*
ACCUMULATOR.

3136
SECONDARY ELECTRONS.
Electrons formed in interaction between primary particle or photon and electrons of medium.
D Sekundärelektronen
F électrons secondaires
P elektrony wtórne
R вторичные электроны

3137
SECONDARY IONIZATION.
Ionization of atoms of matter caused by their collision with fast electrons (δ rays) produced during primary ionization.
D Sekundärionisation
F ionisation secondaire
P jonizacja wtórna
R вторичная ионизация

3138
SECONDARY PHOTOCHEMICAL REACTION.
Reaction which may proceed readily without absorption of any radiative energy because of presence of very active products (e.g. electronically excited molecules, ions or free radicals) formed during primary photochemical process.
D photochemische Sekundärreaktion
F réaction secondaire
P reakcja fotochemiczna wtórna
R вторичная фотохимическая реакция

SECONDARY RADIATION. *See* X-RAY FLUORESCENCE.

SECONDARY REACTION. *See* INDUCED REACTION.

3139
SECONDARY SALT EFFECT.
Influence of electrolytes present in solution on rate constant of proceeding reaction catalyzed by acid or base. Secondary salt effect is expressed as change of equilibrium state of ionization of acid or base.
D sekundärer Salzeffekt
F effet de sel secondaire
P efekt solny wtórny
R вторичный солевой эффект

3140
SECONDARY STANDARD (*in volumetric analysis*).
Substance for which content of active component was defined by comparison with primary standard; it is used in standardization of solutions.
D sekundäre Urtitersubstanz
F étalon secondaire
P wzorzec wtórny
R ...

3141
SECONDARY STANDARD SUBSTANCE (*in thermochemistry*).
Substance used for calibration of calorimetric systems after its previous comparison with any primary standard substance.
D sekundäre Eichsubstanz
F étalon secondaire
P wzorzec wtórny termochemiczny
R втор(ичн)ое стандартное вещество

3142
SECONDARY VALENCY.
According to Werner's theory, valency characteristic for compounds unable to dissociate; corresponds to coordination bond.
D Nebenvalenz
F valence secondaire
P wartościowość poboczna, wartościowość drugorzędowa, wartościowość wtórna
R побочная валентность

3143
SECOND COORDINATION SPHERE (outer coordination sphere).
Surroundings of complex ion composed, among others, of ions beyond first coordination sphere neutralizing charge of complex.
D zweite Sphäre, äußere Sphäre
F ...
P sfera koordynacyjna zewnętrzna
R внешняя координационная сфера

3144
SECOND LAW OF THERMODYNAMICS.
For each c-component phase there is a function of state S, called entropy of this phase

$$S^\alpha = S^\alpha(U^\alpha, V^\alpha, n_1^\alpha, n_2^\alpha, ..., n_c^\alpha)$$

(where U = internal energy, V = volume, $n_1, n_2, ..., n_c$ = numbers of moles of phase components, and α numbers system phases) such that
(1) Gibbs equation is satisfied,
(2) entropy of system is extensive quantity $S = \sum_\alpha S^\alpha$, and

(3) principle of entropy increase is satisfied for all possible adiabatic

transitions between local equilibrium states of system.

D zweiter Hauptsatz der Thermodynamik
F deuxième principe de la thermodynamique, principe de Carnot
P zasada termodynamiki druga
R второе начало термодинамики

3145
SECOND MOMENT.
Mean square of resonance line width computed with respect to centre of gravity of line. Quantity used in structural studies of crystals by nuclear magnetic resonance method.

D zweites Moment
F second moment, deuxième moment
P moment drugiego rzędu, moment drugi
R второй момент, момент второго порядка

SECOND MOMENT OF FLUCTUATING VARIABLE. *See* **MEAN SQUARE DEVIATION.**

3146
SECOND-ORDER ASYMMETRIC TRANSFORMATION.
Process of privileged crystallization of one of the epimers taking part in first-order asymmetric transformation. Constant lack of equilibrium in solution of epimers is caused by precipitation of less soluble epimer. This effect enables almost quantitative conversion of one epimer into another to take place.

D ...
F ...
P przekształcenie asymetryczne drugiego rodzaju
R асимметрическое превращение второго рода

3147
SECOND-ORDER PHASE TRANSITION.
Process accompanied by sudden change of second order derivatives of Gibbs free energy with respect to pressure and temperature, i.e. compressibility and molal heat capacity.

D Phasenumwandlung zweiter Ordnung
F transformation de second ordre
P przemiana (fazowa) drugiego rzędu
R фазовый переход второго рода

3148
SECOND-ORDER REACTION.
Reaction in which kinetic equation is marked by sum of powers of concentrations of reactants being equal to two.

D Reaktion 2. Ordnung, Reaktion zweiter Ordnung
F réaction du deuxième ordre
P reakcja drugiego rzędu
R реакция второго порядка

SECOND TRIAD. *See* **LIGHT PLATINUM METALS.**

3149
SECOND VIRIAL COEFFICIENT.
Coefficient B in virial equation of state. B has the same dimension as molar volume, depends on temperature and is characteristic of a given gas.

D zweiter Virialkoeffizient
F second coefficient viriel
P współczynnik wirialny drugi
R второй вириальный коэффициент

3150
SECULAR EQUILIBRIUM.
Stationary state between radioactive parent nuclide and its daughter for the case when half-life of parent nuclide is very long compared with that of daughter. For this equilibrium, the relation is

$$N_1\lambda_1 = N_2\lambda_2 = ... = N_i\lambda_i$$

where N_i = number of nuclei of i-th member of radioactive series; λ_i = disintegration constant of this member.

D (radioaktives) Dauer-Gleichgewicht, säkulares Gleichgewicht
F équilibre séculaire
P równowaga (promieniotwórcza) trwała, równowaga wiekowa
R вековое равновесие

3151
SEDIMENTATION.
Precipitation and settling of particles of dispersed phase under influence of gravitation or rotation.

D Sedimentation
F sédimentation
P sedymentacja
R седиментация

3152
SEDIMENTATION ANALYSIS.
Method of mechanical analysis based on difference of rate of sedimentation of particles of investigated substance dispersed in a given medium in order to determine concent of particles of different size.

D Sedimentationsanalyse
F analyse de sédimentation
P analiza sedymentacyjna
R седиментационный анализ, седиментометрический анализ

3153
SEDIMENTATION BALANCE.
Balance allowing study rate of gravitational sedimentation of suspensions, e.g. Figurowski balance.

D Sedimentationswaage
F balance de sédimentation
P waga sedymentacyjna
R седиментационные весы

3154
SEDIMENTATION EQUILIBRIUM.
Equilibrium state between sedimentation
and diffusion of particles of dispersed
phase, i.e. equilibrium state in outer field
of forces (gravitational field, centrifugal
field) leading to formation of
concentration gradient.

D Sedimentationsgleichgewicht
F équilibre de sédimentation
P równowaga sedymentacyjna
R седиментационное равновесие

3155
SEDIMENTATION POTENTIAL (Dorn
effect, electrophoretic potential).
Difference of electrical potentials arising
between upper and low layer of vertically
placed column with sedimenting particles.

D Sedimentationspotential, Dorn-Effekt,
elektrophoretisches Potential
F potentiel de sédimentation, effet Dorn
P potencjał sedymentacji, efekt Dorna
R потенциал седиментации

3156
SEDIMENTATION RATE, U.
Rate of sedimentation of particles due to
gravitational or centrifugal forces which,
in case of spherical particles of radius r,
can be expressed as

$$U = \frac{2}{g} \frac{(\gamma - \gamma_0)a}{\eta} r^2$$

where γ and γ_0 = densities of dispersed
phase and dispersion medium respectively,
η = viscosity coefficient of sol,
$a = g$ = constant of gravitation
(sedimentation due to gravitation),
or $a = \omega^2 x$ (where ω = angular velocity
of rotation of centrifuge, x = distance of
particle from axis of rotation for
sedimentation due to rotation).

D Sedimentationsgeschwindigkeit
F vitesse de sédimentation
P szybkość sedymentacji
R скорость седиментации

3157
SEGREGATION.
Differentiation of composition of alloy
which occurs e.g. during solidification in
its various parts.

D Seigerung, Saigerung
F ségrégation
P segregacja
R сегрегация

3158
SELECTION RULES.
Rules which govern possibility of
radiative transitions between quantum
mechanical states of system.

D Auswahlregeln
F règles de sélection
P reguły wyboru
R правила отбора

3159
SELECTIVE ADSORPTION.
Adsorption from multicomponent mixture
which selectively increases concentration
of one of components on surface of
adsorbent; it is due to different adsorptive
powers of components for definite
adsorbent.

D selektive Adsorption
F adsorption sélective
P adsorpcja selektywna
R селективная адсорбция

3160
SELECTIVE REACTION.
Reaction of selective reagent under
optimum conditions.

D selektive Reaktion
F réaction sélective
P reakcja selektywna
R селективная реакция

3161
SELECTIVE REAGENT.
Reagent reacting under specified
conditions with some small number of
ions or compounds.

D selektives Reagens
F réactif sélectif
P odczynnik selektywny
R селективный реагент

3162
SELECTIVE REDUCTION.
Preferential reduction of one (or more)
functional groups from among other
groups, which could be reduced, present
in molecule of organic compound.

D selektive Reduktion
F réduction sélective
P redukcja selektywna
R селективное восстановление

3163
SELECTIVITY (of ion exchanger).
Ability of ion exchanger to prefer some
ions in comparison with others.

D Selektivität
F sélectivité
P selektywność
R селективность, избирательность

3164
SELECTIVITY COEFFICIENT
(equilibrium quotient), $k_{A,B}$.
Product of concentrations of products of
ion exchange reaction (raised to
appropriate powers) divided by analogous
product of substrate concentrations.

D Selektivitätskoeffizient
F coefficient d'échange, constante
apparente d'échange
P współczynnik selektywności
R коэффициент избирательности

SELECTIVITY OF CATALYST. *See*
SPECIFITY OF CATALYST.

3165
SELECTIVITY OF REAGENT.
Property of reagent to react with limited
number of chemical substances under
optimum conditions.
D Selektivität
F sélectivité
P selektywność odczynnika
R селективность

3166
SELENIUM, Se.
Atomic number 34.
D Selen
F sélénium
P selen
R селен

3167
SELF-ABSORPTION.
Absorption of electromagnetic or
corpuscular radiation by emitting medium.
D Selbstabsorption
F auto-absorption
P autoabsorpcja, samopochłanianie
R самопоглощение

3168
SELF-ABSORPTION (*in emission spectral
analysis*).
Decrease of radiation intensity of
wavelength corresponding to centre of
spectral line due to absorption in outer
gaseous layer surrounding excitation
source.
D Selbstabsorption
F réabsorption, self-absorption de raies,
auto-absorption
P autoabsorpcja linii
R самопоглощение

3169
SELF-CONSISTENT FIELD.
Effective potential field acting on
particles in single-particle approximation
for many-particle systems; it follows from
averaging inter-particle interactions.
D Selfconsistentfield
F champ self-consistent
P pole samouzgodnione
R самосогласованное поле

SELF-CONSISTENT FIELD METHOD.
See HARTREE-FOCK METHOD.

3170
SELF-DIFFUSION.
Spontaneous displacement of chemical
components in their own medium caused
by thermal movement of particles, e.g.
self-diffusion of lead ions in crystals of
a lead salt. It is usually observed and
measured by isotopic tracers.
D Selbstdiffusion
F auto-diffusion
P samodyfuzja
R самодиффузия

3171
SELF-DIFFUSION COEFFICIENT.
Proportionality coefficient D in Fick's
laws applied to self-diffusion. The true
self-diffusion coefficient should be
distinguished from that of an isotopic
tracer because of isotopic effects.
D Selbstdiffusionskoeffizient
F coefficient d'auto-diffusion
P współczynnik autodyfuzji
R коэффициент самодиффузии

3172
SELF-IRRADIATION.
Partial (in case of γ-emitters) or
practically complete (in case of soft
radiation from β- and α-emitters)
absorption of energy of ionizing radiation
by the medium in which a radioactive
nuclide is present.
D Selbstbestrahlung
F auto-irradiation
P samonapromienianie
R самооблучение

3173
SELF-REVERSAL (*in emission spectral
analysis*).
Limiting case of self-absorption when
radiation corresponding to spectral line
centre is completely absorbed.
D Selbstumkehr
F renversement (d'une raie)
P odwrócenie linii
R самообращение

3174
SELF-SCATTERING.
Scattering of radiation by material that
emits it.
D Selbststreuung
F autodiffusion
P samorozpraszanie
R саморассеяние

SEMICOLLOID. *See* ASSOCIATION
COLLOID.

3175
SEMICONDUCTOR.
Material in which electrons or electron holes are carriers of electric current, characterized by specific conductance of 10^{-9} to $10^{2}\,\Omega^{-1}\,cm^{-1}$ (at room temperature), increasing with temperature, according to relation (valid in defined temperature ranges):

$$\sigma = \sigma_0 \exp\left(-\frac{E}{kT}\right)$$

where σ = specific conductance at temperature T, E = activation energy, k = Boltzmann constant, σ_0 = empiric constant.

D Halbleiter
F semiconducteur
P półprzewodnik
R полупроводник

3176
SEMICONDUCTOR DETECTOR (solid-state detector).
Detector of ionizing radiation composed of a semiconductor material containing the polarized p-n joint (see: electron-defect semiconductor and electron-excess semiconductor).

D Halbleiterdetektor, Festkörperdetektor
F détecteur semi-conducteur
P detektor półprzewodnikowy
R полупроводниковый детектор

3177
SEMIDINE REARRANGEMENT.
Acid-catalyzed conversion of p-derivatives of hydrazobenzene to corresponding p-derivatives of 4′-amino- or 2′-aminodiphenylamine (semidine derivatives), e.g.

D Semidin-Umlagerung
F transposition semi(benzi)dinique
P przegrupowanie semidynowe
R семидиновая перегруппировка

SEMIMETALS. See METALLOIDS.

3178
SEMIMICRO-ANALYSIS.
Analysis of sample with mass of 0.01—0.1 g and for gases of volume 1—10 cm³.

D Halbmikroanalyse, Semimikroanalyse
F semi-microanalyse

P półmikroanaliza, analiza w skali półmikro
R полумикроанализ

SEMIMICRO METHOD. See CENTIGRAM METHOD.

3179
SEMI-PERMEABLE MEMBRANE.
Partition between system parts which allows for transfer of some components but not others.

D halbdurchlässige Membran, semipermeable Wand, semipermeable Membran
F paroi semi-perméable
P odgraniczenie półprzepuszczalne, przegroda półprzepuszczalna, membrana półprzepuszczalna, błona półprzepuszczalna
R полупроницаемая диафрагма, полупроницаемая мембрана

SEMIPOLAR BOND. See COORDINATE BOND.

3180
SENSITIVITY (of determination or analytical method).
Lowest value of concentration (or content) of constituent determined by a given method, which allows two values to be treated as different at assumed level of probability.

D Empfindlichkeit
F sensibilité
P czułość
R чувствительность

3181
SENSITIVITY TO LIGHT.
Ability of various substances to undergo different chemical changes under influence of light.

D Lichtempfindlichkeit
F sensibilité à la lumière
P światłoczułość
R светочувствительность

3182
SENSITIVITY TO LIGHT (in photography).
Ability of a photographic material to display visual image under action of light which is followed by development. Quantitative measure of sensitivity to light for different materials is given by reciprocal of their exposure which is responsible for quantitative blackening of photographic materials to be investigated comparatively.

D Lichtempfindlichkeit
F sensibilité à la lumière
P światłoczułość
R светочувствительность

3183
SENSITIZATION.
In colloid chemistry, effect of certain substances added in very small quantities to sol on its sensitivity to action of electrolytes.
D Sensibilisierung, Sensibilisation
F sensibilisation, sensitisation
P sensybilizacja
R сенсибилизация

3184
SENSITIZED CHEMILUMINESCENCE.
Emission of light by atoms or molecules excited by energy transferred from other molecules previously excited in chemical reaction.
D sensibilisierte Chemilumineszenz
F chimiluminescence sensibilisée
P chemiluminescencja sensybilizowana
R сенсибилизированная
   хемилюминесценция

3185
SENSITIZED FLUORESCENCE.
Fluorescence of molecules or atoms excited by other molecules or atoms which have previously absorbed photon.
D sensibilisierte Fluoreszenz
F fluorescence sensibilisée
P fluorescencja sensybilizowana
R сенсибилизированная флуоресценция

SENSITIZER. See PHOTOSENSITIZER.

3186
SENSITIZER (in photography).
Synthetic dye (e.g. erythrosin, cyanine dyes) which is adsorbed upon crystals of silver halides when added to photographic emulsion which makes the latter sensitive to light within visible and near-infrared spectral range.
D (spektraler) Sensibilisator
F sensibilisateur
P sensybilizator optyczny
R оптический сенсибилизатор

3187
SENSITOMETRY.
Branch of science which considers quantitatively photosensitive properties of photographic materials.
D Sensitometrie
F sensitométrie
P sensytometria
R сенситометрия

3188
SEPARATION ENERGY.
Energy necessary to detach particle from system in which is bound; e.g. energy needed for separation of electron from atom, or of nucleon from atomic nucleus.

D Trennarbeit
F énergie de séparation
P energia separacji
R энергия отделения

3189
SEPARATION FACTOR.
Quotient of ratios of two substances to be separated in initial material and after separation.
D Trennfaktor
F facteur de séparation
P współczynnik rozdzielenia
R фактор разделения

3190
SEQUENCE.
In vibrational structure of electronic spectrum, group of oscillation bands with the same value of difference of vibrational quantum numbers.
D Sequenz
F séquence
P sekwens
R секвенция

SEQUENCE RULE. See PRIORITY RULE.

SERIES. See SAMPLE.

SERIES OF SPECTRAL LINES. See SPECTRAL SERIES.

3191
SERINI REACTION.
Preparation of ketones from secondary acetate from secondary-tertiary 1,2-glycol with zinc dust. In chemistry of steroids this reaction is applied for removal of residual acetic acid from 17-hydroxy-20-acetoxy-derivatives

$$CH_3CO-O-\underset{\underset{17}{|}}{\overset{\overset{CH_3}{|}}{CH}}\ OH \xrightarrow[toluene]{Zn} \underset{\underset{17}{|}}{\overset{\overset{CH_3}{|}}{C}}{=}O \ H$$

D Serini-Reaktion, Serini Glykol→
   Desoxyketon-Umwandlung
F réaction de Serini
P reakcja Seriniego
R реакция Серини

3192
SHELL MODEL.
Model of atomic nucleus based on concept of similarity between structure of energetic levels in nucleus to structure of electron shells in atom. Energy levels of nucleus are arranged in groups. Levels within a particular group do not differ greatly in energy and belong to one shell.
D Schalen-Modell
F modèle des couches
P model powłokowy
R оболочечная модель

3193
SHIELDING CONSTANT (screening constant), σ.
In theory of NMR chemical shifts dimensionless quantity which defines effective magnetic field at a given nucleus, $H_{ef} = H_{extern}(1-\sigma)$.
D Abschirmungskonstante
F constante d'écran
P stała ekranowania
R константа экранирования, постоянная экранирования

3194
SHIELDING EFFECT.
In S_N1 reaction, effect protecting reaction centre by leaving group; due to this, nucleophile can attack electrophilic centre from opposite side.
D Abschirmungseffekt
F effet d'écran
P efekt osłaniania
R эффект экранирования

3195
SHIELDING OF NUCLEAR CHARGE.
Apparent decrease of nuclear charge interacting with outer electrons due to screening of nucleus by the other electrons.
D Kernabschirmung
F blindage du noyau, effet d'écran
P ekranowanie jądra przez elektrony
R экранирование ядра

3196
SHIELDING OF NUCLEUS (screening of nucleus).
Decrease of magnetic field in vicinity of nucleus by induced currents in surrounding electronic shells.
D Abschirmung des Kerns
F écran du noyau
P ekranowanie jądra (magnetyczne), przesłanianie jądra (magnetyczne)
R экранирование ядра

3197
SHOCK TUBE.
Device for generation of flat shock wave by piercing membrane between compartments with gases at high and low pressures.
D Stoßrohr
F tube de choc
P rura uderzeniowa
R ударная труба

3198
SHOCK WAVE.
Propagation of powerful mechanical disturbance in form of discontinuity moving at ultrasonic speed.
D Stoßwelle, Schockwelle
F onde de choc
P fala uderzeniowa
R ударная волна

3199
SHORT-LIVED RADIOISOTOPE.
Term which conventionally defines radioactive nuclides of half-lives shorter than 100 hours.
D kurzlebiges Isotop
F radioisotope de période courte
P izotop krótkożyciowy
R короткоживущий изотоп

3200
SHORT-RANGE ORDER.
Arrangement of atoms or molecules in zone of close neighbours of particle. Concept used in phase transitions: order-disorder.
D Nahordnung
F ordre à parcour réduit
P uporządkowanie bliskiego zasięgu
R ближний порядок

3201
SIDE CHAIN (branch).
System of chain-like linked atoms branching from main chain of atoms or from cyclic arrangement of atoms, e.g.

$$CH_3-CH_2-CH_2-CH-CH_2-CH_2-CH_3$$
$$CH_2-CH_3$$

D Seitenkette
F chaîne latérale
P łańcuch boczny
R боковая цепь, побочная цепь

3202
SIDGWICK-POWELL THEORY.
Theory which assumes that electrostatic repulsion forces, acting between electron pairs (bonding or non-bonding) of valence shells of atoms, are of decisive character with respect to spatial structure of molecule.
D Sidgwick-Powell-Theorie
F ...
P teoria Sidgwicka i Powella
R теория Сиджвика

3203
SIEVE ACTION (sieve effect).
Exclusion from stationary phase (adsorbent, ion exchanger etc.) of molecules or ions whose dimensions are greater than pore size of stationary phase.
D Siebwirkung
F effet de tamis
P efekt sitowy
R ситовый эффект

3204
SIEVE ANALYSIS (screen analysis).
Method of mechanical analysis based on separation of particles of investigated substance by use of different sieves and classification of sizes of these particles.
D Siebanalyse
F analyse granulométrique par tamisage
P analiza sitowa
R ситовый анализ

SIEVE EFFECT. *See* SIEVE ACTION.

3205
SIGMA BOND (σ-bond).
Chemical bond formed by valence electrons present in molecular σ-bonding orbital, symmetric with respect to bond axis, e.g. C—H bonds in methane.
D σ-Bindung
F liaison σ
P wiązanie σ
R σ-связь

3206
SIGMA COMPLEX (σ-complex).
Product of addition of cation to one of two carbon atoms of double bond system, requiring π-electron pair from that system to form σ-bond. The σ-complex is formed from the π-complex, e.g.

π-complex

σ-complex

D σ-Komplex
F complexe σ
P kompleks σ
R σ-комплекс

3207
SIGMA ELECTRONS (σ-electrons).
Electrons assumed to occupy σ molecular orbitals.
D σ-Elektronen
F électrons σ
P elektrony σ
R σ-электроны

3208
SIGMA HYPERON (Σ-hyperon).
Elementary particle, a hadron, of baryon number $B = 1$, lepton number $L = 0$, spin quantum number $J = 1/2$, strangeness $S = -1$, isospin $I = 1$, unstable. Σ-hyperon occurs in three states of charge (isospin triplet): of positive elementary electric charge (Σ^+-hyperon; mass $m = 1\ 189.42$ MeV, third component of isospin $I_3 = 1$, mean lifetime $0.800 \cdot 10^{-10}$ s), electrically neutral (Σ^0-hyperon; $m = 1\ 192.51$ MeV, $I_3 = 0$, mean lifetime less than $1.0 \cdot 10^{-14}$ s) and of negative elementary electric charge (Σ^--hyperon; $m = 1197.37$ MeV, $I_3 = -1$, of mean lifetime $1.489 \cdot 10^{-10}$ s).
D Σ-Hyperon
F hypéron Σ, hypéron sigma
P hiperon Σ
R Σ-гиперон, сигма-гиперон

3209
SIGMA ORBITAL (σ-orbital).
Molecular orbital symmetric with respect to rotation about bond axis.
D σ-Orbital
F orbitale σ
P orbital molekularny (typu) σ
R σ-орбиталь

3210
SIGNIFICANCE LEVEL.
Presumed probability of making type 1 error during verification of statistical hypothesis.
D Sicherheitswahrscheinlichkeit
F niveau de signification, seuil de signification
P poziom istotności
R уровень значимости

3211
SILICON, Si.
Atomic number 14.
D Silicium, Silizium
F silicium
P krzem
R кремний

3212
SILVER, Ag.
Atomic number 47.
D Silber
F argent
P srebro
R серебро

3213
SIMIRAL CHIRAL CENTRES (identical chiral centres).
Chiral centres identical with respect to chemical structure but different in configuration, e.g. each molecule of the three tartaric acids, $(+)$, $(-)$ and meso, has two equivalent chiral centres.
D gleichwertige Chiralitätszentren
F . . .
P centra chiralności równocenne
R . . .

3214
SIMONINI REACTION.
Formation of ester from two molecules of silver salt of carboxylic acid and one molecule of iodine, e.g.

$$2RCOOAg + I_2 \longrightarrow RCOOR + CO_2 + 2AgI$$

D Simonini Silbersalz-Abbau, Simonini Decarboxylierung
F réaction de Simonini
P reakcja Simoniniego
R реакция Симонини

3215
SIMONIS REACTION.
Preparation of chromone and coumarine from phenols and β-ketoesters in presence of phosphorus pentoxide, e.g.

D Simonis Chromon-Ringschluß
F réaction de Simonis
P reakcja Simonisa
R реакция Симониса

SIMPLE CHAIN REACTION. *See* STRAIGHT CHAIN REACTION.

3216
SIMPLE ELECTRODE.
Electrode (half-cell) at which only one electrode reaction occurs.

D Einfachelektrode
F électrode simple
P elektroda prosta
R простой электрод

3217
SIMPLE ELEMENT (pure element).
Chemical element which is inherently free of its other natural isotopes.

D Reinelement
F élément pur
P pierwiastek czysty, pierwiastek prosty
R чистый элемент

SIMPLE LIGAND. *See* UNIDENTATE LIGAND.

SIMPLE SOLUTION. *See* TRUE SOLUTION.

3218
SIMPLEST FORMULA.
Chemical formula which defines stoichiometry of molecule of chemical compound only.

D...
F...
P wzór najprostszy
R простейшая формула

3219
SIMPLE SUBSTANCE.
Form of existence of chemical element in its free state; some elements may exist in their different (allotropic) forms, e.g. carbon occurs as graphite or diamond.

D...
F corps simple
P substancja prosta
R простое вещество

SIMULTANEOUS REACTIONS. *See* PARALLEL REACTIONS.

3220
SINGLE BEAM SPECTROPHOTOMETER.
Spectrophotometer with both reference material and sample in the same radiation beam.

D Einstrahlspektrometer, Spektralphotometer mit Einstrahlverfahren
F spectrophotomètre monofaisceau
P spektrofotometr jednowiązkowy
R однолучевой спектрофотометр

3221
SINGLE BOND.
Chemical bond formed by pair of valence electrons present in molecular bonding σ-orbital.

D einfache Bindung
F liaison simple
P wiązanie pojedyncze
R простая связь, единичная связь

3222
SINGLE CRYSTAL.
Crystal, very close approximation to perfect crystal (contains negligible number of imperfections and is free from macroscopic defects).

D Einkristall
F monocristal
P monokryształ
R монокристалл

3223
SINGLE ESCAPE PEAK.
Peak in gamma radiation spectrum generated due to formation of electron-positron pair, annihilation of positron formed and escape of one annihilation photon from detector.

D ein-Quant-Escape-Peak, ein-Quant-Escape-Linie

F pic de premier échappement
P pik ucieczki pojedynczej
R пик одиночного вылета, пик вылета
одного кванта

SINGLE PARTICLE APPROXIMATION.
See ONE PARTICLE APPROXIMATION.

SINGLE PARTICLE MODEL. *See* ONE
PARTICLE APPROXIMATION.

3224
SINGLE-SWEEP VOLTAMMETRY
(voltammetry with continuously changing
potential, stationary electrode
voltammetry).
Electrochemical method which consists in
determination of current-potential
relationship during electrolysis of
non-stirred solution or molten salt;
voltage linearly increasing with time is
applied to microelectrode with constant
surface and to large reference electrode.
D Impulspolarographie
F chronoampérométrie linéaire
P chronowoltamperometria
R вольтамперометрия при непрерывно
изменяющемся потенциале

SINGLET. *See* SINGLET STATE.

3225
SINGLET LINE.
In atomic spectroscopy spectral line due
to electron transition between two energy
levels belonging to two atomic singlet
terms.
D Einfachlinie
F simplet
P singlet widmowy
R сингулет, одиночная линия

3226
SINGLET STATE (singlet).
Atomic or molecular state (term) with
total spin quantum number $S = 0$ and
multiplicity equal to 1.
D Singulettzustand, Singulettsystem
F état singulet
P stan singletowy, singlet
R синг(у)летное состояние

3227
SINTERING.
Forming a coherent bonded mass or larger
particles by heating small particles of
solids without melting.
D Sinterung, Zusammensintern
F frittage
P spiekanie
R спекание

3228
SINTERING OF CATALYST.
Changes occurring in solid catalyst due to
keeping it at high temperature. Sintering

of catalyst is usually accompanied by
diminution or loss of its catalytic activity.
D Sinterung des Katalysators,
Zusammensintern der
Katalysatorsubstanz
F frittage du catalyseur
P spiekanie kontaktu
R спекание катализатора

SKEW CONFORMATION. *See* GAUCHE
CONFORMATION.

3229
SKRAUP QUINOLINE SYNTHESIS.
Preparation of quinoline and its
derivatives by heating primary aromatic
amines with glycerol and concentrated
sulfuric acid, in presence of oxidizing
agent (nitrobenzene), e.g.

D Skraup (Chinolin)-Synthese
F réaction de Skraup
P synteza (chinoliny) Skraupa
R реакция Скраупа

3230
SLATER'S ORBITAL.
Atomic orbital of form

$$\Psi = Nr^{n*-1}e^{-Z*r/n*a_0}Y_{lm}(\vartheta, \varphi)$$

where $n*$ = effective quantum number,
$Z*$ = effective nuclear charge,
r = distance from nucleus, ϑ and
φ = angular coordinates of electron,
a_0 = Bohr radius, N = normalization
constant, Y_{lm} = hydrogen-like angular
part of orbital.
D Slatersches Orbital, Slatersche
Wellenfunktion
F orbitale de Slater
P orbital Slatera
R функция Слейтера

3231
SLOW COAGULATION.
Coagulation of colloidal particles partially
discharged when their electrokinetic
potential ζ is between critical potential
value and zero. This coagulation increases
with increasing concentration of
electrolyte used and depends on particular
features of its ions.
D langsame Koagulation
F coagulation lente
P koagulacja powolna
R медленная коагуляция

3232
SLOWING DOWN (moderation).
Decrease in energy of neutrons by scattering through nuclei of atoms of medium.

D Verlangsamung, Abbremsung
F ralentissement, modération
P spowalnianie (neutronów)
R замедление (нейтронов)

3233
SLOW NEUTRONS.
Neutrons having kinetic energies from zero up to about 1000 eV.

D langsame Neutronen
F neutrons lents
P neutrony powolne
R медленные нейтроны

3234
S_E1 MECHANISM (mechanism of unimolecular electrophilic substitution).
Substitution reaction involving formation of carbanion as a result of ionization of organic molecule with subsequent reaction of carbanion with electrophile $Y^{\oplus}$, e.g.

$$-\overset{|}{\underset{|}{C}}:X \rightarrow -\overset{|}{\underset{|}{C}}:^{\ominus} + X^{\oplus}$$

$$-\overset{|}{\underset{|}{C}}:^{\ominus} + Y^{\oplus} \rightarrow -\overset{|}{\underset{|}{C}}Y$$

D S_E1-Mechanismus, Mechanismus der unimolekularen elektrophilen Substitution
F mécanisme de la substitution électrophile S_E1
P mechanizm S_E1, mechanizm jednocząsteczkowego podstawienia elektrofilowego
R механизм S_E1-замещения

3235
S_E2 MECHANISM (mechanism of bimolecular electrophilic substitution).
Substitution reaction involving attack of electrophile $Y^{\oplus}$ on carbon atom from which exchanged group separates off simultaneously as cation $X^{\oplus}$, e.g.

$$Y^{\oplus} + -\overset{|}{\underset{|}{C}}-X \rightarrow \left[\overset{\delta\oplus}{Y}\cdots\overset{|}{\underset{|}{C}}\cdots\overset{\delta\oplus}{X}\right] \rightarrow -\overset{|}{\underset{|}{C}}-Y + X^{\oplus}$$
$$\text{activated complex}$$

D S_E2-Mechanismus, Mechanismus der bimolekularen elektrophilen Substitution
F mécanisme S_E2, mécanisme de la substitution électrophile biparticulaire
P mechanizm S_E2, mechanizm dwucząsteczkowego podstawienia elektrofilowego
R механизм S_E2-замещения, механизм бимолекулярного электрофильного замещения

3236
S_N1 MECHANISM (mechanism of unimolecular nucleophilic substitution).
Substitution involving carbonium ion formation (due to dissociation of RX molecule) as rate-determining step with subsequent reaction of carbonium ion with nucleophile $Y^{\ominus}$, e.g.

$$-\overset{|}{\underset{|}{C}}-X \xrightleftharpoons{\text{slow}} \overset{\diagup}{\underset{|}{C}}{}^{\oplus} + X:^{\ominus}$$

$$\overset{\diagup}{\underset{|}{C}}{}^{\oplus} + Y^{\ominus} \xrightarrow{\text{fast}} -\overset{|}{\underset{|}{C}}-Y$$

D S_N1-Mechanismus, Mechanismus der monomolekularen nucleophilen Substitution
F mécanisme de la réaction S_N1, mécanisme de la substitution nucléophile monomoléculaire
P mechanizm S_N1, mechanizm jednocząsteczkowego podstawienia nukleofilowego
R механизм S_N1-замещения

3237
S_N2 MECHANISM (mechanism of bimolecular nucleophilic substitution).
Substitution involving indirect formation of active complex (the rate-determining step), containing leaving group $X^{\ominus}$ and attacking group $Y^{\ominus}$, both partly bound to carbon atom on which substitution occurs, e.g.

$$Y^{\ominus} + -\overset{|}{\underset{|}{C}}-X \rightarrow \left[\overset{\delta\ominus}{Y}\cdots\overset{|}{\underset{|}{C}}\cdots\overset{\delta\ominus}{X}\right] \rightarrow Y-\overset{|}{\underset{|}{C}}- + X^{\ominus}$$
$$\text{activated complex}$$

D S_N2-Mechanismus, Mechanismus der bimolekularen nucleophilen Substitution
F mécanisme de la réaction S_N2, mécanisme de la substitution nucléophile biparticulaire
P mechanizm S_N2, mechanizm dwucząsteczkowego podstawienia nukleofilowego
R механизм S_N2-замещения, бимолекулярный нуклеофильный механизм замещения

3238
S_N2' MECHANISM (mechanism of bimolecular nucleophilic substitution with allylic rearrangement).
Mechanism of nucleophilic substitution connected with rearrangement observed with allylic derivatives. Nucleophile attacks allylic atom (not carbon atom originally substituted by leaving group). Migration of double bond takes place simultaneously with substitution, e.g.

$$R_3N: + CH_2=CH-\overset{Cl}{\overset{|}{CHCH_3}} \rightarrow$$
$$\overset{\oplus}{R_3N}CH_2-CH=CHCH_3 + Cl^{\ominus}$$

D S_N2'-Mechanismus, Mechanismus der bimolekularen nucleophilen Substitution mit Allylumlagerung
F mécanisme S_N2', mécanisme de la substitution nucléophile biparticulaire avec transposition allylique
P mechanizm S_N2'
R S_N2' механизм нуклеофильного замещения включающий аллильную перегруппировку

3239
S_Ni MECHANISM (mechanism of internal nucleophilic substitution).
Mechanism involving intramolecular nucleophile attack on carbon atom at which substitution occurs. Nucleophile attacks from direction of leaving group. Thus, reaction proceeds with total retention of configuration.

D S_Ni-Mechanismus, Mechanismus der inneren nucleophilen Substitution
F mécanisme S_Ni, mécanisme de la substitution nucléophile interne
P mechanizm S_Ni
R механизм внутримолекулярного нуклеофильного замещения S_Ni

3240
S_Ni' MECHANISM.
Mechanism of internal nucleophilic substitution with rearrangement observed in case of allylic derivatives. Reaction involves intramolecular nucleophilic attack on allylic carbon atom with subsequent migration of double bond, e.g.

D S_Ni'-Mechanismus
F mécanisme S_Ni'
P mechanizm S_Ni'
R механизм S_Ni'-замещения

3241
S_R MECHANISM (mechanism of free-radical substitution).
Mechanism of radical substitution involving homolytic fission of bonds.

Initiated reaction proceeds in chain process, e.g.

$$Cl_2 \xrightarrow{\text{light}} 2Cl\cdot$$
$$Cl\cdot + CH_4 \rightarrow CH_3\cdot + HCl$$
$$CH_3\cdot + Cl_2 \rightarrow CH_3Cl + Cl\cdot \text{ etc.}$$

D S_R-Mechanismus, radikalischer Mechanismus der Substitution
F mécanisme de la substitution radicalaire, mécanisme d'échange homolytique
P mechanizm S_R, mechanizm podstawienia rodnikowego
R радикальный механизм замещения, механизм гомолитического замещения

3242
SODDY-FAJANS DISPLACEMENT LAWS.
Rules which define changes in mass number A and atomic number Z of nuclei produced in radioactive transformations with respect to A and Z values of the parent nuclides. (1) In α-decay, A decreases by 4, Z decreases by 2; (2) in β^--decay, A does not change, Z increases by 1; (3) in β^+-decay, A does not change, Z decreases by 1.

D Soddy-Fajans-Verschiebungssätze
F lois des déplacements radioactifs
P reguły przesunięć Fajansa i Soddy'ego
R правила сдвигов Фаянса и Содди

3243
SODIUM, Na.
Atomic number 11.

D Natrium
F sodium
P sód
R натрий

SODIUM-24. *See* RADIOSODIUM.

3244
SOFT RADIATION.
Ionizing radiation of low energy and poor penetrating power, e.g. β-radiation and X-rays of energy up to several hundred keV.

D weiche Strahlung
F rayonnement mou
P promieniowanie miękkie
R мягкое излучение

3245
SOL.
Colloid in which dispersion medium is liquid.

D Sol
F sol
P zol, roztwór koloidalny
R золь

3246
SOLID.
Body distinguished by crystalline
structure and elasticity of shape.
D fester Körper, Festkörper
F solide
P ciało stałe
R твёрдое тело

3247
SOLID CATALYST.
In heterogeneous catalysis, catalytically
active solid phase.
D Kontakt, Feststoff-Katalysator,
Kontaktkatalysator
F catalyseur solide, catalyseur de contact
P kontakt, masa kontaktowa
R контакт, контактное вещество

3248
SOLID EUTECTIC.
Solid phase which appears during
crystallization of eutectic.
D festes Eutektikum
F eutectique solide
P eutektyk stały
R твёрдая эвтектика

3249
SOLID PHASE.
Phase composed of substances in solid
state.
D feste Phase
F phase solide
P faza stała
R твёрдая фаза

3250
SOLID SOLUTION (mixed crystals).
Solution characterized by miscibility of
constituents in solid phase.
D feste Lösung, Mischkristalle
F solution solide, cristaux mixtes
P roztwór stały, kryształy mieszane
R твёрдый раствор, смешанные
кристаллы

3251
SOLID STATE.
State of aggregation characterized by
resilience (elasticity) of shape and
crystalline structure.
D fester Zustand
F état solide
P stan (skupienia) stały
R твёрдое состояние

SOLID-STATE DETECTOR. See
SEMICONDUCTOR DETECTOR.

SOLID SUPPORT. See SUPPORT.

3252
SOLIDUS (solidus curve).
Plot of temperature at which the last of
liquid phase solidifies, with respect to
composition.

D Soliduskurve, Soliduslinie
F solidus
P solidus
R солидус, линия солидуса

SOLIDUS CURVE. See SOLIDUS.

3253
SOLUBILITY.
Ability of solid, liquid or gaseous
substance to form homogeneous systems
with other substances.
D Löslichkeit
F solubilité
P rozpuszczalność
R растворимость

3254
SOLUBILITY DIAGRAM.
Plot of solubility of substance in solvent,
as function of temperature
(two-component systems) and also as
function of presence of other substances
(multicomponent systems).
D Löslichkeitsbild, Löslichkeitsdiagramm
F diagramme de solubilité
P wykres rozpuszczalności
R диаграмма растворимости

3255
SOLUBILITY PRODUCT (activity
solubility product), K_s.
Ionic product of weakly soluble
electrolyte. Equilibrium constant for
equilibrium between weakly soluble
electrolyte and its saturated solution
(ions) is written as

$$K_s = a_+^{v+} \cdot a_-^{v-}$$

where a_+ and a_- = concentration
activities of cation and anion,
respectively, v_+ and v_- = number of
cations and anions resulting from one
molecule of electrolyte. Value for
solubility product depends on
temperature; in formula for solubility
product for very weakly soluble
electrolytes activities can be substituted
by concentrations.
D (thermodynamisches)
Löslichkeitsprodukt,
Löslichkeitskonstante
F produit de solubilité
P iloczyn rozpuszczalności, stała
rozpuszczalności
R произведение растворимости,
произведение активностей

3256
SOLUTE.
Substance dissolved in a solvent.
D gelöster Stoff, Gelöstes
F soluté
P substancja rozpuszczona
R растворённое вещество

3257
SOLUTION.
Single phase system of many
constituents, in which chemical
composition does not obey laws of
definite proportions and of multiple
proportions and may change continuously
in defined limits.

D Lösung
F solution
P roztwór
R раствор

3258
SOLVATE.
Addition compound formed in solution as
result of solvation.

D Solvat
F solvat
P solwat
R сольват

3259
SOLVATED ELECTRON, e_{solv}^-.
Electron attached to one or several
molecules in liquids.

D solvatisiertes Elektron
F électron solvaté
P elektron solwatowany
R сольватированный электрон

3260
SOLVATION.
Interaction between molecules of solvent
and dissolved ions or molecules, leading to
formation of solvates.

D Solvatation
F solvatation
P solwatacja
R сольватация

3261
SOLVATION NUMBER OF ION.
Number of solvent molecules in first
solvation sphere.

D Solvatationszahl des Ions
F nombre de solvatation d'ion
P liczba solwatacji jonu
R число сольватации иона

3262
SOLVATION OF ION.
Interaction of ion with solvent resulting
in ordering of solvent molecules around
ion.

D Solvatation des Ions
F solvatation d'ion
P solwatacja jonu
R сольватация иона

3263
SOLVATION SHELL.
Zone around ion or molecule formed by
solvent molecules, as result of solvation.

D Solvatationshülle
F sphère de solvatation
P powłoka solwatacyjna
R сольватная оболочка, сольватационная
  сфера

3264
SOLVATOCHROMISM.
Solvent-induced shift of absorption bands
in electronic spectra.

D Solvatochromie
F solvatochromisme
P solwatochromia
R сольватохромия

3265
SOLVOLYSIS.
Metathesis between solvate and solvent,
the latter acting as a nucleophile, e.g.

$$C_2H_5OH + (CH_3)_3CBr \longrightarrow C_2H_5-O-C(CH_3)_3 + HBr$$

D Solvolyse
F solvolyse
P solwoliza
R сольволиз

3266
SOMMELET ALDEHYDE SYNTHESIS.
Preparation of aldehydes by hydrolysis of
adduct of hexamethylenetetramine and
organic halides (usually arylic), e.g.

$$ArCH_2X \xrightarrow{C_6H_{12}N_4} [ArCH_2-C_6H_{12}N_4]^{\oplus} X^{\ominus} \xrightarrow{H_2O} ArCHO$$

D Sommelet (Aldehyd)-Synthese
F réaction de Sommelet
P synteza (aldehydów) Sommeleta
R реакция Соммле

3267
SONOCHEMISTRY (phonochemistry).
Branch of chemistry dealing with
applications of ultrasonic or acoustic
waves for initiation or catalysis of certain
chemical reactions.

D Ultraschallchemie
F phonochimie
P fonochemia, sonochemia
R ...

3268
SORBENT.
Substance able to take up another
substance in process of sorption.

D Sorbens, Sorptionsmittel
F sorbant
P sorbent
R сорбент

SORET EFFECT. See THERMAL
DIFFUSION.

3269
SORPTION.
General term employed for description of ability of taking up given substance (sorptive) from liquid bulk by another condensed phase.
D Sorption
F sorption
P sorpcja
R сорбция

3270
SPACE CHARGE.
Surplus electric change due to presence of uncompensated carriers of current in unit volume.
D Raumladung
F charge d'espace, charge spatiale
P ładunek przestrzenny
R пространственный заряд, объёмный заряд

3271
SPACE-CHARGE POLARIZATION.
Dielectric polarization due to displacement of free charge carriers which are not neutralized at electrodes.
D Zwischenpolarisation
F polarisation de charge d'espace
P polaryzacja migracyjna, polaryzacja ładunku przestrzennego
R миграционная поляризация

SPACE CHEMISTRY. See COSMOCHEMISTRY.

3272
SPACE GROUP.
Group of symmetry elements transforming space lattice into itself (self-coincidence). Except elements of symmetry characteristic for elementary cell, every space group contains elements of translation symmetry, and may also contain screw axis of symmetry and mirror translation planes. There are 230 three-dimensional Fedorov and Schoenflies space groups.
D Raumgruppe
F groupe d'espace, groupe spatial
P grupa przestrzenna
R пространственная группа

SPACE POLYMER. See NETWORK POLYMER.

3273
SPACE QUANTIZATION.
Limitation of allowed directions of spin, orbital or total magnetic moment of atom or molecule in magnetic or electric field to discrete, small number of orientations defined by corresponding magnetic quantum numbers.
D Richtungsquantelung
F quantification spatiale
P kwantowanie przestrzenne
R пространственное квантование

3274
SPALLATION.
Smashing of atomic nucleus into several fragments by high energy particle.
D Spallation
F spallation
P kruszenie, spalacja
R реакция скалывания

SPARK LINE. See ION LINE.

3275
SPARK SPECTRUM.
Emission spectrum of element obtained by electric spark excitation; it contains chiefly emission lines of singly ionized atoms, but also spectral lines of multiply ionized and neutral atoms.
D Funkenspektrum
F spectre d'éntincelles
P widmo iskrowe
R искровой спектр

SPECIFIC ABSORBANCE. See ABSORPTIVITY.

3276
SPECIFIC ACID-BASE CATALYSIS.
Acid-base catalysis due to only one acid or base coupled with solvent.
D spezifische (Säure-Base-)Katalyse
F catalyse (acido-basique) spécifique
P kataliza (kwasowo-zasadowa) specyficzna
R специфический (кислотно-основной) катализ

3277
SPECIFIC ACTIVITY (specific radioactivity).
Activity of a radioactive material with reference to unit mass or unit volume (for solutions).
D spezifische Aktivität
F activité spécifique
P aktywność właściwa
R удельная активность

3278
SPECIFIC COAGULATION TIME.
Time in which initial number of colloidal particles in unit volume decreases due to coagulation to half its initial value.
D Halbwertszeit der Koagulation
F ...
P czas połowicznej koagulacji
R время половинной коагуляции

3279
SPECIFIC CONDUCTANCE (electrolytic conductivity), $\varkappa$.

Parameter which characterizes electrolytic conductor

$$R = \rho \frac{l}{S} \qquad \varkappa = \frac{1}{\rho}$$

where R = resistance of conductor, l and S its length and cross-section area respectively, ρ = specific resistance.

D spezifische Leitfähigkeit
F conductivité spécifique
P przewodnictwo właściwe
R удельная электропроводность

3280
SPECIFIC HEAT (specific heat capacity), c.
Intensive quantity, heat capacity of unit mass of a given substance. Specific heat capacities are defined either at constant pressure, c_p, or at constant volume c_V.

D spezifische Wärme
F chaleur spécifique
P ciepło właściwe
R удельная теплоёмкость

SPECIFIC HEAT CAPACITY. See SPECIFIC HEAT.

3281
SPECIFIC ION EXCHANGER (chelating resin).
Ion exchanger that, due to presence of special functional groups, shows distinct preference for one or several ions.

D spezifischer Ionenaustauscher
F échangeur spécifique
P jonit selektywny, jonit specyficzny, jonit chelatujący
R селективный ионит

3282
SPECIFIC IONIZATION.
Number of ion pairs formed per unit distance along track of ionizing particle passing through matter.

D spezifische Ionisation
F ionisation spécifique
P jonizacja właściwa
R удельная ионизация, линейная плотность ионизации

3283
SPECIFICITY OF CATALYST (selectivity of catalyst).
Ability of catalyst to change rate of one defined type of reaction from among several possible reactions in given system. The higest specifity is exhibited by some enzymes influencing only one reaction.

D spezifische Katalysatorwirkung, lenkende Wirkung des Katalysators, selektive Wirkung des Katalysators
F sélectivité du catalyseur
P specyficzność katalizatora, selektywność katalizatora

R специфичность катализатора, селективность катализатора

3284
SPECIFICITY OF REAGENT.
Property of reagent to react with only one chemical individual under optimum conditions.

D Spezifität
F spécifité
P specyficzność odczynnika
R специфичность реагента

SPECIFIC PHASE. See PHASE POINT.

3285
SPECIFIC POLARIZATION.
Quantity defined by

$$p = \frac{\varepsilon_r - 1}{\varepsilon_r + 2} \frac{1}{\rho}$$

where: ε_r = relative permittivity of medium, ρ = its density.

D spezifische Polarisation
F polarisation spécifique
P polaryzacja właściwa
R удельная поляризация

SPECIFIC RADIOACTIVITY. See SPECIFIC ACTIVITY.

3286
SPECIFIC REACTION.
Reaction of specific reagent under optimum conditions.

D spezifische Reaktion
F réaction spécifique
P reakcja specyficzna
R специфичная реакция, специфическая реакция

3287
SPECIFIC REAGENT.
Reagent reacting under specified conditions only with one ion or compound.

D spezifisches Reagens
F réactif spécifique
P odczynnik specyficzny
R специфический реагент

SPECIFIC REFRACTION. See SPECIFIC REFRACTIVITY.

3288
SPECIFIC REFRACTIVITY (specific refraction), r.
Function of refractive index n of substance, in practice independent of temperature, pressure and phase; usually given by Lorenz-Lorentz formula

$$r = \frac{n^2 - 1}{n^2 + 2} \frac{1}{\rho}$$

D spezifische Refraktion
F réfraction spécifique
P refrakcja właściwa
R удельная рефракция

3289
SPECIFIC ROTATION.
Rotation of polarization plane of monochromatic light by liquid layer of thickness 10 cm and containing 1 g of a given substance in 1 cm³; measured in degrees.
D spezifische Drehung
F rotation spécifique
P skręcalność właściwa
R удельное вращение

3290
SPECIFIC SURFACE OF ADSORBENT.
Total surface area corresponding to defined unit mass of adsorbent.
D ...
F surface spécifique d'adsorbant
P powierzchnia właściwa (adsorbenta)
R удельная поверхность адсорбента

3291
SPECIFIC VISCOSITY (viscosity ratio excess), η_{sp}.
Ratio of difference between viscosity coefficients of solution and pure solvent to viscosity coefficient of pure solvent

$$\eta_{sp} = \frac{\eta - \eta_0}{\eta_0} = \eta_{rel} - 1$$

where η = viscosity coefficient of solution, η_0 = viscosity coefficient of pure solvent, η_{rel} = relative viscosity of solution.
D spezifische Viskosität, relative Viskositätsänderung
F viscosité spécifique
P lepkość właściwa
R удельная вязкость

3292
SPECIFIC VOLUME.
Volume of unit mass of uni- or multicomponent system.
D spezifisches Volumen
F volume massique, volume spécifique
P objętość właściwa
R удельный объём

SPECPURE REAGENT. See
SPECTRALLY PURE REAGENT.

3293
SPECTRAL GHOSTS.
False spectral lines arising from imperfection of diffraction monochromator.
D Geister, falsche Linien
F raies fantômes
P duchy (*w widmie*)
R духи, ложные спектральные линии

3294
SPECTRAL LINE.
Slit image in focal plane of spectrograph or spectrometer created by electromagnetic radiation.

D Spektrallinie
F raie spectrale
P linia spektralna, linia widmowa
R спектральная линия

3295
SPECTRALLY PURE REAGENT
(spectral pure reagent, specpure reagent).
Reagent provided with certificate indicating spectral lines of impurities detectable under specified conditions of excitation.
D spektrographisch reines Reagens, spektralreines Reagens, spektroskopisch reines Reagens
F réactif spectroscopiquement pur
P odczynnik spektralnie czysty
R спектрально чистый реактив

SPECTRAL PURE REAGENT. See
SPECTRALLY PURE REAGENT.

3296
SPECTRAL SERIES (series of spectral lines).
In atomic spectroscopy group of lines corresponding to some fixed value of principal quantum number of either initial or final energetic level.
D Spektralserie
F série spectrale
P seria widmowa
R спектральная серия

3297
SPECTRAL TERM (spectroscopic term).
In hydrogen atom spectroscopy, ratio of Rydberg constant and square of principal quantum number of a given electronic state. Generally, in atomic spectroscopy, energy of a given electronic energy level in kaysers (cm^{-1}).
D Spektralterm
F terme spectroscopique
P term widmowy
R спектральный терм

3298
SPECTROANALYSIS (spectrum analysis, spectroscopic analysis).
Qualitative and quantitative determination of sample's chemical composition by measurement of its spectra.
D Spektralanalyse
F analyse spectrale, analyse spectroscopique
P analiza spektralna, analiza widmowa, analiza spektroskopowa
R спектральный анализ

3299
SPECTROCHEMICAL ANALYSIS.
Spectroscopic determination of sample's qualitative and quantitative chemical composition.

Remark: In Russian and Polish also used to denote emission spectroanalysis with previous chemical treatment of sample.

D spektrochemische Analyse, chemische Spektralanalyse
F analyse spectrochimique
P analiza spektrochemiczna
R спектрохимический анализ

3300
SPECTROCHEMICAL BUFFER.
In emission spectral analysis substance added to sample in order to minimize influence of other elements of sample on spectrum of a given element.

D spektrochemischer Puffer
F tampon spectral
P bufor spektroskopowy
R спектроскопический буфер

3301
SPECTROCHEMICAL CARRIER.
Substance added to spectral sample in order to facilitate its transition to gaseous state.

D Spektralträger
F entraîneur, porteur
P nośnik spektroskopowy
R спектроскопический носитель

3302
SPECTROCHEMICAL SERIES.
Series of ligands arranged by increasing value of energy splitting (Δ or 10 Dq) in octahedral complexes formed by these ligands: $I^- < Br^- < Cl^- < F^- < H_2O < < NH_3 < H_2N(CH_2)_2NH_2 < CN^-$ (cf. crystal field splitting energy).

D spektrochemische Reihe
F série spectrochimique (de Tsuchida)
P szereg spektrochemiczny (Tsuchidy)
R спектрохимический ряд (Цушиды)

3303
SPECTROFLUORIMETER.
(spectrofluorometer).
Instrument for measurement of intensity of fluorescence spectrum (after splitting of radiation in monochromator).

D Spektrofluorimeter
F spectrofluorimètre, spectrofluoromètre
P spektrofluorymetr
R спектрофлуориметр, спектрофлуорометр

3304
SPECTROFLUORIMETRY
(spectrofluorometry).
Measurement of intensity of split fluorescence spectrum.
Remark: This term is frequently applied in context of fluorimetric analysis utilizing spectrofluorimeter.

D Spektrofluorimetrie
F spectrofluorimétrie

P spektrofluorymetria
R спектрофлуориметрия, спектрофлуорометрия

SPECTROFLUOROMETER. *See* SPECTROFLUORIMETER.

SPECTROFLUOROMETRY. *See* SPECTROFLUORIMETRY.

3305
SPECTROGRAM.
Photographic or graphical recording of spectrum.

D Spektrogramm
F spectrogramme
P spektrogram
R спектрограмма

3306
SPECTROGRAPH.
Device splitting electromagnetic radiation beam into monochromatic rays and recording spectrum obtained on photographic plate.

D Spektrograph
F spectrographe
P spektrograf
R спектрограф

3307
SPECTROGRAPHIC ANALYSIS.
Spectroanalysis based on measurement of sample emission spectrum recorded on photographic plate.

D spektrographische Analyse
F analyse spectrographique
P analiza spektrograficzna
R спектрографический анализ

3308
SPECTROMETER.
Device used for measurement of intensity of electromagnetic radiation and composed of entrance slit, monochromator and one or more exit slits connected to (photoelectric) detector; measurements are taken at some chosen wavelengths.

D Spektrometer
F spectromètre
P spektrometr
R спектрометр

SPECTROMETER. *See* SPECTROPHOTOMETER.

3309
SPECTROMETRIC ANALYSIS.
Emission spectral analysis using photoelectric detectors for direct measurement of intensity of spectral lines.

D spektrometrische Analyse
F analyse spectrométrique, analyse spectrale à lecture directe
P analiza spektrometryczna
R спектрометрический анализ

3310
SPECTROPHOTOMETER (spectrometer).
Device for measurement of radiation transmittance and/or absorbancy as function of wavelength. It is composed of radiation source, monochromator, absorption cell (for different phases), detector and measuring unit.
D Spektralphotometer, Spektrophotometer
F spectrophotomètre
P spektrofotometr
R спектрофотометр

3311
SPECTROPHOTOMETRIC TITRATION (photometric titration).
Titration method with end point determined by spectrophotometer from point of inflection on curve representing dependence of absorption coefficient on amount of added reagent.
D photometrische Titration
F titrage (spectro)photométrique, titrimétrie colorimétrique
P miareczkowanie spektrofotometryczne, miareczkowanie foto(kolory)metryczne
R фотометрическое титрование

3312
SPECTROPHOTOMETRY.
Branch of science dealing with investigation of dependence between intensity of electromagnetic radiation interacting with matter (emission, absorption, reflection, scattering) and radiation wavelength; this term usually refers to ultraviolet, visible and infrared region.
D Spektralphotometrie
F spectrophotométrie
P spektrofotometria
R спектрофотометрия

3313
SPECTROPOLARIMETER.
Device used for measurement of dispersion of optical rotation.
D Spektropolarimeter
F spectropolarimètre
P spektropolarymetr, polarymetr spektralny
R спектрополяриметр

3314
SPECTROPOLARIMETRY.
Analytical method based on measurement of dispersion of optical rotation. Used for determination of optically active compounds, structural research and for identification of some optically active groups.
D Messung der optischen Rotationsdispersion
F spectropolarimétrie

P analiza spektropolarymetryczna, spektropolarymetria
R спектрополяриметрический анализ, спектрополяриметрия

3315
SPECTROSCOPE.
Device splitting electromagnetic radiation beam into monochromatic rays and enabling obtained spectrum to be observed visually.
D Spektroskop
F spectroscope
P spektroskop
R спектроскоп

SPECTROSCOPIC ANALYSIS. *See* SPECTROANALYSIS.

SPECTROSCOPIC ANALYSIS. *See* VISUAL SPECTROSCOPIC ANALYSIS.

SPECTROSCOPIC SPLITTING FACTOR g_J. *See* LANDÉ FACTOR.

SPECTROSCOPIC TERM. *See* SPECTRAL TERM.

3316
SPECTROSCOPY.
Branch of physics and chemistry dealing with investigation of structure and properties of atoms, molecules and nuclei by absorption, emission and scattering of electromagnetic radiation.
D Spektroskopie
F spectroscopie
P spektroskopia
R спектроскопия

3317
SPECTRUM.
All frequencies absorbed or emitted by a given system. Ordered correspondence between intensity of absorbed or emitted radiation and its wavelength, wavenumber or frequency, usually represented graphically.
D Spektrum
F spectre
P widmo
R спектр

SPECTRUM ANALYSIS. *See* SPECTROANALYSIS.

3318
SPECTRUM PROJECTOR.
Device for observation on screen of enlarged spectra recorded on photographic plate.
D Spektrenprojektor
F projecteur de spectres
P spektroprojektor
R спектропроектор

SPENT FUEL ELEMENTS. *See*
BURNT-UP FUEL ELEMENTS.

3319
SPHERICAL TOP MOLECULE.
Molecule with the same value of three
principal momenta of inertia, e.g. CH_4.
D Kugelkreiselmolekül
F molécule cuspidale sphérique
P cząsteczka typu bąka sferycznego,
   cząsteczka-rotator sferyczny
R молекула типа сферического волчка

sp HYBRID. *See* DIGONAL HYBRID.

sp^2 HYBRID. *See* TRIGONAL HYBRID.

sp^3 HYBRID. *See* TETRAGONAL
HYBRID.

3320
SPIN (spin angular momentum).
Intrinsic particle angular momentum;
vector quantity of no classical
counterpart.
Remark: Term spin also is used to denote spin
quantum number.
D Spin
F spin
P moment pędu spinowy, spin
R спин

SPIN ANGULAR MOMENTUM. *See*
SPIN.

3321
SPIN DECOUPLING.
Elimination of coupling between two
nuclear spins I_1 and I_2 which had formed
multiplet structure of nuclear magnetic
resonance line during observation of
resonance signal of nuclei of spin I_1.
Achieved by irradiating sample with
alternating magnetic field of frequency
corresponding to resonance frequency of
nuclei of spin I_2; technique used for
analysing complex NMR spectra.
D Spin-Entkopplung
F découplage de spin
P rozsprzężenie spinowe
R развязка спинов

3322
SPIN DENSITY (unpaired electron
density).
Difference of electron densities for
electrons associated with α and β spin
functions.
D Spindichte
F densité de spin
P gęstość spinowa
R спиновая плотность

3323
SPIN ECHO.
Additional signal observed in pulse
techniques of paramagnetic resonance
when high-frequency electromagnetic
field has been switched off.
D Spin-Echo
F écho de spin
P echo spinowe
R спиновое эхо

3324
SPIN-FREE COMPLEX (high-spin
complex).
Complex in which number of unpaired
electrons is maximal (from view-point of
their possible distribution on t_{2g} and e_g
orbitals).
D magnetischer normaler Komplex
F complexe à spins non appariés,
   complexe à spins élevés
P kompleks wysokospinowy
R высокоспиновый комплекс

3325
SPIN-LATTICE RELAXATION
(longitudinal relaxation).
Relaxation in nuclear magnetic resonance
arising from energy exchange between
spin system and crystal lattice.
D Spin-Gitter-Relaxation, longitudinale
   Relaxation
F relaxation spin-réseau, relaxation
   longitudinale
P relaksacja spinowo-sieciowa, relaksacja
   podłużna
R спин-решёточная релаксация,
   продольная релаксация

3326
SPIN MOMENT, p_S.
Vector quantity defined by
$p_S = \hbar \sqrt{S(S+1)}$, where S = spin quantum
number. Spin magnetic moment
$\mu_S = \sqrt{S(S+1)}\, \mu_B$ corresponds to spin
moment, where μ_B = Bohr magneton.
D Spinmoment
F moment de spin
P moment spinowy atomu
R спиновой момент

3327
SPINORBITAL.
One-electron wave function in atom or
molecule given as product of spatial part
(orbital) and spin function of electron.
D Teilchenfunktion
F spin-orbitale
P spinorbital
R спин-орбиталь

24*

3328
SPIN-ORBIT COUPLING (spin-orbit interaction).
Interaction between spin of electron (or total spin of electrons) with angular momentum of electron (or total angular momentum of electrons) giving total angular momentum of system.
D Spin-Bahn-Kopplung, Spin-Bahn-Wechselwirkung
F couplage spin-orbite, interaction spin-orbite
P sprzężenie spinowo-orbitalne, oddziaływanie spinowo-orbitalne, sprzężenie spin-orbita
R спин-орбитальная связь, спин-орбитальное взаимодействие

SPIN-ORBIT INTERACTION. *See* SPIN-ORBIT COUPLING.

3329
SPIN QUANTUM NUMBER, J (s).
Number which defines allowed quantized values of spin angular momentum. For electron spin quantum number assumes only one value $J = 1/2$.

D Spinquantenzahl
F nombre quantique de spin
P liczba kwantowa spinowa
R спиновое квантовое число

3330
SPIN-SPIN COUPLING (spin-spin interaction).
Isotropic interaction of nuclear spins through molecular bonding electrons producing multiplet structure of NMR lines.
D Spin-Spin-Kopplung, Spin-Spin-Wechselwirkung
F couplage spin-spin, interaction spin-spin
P sprzężenie spinowo-spinowe, oddziaływanie spinowo-spinowe, sprzężenie spin-spin
R спин-спиновая связь, спин-спиновое взаимодействие

3331
SPIN-SPIN COUPLING CONSTANT, J.
Constant, in Hz, which determines isotropic indirect spin-spin interaction between nuclear spins leading to formation of multiplet structure of NMR lines.
D Spin-Kopplungskonstante J
F effet J de couplage spin-spin
P stała sprzężenia spinowo-spinowego J
R константа спин-спиновой связи J

SPIN-SPIN INTERACTION. *See* SPIN-SPIN COUPLING.

3332
SPIN-SPIN RELAXATION (transverse relaxation).

Relaxation in nuclear magnetic resonance arising from spin-spin dipole-dipole interaction.
D Spin-Spin Relaxation, transversale Relaxation
F relaxation spin-spin, relaxation transversale
P relaksacja spinowo-spinowa, relaksacja poprzeczna
R спин-спиновая релаксация, поперечная релаксация

3333
SPIN TEMPERATURE, T_s.
Concept introduced in consideration of the population of Zeeman levels in spin-lattice relaxation. For two levels with $M = +1/2$ and $M = -1/2$ spin temperature is defined by ratio of their populations

$$\frac{n_{+1/2}}{n_{-1/2}} = \exp\left(2\mu H/kT_s\right)$$

where μ = magnetic moment, H = magnetic field strength, k = Boltzmann constant, M = projection of spin onto direction of external magnetic field.

D Spin-Temperatur
F température des spins
P temperatura spinowa
R спиновая температура

3334
SPIN WAVES.
Concept utilized in quantum mechanical theory of magnetism in order to describe propagation of spin-excited states in magnetically ordered crystals (e.g. in ferromagnets).

D Spinwellen
F ondes de spin
P fale spinowe
R спиновые волны

3335
SPIN WAVE THEORY.
Quantum theory of magnetically ordered crystals of ferro-, ferri- and antiferromagnets.

D Spinwellentheorie
F théorie des ondes de spin
P teoria fal spinowych
R теория спиновых волн

3336
SPIRO ATOM.
Atom shared by two ring systems, e.g. carbon atom in spirobicyclobutane or nitrogen atom in 1,1'-spirobipiperidinium bromide

D Spiroatom
F atome spiro
P atom spiro
R спирановый атом

3337
SPONTANEOUS FISSION.
Radioactive decay of the heaviest nuclei,
consisting in their spontaneous decay to
two (rarely three or four) fragments
which are nuclei of elements appearing
in centre of periodic system.

D spontane Spaltung
F fission spontanée
P rozszczepienie samorzutne
R спонтанное деление

3338
SPONTANEOUS MAGNETIZATION.
Saturation magnetization of ferro- or
ferrimagnet domains existing also in
absence of external magnetic field.

D spontane Magnetisierung
F aimantation spontanée
P namagnesowanie spontaniczne,
namagnesowanie samorzutne
R спонтанная намагниченность,
самопроизвольная намагниченность

3339
SPONTANEOUS POLARIZATION.
Orientation polarization in ferroelectrics
below Curie point exhibited in absence of
external electric field.

D spontanische Polarisation
F polarisation spontanée
P polaryzacja spontaniczna
R спонтанная поляризация

3340
SPOT TEST ANALYSIS.
Technique in trace analysis; it is based on
reaction carried out with one drop of
solution on surface of sample, on porcelain
plate or paper. Appearance of appropriate
colour, precipitate or gas bubbles indicates
presence of suspected component.

D Tüpfelanalyse
F analyse à la goutte, analyse à la touche
P analiza kroplowa
R капельный анализ

3341
SPUR.
Group of ionic and free radical products
of radiolysis located close to one another
in zone of primary ionization of high
energy particle or photon. Spur region
diffuses with time and eventually
becomes indistinguishable with bulk of
liquid.

D Spur
F grappe, spur
P gniazdo
R шпора

3342
SPUR REACTIONS.
Reactions occurring inside spurs, i.e.
reactions of intermediate products of
radiolysis in places where their initial
concentrations are high.

D Spur-Reaktionen
F . . .
P reakcje wewnątrzśladowe, reakcje
wewnątrz gniazd
R внутритрековые реакции

3343
SQUARE PLANAR COMPLEX.
Complex of coordination number 4
in which donor atoms of ligands are
distributed at corners of square
(hybridization of dsp^2 type).

D quadratischer Komplex
F complexe plan
P kompleks płaski kwadratowy
R плоский квадратный комплекс

3344
SQUARE-WAVE POLAROGRAPHY.
Method based on recording and analysis
of square-wave alternating current which
flows in polarographic circuit when
voltage linearly increasing with time
applied to electrodes is modulated with
square-wave voltage of frequency
200—250 Hz and amplitude 5—50 mV.

D square-wave-Polarographie
F polarographie à tension carée
P polarografia zmiennoprądowa
prostokątna
R полярография с наложением
прямоугольного напряжения

3345
STABILITY CONSTANT (overall stability
constant, cumulative formation
constant), β_n.
Equilibrium constant of formation
reaction of complex $M + nL \rightleftarrows ML_n$.
Under conditions guaranteeing constancy
of activation coefficients

$$\beta_n = \frac{[ML_n]}{[M][L]^n}$$

where $[ML_n]$, $[M]$ and $[L]$ = equilibrium
concentrations of complex, metal and
ligand respectively.

D Bruttostabilitätskonstante
F constante de stabilité globale, constante
cumulative
P stała trwałości (skumulowana), stała
trwałości pełna, stała trwałości
całkowita
R общая константа устойчивости

3346
STABLE COMPLEX.
Complex which practically does not
dissociate in solution, and is characterized
by large value of stability constant
($> 10^8$).
D starker Komplex
F complexe stable
P kompleks trwały
R прочный комплекс

3347
STABLE NUCLIDE.
Nuclide not undergoing spontaneous
measurable radioactive decay.
D stabiles Nuklid
F nucléide stable
P nuklid trwały
R стабильный изотоп

3348
STABLE STATE.
State of system in which all conditions of
diffusion, mechanical and thermal
stability are fulfilled.
D ...
F état stable
P stan stabilny, stan trwały
R устойчивое состояние, стабильное
    состояние

3349
STALAGMOMETER.
Device for comparative determination of
surface or interfacial tension of liquids
by measuring masses of drops falling
from vertical vessel resembling pipette
and ending in capillary tube (which has
flattened tip) with known inner diameter.
D Stalagmometer
F stalagmomètre
P stalagmometr
R сталагмометр

3350
STANDARD CELL.
Galvanic cell used as standard in
measurements of EMF of cells; currently
exclusively Weston normal cell.
D Normalelement
F élément étalon, pile étalon
P ogniwo wzorcowe
R стандартная цепь, стандартный
    элемент, эталонный элемент

3351
STANDARD CONDITIONS (normal
conditions).
Temperature 0°C and pressure
$1.013\,25 \cdot 10^{-1}$ MPa (formerly $p = 1$ atm).
D Norm(al)bedingungen
F conditions normales
P warunki normalne
R нормальные условия

3352
STANDARD CURVE METHOD.
Method of colorimetric analysis based on
determination of concentration dependence
of absorption of standard solutions.
Unknown concentration of investigated
substance is then determined by
absorption measurement.
D ...
F méthode de courbe d'étalonage
P metoda krzywej wzorcowej
R ...

3353
STANDARD DEVIATION (root mean
square deviation).
Absolute measure of distribution of actual
values of quantity A with respect to its
statistical average $\langle A \rangle$:

$$\sqrt{\langle (A - \langle A \rangle)^2 \rangle}.$$

D Standardabweichung
F fluctuation absolue
P fluktuacja bezwzględna, odchylenie
    standardowe
R абсолютная флуктуация

3354
STANDARD DEVIATION (root mean
square deviation *in series*), s.
Statistic which determines for series of
measurements scattering of results around
corresponding mean value, given by
square root of variance and computed
according to

$$s = \left[\frac{1}{n-1} \sum_{i=1}^{n} (x_i - \overline{x})^2 \right]^{1/2}$$

where x_i = value of i-th element in series,
$\overline{x}$ = arithmetic mean and n = number of
elements in series.
D Standardabweichung (*der Stichprobe*)
F écart-type, écart quadratique moyen,
    déviation standard
P odchylenie standardowe, błąd
    kwadratowy średni
R среднее квадратичное отклонение,
    стандартное отклонение

3355
STANDARD DEVIATION OF THE
MEAN, $\overline{s}$.
Quantity characterizing scattering of
results from a given sample, defined by
$\overline{s} = s/\sqrt{n}$, where s = standard deviation
for single measurement and n = number
of measurements in sample.
D Standardabweichung des Mittelwertes
F erreur standard de moyenne, écart-type
    de moyenne
P błąd średni średniej
R среднее квадратичное отклонение

STANDARD ELECTRODE POTENTIAL.
See NORMAL ELECTRODE POTENTIAL.

3356
STANDARD ELECTROMOTIVE FORCE
(normal electromotive force).
Electromotive force cell in which all
substances participating in reactions
occurred in cell are in standard state.

D elektromotorische Standard-Kraft
F force électromotrice standard
P siła elektromotoryczna normalna, siła
elektromotoryczna standardowa
R стандартная электродвижущая сила

3357
STANDARD ENTHALPY OF REACTION
(standard heat of reaction), $\Delta H^{\ominus}_{T,p}$.

Extensive quantity, measure of heat effect
in isothermal-isobaric chemical reaction

$$\Delta H^{\ominus}_{T,p} = \sum_B \nu_B H^{\ominus}_B$$

where ν_B = stoichiometric coefficient of
reagent B (positive for products and
negative for substrates), $H^{\ominus}_B$ = molar
enthalpy of reagent B in standard state
(of temperature and pressure identical
for all reactants), $\sum_B$ denotes summation

over all reactants.

D Standardreaktionsenthalpie,
Standardwert der Reaktionsenthalpie,
Reaktionsenthalpie für den
Standardzustand
F enthalpie standard de réaction,
enthalpie de réaction à l'état standard,
chaleur de réaction (à pression
constante)
P entalpia reakcji chemicznej
standardowa, ciepło izobarycznej reakcji
chemicznej standardowe
R стандартная энтальпия реакции,
(стандартная) теплота реакции при
постоянном давлении, скрытая теплота
изобарно-изотермической реакции

STANDARD HEAT OF REACTION. *See*
STANDARD ENTHALPY OF REACTION.

STANDARD HYDROGEN ELECTRODE.
See NORMAL HYDROGEN ELECTRODE.

3358
STANDARD SAMPLE.
Substance with strictly defined chemical
composition given in producer's certificate,
determined by large number of analyses
carried out in various laboratories.
Standard sample is used in comparative
analyses, estimation of analytical methods
etc.

D genormtes Muster
F substance de base
P wzorzec analityczny, normalka
R стандартный образец

STANDARD SERIES METHOD. *See*
VISUAL COMPARISON METHOD.

3359
STANDARD SOLUTION.
Solution with strictly described
properties, e.g. pH concentration of
analysed ions, colour, electrical
conductivity.

D Standardlösung
F solution étalon
P roztwór wzorcowy
R стандартный раствор

3360
STANDARD STATE (*of component in
a given phase*).
Arbitrarily chosen thermodynamic state
with respect to which thermodynamic
quantities of component in a given phase
are determined. In simplest case standard
state can be chosen as state of pure
component or state of component in
infinitely diluted solution.

D Standardzustand
F état standard
P stan standardowy
R стандартное состояние

3361
STANDARD STATES OF UNIT
ACTIVITY.
Reference states for calculating
thermodynamic activity such that: (1) in
solutions diluted with respect to component
B, activity of this component a_B tends
to concentration value c_B, when this
concentration tends to zero

$$\lim_{c_B \to 0} \frac{a_B}{c_B} = 1$$

(2) activity of pure substances under
pressure of one atmosphere is equal to
unity.

D . . .
F . . .
P stany odniesienia aktywności
R . . .

3362
STARK EFFECT.
Splitting of atomic or molecular energy
levels or splitting of spectral lines due to
external electric field.

D Stark-Effekt
F effet Stark
P efekt Starka
R явление Штарка

3363
STARK-EINSTEIN LAW (*of photochemical equivalence*).
Each molecule engaged in primary photochemical process absorbs only one quantum of radiation initiating this reaction.
D Stark-Einsteinsches Äquivalenzgesetz, Quantenäquivalenzgesetz
F loi de l'équivalent photochimique d'Einstein
P prawo równoważności fotochemicznej Einsteina
R закон квантовой эквивалентности Эйнштейна

3364
STATE OF AGGREGATION (physical state).
Physical state in which matter exists at defined conditions of pressure and temperature. Differences between particular states of aggregation are determined by interactions between particles and their mutual distances.
D Aggregatzustand
F état d'agrégat, état physique
P stan skupienia
R агрегатное состояние

STATE PARAMETER. *See* STATE VARIABLE.

3365
STATE QUANTITIES (thermodynamic quantities, thermodynamic properties, functions of state, macroscopic variables).
Quantities which depend only on state of system. Change of state quantity depends solely on initial and final state of system and does not depend on way in which state is changed.
D Zustandsfunktionen, Zustandsgrößen
F fonctions d'état, grandeurs macroscopiques
P funkcje stanu, funkcje termodynamiczne, wielkości termodynamiczne
R функции состояния, термодинамические величины

3366
STATE VARIABLE (thermodynamic parameter, coordinate of thermodynamic system, macroscopic parameter, state parameter).
Quantity used to characterize state of thermodynamic system.
D Zustandsvariable
F paramètre du système, variable d'état, grandeur d'état, paramètre indépendant, variable indépendante

P parametr stanu, zmienna stanu, parametr termodynamiczny
R параметр состояния

3367
STATIC METHOD OF RATE MEASUREMENTS.
Execution of measurements of chemical reaction rate in system without enforced flow of reactants.
D statische Methode
F méthode statique
P metoda statyczna pomiarów kinetycznych
R статический метод исследования кинетики реакции

3368
STATIC PERMITTIVITY, ε_s.
Electric permittivity in static electric field.
D statische Dielektrizitätszahl
F permittivité statique
P przenikalność elektryczna statyczna
R статическая диэлектрическая проницаемость

3369
STATIC REACTIVITY INDICES.
Reactivity indices following from consideration of electronic structure of molecule in preliminary stage of chemical reaction. At this stage changes in molecular electronic structure due to approach of reagent are rather small (polarization of electronic cloud). One of these indices is given by π-electron density.
D statische Indizes chemischer Reaktionsfähigkeit
F indices statiques de réactivité (chimique)
P indeksy reaktywności statyczne
R статические индексы реакционной способности

STATIONARY ELECTRODE VOLTAMMETRY. See SINGLE-SWEEP VOLTAMMETRY.

3370
STATIONARY PHASE (*in chromatography*).
Adsorbent, ion exchanger or liquid on solid support that is able to retain components being separated, and through which flows stream of mobile phase.
D stationäre Phase, unbewegliche Phase
F phase stationnaire
P faza nieruchoma
R неподвижная фаза

3371
STATIONARY STATE (steady state).
State in which all properties have time-independent values.

D stationärer Zustand
F état stationnaire
P stan stacjonarny
R стационарное состояние,
   установившееся состояние

3372
STATISTIC.
Characteristic function describing a given series.
D Stichprobenfunktion
F statistique
P statystyka, parametr próby
R статистика

STATISTICAL ASSEMBLY. *See* STATISTICAL ENSEMBLE.

3373
STATISTICAL AVERAGE (ensemble average, expectation value), $\langle A \rangle$.
Expectation value of given quantity A (e.g., energy of system)

$$\langle A \rangle = \sum_{A'} A' P(A')$$

where $P(A') = $ probability that A has value A'. For continuous distribution of values of A', summation has to be replaced by integration.
D gemittelter Wert, Gesamtheitsmittel, Scharmittelwert, Scharmittel, Phasenmittelwert
F moyenne d'ensemble, valeur moyenne en phase
P średnia statystyczna, wartość średnia w zespole
R статистическое среднее значение, среднее по совокупности

3374
STATISTICAL ENSEMBLE (statistical assembly, representative ensemble, Gibbs ensemble).
Virtual set of statistically independent physical systems satisfying following conditions: (1) particles of systems are all of the same type and interact through the same interaction potential, (2) distribution of probability of microstates of these systems is given. In statistical mechanics it is convenient to consider given system as randomly chosen from some appropriate statistical ensemble.
D statistische Gesamtheit, virtuelle Gesamtheit, statistisches Ensemble, repräsentatives Ensemble
F ensemble statistique, ensemble représentatif, ensemble virtuel
P zespół statystyczny
R статистический (фазовый) ансамбль, статистический ансамбль Гиббса

3375
STATISTICAL MECHANICS.
Branch of theoretical physics which deals with description and interpretation of macroscopic properties of matter by using known particle properties (i.e. properties of atoms, molecules, ions, etc.) and interactions between them.
D statistische Mechanik
F mécanique statistique
P mechanika statystyczna
R статистическая механика

3376
STATISTICAL OPERATOR (density matrix, density operator).
Operator which in statistical quantum mechanics determines probability of result of measurement of any quantity. Statistical operator allows determination of all macroscopic properties of given system.
D statistischer Operator, Dichtematrix
F opérateur statistique, matrice densité, operateur densité
P operator statystyczny, operator gęstości, macierz gęstości
R статистический оператор, матрица плотности, статистическая матрица

3377
STATISTICAL PROBABILITY.
Measure of relative frequency of events (features) in a given statistical material; approximates to probability a priori.
D statistische Wahrscheinlichkeit
F prpobabilité statistique
P prawdopodobieństwo statystyczne
R статистическая вероятность

3378
STATISTICAL TEST.
Method of verification of hypotheses about either unknown distribution of random variable or value of unknown parameter.
D statistischer Test
F test statistique
P test statystyczny
R статистическая проверка

3379
STATISTICAL THERMODYNAMICS.
Branch of statistical mechanics dealing with dependence of thermodynamic functions on structure and mutual interactions of particles composing macroscopic systems.
D statistische Thermodynamik
F thermodynamique statistique
P termodynamika statystyczna
R статистическая термодинамика

3380
STATISTICAL WEIGHT (statistical
weight factor).
Ratio of probability of event from group
of events with some common feature to
probability of any event with assumption
of a priori equivalence of all events.
D statistisches Gewicht
F poids statistique
P waga statystyczna
R статистический вес

STATISTICAL WEIGHT FACTOR. *See*
STATISTICAL WEIGHT.

3381
STATISTICS.
Branch of science based on probability
theory and used for making decisions in
indeterminate circumstances, i.e. methods
which allow determination of objective
regularities in investigated phenomena
when precise determination of
experimental result is impossible because
of measurement random errors.
D mathematische Statistik
F statistique mathématique
P statystyka matematyczna
R математическая статистика

STAUDINGER FUNCTION. *See*
REDUCED VISCOSITY.

STEADY STATE. *See* STATIONARY
STATE.

3382
STEP FILTER (step weakener).
Quartz, occasionally glass, plate which
enables intensity of passing radiation to
be weakened in controlled way, different
for distinct weakener steps.
D Stufenfilter
F filtre à échelons
P osłabiacz stopniowy, filtr stopniowy
R ступенчатый ослабитель

3383
STEPHEN REACTION.
Preparation of aldehydes from nitriles by
action of anhydrous stannous chloride in
diethyl ether saturated with hydrogen
chloride, e.g.

$$RCN + HCl \longrightarrow RC(Cl){=}NH \xrightarrow{SnCl_2}$$

$$RCH{=}NH{\cdot}HCl \xrightarrow[-NH_4Cl]{H_2O} RCHO$$

D Stephen-Reaktion, Stephen
Nitril-Reduktion
F synthèse de Stephen
P reakcja Stephena
R реакция Стефена

STEP WEAKENER. *See* STEP FILTER.

STEPWISE DISSOCIATION CONSTANT.
See CONSECUTIVE INSTABILITY
CONSTANT.

3384
STEPWISE ELUTION.
Elution of mixture with aid of several
successively introduced eluents, each of
which selectively elutes single substances
(ions) or their groups.
D stufenweise Elution
F élution par éluants successifs
P elucja stopniowana
R ступенчатое элюирование

STEPWISE FORMATION CONSTANT.
See CONSECUTIVE STABILITY
CONSTANT.

3385
STEREOBASE UNIT.
Smallest set of one, two or more
successive configurational base units, that
prescribes repetition in polymer
macromolecule.
D ...
F ...
P jednostka powtarzalna przestrzennie,
jednostka przestrzennie podstawowa
R стереоблок

3386
STEREOCHEMICAL EFFECT (*in
coordination chemistry*).
Influence of size and spatial structure
of ligand on stability of complex. In
chelate complexes this effect is manifested
by influence of substituents in chelating
agent on formation and stability of these
complexes (*cf.* steric hindrance in chelate
complexes).
D sterischer Effekt
F effet stérique
P efekt steryczny, efekt stereochemiczny
R стерический эффект

3387
STEREOCHEMISTRY.
Branch of chemistry dealing with steric
arrangement of atoms in molecules.
D Stereochemie, Raumchemie
F stéréochimie
P stereochemia
R стереохимия

STEREO-FORMULA. *See*
STEREOMETRIC FORMULA.

STEREOISOMERIC FORMULA.
See STEREOMETRIC FORMULA.

3388
STEREOISOMERISM.
Type of isomerism caused by different
configuration of molecules (stereoisomers)
which are not structural isomers.
D Stereoisomerie, Raumisomerie
F stéréo-isomérie
P stereoizomeria, izomeria przestrzenna
R стереоизомерия, пространственная
изомерия

3389
STEREOMETRIC FORMULA
(stereoisomeric formula, stereo-formula).
Chemical formula portraying spatial
arrangements of atoms in molecule, thus
enabling differentiation between
stereoisomers, e.g.

D-(+)-glyceric aldehyde   L-(−)-glyceric aldehyde

D Stereoformel, Projektionsformel
F formule stérique
P wzór przestrzenny
R пространственная формула

3390
STEREOREGULAR POLYMER.
Regular polymer composed of
macromolecules having regular repetition
of identical stereobase units at all
stereoisomeric sites of main chain.

D stereoreguläres Polymer
F polymère stéréorégulier
P polimer stereoregularny
R стереорегулярный полимер

3391
STEREOSELECTIVE
POLYMERIZATION.
Polymerization in which polymer
macromolecule is formed from mixture of
stereoisomeric monomer molecules by
preferential incorporation of one species
into growing polymer chain.

D stereoselektive Polymerisation
F ...
P polimeryzacja stereoselektywna
R стереоселективная полимеризация

3392
STEREOSELECTIVE REACTION.
Formation of one of possible stereoisomers
in predominant amount; such isomer
possesses defined configuration, e.g.

$$(\pm) \; C_6H_5-CH-CH_2-C_6H_5 \; \xrightarrow[-HCl]{^{\ominus}OH}$$
$$| \atop Cl$$

$$C_6H_5 \diagdown \quad \diagup H \qquad C_6H_5 \diagdown \quad \diagup C_6H_5$$
$$C=C \qquad + \qquad C=C$$
$$H \diagup \quad \diagdown C_6H_5 \qquad H \diagup \quad \diagdown H$$
$$70\% \qquad\qquad 30\%$$

D stereoselektive Reaktion
F réaction stéréosélective
P reakcja stereoselektywna
R стереонаправленная реакция

3393
STEREOSPECIFIC POLYMERIZATION.
Polymerization in which tactic polymer
is formed.

D stereoregulierte Polymerisation,
stereospezifische Polymerisation
F polymérisation stéréospécifique
P polimeryzacja stereospecyficzna,
polimeryzacja stereoregularna
R стереоспецифическая полимеризация

3394
STEREOSPECIFIC REACTION.
Formation of product of strictly defined
steric structure from substrate of known
configuration, e.g. trans-but-2-ene on
treatment with bromine forms only
meso-2,3-dibromobutane.

D stereospezifische Reaktion
F réaction stéréospécifique
P reakcja stereospecyficzna
R стереоспецифическая реакция

3395
STERIC EFFECT.
Effect exerted by steric system of atoms
in molecule on its physical properties and
chemical reactivity (or conformation).
Remarks: Term has wide application, see
shielding effect, hindering effect, transannular
effect, internal strain.

D sterischer Effekt
F effet stérique
P efekt przestrzenny, efekt steryczny
R пространственный эффект,
стерический фактор

3396
STERIC FACTOR (probability factor), *P*.
Fraction of active collisions which occur
when molecule assumes geometrical
orientation necessary for reaction. Steric
factor is smaller than one in majority of
reactions.

D sterischer Faktor,
Wahrscheinlichkeitsfaktor
F facteur stérique, facteur de probabilité
P współczynnik steryczny, współczynnik
orientacji, współczynnik
prawdopodobieństwa
R стерический фактор, стерический
множитель, вероятностный фактор

3397
STERIC HINDRANCE.
Specific interaction between atoms or
groups of atoms in molecule, influencing
reactivity of a site in compound, for
instance, by interfering in coupling, in
acquiring given conformation or by
decreasing reaction rate.

D sterische Hinderung
F encombrement stérique
P przeszkoda przestrzenna, przeszkoda
steryczna, zawada przestrzenna
R пространственное затруднение,
стерическое препятствие

3398
STERIC HINDRANCE (blocking group).
In coordination chemistry, chelating agent
group which decreases stability of
complex by steric hindrance,
e.g. derivatives of 8-hydroxyquinoline
(in position 2) form less stable chelate
complexes than parent reagent alone
(cf. steric effect in coordination
chemistry).
D sterische Hinderung
F obstacle stérique
P przeszkoda przestrzenna, zawada
przestrzenna, grupa blokująca
R пространственное затруднение

3399
STERN'S THEORY OF DOUBLE LAYER.
Theory which unites basic ideas of
theories of structure of double layer of
Helmholtz and Gouy and Chapman at
electrode-electrolyte interface: it assumes
existence on solution side next to metal
surface of compact layer (Helmholtz) and
further away of diffuse layer (Gouy and
Chapman).
D Sternsche Theorie der Doppelschicht
F théorie de la couche double de Stern
P teoria budowy warstwy podwójnej
Sterna
R теория двойного (электрического) слоя
Штерна, адсорбционная теория
двойного электрического слоя Штерна

3400
STERN-VOLMER REACTIONS.
Exchange of atoms between two molecules,
at least one of which is in excited state.
D Stern-Volmer Reaktionen
F ...
P reakcje Sterna i Volmera
R реакции Штерна-Фольмера

3401
STEVENS REARRANGEMENT.
Migration of benzyl group, for instance,
in phenacylbenzyldimethylammonium
bromide, caused by heating it with dilute
solution of sodium hydroxide or by action
of sodium amalgam on aqueous bromide
solution

$$C_6H_5COCH_2-\overset{CH_3}{\underset{CH_3}{\overset{|}{\underset{|}{N}}}}-CH_2-C_6H_5 \quad \overset{Br^\ominus}{\underset{-HBr}{\overset{\overset{\oplus}{OH}}{\longrightarrow}}}$$

$$C_6H_5COCH-N(CH_3)_2$$
$$\underset{CH_2-C_6H_5}{|}$$

D Stevens-Umlagerung, (Stevens)
Ylid-Amin-Isomerisation
F transposition de Stevens
P przegrupowanie Stevensa
R перегруппировка Стивенса

3402
STOBBE CONDENSATION.
Condensation of aldehydes and ketones
with succinic acid esters in presence of
sodium ethanolate; products are
monoesters of alkylidenosuccinic acids,
which upon hydrolysis yield alkylidenosuccinic acids, e.g.

$$\overset{CH_3}{\underset{CH_3}{>}}CO + \underset{CH_2-COOC_2H_5}{\overset{CH_2-COOC_2H_5}{|}} \quad \overset{C_2H_5ONa}{\underset{H^\oplus}{\longrightarrow}}$$

$$\overset{CH_3}{\underset{CH_3}{>}}C=\overset{CH_2-COOC_2H_5}{\underset{}{\overset{|}{C}}}-COOH \quad \overset{hydrolysis}{\longrightarrow} \quad \overset{CH_3}{\underset{CH_3}{>}}C=\overset{CH_2-COOH}{\underset{}{\overset{|}{C}}}-COOH$$

D Stobbe (Bernsteinsäureester)-
Kondensation
F condensation de Stobbe
P kondensacja Stobbego
R конденсация Штоббе

3403
STOICHIOMETRIC COEFFICIENT, ν.
Number in front of symbol (formula) of
reactant in chemical equation.
Stoichiometric coefficient determines
number of moles of reactant taking part
in a given reaction.
D stöchiometrischer Koeffizient,
stöchiometrische Zahl, stöchiometrischer
Faktor
F coefficient stoechiométrique
P współczynnik stechiometryczny
R стехиометрический коэффициент

3404
STOICHIOMETRIC COMPOUNDS
(daltonides).
Chemical compounds fulfilling law of
definite proportions.
D stöchiometrische Verbindungen,
Daltonide
F daltonides
P związki stechiometryczne, daltonidy
R дальтониды

STOICHIOMETRIC EQUATION.
See CHEMICAL EQUATION.

STOICHIOMETRIC FACTOR.
See ANALYTICAL FACTOR.

3405
STOICHIOMETRY.
Science dealing with proportions by
weight in which chemical elements or
compounds react; it includes fundamental
chemical laws pertaining to quantitative
description of composition of chemical
compounds.
D Stöchiometrie
F stoechiométrie
P stechiometria
R стехиометрия

3406
STOKES LAW.
Wave length of luminescent light is longer than that of corresponding exciting radiation.
D Stokessche Regel
F loi de Stokes
P prawo Stokesa
R правило Стокса

3407
STOKES LINE.
Raman line with frequency decreased with respect to that of excitation line.
D Stokessche Linie
F ligne de Stokes
P linia stokesowska
R стоксова линия

3408
STOPPING POWER.
Measure of effect of a substance upon kinetic energy of charged particle passing through it. It is equal to energy loss of the particle per unit length.
D Bremsvermögen
F pouvoir d'arrêt (linéique)
P zdolność hamowania (bezwzględna)
R тормозная способность

STORAGE CELL. *See* ACCUMULATOR.

STRAIGHT CHAIN. *See* NORMAL CHAIN.

3409
STRAIGHT CHAIN REACTION (simple chain reaction).
Chain reaction which proceeds without chain branching.
D unverzweigte Kettenreaktion
F réaction en chaîne droite
P reakcja łańcuchowa prosta
R простая цепная реакция, неразветвлённая цепная реакция

3410
STRANGENESS, *S.*
Additive quantum number attributed to hadrons and photons. Sum of strangeness is preserved in strong and electromagnetic interactions, but changes in weak interactions. Strangeness has integral values.
D Fremdartigkeit, Seltsamkeit, Strangeness
F étrangeté
P dziwność
R странность

3411
STRANGENESS CONSERVATION LAW.
In strong and electromagnetic interactions, sums of strangeness of initial and final particles are equal. Strangeness conservation law is not kept in weak interactions.
D Erhaltungssatz der Strangeness
F loi de conservation d'étrangeté
P prawo zachowania dziwności
R закон сохранения странности

3412
STRANGE PARTICLES.
Group of unstable particles of non-zero strangeness S (K-mesons, Λ-, Σ-, Ξ-, Ω-hyperons and a series of resonances).
D fremdartige Teilchen, seltsame Teilchen
F particules étranges
P cząstki dziwne
R странные частицы

3413
STREAMING MERCURY ELECTRODE.
Electrode consisting of jet of mercury flowing from capillary under pressure. In practice its surface is constant in time.
D Quecksilberstrahlelektrode
F électrode à jet de mercure
P elektroda rtęciowa strumieniowa
R струйчатый ртутный электрод

3414
STREAMING POTENTIAL.
Difference of electrical potentials arising on both sides of porous solid body or between tips of capillary as result of forced flow of liquid.
D Strömungspotential
F potentiel d'écoulement
P potencjał przepływu
R потенциал течения

3415
STRESS RELAXATION.
Decrease of internal stress in system affected by constant long-time deformation.
D Spannungsrelaxation
F relaxation de tension
P relaksacja naprężeń
R релаксация напряжения

3416
STRESS RELAXATION TIME.
Time in which internal stresses in deformed system decrease by a factor of e.
D Spannungsrelaxationszeit
F temps de relaxation de tension
P czas relaksacji naprężeń
R время релаксации напряжения

STRETCHING FORCE CONSTANT.
See BCND FORCE CONSTANT.

3417
STRETCHING VIBRATIONS (valence vibrations).
Normal vibrations in molecules which approximately correspond solely to change of bond lengths.

D Valenzschwingungen
F vibrations de valence
P drgania rozciągające, drgania walencyjne
R валентные колебания

3418
STRIPPING.
Nuclear reaction of direct interaction, consisting in addition of one or several nucleons of incident particle to target nucleus. Remaining part of incident particle continues its route with slight deflection.

D Stripping-Reaktion, Strippingprozeß
F stripage, cassure en vol
P stripping, reakcja zdarcia
R реакция срыва

3419
STRIPPING ANALYSIS.
Method of determination of low concentrations of metal ions based on preliminary electro-deposition of determined metals on hanging mercury drop electrode from intensely stirred solution during controlled time. Resulting amalgam is oxidized (usually in stationary electrode voltammetry conditions) and concentration of ions in solution, from which amalgam was formed is determined from observed current value.

D inverse Polarographie
F chronoampérométrie linéaire par redissolution anodique
P metoda wiszącej kropli
R метод амальгамной полярографии с накоплением

3420
STRONG ELECTROLYTE.
Electrolyte with degree of dissociation approximately equal to one, regardless of concentration (complete dissociation).

D starker Elektrolyt
F électrolyte fort
P elektrolit mocny
R сильный электролит

3421
STRONG INTERACTION (nuclear interaction).
Interaction of particles resulting in processes which take place in time range of $10^{-24} - 10^{-21}$ s (rapid processes). Strong interaction fullfils the greatest number of conservation laws.

D starke Wechselwirkung
F interaction forte, interaction nucléaire
P oddziaływanie silne, oddziaływanie jądrowe
R сильное взаимодействие

3422
STRONG LIGAND FIELD.
Crystal ligand field causing splitting of energy levels (terms) of central ion, large compared with distances between energy levels (terms) of free ion. In complex of such field, number of unpaired electrons may be decreased compared with free ion (cf. crystal field splitting energy, low-spin complex).

D starkes Ligandenfeld
F champ intense des coordinats
P pole (krystaliczne) ligandów silne
R сильное поле лигандов

3423
STRONTIUM, Sr.
Atomic number 38.

D Strontium
F strontium
P stront
R стронций

STRONTIUM-90. *See* RADIOSTRONTIUM.

3424
STRUCTURAL FORMULA.
Chemical formula which defines, apart from type and number of atoms, also bonds between them, e.g. for water molecule H—O—H, for methane

$$H-\overset{\displaystyle H}{\underset{\displaystyle H}{C}}-H$$

D Strukturformel, Konstitutionsformel, Valenzstrichformel
F formule développée, formule de structure
P wzór kreskowy, wzór budowy, wzór strukturalny
R структурная формула

3425
STRUCTURAL ISOMERISM.
Type of isomerism caused by different sequence in linkage of atoms in isomeric molecules.

D Strukturisomerie
F isomérie de structure
P izomeria strukturalna
R структурная изомерия

STRUCTURAL ISOMERISM.
See LINKAGE ISOMERISM (*in coordination chemistry*).

3426
STRUCTURAL VISCOSITY (non-Newtonian viscosity).
Excess of viscosity coefficient of sol over Newtonian viscosity, i.e. difference

between viscosity coefficients of sol in the case of small and very great velocity gradients. Change in viscosity coefficient with changing velocity gradient is most frequently caused by formation of internal structures.

D Strukturviskosität
F viscosité de structure
P lepkość strukturalna, lepkość anormalna
R структурная вязкость, аномалия вязкости

3427
STUDENT'S DISTRIBUTION
(t-distribution).
Probability distribution of t statistic with probability density function given by

$$f(t) = \frac{\Gamma[(k+1)/2]}{\Gamma(k/2)\sqrt{\pi k}} \left(1+\frac{t^2}{k}\right)^{-(k+1)/2}$$

for $-\infty < t < +\infty$ and for number of degrees of freedom $k = n-1$. When population is large enough $(n > 30)$, Student's distribution can be practically approximated by normal distribution.

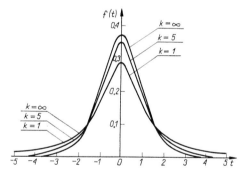

D Studentsche $(t-)$Verteilung, t-Verteilung
F loi t de Student
P rozkład (Studenta) t
R распределение t Стьюдента

3428
STUDENT'S t-TEST.
Test of agreement which allows decision to be made about significance of deviation from mean or difference between two mean values. Value of function t is equal to ratio of observed deviation or difference and standard deviation and is compared with tabular value of t, which shows frequency of event at which deviation of this value can occur for series with a given number of degrees of freedom.

D t-Test
F test t
P test t-Studenta
R t-критерий, критерий Стьюдента

3429
SUBEXCITATION ELECTRONS.
Final generations of secondary electrons having energies lower than minimum energy of electron excitations in a given system.

D Subexzitation-Elektronen, sehr langsame Elektronen
F électrons de sous-excitation
P elektrony podwzbudzeniowe
R слабовозбуждённые электроны

3430
SUB-GROUP (odd series, B family).
Part of group of periodic system composed of transition elements.

D Nebengruppe
F famille secondaire, sous-groupe, groupe auxiliaire
P grupa poboczna, grupa dodatkowa, podgrupa
R побочная группа, подгруппа

3431
SUBLIMATION.
Transition from solid to vapour phase.

D Sublimation
F sublimation
P sublimacja
R сублимация

3432
SUBLIMATION TEMPERATURE.
Temperature at which substance is converted from solid into gaseous state.

D Sublimation-Temperatur
F point de sublimation
P temperatura sublimacji
R температура сублимации, температура возгонки

SUBMICRO METHOD. *See* NANOGRAM METHOD.

3433
SUBSTANCE (matter, body).

D Substanz, Materie, Stoff
F substance, matière, corps
P substancja, materia, ciało
R вещество, материя, тело

3434
SUBSTANCE (pure substance, chemical individual).
Every definite chemical compound (complex substance), or chemical element in its free state (simple substance).

D (reine) Substanz, chemisches Individuum
F substance, individu chimique
P substancja (chemiczna), indywiduum chemiczne
R (чистое) вещество, химический индивид, химический индивидуум

3435
SUBSTITUENT (substituent group).
Atom (or group of atoms) replacing
mainly hydrogen atom on carbon atom in
organic compound molecule.
D Substituent
F substituant
P podstawnik
R заместитель, замещающая группа

SUBSTITUENT GROUP. *See*
SUBSTITUENT.

3436
SUBSTITUTION, S.
Replacement of atom (or group of atoms)
in molecule by another atom (or group of
atoms). The following types of substitution
are distinguished in organic chemistry:
nucleophilic S_N, electrophilic S_E and free
radical S_R, depending on mechanism of
particular reaction.
D Substitution
F substitution
P podstawienie, substytucja
R замещение

3437
SUBSTITUTIONAL SOLID SOLUTION.
Solid solution in which atoms or ions of
dissolved substance substitute atoms or
ions in matrix substance; crystallographic
conditions are defined by Grimm's law.
D Substitutionsmischkristalle
F solution solide de substitution
P roztwór stały substytucyjny
R твёрдый раствор замещения

3438
SUBSTOICHIOMETRIC ISOTOPE
DILUTION.
Isotope dilution in which for separation
of the constituent to be determined
a substoichiometric quantity of reagent is
used.
D substöchiometrische
  Isotopenverdünnung
F dilution isotopique substœchiométrique
P rozcieńczenie izotopowe
  substechiometryczne
R субстехиометрический метод
  в изотопном разбавлении

3439
SUBSTOICHIOMETRIC SEPARATION.
Separation of component in sample by
means of reagent added in a quantity
smaller than the stoichiometric one.
D unterstöchiometrische Trennung
F séparation sous-stœchiométrique
P wydzielanie substechiometryczne
R субстехиометрический метод
  выделения

3440
SUBSTRATE (reactant).
Substance undergoing chemical changes;
in stoichiometric equation of reaction
symbols (formulae) of substrate are placed
on left.
D Reaktand, Ausgangsstoff, Substrat
F substrat
P substrat reakcji
R реагирующее вещество, реактант,
  субстрат реакции

3441
SUBTRACTIONAL SOLID SOLUTION.
Solid solution with excess of one of
constituents, due to fact that some nodes
in lattice are left unoccupied by second
constituent.
D Defektmischkristalle, Mischkristalle mit
  Leerstellen
F ...
P roztwór stały subtrakcyjny
R твёрдый раствор вычитания

SUBULTRAMICRO METHOD. *See*
PICOGRAM METHOD.

SULFOCHLORINATION. *See*
CHLOROSULFONATION.

3442
SULFONATION.
Substitution of hydrogen atom in organic
compound molecule by sulfonic group
—SO_3H, effected by sulfuric acid or sulfur
trioxide, e.g.

$$C_6H_6 + HO—SO_3H \xrightarrow{SO_3} C_6H_5—SO_3H + H_2O$$

D Sulfonierung
F sulfonation
P sulfonowanie
R сульфирование

3443
SULFUR, S.
Atomic number 16.
D Schwefel
F soufre
P siarka
R сера

SULFUR-35. *See* RADIOSULFUR.

3444
SULFUR GROUP.
Elements of sixth main group of periodic
system, except for oxygen (sulfur,
selenium, tellurium, polonium).
Remark: Term previously used to denote
present oxygen family.
D Schwefelgruppe
F groupe du soufre
P siarkowce
R группа серы, элементы группы серы

3445
SULFURIZATION.
Introduction of sulfur into molecule
of chemical compound, e.g.

$$2C_6H_6 + S_2Cl_2 \rightarrow C_6H_5-S-C_6H_5 + S + 2HCl$$

D Sulfurierung
F sulfur(is)ation
P sulfuryzowanie
R ...

SULPHUR. *See* SULFUR.

3446
SUM PEAK.
Peak in gamma radiation spectrum
formed as a result of simultaneous
photoelectric absorption of two γ-photons
of different energies.

D Summen-Peak
F pic de somme
P pik sum(aryczn)y
R пик суммарной энергии, суммарный
пик

3447
SUPERCONDUCTIVITY.
Disappearence of electric resistance below
certain temperature (usually in range
15—0.4 K). At the same time material
becomes perfectly diamagnetic.

D Supraleitung, Supraleitfähigkeit
F supraconductivité
P nadprzewodnictwo
R сверхпроводимость

3448
SUPERCONDUCTIVITY MODEL
(superfluid model, pairing correlations
model).
Model of atomic nucleus which takes into
account residual interaction between
nucleons. This interaction causes
association of protons and neutrons into
pairs of zero-spin and gives rise to effects
similar to the superconductivity which
occurs in conductor due to strong
correlation of electron pairs.

D Supraleitfähigkeitsmodell
F modèle de superconductivité
P model nadprzewodnikowy, model
skorelowanych par
R сверхтекучая модель, модель парных
корреляций

3449
SUPEREXCHANGE.
In magnetochemistry, mechanism of
strong indirect exchange interactions
between paramagnetic ions through
diamagnetic ion positioned between them
(e.g. oxide ion). Exhibited by magnetically
ordered crystals (ferro-, antiferro- and
ferrimagnetic).

D Superexchange
F superéchange
P nadwymiana
R косвенный обмен, сверхобмен

3450
SUPEREXCITED MOLECULE.
Molecule with about 10 eV excess energy,
sufficient usually for scission to free
radicals and for reactions with zero
activation energy.

D hochangeregtes Molekül,
superangeregtes Molekül
F ...
P cząsteczka nadwzbudzona
R сверхвозбуждённая молекула

3451
SUPERFLUIDITY.
Property of helium II to flow through
narrow capillary tubes without friction.

D Supraflüssigkeit, Überflüssigkeit,
Superfluidität
F superfluidité
P nadciekłość
R сверхтекучесть

SUPERFLUID MODEL. *See*
SUPERCONDUCTIVITY MODEL.

3452
SUPERHEATED LIQUID.
Fluid in metastable state, heated at
constant pressure to temperature higher
than boiling temperature or subjected
isothermally to pressure lower than
pressure of saturated vapour. Disturbance
of this metastable state causes violent
change of liquid into gaseous phase.

D überhitzte Flüssigkeit
F liquide surchauffé
P ciecz przegrzana
R перегретая жидкость

3453
SUPERHEATED VAPOUR.
Vapour of pressure which is lower than
pressure of saturated vapour at the same
temperature.

D ungesättigter Dampf, überhitzter Dampf
F vapeur surchauffée
P para nienasycona, para przegrzana
R ненасыщенный пар, перегретый пар

SUPERLATTICE STRUCTURE. *See*
SUPERSTRUCTURE.

3454
SUPERPARAMAGNETISM.
Quasi-paramagnetism displayed by highly
disintegrated ferromagnets (particle
volume of 10^{-18} cm^3 or lower).

D Superparamagnetismus
F superparamagnétisme
P nadparamagnetyzm
R сверхпарамагнетизм

3455
SUPERPOSITION PRINCIPLE (principle
of superposition).
If quantum mechanical system is allowed
to have states Ψ_1, Ψ_2, Ψ_3, ..., then in state
given by $\Psi = \sum_i c_i \Psi_i$ probability that
measurement of some quantity will give
value characteristic for state Ψ_j is
equal c_j^2.

D Superpositionsprinzip
F principe de superposition
P zasada superpozycji stanów
R принцип суперпозиции

3456
SUPERSATURATED SOLUTION.
Solution in which concentration of
dissolved constituent is higher than
concentration of saturated solution in the
same conditions. Supersaturated solution
is metastable system.

D übersättigte Lösung
F solution sursaturée
P roztwór przesycony
R пересыщенный раствор

3457
SUPERSATURATED VAPOUR.
Vapour in unstable state, isobarically
cooled below boiling point, or compressed
isothermally to pressure larger than
pressure of saturated vapour;
supersaturation disappears in presence of
condensation seeds and vapour transforms
into saturated state.

D übersättigter Dampf
F vapeur sursaturée
P para przesycona
R пересыщенный пар

3458
SUPERSTRUCTURE (superlattice
structure).
Additional long-range order among
constituents of solid solution, consisting
either in splitting of equivalent atomic
positions into two non-equivalent
subpopulations occupied by different
kinds of atoms, or in multiplication of
periods due to occupation of nodes
of straight lattice by different atoms,
sequence of which is repeated
periodically.

D Überstruktur, übergeordnetes Gitter
F superstructure, super-réseau
P nadstruktura
R сверхструктура, сверхрешётка

3459
SUPPLIED ENTROPY (exchanged
entropy).
Part of entropy increase which (1) is due
to exchange of heat and mass with
surroundings, and (2) changes sign if
direction of exchange is reversed.

D Entropieströmung, zugeströmte Entropie
F entropie échangée
P entropia przekazana
R притекнувшая энтропия

3460
SUPPORT (solid support in
chromatography).
Porous solid (usually inert) which can sorb
liquid phase.

D fester Träger
F support (solide)
P nośnik
R (твёрдый) носитель

3461
SUPPORTING ELECTRODE.
In spark and arc spectroscopy electrode
prepared of material different from that
of sample; sample is then placed on/in
this electrode.

D Trägerelektrode, Hilfselektrode
F électrode auxiliaire
P elektroda (spektralna) pomocnicza
R посторонний электрод,
   вспомагательный электрод

3462
SUPPORTING ELECTROLYTE.
In polarography or other electrochemical
techniques, electrolyte introduced to
examined solution in large concentration
in order to increase its conductivity; ions
of supporting electrolyte should not react
with electrode over large potential range.

D Zusatzelektrolyt
F électrolyte indifférent
P elektrolit podstawowy
R посторонний электролит

3463
SURFACE-ACTIVE AGENT (surfactant).
Substance which lowers surface tension
of medium in which it is dissolved, and/or
interfacial tension with other phases.

D grenzflächenaktiver Stoff,
   oberflächenaktiver Stoff, Tensid,
   Surfactant
F agent tensio-actif, surfactif
P substancja powierzchniowo-czynna
R поверхностно-активное вещество

3464
SURFACE-ACTIVE FILM (adsorbed film).
Thin layer of insoluble (or hardly soluble)
substance at surface of liquid which
changes its surface tension.

D oberflächenaktive Schicht
F pellicule surfactive
P błonka adsorpcyjna, warstwa
   adsorpcyjna
R адсорбционная плёнка

3465
SURFACE ACTIVITY.
Ability of some solutes to accumulate on

(or near) solvent surface, thus lowering its surface tension.

D Oberflächenaktivität
F activité superficielle
P aktywność powierzchniowa
R поверхностная активность

3466
SURFACE CHARGE DENSITY.
Electric charge of unit surface area.

D Oberflächendichte der Ladung
F densité superficielle de charge
P gęstość ładunku powierzchniowa
R поверхностная плотность заряда

3467
SURFACE COMPOUNDS THEORY.
Theory of heterogeneous catalysis, explaining mechanism of catalysis by formation of intermediate, unstable compounds of substrates with atoms or ions present in surface layers of catalyst.

D Theorie der
    Oberflächenzwischengebilden
F ...
P teoria związków powierzchniowych
R теория активных ансамблей

3468
SURFACE DIFFUSION.
Diffusion occurring in two-dimensional layer of adsorbate along surface of adsorbent.

D Oberflächendiffusion
F diffusion superficielle
P dyfuzja powierzchniowa
R поверхностная диффузия

SURFACE ELECTRIC POTENTIAL
DIFFERENCE. *See* SURFACE
POTENTIAL.

3469
SURFACE FILM (condensed surface film).
Thin layer of closely-packed molecules of insoluble substance at surface of liquid.

D Oberflächenschicht
F pellicule superficielle
P błonka powierzchniowa, warstwa
    powierzchniowa
R поверхностная плёнка

3470
SURFACE PHENOMENA.
Group of all physico-chemical phenomena observed at interfaces between different phases.

D Oberflächenerscheinungen,
    Grenzflächenerscheinungen
F phénomènes de surface
P zjawiska powierzchniowe
R поверхностные явления

3471
SURFACE POTENTIAL (surface electric potential difference), χ.

Difference of electrical potentials between point inside phase and point separated 10^{-4} cm from phase surface.

D Oberflächenpotential
F tension électrique de surface
P potencjał powierzchniowy fazy
R поверхностный потенциал

3472
SURFACE PRESSURE.
Two-dimensional pressure which occurs in surface film due to random thermal movement of adsorbed molecules along surface.

D Oberflächendruck
F pression superficielle
P ciśnienie powierzchniowe
R поверхностное давление

3473
SURFACE TENSION, σ.
Free surface energy required to develop defined unit of free surface area.

D Oberflächenspannung
F tension superficielle
P napięcie powierzchniowe
R поверхностное натяжение

SURFACTANT. *See* SURFACE-ACTIVE
AGENT.

3474
SUSPENSION.
Disperse system in which particles of solid dispersed phase have dimensions greater than 500 nm.

D (grobe) Suspension, grobdisperses
    System, mechanische Zerteilung
F suspension
P zawiesina (mechaniczna), suspensja,
    układ o rozdrobnieniu mechanicznym
R суспензия

3475
SUSPENSOID.
Historical term denoting an irreversible lyophobic sol.
Remark: Do not confuse with suspension, i.e. mechanical suspension.

D Suspensoid, Suspensionskolloid
F suspensoïde
P suspensoid
R суспенсоид

3476
SWELLING.
Process of binding liquid or gas by gel or solid accompanied by increase in volume (shape of material that swells does not change in principle).

D Quellung
F gonflement
P pęcznienie
R набухание

25*

3477
SWELLING PRESSURE.
Pressure arising when gel or solid swells and increases its volume.

D Quellungsdruck
F pression de gonflement
P ciśnienie pęcznienia
R давление набухания

3478
SYMMETRIC TOP MOLECULE.
Molecule, all principal momenta of inertia of which are finite and two of them are different, e.g. CH_3Cl.

D symmetrisches Kreiselmolekül
F molécule cuspidale symétrique
P cząsteczka typu bąka symetrycznego, cząsteczka-rotator symetryczny
R молекула типа симметричного волчка

3479
SYMMETRY (in physics).
Invariance of physical systems or interactions with respect to a certain transformation (leads usually to corresponding conservation law).

D Symmetrie
F symétrie
P symetria
R симметрия

3480
SYMMETRY (in crystallography).
Attribute of physical object (e.g. crystal) or abstract object (e.g. geometrical structure) consisting in equivalency i.e. indistinguishability of its properties in defined directions. These directions are called equiponderant with respect to symmetry of considered property.

D Symmetrie
F symétrie
P symetria
R симметрия

3481
SYMMETRY CLASS.
Population of objects (both physical like crystals and abstract, e.g. vectors, functions etc.) undergoing identical transformation into self-coincidence with respect to all symmetry elements belonging to the same point group.
Remark: Symbol of this point group is used to denote symmetry class, sometimes leading to confusion of class of objects with group of operations which may be performed on these objects.

D Symmetrie-Klasse
F classe de symétrie
P klasa symetrii
R класс симметрии

3482
SYMMETRY ELEMENT.
Operator which transforms defined mathematical or physical object into repetition of itself. Basic symmetry elements are: centre of symmetry, rotation axis, and symmetry plane.

D Symmetrieelement
F élément de symétrie
P element symetrii
R элемент симметрии

3483
SYMMETRY NUMBER (of molecule).
Number of identical molecular configurations obtained by rotation of molecule with identical nuclei.

D Symmetriezahl
F nombre de symétrie
P liczba symetrii
R число симметрии, фактор симметрии

3484
SYMMETRY OPERATIONS.
Displacement of geometric objects bringing them to self-coincidence. Principal symmetry operations for finite objects are: revolution around n-fold axis by $360/n°$ angle where n is a natural number, reflection in plane and reflection at point; for infinite objects (space lattices): translations and their combinations with revolutions, where number of axes is limited: $n = 2, 3, 4$ and 6.

D Symmetrieoperationen, Deckoperationen
F opérations de symétrie
P operacje symetrii
R операции симметрии

3485
SYN-ANTI ISOMERISM.
Obsolete term formerly denoting cis-trans isomerism in compounds containing
$\diagdown$C=N— or —N=N— groups

D syn-anti-Isomerie
F isomérie syn-anti
P izomeria syn-anti
R син-анти изомерия

3486
SYNDIOTACTIC POLYMER.
Regular polymer composed of molecules having regular repetition of alternating enantiomeric configurational base units at least at one main-chain chiral or prochiral atom in configurational base unit.
In syndiotactic polymer stereobase unit consist of two enantiomeric configurational base units

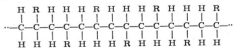

D syndiotaktisches Polymer
F polymère syndiotactique
P polimer syndiotaktyczny
R синдиотактический полимер

3487
SYNERESIS.
Process of exuding part of dispersion phase from gelatin.

D Synärese, Synäresis
F synérèse
P synereza
R синерезис

3488
SYNERGISTIC COAGULATION.
Coagulation of sol by mixture of electrolytes, when presence of one coagulant makes sol sensitive to action of another coagulant; hence sum concentration of electrolytes required for coagulation is considerably lower than would seem necessary from additive action.

D Sensibilisierung (mit Elektrolyten)
F sensitisation
P synergizm
R синергизм

3489
SYNTHESIS.
Production of chemical compound by union of elements or simple compounds or by degradation of complex compound, e.g.

$$CaO + CO_2 \rightarrow CaCO_3$$

D Synthese
F synthèse
P synteza
R синтез

3490
SYSTEMATIC ERROR (bias).
Deviation of mean value of population of results of measurement (determination) from value assumed as true, which occurs systematically when using a given method or instrument.

D systematischer Fehler
F erreur systématique
P błąd systematyczny
R систематическая ошибка

3491
SZILARD-CHALMERS EFFECT (recoil effect).
Rupture of chemical bond between recoiling atom and molecule in which it was incorporated due to nuclear transformation or reaction.

D Szilard-Chalmers-Effekt,
   Rückstoßeffekt
F effet Szilard-Chalmers, effet de recul
P efekt Szilarda i Chalmersa, efekt odrzutu
R эффект Сциларда-Чалмерса, эффект отдачи

3492
SZYSZKOWSKI'S EQUATION.
Equation giving dependence of surface tension σ of aqueous solution of surface-active substance (surfactant) on its concentration c, viz.

$$\sigma_0 - \sigma = a \, \log (1 + bc)$$

where σ_0 = surface tension of water, a, b = constants characteristic of a given surfactant; equation is true for solutions of substances which do not form micelles.

D Szyszkowski-Gleichung
F équation de Szyszkowski
P równanie Szyszkowskiego
R уравнение Шишковского

3493
TACTICITY.
Steric order in main polymeric chain.

D Taktizität
F tacticité
P taktyczność
R тактичность

3494
TACTIC POLYMER.
Regular polymer in whose macromolecules there is an ordered structure with respect to configurations around at least one main-chain site of stereoisomerism per constitutional base unit.

D taktisches Polymer
F polymère tactique
P polimer taktyczny
R тактический полимер

3495
TAFEL'S EQUATION.
Relation between overvoltage and density of polarizing current j

$$\eta = a - b \lg j$$

where a and b = constants; equation is valid if in process transition polarization predominates and anodic (cathodic) current is so large that one may neglect cathodic (anodic) current.

D Tafelsche Gleichung
F formule de Tafel, loi de Tafel
P wzór Tafela
R формула Тафеля

TAGGED ATOMS. See LABELLED ATOMS.

3496
TAILING (*in chromatography*).
Asymmetric spreading of rear edge of
chromatographic band during
development.
D Schwanzbildung
F effet de traînée
P tworzenie się ogonów
R образование хвостов

3497
TANTALUM, Ta.
Atomic number 73.
D Tantal
F tantale
P tantal
R тантал

3498
TARGET.
Substance or object exposed to action of
nuclear radiation or bombarded with
accelerated ions or elementary particles.
During these processes specific nuclear
reactions occur in the target.
D Target, Auffänger, Treffplatte
F cible (nucléaire)
P tarcza
R мишень

3499
TAST POLAROGRAPHY.
Variation of polarography consisting in
recording of current at end of drop life;
due to this influence of capacity current
is eliminated to some extent and limiting
currents are larger than those recorded in
classical polarography.
D Tastpolarographie
F ...
P polarografia prądu maksymalnego,
polarografia z bramkowaniem
R таст-полярография, полярография
с регистрацией максимального тока

3500
TAUTOMERIC EQUILIBRIUM.
Thermodynamic equilibrium of mutual
conversion of tautomeric forms, e.g. for
keto-enolic tautomerism, this equilibrium
is expressed by equation

$$K_T = \frac{[EH]}{[KH]} = \frac{K_{KH}}{K_{EH}}$$

where K_T = tautomeric equilibrium
constant, [EH] and [KH] = concentrations
of enol and ketone forms, K_{KH} and
K_{EH} = acid constants of enol and ketone
forms. Content of either form in mixture
is expressed in percent.
D Tautomeriegleichgewicht
F équilibre tautomérique
P równowaga tautomeryczna
R таутомерное равновесие

3501
TAUTOMERISM.
Type of structural isomerism caused by
existence of thermodynamic equilibrium
between isomers (tautomers) whose
molecules differ by arrangement of atoms
and electron density distribution.
D Tautomerie
F tautomérie
P tautomeria
R таутомерия

3502
TAU VALUE (τ-value).
Number characterizing proton chemical
shift in NMR spectra: $\tau = 10-\delta$, where
δ = chemical shift in ppm, measured with
respect to tetramethylsilane standard.
D τ-Wert
F ...
P wartość τ
R значение τ

t-DISTRIBUTION. *See* STUDENT'S
DISTRIBUTION.

3503
TECHNETIUM, Tc.
Atomic number 43.
D Technetium
F technétium
P technet
R технеций

3504
TELLURIUM, Te.
Atomic number 52.
D Tellur
F tellure
P tellur
R теллур

3505
TELOMER.
Product of telomerization, oligomer
composed of molecules containing
terminal groups incapable (under
conditions of synthesis) to react with
added monomer to give larger molecules
of identical chemical composition.
D Telomer
F télomère
P telomer
R теломер

3506
TELOMERIZATION.
Polymerization of telogens; these are not
monomers, but are capable to initiate,
transfer and end the chain.
D Telomerisation
F télomérisation
P telomeryzacja
R теломеризация

3507
TEMPERATURE (empirical temperature), t.
Intensive function of state which has following property: if system in thermodynamic equilibrium is composed of two diathermally partitioned parts A and B then their temperatures are equal, $t_A = t_B$.
D (empirische) Temperatur
F température
P temperatura (empiryczna)
R температура

3508
TEMPERATURE COEFFICIENT OF REACTION RATE.
Ratio of reaction rate constants at two temperatures; in particular, quotient n, of rate constant k at temperature t divided by k measured at temperature 10 degrees lower.

$$n = \frac{k_t}{k_{t-10°}}$$

for many reactions $2 < n < 3$.
D Temperaturkoeffizient der Reaktionsgeschwindigkeit
F coefficient de van't Hoff, coefficient thermique (des vitesses de réactions)
P współczynnik temperaturowy (szybkości) reakcji
R температурный коэффициент (скорости) реакции

3509
TEMPERATURE-PROGRAMMED CHROMATOGRAPHY.
Chromatographic technique in which column temperature is changed in course of analysis according to predetermined programme.
D temperaturprogrammierte Chromatographie
F chromatographie à température programmée
P chromatografia z programowaniem temperatury
R хроматография с программированием температуры

3510
TENSAMMETRY.
Alternating-current polarography applied to study of adsorption and desorption phenomena of various substances on dropping mercury electrode.
D Tensammetrie
F ...
P tensammetria
R тенсамметрия

3511
TERBIUM, Tb.
Atomic number 65.
D Terbium
F terbium
P terb
R тербий

TERM MULTIPLICITY. *See* MULTIPLICITY.

TERMOLECULAR REACTION. *See* TRIMOLECULAR REACTION.

3512
TERPOLYMER.
Product of copolymerization of mixture of three species of monomer.
D Terpolymer
F terpolymère
P terpolimer
R терполимер, тройной сополимер

TEST OF AGREEMENT. *See* TEST OF GOODNESS OF FIT.

3513
TEST OF GOODNESS OF FIT (test of agreement).
Statistical test used for verification of hypotheses about form of probability distribution.
D Test der Anpassung
F test de conformité
P test zgodności
R критерий соответствия

3514
TEST OF SIGNIFICANCE.
Statistical test which allows decision to be made whether, after trial, a given hypothesis should be rejected or whether there is no reason for its rejection, to small risk of error (measured by significance level).
D Test der Sicherheit
F test de signification
P test istotności
R критерий значимости

TEST-PAPER. *See* INDICATOR PAPER.

3515
TETRAGONAL HYBRID (sp³ hybrid).
One of four equivalent hybridized atomic orbitals formed from one s and three p orbitals. Angle between two sp³ hybrids is 109°27′; used for instance to explain structure of methane molecule.
D tetraedrische Hybride, tetraedrisches Hybridorbital, te-Hybride
F hybride tétraédrique, hybride sp³
P hybryd tetraedryczny, hybryd sp³
R тетраэдрическая гибридная орбиталь

3516
TETRAHEDRAL COMPLEX.
Complex of coordination number 4, in
which donor atoms of ligands are
distributed at vertices of regular or
distorted tetrahedron. In this complex
hybridization sp^3, d^3s is involved.
D tetraedrischer Komplex
F complexe tétraédrique
P kompleks tetraedryczny
R тетраэдрический комплекс

3517
THALLIUM, Tl.
Atomic number 81.
D Thallium
F thallium
P tal
R таллий

3518
THEORETICAL CHEMISTRY.
Branch of physical chemistry dealing with
applications of theory for description of
various physico-chemical phenomena.
D theoretische Chemie
F chimie théorique
P chemia teoretyczna
R теоретическая химия

THEORETICAL END-POINT OF
TITRATION. See EQUIVALENCE POINT
OF TITRATION.

3519
THEORETICAL PLATE NUMBER (in
chromatography), n.
Parameter characterizing column
performance
$$n = 16\left(\frac{\text{retention volume}}{\text{peak width}}\right)^2$$
where peak width is denoted by base of
triangle with sides tangent to inflection
points of peak.
D Bodenzahl, effektive Bodenzahl der
Säule, Zahl der theoretischen Stufen der
Kolonne
F nombre de plateaux théoriques
P liczba półek teoretycznych
R число теоретических тарелок

3520
THEORY OF ACTIVE CENTRES.
Theory of heterogeneous catalysis which
implies progress of reaction on active
centres of catalyst.
D Theorie der aktiven Zentren
F théorie des centres actifs
P teoria centrów aktywnych
R теория активных центров

THEORY OF MULTIPLETS. See
BALANDIN THEORY OF MULTIPLETS.

THEORY OF STRAINLESS RINGS. See
SACHSE-MOHR THEORY.

3521
THERMAL ANALYSIS.
Method based on investigation of
temperature changes of investigated
system in time of its heating or cooling
in order to determine its chemical
composition and/or phase change of
substance.
D Thermoanalyse, thermische Analyse
F analyse thermique
P analiza termiczna
R термический анализ

3522
THERMAL COLUMN.
Block of graphite, several metres thick,
leading from nuclear reactor core, through
reflector and shield, used for
thermalization of neutrons.
D thermische Säule
F colonne thermique
P kolumna termiczna
R тепловая колонна

3523
THERMAL CONDUCTIVITY CELL
(katharometer).
Device that records composition of gas
leaving chromatographic column by
measuring its thermal conductivity.
D Wärmeleitfähigkeits-Meßzelle
F cellule de conductivité thermique,
catharomètre
P detektor termokonduktometryczny,
katarometr
R термокондуктометрический детектор,
катарометр

3524
THERMAL DE BROGLIE WAVELENGTH
(quantum length).
Parameter $\lambda = \sqrt{\dfrac{h^2}{2\pi mkT}}$ (m = particle
mass, k = Boltzmann constant, and
h = Planck constant) which represents
order of magnitude of de Broglie
wavelength for typical particle in
thermodynamic system at temperature T;
if $\lambda \ll \sqrt[3]{v}$, where v denotes average
volume per particle, quantum effects are
negligible; they become important for λ
comparable to or greater than $\sqrt[3]{v}$.
D thermische (de Broglie-)Wellenlänge
F longueur d'onde thermique de
de Broglie
P długość fali de Broglie'a termiczna
R тепловая длина волны (де Бройля)

3525
THERMAL DIFFUSION (thermodiffusion,
Soret effect).
Transport of matter due to temperature
gradient.

D Thermodiffusion, (Ludwig-)Soret-Effekt
F diffusion thermique
P termodyfuzja, termotransport, efekt Soreta
R термодиффузия, эффект Соре

3526
THERMAL DISSOCIATION.
Dissociation at elevated temperature.

D thermische Dissoziation
F dissociation thermique
P dysocjacja termiczna
R термическая диссоциация

3527
THERMAL EQUATION OF STATE.
Equation which gives dependence of phase pressure p on its temperature T, volume V and number of moles of its constituents n_1, n_2, ..., n_c

$$p = p(T, V, n_1, n_2, ..., n_c)$$

D thermische Zustandsgleichung
F équation (thermique) d'état
P równanie stanu termiczne
R термическое уравнение состояния

3528
THERMAL EQUILIBRIUM.
Thermodynamic state of system with the same temperature in all coexisting phases.

D thermisches Gleichgewicht
F équilibre thermique
P równowaga termiczna
R термическое равновесие

3529
THERMAL EXPLOSION.
Transition of steady state of exothermal reaction into non-stationary state in which reaction gradually undergoes acceleration due to temperature increase.

D Wärmeexplosion
F explosion thermique
P wybuch termiczny
R тепловой взрыв

THERMALLY CONDUCTING WALL. See DIATHERMIC PARTITION.

3530
THERMAL NEUTRONS.
Neutrons in thermal equilibrium with substance in which they exist. In normal conditions energy of thermal neutrons is 0.01—0.1 eV: most probable speed of thermal neutrons at 20°C is 2200 m/s.

D thermische Neutronen
F neutrons thermiques
P neutrony termiczne
R тепловые нейтроны

3531
THERMAL SPIKE.
Local heating of liquid or solid medium, confined to a molecular scale, due to the stopping of ion or neutral heavy particle of high kinetic energy.

D Wärmestörbereich
F zone lacunaire thermique
P kolec termiczny
R тепловой пик, тепловой клин

3532
THERMOCHEMICAL EQUATION.
Stoichiometric representation of chemical reaction or physical process which takes into account both amounts and nature of all reactants or constituents, their aggregation state, value of heat effect as well as temperature and pressure i.e. conditions of the transformation.

D thermochemische Gleichung
F équation thermochimique
P równanie termochemiczne
R термохимическое уравнение

3533
THERMOCHEMISTRY.
Branch of physical chemistry dealing with studies on heat effects of chemical reactions as well as their dependence on physical parameters.

D Thermochemie
F thermochimie
P termochemia
R термохимия

3534
THERMOCHROMISM.
Reversible change of colour of chemical compound with temperature; usually high-temperature form absorbs at longer wavelength than the low-temperature one. Bianthron is typical thermochromic compound.

D Thermochromie
F thermochromisme
P termochromia
R термохромизм

THERMODIFFUSION. See THERMAL DIFFUSION.

3535
THERMODIFFUSION POTENTIAL.
Difference of inner potentials resulting from temperature gradient in phase.

D Thermodiffusionspotential
F potentiel de thermodiffusion
P potencjał termodyfuzyjny
R термодиффузионный потенциал

THERMODYNAMIC ACTIVITY. See ACTIVITY.

THERMODYNAMICALLY UNSTABLE COMPLEX. See UNSTABLE COMPLEX.

3536
THERMODYNAMIC CALORIE, cal_{th}.
Obsolete energy unit; in SI system
$1\ cal_{th} = 4.1840$ J.
D thermochemische Kalorie
F calorie thermochimique
P kaloria termochemiczna
R термохимическая калория

THERMODYNAMIC CHANGE. *See*
THERMODYNAMIC PROCESS.

THERMODYNAMIC ENGINE. *See* HEAT
ENGINE.

3537
THERMODYNAMIC EQUILIBRIUM
(complete internal equilibrium,
macroscopic equilibrium state).
Thermodynamic state in which entropy
source and all thermodynamic forces
vanish; in thermodynamic equilibrium no
irreversible process is possible.
D thermodynamisches Gleichgewicht,
(inneres) Gleichgewicht
F équilibre (thermodynamique)
P równowaga termodynamiczna
R термодинамическое равновесие,
термодинамические равновесное
состояние, статистическое равновесие

3538
THERMODYNAMIC EQUILIBRIUM
CONSTANT, K_a.
Ratio of product of equilibrium activities
of products and product of equilibrium
activities of substrates. Activity of each
reactant is raised to power equal to
corresponding stoichiometric coefficient.
Thermodynamic equilibrium constant
depends on temperature and pressure.
D thermodynamische
Gleichgewichtskonstante
F constante d'équilibre (thermodynamique)
P stała równowagi (chemicznej)
termodynamiczna
R термодинамическая константа
равновесия

THERMODYNAMIC FLOWS. *See*
THERMODYNAMIC FLUXES.

THERMODYNAMIC FLUCTUATIONS.
See EQUILIBRIUM FLUCTUATIONS.

3539
THERMODYNAMIC FLUXES (fluxes,
thermodynamic flows, generalized fluxes,
rates, currents).
Quantities characterizing velocities of
individual irreversible processes; they
vanish at thermodynamic equilibrium.
D generalisierte Ströme
F flux

P przepływy termodynamiczne
R потоки

3540
THERMODYNAMIC FORCES (affinities).
Gradients of intensive quantities
characterizing distance from
thermodynamic equilibrium.
D generalisierte Kräfte, (treibende) Kräfte
F forces (généralisées)
P bodźce termodynamiczne, siły
termodynamiczne
R термодинамические силы,
(термодинамические движущие) силы

THERMODYNAMIC FUNCTIONS. *See*
CHARACTERISTIC FUNCTIONS.

3541
THERMODYNAMIC LIMIT (infinite
volume limit).
Limit for which volume of system
approaches infinity, $V \to \infty$; particle
number density remains constant,
$\dfrac{\langle N \rangle}{V} = $ constant. Average values in
assembly composed of finite volume
systems become extensive or intensive as
characteristic for thermodynamic
quantities only in thermodynamic limit.
D thermodynamischer Grenzfall,
thermodynamische Grenze
F limite thermodynamique
P granica termodynamiczna
R термодинамический предел

THERMODYNAMIC PARAMETER. *See*
STATE VARIABLE.

THERMODYNAMIC POTENTIAL. *See*
GIBBS FREE ENERGY.

THERMODYNAMIC POTENTIALS. *See*
CHARACTERISTIC FUNCTIONS.

3542
THERMODYNAMIC PROBABILITY (*of
macroscopic state*).
Number of microscopic states constituting
a given macroscopic state.
D thermodynamische Wahrscheinlichkeit
F probabilité thermodynamique
P prawdopodobieństwo termodynamiczne
R термодинамическая вероятность

3543
THERMODYNAMIC PROCESS
(thermodynamic change, thermodynamic
transition).
Transition of system from one
thermodynamic state to another (which
can also be identical with initial one)
accompanied macroscopically by chemical
reaction, mass transport, or energy
transfer within system or between system
and surroundings.

D Zustandsänderung, thermodynamischer
  Vorgang, thermodynamischer Prozeß
F transformation thermodynamique
P przemiana termodynamiczna, proces
  termodynamiczny
R термодинамический процесс, изменение
  состояния

THERMODYNAMIC PROPERTIES. *See*
STATE QUANTITIES.

THERMODYNAMIC QUANTITIES. *See*
STATE QUANTITIES.

3544
**THERMODYNAMICS OF
IRREVERSIBLE PROCESSES**
(irreversible thermodynamics,
non-equilibrium thermodynamics).
Phenomenological theory of dissipative
processes in macroscopic systems in
thermodynamic non-equilibrium states.

D Thermodynamik der irreversiblen
  Prozesse
F thermodynamique des phénomènes
  irréversibles
P termodynamika procesów
  nieodwracalnych, termodynamika
  stanów nierównowagi
R термодинамика необратимых
  процессов, термодинамика
  неравновесных состояний

THERMODYNAMIC STATE.
See MACROSCOPIC STATE.

3545
THERMODYNAMIC SYSTEM
(macroscopic system).
System composed of large number of
particles (most frequently molecules),
which behaves according to laws of
macroscopic thermodynamics.

D thermodynamisches System
F système thermodynamique
P układ termodynamiczny, układ
  makroskopowy
R термодинамическая система,
  макроскопическая система

3546
THERMODYNAMIC TEMPERATURE
(absolute temperature), T.
Intensive thermodynamic quantity

$$T = \left(\frac{\partial U}{\partial S}\right)_V$$

where U = internal energy, S = entropy,
and V = phase volume. Thermodynamic
temperature is equivalent to temperature
determined by gas thermometer.

D absolute Temperatur
F température thermodynamique,
  température absolue, température de
  Carnot
P temperatura termodynamiczna,
  temperatura bezwzględna
R термодинамическая температура,
  абсолютная температура

THERMODYNAMIC TRANSITION. *See*
THERMODYNAMIC PROCESS.

3547
THERMOGRAVIMETRY.
Method of thermal analysis consisting in
use of thermobalance to determine change
of mass of investigated substance in
relation to temperature or time.

D Thermogravimetrie
F thermogravimétrie
P termograwimetria, analiza
  termograwimetryczna
R термогравиметрия

3548
THERMOLUMINESCENCE.
Emission of radiation by increasing
temperature of system excited at lower
temperatures.

D Thermolumineszenz
F thermoluminescence
P termoluminescencja
R термолюминесценция

3549
**THERMOLUMINESCENT DOSIMETER,
TLD.**
Device with solid-state or rigid substance
(there exist several different models,
according to shape and size), which after
being irradiated with high energy
radiation, produces proportional amount
of light to ionizing energy received by it.

D Thermolumineszenzdosimeter
F dosimètre thermoluminescent
P dawkomierz termoluminescencyjny
R термолюминесцентный дозиметр

3550
THERMOMAGNETIC ANALYSIS.
Method of analysis and determination of
phase components of materials containing
ferromagnets (e.g. catalysts) and method
of investigation of appearance of
ferromagnetic phases during
recrystallization of metal alloys; it
utilizes measurements of magnetization
as function of temperature.

D thermomagnetische Analyse
F analyse thermo-magnétique
P analiza termomagnetyczna
R термомагнитный анализ

3551
THERMOMETRIC TITRATION
(enthalpimetric titration).
Titration for which end point is
determined on basis of changes of
temperature of analysed solution (liquid
or gas) resulting from heat of reaction.
D thermometrische Titration,
enthalpometrische Titration,
Enthalpie-Titration
F titrage thermochimique
P miareczkowanie termometryczne
R термометрическое титрование

3552
THERMONUCLEAR REACTION (nuclear
fusion).
Combination of two light nuclei to form
heavier nucleus with release of energy
equal to difference between nuclear
binding energy of products and sum of
binding energies of the two light nuclei.
D thermonukleare Reaktion
F réaction thermonucléaire
P reakcja termojądrowa, reakcja
termonuklearna, reakcja syntezy
jądrowej
R термоядерная реакция, реакция
синтеза

3553
THERMOSTAT.
Device maintaining constant temperature
automatically.
D Thermostat
F thermostat
P termostat
R термостат

THERMOSTATICS. *See* CLASSICAL
THERMODYNAMICS.

3554
THICK TARGET.
Target of such thickness that there is
appreciable energy loss or absorption of
incident particles or photons traversing it.
D dicker Target, dicke Treffplatte
F cible épaisse
P tarcza gruba
R толстая мишень

3555
THIN-LAYER CHROMATOGRAPHY,
TLC.
Chromatography that employs stationary
phase consisting of thin layer of
fine-grained solid (adsorbent) spread on
glass plate or pressed into plastic foil.
D Dünnschicht-Chromatographie
F chromatographie en couche mince
P chromatografia cienkowarstwowa,
chromatografia w cienkich warstwach
R хроматография в тонких слоях,
тонкослойная хроматография

3556
THIN-LAYER ELECTROCHEMISTRY.
Electrolysis of thin layer (1—100 μm) of
electrolyte solutions carried out under
various conditions (chronoamperometry,
chronopotentiometry, stationary electrode
voltammetry).
D ...
F ...
P elektrochemia cienkowarstwowa
R тонкослойная электрохимия

3557
THIN TARGET.
Target of such small thickness that there
is negligible energy loss or absorption of
incident particles or photons traversing it.
D dünner Target
F cible mince
P tarcza cienka
R тонкая мишень

3558
THIRD LAW OF THERMODYNAMICS
(Nernst-Planck theorem).
Only finite amount of energy can be taken
out of equilibrium system in form of
heat and during this process
thermodynamic temperature tends to zero
and entropy to some value S_0, which
(1) is equal to zero for one-component
phases and for ordered multicomponent
phases,
(2) in case of disordered multicomponent
phases is positive and depends only on
number of components.
D dritter Hauptsatz der Thermodynamik,
Nernstscher Wärmesatz
F troisième principe de la
thermodynamique, postulat de
Nernst-Planck, principe
de Nernst-Planck
P zasada termodynamiki trzecia, teoremat
Nernsta i postulat Plancka
R третье начало термодинамики, третий
принцип термодинамики

3559
THIRD-ORDER REACTION.
Reaction with kinetic equation
characterized by sum of powers in which
concentrations of reactants occur being
equal to three.
D Reaktion 3. Ordnung, Reaktion dritter
Ordnung
F réaction du troisième ordre
P reakcja trzeciego rzędu
R реакция третьего порядка

THIRD TRIAD. *See* HEAVY PLATINUM
METALS.

3560
THIXOTROPY.
Isothermal reversible conversion of gel

into sol under influence of mechanical shaking.

D Thixotropie
F thixotropie
P tiksotropia
R тиксотропия

3561
THORIUM, Th.
Atomic number 90.

D Thorium
F thorium
P tor
R торий

3562
THORIUM D, ThD.
Isotope of lead. Atomic number 82. Mass number 208.

D Thoriumblei, Thorium D
F plomb du thorium, thorium D
P toroołów, tor D, ThD
R торий D

3563
THORIUM SERIES.
Naturally occurring series with first and last members thorium, ^{232}Th, and lead, ^{208}Pb, respectively. The mass number of each member of this series is $A = 4n$, where n = natural number.

D Thorium-Reihe
F famille du thorium
P szereg torowy
R ряд тория

3564
THORON (thorium emanation), Tn.
Isotope of radon. Atomic number 86. Mass number 220.

D Thoron, Thorium-Emanation
F thoron
P toron, emanacja torowa
R торон

3565
THREE CARBON TAUTOMERISM.
Type of prototropic tautomerism connected with reversible rearrangement of α,β-ethylenic compounds into corresponding β,γ-derivatives, e.g.

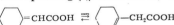

D Drei-Kohlenstoff-Tautomerie
F tautomérie tricentrique carboné
P tautomeria trójwęglowa
R трёхуглеродная таутомерия

3566
THREE-CENTRE BOND.
Conception of bond present in molecule with electron deficiency, postulated by Longuet-Higgins, e.g. B—B bond in B_2H_6. According to it, B—B bond is formed by electrons present in three-centred molecular orbitals overlapping hydrogen atom and two boron atoms.

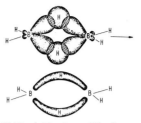

D Drei-Zentren-Bindung
F liaison tricentrique, liaison par point anionique
P wiązanie trójatomowe, wiązanie trójcentrowe
R трёхцентровая связь

3567
THREE-SIGMA RULE.
If distribution in population of results of measurements (determinations) is unknown because of their small number, then 0.99 confidence level is accepted for confidence interval $\bar{x} \pm 3s$, where s = standard deviation in series. Nowadays this rule is rarely used and may be applied only in case of distributions close to normal.

D Zusammenhang von 3σ-Grenze
F règle des trois sigmas
P reguła trzech sigm
R критерий трёх сигм

3568
THREO FORM.
Diastereoisomer having two adjacent non-equivalent chiral centres of specific configuration: two identical or similar atoms (or groups of atoms) are situated on opposite sides of Fischer projection formula, e.g.

$$
\begin{array}{c}
\mathrm{CH_3} \\
\mathrm{Br-\!\!\!-H} \\
\mathrm{H-\!\!\!-OH} \\
\mathrm{CH_3}
\end{array}
$$

D threo-Form
F forme thréo
P odmiana treo
R трео-форма

3569
THRESHOLD DOSIMETER (go-no-go dosimeter).
Ionizing radiation dosimeter recording abrupt change of properties in effect of absorption of defined dose of radiation.

D Schwellwert-Dosimeter
F dosimètre (du seuil) limite
P dawkomierz progowy
R пороговый дозиметр

3570
THRESHOLD ENERGY.
The smallest energy that a particle or quantum must have to cause, for instance, ionization of a molecule, displacement of an atom in crystal lattice, nuclear activation by ionizing radiation or nuclear reaction.
D Schwellenenergie
F énergie à seuil
P energia progowa
R пороговая энергия

3571
THULIUM, Tm.
Atomic number 69.
D Thulium
F thulium
P tul
R тулий

TIFFENEAU CONVERSION. *See*
TIFFENEAU REACTION.

3572
TIFFENEAU REACTION (Tiffeneau conversion).
The rearrangement of cyclic amino-alcohols (obtained from reduction of cyanohydrins) to cyclic ketones with extension of ring, effected by nitrous acid, e.g.

D Tiffeneau-Ringerweiterung
F réaction de Tiffeneau
P reakcja powiększania pierścieni Tiffeneau
R перегруппировка Тиффено

3573
TIME AVERAGE, $\overline{A}$.
Time-averaged value of quantity A

$$\overline{A} = \lim_{t^* \to \infty} \frac{1}{t^*} \int\limits_0^{t^*} A(t)\,dt$$

where $A(t)$ = value of function A at the time t, t^* = a fairly long time-averaging interval.
D Zeitmittel
F (valeur) moyenne temporelle
P średnia czasowa
R временное среднее значение, среднее по времени

3574
TIME-DEPENDENT SCHRÖDINGER EQUATION (Schrödinger equation with time).
Equation which determines time evolution of quantum mechanical system;

$$H\varphi = i\hbar\frac{\partial\varphi}{\partial t}$$

where H = hamiltonian of system,
φ = wave function, which depends on spatial and spin coordinates and time,
$i = \sqrt{-1}$, $\hbar$ = Planck's constant divided by 2π.
D zeitabhängige Schrödinger-Gleichung
F équation d'ondes contenant le temps
P równanie Schrödingera zawierające czas
R уравнение Шредингера

3575
TIME-INDEPENDENT SCHRÖDINGER EQUATION.
Equation which determines stationary states of quantum mechanical system; $H\varphi = E\varphi$, where H = hamiltonian of system, φ and E = wave function and corresponding stationary state energy respectively.
D zeitunabhängige Schrödinger-Gleichung
F équation de Schrödinger
P równanie Schrödingera nie zawierające czasu
R уравнение Шредингера для стационарных состояний

3576
TIME-RESOLVED SPECTROSCOPY.
Technique of spectrophotometric measurements with recording of absorption spectra in time intervals of 5 μs to 0.01 s. Most frequently used for study of kinetics of fast processes.
D Kurzzeitspektroskopie
F ...
P spektrofotometria szybka
R ...

3577
TIME-RESOLVED SPECTRUM.
Spectrum obtained by using cyclic excitation source. Subsequent parts along spectrum correspond to subsequent time-ordered excitation phases.
D zeitlaufgelöstes Spektrum
F spectre résolu dans le temps
P widmo rozdzielone w czasie
R спектр разрешённый по времени

3578
TIN, Sn.
Atomic number 50.
D Zinn
F étain
P cyna
R олово

3579
TISHCHENKO PREPARATION OF ESTERS.
Preparation of esters by intermolecular condensation of aldehydes in presence of alcoholates of aluminium, magnesium or sodium, e.g.

$$2\,RCH_2CHO \xrightarrow{Al(OC_2H_5)_3}$$

$$\left[\begin{array}{c} RCH_2CH-O-Al(OC_2H_5)_2 \\ | \oplus \\ O-CHCH_2R \end{array}\right] \rightarrow RCH_2COOCH_2CH_2R$$

D Tischtschenko-Claisen
  Aldehyd→Ester-Hydridanionotropie
F réaction de Tischtschenko
P reakcja Tiszczenki
R реакция Тищенко

3580
TITANIUM, Ti.
Atomic number 22.

D Titan
F titane
P tytan
R титан

TITANIUM FAMILY. *See* TITANIUM GROUP.

3581
TITANIUM GROUP (titanium family).
Elements in fourth sub-group of periodic system with structure of outer electronic shells of atoms $(n-1)d^2ns^2$: titanium, zirconium, hafnium (thorium).

D . . .
F groupe du titane
P tytanowce
R подгруппа титана, элементы подгруппы титана

3582
TITRANT.
Solution with precisely known titre used in volumetric analysis.

D Maßlösung, Titrans, Reagenzlösung, Titrierflüssigkeit
F solution titrée
P roztwór mianowany
R стандартный титрованный раствор

3583
TITRATION.
Addition of standardized solution of reagent to solution of analysed substance in order to determine it using volume of added reagent.

D Titration, Titrierung
F titrage
P miareczkowanie
R титрование

3584
TITRATION CURVE.
Plot which shows changes of measured quantity in titration according to volume of titrant.

D Titrationskurve
F courbe de titrage
P krzywa miareczkowania
R кривая титрования

3585
TITRATION ERROR (volumetric error).
Difference between volume of titration solution necessary for end-point and volume sufficient for equivalence point to be reached.

D Titrierfehler, Titrationsfehler, Indikationsfehler
F erreur de titrage
P błąd miareczkowania
R ошибка титрования

3586
TITRE.
Number of grams of analysed component which react with 1 cm³ of titrant in strickly determined conditions.

D Titer
F titre de la solution
P miano
R титр

3587
TITRIMETRIC ANALYSIS (volumetric analysis).
Technique of quantitative analysis based on determination of solution component from measured volume of titrant used for quantitative reaction with this component.

D Maßanalyse, Titrieranalyse, titrimetrische Analyse, volumetrische Analyse
F volumétrie, analyse volumétrique, titrimétrie
P analiza miareczkowa, analiza objętościowa
R объёмный анализ, титриметрический анализ

TORSIONAL STRAIN. *See* PITZER STRAIN.

TOTAL ABSORPTION PEAK. *See* PHOTOPEAK.

TOTAL ANALYSIS. *See* COMPLETE ANALYSIS.

3588
TOTAL ANGULAR MOMENTUM.
Vector quantity defined as sum of orbital and spin angular momentum of system.

D Gesamtdrehimpuls
F moment cinétique total, moment angulaire total
P moment pędu całkowity, kręt całkowity
R полный момент импульса, полный момент количества движения

3589
TOTAL CROSS-SECTION.
Differential cross-section integrated over full solid angle Ω and over total energy.

D integraler Wirkungsquerschnitt
F section efficace totale
P przekrój czynny całkowy
R полное (эффективное) сечение

3590
TOTAL HEAT OF SOLUTION.
Enthalpy change of system in process of dissolving one mole of a given substance in such an amount of pure solvent as finally to reach saturation point.

D ganze Lösungswärme
F chaleur totale de dissolution
P ciepło rozpuszczania pełne
R полная теплота растворения

3591
TOTAL ION EXCHANGE CAPACITY
(ion exchange capacity).
Number of equivalents of ionogenic groups per unit of dry mass (mass capacity), or bed volume (volume capacity) of exchanger.

D Gesamtkapazität, Kapazität (*des Ionenaustauschers*)
F capacité d'échange
P zdolność wymienna jonitu całkowita, pojemność właściwa jonitu całkowita
R ионообменная ёмкость

3592
TOTAL POLARIZABILITY, α.
Sum of atomic, electronic and orientation polarizabilities given by

$$\alpha = \frac{P}{N_L E}$$

where: P = dielectric polarization of medium, N_L = number of molecules in unit volume, E = strength of local field acting on molecule.

D Gesamtpolarisierbarkeit
F polarisabilité totale
P polaryzowalność całkowita
R полная поляризуемость

TOTAL RETENTION VOLUME. *See* RETENTION VOLUME.

3593
TRACE ANALYSIS.
Identification or determination of trace components of sample.

D Spurenanalyse
F analyse de traces
P analiza śladowa
R определение следов

3594
TRANSAMINATION.
Reaction between amino acid and α-ketoacid, involving migration of amino group, e.g.

$$\underset{\text{H}_2\text{N}-\overset{\displaystyle R}{\underset{|}{\text{C}}\text{H}-\text{COOH}}}{} + \underset{\text{O}=\overset{\displaystyle R'}{\underset{|}{\text{C}}-\text{COOH}}}{} \rightleftarrows \underset{\text{O}=\overset{\displaystyle R}{\underset{|}{\text{C}}-\text{COOH}}}{}$$

$$+ \underset{\text{H}_2\text{N}-\overset{\displaystyle R'}{\underset{|}{\text{C}}\text{H}-\text{COOH}}}{}$$

D Transaminierung
F transamination
P transaminacja
R переаминирование

3595
TRANSANNULAR EFFECT.
Mutual influence exerted by atoms (or groups of atoms) located in remote points of medium-size ring, for instance, at 1,5- or 1,6-position of 8—12-membered ring. This effect is accompanied by specific physical constants and very often by specific chemical reactions (*see* transannular migration).

D transannularer Effekt
F effet transannulaire
P efekt transanularny
R трансаннулярный эффект

3596
TRANSANNULAR MIGRATION.
Type of rearrangement characteristic for multi-membered alicyclic compounds, (8—12 carbon atoms) involving hydrogen atom (or group of atoms) migration to carbon atom sterically close, but situated further in the ring (3 to 4 carbon atoms away), e.g.

D Transannularumlagerung
F transposition transannulaire, interaction transannulaire
P przegrupowanie transanularne
R трансаннулярная реакция

3597
TRANS EFFECT.
Effect of weakening of bond of each group in trans position with respect to coordinating ligand (cf. trans effect series).
D *Trans*-Effekt
F effet *trans*
P efekt *trans*
R транс-влияние (Черняева)

3598
TRANS EFFECT SERIES (order of increasing trans effect).
Ligands arranged by increasing trans effect, e.g. $H_2O < OH^- < NH_3 <$ $< $ pyridine $< Cl^- < Br^- < NCS^- \sim I^- \sim$ $\sim NO_2^- \sim SO_3H^- \sim PR_3 \sim R_2S \sim$ $\sim SC(NH_2)_2 < NO < CO \sim C_2H_4 \sim CN^-$.
D Reihe des *Trans*-Effektes
F série d'effet *trans*
P szereg efektu *trans*
R ряд изменения трансвлияния аддендов

3599
TRANSESTERIFICATION.
Reaction of ester with alcohol (alcoholysis), ester with acid (acidolysis) or between two different esters, leading to exchange of alkoxyl or acyl group of ester, e.g.

$CH_3COOC_2H_5 + C_4H_9OH \overset{H^{\oplus}}{\rightleftharpoons}$
$CH_3COOC_4H_9 + C_2H_5OH$

$CH_2{=}CHCOOCH_3 + HCOOH \overset{H^{\oplus}}{\longrightarrow}$
$CH_2{=}CHCOOH + HCOOCH_3$

$C_6H_5COOC_2H_5 + CH_3COOC_4H_9 \overset{H^{\oplus}}{\rightleftharpoons}$
$C_6H_5COOC_4H_9 + CH_3COOC_2H_5$

D Umesterung
F transestérification
P transestryfikacja
R переэтерификация

3600
TRANSFERENCE NUMBER OF ION.
Ratio of electrical charge carried by one type of ions to total charge flowing across electrolytic conductor.
D Überführungszahl des Ions
F nombre de transport d'ion
P liczba przenoszenia jonu
R число переноса иона

TRANSFER EQUILIBRIUM. *See* PHASE EQUILIBRIUM.

3601
TRANSFER REACTION.
Nuclear reaction in which bombarding nucleus transfers to target nucleus one or more nucleons.
D Übertragungs-Reaktion
F réaction de transfert
P reakcja przekazu, reakcja transferu
R реакция передачи

TRANS FORM. *See* TRANS ISOMER.

3602
TRANSIENT EQUILIBRIUM (transient radioactive equilibrium).
Stationary state between radioactive parent nuclide and its daughter for the case when parent nuclide's half-life is up to a hundred times longer than that of daughter.
D (radioaktives) Übergangsgleichgewicht, laufendes (radioaktives) Gleichgewicht
F équilibre (radioactif) transitoire
P równowaga (promieniotwórcza) przejściowa
R переходное (радиоактивное) равновесие

TRANSIENT RADIOACTIVE EQUILIBRIUM. *See* TRANSIENT EQUILIBRIUM.

3603
TRANS ISOMER (*trans* form).
Diastereoisomer in which identical or similar groups are situated on opposite sides of plane passing through double bond or through ring surface, e.g. *trans*-2-butene

D *trans*-Isomer, *trans*-Form
F isomère *trans*
P izomer *trans*, odmiana *trans*
R транс-изомер, транс-форма

3604
TRANS ISOMER (*trans* form).
In coordination chemistry, isomer in which two similar ligands are in opposite positions, i.e. in square complexes $[Ma_2b_2]$ ligands occupy opposite vertices of square, and in octahedral complexes $[Ma_4b_2]$ opposite corners of octahedron.
D *trans*-Form
F isomère *trans*
P izomer *trans*
R изомер *транс*

TRANSITION COMPLEX. *See* ACTIVATED COMPLEX.

3605
TRANSITION ELEMENTS (transition metals).
Elements the atoms of which have partly-filled outermost electronic d sub-shell. Structure of outer electronic shells of atoms of these elements can, in general, be represented by $(n-1)d^{1-9}ns^2$, although there exist elements in which number of electrons in ns sub-shell is smaller than 2. Four series of transition elements are distinguished which correspond to filling of 3d, 4d, 5d, and 6d sub-shells.
D Übergangselemente, Übergangsmetalle
F éléments de transition
P pierwiastki przejściowe, metale przejściowe
R переходные элементы, переходные металлы

TRANSITION METALS. *See* **TRANSITION ELEMENTS.**

3606
TRANSITION MOMENT.
Matrix element of dipole moment operator between wave functions of two energetic states of the system. Direction of transition moment determines polarization of radiative transition. Squared absolute value of transition moment is proportional to dipole transition probability.
D Übergangsmoment
F moment de transition
P moment przejścia
R момент перехода

3607
TRANSITION POLARIZATION.
Type of electrolytic polarization consisting in transfer of electrons or ions across interphase, slow in comparison with other steps of electrode reaction. Transition polarization is expressed by transition overvoltage.
D Durchtrittspolarisation
F polarisation d'activation
P polaryzacja przejścia
R переходная поляризация

3608
TRANSITION PROBABILITY.
Number which expresses fraction of system of quantum mechanical statistical ensemble undergoes transition from one stationary state φ_1 to other stationary state φ_2 in unit time interval due to some perturbation.
D Übergangswahrscheinlichkeit
F probabilité de transition
P prawdopodobieństwo przejścia
R вероятность перехода

TRANSITION STATE. *See* **ACTIVATED COMPLEX.**

TRANSITION-STATE THEORY. *See* **ABSOLUTE REACTION RATE THEORY.**

3609
TRANSITION TIME.
In chronopotentiometry, time which elapses from beginning of electrolysis to moment at which concentration of electrolyzed substance at electrode surface drops to zero.
D Transitionszeit
F temps de transition
P czas przejścia
R переходное время

3610
TRANSLATION.
Operation consisting in parallel shift by a given displacement of all points belonging to defined geometrical structure. Translation is described explicitly by direction, sense of vector, and length of section through which operation was conducted.
D Translation
F translation
P translacja
R трансляция

3611
TRANSLATIONAL ENERGY.
Kinetic energy of translation of molecule.
D Translationsenergie
F énergie de translation
P energia translacyjna
R трансляционная энергия

3612
TRANSLATIONAL PARTITION FUNCTION, f_{tr}.
Function defined by $f_{tr} = V/\lambda^3$, where V = volume of gas and λ = thermal de Broglie wavelength. Translational partition function determines influence of translational degrees of freedom of molecule on thermodynamic properties of corresponding ideal gas.
D Translationszustandssumme, Verteilungsfunktion der Schwerpunktstranslation
F fonction de partition de translation, somme d'états de translation
P funkcja rozdziału translacyjna
R поступательная статистическая сумма, поступательная сумма состояний

3613
TRANSLATIONAL SYMMETRY.
Symmetry of system in which surroundings of two points of this system are congruent with respect to the same translation.
D Translationssymmetrie
F symétrie de translation
P symetria translacyjna
R трансляционная симметрия

3614
TRANSMETHYLATION.
Migration of methyl group —CH_3 from one molecule to another, stimulated by chemical or biochemical catalyst.
D Transmethylierung
F transméthylation
P transmetylowanie
R трансметилирование

TRANSMISSION. *See* TRANSMITTANCE.

3615
TRANSMISSION COEFFICIENT.
Fraction of activated complexes which decay into products. Transmission coefficient is almost invariably equal to one and differs from one only in reactions accompanied by changes in electronic energy.
D Transmissionskoeffizient, Durchgangsfaktor
F coefficient de transmission
P współczynnik przejścia, współczynnik transmisji
R коэффициент прохождения, трансмиссионный коэффициент

3616
TRANSMITTANCE (transmittancy, transmission), *T*.
Ratio of radiation intensity passing through investigated sample (I) to that of radiation passing through reference (I_0).
D (optische) Durchlässigkeit, Transmission, Durchlässigkeitsgrad
F (facteur de) transmission
P transmitancja, przepuszczalność, wartość transmisji
R прозрачность, пропускание, коэффициент пропускания, трансмиссия

TRANSMITTANCY. *See* TRANSMITTANCE.

3617
TRANSPORT COEFFICIENTS.
Proportionality constants between flux (e.g. of heat, electric charge, particles) and corresponding force responsible for it (e.g. temperature gradient, electrostatic potential gradient, concentration gradient).
D Transportkoeffizienten
F coefficients de transport
P współczynniki transportu
R коэффициенты переноса

3618
TRANSURANIC ELEMENTS (posturanic elements, transuranium elements).
Artificially obtained radioactive elements belonging to actinoids of atomic number Z > 92: neptunium, plutonium, americium, curium, berkelium, californium, einsteinium, fermium,

mendelevium, nobelium, lawrencium, 104, 105.
D Transurane
F transuraniens, éléments transuraniques
P pierwiastki transuranowe, pierwiastki pozauranowe, transuranowce, transurany
R трансурановые элементы, заурановые элементы

TRANSURANIUM ELEMENTS. *See* TRANSURANIC ELEMENTS.

TRANSVERSE RELAXATION. *See* SPIN-SPIN RELAXATION.

3619
TRAP (trapping state).
Localized energetic state in solids or liquids able to catch electrons (electron trap), to give up electrons (hole-trap), to participate in both processes with equal probability (recombination centre), or to catch specified free radicals (radical trap).
D Haftstelle
F piège
P pułapka, poziom pułapkowy, stan pułapkowy
R ловушка

3620
TRAPPED ELECTRON, e_{tr}^-.
Electron, knocked out in absorption of high energy radiation, and after being slowed down to appropriate energy, caught in specific electron trap.
D gefangenes Elektron, fixiertes Elektron
F électron piégé
P elektron spułapkowany
R захваченный электрон

3621
TRAPPING OF RADICALS.
Stabilization of radicals for period of time as long as needed, in medium preventing their diffusion and recombination.
D Stabilisierung der Radikale
F piégeage des radicaux
P stabilizacja rodników, zamrażanie rodników
R стабилизация радикалов

TRAPPING STATE. *See* TRAP.

3622
TRAUBE'S RULE.
In dilute aqueous solutions of surface-active substances (surfactants) whose hydrophobic groups are long aliphatic carbon chains, surface activity increases with increase in length of chains.
D Traubesche Regel
F règle de Traube
P reguła Traubego
R правило Траубе

3623
TRIGONAL HYBRID (sp² hybrid).
One of three equivalent hybridized atomic orbitals formed from one s and two p atomic orbitals. Angle between two sp² hybrids in 120°; used for instance to explain structure of ethylene molecule.
D trigonale Hybride, trigonales Hybridorbital, tr-Hybride
F hybride trigonal, hybride sp²
P hybryd trygonalny, hybryd sp²
R тригональная гибридная орбиталь

3624
TRIMER.
Oligomer whose molecule is composed of three units of the same monomer.
D Trimer
F trimère
P trimer
R тример

3625
TRIMOLECULAR REACTION (termolecular reaction).
Elementary reaction of molecularity equal to three.
D trimolekulare Reaktion
F réaction trimoléculaire
P reakcja trójcząsteczkowa
R тримолекулярная реакция

3626
TRIPLE BOND.
Chemical bond formed by three pairs of valence electrons; one pair is present in molecular bonding σ-orbital and two others occupy molecular bonding π-orbitals with nodal planes perpendicular to one another, e.g., $C \equiv C$ bond in acetylene.
D Dreifachbindung
F liaison triple
P wiązanie potrójne
R тройная связь

3627
TRIPLE POINT.
Point on phase diagram determining equilibrium coexistence of three phases.
D Tripelpunkt
F point triple
P punkt potrójny
R тройная точка

TRIPLET. See TRIPLET STATE.

3628
TRIPLET (in atomic spectroscopy).
Group of spectral lines arising due to transition between energy levels corresponding to two atomic triplet terms.
D Triplett
F triplet
P tryplet widmowy
R триплет

3629
TRIPLET STATE (triplet).
Atomic or molecular state (term) with total spin quantum number $S = 1$ and three possibilities of projection of spin angular momentum onto direction of the external magnetic field, corresponding to three values of magnetic spin quantum number $M_s = \pm 1$ and 0; triplet multiplicity is equal to 3.
D Triplettzustand, Triplettsystem
F état triplet
P stan trypletowy, tryplet
R триплетное состояние

3630
TRIPLET-TRIPLET ANNIHILATION.
Transition of two colliding molecules from their triplet state to corresponding ground and excited singlet states due to their intermolecular energy exchange. Molecule raised to singlet state subsequently displays delayed fluorescence.
D ...
F ...
P anihilacja tryplet-tryplet
R триплет-триплетная аннигиляция

3631
TRITIUM, T.
Isotope of hydrogen. Atomic number 1. Mass number 3.
D Tritium, überschwerer Wasserstoff, schweres Deuterium
F tritium
P tryt, wodór superciężki, radiowodór
R тритий, сверхтяжёлый водород

3632
TROUTON'S RULE.
Molal latent heat of vaporization of liquid (at normal boiling point under atmospheric pressure) divided by this boiling temperature on absolute scale is nearly the same for a great number of substances and approximately equal to 21 cal mole⁻¹ K⁻¹.
D Pictet-Troutonsche Regel
F règle de Trouton
P reguła Troutona i Picteta
R правило Трутона

3633
TRUE ACTIVATION ENERGY.
Apparent activation energy of reaction in multiphase system E^x increased by value of adsorption heat of substrate λ_A and diminished by value of adsorption heat of product λ_B

$$E = E^x + \lambda_A - \lambda_B$$

D wahre Aktivierungsenergie
F énergie d'activation réelle
P energia aktywacji rzeczywista
R истинная энергия активации

3634
TRUE DEGREE OF DISSOCIATION.
Fraction of molecules of electrolyte which
dissociate into ions; this value is
determined by measurement of light
absorption or from measurements of
conductivity of electrolytic solutions (*see*
degree of dissociation).

D wahrer Dissoziationsgrad
F taux d'ionisation (vrai), coefficient
  d'ionisation (vrai), degré de dissociation
  électrolytique (vrai)
P stopień dysocjacji rzeczywisty
R степень электролитической
  диссоциации (истинная)

3635
TRUE HALF-WIDTH (true width).
Spectral line width as given by features
of emitting or absorbing species, free of
instrument factors; it is usually given as
true half-width $\Delta\nu_{1/2}^{t}$.

D wahre Halbwertsbreite $\Delta\nu_{1/2}$
F largeur vraie (à mi-absorption)
P szerokość linii widmowej rzeczywista
R истинная ширина спектральной линии

TRUE MOLAL HEAT CAPACITY. *See*
MOLAL HEAT CAPACITY.

3636
TRUE SOLUTION (molecular solution,
simple solution).
Limiting case of dispersed system in
which particles of dispersed phase have
dimensions smaller than 1 nm, therefore
of order of simple molecules.

D molekulardisperses System, wahre
  Lösung, molekulare Lösung
F dispersoïde moléculaire, solution vraie,
  solution moléculaire, solution ordinaire
P układ o rozdrobnieniu cząsteczkowym,
  układ o rozdrobnieniu granicznym,
  roztwór właściwy
R молекулярно-дисперсная система,
  высокодисперсная система, истинный
  раствор

TRUE WIDTH. *See* TRUE HALF-WIDTH.

TRUNK CHAIN. *See* MAIN CHAIN.

TSCHUGAEFF DEHYDRATION. *See*
CHUGAEV REACTION.

3637
t STATISTIC.
Statistic which determines degree of
agreement between mean in series $\bar{x}$ and
mean of population s; given by equation

$$t = \frac{\bar{x} - \mu}{s/\sqrt{n}}$$

where independent random variables x_i
have normal distribution with mean value
μ and with variances σ_i^2; $\bar{x} = \sum x_i/n$;
$s^2 = \sum (x_i - \bar{x})^2/(n-1)$; t statistic has
t-distribution.

D ...
F variable aléatoire t
P statystyka t
R t-статистика

3638
TUNGSTEN (wolfram), W.
Atomic number 74.
D Wolfram
F tungstène
P wolfram
R вольфрам

3639
TUNNEL EFFECT (tunneling).
Passage of particle through region in
which its potential energy is higher than
its initial kinetic energy. Tunnel effect is
governed by laws of quantum mechanics
(i.e. it is inconsistent with laws of classical
mechanics).
D Tunneleffekt
F effet de tunnel
P efekt tunelowy
R туннельный эффект

TUNNELING. *See* TUNNEL EFFECT.

3640
TURBIDIMETER.
Device for measurement of decrease of
light transmittance due to solution
containing suspension of solid particles
placed in light path.
D Turbidimeter
F turbidimètre
P turbidymetr
R турбидиметр

TURBIDIMETRIC TITRATION. *See*
HETEROMETRIC TITRATION.

3641
TURBIDIMETRY.
Analytical method based on light
transmittance measurements related to
changes of solution opacity due to
isolation of determined component in form
of deposit or emulsion.
D Turbidimetrie
F turbidimétrie
P analiza turbidymetryczna,
  turbidymetria
R турбидиметрия

3642
TWIN CALORIMETERS (double adiabatic calorimeter).
Calorimetric system consisted of two identical calorimeters symmetrically located inside common calorimetric jacket. Process investigated is carried out in one of the two calorimeters ("measuring" one) whereas the second ("standard") is taken as reference.
D Zwillingskalorimeter
F calorimètres semblables, calorimètres jumelés
P kalorymetry bliźniacze
R двойные калориметры

TWIN COMPARATOR. *See* DOUBLE COMPARATOR.

TWO-COMPONENT SYSTEM. *See* BINARY SYSTEM.

3643
TWO-DIMENSIONAL CHROMATOGRAPHY.
Chromatographic technique used in paper or thin-layer chromatography in which development is carried out successively in two perpendicular directions.
D zweidimensionale Chromatographie
F chromatographie à deux dimensions
P chromatografia dwukierunkowa, technika dwukierunkowa
R двухмерная хроматография

TWO-PHOTON TRANSITION. *See* TWO-QUANTUM TRANSITION.

3644
TWO-QUANTUM TRANSITION (two-photon transition).
Simultaneous absorption of two photons by atom or molecule.
D zwei-Photonen-Übergänge
F excitation à deux photons
P przejścia dwufotonowe
R двухквантовые процессы

3645
TYNDALL EFFECT.
Dispersion of light by colloidal particles of dimensions smaller than wavelength of the light, commonly used in detection of colloidal inhomogeneity.
D Tyndall-Effekt, Tyndallphänomen
F effet de Tyndall
P efekt Tyndalla
R эффект Тиндаля

3646
TYPE 1 ERROR.
Possible error in verifying statistical hypothesis due to rejection of correct hypothesis.
D Fehler erster Art
F risque de première espèce
P błąd pierwszego rodzaju
R ошибка первого рода

3647
TYPE 2 ERROR.
Possible error in verifying statistical hypothesis due to assumption of false hypothesis.
D Fehler zweiter Art
F risque de seconde espèce
P błąd drugiego rodzaju
R ошибка второго рода

3648
ULLMANN REACTION.
Condensation of two aryl groups in reaction of haloarenes in presence of powdered copper, at 100—300°C, resulting in formation of biaryls, e.g.

$$2\,C_6H_5I + Cu \xrightarrow{\Delta} C_6H_5\!-\!C_6H_5 + CuI_2$$

This reaction is also applied to synthesis of aryl ethers and derivatives of phenylanthranilic acid, e.g.

D Ullmann-Reaktion
F réaction d'Ullmann
P reakcja Ullmanna
R конденсация Ульмана

3649
ULTRAFILTRATION.
Filtration through membranes whose pores are small enough to retain colloidal particles.
D Ultrafiltration
F ultrafiltration
P ultrafiltracja
R ультрафильтрация

ULTRAMICRO METHOD. *See* MICROGRAM METHOD.

3650
ULTRAMICROSCOPE.
Microscope in which use is made of Tyndall effect to observe colloidal particles.
D Ultramikroskop
F ultramicroscope
P ultramikroskop
R ультрамикроскоп

3651
ULTRAMICROSPECTROPHOTOMETRY.
Spectroscopic technique used for
absorption measurements for
microgramme samples by utilization of
so-called microvessels made of capillary
tubes which enable spectrophotometric
studies to be conducted on 20—30 nm^3 of
liquids.
D *(mikrochemische Arbeitstechnik für*
Photometrie)
F ...
P ultramikrospektrofotometria
R ультрамикроспектрофотометрия

ULTRAULTRAMICRO METHOD. *See*
NANOGRAM METHOD.

3652
ULTRAVIOLET, UV.
Electromagnetic radiation range of
wavelength between 1 nm to ca. 400 nm.
D Ultraviolett
F ultraviolet, région ultraviolette
P nadfiolet, ultrafiolet
R ультрафиолетовая область

ULTRAVIOLET. *See* NEAR
ULTRAVIOLET.

3653
ULTRAVIOLET SPECTRUM (UV
spectrum).
Spectrum in ultraviolet wavelength region
with bands corresponding to transitions
between different electronic states of
atoms or molecules.
D Ultraviolett-Spektrum
F spectre ultraviolet
P widmo w nadfiolecie, widmo UV
R ультрафиолетовый спектр

3654
UNCERTAINTY PRINCIPLE (Heisenberg
relation, indeterminancy principle).
Principle claiming that simultaneous,
infinitely accurate measurement of
particle position and momentum, or
energy and time is impossible.
D Unbestimmtheitsbeziehung
F principe d'incertitude, relation
d'incertitude, relation de Heisenberg
P zasada nieoznaczoności, relacja
nieoznaczoności Heisenberga
R принцип неопределённости

3655
UNCOMPENSATED HEAT, Q_n.
Quantity defined as:

$$Q_n = TdS - \delta Q$$

where δQ = elementary heat, S = entropy
of system, T = thermodynamic

temperature. In irreversible processes Q_n
is never equal to zero and always positive.
D nichtkompensierte Wärme
F chaleur non compensée
P ciepło nieskompensowane
R некомпенсированная теплота

3656
UNDER-EXPOSURE REGION.
Curved part of emulsion calibration curve
of photographic emulsion corresponding to
region of insufficient exposure.
D Unterexpositionsgebiet
F région de sous-exposition
P obszar niedoświetleń
R область недодержек

UNEQUAL DISSYMMETRY CENTRES.
See DISSIMILAR CHIRAL CENTRES.

3657
UNIDENTATE LIGAND (simple ligand,
monofunctional ligand).
Ligand which occupies only one
coordination position around central
atom.
D einzähniger Ligand
F coordinant unidenté
P ligand jednodonorowy, ligand
jednofunkcyjny, ligand jednomiejscowy
R однокоординационный адденд

3658
UNIFIED MODEL (*of nucleus*).
Model of atomic nucleus as system
consisting of a core, described by
collective model and one or several
external (valency) nucleons. Unified model
is combination of collective and shell
models.
D kombiniertes Modell
F modèle unifié
P model uogólniony
R обобщенная модель

UNIMOLECULAR ADSORPTION.
See MONOLAYER ADSORPTION.

UNIMOLECULAR REACTION.
See MONOMOLECULAR REACTION.

3659
UNIT CELL.
The smallest parallelepiped which exhibits
all characteristic geometric features of
space lattice (besides translation
symmetry) and enables reconstruction of
it by simple repetition of cell in
directions of its three edges.
D Elementarzelle, Einheitszelle
F maille élémentaire, forme fondamentale
P komórka elementarna
R элементарная ячейка

3660
UNITED ATOM.
Atom obtained by artificial process of
uniting nuclei of diatomic molecule.
D vereinigtes Atom
F atome unifié
P atom zjednoczony
R объединённый атом

3661
UNIT SAMPLE.
Primary sample taken from one package.
D Einzelprobe
F ...
P próbka jednostkowa
R индивидуальная проба

3662
UNIVARIANT PHASE SYSTEM
(univariant system).
System with one thermodynamic degree
of freedom.
D univariantes Phasensystem, univariantes
System
F système monovariant, monovariant
système de phases
P układ (fazowy) jednozmienny
R одновариантная (фазовая) система

UNIVARIANT SYSTEM. *See*
UNIVARIANT PHASE SYSTEM.

3663
UNIVERSAL CONSTANTS (absolute
constants, fundamental constants).
Physical constants whose numerical
values (according to present status of
knowledge) do not depend on such
parameters as pressure, volume,
temperature and properties of substance
or systems which are used for their
measurement but on system of units;
examples of universal constants are:
velocity of light in vacuum, absolute zero
point temperature, gas constant, Avogadro
constant, gravitational constant, charge
and rest mass of electron, etc.
D universelle Konstanten, absolute
Konstanten, Naturkonstanten,
Grundkonstanten
F constantes universelles, constantes
fondamentales, constantes physiques
P stałe uniwersalne
R универсальные постоянные,
универсальные константы, физические
константы

3664
UNIVERSAL FREQUENCY FACTOR,
kT/h.
Coefficient characterizing effective rate of
transition of energy barrier by activated
complex. This factor depends only on
thermodynamic temperature T and not on
substrates or reaction type (k = reaction-
rate constant, h = Planck constant).

D Übergangsfaktor kT/h
F facteur kT/h
P współczynnik częstościowy uniwersalny,
współczynnik kT/h
R универсальный фактор частоты,
множитель kT/h

UNIVERSAL GAS CONSTANT. *See* GAS
CONSTANT.

3665
UNIVERSAL INDICATOR (multiple range
indicator).
Mixture of several pH indicators which
changes continuously in colour over large
pH range; it is used for approximate
determination of pH.
D Universalindikator
F indicateur universel
P wskaźnik uniwersalny
R универсальный индикатор,
комбинированный индикатор

3666
UNIVERSAL INDICATOR PAPER.
Indicator paper saturated with suitable
mixture of pH indicators for approximate
estimation of wide range of pH of
solutions.
D Universalindikatorpapier
F papier universel
P papierek uniwersalny
R универсальная реактивная бумага

UNLIMITED COMPATIBILITY. *See*
COMPLETE MISCIBILITY.

UNPAIRED ELECTRON DENSITY. *See*
SPIN DENSITY.

3667
UNPAIRED ELECTRONS.
Electrons on singly occupied atomic or
molecular orbitals; they are responsible
for instance for paramagnetism of atoms
and molecules.
D ungepaarte Elektronen
F électrons non appariés, électrons non
couplés
P elektrony niesparowane
R неспаренные электроны

3668
UNSATURATED BOND (unsaturated
linkage).
Multiple bond characterized by its special
reactivity, e.g. double bond in alkenes or
triple bond in alkynes.
D ungesättigte Bindung
F liaison non saturée
P wiązanie nienasycone
R непредельная связь

UNSATURATED LINKAGE. *See*
UNSATURATED BOND.

3669

UNSATURATED WESTON CELL.
Weston cell with cadmium sulfate solution
saturated at 4°C (unsaturated at room
temperature); it has very low temperature
coefficient of EMF.

D ungesättigtes Weston-Element
F élément Weston insaturé
P ogniwo Westona nienasycone
R ненасыщенный элемент Вестона

3670

UNSEALED SOURCE.
Any radioactive source of such a
construction, that the radioactive
substance contained in it is in direct
contact with surroundings. Examples of
such sources are radioactive substances
undergoing radiochemical operations.

D offenes radioaktives Präparat
F source non scellée
P źródło (promieniowania) otwarte
R открытый источник

UNSTABLE CHEMICAL EQUILIBRIUM.
See FALSE CHEMICAL EQUILIBRIUM.

3671

UNSTABLE COMPLEX
(thermodynamically unstable complex).
Complex which readily undergoes
dissociation in solution, i.e. is characterized
by small value of stability constant.

D schwacher Komplex
F complexe instable
P kompleks nietrwały
R малоустойчивый комплекс, нестойкий
   комплекс

3672

UNSTABLE STATE.
State of system in which at least one
condition of stability is not fulfilled
(*cf.* condition of diffusion, mechanical and
thermal stability of phase).

D instabiler Zustand
F état instable
P stan niestabilny, stan nietrwały
R лабильное состояние, (абсолютно)
   неустойчивое состояние

URANIDES. *See* URANOIDS.

3673

URANIUM, U.
Atomic number 92.

D Uran
F uranium
P uran
R уран

3674

URANIUM SERIES.
Naturally occurring series with first and
last members uranium, ^{238}U, and lead,
^{206}Pb, respectively. The mass number of

each member of this series is $A = 4n+2$,
where n = natural number.

D Uran-Radium-Reihe
F famille de l'uranium
P szereg urano-radowy
R ряд урана

3675

URANOIDS (uranides).
Elements from actinide family starting
with uranium (uranium, neptunium,
plutonium, americium, curium, berkelium,
californium, einsteinium, fermium,
mendelevium, nobelium, lawrencium).

D Uran(o)ide
F uranides
P uranowce
R ураниды

3676

USANOVICH-EBERT-KONOPIK'S
THEORY OF ACIDS AND BASES.
The most general contemporary theory of
acids and bases; acids are species having
tendency to lose protons or arbitrary
cations or to add on electrons or arbitrary
anions, bases are species having tendency
to add on protons or arbitrary cations or
to lose electrons or arbitrary anions.

D Usanowitsch-Ebert-Konopiksche
   Säuren-Basen Theorie
F théorie des acides et des bases de
   Usanovich, Ebert et Konopik
P teoria kwasów i zasad Usanowicza,
   Eberta i Konopika
R теория кислот и оснований
   Усановича-Эберта-Конопика

UV SPECTRUM. *See* ULTRAVIOLET
SPECTRUM.

3677

VACANCY (vacant lattice site *atomic or
ionic*).
Vacant (unoccupied) node in crystal lattice
of atomic or ionic crystal.

D Leerstelle, Gitterloch, Gitterlücke
F vacance, lacune
P luka, wakans
R вакансия, вакантный узел

VACANT LATTICE SITE.
See VACANCY.

3678

VACUUM ULTRAVIOLET.
Electromagnetic radiation range of
wavelength between ca. 1 nm to
ca. 170 nm.

D Schumann-Ultraviolett, Schumann-UV
F . . .
P nadfiolet próżniowy, nadfiolet
   Schumanna
R вакуумная ультрафиолетовая область

VALENCE. *See* VALENCY.

3679
VALENCE BAND (*in band theory of solids*).
Energy band originating from overlap of highest completely filled electronic levels of all atoms or ions forming crystal lattice.
D Valenzband, Grundband
F bande de valence
P pasmo podstawowe, pasmo walencyjne
R валентная зона

3680
VALENCE-BOND METHOD (VB method).
Quantum chemistry method used to compute approximate wave functions of many-electron systems in form of linear combinations of known functions corresponding to some preselected electron distributions in molecule (viz. so-called resonance structures).
D Valenzstrukturverfahren
F méthode de la liaison de valence
P metoda wiązań walencyjnych, metoda VB
R метод валентных связей

3681
VALENCE-BOND THEORY.
Theory which assumes that formation of complex compound is accompanied by transfer of electrons of ligands into hybridized orbitals of central ion. Characteristic features of this theory is preservation, in general, of individuality of atoms within complex and action of exchange forces between ligand and central ion, which corresponds to complete covalence of bonds. Considering assumed hybridization, theory makes possible easy explanation of directed valence bonds and stereochemistry of complexes.
D Valency-Bond-Theorie
F théorie de la liaison de valence
P teoria wiązań walencyjnych
R теория валентных связей

3682
VALENCE ELECTRONS (valency electrons).
Electrons of outer electronic shell in atom. They play decisive role in formation of chemical bonds.
D Valenzelektronen, Aussenelektronen
F électrons de valence, électrons périphériques, électrons de la couche extérieure
P elektrony walencyjne, elektrony wartościowości, elektrony zewnętrznej powłoki
R валентные электроны, внешние электроны, электроны внешней оболочки

3683
VALENCE SHELL.
Outer electron shell of atom, occupied in its ground state.
D Valenzschale
F couche de valence
P powłoka walencyjna, warstwa walencyjna
R валентная оболочка

VALENCE VIBRATIONS. *See* STRETCHING VIBRATIONS.

3684
VALENCY (valence).
Ability of atom or free radical to join with other atoms or radicals in constant ratios.
D Wertigkeit, Valenz
F valence
P wartościowość, walencyjność
R валентность

VALENCY ELECTRONS. *See* VALENCE ELECTRONS.

3685
VANADIUM, V.
Atomic number 23.
D Vanadium, Vanadin
F vanadium
P wanad
R ванадий

VANADIUM FAMILY. *See* VANADIUM GROUP.

3686
VANADIUM GROUP (vanadium family).
Elements in fifth sub-group of periodic system with structure of outer electronic shells of atoms $(n-1)d^3ns^2$: vanadium, niobium, tantalum, (protactinium).
D Gruppe der Erdsäuren, Vanadingruppe
F sous-groupe du vanadium
P wanadowce
R подгруппа ванадия, элементы подгруппы ванадия

VAN DER WAALS ADSORPTION. *See* PHYSICAL ADSORPTION.

3687
VAN DER WAALS BOND (molecular bond).
This bond, considerably weaker than ordinary chemical linkage, is caused by van der Waals forces. It occurs e.g. between atoms of noble gases and between molecules in molecular crystals.
D van der Waals-Molekularattraktion, Molekular-Bindung, Edelgasbindung
F liaison (de) Van der Waals, liaison moléculaire
P wiązanie van der Waalsa
R ван-дер-ваальсова связь, молекулярная связь

VAN DER WAALS CRYSTALS.
See MOLECULAR CRYSTALS.

3688
VAN DER WAALS EQUATION OF
STATE.
Empirical (thermal) equation of state
of real gas

$$\left(p+\frac{a}{V_{\mathrm{m}}^{2}}\right)(V_{\mathrm{m}}-b) = RT$$

where p = pressure, V_{m} = molar gas
volume, T = thermodynamic temperature,
R = universal gas constant, a, b =
constants characteristic for the given gas;
value of a/V_{m}^{2} represents correction for
forces of mutual attraction of molecules,
constant b is correction for molecules' own
molar volume.

D van der Waalssche Zustandsgleichung
F équation (d'état) de Van der Waals
P równanie (stanu) van der Waalsa
R уравнение (состояния)
　Ван-дер-Ваальса

3689
VAN DER WAALS FORCES.
Attractive part of intermolecular
interaction. Van der Waals forces include:
interaction of temporary fluctuating
multipole moments (dispersion effect),
interaction of permanent multipoles with
induced multipoles (induction effect),
and interaction of permanent multipoles
(orientation effect).

D van der Waals-Kräfte, van der Waalsche
　Kräfte
F forces de Van der Waals
P siły van der Waalsa
R силы Ван-дер-Ваальса,
　ван-дер-ваальсовы силы

3690
VAN DER WAALS ISOTHERM.
p-V isotherm described by van der Waals
equation.

D van der Waalssche Isotherme
F isotherme de Van der Waals
P izoterma van der Waalsa
R изотерма Ван-дер-Ваальса

3691
VAN DER WAALS RADIUS.
One half of distance between two identical
atoms when repulsive and attractive
forces are completely balanced. Sum of
van der Waals radii of two atoms is called
van der Waals distance.

D van der Waals-Radius
F rayon de Van der Waals
P promień van der Waalsa
R ван-дер-ваальсов радиус

3692
VAN'T HOFF COMPLEX.
In homogeneous catalysis, rate of decay
of substrate — catalyst complex into
starting compounds is considerably lower
than rate of decay into reaction products.

D van't Hoffscher Zwischenkörper, van't
　Hoffscher Komplex
F complexe de Van't Hoff, intermédiaire
　de Van't Hoff
P substancja przejściowa typu van't Hoffa,
　kompleks van't Hoffa
R промежуточное соединение
　Вант-Гоффа, промежуточное вещество
　Вант-Гоффа

3693
VAN'T HOFF EQUATION (van't Hoff
law).
Formula describing dependence of osmotic
pressure π of infinitely diluted solution
of non-electrolyte on thermodynamic
temperature T and molar concentration
of solution c

$$\pi = cRT$$

where R = molar gas constant.

D van't Hoffsches Gesetz
F loi de Van't Hoff
P wzór van't Hoffa
R формула Вант-Гоффа, закон
　Вант-Гоффа

VAN'T HOFF EQUATION. *See* VAN'T
HOFF ISOBAR.

3694
VAN'T HOFF ISOBAR (van't Hoff
equation, van't Hoff relation).
Equation which determines influence of
isobaric temperature changes on
thermodynamic constant of chemical
equilibrium

$$\left(\frac{\partial \ln K_a}{\partial T}\right)_p = \frac{\Delta H^{\ominus}}{RT^2}$$

where K_a = thermodynamic chemical
equilibrium constant, T = thermodynamic
temperature, p = pressure, $\Delta H^{\ominus}$ =
standard enthalpy of chemical reaction,
and R = molar gas constant.

D van't Hoffsche Isobare, van't Hoffsche
　Reaktionsisobare, van't Hoffsche
　Gleichung
F formule de Van't Hoff, Van't Hoff
　isobare
P równanie izobary van't Hoffa, izobara
　van't Hoffa, izobara reakcji
R уравнение изобары Вант-Гоффа,
　изобара Вант-Гоффа, изобара реакции

3695
VAN'T HOFF ISOCHORE (reaction isochore).

Equation which determines influence of isochoric temperature changes on thermodynamic constant of chemical equilibrium

$$\left(\frac{\partial \ln K_a}{\partial T}\right)_V = \frac{\Delta U^{\ominus}}{RT^2}$$

where K_a = thermodynamic chemical equilibrium constant, T = thermodynamic temperature, V = volume, $\Delta U^{\ominus}$ = standard internal energy of chemical reaction, and R = molar gas constant.

D van't Hoffsche Reaktionsisochore, Reaktionsisochore, van't Hoffsche Gleichung
F Van't Hoff isochore
P równanie izochory van't Hoffa, izochora van't Hoffa, izochora reakcji
R уравнение изохоры Вант-Гоффа, изохора Вант-Гоффа

VAN'T HOFF LAW. *See* VAN'T HOFF EQUATION.

VAN'T HOFF RELATION. *See* VAN'T HOFF ISOBAR.

3696
VAN'T HOFF RULE.
Increase of temperature by 10 degrees (kelvins) doubles (roughly) rate of reaction.

D van't Hoffsche Regel
F règle de Van't Hoff
P reguła van't Hoffa
R правило Вант-Гоффа

3697
VAN'T HOFF RULE OF OPTICAL SUPERPOSITION.
The molar rotation of compound containing several chiral centres is the sum of rotations of particular chiral centres in molecule. Rule does not hold when chiral centres are located in direct neighbourhood.

D van't Hoffsches Prinzip, Superpositionsprinzip
F règle de superposition optique de Van't Hoff
P reguła superpozycji optycznej van't Hoffa, zasada addytywności van't Hoffa
R принцип оптической суперпозиции, принцип оптической аддитивности Вант-Гоффа

3698
VAPORIZATION.
Transition from liquid to vapour state.

D Verdampfung
F vaporisation
P parowanie
R парообразование

3699
VAPOUR PRESSURE.
Pressure of vapour in equilibrium with corresponding condensed phase.

D Sättigungsdampfdruck, Dampfdruck
F pression de vapeur, tension de vapeur (saturante)
P prężność pary nasyconej
R давление (насыщенного) пара

VARIABILITY. *See* NUMBER OF THERMODYNAMIC DEGREES OF FREEDOM.

3700
VARIANCE (*in series*), s^2.
Statistic characterizing dispersion and given by equation

$$s^2 = \left[\sum_{i=1}^{n} (x_i - \bar{x})^2\right] / (n-1)$$

where x_i = value of series element, $\bar{x}$ = arithmetic mean, and n = number of elements in series.

D Varianz
F variance (empirique), carré des écarts quadratiques moyens
P wariancja
R дисперсия

VARIANCE. *See* MEAN SQUARE DEVIATION.

VARIANCE. *See* NUMBER OF THERMADYNAMIC DEGREES OF FREEDOM.

3701
VARIANCY PROPAGATION.
Principle of calculation of variance of indirectly measured quantity x, which is function of number of independent stochastic variables $x = f(x_1, x_2, ..., x_n)$, according to equation

$$\sigma^2(x) = \sum_{i=1}^{n} (\partial x / \partial x_i)^2 \sigma^2(x_i)$$

This principle is used for estimation of maximum absolute error.

D ...
F ...
P propagacja wariancji
R распространение дисперсии

VARIATE. *See* RANDOM VARIABLE.

3702
VARIATIONAL METHOD.
Computational method used to find best possible approximation to solution of given problem within given class of functions.

D Variationsmethode
F méthode des variations
P metoda wariacyjna
R вариационный метод

VB METHOD. *See* VALENCE BOND METHOD.

VESSEL CONSTANT. *See* CELL CONSTANT OF CONDUCTOMETRIC VESSEL.

3703
VIBRATIONAL CHARACTERISTIC TEMPERATURE (*of a molecule*), Θ_v.
Parameter expressed in units of temperature $\Theta_v = \dfrac{h\nu}{k}$, where h = Planck constant, k = Boltzmann constant, and ν = classical nuclear vibration frequency in molecule. Vibrational characteristic temperature enters expression for molecular vibrational partition function.

D charakteristische Temperatur der Schwingung
F température caractéristique de vibration
P temperatura oscylacji (cząsteczki) charakterystyczna
R характеристическая колебательная температура

3704
VIBRATIONAL ENERGY.
Kinetic and potential energy of vibrations of molecule.

D Schwingungsenergie
F énergie de vibration
P energia oscylacyjna
R колебательная энергия

VIBRATIONAL ENERGY LEVELS OF A MOLECULE. See MOLECULAR VIBRATIONAL ENERGY LEVELS.

3705
VIBRATIONAL PARTITION FUNCTION, $v(T)$.
Function defined by
$$v(T) = \left[2 \sin h \, \frac{\Theta_v}{2T} \right]^{-1}$$
where T = thermodynamic temperature, and Θ_v = characteristic vibration temperature; vibrational partition function determines influence of vibrations of nuclei in diatomic molecule on thermodynamic properties of corresponding gas.

D Schwingungszustandssumme, Verteilungsfunktion der Schwingung
F fonction de partition de vibration, somme d'états de vibration
P funkcja rozdziału oscylacyjna
R колебательная статистическая сумма, колебательная сумма состояний

3706
VIBRATIONAL QUANTUM NUMBER, n.
In quantum mechanics number which determines energy E_n of harmonic oscillator according to $E_n = h\nu(n+1/2)$; where h = Planck constant, ν = vibration frequency; n assumes non-negative integral values.

D Schwingungsquantenzahl
F nombre quantique de vibration
P liczba kwantowa oscylacji
R колебательное квантовое число

3707
VIBRATION BAND.
In IR spectroscopy band corresponding to vibrational transition. There is also simultaneous change of rotational energy.

D Schwingungsband
F bande de vibration
P pasmo oscylacyjne, pasmo wibracyjne
R колебательная полоса

VIBRATION-ROTATIONAL SPECTRUM. *See* VIBRATION-ROTATION SPECTRUM.

3708
VIBRATION-ROTATION SPECTRUM (vibration-rotational spectrum).
Spectrum corresponding to transitions between two vibrational levels of the same electronic state with simultaneous change of rotational energy.

D Rotationsschwingungsspektrum
F spectre de vibration-rotation
P widmo rotacyjno-oscylacyjne
R колебательно-вращательный спектр

3709
VIBRATION SPECTRUM (oscillation spectrum).
Spectrum in infrared region arising from transitions between vibrational energy levels of the same electronic state.

D Schwingungsspektrum
F spectre de vibration
P widmo oscylacyjne
R колебательный спектр

3710
VIBRATIONS (oscillations).
In general any periodic movements. In molecular spectroscopy periodic movements of molecular fragments (atoms or atomic groups); also oscillations of whole molecule with respect to its equilibrium position.

D Schwingungen
F vibrations
P drgania, oscylacje, wibracje
R колебания

3711
VIBRONIC LEVEL.
Vibrational-electronic molecular energy
level.
D vibronischer Zustand
F état vibronique, état de
vibration-rotation
P stan wibroniczny, stan wibronowy
R электронно-колебательный уровень

3712
VINYLATION.
Introduction of vinyl group —CH=CH₂
into molecule of organic compound,
usually by addition of compound with
active hydrogen atoms to acetylene or its
derivative, e.g.

$$HC{\equiv}CH + CH_3OH \xrightarrow{KOH} H_2C{=}CH{-}OCH_3$$

D Vinylierung
F vinylation
P winylowanie
R винилирование

3713
VINYLOGY RULE.
Influence of functional group on certain
other group of atoms in molecule may
spread beyond α-position, especially if the
latter group is separated from the former
by one or more conjugated double bonds.
E.g. aldehyde group activates methyl
group in crotonaldehyde (vinylogue of
acetaldehyde) similary to acetaldehyde.
D Vinylogieprinzip
F principe de vinylogie
P reguła winylogii
R принцип винилогии

3714
VIRIAL (of gas).
Statistical average

$$\left\langle \sum_{\nu} \boldsymbol{x}_\nu \cdot \frac{\partial U}{\partial \boldsymbol{x}_\nu} \right\rangle$$

where ν enumerates gas molecules,
$\boldsymbol{x}$ = position vector of centre of mass
of molecule, and U = potential energy of
intermolecular interactions. Virial
determines influence of intermolecular
interactions on gas pressure.
D Virial
F viriel (du gaz)
P wiriał
R вириал

3715
VIRIAL COEFFICIENTS.
Coefficients B, C, ... in virial equation
of state (respectively called: second, third
etc. virial coefficient); values
characteristic for the given gas and
depending on temperature.

D Virialkoeffizienten
F coefficients viriels
P współczynniki wirialne
R вириальные коэффициенты

3716
VIRIAL EQUATION OF STATE.
Thermal equation of state of real gas in
the form of power series with respect to
molar volume V_m or pressure p

$$pV_m = RT \left(1 + \frac{B}{V_m} + \frac{C}{V_m^2} + ... \right)$$

or

$$pV_m = RT(1 + B'p + C'p^2 + ...)$$

where T = thermodynamic temperature,
R = universal gas constant, B, C ... or
B', C' ... = virial coefficients.
D Virialgleichung
F équation d'état virielle
P równanie stanu wirialne
R вириальное уравнение состояния

VISCOELASTIC BEHAVIOUR. *See*
VISCOELASTICITY.

3717
VISCOELASTICITY (viscoelastic
behaviour, viscoelastic properties).
Rheological properties of such amorphous
plastic polymers which, with respect to
violent and brief forces, behave like an
elastic body, and with respect to
long-acting forces behave like a viscous
liquid displaying increasing deformation
with time. This phenomenon is caused by
rates of relaxation much faster than in
typical solids, but slower than in typical
liquids.
D Viskoelastizität, viskoelastische
Eigenschaften, viskoelastisches
Verhalten
F viscoélasticité, comportement
visco-élastique
P lepkosprężystość
R вязкоупругость

VISCOELASTIC PROPERTIES. *See*
VISCOELASTICITY.

3718
VISCOMETER.
Apparatus designed to measure viscosity.
D Viskosimeter
F viscosimètre
P lepkościomierz, wiskozymetr
R вискозиметр

3719
VISCOMETRY.
Branch of physical chemistry concerned
with study of viscosity of different
systems.

D Viskosimetrie
F viscosimétrie
P wiskozymetria
R вискозиметрия

3720
VISCOSITY (internal friction).
Resistance exerted by medium during
movement of one part of it against
another.
Remark: Expression used also in meaning of
viscosity coefficient.

D Viskosität, Zähigkeit, innere Reibung
F viscosité, frottement intérieur
P lepkość, tarcie wewnętrzne
R вязкость, внутреннее трение

3721
VISCOSITY (viscosity coefficient), η.
Measure of viscosity, coefficient
determined by formula

$$\eta = \frac{F}{A\dfrac{\mathrm{d}v}{\mathrm{d}x}}$$

where F = tangent force necessary to
displace liquid layer of surface area A
with velocity $\mathrm{d}v$ with respect to second
layer at distance $\mathrm{d}x$. Unit of viscosity in
SI system is pascal-second.

D dynamische Viskosität,
Viskositätskoeffizient,
Viskositätskonstante
F coefficient de viscosité, coefficient
de frottement intérieur
P lepkość dynamiczna, współczynnik
lepkości (dynamicznej), lepkość
(bezwzględna)
R (динамическая) вязкость, коэффициент
(динамической) вязкости

VISCOSITY COEFFICIENT. *See*
VISCOSITY.

VISCOSITY RATIO. *See* RELATIVE
VISCOSITY.

VISCOSITY RATIO EXCESS.
See SPECIFIC VISCOSITY.

3722
VISIBLE (visible region), VIS.
Electromagnetic radiation range of
wavelength between ca. 400 nm and
ca. 750 nm to which human eye is
sensitive.

D Sichtbare, sichtbares Gebiet
F région visible
P zakres widzialny
R видимая область (спектра)

VISIBLE REGION. *See* VISIBLE.

3723
VISIBLE SPECTRUM.
Spectrum in visible wavelength region
with lines or bands corresponding to
transitions between different electronic
energy levels of atoms or molecules.

D sichtbares Spektrum
F spectre visible
P widmo widzialne
R видимый спектр

3724
VISUAL COMPARISON METHOD
(standard series method).
Colorimetric analysis method for
determination of concentration in
investigated solution by comparison with
series of coloured standards (solutions or
solids) corresponding to known
concentrations. Most frequently carried
out by using comparators.

D Methode von Serie der Standarden
F méthode de comparaison visuelle,
méthode de comparaison à série
d'étalons
P metoda serii wzorców, metoda skali
barw, metoda skali kolorymetrycznej
R метод стандартных серий, метод
стандартной шкалы

VISUAL INDICATOR. *See* INDICATOR.

3725
VISUAL PHOTOMETRY.
Radiation intensity measurement method
by visual equalization of illumination of
two surfaces, one of which is illuminated
by comparative light beam.

D visuelle Photometrie
F photométrie visuelle
P fotometria wizualna
R визуальная фотометрия

3726
VISUAL SPECTROSCOPIC ANALYSIS
(spectroscopic analysis).
Spectroanalysis using the human eye as
radiation detector.

D visuelle Spektralanalyse
F analyse spectrale visuelle
P analiza spektroskopowa
R визуальный спектральный анализ

3727
VISUAL TITRATION.
Titration in which end point is determined
visually (on basis of colour changes,
appearance or disappearance of
fluorescence etc.)

D visuelle Titration
F . . .
P miareczkowanie wizualne
R визуальное титрование

3728
VITREOUS STATE (glassy state).
Amorphous state characterized by rigid
arrangement of molecules, occupying
defined, unchangeable positions.
Deformation caused by external force is
elastic and is described by Hook's law.
D glasiger Zustand, Glaszustand,
glasartiger Zustand
F état vitreux
P stan szklisty
R стеклообразное состояние

VOID VOLUME. *See* FREE VOLUME.

3729
VOLATILITY (fugacity), a_p.
Effective gas pressure in uni- or
multi-component system defined by

$$\mu = \mu^{\ominus} + RT \ln a_p$$

where μ = molar thermodynamic potential
(in uni-component systems) or chemical
potential of component (in
multi-component systems), $\mu^{\ominus}$ = molar
thermodynamic potential or chemical
potential of component in standard state,
respectively, R = molar gas constant and
T = thermodynamic temperature.
D Fugazität, Flüchtigkeit
F fugacité
P lotność, aktywność ciśnieniowa
R фугитивность, активное давление,
летучесть

3730
VOLATILITY COEFFICIENT, f_p.
Measure of deviation of real gas
properties from those of ideal gas,
defined by ratio of volatility of a given
gas a_p to pressure p of ideal gas in the
same conditions

$$f_p = a_p/p$$

D Aktivitätskoeffizient
F coefficient d'activité
P współczynnik aktywności ciśnieniowej
R коэффициент фугитивности

VOLTAMETER. *See* COULOMETER.

3731
VOLTAMMETRY.
Electrochemical method consisting in
recording and analysis of dependence of
electric current on potential of working
electrode.
D Voltammetrie
F voltamétrie
P woltamperometria
R вольтамперометрия

VOLTAMMETRY WITH
CONTINUOUSLY CHANGING
POTENTIAL. *See* SINGLE-SWEEP
VOLTAMMETRY.

VOLTA POTENTIAL DIFFERENCE. *See*
CONTACT POTENTIAL DIFFERENCE.

3732
VOLUME FRACTION, Φ_B.
Ratio of volume of dissolved substance B
and volume of solution

$$\Phi_B = \frac{n_B V_B}{\sum_i n_i V_i}$$

where n_B = number of moles of dissolved
substance B, V_B = molar volume of
dissolved substance B, $\sum_i n_i V_i$ = sum of
volume of all constituents in solution.
D Volumenbruch
F fraction en volume
P ułamek objętościowy, stężenie
objętościowe
R объёмная доля

3733
VOLUME OF ACTIVATION, $\Delta V^{\ne}$.
Difference of volume of initial and
transition state experimentally
determined by measurement of pressure
dependence of rate constant of a given
process (chemical reaction, diffusion,
viscosity) and by using formula

$$\left(\frac{\partial \ln k}{\partial p}\right)_T = \frac{\Delta V^{\ne}}{RT}$$

where k = rate constant of a given
process, p = pressure, R = molar gas
constant and T = thermodynamic
temperature.
D Aktivierungsvolumen
F volume d'activation
P objętość aktywacji
R объём активации

VOLUMETRIC ANALYSIS. *See*
TITRIMETRIC ANALYSIS.

VOLUMETRIC ERROR. *See* TITRATION
ERROR.

3734
VON AUWERS-SKITA RULE.
Cis-isomers of disubstituted cycloalkane
derivatives (mainly C_5 and C_6) usually
have higher boiling points, higher
densities and higher refractive indices
than *trans*-isomers.
D von Auwers-Skita-Regel
F règle d'Auwers et Skita
P reguła Auwersa i Skity
R правило Ауверса-Скита

3735
VON BRAUN CYANOGEN BROMIDE
REACTION.
Breaking of C—N bond in tertiary amines
with cyanogen bromide, e.g.

$R_3N + BrCN \rightarrow [R_3\overset{\oplus}{N}-C\equiv N]^{\oplus} \; Br^{\ominus} \xrightarrow{\Delta}$

$R_2N-C\equiv N + RBr$

D von Braunsche Bromcyan-Reaktion, von Braun tert. Amin-Abbau
F réaction de von Braun
P reakcja (bromocyjanowa) von Brauna
R расщепление аминов по Брауну

3736
VON BRAUN REACTION.
Decomposition of acylated aliphatic amine derivatives to alkyl halides and nitriles, effected by heating with PBr_5 or PCl_5, e.g.

$\overset{O}{\overset{\|}{RNHCC_6H_5}} + PBr_5 \xrightarrow{\Delta}$

$RBr + HBr + C_6H_5C\equiv N + POBr_3$

Von Braun reaction causes also decyclization of secondary derivatives of aromatic amines, e.g.

$C_6H_5C=O$
$C_6H_5C\equiv N + BrCH_2(CH_2)_3CH_2Br + POBr_3$

D von Braun (Amin)-Abbau
F réaction de von Braun
P degradacja (amidów) von Brauna
R реакция Брауна

3737
VON NEUMANN'S EQUATION
(Liouville-von Neumann equation, quantum-mechanical Liouville equation).
Conservation equation for statistical operator

$$\frac{\partial\rho}{\partial t} + \frac{i}{\hbar}(H\rho - \rho H) = 0$$

where ρ = statistical operator, H = Hamiltonian of system, t = time, $i = \sqrt{-1}$ and $\hbar$ = Planck constant divided by 2π. In quantum statistical mechanics von Neumann's equation determines time evolution of isolated system.

D von Neumann-Gleichung
F équation de von Neumann, équation opératorielle de Schrödinger
P równanie von Neumanna
R уравнение фон Неймана

3738
WAGNER-MEERWEIN REARRANGEMENT.
Acid-catalyzed conversion of pinacoline alcohol (and its analogues) into alkenes, involving anionic migration (in 1,2-rearrangement) from alkyl or aryl

group and elimination of water molecule, e.g.

$\underset{R}{\overset{R}{R-C-CH-R}} \underset{OH}{\overset{H^{\oplus}}{\xrightarrow{-H_2O}}} \underset{R}{\overset{R}{C=C}}\overset{R}{\underset{R}{}}$

In case of bicyclic terpenes, this reaction results in expansion or diminishing of rings, e.g.

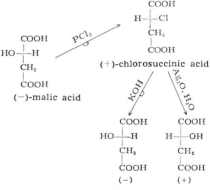

D Retropinakolin-Umlagerung, Wagner-Meerwein-Umlagerung
F transposition rétropinacolique, transposition de Wagner-Meerwein
P przegrupowanie retropinakolinowe, przegrupowanie Wagnera i Meerweina
R ретропинаколиновая перегруппировка, перегруппировка Вагнера-Меервейна

3739
WALDEN INVERSION.
Inversion of configuration of molecule taking place in S_N2 reactions at atom forming chirality centre. Attacking substituent approaches from opposite side to group leaving molecule, e.g.

(−)-malic acid; (+)-chlorosuccinic acid; (−); (+)

D Waldensche Umkehrung
F inversion de Walden
P inwersja Waldena
R Вальденовское обращение

3740
WALDEN'S RULE.
Independently of nature of solvent, $\Lambda_0\eta_0$ = const., where Λ_0 = limiting equivalent conductance of electrolyte in a given solvent, η_0 = viscosity of solvent; empirical rule holds for small ions.

D Waldensche Regel
F règle de Walden
P reguła Waldena
R правило Писаржевского-Вальдена

3741
WAVE FUNCTION.
Function of coordinates or momenta of all particles of given system describing its quantum mechanical state; its modulus square determines probability of finding system in coordinate or momentum space point given by function arguments. Wave function is given by solution of Schrödinger equation.
D Wellenfunktion
F fonction d'onde
P funkcja falowa, amplituda prawdopodobieństwa
R волновая функция

WAVE MECHANICS. *See* QUANTUM MECHANICS.

3742
WAVENUMBER, $\tilde{\nu}$ (σ).
Number of waves of electromagnetic radiation in 1 cm.
Remark: Term frequency (ν) is often misused for wavenumber ($\tilde{\nu}$). They are related by equation: $\tilde{\nu} = \nu/c$ where ν = frequency, c = speed of light.
D Wellenzahl
F nombre d'ondes
P liczba falowa
R волновое число

3743
WAVE PACKET.
Wave function different from zero only in very limited part of space.
D Wellenpaket
F paquet d'ondes
P pakiet falowy, paczka falowa
R волновой пакет

3744
WAVE VECTOR.
Vector perpendicular to head of wave, of absolute value $k = 2\pi/\lambda$, where λ = wavelength (e.g. of elastic wave in crystal).
D Wellenvektor
F vecteur d'onde
P wektor falowy
R волновой вектор

3745
WEAK ELECTROLYTE.
Electrolyte with degree of dissociation changing from value close to one for very diluted solutions to zero for very concentrated ones (partial dissociation).
D schwacher Elektrolyt
F électrolyte faible
P elektrolit słaby
R слабый электролит

3746
WEAK INTERACTION.
Interaction of particles resulting in processes which take place in time range of 10^{-10}—10^{-6} s (slow processes). Weak interaction satisfies only a few laws of conservation.
D schwache Wechselwirkung
F interaction faible
P oddziaływanie słabe
R слабое взаимодействие

3747
WEAK LIGAND FIELD.
Crystal ligand field causing splitting of energy levels (terms) of central ion, small compared with distances between energy levels (terms) of free ion. In complex of such field, number of unpaired electrons remains the same as in free ion (*cf.* crystal field splitting energy, spin-free complex).
D schwaches Ligandenfeld
F champ faible des coordinats
P pole (krystaliczne) ligandów słabe
R слабое поле лигандов

3748
WEERMAN DEGRADATION (Weerman synthesis of aldehydes).
Preparation of aldehydes impoverished by one carbon atom by action of aqueous solution of sodium hypochlorite on appropriate α-hydroxyacid amides. This reaction is applied for shortening carbon chain in aldoses

$$
\begin{array}{c}
\text{CHO} \\
| \\
\text{CHOH} \\
| \\
\text{CHOH} \\
|
\end{array}
\xrightarrow[\substack{\text{1. oxidation} \\ \text{2. lactonization} \\ \text{3. +NH}_3}]{}
\begin{array}{c}
\text{CONH}_2 \\
| \\
\text{CHOH} \\
| \\
\text{CHOH} \\
|
\end{array}
\xrightarrow[\text{NaOH}]{\text{NaOCl}}
$$

$$
\begin{bmatrix}
\text{N=C=O} \\
| \\
\text{CHO\!-\!H} \\
| \\
\text{CHOH} \\
|
\end{bmatrix}
\xrightarrow{\text{NaOH}}
\begin{array}{c}
\text{NaNCO} \\
\text{CHO} \\
| \\
\text{CHOH} \\
|
\end{array}
$$

D Weerman Aldonsäureamid-Abbau
F dégradation de Weerman
P degradacja Weermana
R расщепление по Веерману

WEERMAN SYNTHESIS OF ALDEHYDES. *See* WEERMAN DEGRADATION.

3749
WEIGHED AMOUNT (weighed sample).
Weighed amount of substance.
D Einwaage
F pesée
P odważka
R навеска

WEIGHED SAMPLE. *See* WEIGHED AMOUNT.

WEIGHT AVERAGE MOLECULAR WEIGHT. *See* MASS AVERAGE RELATIVE MOLECULAR MASS.

WEISSENBERG CAMERA. *See* WEISSENBERG GONIOMETER.

3750
WEISSENBERG GONIOMETER (Weissenberg camera).
X-ray camera in which revolution of crystal is synchronized with shift of photographic film in direction parallel to axis of revolution. This method simplifies indexing of reflections.
D Weissenberg-Goniometer
F goniometre de Weissenberg, chambre de Weissenberg
P goniometr Weissenberga
R (рентген)гониометр Вейсенберга

3751
WEIZSÄCKER'S MASS FORMULA.
Semiempirical formula for mass or atomic nucleus binding energy as function of its mass number A and atomic number Z, based on liquid-drop model of nucleus.
D (Weizsäcker-)Massenformel
F formule massique de Weizsäcker
P formuła masowa Weizsäckera
R массовая формула Вейцзеккера

3752
WELL-SCINTILLATION COUNTER (well-type counter).
Scintillation detector in which scintillator is shaped like a well, enabling small samples to be placed there.
D Bohrloch-Szintillationszähler
F scintillateur à puits
P licznik studzienkowy, licznik wnękowy
R счётчик с сцинтиллятором колодезного типа

WELL-TYPE COUNTER. *See* WELL-SCINTILLATION COUNTER.

3753
WERNER THEORY.
Theory of structure of complex compounds which assumes coordination of ligands around central atom. Assumed presence of principal and auxiliary valences is characteristic for this theory.
D Werners Koordinationslehre
F théorie de Werner
P teoria Wernera
R теория Вернера

WESTON CADMIUM CELL. *See* WESTON NORMAL CELL.

3754
WESTON NORMAL CELL (Weston standard cell, Weston cadmium cell).

Cell represented by scheme:

Hg, $HgSO_{4(s)}$ | $CdSO_4 \cdot \frac{8}{3} H_2O$ (sat. solution)| Cd(Hg) (12.5%)

EMF of this cell at 20°C is 1.01830 V and is standard of international volt.
D Normalelement, Normal-Weston-Element, Cadmium-Normalelement
F élément Weston normal, pile étalon Weston
P ogniwo Westona normalne
R стандартный элемент Вестона, нормальный элемент Вестона, эталонный элемент Вестона

WESTON STANDARD CELL. *See* WESTON NORMAL CELL.

3755
WET METHOD.
Method of investigation of solid samples after solution.
D nasses Verfahren
F méthode par voie humide
P metoda mokra
R анализ мокрым путём

3756
WETTING.
Establishment of certain definite wetting angle Θ in gas-liquid-solid three-phase system; when $\Theta > 90°$ liquid is said in practice not to wet solid, whereas complete (ideal) wetting implies contact angle of 0°.
D Benetzung
F mouillage
P zwilżanie
R смачивание

WETTING ANGLE. *See* CONTACT ANGLE.

WHITE RADIATION. *See* X-RAY BACKGROUND RADIATION.

3757
WIEDEMANN-FRANZ LAW (Lorentz relation).
Ratio of thermal conductivity λ and electric conductivity σ is constant at defined temperature T for all metals and is described by formula

$$\frac{\lambda}{\sigma} = -LT$$

where $L = 2.45 \cdot 10^{-8}$ V²/K² is Lorentz number.
D Wiedemann-Franzsches Gesetz
F loi de Wiedemann et Franz
P prawo Wiedemanna i Franza
R закон Видемана и Франца

WIEN'S EFFECT. *See* FIELD STRENGTH EFFECT.

3758
WILLGERODT REACTION.
Conversion of alkyl aryl ketones into amides containing the same number of carbon atoms, by heating ketones with aqueous solution of ammonium polysulfide, e.g.

$$C_6H_5COCH_3 \xrightarrow[\Delta]{(NH_4)_2S_x} C_6H_5CH_2CONH_2$$

D Willgerodt-Reaktion, Willgerodt-Kindler Redox-Amidierung
F réaction de Willgerodt
P reakcja Willgerodta
R реакция Вильгеродта

3759
WILLIAMSON SYNTHESIS OF ETHERS.
Preparation of ethers by treating alcoholates with alkyl halides, e.g.

$$RONa + R'X \rightarrow R-O-R' + NaX$$

D Williamson-Synthese, Williamson Alkoholat-Alkylierung
F synthèse de Williamson
P synteza Williamsona
R синтез Вильямсона

3760
WINDOW COUNTER.
Counter tube of the G-M type with window of low absorption permitting detection of β-particles of low penetrating power.

D Fensterzählrohr, Glockenzählrohr
F compteur à fenêtre, compteur type cloche
P licznik okienkowy
R торцевой счётчик

3761
WITTIG REACTION.
Preparation of alkenes from aldehydes and ketones and phosphoric ylids (alkylidenetriphenylphosphates), e.g.

D Wittig-Reaktion, Wittig aufbauende Carbonyl-Olefinierung, Wittig Phosphin-Methylen-Reaktion
F réaction de Wittig
P reakcja Wittiga
R реакция Виттига

3762
WOHL-ZEMPLÉN DEGRADATION OF SUGARS.
Shortening of carbon chain in aldoses; acetylation of aldose oxime with acetic anhydride in presence of sodium acetate leads to formation of acetylated nitrile of aldonic acid. The latter on treatment with sodium methylate releases hydrogen cyanide, forming aldose having one carbon atom less.

D Wohl-Zemplénscher Abbau, Wohl Zuckkernitril-Abbau
F dégradation de Wohl
P degradacja Wohla i Zempléna
R расщепление по методу Воля-Цемплена

3763
WOHL-ZIEGLER BROMINATION.
Bromination of alkenes in allyl position with N-bromosuccinoimide

D Wohl-Ziegler Allyl-Bromierung
F réaction de Wohl-Ziegler
P reakcja (bromowania) Wohla i Zieglera
R реакция Воля-Циглера

3764
WOLFF-KISHNER REDUCTION.
Reduction of carbonyl compound to hydrocarbon involving hydrazone formation and heating the latter to 150—200°C in presence of strong bases, e.g.

D Wolff-Kishner-Reaktion, Kishner-Wolff Carbonyl→Methylen-Reduktion
F réaction de Wolff-Kishner
P redukcja Wolffa i Kiżnera
R реакция Вольфа-Кижнера

3765
WOLFF REARRANGEMENT OF DIAZOKETONES.
Rearrangement of α-diazoketones into ketenes (with elimination of nitrogen molecule) and further into carboxylic acids or their derivatives in presence of

catalysts (silver oxide, Raney nickel, electromagnetic radiation), e.g.

$$RCOCHN_2 \xrightarrow[-N_2]{Ag_2O} [RCH=C=O] \xrightarrow{H_2O} RCH_2COOH$$

D Wolff (Diazoketon→Keten)-Umlagerung
F transposition de Wolff
P przegrupowanie (dwuazoketonów) Wolffa
R перегруппировка Вольфа

WOLFRAM. *See* TUNGSTEN

3766
WORK FUNCTION (emission work).
Work needed to transfer particle from solid state to vacuum; it corresponds to binding energy of particle in solid state; e.g. work needed for removing electron from metal.

D Austrittsarbeit
F travail d'extraction
P praca wyjścia
R работа выхода

WORK FUNCTION. *See* HELMHOLTZ FREE ENERGY.

3767
WURTZ REACTION (Wurtz synthesis).
Preparation of alkanes by action of sodium on alkyl halides

$$2RX + 2Na \longrightarrow R—R + 2NaX$$

D Wurtz-Reaktion, Wurtz Alkylhalogenid-Kondensation
F synthèse de Wurtz
P reakcja Wurtza
R реакция Вюрца

WURTZ SYNTHESIS. *See* WURTZ REACTION.

3768
XANTHO-PROTEIN REACTION.
Coloured (yellow) reaction of amino acids containing aromatic nuclei with concentrated nitric acid; applied for detection of protein.

D Xanthoproteinreaktion
F réaction xanthoprotéique
P reakcja ksantoproteinowa
R ксантопротеиновая реакция

3769
XENON, Xe.
Atomic number 54.

D Xenon
F xénon
P ksenon
R ксенон

3770
XEROGEL.
Dry gel formed from lyogel after separation of dispersion medium with simultaneous decrease in volume.

D Xerogel
F xérogel
P kserożel
R ксерогель

3771
XI HYPERON (Ξ-hyperon, cascade particle).
Elementary particle, a hadron, of baryon number $B = 1$, lepton number $L = 0$, quantum spin number $J = 1/2$, strangeness $S = -2$, isospin $I = 1/2$, unstable. Ξ-hyperon occurs in two states of charge (isospin doublet): electrically neutral (Ξ⁰-hyperon; mass $m = 1\,314.7$ MeV, third component of isospin $I_3 = 1/2$, mean lifetime $3.03 \cdot 10^{-10}$ s) and of negative elementary electric charge (Ξ⁻-hyperon; $m = 1\,321.31$ MeV, $I_3 = -1/2$, mean lifetime $1.660 \cdot 10^{-10}$ s).

D Ξ-Hyperon
F hypéron Ξ, hypéron xi
P hiperon Ξ, hiperon kaskadowy
R Ξ-гиперон, кси-гиперон

3772
X-RADIATION (X-rays).
Electromagnetic radiation of wavelength (approximately) from 5 to 2000 pm.

D Röntgenstrahlung
F rayonnement X
P promieniowanie rentgenowskie, promieniowanie X, promienie Roentgena
R рентгеновские лучи

3773
X-RAY ABSORPTION ANALYSIS.
Spectroanalysis based on investigation of X-ray absorption spectrum.

D Röntgenabsorptionsanalyse
F analyse par absorption de rayons X
P analiza rentgenowska absorpcyjna
R абсорбционный рентгеновский анализ

3774
X-RAY BACKGROUND RADIATION (white radiation, bremsstrahlung).
X-ray continuous radiation produced by rapid hindrance of movement of charged particles (e.g. electrons) in Coulomb field of atomic nuclei.

D Bremsstrahlung
F rayonnement de freinage
P promieniowanie hamowania
R тормозное излучение, белое рентгеновское излучение

3775
X-RAY DIFFRACTION.
Scattering and interference of X-rays passing through matter.

D Röntgenstrahlenbeugung
F diffraction des rayons X
P dyfrakcja promieni rentgenowskich
R дифракция рентгеновских лучей

3776
X-RAY DIFFRACTION ANALYSIS.
Qualitative and quantitative determination
of composition of crystalline substances
by X-ray diffraction measurement.
D Röntgendiffraktionsanalyse
F analyse par diffraction aux rayons X,
analyse diffractométrique aux rayons X
P analiza rentgenowska dyfrakcyjna
R рентгеновский дифракционный анализ

3777
X-RAY DIFFRACTOMETER.
Instrument used for structural and
analytical research utilizing X-ray
diffraction on a given material.
D Röntgendiffraktometer
F diffractomètre de rayons X
P dyfraktometr rentgenowski
R рентгеновский дифрактометр

3778
X-RAY ESCAPE PEAK.
Peak in gamma radiation or X-ray
spectrum caused by escape of X-rays
emitted by an atom of the detector
material after photoelectron emission.
D Röntgen-Escape-Peak
F pic de fuite des photons
P pik ucieczki promieniowania
rentgenowskiego
R рентгеновский пик утечки

3779
X-RAY FLUORESCENCE (secondary
radiation).
X radiation emitted when atoms are
ionized by incident X-ray photons which
eject electrons from their inner shells
(mainly K and L).
D Röntgenfluoreszenzstrahlung
F rayonnement X secondaire
P fluorescencja rentgenowska,
promieniowanie wtórne
R вторичное рентгеновское излучение,
флуоресцентное рентгеновское
излучение

3780
X-RAY FLUORESCENCE ANALYSIS.
Spectroanalysis based on investigation of
X-ray fluorescence spectra, i.e. analysis of
secondary X-ray radiation spectrum
excited by primary one.
D Röntgenfluoreszenzspektralanalyse
F analyse spectrométrique de
fluorescence X, analyse par
(spectrométrie de) fluorescence X
P analiza fluorescencyjna rentgenowska,
analiza rentgenofluorescencyjna
R флуоресцентный рентгеновский
анализ, анализ по вторичным
рентгеновским спектрам

3781
X-RAY PHOTOGRAPH.
Photographic image obtained with X rays
by different methods (Debye, Scherrer
and Hull, Laue, Weissenberg, rotating
crystal etc.), revealing structural features
of solid, investigated as crystal or powder
(symmetry and dimensions of elementary
cell, location of atoms or ions in
elementary cell).
D Röntgenaufnahme, Röntgenbild
F radiogramme
P rentgenogram
R рентгенограмма, рентгеновская
диаграмма

X-RAYS. See X-RADIATION.

3782
X-RAY SPECTROCHEMICAL
ANALYSIS.
Spectroanalysis based on investigation of
X-ray spectrum.
D Röntgenspektralanalyse
F analyse par spectrométrie X
P analiza rentgenowska spektralna
R рентгеновский спектральный анализ,
рентгеноспектральный анализ

3783
X-RAY SPECTROMETER.
Spectrometer for study of X-ray radiation.
D Röntgenspektrometer
F spectromètre à rayons X
P spektrometr rentgenowski
R рентгеновский спектрометр

3784
X-RAY TUBE.
Vacuum tube generating X-ray radiation
due to anti-cathode bombardment by
electron beam.
D Röntgenröhre
F tube à rayons X
P lampa rentgenowska
R рентгеновская трубка

3785
YIELD (of reaction).
Ratio of amount of product obtained in
chemical reaction and theoretical amount
of product following from stoichiometry
for the same amount of substrates;
expressed as percentage.
D Ausbeute, Reaktionsausbeute
F rendement
P wydajność reakcji
R выход

3786
YLID.
Neutral mesomeric molecule in which the
carbanion is linked directly with the
electropositive heteroatom

$$\overset{\oplus}{Z}-\overset{\ominus}{C}\diagup \leftrightarrow Z=C\diagup$$

where Z denotes nitrogen, phosphorus, sulfur, arsenic and antimony; ylids are strong nucleophiles.

D Ylid
F ylide
P ylid
R илид

3787
YTTERBIUM, Yb.
Atomic number 70.

D Ytterbium
F ytterbium
P iterb
R иттербий

3788
YTTRIUM, Y.
Atomic number 39.

D Yttrium
F yttrium
P itr
R иттрий

3789
YTTRIUM EARTHS.
Oxides of yttrium and lanthanoids from europium (Z = 63) to lutetium (Z = 71).

D Yttererden
F terres yttriques
P ziemie itrowe
R иттриевые земли

3790
ZEEMAN EFFECT.
Splitting of atomic or molecular energy levels or splitting of spectral lines due to external magnetic field.

D Zeeman-Effekt
F effet Zeeman
P efekt Zeemana
R явление Зеемана

3791
ZEEMAN LEVELS.
Sublevels arising from splitting of energetic level of atom, ion, molecule or paramagnetic nucleus in magnetic field.

D Zeeman-Komponenten
F niveaux Zeeman
P poziomy zeemanowskie
R зеемановские уровни

3792
ZEOLITES.
Crystalline aluminosilicates with ion exchange properties.

D Zeolithe
F zéolit(h)es
P zeolity
R цеолиты

3793
ZEOTROPIC SYSTEM.
System with no extremal vapour pressure or extreme of boiling point.

D zeotropes System
F système zéotrope
P układ zeotropowy
R зеотропная система

3794
ZEOTROPY.
Property of systems characterized by non-existence of either extremal vapour pressure or extreme of boiling point.

D Zeotropie
F zéotropie
P zeotropia
R зеотропия

3795
ZERO CHARGE POTENTIAL (null potential).
Potential of electrode (half-cell) measured with respect to a given reference electrode, at which electrical charge of metal surface is zero, e.g. for mercury is $+0.19$ V as compared to normal hydrogen electrode in absence of substance which may be specifically adsorbed on electrode.

D Nullpunktspotential
F point de charge zéro
P potencjał ładunku zerowego, punkt zerowy elektrody
R нулевая точка (электрода), потенциал нулевого заряда

3796
ZERO-ORDER REACTION.
Reaction with kinetic equation characterized by sum of exponents in which concentrations of reactants occur being equal to zero; i.e. reaction of rate which does not change with change of concentration of reactants.

D Reaktion 0. Ordnung, Reaktion nullter Ordnung
F réaction d'ordre zéro
P reakcja zerowego rzędu
R реакция нулевого порядка

ZERO-POINT ENERGY. *See* ZERO-TEMPERATURE ENERGY.

3797
ZERO TEMPERATURE (absolute zero of temperature).
Value of thermodynamic temperature equal to 0 K characteristic for states which according to third law of thermodynamics cannot lose any more energy in form of heat.

D absoluter Temperaturnullpunkt, absoluter Nullpunkt
F zéro absolu, zéro de Carnot
P zero bezwzględne (temperatury)
R абсолютный нуль (температуры)

3798
ZERO-TEMPERATURE ENERGY (zero-point energy).
Energy of ground state of oscillator or population of oscillators. It is independent of temperature.

D Nullpunktsenergie
F énergie nulle
P energia zerowa
R нулевая энергия

3799
ZEROTH LAW OF THERMODYNAMICS.
If two systems are in thermal equilibrium with a third, then all of them are in mutual thermal equilibrium. It is assumed that partitions between these systems are all diathermal.

D nullter Hauptsatz der Thermodynamik
F principe zéro de la thermodynamique
P zasada termodynamiki zerowa
R нулевой принцип термодинамики

ZETA POTENTIAL. *See*
ELECTROKINETIC POTENTIAL.

3800
ZINC, Zn.
Atomic number 30.

D Zink
F zinc
P cynk
R цинк

ZINC FAMILY. *See* ZINC GROUP.

3801
ZINC GROUP (zinc family).
Elements in second sub-group of periodic system with structure of outer electronic shells of atoms $(n-1)d^{10}ns^2$: zinc, cadmium, mercury.

D Zinkgruppe
F famille du zinc, sous-groupe du zinc
P cynkowce
R подгруппа цинка, элементы подгруппы цинка

3802
ZINCKE PREPARATION OF SULFENYL HALIDES.
Preparation of halides, mainly arylsulfenylic, by action of chlorine or bromine on aryl disulfides, thiophenols or arylbenzylic sulfides, e.g.

$$ArS-SAr + X_2 \rightarrow 2ArSX$$
$$C_6H_5SH + Cl_2 \rightarrow C_6H_5SCl + HCl$$
$$ArSCH_2C_6H_5 + 2Cl_2 \rightarrow ArSCl + C_6H_5CHCl_2 + HCl$$

D ...
F méthode de Zincke
P synteza halogenków sulfenylowych Zinckego
R ...

3803
ZIRCONIUM, Zr.
Atomic number 40.

D Zirkon(ium)
F zirconium
P cyrkon
R цирконий

3804
ZONAL CRYSTALS.
Crystals characterized by their stratified (zonal) structure which is due to changes in their chemical composition during growth of crystals.

D Zonenkristalle
F ...
P kryształy zonalne
R ...

3805
ZONE LAW.
Every face of crystal belongs to at least two intersecting zones.

D Zonengesetz
F loi de zone
P prawo pasowe
R правило зоны, закон поясов

ZWITTERION. *See* AMPHOTERIC ION.

NAME INDEX

DEUTSCHES WÖRTERVERZEICHNIS

Alkylation *f* 119
Alkylierung *f* 119
α-Amidoalkylierung 143
Leuckart Amin-~ 2000
Williamson Alkoholat-~ 3759
allgemeine Chemie *f* 1474
allgemeine Säure-Base-Katalyse *f* 1473
Allingersche Regel *f* 120
Allotropie *f* 121
Allylumlagerung *f* 125
Alpha-Teilchen-Modell *n* 129
Altern *n* 107
alternierendes Copolymer *n* 131
Altersbestimmung *f*
radioaktive ~ 2879
Alterung *f* 107
~ der Sole 108
~ des Katalysators 106
Aluminium *n* 133
Amadori Aldose-N-Glykosid →
   Isoglykosamin-Umlagerung *f* 135
Amadori-Umlagerung *f* 135
Amagatsches Gesetz *n* 136
Amalgampolarographie *f* 137
Americium *n* 141
Amerizium *n* 141
Amidierung *f*
Willgerodt-Kindler Redox-~ 3758
α-Amidoalkylierung *f* 143
Aminierung *f* 144
Leuckart reduktive Carbonyl-~ 2000
Transaminierung 3594
Tschitschibabin Pyridin-~ 595
Aminolyse *f* 145
Aminomethylierung *f* 146
Amminokomplex *n* 147
Ammonolyse *f* 148
amorpher Körper *m* 149
amorpher Stoff *m* 149
amorpher Zustand *m* 150
Amperometrie *f* 152
amperometrische dead-stop-Methode *f* 343
amperometrische Titration *f* 152
Ampholyt *m* 154
ampholytisches Lösungsmittel *n* 153
amphoterer Elektrolyt *m* 154
amphoterer Ionenaustauscher *m* 156
Amphoterie *f* 157
Analyse *f*
Aktivierungsanalyse 61 (*vide*)
Atomabsorptionsanalyse 241
Atomfluoreszenzanalyse 246
chemische ~ 565
chromatographische ~ 607
differentiale thermische ~ 995
Elektroanalyse 1102, 1118
Elementaranalyse 1191
flammenphotometrische ~ 1381
flammenspektrophotometrische ~ 159
Fluoreszenzanalyse 1394
fluorimetrische ~ 1394
Frontalanalyse 1433
funktionelle ~ 1438
Gasanalyse 1456
Gesamtanalyse 681
Gewichtsanalyse 1507
Halbmikroanalyse 3178
Instrumentalanalyse 1744
Instrumentenanalyse 1744
interferometrische ~ 1759
Isotopen-Verdünnungsanalyse 1883
Isotopenanalyse 1892
Konformationsanalyse 723
Lumineszenzanalyse 2074
Makroanalyse 2084

Analyse
Maßanalyse 3587
massenspektrometrische ~ 2131
Mikroanalyse 2216
Neutralisationsanalyse 41
polarographische ~ 2709
qualitative ~ 2830
quantitative ~ 2833
radiochemische ~ 2895
radiometrische ~ 2909
Reduktionsanalyse 2989
Röntgenabsorptionsanalyse 3773
Röntgendiffraktionsanalyse 3776
Röntgenfluoreszenzspektralanalyse 3780
Sedimentationsanalyse 3152
Semimikroanalyse 3178
Siebanalyse 3204
Spektralanalyse 3298
spektrochemische ~ 3299
spektrographische ~ 3307
spektrometrische ~ 3309
Spurenanalyse 3593
thermische ~ 3521
Thermoanalyse 3521
thermomagnetische ~ 3550
Titrieranalyse 3587
titrimetrische ~ 3587
Tüpfelanalyse 3340
Varianzanalyse 160
volumetrische ~ 3587
Analyseformel *f* 1215
Analysenfehler *m* 164
Analysenlinie *f* 168
analytische Chemie *f* 162
analytischer Faktor *m* 165
Anaphorese *f* 171
anchimere Beschleunigung *f* 172
Anfangsstadium *n* 1735
angeregter Kern *m* 1300
angeregter Zustand *m* 1301
angeregtes Molekül *n* 1299
angereichertes Material *n* 1238
angewandte Chemie *f* 207
anharmonischer Oszillator *m* 176
Anion *n* 177
Carbanion 474
Anionenaustauscher *m* 178
Anionotropie *f* 179
Anisotropie *f* 180
diamagnetische ~ 963
magnetische ~ 2096
Anlagerungskomplex *m* 2399
Anlassen *n* 181
Annihilation *f* 182
Annihilationsstrahlung *f* 184
Anode *f* 185
Anodenflüssigkeit *f* 187
Anodenlösung *f* 187
anodische Teilstromdichte *f* 186
Anolyt *m* 187
Anomere *npl* 188
anorganische Chemie *f* 1740
anorganischer Ionenaustauscher *m* 1741
Anregung *f* 1295
Einelektronanregung 2459
Anregungsenergie *f* 1296
Anregungsfunktion *f* 1297
Anreicherung *f*
Isotopenanreicherung 1895
Anreicherungsfaktor *m* 1239
Ansa-Verbindungen *fpl* 189
Ansätze *mpl*
phänomenologische ~ 2619
Ansprechwahrscheinlichkeit *f*
   des Detektors 1787

Bahndrehimpulsquantenzahl f 2482
Baker-Nathan-Effekt m 1682
Balandin-Multiplett-Theorie f 302
Bananen-Bindung f 329
Band n
B-~ 321
CT-~ 554
Eichband 2991
Elektronenband 1154
Energieband 122 (vide)
Grundband 3679
Grundschwingungsband 1441
K-~ 1913
Kombinationsband 676
Ladungsaustauschband 554
Leitfähigkeitsband 715
Leitungsband 715
Parallel-~ 2545
R-~ 2952
Rotation-Schwingungsband 3081
Schwingungsband 3707
senkrechtes ~ 2601
Valenzband 3679
verbotenes ~ 1405
Bandenkante f 303
Bandenkopf m 303
Bandenspektrum n 304
Bändermodell n der Festkörpertheorie 305
Bändertheorie f von Festkörper 305
Barbier-Wieland Carbonsäure-Abbau m 306
Barbier-Wieland-Reaktion f 306
Barium n 307
Barkhausen-Effekt m 308
Barn n 309
Barriere f
Coulomb-~ 786
Energiebarriere 2740
Rotationsbarriere 3075
Bart Arsenit-Arylierung f 311
Bart-Reaktion f 311
Bartonscher Effekt m 2061
Baryonenresonanzen fpl 314
Baryonenzahl f 312
Base f 315
Basis-Linien-Verfahren n 316
basisches Lösungsmittel n 2808
Basizität f 317
~ der Säure 318
Bastardisierung f 1651
Bastardorbital n 1652
bathmochrome Gruppe f 319
bathochrome Absorptionsverschiebung f 320
bathochrome Gruppe f 319
Baufehler mpl
chemische ~ 1701
Kristallbaufehler 1974
Béchamp Nitroaryl-Reduktion f 325
Béchamp-Reduktion f 325
Beckmann Oxim→Amid-Umlagerung f 326
Beckmann-Umlagerung f 326
Bedingung f
~ für die mechanische Stabilität 712
~ für die Stabilität in Bezug auf Diffusion 713
~ für die thermische Stabilität 714
Bedingungen fpl
Normalbedingungen 3351
Normbedingungen 3351
begrenzte Mischbarkeit f 2562
Begrenzungsfläche f 1754
Behälter m 743
Beibehalten n der Konfiguration 3047

Benetzung f 3756
Benetzungswärme f 1571
Benzidinumlagerung f 330
Benzilsäureumlagerung f 331
Benzoinkondensation f 332
Benzoylierung f 333
Benzylierung f 334
Berechnungsmethode f
Hendersonsche ~ der Diffusionspotential 1582
Plancksche ~ des Diffusionspotentials 2677
Bereich m
azeotroper ~ 289
Bereiche mpl 1043
Bereichsstruktur f 1044
Berkelium n 335
Berthelot-Nernst-Verteilungsgesetz n 337
Berthelotsche Zustandgleichung f 336
Berthollidverbindungen fpl 2396
Beryllium n 338
Beschleuniger m 551
Linearbeschleuniger von Elektronen 2023
Teilchenbeschleuniger 551
beschleunigte Elektronen npl 27
Beschleunigung f
anchimere ~ 172
Besetzungszahl f 2447
mittlere ~ 281
Beständigkeit f
Strahlenbeständigkeit 2914
Bestimmung f 952
~ nach den funktionellen Gruppen 1438
Korngrößen-~ 2569
Bestimmungsmethode f
Hittorfsche ~ der Überführungszahlen 1606
Bestrahlung f 1848
Pulsbestrahlung 2816
Selbstbestrahlung 3172
Bestrahlungskammer f 1849
Bestrahlungsraum m 1849
Bestrahlungsschäden mpl 2861
Bestrahlungsschleife f 1850
BET-Adsorptionsisotherme f 347
BET-Gleichung f 346
Beta-Rückstreuverfahren n 343
Betatron n 345
Bethe-Weizsäcker-Zyklus m 484
Bethe-Zyklus m 484
Beugung f
Atom- und Molekülbeugung 997
Elektronenbeugung 1146
Lichtbeugung 2010
Neutronenbeugung 2365
Röntgenstrahlenbeugung 3775
Beugungsordnung f 2484
bewegliche Phase f 2247
Beweglichkeit f
Ionenbeweglichkeit 1815
Bewegung f
Brownsche ~ 440
Beziehung f
Brönsted-~ 433
Brönstedsche ~ 438
Duhem-Margulessche ~ 1070
Gibbs-Duhemsche ~ 1484
Ionenbeziehung 1811
Masse-Energie-~ 1088
Unbestimmtheitsbeziehung 3654
Beziehungen fpl
Maxwellsche ~ 2144
Bezirke mpl 1043
Bezirksstruktur f 1044
Bezugs-EMK f 3010
Bezugselektrode f 2992

Bouveault-Formylierung *f* 413
Boyle-Mariottesches Gesetz *n* 415
Boyle-Punkt *m* 416
Boyle-Temperatur *f* 416
Bragg-Gray-Kammer *f* 418
Bragg-Spektrometer *n* 419
Braggsche Gleichung *f* 420
Braggsche Kurve *f* 417
Bravaisgitter *n* 426
Brechung *f*
Doppelbrechung 362, 1061
Lichtbrechung 2997
Brechungsexponent *m* 2998
Brechungsindex *n* 9
Brechungskoeffizient *m* 9
Brechungsverhältnis *n* 2998
Brechzahl *f* 9
Bredtsche Regel *f* 429
Breit-Wigner-Formel *f* 431
Bremsstrahlung *f* 3774
Bremssubstanz *f* 2249
Bremsvermögen *n* 3408
lineares ~ 2029
Massenbremsvermögen 2136
Brennelemente *npl*
abgestellte ~ 448
ausgebrannte ~ 448
Brennstoff *m*
Kernbrennstoff 2412
Brennstoffelement *n* 1435
Brennstoffkreislauf *m* 2413
Brennstoffzyklus *m* 2413
Briefumschlag-Konformation *f* 1248
Brillaun-Zonen *fpl* 434
Brom *n* 436
Bromierung *f* 435
Wohl-Ziegler Allyl-~ 3763
Brönsted-Beziehung *f* 438
Brönsted-Bjerrum Gleichung *f* 437
Brönstedsche Beziehung *f* 438
Brönstedsche Säuren-Basen-Theorie *f* 439
Brownsche Bewegung *f* 440
Bruch *m*
Massenbruch 2128
Molenbruch 2289
Volumenbruch 3732
Brücken-Gruppe *f* 433
Brückenatom *n* 432
bruckenbildende Gruppe *f* 433
Brüter *m* 430
Brutreaktor *m* 430
Bruttodissoziationskonstante *f* 2516
~ des Komplexes 2516
Bruttoformel *f* 2277
Bruttostabilitätskonstante *f* 3345
Bucherer-Bergs Hydantoin-Ringschluß *m* 441
Bucherer-Lepetit Naphthol→
Naphthylamin-Umwandlung *f* 442
Bucherer-Lepetit-Reaktion *f* 442
Bunsen-Roscoesches Gesetz *n* 2976
Bunsensches Eiskalorimeter *n* 447
Büschelmodell *n* 640
Buttersäuregärung *f* 453

C

Cadiot-Chodkiewicz-Reaktion *f* 454
Cadmium *n* 455

Cadmium-Normalelement *n* 3754
Caesium *n* 456
Cahn-Ingold-Prelogsche Konvention *f* 3083
Calcium *n* 459
Californium *n* 461
calorische Zustandsgleichung *f* 463
Cannizzaro Aldehyd-Dismutation *f* 467
Cannizzaro-Reaktion *f* 467
Carathéodory Unerreichbarkeits-Axiom *n* 473
Carathéodorysches Postulat *n* 473
Carbanion *n* 474
Carben *n* 475
Carbeniat-Ion *n* 474
Carbeniumion *n* 477
carbocyclischer Ring *m* 478
Carboniumion *n* 477
Carbonylierung *f* 485
Carboxylierung *f* 486
Kolbe-Schmitt Phenolat-~ 1943
Carnot-Prozeß *m* 489
Carnotscher Kreisprozeß *m* 489
Carnotscher Satz *m* 490
Cäsium *n* 456
Catenane *npl* 506
Centigramm-Methode *f* 515
Cer *n* 523
Čerenkov-Strahlung *f* 522
Čerenkov-Zähler *m* 521
Ceriterden *fpl* 524
Cerium *n* 523
Chapman-Enskog-Methode *f* 540
Chapman Imidoester→Amid-Umlagerung *f* 541
Chapman-Umlagerung *f* 541
Charakter *m*
aromatischer ~ 218
charakteristische Absorptionsfrequenz *f* 544
charakteristische Frequenz *f* 543
charakteristische Funktionen *fpl* 545
charakteristische Koordinationszahl *f* 542
charakteristische Röntgenstrahlung *f* 547
charakteristische Temperatur *f* 546
charakteristische Temperatur *f* der Rotation 3076
charakteristische Temperatur *f* der Schwingung 3703
charakteristisches Röntgenspektrum *n* 548
Chelatbildungsreaktion *f* 562
Chelatbindungen *fpl* 558
Chelateffekt *m* 559
Chelatkomplex *m* 557
Chelatligand *m* 563
Chelatometrie *f* 564
Chelatring *m* 561
Chelatverbindung *f* 557
Chemie *f* 593
allgemeine ~ 1474
analytische ~ 162
angewandte ~ 207
anorganische ~ 1740
Astrochemie 782
Biochemie 354
biologische ~ 354
~ der heißen Atome 1634
Elektrochemie 1106
Geochemie 1477
Kernchemie 2407

Detonation *f* 953
Detonationswelle *f* 954
Deuterierung *f* 955
Deuterium *n* 956
schweres ~ 3631
Deuteron *n* 957
Deuton *n* 957
Diagramm *n*
Laue-~ 1977
Löslichkeitsdiagramm 3254
Molekulardiagramm 2272
Phasendiagramm 2611
Schichtliniendiagramm 3012
Zustandsdiagramm 2611
Diagramme *npl*
Feynman-~ 1353
Dialysator *m* von Graham 1500
Dialyse *f* 962
Elektrodialyse 1116
Diamagnetika *npl* 964
diamagnetische Abschirmung *f* des Kerns 965
diamagnetische Anisotropie *f* 963
diamagnetische Körper *mpl* 964
diamagnetische Stoffe *mpl* 964
diamagnetische Suszeptibilität *f* 966
Diamagnetismus *m* 967
Diastereoisomere *npl* 969
Diastereoisomerie *f* 968
Diastereomerie *f* 968
diathermische Wand *f* 970
Diazotierung *f* 971
Dichroismus *m* 972
Zirkulardichroismus 621
Dichte *f* 938
Austauschstromdichte 1286
~ der Leitungselektronen 1415
~ der Zustände 940
Elektronendichte 1145, 1415
Entropiedichte 1243
Flußdichte 2857
Ionisationsdichte 1830
Ladungsdichte 550
Leuchtdichte 2072
molekulare ~ 2567
Moleküldichte 2567
Oberflächendichte (vide)
Spindichte 3322
Strahlungsflußdichte 2357
Stromdichte 859
Teilchendichte 2567
Teilstromdichte (vide)
Verteilungsdichte 2787
Wahrscheinlichkeitsdichte 939
Dichtematrix *f* 3376
Dicke *f*
Halbwertdicke 1529
Schichtdicke 3102
dicke Treffplatte *f* 3554
dicker Target *m* 3554
Dieckmann intermolekulare Esterkondensation *f* 973
Dieckmann-Reaktion *f* 973
Dielektrikum *n* 974
dielektrische Absorption *f* 975
dielektrische Dispersion *f* 976
dielektrische Polarisation *f* 978, 979
dielektrische Relaxation *f* 980
dielektrische Relaxationszeit *f* 981
dielektrische Sättigung *f* 982
dielektrische Verluste *mpl* 977
dielektrischer Verlustfaktor *m* 1033
Dielektrizitätskonstante *f* 2599
absolute ~ 2599

Dielektrizitätskonstante
~ des Vakuums 2600
komplexe ~ 687
relative ~ 3014
Dielektrizitätszahl *f* 2599, 3014
~ des Vakuums 2600
komplexe ~ 687
optische ~ 2473
statische ~ 3368
Diels-Alder-Reaktion *f* 985
Dienanaloga *npl* 984
Dienon→Phenol-Umlagerung *f* 986
Diensynthese *f* 985
Differentialdetektor *m* 990
differentiale thermische Analyse *f* 995
Differentialthermoanalyse *f* 995
differentielle Adsorptionswärme *f* 991
differentielle Kapazität *f* der Doppelschicht 988
differentielle Lösungswärme *f* 2559
differentielle Quellungswärme *f* 2560
differentielle Spektralphotometrie *f* 994
differentielle Verdünnungswärme *f* 2558
differentieller Wirkungsquerschnitt *m* 989
differenzierendes Lösungsmittel *n* 996
Diffraktion *f*
~ des Lichtes 2010
Elektronendiffraktion 1146
Diffraktometer *n*
Röntgendiffraktometer 3777
diffuse Doppelschicht *f* 998
diffuse Reflexion *f* 999
diffuse Schicht *f* 998
Diffusion *f* 1000
Oberflächendiffusion 3468
Selbstdiffusion 3170
Thermodiffusion 3525
Diffusionsgebiet *n* des Prozesses 1005
Diffusionsgesetz *n*
erster Ficksches ~ 1354
zweites Ficksches ~ 1355
Diffusionsgrenzstrom *m* 2015
Diffusionskoeffizient *m* 1001
Diffusionsmodell *n* der Radiolyse 1003
Diffusionspolarisation *f* 1004
Diffusionspotential *n* 2045
Diffusionsschicht *f* 1002
Diffusionsthermoeffekt *m* 1069
digonale Hybride *f* 1006
Dilatanz *f* 1007
Dimerisation *f* 1010
dimolekulare Reaktion *f* 349
Dipol *m* 1098
elektrischer ~ 1098
induzierter ~ 1715
magnetischer ~ 2098
permanenter ~ 2597
Dipol-Wechselwirkung *f* 1012
1,3-dipolare Addition *f* 1011
Dipolmolekül *n* 2706
Dipolmoment *n* 1099
elektrisches ~ 1099
induziertes ~ 1716
permanentes ~ 2598
Diradikale *npl* 360
direkte Isotopenverdünnung *f* 1013
direkte Kernreaktion *f* 1014
diskontinuierliches Spektrum *n* 1017
diskontinuierliches System *n* 1016
Dismutation *f* 1029
Aldehyd-~ 467

Elektronen
δ-∼ 934
π-∼ 2667
σ-∼ 3207
gepaarte ∼ 2535
Hüllenelektronen 2480
innere ∼ 1737
Konversionselektronen 755
Leuchtelektronen 2468
lockernde ∼ 192
lokalisierte ∼ 2056
n-∼ 2388
nichtbindende ∼ 2388
nonbonding ∼ 2388
sehr langsame ∼ 3429
Sekundärelektronen 3136
Subexzitation-∼ 3429
ungepaarte ∼ 3667
Valenzelektronen 2468, 3682
Elektronen-Donator-Akzeptor-Komplex m 1045
Elektronen-Leitung f 1140
Elektronenaffinität f 1136
Elektronenakzeptor m 1135
Elektronenaustauscher mpl 1151
Elektronenband n 1154
Elektronenbeugung f 1146
Elektronendichte f 1145, 1415
Elektronendiffraktion f 1146
Elektronendonator m 1147
Elektroneneinfang m 1138
Elektronenformel f 1156
Elektronengas n 1152
entartetes ∼ 903
Elektronenisomerie f 1157
Elektronenkonfiguration f 1141
Elektronenleiter m 2192
Elektronenleptonenzahl f 1158
Elektronenmangel m 1144
Elektronenniveaus npl des Moleküls 2273
Elektronenoktett n 1176
Elektronenpaar n 1177
bindendes ∼ 398
Elektronenpaarbindung f 804
Elektronenresonanz f
paramagnetische ∼ 1178
Elektronenschale f 1164
Elektronensextett n 220
Elektronensonde-Röntgenmikroanalysator m 1175
Elektronenspektroskopie f 1166
Elektronenspektrum n 1167, 1174
Elektronenspin-Resonanz f 1178
Elektronenstrahl m 1137
Elektronentheorie f
∼ der Katalyse 1169
∼ der Metalle 1179
Elektronenübergang m 2478
Elektronenunterschale f 1168
Elektronenvolt f 1180
Elektronenwärme f 1165
Elektronenwolke f 1139
Elektronenzustandssumme f 1160
elektronische Fehlordnung f 1155
elektronische Polarisierbarkeit f 1161
Elektronpolarisation f 1162, 1163
Elektronverschiebungspolarisation f 1162
Elektroosmose f 1181
elektrophiler Mechanismus m der Addition an Kohlenstoff-Kohlenstoff-Mehrfach-bindungen 2159

elektrophiles Agens n 1182
Elektrophorese f 1183
elektrophoretischer Effekt m 1184
elektrophoretisches Potential n 3155
Elektrophotometer n 2631
Elektroselektivität f 1185
elektrostatische Bindung f 1186
Elektrovalenz f 1187
elektroviskoser Effekt m 1188
Element n 573
Brennstoffelement 1435
chemisches ∼ 573
Daniell-∼ 875
∼ 104 1189
∼ 105 1190
Konzentrationselement 698
Leclanché-∼ 1994
Mischelement 2244
Normal-Weston-∼ 3754
Normalelement 3350, 3754 (vide)
Radioelement 2902
Reinelement 3217
Sekundärelement 31
Symmetrieelement 3482
ungesättigtes Weston-∼ 3669
Elementaranalyse f 1191
elementare Arbeit f 1197
Elementarprozesse mpl der Radiolyse 1195
Elementarquantum n 2675
Elementarreaktion f 1196
Elementarteilchen npl 1194
Elementarzelle f 3659
einfache primitive ∼ 2776
Elemente npl
∼ der Lanthanreihe 1966
Übergangselemente 3605
Eliminierung f 1199
Eluat n 1200
Eluat-Volumen n 1084
Eluens n 1201
Elution f 1202
stufenweise ∼ 3384
Elutionskurve f 1203
Elutionsmittel n 1201
Emanation f
Actinium-∼ 55
Aktinium-∼ 55
Radium-∼ 2926
Thorium-∼ 3564
Emanationen fpl 1206
Emanationsmethode f 1205
Emaniervermögen n 1204
Emde-Abbau m 1207
Emde-Abbau m quartärer Ammoniumsalze 1207
Emissionsspektralanalyse f 1210
Emissionsspektroskopie f 1211
Emissionsspektrum n 1212
Empfindlichkeit f 3180
Lichtempfindlichkeit 3181, 3182
Strahlenempfindlichkeit 2915
empirische Formel f 1215
empirische Temperatur f 3507
empirische Verteilung f 1214
Emulsion f 1216
Emulsionpolymerisation f 1218
Emulsionskolloid n 1219
Emulsoid n 1219
Enantiomere npl 1221
enantiotrope Phasen fpl 1222
Enantiotropie f 1223
endo-exo-Isomerie f 1224

Fermi-Fläche f 1343
Fermi-Resonanz f 1342
Fermi-Temperatur f 1344
Ferminiveau n 1340
Fermionen npl 1341
Fermium n 1345
fernes Ultraviolett n 1327
fernes UV 1327
Fernordnung f 2063
Ferrimagnetika npl 1347
ferrimagnetische Stoffe mpl 1347
Ferrimagnetismus m 1346
Ferrite npl 1348
Ferroelektrikum n 1349
Ferromagnetika npl 1351
ferromagnetische Stoffe mpl 1351
Ferromagnetismus m 1350
feste Lösung f 3250
feste Phase f 3249
fester Körper m 3246
fester Träger m 3460
fester Zustand m 3251
festes Eutektikum n 3248
Festigkeit f
Bindungsfestigkeit 400
Festkörper m 3246
Festkörperdetektor m 3176
Feststoff-Katalysator m 3247
Feynman-Diagramme npl 1353
Feynman-Graphen mpl 1353
Film m
monomolekularer ~ 2299
Filmdosimeter n 1358
Filter m 2012
Absorptionsfilter 18
Graufilter 2356
Interferenz-Verlauffilter 1756
Interferenzfilter 1756
Lichtfilter 2012
Neutralfilter 2356
Stufenfilter 3382
Filtrat n 1359
Filtration f 1360
Filtrieren n 1360
Fischer-Hepp Nitrosamin-Umlagerung f 1368
Fischer-Hepp-Umlagerung f 1368
Fischer Indol-Synthese f 1369
Fischer-Tropsch Kohlenoxid-Druck-hydrierung f 1371
Fischer-Tropsch-Verfahren n 1371
Fischersche Projektionsformel f 1370
Fishersche F-Verteilung f 1333
Fittig-Reaktion f 1379
Fittig-Synthese f 1379
fixiertes Elektron n 3620
Fixierungspaar n 751
Fläche f
Energiefläche 1234
Fermi-~ 1343
Flüssigkeitsoberfläche (vide)
Oberfläche (vide)
Potentialfläche 2742
Flammenphotometer n 1380
Flammenphotometrie f 1381
flammenphotometrische Analyse f 1381
Flammenspektralphotometer n 1382
Flammenspektrophotometer n 1382
Flammenspektrophotometrie f 159

flammenspektrophotometrische Analyse f 159
Flash-Technik f 1384
Flockung f 1386
Flockungschwellenwert m 644
Flockungskonzentration f 644
Flockungsregel f
Hardy-Schulze-~ 3128
Flockungswert m 644
Flotation f 1387
Flüchtigkeit f 3729
Fluid n 1391
Fluidität f 1392
Superfluidität 3451
Fluidum n 1391
Fluor n 1400
Fluoreszenz f 1393
sensibilisierte ~ 3185
verzögerte ~ 928
Fluoreszenzanalyse f 1394
Fluoreszenzspektrum n 1396
Fluorgruppe f 1536
Fluorieren n 1399
Fluorierung f 1399
Fluorimeter n 1398, 3303
Fluorimetrie f 1394
fluorimetrische Analyse f 1394
Fluorometrie f 1394
Fluß m 1403
Entropiefluß 1245
Induktionsfluß 2101
magnetischer ~ 2101
Strahlungsfluß 1403
Flußdichte f 2857
Flüssig-Flüssig Extraktion f 1315
flüssige Kristalle mpl 2042
flüssige Phase f 2046
flüssiger Aggregatzustand m 2047
Flüssiger Ionenaustauscher m 2044
flüssiger Zustand m 2047
Flüssigkeit f 2039
Anodenflüssigkeit 187
Kathodenflüssigkeit 509
Supraflüssigkeit 3451
Titrierflüssigkeit 3582
Überflüssigkeit 3451
überhitzte ~ 3452
Flüssigkeit-Chromatographie f 2040
Flüssigkeiten fpl
kristalline ~ 2042
Flüssigkeitsoberfläche f
freie ~ 1420
Flüssigkeitszählrohr n 2041
Flußkonverter m 1402
Flußmittel n 1401
Flußumwandler m 1402
Folgenreaktionen fpl 734
Folgenreaktionsystem n 734
Form f
aci-~ 50
Bootsform 379
cis-~ 623, 624
cisoide und transoide ~ 3132
Enol-~ 1236
erythro-~ 1268
keto-~ 1923
meso-~ 2179
polymorphische ~ 2725
Ringform 866
Sesselform 537
threo-~ 3568
trans-~ 3603, 3604
Wannenform 379

Grenze
Erfassungsgrenze 2018
Nachweisgrenze 2018
thermodynamische ~ 3541
Grenzen *fpl*
Explosionsgrenzen 1307
Zündgrenzen 2020
Grenzfall *m*
thermodynamischer ~ 3541
Grenzfläche *f*
Phasengrenzfläche 1754
grenzflächenaktiver Stoff *m* 3463
Grenzflächenerscheinungen *fpl* 3470
Grenzflächenspannung *f* 1755
Grenzgesetz *n*
Debye-Hückel-Onsagersches ~ 887
Debye-Hückelsches ~ 889
Grenzgleichung *f*
Debye-Hückelsche ~ 889
Grenzkonzentration *f* 2016
Grenzleitfähigkeit *f* 2017
Grenzstrom *m* 2014
Grenzstrukturen *fpl* 3044
Grenzviskosität *f* 1789
Grignard-Addition *f* 1510
Grignard Organomagnesium-Addition *f*
1510
Grimmsches Gesetz *n* 1511
grobdisperses System *n* 3474
grobe Suspension *f* 3474
grober Fehler *m* 1512
große kanonische Gesamtheit *f* 1504
große Verteilungsfunktion *f* 1505
große Zustandssumme *f* 1505
Größen *fpl*
extensive ~ 1309
intensive ~ 1752
Konfigurationsgrößen 725
kritische ~ 813
partielle molare ~ 2564
Qualitätsgrößen 1752
Quantitätsgrößen 1309
reduzierte ~ 2982
Zustandsgrößen (*vide*) 3365
großes Potential *n* 1506
großes thermodynamisches Potential *n*
1506
großkanonische Gesamtheit *f* 1504
Grotthuss-Drapersches Gesetz *n* 1514
Grundband *n* 3679
Grundbaustein *m* 739
~ des Polymers 739
Grundgesamtheit *f* 2730
Grundkonstanten *fpl* 3663
Grundschwingungsband *n* 1441
Grundviskosität *f* 1789
Grundzustand *m* 1515
Gruppe *f* 167, 1516
antiauxochrome ~ 191
auxochrome ~ 280
bathmochrome ~ 319
bathochrome ~ 319
Bor-Alluminium-~ 406
Borgruppe 406
Brücken-~433
brückenbildende ~ 433
chromophore ~ 615
Eisengruppe 1352
Erdalkaligruppe 339
Fluorgruppe 1536
funktionelle ~ 1437
~ der Erdsäuren 3636
~ des Chroms 613
~ von Elementen 167, 1321, 1322
Hauptgruppe 2119

Gruppe
hypsochrome ~ 1689
Kohlenstoff-Silicium-~ 482
koordinierende ~ 760
koordinierte ~ 2006
Kupfergruppe 775
Mangangruppe 2120
Nebengruppe 3430
Nukleonengruppe 639
polare ~ 2695
Punktgruppe 2693
Raumgruppe 3272
Sauerstoff-Schwefel-~ 538
saure ~ 43
Scandiumgruppe 3115
Schutzgruppe 2796
Schwefelgruppe 3444
Stickstoff-Phosphor-~ 2383
Stickstoffgruppe 2383
Vanadingruppe 3686
Zinkgruppe 3801
Gruppenmoment *n* 1517
Gruppenreagens *n* 1518
Gruppenschwingungen *fpl* 1519
gyromagnetische Effekte *mpl* 1521
gyromagnetisches Verhältnis *n* 1522

H

H-Theorem *n* 1643
Hadronen *npl* 1523
Hafnium *n* 1524
Haftstelle *f* 3619
Haftvermögen *n* 83
Hahnsche Fällungsregel *f* 1525
halbdurchlässige Membran *f* 3179
Halbleiter *m* 3175
Defekthalbleiter 1143
Eigenhalbleiter 1788
lichtelektrischer ~ 2633
n-~ 1150
organischer ~ 2488
p-~ 1143
Halbleiterdetektor *m* 3176
Halbmetalle *npl* 2194
Halbmikro-Methode *f* 515
Halbmikroanalyse *f* 3178
halbpolare Bindung *f* 758
Halbstufenpotential *n* 1531
Halbwertsbreite *f* 1526, 1530
scheinbare ~ $\Delta v_{1/2}^s$ 204
wahre ~ $\Delta v_{1/2}$ 3635
Halbwertsdicke *f* 1529
Halbwertszeit *f* 1527, 1528
~ der Koagulation 3278
Halbzelle *f* 1109
Hall-Effekt *m* 1533
Haloformreaktion *f* 1534
Halogene *npl* 1536
Pseudo-~ 2813
Halogenierung *f* 1535
Hell-Volhard-Zelinsky α-~ 1578
Halogenokomplex *m* 1532
Halogenwanderung *f*
Orton ~ 2500
Halogenwasserstoffabspaltung *f* 926
Hamilton-Operator *m* 1537
Hammett-Gleichung *f* 1538
Hammettsche Aciditäts-Funktion *f* 46
Händigkeit *f* 596
hängende Hg-Tropfen-Elektrode *f* 1539
Hantzsch Pyridin-Synthese *f* 1540
Hardy-Schulze-Flockungsregel *f* 3128
harminischer Oszillator *m* 1543
harte Strahlung *f* 1541

Hartreesche Methode *f* des
  Selfconsistentfield 1544
Häufigkeit *f*
  relative ~ 3012
  Isotopenhäufigkeit 26
Häufigkeitsfaktor *m* 2761
Hauptgleichung *f*
  Gibbssche ~ 1488
Hauptgruppe *f* 2119
Hauptkette *f* 2117
Hauptpolarisierbarkeiten *fpl* 2777
Hauptquantenzahl *f* 2778
Hauptsatz *m*
  dritter ~ der Thermodynamik 3558
  erster ~ der Thermodynamik 1364
  nullter ~ der Thermodynamik 3799
  zweiter ~ der Thermodynamik 3144
Hauptvalenz *f* 2775
Hedval-Effekt *m* 1575
heiße Atome *npl* 1635
heiße Reaktionen *fpl* 1640
heiße Zelle *f* 1636
heißes Laboratorium *n* 1637
heißes Plasma *n* 1638
heißes Radikal *n* 1639
Helium *n* 1577
Helizität *f* 1576
Hell-Volhard-Zelinsky α-Halogenierung *f*
  1578
Helmholtz-Schicht *f* 678
Helmholtzsche Doppelschicht *f* 678
Helmholtzsche Theorie *f* der Doppelschicht
  1580
Hemmstoff *m* 1733
Hendersonsche Berechnungsmethode *f* des
  Diffusionspotentials 1582
Hendersonsche Gleichung *f* 1581
Henrysches Gesetz *n* 1583
Herman-Mauguin-Symbole *npl* 1584
Hesssches Gesetz *n* 1585
Heteroatom *n* 1586
Heteroazeotrop *n* 1587
heteroazeotrope Mischung *f* 1587
heteroazeotroper Punkt *m* 1588
heteroazeotropes System *n* 1589
heterocyclischer Ring *m* 1590
heterodisperses System *n* 2717
heterogene Katalyse *f* 1592
heterogener Isotopenaustausch *m* 1593
heterogenes System *n* 1594
Heterolyse *f* 1595
heterolytische Dissoziation *f* 1595
heterometrische Titration *f* 1596
heteronukleares zweiatomiges Molekül *n*
  1597
heteropolare Bindung *f* 1811
Heteropolymerisation *f* 1599
Heteropolysäure *f* 1598
heterozeotropes System *n* 1600
Hevesy-Paneth-Methode *f* 2540
Hilfselektrode *f* 3461
Hinderung *f*
  sterische ~ 3397, 3398
Hittorfsche Bestimmungsmethode *f* der
  Überführungszahlen 1606
hochangeregtes Molekül *n* 3450
hochelastischer Zustand *m* 1601
Hochfrequenzspektroskopie *f* 2918

Hochfrequenztitration *f* 2505
höchstzulässige Dosis *f* 2140
Hoesch-Houben Phenolketon-Synthese *f*
  1642
Hoesch-Synthese *f* 1642
Hofmann-Abbau *m* 1607
Hofmann-Abbau *m* quartärer
  Ammoniumhydroxide 1607
Hofmann Carbonsäure-Amid →
  → Amin-Abbau *m*
Hofmann-Martius N-Alkylanilin →
  → C-Alkylanilin-Umlagerung *f* 1608
Hofmann-Martius-Umlagerung *f* 1608
Hofmann-Regel *f* 1610
Hofmeistersche Reihen *fpl* 1611
Höhe *f* des effektiven Bodens 1080
Hohlkathodenlampe *f* 1615
Holmium *n* 1616
Homoazeotrop *n* 1617
homoazeotropes System *n* 1618
homocyclischer Ring *m* 1619
homodisperses System *n* 2295
homogene Katalyse *f* 1620
homogener Isotopenaustausch *m* 1622
homogenes System *n* 1624
Homogenisation *f* der Emulsionen 1625
Homogenisierung *f* der Emulsionen 1625
Homologe *npl* 1628
homologe Linien *fpl* 1626
homologe Reihe *f* 1627
Homologen *npl* 1628
Homologie *f* 1629
Homolyse *f* 1630
homolytische Dissoziation *f* 1630
homonukleares zweiatomiges Molekül 1631
homöopolare Bindung *f* 804
Homopolymer *n* 1632
Homopolymerisat *n* 1632
Homopolymerisation *f* 1633
Houben-Fischer Nitril-Synthese *f* 1641
Hückel-Regel *f* 1645
Hückelsche Gleichung *f* 1644
Hugoniot-Gleichung *f* 1647
Hugoniot-Relation *f* 1647
Hüllenelektronen *npl* 2480
Hume-Rothery Phasen *fpl* 1648
Hundsche Regeln *fpl* 1649
Hunsdiecker-Borodin Silbersalz-
  -Decarboxylierung *f* 1650
Hunsdiecker-Reaktion *f* 1650
Hybride *f*
  digonale ~ 1006
  sp ~ 1006
  te-~ 3515
  tetraedrische ~ 3515
  tr-~ 3623
  trigonale ~ 3623
Hybridisierung *f* 1651
Hybridorbital *n* 1652
  tetraedrisches ~ 3515
  trigonales ~ 3623
Hydratation *f* 1654, 1655
  ~ des Ions 1658
Hydratationswärme *f* 1563
Hydratationszahl *f* des Ions 1657
hydratisiertes Elektron *n* 1653
Hydratisierung *f* 1654
  Kutscheroff Acetylen-~ 1951

Hydratisomerie f 1656
Hydrierung f 1665
Birch-Hückel-~ 361
destruktive ~ 1672
hydroaromatischer Ring m 1659
Hydroborierung f 1660
hydrodynamisches Stadium n 1661
Hydroformylierung f 1662
Hydrogel n 1663
Hydrogenation f 1665
Hydrogenolyse f 1672
Hydrogenolysis f 1672
Hydrolyse f 1673, 1674
Hydrolysengrad m 917
Hydrolysenkonstante f 1675
hydrolytische Adsorption f 1676
Hydrosol n 1677
Hydroxamsäure → Isocyanat-Abbau m 2066
Hydroxokomplex m 1678
Hydroxylierung f 1679
Hydroxymethylierung f 1680
Hyperfeinstruktur f 1685, 1686
Hyperfeinstruktur-Aufspaltung f 1684
Hyperfeinstrukturwechselwirkung f 1683
Hyperkonjugation f 1682
Hyperladung f 1681
Hyperon n
Λ-~ 1960
Ξ-~ 3771
Σ-~ 3208
Ω-~ 2456
Hyperons npl 1687
hypochromer Effekt m 1688
Hypothese f
Nullhypothese 2440
hypsochrome Absorptionsverschiebung f 1690
hypsochrome Gruppe f 1689
hypsometrische Formel f 310
Hysterese f 1691
Adsorptionshysterese 96
magnetische ~ 2102
Hysteresis f 1691

I

I-Spannung f 1781
ideal verdünnte Lösung f 1692
ideale Gasgesetze npl 2588
ideale Mischung f 1695
idealer Kristall m 2586
ideales Gas n 1693
ideales Gasgemisch n 2587
Idealkristall m 2586
Identifizierung f 1696
Identitätsreaktion f 1697
Ilkowič Gleichung f 1700
Impulsmethode f
coulostatische ~ 791
Impulspolarographie f 3224
Impulsreaktor m 2815
Impulssatz m 2291
Index n
Brechungsindex 9
Indexe mpl
kritische ~ 807
indifferentes Lösungsmittel n 208
Indikationsfehler m 3585

Indikator m 1708
Adsorptionsindikator 97
Außen-~ 1311
äußerer ~ 1311
Innen-~ 1772
innerer ~ 1772
isotoper ~ 1893
pH-~ 40
radioaktiver ~ 2889
Säure-Basen-~ 40
Universalindikator 3665
Indikatorelektrode f 1709
Indikatorfehler m 1710
Indikatormethode f 1899
radioaktive ~ 2890
Indikatorpapier n 1711
Universalindikatorpapier 3666
Indium n 1712
individuelle Bildungskonstante f 735
individuelle Dissoziationskonstante f 733
individuelle Dissoziationskonstante f des Komplexes 733
individuelle Stabilitätskonstante f 735
Individuum n 588
chemisches ~ 3434
Indizes mpl
dynamische ~ chemischer Reaktionsfähigkeit 1071
~ chemischer Rekationsfähigkeit 2963
kristallographische ~ 2235
Millersche ~ 2235
statische ~ chemischer Reaktionsfähigkeit 3369
Indopheninreaktion f 1713
Induktion f
chemische ~ 578
Kerninduktion 2415
magnetische ~ 2103
Induktionseffekt m 1720, 1723
Induktionsfaktor m 1721
Induktionsfluß m 2101
Induktionsperiode f 1722
Induktor m 1724
~ der induzierten Reaktion 1724
induzierte Aktivität f 1714
induzierte Polarisation f 1717
induzierte Radioaktivität f 1718
induzierte Reaktion f 1719
induzierte Reaktionen fpl 800
induzierter Dipol m 1715
induziertes Dipolmoment n 1716
inerter Komplex m 1726
infinitesimale Zustandsänderung f 1730
Infrarot n 1731. 2232
kurzwelliges ~ 2336
langwelliges ~ 1326
nahes ~ 2336
Infrarot-Spektrum n 1732
Inhibitor m 1733
~ der Kettenreaktion 1734
Initiator m der Kettenreaktion 529
inkohärente Streuung f 1703
Innen-Indikator m 1772
Innenwiderstand m des Elements 1775
innere Adsorption f 1766
innere Elektrolyse f 1770
innere Elektronen npl 1737
innere Energie f 1771
innere Konversion f 1767
innere Reibung f 3720
innere Rückkehr f 1776
innere Spannung f 1781
innere Strahlungsquelle f 1777
innere Umwandlung f 1768

innere Zustandssumme *f* 1773
innerer Druck *m* 1774
innerer Indikator *m* 1772
innerer orbitaler Komplex *m* 1738
innerer Photoeffekt *m* 2632
innerer Standard *m* 1778, 1779
inneres elektrisches Potential *n* 1736
inneres Feld *n* 2051
inneres Gleichgewicht *n* 3537
inneres Komplexsalz *n* 2355
instabiler Zustand *m* 3672
instabiles chemisches Gleichgewicht *n* 1320
Instrumentalanalyse *f* 1744
Instrumentenanalyse *f* 1744
Integral *n*
Konfigurationsintegral 2572
Stoßintegral 659
Integraldetektor *m* 1748
Integraldosis *f* 1749
integrale Adsorptionswärme *f* 1750
integrale Kapazität *f* der Doppelschicht 1747
integrale Lösungswärme *f* 1568
integrale Quellungswärme *f* 1751
integrale Verdünnungswärme *f* 1557
integraler Wirkungsquerschnitt *m* 3589
Intensität *f*
Magnetisierungsintesität 2112
intensive Eigenschaften *fpl* 1752
intensive Größen *fpl* 1752
intensive Parameter *mpl* 1752
intensive Zustandsvariablen *fpl* 1752
Interferenz-Ordnungszahl *f* 2484
Interferenz-Verlauffilter *m* 1756
Interferenzfilter *m* 1756
Interferenzspektroskop *n* von Fabry-Perot 1319
Interferometer *n* 2470
optisches ∼ 2470
Interferometrie *f* 1759
interferometrische Analyse *f* 1759
Interkombination *f*
strahlungslose ∼ 1784
Interkombinationsübergang *m* 1753
intermetallische Phasen *fpl* 1761
intermolekulare Wasserstoffbindung *f* 1764
intermolekulare Wasserstoffbrücke *f* 1764
intermolekulare
    Wasserstoffbrückenbindung *f* 1764
internationale Kalorie *f* 1901
Intervall *n*
Konfidenzintervall 720
Vertrauensintervall 720
intramolekulare Reaktion *f* 1786
intramolekulare Wasserstoffbrücke *f* 1785
invariantes Phasensystem *n* 1790
invariantes System *n* 1790
inverse Polarographie *f* 3419
Inversion *f* 1791
∼ der Konfiguration 1794
Inversionsachse *f* 1792
Inversionsdrehachse *f* 1792
Inversionslinie *f* 1793
Inversionspunkt *m* 1796
Inversionsspektrum *n* 1795
Inversionstemperatur *f* 1796
Ion *n* 1799
Carbeniat-∼ 474
Carbeniumion 477

Ion
Carboniumion 477
Komplexion 686
Lyat-∼ 2079
Lyonium-∼ 2081
Phenonium-∼ 2620
primäres ∼ 2554
Radikalion 1841
wasserstoffähnliches ∼ 1670
Zentralion 513
Zwitterion 155
Ion-Molekül-Reaktion *f* 1838
Ionen *mpl*
Coionen 655
Gegenionen 797
nichtklassische ∼ 2390
Ionen-Paar *n* 1800
Ionen-Wertigkeit *f* 1825
Ionenaktivitätskoeffizient *m* 73
mittlerer ∼ 2147
Ionenäquivalentleitfähigkeit *f* 1812
Ionenassoziationskomplex *m* 1800
Ionenatmosphäre *f* 1810
Ionenausbeute *f* 1826
Ionenaustauschchromatographie *f* 1803
Ionenaustauscher *m* 1806
amphoterer ∼ 156
anorganischer ∼ 1741
flüssiger ∼ 2044
Kunstharz-∼ 1807
makro-retikularer ∼ 2088
spezifischer ∼ 3281
Ionenaustauscher-Membrane *f* 1808
Ionenaustauscherharz *n* 1807
Ionenaustauschersäule *f* 1804
Ionenbeweglichkeit *f* 1815
Ionenbeziehung *f* 1811
Ionenbindung *f* 1811
Ionencluster *n* 1801
Ionendosis *f* 1308
Ionendurchmesser *m*
mittlerer ∼ 2151
Ionenentladung *f* 1015
Ionenkonzentration *f*
mittlere ∼ 2150
Ionenkristalle *mpl* 1814
Ionenladungszahl *f* 1825
Ionenleiter *m* 1813
Ionenleitfähigkeit *f* 1812
Ionenlinie *f* 1837
Ionenpaar *n* 1800, 1840
Ionenprodukt *n* 1817
∼ des Lösungsmittels 1818
∼ des Wassers 1819
Ionenradius *m* 1820
Ionenreaktion *f* 1821
Ionenrefraktion *f* 1822
ionensensitive Elektrode *f* 1843
Ionensiebe *npl* 1823
Ionenspektrum *n* 1844
Ionenstärke *f* 1824
Ionenstärke-Gesetz *n* von Lewis und Randall 2002
Ionenverzögerungs-Verfahren *n* 1842
Ionenwanderung *f*
∼ im Feld 2234
Ionenwolke *f* 1810
Ionisation *f* 1828
Autoionisation 274
Photoionisation 2640
Präionisation 274
Primärionisation 2769
Sekundärionisation 3137
spezifische ∼ 3292

Ionisationsdetektor *m* 1831
Ionisationsdichte *f* 1830
Ionisationsisomerie *f* 1833
Ionisationskammer *f* 1829
Ionisationspotential *n* 1834
ionische Polymerisation *f* 1816
ionisierende Strahlenpartikel *f* 1835
ionisierende Strahlung *f* 1836
Ionisierung *f* 1828
Ionisierungsenergie *f* 1832
Ionisierungsspannung *f* 1834
Ionium *n* 1827
ionogene Bindung *f* 1811
Ionophorese *f* 1839
Ionradikal *n* 1841
Iridium *n* 1845
irreversible Koagulation *f* 1852
irreversible Reaktion *f* 1855
irreversibler Prozeß *m* 1854
irreversibler Vorgang *m* 1854
irreversibles Kolloid *n* 1853
Irving-Williams-Reihe *f* 1856
isentrope Zustandsänderung *f* 1857
Ising-Modell *n* 1858
iso-elektronische Moleküle *npl* 1868
Isobare *f* 1860
van't Hoffsche ~ 3694
Isobare *npl* 1862
isobare Zustandsänderung *f* 1861
Isobarenspin *m* 1880
isobarer Prozeß *m* 1861
Isochore *f* 1863
isochorer Prozeß *m* 1864
Isodimorphie *f* 1865
isodisperses System *n* 2295
Isodosis *f* 1866
isoelektrischer Punkt *m* 1867
isoelektrisches Prinzip *n* 1869
Isolator *m* 1746
isolierte Doppelbindungen *fpl* 1870
isoliertes System *n* 1871
Isomer *n*
cis-~ 623
trans-~ 3603
Isomere *npl* 1876
Kernisomere 2417
Konformationsisomere 729
Konstellationsisomere 729
Spiegelbildisomere 1221
Isomerenübergang *m* 1873
isomerer Kern *m* 2199
Isomerie *f* 1874
Atrop-Isomerie 264
Bindungsisomerie 2034
cis-trans-~ 625, 626
Elektronenisomerie 1157
endo-exo-~ 1224
geometrische ~ 625, 626
Hydratisomerie 1656
Ionisationsisomerie 1833
Kernisomerie 2416
Kettenisomerie 530
Koordinationsisomerie 764
optische ~ 2471
Polymerisationsisomerie 2723
Raumisomerie 3388
Salzisomerie 2034
Spiegelbildisomerie 2471
Stellungsisomerie 2731 (vide)
Stereoisomerie 3388
Strukturisomerie 3425
syn-anti-~ 3485

Isomerisation *f* 1875
Stevens Ylid-Amin-~ 3401
Ylid-Amin-~ 3401
Isomerisierung *f*
cis-trans-~ 627
Photoisomerisierung 2641
Isomorphie *f* 1877
Isomorphismus *m* 1877
Isonitrilreaktion *f* 488
Isopolysäure *f* 1878
Isorotations-Regel *f* von Hudson 1646
isosbestischer Punkt *m* 1879
Isospin *m* 1880
Isospin-Multipletts *npl* 553
Isostere *f*
Adsorptionsisostere 98
isostrukturelle Verbindungen *fpl* 1882
isotaktisches Polymer *n* 1883
Isotherme *f* 1884
Adsorptionsisotherme 99 (vide)
Austauschisotherme 1805
~ von Freundlich 1428
kritische ~ 808
Langmuirsche ~ 1965
lineare ~ 2025
Reaktionsisotherme (vide)
van der Waalssche ~ 3690
Verteilungsisotherme 1040
isotherme Zustandsänderung *f* 1886
isothermer Vorgang *m* 1886
isothermes Kalorimeter *n* 1885
Isotonen *npl* 1887
Isotop *n*
kurzlebiges ~ 3199
langlebiges ~ 2062
Leitisotop 1898 (vide)
Sr-~ 90
Isotope *npl* 1891
isotope Strahlenquelle *f* 2887
Isotopen-Verdünnungsanalyse *f* 1888
Isotopenanalyse *f* 1892
Isotopenanreicherung *f* 1895
Isotopenaustausch *m* 1896
heterogener ~ 1593
homogener ~ 1622
Isotopeneffekt *m*
kinetischer ~ 1889
Isotopeneffekte *mpl* 1890
Isotopenhäufigkeit *f* 26
Isotopenspin *m* 1880
Isotopenverdünnung *f* 1888
direkte ~ 1013
einfache ~ 1013
substöchiometrische ~ 3438
umgekehrte ~ 3050
isotoper Indikator *m* 1898
isotoper Träger *m* 1893
isotoper Wirkungsquerschnitt *m* 1894
isotopischer Träger *m* 1893
Isotropie *f* 1900
isotype Verbindungen *fpl* 1882
Ivanoff Hydroxycarbonsäure-Synthese *f* 1902
Iwanow-Reaktion *f* 1902

J

Jacobsen Alkylwanderung *f* 1903
Jacobsen-Reaktion *f* 1903
Jahn-Teller-Effekt *m* 1904, 1905
Jahn-Teller-Kopplung *f* 1904
Jahn-Teller-Theorem *n* 1906

Kerninduktion *f* 2415
Kernisomere *npl* 2417
Kernisomerie *f* 2616
Kernkräfte *fpl* 2411
kernmagnetische Resonanz *f* 2419
kernmagnetische Resonanzspektroskopie
  f 2421
kernmagnetisches Resonanz-Spektrometer
  n 2420
kernmagnetisches Resonanzspektrum
  n 2422
Kernmagneton *n* 2423
Kernmodell *n*
optisches ∼ 2472
Kernmodelle *npl*
kollektive ∼ 657
Kernmoment *n*
magnetisches ∼ 2418
Kernpotential *n* 2424
Kernquadrupolmoment *n* 2426
Kernquadrupolresonanz *f* 2427
Kernreaktion *f* 2430
direkte ∼ 1014
störende ∼ 1757
Kernreaktionsenergie *f* 1232
Kernreaktionsenergiebilanz *f* 2432
Kernreaktionskanal *m* 2431
Kernreaktorkühlmittel *n* 2433
Kernreinheit *f* 2425, 2885
Kernspaltung *f* 2410
Kernspaltungskettenreaktion *f* 1373
Kernspektroskopie *f* 2434
Kernspin *m* 2435
Kernstrahlung *f* 2428
Kernstrahlungsspektrum *m* 2429
Kerr-Effekt *m* 1921
Kerr-Konstante *f* 1920
Keto-Enol-Tautomerie *f* 1922
Keto-Form *f* 1923
Ketonspaltung *f* 1924
Kette *f*
chemische ∼ 568
galvanische ∼ 1445
gerade ∼ 2398
geradelinige ∼ 2398
Hauptkette 2117
Kohlenstoffkette 481
Konzentrationskette 698
lineare ∼ 2398
normale ∼ 2398
Reaktionskette 533
reversible ∼ 3051
Seitenkette 3201
Stammkette 2117
verzweigte ∼ 1407
Kettenabbruch *m* 536
Kettenabbruchreaktion *f* 536
Kettenentwicklung *f* 534
Kettenisomerie *f* 530
Kettenlänge *f* 531, 532
∼ des Polymers 532
Kettenreaktion *f* 535
unverzweigte ∼ 3409
verzweigte ∼ 424
Kettenreaktionsträger *m* 526
Kettenstart *m* 528
Kettenstartreaktion *f* 528
Kettenverzweigung *f* 525
Kiliani-Fischer Cyanhydrin-Synthese
  f 1925

kinematische Viskosität *f* 1926
Kinetik *f*
chemische ∼ 579
Reaktionskinetik 579
kinetische Einzelordnung *f* 2486
kinetische Gastheorie *f* 1933
kinetische Reaktionsordnung *f* 2485
kinetische Spektroskopie *f* 1931
kinetischer Isotopeneffekt *m* 1889
kinetischer Strom *m* 1927
kinetisches Gebiet *n* des Prozesses 1930
kinetisches Stadium *n* 1932
Kirchhoffsche Gleichung *f* 1934
Kirchhoffsches Gesetz *n* 1934, 1935
Kishner-Wolff Carbonyl→Methylen-
  -Reduktion *f* 3764
Klasse *f*
Symmetrie-∼ 3431
Klassen *fpl*
Kristallklassen 830 (*vide*)
klassische Dissoziationskonstante *f* des
  Elektrolyts 1123
klassische Thermodynamik *f* 632
Knight-Shift 1937
Knight-Verschiebung *f* 1937
Knock-out-Reaktion *f* 1938
Knoevenagel Aldol-Kondensation *f* 1939
Knoevenagel Crotonisierung *f* 1939
Knoevenagel-Reaktion *f* 1939
Knorr Pyrrol-Synthese *f* 1940
Koagel *n* 642
Koagulation *f* 643
irreversible ∼ 1852
langsame ∼ 3231
orthokinetische ∼ 2499
perikinetische ∼ 2589
reversible ∼ 3052
schnelle ∼ 2948
Koaleszenz *f* 645
Koazervation *f* 641
Kobalt *n* 646
∼-60 2900
Kobalteinheit *f*
Koeffizient *m*
Absorptionskoeffizient 14 (*vide*) 25
Adsorptionskoeffizient 93
Aktivitätskoeffizient 72, 3730
Ausdehnungskoeffizient 650 (*vide*)
Brechungskoeffizient 9
Diffusionskoeffizient 1001
Extinktionskoeffizient (*vide*)
Ionenaktivitätskoeffizient 73 (*vide*)
Joule-Thomsonscher-∼ 1910
Katalysekoeffizient 502
∼ der inneren Konversion 1769
Kompressibilitätskoeffizient 649
Konfidenzkoeffizient 721
Kristallisationskoeffizient 1621
osmotischer ∼ 2510
Schwächungskoeffizient (*vide*)
Selbstdiffusionskoeffizient 3171
Selektivitätskoeffizient 3164 (*vide*)
Spannungskoeffizient 2765 (*vide*)
stöchiometrischer ∼ 3403
Temperaturkoeffizient (*vide*)
Transmissionskoeffizient 3615
Verteilungskoeffizient 1039
Virialkoeffizient (*vide*)
Viskositätskoeffizient 3721
Koeffizienten *mpl*
phänomenologische ∼ 2618
Transportkoeffizienten 3617
Virialkoeffizienten 3715
Koenigs-Knorr β-Glykosidierung *f* 1941

Kraft
elektromotorische Standard-~ 3356
Kräfte *fpl* 3540
Adsorptionskräfte 95
Coulomb-~ 785
Dispersionskräfte 1024
generalisierte ~ 3540
Kernkräfte 2411
treibende ~ 3540
van der Waals-~ 3689
van der Waalssche ~ 3689
zwischenmolekulare ~ 1763
Kramersches Theorem *n* 1949
Kreiselmolekül *n*
asymmetrisches ~ 237
Kugelkreiselmolekül 3319
symmetrisches ~ 3478
Kreisprozeß *m* 865
Born-Haberscher ~ 402
Carnotscher ~ 489
reversibler ~ 3054
Kreuzfeuermethode *f* 818
Kriewitz-Prins Olefin-Formaldehyd-
   -Addition *f* 2784
Kristall *m* 827
Einkristall 3222
idealer ~ 2586
Idealkristall 2586
Kristall-Lumineszenz *f* 845
Kristallachsen *fpl* 828
Kristallbaufehler *mpl* 1974
Kristallchemie *f* 829
Kristalle *mpl*
Atomkristalle 245
flüssige ~ 2042
Ionenkristalle 1814
~ mit Wasserstoffbrücken 1668
Mischkristalle 3250 (*vide*)
Molekularkristalle 2271
Molekülkristalle 2271
Mosaikkristalle 2305
Zonenkristalle 3804
Kristallfeld *n* 831
Kristallfeldstabilisierungsenergie *f* 833
Kristallfeldtheorie *f* 834
Kristallgitter *n* 835
kristallin-flüssiger Zustand *m* 2183
kristalline Flüssigkeiten *fpl* 2042
kristalliner Zustand *m* 837
kristallines Aggregat *n* 2716
kristallines Polymer *n* 836
kristallinischer Zustand *m* 837
Kristallisation *f* 839
~ aus der Gasphase 840
Kristallisationskoeffizient *m* 1621
Kristallisationspolarisation *f* 841
Kristallisationswärme *f* 1556
Kristallit *m* 838
Kristallklassen *fpl* 830
natürliche ~ 830
kristallographische Achsen *fpl* 828
kristallographische Indizes *mpl* 2235
Kristalloid *n* 844
Kristallphosphoren *npl* 846
Kristallstruktur *f* 847
Kristallsysteme *npl* 843
Kristallzone *f* 848
Kriterium *n*
magnetisches ~ 2097
kritische Daten *pl* 813
kritische Größen *fpl* 813
kritische Indexe *mpl* 807
kritische Isotherme *f* 808
kritische Lösungstemperatur *f* 814

kritische Masse *f* 809
kritische Mischungstemperatur *f* 814
kritische Opaleszenz *f* 810
kritische Phase *f* 811
kritische Temperatur *f* 816
kritischer Druck *m* 812
kritischer Komplex *m* 58
kritischer Zustand *m* 815
kritisches Volumen *n* 817
Krone-Konformation *f* 821
Kryohydrate *n* 822
kryohydratischer Punkt *m* 823
Kryohydratpunkt *m* 823
Kryometrie *f* 824
Kryoskopie *f* 824
kryoskopische Konstante *f* 825
Kryosol *n* 826
Krypton *n* 1950
Kryptonat *n*
radioaktives ~ 2884
Kugelkreiselmolekül *n* 3319
kumulierte Doppelbindungen *fpl* 849
Kunstharz-Ionenaustauscher *m* 1807
künstliche Radioaktivität *f* 1718
Kupellation *f* 850
Kupfer *n* 774
Kupfergruppe *f* 775
Kupplung *f* 801
Kurve *f* 417
Absorptionskurve 16, 17
Bildungskurve 1408
Braggsche ~ 417
Dispersionskurve 1022
Durchbruchskurve 428
Eichkurve 163
Elektrokapillarkurve 1100
Elutionskurve 1203
Energiekurve (*vide*)
Liquiduskurve 2048
Morse-~ 2304
Phasenkurve 2616
Potential-Zeitkurve 2744
Potentialkurve 2741
Schwärzungskurve 1217
Soliduskurve 3252
Stromspannungskurve (*vide*)
Titrationskurve 3584
kurzlebiges Isotop *n* 3199
kurzwelliges IR 2336
Kurzzeitspektroskopie *f* 3576
Kutscheroff Acetylen-Hydratisierung *f*
   1951
Küvette *f* 13
Absorptionsküvette 13

L

labiler Komplex *m* 1955
labiles chemisches Gleichgewicht *n* 1320
Laboratorium *n*
heißes ~ 1637
Ladung *f*
Hyperladung 1681
Raumladung 3270
Ladungsaustauschband *n* 554
Ladungsdichte *f* 550
Ladungsträger *m* 1097
Ladungsübertragungskomplex *m* 1045
Ladungsverteilung *f* 549
Ladungswert *m* 2522
Ladungswolke *f* 549

Laktam-Laktim Tautomerie f 142
Lambert-Beersches Gesetz n 1961
Lampe f
Hohlkathodenlampe 1615
Landé-Faktor m 1962
Landéscher Aufspaltungsfaktor m 1962
Länge f
Bindungslänge 396
Kettenlänge 531, 532 (vide)
Wellenlänge (vide)
Langevin-Theorie f 1963
Langevinsche Theorie f 1963
langlebiges Isotop n 2062
Langmuirsche Adsorptionsisotherme f 1965
Langmuirsche Gleichung f 1964
Langmuirsche Isotherme f 1965
langsame Koagulation f 3231
langsame Neutronen npl 3233
langwelliges Infrarot n 1326
Lanthan n 1967
Lanthanide npl 1966
Lanthanoide npl 1966
Larmor-Frequenz f 1968
Larmor-Präzession f 1969
latente Oberflächenwärme f 1971
latente Wärme f 1553
latentes photographisches Bild n 1972
Latimersche Gleichung f 1973
Laue-Aufnahme f 1977
Laue-Diagramm n 1977
Laue-Photogramm n 1977
laufendes Gleichgewicht n 3602
laufendes radioaktives Gleichgewicht n 3602
Laves-Phasen fpl 1978
Lawrencium n 1988
Lawrentium n 1988
LCAO-Methode f 1989
Le Châtelier-Braunsches Prinzip n 1993
Lebensdauer f
mittlere ~ 2152, 2153, 2446
Leclanché-Element n 1994
Leerstelle f 3677
Leerversuch m 373
Legierung f 124
leichte Platinmetalle npl 2013
leichter Wasserstoff m 2802
Leiter m 719
Defektleiter 1143
elektrolytischer ~ 1813
Elektronenleiter 2192
Halbleiter 3175 (vide)
Ionenleiter 1813
metallischer ~ 2192
Photoleiter 2633
Störstellenhalbleiter 1317
Überschußleiter 1150
Leiterpolymer m 1958
Leitfähigkeit f
Grenzleitfähigkeit 2017
Ionenäquivalentleitfähigkeit 1812
Ionenleitfähigkeit 1812
molare ~ 2257
spezifische ~ 3279
Supraleitfähigkeit 3447
Leitfähigkeitsband n 715
Leitfähigkeitsdispersion f 1027
Leitfähigkeitsmesser m 716
Leitfähigkeitsmessungen f 718
Leitfähigkeitstitration f 717

Leitisotop n 1898
radioaktives ~ 2889
Leitisotopenmethode f 1899
Leitung f
Defektelektronen-~ 1614
Elektronen-~ 1140
n-~ 1140
Photoleitung 2632
Supraleitung 3447
Wärmeleitung 1550
Leitungsband n 715
Leitungsfaktor m 2752
lenkende Wirkung f des Katalysators 3283
Lennard-Jones-Potential n 1995
µ-Lepton n 2313
Leptonen npl 1998
Leptonenzahl f 1996
Elektronenleptonenzahl 1158
Lethaldosis f 1999
mittlere ~ 2168
letzte Linien fpl 2932
letzte Lösungswärme f 1970
Leuchtdichte f 2072
Leuchtelektronen npl 2468
Leuchtstoffe mpl 2076
Leuckart Amin-Alkylierung f 2000
Leuckart-Reaktion f 2000
Leuckart reduktive Carbonyl-Aminierung f 2000
Leuckart-Wallach-Reaktion f 2000
Lewis-Sargentsche Formel f 2003
Lewis-Sargentsche Gleichung f 2003
Lewissche Säuren-Basen-Theorie f 2004
Licht n
polarisiertes ~ 2704
Lichtabsorption f 20
Lichtbeugung f 2010
Lichtblitzphotolyse f 1383
Lichtblitzspektroskopie f 1384
Lichtbrechung f 2997
lichtelektrischer Halbleiter m 2633
Lichtempfindlichkeit f 3181, 3182
Lichtfilter m 2012
Lichtquant n 2646
Lichtquelle f 1298
Lichtstärke f 2077
Liesegangsche Ringe mpl 2005
Ligand m 2006
Chelatligand 563
einzähniger ~ 3657
mehrwertiger ~ 2317
mehrzähniger ~ 2317
Ligandatom n 759
Ligandenfeld n 2007
schwaches ~ 3747
starkes ~ 3422
Ligandenfeldtheorie f 2008
Ligandenzahl f
durchschnittliche ~ 2009
Linearbeschleuniger m von Elektronen 2023
lineare Dispersion f 2022
lineare Energieübertragung f 2024
lineare Isotherme f 2025
lineare Kette f 2398
lineare Regression f 2028
linearer Schwächungskoeffizient m 2021
lineares Bremsvermögen n 2029
lineares Molekül n 2026
lineares Polymer n 2027

Linie *f*
Analysenlinie 168
Antistokessche ~ 202
Atomlinie 263
Bogenlinie 263
ein-Quant-Escape-~ 3223
Einfachlinie 3225
Funkenlinie 1837
Inversionslinie 1793
Ionenlinie 1837
Phasengleichgewichtslinie 2613
Phasenlinie 2616
Photolinie 2649
Raman-~ 2935
Rayleigh-~ 2950
Resonanzlinie 3042
Soliduslinie 3252
Spektrallinie 3294
Stokessche ~ 3407
zwei-Quanten-Escape-~ 1060
Linien *fpl*
falsche ~ 3293
homologe ~ 1626
letzte ~ 2932
Paarlinien 2537
Linienbreite *f* 2033
natürliche ~ 2332
Liniendefekte *mpl* 2030
Linienkoinzidenz *f* 2031
Linienpaar *n* 2032
Linksdrehung *f* 1959
Liouville-Gleichung *f* 2035
Liouvillescher Satz *m* 2036
Lippmann-Gleichung *f* 2037
Liquiduskurve *f* 2048
liquokristalliner Zustand *m* 2183
Lithium *n* 2049
lithiumgedrifteter Germaniumdetektor *m* 2050
lockernde Elektronen *npl* 192
logarithmische Normalverteilung *f* 2060
logarithmisches Verteilungsgesetz *n* 2059
lokale Entropieerzeugung *f* 2052
lakales elektrisches Feld *n* 2051
lokales Feld *n* 2051
lokales Gleichgewicht *n* 2053
lokales Potential *n* 2058
lokalisierte Adsorption *f* 2055
lokalisierte Bindung *f* 2057
lokalisierte Elektronen *npl* 2056
lokalisiertes Orbital *n* 2057
Lokalisierungsenergie *f* 2054
longitudinale Relaxation *f* 3325
Lorenz-Feld *n* 2065
Löschung *f* der Lumineszenz 2847
Löslichkeit *f* 3253
Löslichkeitsbild *n* 3254
Löslichkeitsdiagramm *n* 3254
Löslichkeitskonstante *f* 3255
Löslichkeitsprodukt *n* 3255
thermodynamisches ~ 3255
Lossen Hydroxamsäure→Isocyanat-Abbau *m* 2066
Lossen-Reaktion *f* 2066
Lösung *f* 3257
Anodenlösung 187
Bezugslösung 2993
feste ~ 3250
gesättigte ~ 3107
ideal verdünnte ~ 1692
Kathodenlösung 509
Maßlösung 3582
molekulare ~ 3636
nichtideale ~ 2972

Lösung
Pufferlösung 443
Reagenzlösung 3582
reale ~ 2972
Standardlösung 3359
übersättigte ~ 3456
Vergleichslösung 680, 2993
wahre ~ 3636
Lösungsdruck *m*
elektrolytischer ~ 1125
Lösungsmittel *n*
Akzeptorlösungsmittel 2808
ampholytisches ~ 153
aprotisches ~ 208
basisches ~ 2808
differenzierendes ~ 996
Donorlösungsmittel 2603
indifferentes ~ 208
nivellierendes ~ 2001
saures ~ 2803
Lösungstemperatur *f*
kritische ~ 814
Lösungswärme *f* 1568
differentielle ~ 2559
erste ~ 1363
ganze ~ 3590
integrale ~ 1568
letzte ~ 1970
Ludwig-Soret-Effekt *m* 3525
Luggin-Kapillare *f* 2071
Lumineszenz *f* 2073
Biolumineszenz 357
Chemilumineszenz 591
Kristall-~ 845
Photolumineszenz 2642
Radiolumineszenz 2907
Radiophotolumineszenz 2907
Radiothermolumineszenz 2921
Thermolumineszenz 3548
Lumineszenzanalyse *f* 2074
Lumineszenzdosimeter *n* 2075
Luminophore *mpl* 2076
Lutetium *n* 2078
Lyat-Ion *m* 2079
Lyogel *n* 2080
Lyonium-Ion *m* 2081
lyophiles Kolloid *n* 2082
lyophobes Kolloid *n* 2083
lyotrope Reihen *fpl* 1611

M

Madelungsche Konstante *f* 2092
magische Nukleonenzahlen *fpl* 2094
magische Zahlen *fpl* 2094
magischer Kern *m* 2093
Magnesium *n* 2095
Magnetfeld *n* 2099
magnetische Anisotropie *f* 2096
magnetische Feldstärke *f* 2100
magnetische Hysterese *f* 2102
magnetische Induktion *f* 2103
magnetische Permeabilität *f* 2106
magnetische Quantenzahl *f* 2107
magnetische Resonanz *f* 2108
magnetische Spinquantenzahl *f* 2109
magnetische Struktur *f* 2110
magnetische Suszeptibilität *f* 2111
magnetische Waage *f* nach Gouy 1495
magnetischer anomaler Komplex *m* 2069
magnetischer Dipol *m* 2098
magnetischer Fluß *m* 2101
magnetischer normaler Komplex *m* 3324

Molekül
lineares ~ 2026
Makromolekül 2087
polares ~ 2706
superangeregtes ~ 3450
Molekular-Bindung f 3687
Molekularattraktion f
van der Waals-~ 3687
Molekulardiagramm n 2272
molekulardisperses System n 3636
molekulare Dichte f 2567
molekulare Lösung f 3636
n-molekulare Verteilungsfunktion 1476
Molekularfeld n 2275
Molekularfeldtheorie f 2276
Molekulargewicht n 3013 (vide)
relatives ~ 3013
Zahlenmittel-~ 2441
Molekularität f
Reaktionsmolekularität 2278
Molekularkristalle mpl 2271
Molekularorbital n 2279
Molekularpolarisation f 2261
Molekularprodukte npl der Radiolyse 2281
Molekularstrahlmethode f 2268
Molekularstrahlverfahren n 2268
Moleküldichte f 2567
Moleküle npl
iso-elektronische ~ 1868
Molekülkolloid n 2270
Molekülkristalle mpl 2271
Molekülorbital n
antibindendes 193
bindendes ~ 395
nichtbindendes ~ 2399
Molekülsiebe npl 2283
Molekülspektralanalyse f 2284
Molekülspektroskopie f 2285
Molekülspektrum n 2286
Molekülverbindungen fpl 78
Molenbruch m 2289
Molgewicht n 3013
Molpolarisation f
komplexe ~ 684
Molrefraktion f 2263
Molvolumen n 2265
partielles ~ 2563
Molwärme f 2251
mittlere ~ 2154
bei konstantem Druck 2252
~ bei konstanten Volumen 2253
scheinbare ~ 205
wahre ~ 2251
Molybdän n 2290
Moment n
Atommoment (vide)
Bahnmoment 2481
Bindungsmoment 392
Dipolmoment 1099 (vide)
effektives magnetisches ~ 1081
Gruppenmoment 1517
Kernmoment (vide)
Kernquadrupolmoment 2426
magnetisches ~ 2104
Quadrupolmoment 2828
Spinmoment 3326
Übergangsmoment 3606
zweites ~ 3145
Monitor m 2292
monochromatische Strahlung f 2293
Monochromator m 2294
monodisperses System n 2295
Monomer n 2297
monomere Einheit f 2298

Monomereinheit f 2298
monomolekulare Reaktion f 2300
monomolekulare Schicht f 2299
monomolekulare Schicht-Adsorption f 2296
monomolekularer Film m 2299
monomolekularer Mechanismus m der Eliminierung 1208
monotrope Phasen fpl 2302
Monotropie f 2303
Morse-Funktion f 2304
Morse-Kurve f 2304
Mosaikkristalle mpl 2305
Mosaikstruktur f 2306
Moseleysches Gesetz n 2307
Mössbauer-Effekt m 2308
Mössbauer-Spektroskopie f 2309
Mullikensche Elektronegativitätsskala f der Elemente 2314
Multiplett n 2321, 2322
Multiplett-Theorie f 302
Multipletts npl
Isospin-~ 553
Multiplizität f 2323, 2324
Multipolymerisation f 773
Muon n 2313
Murexidreaktion f 2328
Muster n
genormtes ~ 3353
Mutarotation f 2329
Mutation f
strahleninduzierte ~ 2868
Mutternuklid n 2555
My-Neutrino n 2326
Myon n 2313
Myonenzahl f 2327

N

n-Elektronen npl 2388
n-Halbleiter m 1150
n-Leitung f 1140
Nachbargruppenbeteiligung f 2343
Nacheffekt m 105
Nachweis m der Kernstrahlung 951
Nachweisgrenze f 2018
Nachweisreaktion f 1697
mikrochemische ~ 2223
Nachwirkungseffekt m 105
Näherung f
Born-Oppenheimer ~ 403
nahes Infrarot n 2336
Nahordnung f 3200
Nametkin Retropinakolin-Umlagerung f 2330
Nametkin-Umlagerung f 2330
Nanogrammethode f 2331
Naphthol→Naphthylamin-Umwandlung f 442
nasses Verfahren n 3755
Natrium n 3243
~-24 n 2917
Naturkonstanten fpl 3663
natürliche Kristallklassen fpl 830
natürliche Linienbreite f 2332
natürliche Radioaktivität f 2334
natürliche Strahlung f 2333
natürliches Radionuklid n 2335

natürliches System *n* der Elemente 2592
Nebengruppe *f* 3430
Nebenquantenzahl *f* 2482
Nebenvalenz *f* 3142
Néel-Punkt *m* 2338
Néel-Temperatur *f* 2338
Nef Acinitroalkan-Spaltung *f* 2339
negative Katalyse *f* 2341
negative Wertigkeit *f* 2342
negativer Katalysator *m* 1733
negatives Azeotrop *n* 2340
Nencki C-Acylierung *f* 2344
Nenitzescu hydrierende Acylierung *f* 2345
Neodym *n* 2346
Neon *n* 2347
nephelauxetische Reihe *f* der Liganden 2348
nephelauxetische Serie *f* der Liganden 2348
Nephelometer *n* 2349
Nephelometrie *f* 2350
Neptunium *n* 2351
Nernstscher Verteilungssatz *m* 2352
Nernstscher Wärmesatz *m* 3558
Netzebene *f* 1976
Neutralfilter *m* 2356
Neutralisation *f* 2357
Neutralisationsanalyse *f* 41
Neutralisationstitration *f* 41
Neutralisationswärme *f* 1565
Neutrino *n* 2358
Antineutrino 198
e-~ 1159
Elektron-~ 1159
My-~ 2326
μ-~ 2326
Neutron *n* 2359
Antineutron 199
Neutronen *npl*
epithermische ~ 1255
langsame ~ 3233
mittelschnelle ~ 1760
prompte ~ 2793
schnelle ~ 1328
Spaltneutronen 1375
thermische ~ 3530
verzögerte ~ 929
Neutronenabsorptionsmethode *f* 2360
Neutronenabsorptionsverfahren *n* 2360
Neutronenaktivierung *f* 2361
Neutronenaktivierungsanalyse *f* 2362
Neutronenausbeute *f* 2367
Neutronenbeugung *f* 2365
Neutronendetektor *m* 2364
Neutronengenerator *m* 2366
Neutronenquelle *f*
radioaktive ~ 1897
Neutronenspektrum *n* 2369
Neutronenstrahlungseinfang *m* 2368
Neutronenthermalisierung *f* 2370
Neutronenzähler *m* 2363
Neutronenzählrohr *n* 2363
Newmansche Formel *f* 2371
Newtonsche Viskosität *f* 2372
nicht umkehrbare Reaktion *f* 1855
nichtbindende Elektronen *npl* 2388
nichtbindendes Molekülorbital *n* 2389
nichtideale Lösung *f* 2972
nichtisotoper Träger *m* 2392

nichtisotopischer Träger *m* 2392
nichtklassische Ionen *npl* 2390
nichtkompensierte Wärme *f* 3655
Nichtmetalle *npl* 2393
nichtstöchiometrische Verbindungen *fpl* 2396
nichtumkehrbarer Vorgang *m* 1854
Nickel *n* 2373
Niederschlag *m*
aktiver ~ 68
radioaktiver ~ 2883
Niementowski Chinolin-Ringschluß *m* 2375
Nilsson-Modell *n* 2377
Ninhydrinreaktion *f* 2378
Niob *n* 2379
Nitrierung *f* 2380
Nitro-Acinitro-Tautomerie *f* 2381
Nitrosierung *f* 2384
Niveau *n*
Azeptorniveau 30
Donatorniveau 1047
Ferminiveau 1340
Kernenergieniveau 2409
Niveaus *npl*
Elektronenniveaus (*vide*)
Energieniveaus (*vide*)
Rotationsniveaus (*vide*)
Schwingungsniveaus (*vide*)
nivellierendes Lösungsmittel *n* 2001
Nobelium *n* 2386
nonbonding Elektronen *npl* 2388
Normal-Wasserstoffelektrode *f* 2403
Normal-Weston-Element *n* 3754
Normalbedingungen *fpl* 3351
normale Kette *f* 2398
normale Konzentration *f* 2400
normale Siedetemperatur *f* 2397
Normalelement *n* 3350, 3754
Cadmium-~ 3754
normaler Siedepunkt *m* 2397
Normalität *f* 2400
Normalkomplex *m* 2399
Normalpotential *n* 2402
~ der Elektrode 2402
Normalschwingungen *fpl* 2405
Normalverteilung *f* 2401
logarithmische ~ 2060
Normbedingungen *fpl* 3351
Normierung *f* der Wellenfunktion 2404
nucleophiles Agens *n* 2438
Nuklearreinheit *f* 2425
Nukleon *n* 2437
Nukleonengruppe *f* 639
Nukleonenzahlen *fpl*
magische ~ 2094
Nuklid *n* 2439
Mutternuklid 2555
Radionuklid 2912 (*vide*)
stabiles ~ 3347
Tochternuklid 878
Nulleffekt *m* 292, 294
Nullhypothese *f* 2400
Nullpunkt *m*
absoluter ~ 3797
Nullpunktsenergie *f* 3798
Nullpunktspotential *n* 3795
nullter Hauptsatz *m* der Thermodynamik 3799
Nullwert *m* 292, 294
Nummer *f*
Atomnummer 250 (*vide*)

S_R-Mechanismus m 3241

S_R-Mechanismus m 3241
Sachse-Mohr-Theorie f 3096
Saigerung f 3157
Saizew-Regel f 3113
säkulares Gleichgewicht n 3150
Salz n
Komplexsalz 689
Salzbrücke f 3097
Salzeffekt m
primärer $\sim$ 2772
sekundärer $\sim$ 3139
Salzfehler m 3098
Salzisomerie f 2034
Salzspaltkapazität f 1802
Samarium n 3100
Sammelprobe f 1513
Sammler m 31
Fraktionssammler 1411
Sandmeyer Diazonium-Austausch m 3104
Sandmeyer-Reaktion f 3104
Sandwich-Struktur f 3105
Sandwich-Verbindung f 3105
Sättigung f
dielektrische $\sim$ 982
Sättigungsaktivität f 3109
Sättigungsdampfdruck m 3699
Sättigungsmagnetisierung f 3111
Satz m
Carnotscher $\sim$ 490
Drehimpulssatz 175
Energiesatz 1227
Erhaltungssatz (vide)
Gleichverteilungssatz 1229
Hauptsatz (vide)
Impulssatz 2291 (vide)
Liouvillescher $\sim$ 2036
$\sim$ von Cailletet und Mathias 458
$\sim$ von der Erhaltung der Energie 1227
Wärmesatz (vide)
Sauerstoff m 2530
schwerer $\sim$ 1572
Sauerstoff-Schwefel-Gruppe f 538
Sauerstoffelektrode f 2531
Säule f
chromatographische $\sim$ 609
Golay-$\sim$ 470
Ionenaustauschersäule 1804
Kapillarsäule 470
thermische $\sim$ 3522
Säure f 38
Säure-Base-Katalyse f 39
Säure-Base-Titration f 41
Säure-Basen-Indikator m 40
Säure-Rest m 49
saure Gruppe f 43
Säureamid-Imid-Tautomerie f 142
saures Lösungsmittel n 2803
Säurespaltung f 42
Scandium n 3114
Scandiumgruppe f 3115
Scavengen n 3120
Scavenger m 3119
Scavenging n 3120
Schäden mpl
Bestrahlungsschäden 2861
Strahlenschäden 2861
Schale f
Elektronenschale 1164
Elektronenunterschale 1168
Valenzschale 3683
Schalen-Modell n 3192
Scharmittel n 3373

Scharmittelwert m 3373
Schaum m 1404
Scheibenelektrode f
rotierende $\sim$ mit Ring 3070
scheinbare Aktivierungsenergie f 203
scheinbare Halbwertsbreite f $\Delta\nu_{1/2}$ 204
scheinbare Molwärme f 205
scheinbare Reaktionsordnung f 206
Schema n
Zerfallschema 897
Schicht f
diffuse $\sim$ 998
Diffusionsschicht 1002
Doppelschicht (vide)
Helmholtz-$\sim$ 678
monomolekulare $\sim$ 2299
oberflächenaktive $\sim$ 3464
Oberflächenschicht 3469
unendlich dicke $\sim$ 1728
unendlich dünne $\sim$ 1729
Schichtdicke f 3102
Schichtlinienaufnahme f 3072
Schichtliniendiagramm n 3072
Schiemann-Kernfluorierung f 3122
Schiemann-Reaktion f 3122
Schild m
biologischer $\sim$ 356
Schmelzen n 2171
Schmelztemperatur f 2172
Schmelzwärme f 1561
Schmidt Carbonyl-Abbau m 3123
Schmidt-Reaktion f 3123
Schmidt-Regel f 3124
schnelle Koagulation f 2948
schnelle Neutronen npl 1328
Schockwelle f 3198
Schoenfliessche Symbole npl 3125
Schotten-Baumann Acylierung f 3126
Schottkysche Fehlordnung f 3127
Schraubenachse f 3133
Schraubenversetzungen fpl 3134
Schumann-Ultraviolett n 3678
Schumann UV 3678
Schutz m 2797
Strahlenschutz 2871
Schutzeffekt m 2799
Schutzgruppe f 2796
Schutzkolloid n 2798
schwache Wechselwirkung f 3746
schwacher Elektrolyt m 3745
schwacher Komplex m 3671
schwaches Ligandenfeld n 3747
Schwächungskoeffizient m
linearer $\sim$ 2021
Massenschwächungskoeffizient 2123
Schwankung f
relative mittlere $\sim$ 3017
Schwankungen fpl
$\sim$ in der Nähe des Gleichgewichtes 1261
statistische $\sim$ 1389
Schwankungsquadrat n
mittleres $\sim$ 2155
Schwanzbildung f 3496
schwarzer Körper m 369
Schwärzung f 370
photographische $\sim$ 370
Schwärzungskurve f 1217
Schwebungsmethode f 1591
Schwefel m 3443
$\sim$-35 m 2920

Schwefelgruppe f 3444
Schwellenenergie f 3570
Schwellwert-Dosimeter n 3569
Schwenkaufnahme f 2502
Schwenkverfahren n 2501
schwere Platinmetalle npl 1573
schwerer Sauerstoff m 1572
schwerer Wasserstoff m 956
schweres Deuterium n 3631
schweres Wasser n 1574
Schwingungen fpl 3710
Deformationsschwingungen 328
entartete ~ 905
Gruppenschwingungen 1519
Normalschwingungen 2405
Valenzschwingungen 3417
Schwingungsband n 3707
Schwingungsenergie f 3704
Schwingungsniveaus npl des Moleküls
2287
Schwingungsquantenzahl f 3706
Schwingungsspektrum n 3709
Schwingungszustandssumme f 3705
Sedimentation f 3151
Sedimentationsanalyse f 3152
Sedimentationsgeschwindigkeit f 3156
Sedimentationsgleichgewicht n 3154
Sedimentationspotential n 3155
Sedimentationswaage f 3153
sehr langsame Elektronen npl 3429
Seigerung f 3157
Seitenkette f 3201
sekundäre Eichsubstanz f 3141
sekundäre Urtitersubstanz f 3140
Sekundärelelektronen npl 3136
Sekundärelement n 31
sekundärer Salzeffekt m 3139
Sekundärionisation f 3137
Sekundärreaktion f 1719
photochemische ~ 3138
Selbstabsorption f 3167, 3168
Selbstbestrahlung f 3172
Selbstdiffusion f 3170
Selbstdiffusionskoeffizient m 3171
Selbstentzündung f 271
Selbstentzündungstemperatur f 272
Selbststreuung f 3174
Selbstumkehr f 3173
Selbstersetzung f
strahleninduzierte ~ 278
Selbstzündung f 271
selektive Adsorption f 3159
selektive Reaktion f 3160
selektive Reduktion f 3162
selektive Wirkung f des Katalysators 3283
selektives Reagens n 3161
Selektivität f 3163, 3165
Selektivitätskoeffizient m 3164
korrigierter ~ 778
Selen n 3166
Selfconsistentfield n 3169
Seltenerdmetalle npl 1966
Seltsamkeit f 3410
seltsame Teilchen npl 3412
Semidin-Umlagerung f 3177
Semikolloid n 232
Semimikroanalyse f 3178
semipermeable Membran f 3179

semipermeable Wand f 3179
semipolare Bindung f 758
senkrechtes Band n 2601
sensibilisierte Chemilumineszenz f 3184
sensibilisierte Fluoreszenz f 3185
Sensibilisation f 3183
Sensibilisator m 2656, 2916, 3186
spektraler ~ 3186
Sensibilisierung f 2476, 3183
~ mit Elektrolyten 3488
Photosensibilisierung 2654
Sensitometrie f 3187
Sequenz f 3190
Serie f
nephelauxetische ~ der Liganden 2348
Spektralserie 3296
Serini Glykol→Desoxyketon-Umwandlung
f 3191
Serini-Reaktion f 3191
Sesselform f 537
Sextett n
Elektronensextett 220
Sicherheitswahrscheinlichkeit f 3210
Sichtbare n 3722
sichtbares Gebiet n 3722
sichtbares Spektrum n 3723
Sidgwick-Powell-Theorie f 3202
Siebanalyse f 3204
Siebe npl
Ionensiebe 1823
Molekülsiebe 2283
Siebwirkung f 3203
Sieden n 385
Siedepunkt m
normaler ~ 2397
Siedepunktserhöhung f
molare ~ 1074
Siedetemperatur f 386
normale ~ 2397
Signal n
Absorptionssignal 17
Dispersionssignal 1022
Silber n 3212
Silbersalz-Decarboxylierung f 1650
Silicium n 3211
Silizium n 3211
Simonini Decarboxylierung f 3214
Simonini Silbersalz-Abbau m 3214
Simonis Chromon-Ringschluß m 3215
Singulettsystem n 3226
Singulettzustand n 3226
Sinnenprüfung 2491
Sinterung f 3227
~ des Katalysators 3228
Skala f
Atomgewichtsskala 248 (vide)
Atommassenskala 248 (vide)
Elektronegativitätsskala 1149
Massenskala (vide)
Skandium n 3114
Skraup Chinolin-Synthese f 3229
Skraup-Synthese f 3229
Slatersche Wellenfunktion 3230
Slatersches Orbital n 3230
Soddy-Fajans-Verschiebungssätze mpl
3242
Sol n 3245
Aerosol 103
Hydrosol 1677
Kryosol 826
Organosol 2492

Synthese
Kolbe-Schmitt-~ 1943
Kostanecki-Robinson Chromon-~ 1947
McFadyen-Stevens-~ 2146
Oxosynthese 1662
Passerini α-Hydroxy-N-Arylamid-~ 2574
Passerini-~ 2574
Pechmann-~ 2581
Pechmann-~ von Cumarin 2581
Perkin Zimtsäure-~ 2595
Photosynthese 2657
Reformatsky β-Hydroxycarbonsäureester-~ 2996
Reformatsky-~ 2996
Reimer-Tiemann Aldehyd-~ 3006
Skraup Chinolin-~ 3229
Skraup-~ 3229
Sommelet Aldehyd-~ 3266
Sommelet-~ 3266
Williamson-~ 3759
Synthesen *fpl*
Reppe-~ 3028
System *n*
abgeschlossenes ~ 1871
azeotropes ~ 290
binäres ~ 351
diskontinuierliches ~ 1016
disperses ~ 1020
divariantes ~ 366
Einkomponentensystem 2457
Einstoffsystem 2457
Folgereaktionsystem 734
geschlossenes ~ 638
grobdisperses ~ 3474
heteroazeotropes ~ 1589
heterodisperses ~ 2717
heterogenes ~ 1594
heteroazeotropes ~ 1600
homoazeotropes ~ 1618
homodisperses ~ 2295
homogenes ~ 1624
invariantes ~ 1790
isodisperses ~ 2295
isoliertes ~ 1871
kolloides ~ 662
kondensiertes ~ 711
kontinuierliches ~ 747
Mehrstoffsystem 2316
molekulardisperses ~ 3636
monodisperses ~ 2295
natürliches ~ der Elemente 2592
offenes ~ 2463
Periodensystem 2592
Phasensystem (*vide*)
polydisperses ~ 2717
Singulettsystem 3226
thermodynamisches ~ 3545
Triplettsystem 3629
univariantes ~ 3662
zeotropes ~ 3793
Zweikomponentensystem 351
systematischer Fehler *m* 3490
Systeme *npl*
Ergodensysteme 1265
Kristallsysteme 843
Szilard-Chalmers-Effekt *m* 3491
Szintillationsspektrometer *n* 3130
Szintillationszähler *m* 3129
Szintillator *m* 3131
Szyszkowski-Gleichung *f* 3492

Ś

Świętosławski und Dorabialska
   adiabatisches Mikrokalorimeter *n* 1049

T

t-Test *m* 3428
t-Verteilung *f* 3427
Tafelsche Gleichung *f* 3495

taktisches Polymer *n* 3494
Taktizität *f* 3493
Tantal *n* 3497
Target *m* 3498
dicker ~ 3554
dünner ~ 3557
Tastpolarographie *f* 3499
Taupunkt *m* 960
Tautomerie *f* 3501
Drei-Kohlenstoff-~ 3565
Keto-Enol-~ 1922
Laktam-Laktim-~ 142
Nitro-Acinitro-~ 2381
Oxo-Cyclo-~ 3064
Protonentautomerie 2809
prototrope ~ 2809
Ring-Ketten-~ 3064
Säureamid-Imid-~ 142
Tautomeriegleichgewicht *n* 3500
Technetium *n* 3503
Technik *f*
Flash-~ 1384
Frontaltechnik 1433
Gradienten-Elutions-~ 1496
te-Hybride *f* 3515
Teilchen *n* 1192
Antiteilchen 200
α-~ 128
β-~ 341
Teilchen *npl*
Elementarteilchen 1194
fremdartige ~ 3412
Kolloidteilchen 663
seltsame ~ 3412
α-~ 130
β-~ 344
Teilchenbahn *f* 2570
Teilchenbeschleuniger *m* 551
Teilchendichte *f* 2567
Teilchenenergie *f* 2566
Teilchenfunktion *f* 3327
Teilchenmenge *f* 151
Teilstromdichte *f*
anodische ~ 186
kathodische ~ 508
teilweise Mischbarkeit *f* 2562
teilweise verdeckte Konformation *f* 2561
Telomer *n* 3505
Telomerisation *f* 3506
Tellur *n* 3504
Temperatur *f* 3507
absolute ~ 3546
antiferromagnetische Curie-~ 2333
Boyle-~ 416
charakteristische ~ 546
charakteristische ~ der Rotation 3076
charakteristische ~ der Schwingung 3703
Curie-~ 854
empirische ~ 3507
Entzündungstemperatur 1699
Erstarrungstemperatur 2172
eutektische ~ 1280
Farbtemperatur 672
Fermi-~ 1344
Inversionstemperatur 1796
Kondensationstemperatur 708
kritische ~ 816
Lösungstemperatur (*vide*)
Mischungstemperatur (*vide*)
Néel-~ 2338
paramagnetische Curie-~ 2547
reduzierte ~ 2984
Schmelztemperatur 2172
Selbstentzündungstemperatur 272
Siedetemperatur 386 (*vide*)
Spin-~ 3333
Sublimation-~ 3432
Temperaturkoeffizient *m* der
   Reaktionsgeschwindigkeit 3508

Wolff Diazoketon → Keten-Umlagerung f 3765
Wolff-Kishner-Reaktion f 3764
Wolff-Umlagerung f 3765
Wolfram n 3638
Wolke f
Elektronenwolke 1139
Ionenwolke 1810
Ladungswolke 549
Wurtz Alkylhalogenid-Kondensation f 3767
Wurtz-Reaktion f 3767
Wurzel f aus dem mittleren Geschwindig-keitsquadrat 3068

X

Xanthoproteinreaktion f 3768
Xenon n 3769
Xerogel n 3770

Y

Ylid m 3786
Ylid-Amin-Isomerisation f 3401
Ytterbium n 3787
Yttererden fpl 3789
Yttrium n 3788

Z

Zähigkeit f 3720
Zahl f
Avogadro-~ 282
Baryonezahl 312
Besetzungszahl 2447
Bodenzahl 3519 (vide)
Brechzahl 9
Dielektrizitätszahl 2599
Elektronenleptonenzahl 1153
Goldzahl 1492
Hydratationszahl (vide)
Ionenladungszahl 1825
Koordinationszahl 765
    (vide) 766
Leptonenzahl 1996
Ligandenzahl (vide)
Massenzahl 2130
Myonenzahl 2327
Ordnungszahl 250
Oxydationszahl 2522 (vide) 2526
Quantenzahl (vide)
Reaktionslaufzahl 1310
Rubinzahl 3085
Solvatationszahl (vide)
stöchiometrische ~ 3403
Stoßzahl 660
Symmetriezahl 3483
Überführungszahl (vide) 3600
Viskositätszahl 2985
Wellenzahl 3742
~ der theoretischen Stufen der Kolonne 3719
~ der Zweierstoße 660
Zählausbeute f 799
Zahlen fpl
magische ~ 2094
Nukleonenzahlen (vide)
Zufallszahlen 2940
Zahlenmittel n 2441
Zahlenmittel-Molekulargewicht n 2441
Zahlenmittelwert m 2441
Zähler m 2860
Bohrloch-Szintillationszähler 3752

Zähler
Čerenkov-~ 521
Geiger-Müller-~ 1469
Neutronenzähler 2363
Proportionalzähler 2794
Szintillationszähler 3129
Tscherenkow ~ 521
2π-~ 792
4π-~ 793
Zählrate f 798
Zählrohr n
Auslösezählrohr 1469
Fensterzählrohr 3760
Flüssigkeitszählrohr 2041
Glockenzählrohr 3760
Geiger-Müller ~ 1469
Neutronenzählrohr 2363
Proportionalzählrohr 2794
Zählrohrcharakteristik f eines Auslöse-zählrohres 794
Zäsium n 456
Zeeman-Effekt m 3790
Zeeman-Komponenten fpl 3791
Zeit f
Freibewegungszeit (vide)
Halbwertzeit (vide) 1527, 1528
Relaxationszeit 3023 (vide)
Totzeit 882
Transitionszeit 3609
Tropfzeit 1067
Vorbrennungszeit 2754
Vorfunkenzeit 2764
zeitabhängige Schrödinger-Gleichung f 3574
Zeitdauer f eines Stoßes 2156
Zeitfaktor m 3110
zeitlaufgelöstes Spektrum n 3577
Zeitmittel n 3573
zeitunabhängige Schrödinger-Gleichung f 3575
Zelle f
chemische ~ 568
Einheitszelle 3659
Elementarzelle 3659 (vide)
galvanische ~ 1445
Halbzelle 1109
heiße ~ 1636
Konzentrationszelle 698 (vide)
Phasenzelle 2610
reversible ~ 3051
Zellkonstante f des Leitfähigkeitgefäßes 514
Zentralatom n 516
Zentralfeld n 517
Zentralion n 518
Zentralkraftfeld n 517
Zentren npl
aktive ~ 67
F-~ 1331
Farbzentren 671
Zentrifugalpotentialwall m 520
Zentrum n
Adsorptionszentrum 91
Akzeptor-~ 29
Chiralitätszentrum 596
Donator-~ 1046
Koordinationszentrum 519
Zeolithe mpl 3792
zeotropes System n 3793
Zeotropie f 3794
Zer n 523
Zerfall m
α-~ 126
β-~ 340
radioaktiver ~ 2380
Zerfallsenergie f
radioaktive ~ 1233

INDEX FRANÇAIS

analyse
~ à la goutte 3340
~ à la touche 3340
~ chimique 565
~ chromatographique 607
~ chronopotentiométrique 618
~ colorimétrique 668
~ conformationelle 728
~ de gaz 1456
~ de sédimentation 3152
~ de traces 3593
~ de variance 160
~ destructive par activation 948
~ diffractométrique aux rayons X 3776
~ électrochimique 1102
~ électrographique 1117
~ élémentaire 1191
~ fluorimétrique 1394
~ fonctionelle 1438
~ frontale 1433
~ granulométrique 2569
~ ~ par tamisage 3204
~ instrumentale 1744
~ interférométrique 1759
~ isotopique 1892
~ non destructive par activation 2391
~ organoleptique 2491
~ par absorption atomique 241
~ par absorption de rayons X 3773
~ par absorption des neutrons 2630
~ par absorption des rayons bêta 342
~ par absorption des rayons gamma 1452
~ par activation 61
~ par activation aux particules chargées 552
~ par activation aux rayons γ 2627
~ par activations dans les photons γ 2627
~ par activation instrumentale 2391
~ par activation neutronique 2362
~ par activation photonucléaire 2627
~ par diffraction aux rayons X 3776
~ par dilution isotopique 1888
~ par fluorescence 1394
~ par fluorescence atomique 246
~ par fluorescence X 3780
~ par luminescence 2074
~ par photométrie de flamme 1381
~ par radioactivation 61
~ par rétrodiffusion du rayonnement bêta 343
~ par spectrométrie X 3782
~ par spectrométrie de fluorescence X 3780
~ par spectrométrie de masse 2131
~ par spectrophotométrie d'absorption 21
~ par spectrophotométrie de flamme 159
~ par spectroscopie atomique 257
~ par spectroscopie d'émission 1210
~ par spectroscopie de masse 2131
~ polarographique 2709
~ pondérale 1507
~ qualitative 2830
~ quantitative 2833
~ radiochimique 2895
~ radiométrique 2909
~ spectrale 3298
~ ~ à lecture directe 3309
~ ~ visuelle 3726
~ spectrochimique 3299
~ spectrographique 3307
~ spectrométrique 3309
~ ~ de fluorescence X 3780
~ spectrophotométrique 161
~ spectroscopique 3298
~ thermique 3521
~ ~ différentielle 995
~ thermo-magnétique 3550
~ totale 681
~ volumétrique 3587
électro-~ 1118
macroanalyse 2074
microanalyse 2216
spectranalyse (vide)
anaphorèse f 171
angle m
~ de contact 741
~ de mouillage 741
~ de perte 2067
anion m 177

anionotropie f 179
anisotropie f 180
~ diamagnétique 963
~ magnétique 2096
anneaux mpl de Liesegang 2005
annihilation f 182
anode f 185
anolyte m 187
anomères mpl 188
antagonisme m de deux électrolytes 190
anti-auxochrome m 191
anticatalyseur m 1733
anti-entraîneur m 1612
antiferromagnétiques mpl 195
antiferromagnétisme m 194
antimatière f 196
antimères mpl 1221
antimoine m 197
antineutretto m 2325
antineutrino m 198
~ ν_μ 2325
~ ν_e 1153
antineutron m 199
antiparticule f 200
antipodes mpl optiques 1221
antiproton m 201
approximation f
~ de Born-Oppenheimer 403
~ en T^3 de Debye 893
aquamétrie f 209
argent m 3212
argon m 213
aromaticité f 218
aromatisation f 221
arsenic m 226
arsonation f 227
arylation f 228
associate m 229
association f 231
astate m 234
astrochimie f 782
atmosphère f ionique 1810
atome m 240
~ central 516
~ donneur du coordinat 759
~ du coordinat 759
~-gramme 1501
~ pont 432
~ spiro 3336
~ unifié 3660
atomes mpl
~ chauds 1635
~ de recul 1635
~ épithermiques 1254
~ marqués 1952
atropo-isomérie f 264
auto-absorption f 3167, 3168
auto-allumage m 271
autocatalyse f 267
autocatalyseur m 268
auto-complexe m 270
auto-diffusion f 3170, 3174
auto-ignition f 271
auto-irradiation f 3172
auto-oxydation f 279
autoprotolyse f 275
autoradiogramme m 276
autoradiographie f 277
autoradiolyse f 278
auxochrome m 280

auxochrome
anti-~ 191
avancement *m* de réaction 1310
axe *m*
~ de symétrie d'inversion 1792
~ hélicoïdal 3133
~ optique 2467
~ spiral 3133
axes *mpl*
~ cristallins 828
~ cristallographiques 828
azéotrope *m* 286
~ négatif 2340
~ positif 2732
azéotropisme *m* 291
azote *m* 2382
azotides *mpl* 2383

B

balance *f*
~ de sédimentation 3153
~ magnétique de Gouy 1495
~ osmotique 2512
balayage *m* 3120
bande *f*
~ B 321
~ d'énergie 122
~ d'énergie permise 122
~ de combinaison 676
~ de conduction 715
~ de conductivité 715
~ de référence 2991
~ de rotation-vibration 3081
~ de transfert de charge 554
~ de valence 3679
~ de vibration 3707
~ du type parallèle 2545
~ du type perpendiculaire 2601
~ électronique 1154
~ fondamentale 1441
~ interdite 1405
~ K 1913
~ R 2952
barn *m* 309
barrière *f*
~ centrifuge 520
~ coulombienne 786
~ d'énergie 2740
~ de potentiel 2739
~ de potentiel coulombienne 786
~ de rotation 3075
baryum *m* 307
base *f* 315
basicité *f* 317
~ d'un acide 318
bateau *m* 379
benzoylation *f* 333
berkélium *m* 335
berthollides *mpl* 2396
béryllium *m* 338
bêtatron *m* 345
bilan *m*
~ énergétique de réaction nucléaire 2432
~ thermique 1546
biocatalyseur *m* 353
biochimie *f* 354
bioluminescence *f* 357
biosynthèse *f* 358
bipolymère *m* 359
biradicaux *mpl* 360
biréfringence *f* 362
~ électrique 1921
bismuth *m* 264
blindage *m* du noyau 3195
bloc *m* 374

bombe *f*
~ à hydrogène 1666
~ atomique 242
~ au cobalt 647
~ calorimétrique 391
bore *m* 405
bosons *mpl* 410
boucle *f* d'irradiation 1850
bouclier *m* biologique 356
branche *f*
~ P 2578
~ Q 2826
~ R 2993
branchement *m* de chaîne 525
breeder *m* 430
bromation *f* 435
brome *m* 436
bromuration *f* 435
broyeur *m* colloïdal 666
bruit *m* de fond 292, 294

C

cadmium *m* 455
cæsium *m* 456
calcium *m* 459
radiocalcium *m* 2893
californium *m* 461
calorie *f*
~ de 14,5 à 15,5°C 464
~ internationale 1901
~ I.T. 1901
~ thermochimique 3536
calorimètre *m* 465
~ à glace de Bunsen 447
~ adiabatique 85
~ isotherme 1885
microcalorimètre 2218 (*vide*)
calorimètres *mpl*
~ jumelés 3642
~ semblables 3642
calorimétrie *f* 466
microcalorimétrie 2219
canal *m*
~ d'expérimentation 322
~ de réaction nucléaire 2431
capacité *f*
~ calorifique 1547
~ ~ à pression constante 1548
~ ~ à volume constant 1549
~ ~ du système calorimétrique 1230
~ d'adsorption 102
~ d'échange 3591
~ d'échange des groupes de fortes acidités 1802
~ d'échange des groupes de fortes basicités 1802
~ différentielle de la couche double 938
~ dynamique limitée à la percée 427
~ intégrale de la couche double 1747
~ utile 427
capillaire *f* de Luggin 2071
capillarité *f* 469
capture *f*
~ de résonance 3036
~ électronique 1138
~ K 1914
~ radiative du neutron 2368
caractère *m* aromatique 218
carbanion *m* 474
carbène *m* 475
carbocation *m* 477
carbonation *f* des phénols selon Kolbe 1943
carbone *m* 479
~ asymétrique 235
radiocarbone 2894
carboxylation *f* 486

carré *m* des écarts quadratiques moyens 3700
carte *f* de contrôle 749
cartouche *f* 2849
cassure *f* en vol 3418
catalyse *f* 495
autocatalyse 267
~ acido-basique 39
~ ~-~ généralisée 1473
~ ~-~ spécifique 3276
~ d'oxydo-réduction 2524
~ enzymatique 1250
~ hétérogène 1592
~ homogène 1620
~ microhétérogène 2225
~ négative 2341
~ positive 2733
~ sous irradiation 2858
~ spécifique 3276
catalyseur *m* 496
anticatalyseur 1733
autocatalyseur 268
biocatalyseur 253
~ de contact 3247
~ mélangé 2242
~ négatif 1733
~ solide 3247
cataphorèse *f* 505
caténanes *fpl* 506
catharomètre *m* 3523
cathode *f* 507
~ de mercure 2178
catholyte *m* 509
cation *m* 510
carbocation 477
cationotropie *f* 512
cellule *f*
~ à haut activité 1636
~ chimique 568
~ de conductivité thermique 3523
~ de l'espace des phases 2610
~ galvanique 1445
~ thermoélectrique 1129
demi-~ 1109
celtium *m* 1524
centre *m*
~ coordinateur 519
~ d'adsorption 91
~ de chiralité 596
centres *mpl*
~ actifs 67
~ d'activité 67
~ de couleur 671
~ F 1331
cérium *m* 523
césium *m* 456
chaîne *f*
~ carbonée 481
~ cinétique de réaction 533
~ de carbone 481
~ de concentration 698
~ de réaction 533
~ droite 2398
~ latérale 3201
~ linéaire 2398
~ ramifiée 1407
~ unique 2117
chalcogènes *mpl* 538
chaleur *f* 1545
~ atomique 247
~ d'adsorption 991, 1552
~ d'hydratation 1563
~ de changement de phase 1553
~ de combustion 1554
~ de condensation 1555
~ de cristallisation 1556
~ de dilution 1557
~ de dissociation 1558
~ de dissolution 1568
~ de formation 1560

chaleur
~ de fusion 1561
~ de la gélification 1562
~ de mélange 1564
~ de mouillage 1571
~ de neutralisation 1565
~ de réaction 1567, 3357
~ de réaction à pression constante 3357
~ de solvatation 1569
~ de sublimation 1570
~ de vaporisation 1559
~ dernière de dissolution 1970
~ intégrale de dilution 1557
~ ~ de dissolution 1568
~ latente 1553
~ ~ superficielle 1971
~ molaire partielle de dilution 2558
~ ~ ~ de dissolution 2559
~ non compensée 3655
~ partielle d'adsorption 991
~ primaire de dissolution 1363
~ spécifique 3280
~ ~ atomique 247
~ ~ électronique 1165
~ ~ molaire 2251
~ ~ ~ à pression constante 2252
~ ~ ~ à volume constant 2253
~ ~ ~ apparente 205
~ ~ ~ moyenne 2154
~ ~ vraie 2251
~ totale d'adsorption 1750
~ ~ de dissolution 3590
chambre *f*
~ à cavité Bragg-Gray 418
~ d'ionisation 1829
~ d'irradiation 1849
~ de Weissenberg 3750
champ *m*
~ coercitif 652
~ cristallin 831
~ d'irradiation 2866
~ de cavité 513
~ de Lorenz 2065
~ de rayonnement 2866
~ de réaction 2958
~ des coordinats 2007
~ des forces centrales 517
~ faible des coordinats 3747
~ intense des coordinats 3422
~ interne 2051
~ local 2051
~ magnétique 2099
~ moléculaire 2275
~ potentiel 2743
~ self-consistent 3169
changement *m* de phase 2617
chaos *m* moléculaire 2269
charge *f*
~ d'espace 3270
~ spatiale 3270
chélate *m* 557
chélation *f* 562
chélatiométrie *f* 564
chemin *m* de réaction 2962
chiffre *m* de coordination 765
chimie *f* 593
astrochimie 782
biochimie 354
~ analytique 162
~ appliquée 207
~ biologique 354
~ de coordination 762
~ des atomes chauds 1634
~ des colloïdes 665
~ des radiations 2859
~ générale 1474
~ inorganique 1740
~ minérale 1740
~ nucléaire 2407
~ organique 2487
~ physique 2659
~ pure 2819
~ quantique 2836

chimie
~ radioactive 2898
~ théorique 3518
cristallochimie 829
électrochimie 1106
géochimie 1447
magnétochimie 2114
microchimie 2221
phonochimie 3267
photochimie 2630
stéréochimie 3387
thermochimie 3533
chimiluminescence *f* 591
~ sensibilisée 3184
chimisorption *f* 592
chiralité *f* 597
prochiralité 2789
chloration *f* 601
chlore *m* 602
chlorosulfonation *f* 605
chloruration *f* 601
choc *m* efficace 1078
chromatogramme *m* 606
chromatographie *f* 610
~ à deux dimensions 3643
~ à température programmée 3509
~ circulaire 2855
~ d'adsorption 92
~ d'échange d'ions 1803
~ de partage 2571
~ des gaz 1457
~ en couche mince 3555
~ en phase gazeuse 1457
~ en phase inversée 1316
~ en phase liquide 2040
~ sur papier 2541
radiochromatographie 2899
chromatopolarographie *f* 611
chrome *m* 612
chromogène *m* 614
chromophore *m* 615
chromophores *mpl* conjugués 691
chronoampérométrie *f* 616
~ linéaire 3224
~ ~ par redissolution anodique 3419
chronopotentiométrie *f* 619
chute *f* ohmique 2452
cible *f* 3498
~ épaisse 3554
~ mince 3557
~ nucléaire 3498
cinétique *f*
~ chimique 579
~ des réactions 579
classe *f* de symétrie 3481
classes *fpl* cristallographiques 830
classification *f* périodique des éléments 2592
clathrates *mpl* 633
coacervation *f* 641
coagel *m* 642
coagulation *f* 643
~ irréversible 1852
~ lente 3231
~ orthocinétique 2499
~ péricinétique 2589
~ rapide 2948
~ réversible 3052
coalescence *f* 645
cobalt *m* 646
radiocobalt 2900
coefficient *m*
~ d'absorption 14, 25
~ d'activité 72, 3730
~ d'activité ionique 73
~ d'adsorption 93
~ d'analyse 165

coefficient
~ d'association 233
~ d'atténuation linéique 2021
~ d'atténuation massique 2123
~ d'auto-diffusion 3171
~ d'échange 3164
~ d'extinction moléculaire 2255
~ d'ionisation 913, 3634
~ d'ionisation vrai 3634
~ d'osmose 2510
~ de compressibilité 649
~ de conversion interne 1769
~ de diffusion 1001
~ de dilatation thermique 650
~ de dissipation 1033
~ de frottement intérieur 3721
~ de Joule-Thomson 1910
~ de partage 1039
~ de puissance 2752
~ de recristallisation 1621
~ de sécurité 721
~ de transmission 3615
~ de Van't Hoff 3508
~ de variation 651
~ de viscosité 3721
~ moyen d'activité ionique 2147
~ R_f 3091
~ stœchiométrique 3403
~ thermique 3508
~ ~ des vitesses de réactions 3508
second ~ viriel 3149
coefficients *mpl*
~ de transport 3617
~ phénoménologiques 2618
~ viriels 3715
cohésion *f* 654
co-ions *mpl* 655
coïncidence *f* de raies 2031
collecteur *m* de fractions 1411
collision *f* efficace 1078
colloïde *m* 662
~ irréversible 1853
~ lyophile 2082
~ lyophobe 2083
~ macromoléculaire 2270
~ micellaire 232
~ protecteur 2798
~ réversible 3053
cologarithme *m* de l'activité des ions hydrogènes 2608
colonne *f*
~ capillaire 470
~ chromatographique 609
~ d'échangeur d'ions 1804
~ thermique 3522
colorimètre *m* 667
~ photoélectrique 2631
colorimétrie *f* 668, 670
~ différentielle 994
combinaisons *fpl* moléculaires 78
combustible *m*
~ nucléaire 2412
~ ~ usé 448
combustion *f* 449, 450, 677
comparateur *m* 679
~ de spectres 1059
complexe *m* 763
auto-~ 270
~ π 2666
~ σ 3206
~ à orbitales externes 2515
~ à orbitales internes 1738
~ à spins appariés 2069
~ à spins élevés 3324
~ à spins faibles 2069
~ à spins non appariés 3324
~ activé 58
~ alcène 2453
~ amminé 147
~ aquo 210
~ d'Arrhenius 222

constitution f 570
container m 743
contamination f 744
conteneur m 743
contraste m 748
contre-électrode f 795
contre-ions mpl 797
conversion f interne 1767, 1768
convertisseur m de neutrons 1402
coordinant m
~ multidenté 2317
~ unidenté 3657
coordination f 761
coordinence f 765
coplanarité f 771
copolymère m 772
~ bloc 375
~ greffé 1498
~ séquencé 375
copolymérisation f 773
coprécipitation f 776
copulation f 801
corps m 3433
~ amorphe 149
~ composé 690
~ gris 1509
~ noir 369
~ parfaitement noir 369
~ simple 3219
corps mpl
~ condensés 709
~ diamagnétiques 964
~ paramagnétiques 2551
corrélation f 779
~ des configurations 780
~ des électrons 1142
couche f
~ d'Helmholtz 678
~ de demi-atténuation 1529
~ de diffusion 1002
~ de valence 3683
~ diffuse 998
~ double 1096
~ ~ électrochimique 1096
~ électronique 1164
~ épaisse 1728
~ infiniment mince 1729
~ monomoléculaire 2299
~ rigide d'Helmholtz 678
double ~ électrochimique 1096
sous-~ électronique 1168
coulomètre m 787
coulométrie f 789
~ à intensité constante 736
~ à potentiel contrôlé 790
coupellation f 850
couplage m
~ de Russel-Saunders 2070
~ jj 1908
~ spin-orbite 3328
~ spin-spin 3330
couple f de raies 2032
courant m
~ capacitif 555
~ catalytique 503
~ cinétique 1927
~ d'adsorption 94
~ d'entropie 1245
~ de diffusion limite 2015
~ de migration 2233
~ limité 2014
~ résiduel 3032
courbe f
~ BET 347
~ caractéristique 1217
~ ~ d'une émulsion 1217
~ ~ du compteur 794
~ d'absorption 16, 17

courbe
~ d'élution 1203
~ d'énergie potentielle 2741
~ d'étalonnage 163
~ de Bragg 417
~ de dispersion 1022
~ de Morse 2304
~ de noircissement 1217
~ de noircissement d'une émulsion 1217
~ de répartition des ions dans l'effluent 428
~ de titrage 3584
~ de vitesse 1929
~ électrocapillaire 1100
~ polarographique 2707
~ potentiel-temps 2744
cours m de réaction 2962
covalence f 803
création f locale d'entropie 2052
cristal m 827
~ idéal 2586
~ parfait 2586
cristallisation f 839, 1423
~ de la phase gazeuse 840
cristallite f 838
cristallochimie f 829
cristalloïde m 844
cristalloluminescence f 845
cristaux mpl
~ avec liaison hydrogène 1668
~ covalents 245
~ d'ions 1814
~ ioniques 1814
~ liquides 2042
~ mixtes 3250
~ moléculaires 2271
~ mosaïques 2305
~ phosphorescents 845
cryohydrate m 822
cryométrie f 824
cryoscopie f 824
cryosol m 826
cuivre m 774
curides mpl 858
curie m 851
curium m 856
cuve f absorbante 13
cyanoéthylation f 863
cyanométhylation f 864
cycle m
~ chélaté 561
~ de Bethe 484
~ de Born-Haber 402
~ de Carnot 489
~ de combustible 2413
~ de l'hydrogène 2807
~ du carbone 484
~ fermé 865
~ hydroaromatique 1659
~ réversible 3054
cyclisation f 868
cycloaddition f
~ 1,2 870
cyclophanes mpl 2543
cyclotron m

D

daltonides mpl 3404
déalcoylation f 884
déamination f 885
débit m
~ de chaleur du rayonnement 1566
~ de dose 1053
~ de neutrons 2367
debye m 894

décarbonylation f 895
décarboxylation f 896
décharge f d'ion 1015
déchets mpl radioactifs 2891
décomposition f 898
~ selon Curtius 862
décontamination f 900
découplage m de spin 3321
décroissance f radioactive 2880
décyclisation f 3065
dédoublement m cétonique 1924
défaut m
~ de masse 2127
~ dû à l'irradiation 2861
défauts mpl
~ dans les cristaux 1974
~ de Frenkel 1424
~ de Schottky 3127
~ linéaires 2030
~ ponctuels 2692
déficit m d'électrons 1144
déformation f des noyaux atomiques 902
dégat m par rayonnement 2861
dégénérescence f 906
~ du niveau énergétique 906
~ quantique 1459
~ ~ de gaz 1459
dégradation f 907, 908
~ d'Emde 1207
~ d'Hofmann 1609
~ de Barbier-Wieland 306
~ de Ruff-Fenton 3086
~ de Weerman 3748
~ de Wohl 3762
~ du polymère 908
~ du polymère par radiation 2862
degré m
~ absolu Kelvin 1916
~ d'association 233, 909
~ d'avancement de réaction 1310
~ d'hydrolyse 917
~ d'ionisation 918
~ d'oxydation 2522
~ de dégénérescence 910
~ de dispersion 912
~ de dissociation électrolytique 913, 3634
~ de dissociation électrolytique vrai 3634
~ de liberté 915
~ de liberté de mouvement 915
~ de liberté du système thermodynamique 916
~ de polymérisation 919
~ de polymérisation de la macromolécule 919
~ de polymérisation moyen 920
~ Kelvin 1916
déhalogénation f 922
dehydrocyclisation f 924
déméthylation f 935
demi-cellule f 1109
demi-largeur f 1526
déminéralisation f 927
dénaturation f des protéines 937
densité f 938
~ d'électrons 1415
~ d'électrons libres 1415
~ d'entropie 1243
~ d'ionisation 1830
~ de charge 550
~ de flux 2857
~ de flux de radiation 2857
~ de probabilité 939, 2787
~ de spin 3320
~ des états 940
~ des particules 2567
~ du courant 859
~ du courant anodique 186
~ du courant cathodique 508
~ du courant d'échange 1286
~ électronique 1145, 1415
~ ionique 1830

densité
~ optique 370
~ superficielle de charge 3466
densitomètre m 2226
déplacement m 1028
~ 1,2 2974
~ chimique 587
~ de Knight 1937
dépolarisant m 941
dépolymérisation f 942
dépôt m actif 68
désactivation f
~ d'état excité 880
~ de la molécule 881
désalcoylation f 884
désamination f 885
descendant m 878
~ radioactif 878
déshalogénation f 922
déshydratation f 923
déshydrogénation f 925
déshydrohalogénation f 926
désintégration f
~ α 126
~ β 340
~ radioactive 2880
photodésintégration 2647
desmotropie f 946
désorption f 947
désoxydant m 2987
désulfuration f 949
détecteur m
~ au germanium compensé au lithium 2050
~ de neutrons 2364
~ de neutrons à résonance 3041
~ de rayonnement 2863
~ différentiel 990
~ intégral 1748
~ par ionisation 1831
~ semiconducteur 3176
~ ~ Ge(Li) 2050
détection f
~ d'un rayonnement 951
~ d'un rayonnement radioactif 951
détermination f de l'âge radioactif 2879
détonation f 953
deutération f 955
deutérium m 956
deutéron m 957
deuton m 957
deuxième loi f de Finck 1355
deuxième moment m 3145
deuxième principe m de la thermodynamique 3144
développement m 958
~ de chromatogramme 958
déviation f 959
~ standard 3354
dextrorotation f 961
diagramme m
~ de Laue 1977
~ de phases 2611
~ de solubilité 3254
~ du cristal tournant 3072
~ moléculaire 2272
diagrammes mpl de Feynman 1353
dialyse f 962
dialyseur m de Graham 1500
diamagnétiques mpl 964
diamagnétisme m 967
diastéréo-isomères mpl 969
diastéréo-isomérie f 968
diazo-réaction f 971

diazotation *f* 971
dichroïsme *m* 972
~ circulaire 621
diélectrique *m* 974
diffraction *f* 2010
~ des atomes et molécules 997
~ des électrons 1146
~ des neutrons 2365
~ des rayons X 3775
~ électronique 1146
diffractomètre *m* de rayons X 3777
diffusion *f* 1000
auto-~ 3170, 3174
~ cohérente 653
~ Compton 697
~ d'échange 1293
~ de Rayleigh 2951
~ des particules 3116
~ du rayonnement 3117
~ élastique 1091
~ incohérente 1703
~ inélastique 1725
~ superficielle 3468
~ thermique 3525
rétrodiffusion 295
dilatance *f* 1007
dilution *f*
~ isotopique 1883
~ ~ inverse 3050
~ ~ simple 1013
~ ~ substœchiométrique 3438
dimérisation *f* 1010
dipôle *m* 1098
~ induit 1715
~ magnétique 2098
~ permanent 2597
diradicaux *mpl* 360
dismutation *f* 1029
~ de radicaux libres 1030
~ selon Cannizzaro 467
dispergation *f* 1018
dispersion *f* 1021
~ de la conductibilité 1027
~ de la lumière 2011
~ diélectrique 976, 983
~ inverse 2975
~ linéaire 2022
~ ~ réciproque 2975
~ rotatoire 2475
dispersité *f* 912
polydispersité 2718, 2724
dispersoïde *m* 1020
~ moléculaire 3636
disproportionation *f* 1029
dissipation *f* d'énergie 1032
dissociation *f* 1036
~ électrolytique 1122
~ hétérolytique 1595
~ homolytique 1630
~ thermique 3526
prédissociation 2760
distance *f* interatomique 396
distillation *f* avec un entraîneur 492
distribution *f*
~ binomiale 352
~ Compton 695
~ de Fischer-Snedecor 1333
~ de la variable aléatoire χ^2 598
~ de Laplace-Gauss 2401
~ de Poisson 2694
~ des doses 1050
~ maxwellienne 2143
~ normale 2401
~ réelle 1214
divariant système *m* de phases 366
domaine *m* azéotropique 289
domaines *mpl* 1043
donneur *m* 1046

donneur
~ d'électrons 1147
dosage *m* 952
~ des éléments 1191
dose *f* 2864
~ absorbée 11
~ d'exposition 1308
~ de rayonnement 2864
~ en profondeur 944
~ intégrale 1749
~ léthale 1999
~ ~ 50% 2168
~ ~ moyenne 2168
~ maximale admissible 2140
~ mortelle 1999
dosimètre *m* 1052, 1055
~ de film 1358
~ de Fricke 1429
~ du seuil limite 3569
~ individuel 2604
~ luminescent 2075
~ photographique 1353
~ thermoluminescent 3549
luminodosimètre 2075
dosimétrie *f* 1056
~ chimique 571
~ de radiation ionisante 1056
~ du rayonnement ionisant 1056
double couche *f* électrochimique 1096
double précipitation *f* 3029
double résonance *f* 1062
doublet *m* 1063
~ électronique 1177
~ liant 398
durée *f*
~ d'existence 2446
~ de vie 2446
~ de vie moyenne 2875
~ moyenne d'une collision 2156
dysprosium *m* 1072

E

eau *f* lourde 1574
ébulliométrie *f* 1073
ébullioscopie *f* 1073
ébullition *f* 385
écart *m* 959
~ hyperfin 1684
~ quadratique moyen 3354
~ ~ ~ relatif 3018
~-type 3354
~-~ de moyenne 3355
~-~ empirique relatif 3018
échange *m*
~ isotopique 1896
~ ~ hétérogène 1593
~ ~ homogène 1622
superéchange 3449
échangeur *m*
~ d'anions 178
~ d'ions 1806
~ d'ions liquide 2044
~ de cations 511
~ macroporeux 2088
~ minéral 1741
~ organique 1907
~ spécifique 3281
échantillon *m* 3101
~ brut 1705
~ global 1513
~ moyen 1956
~ pour laboratoire 1956
~ représentatif 3030
échantillonnage *m*
~ au hasard 2941
~ simple 2941
échelle *f*
~ chimique des poids atomiques 586
~ d'électronégativité 1149

électrons
~ de valence 3682
~ délocalisés 932
~ internes 1737
~ liants 394
~ localisés 2056
~ non appariés 3667
~ non couplés 3667
~ non liants 2388
~ optiques 2468
~ orbitaux 2480
~ périphériques 3632
~ planétaires 2480
~ secondaires 3136
électro-osmose *f* 1181
électrophorèse *f* 1183
électrosélectivité *f* 1185
électrovalence *f* 1187
élément *m* 573, 1445
~ 104 1189
~ 105 1190
~ d'une population 1198
~ Daniell 873
~ de référence 1778
~ de symétrie 3482
~ étalon 3350
~ pur 3217
~ secondaire 31
~ Weston insaturé 3669
~ ~ normal 3754
radioélément 2902
éléments *mpl*
~ de transition 3605
~ de transition interne 1739
~ non-métalliques 2393
~ transuraniques 3618
élimination *f* 1199
éluant *m* 1201
éluat *m* 1200
élution *f* 1202
~ par éluants successifs 3384
~ par gradient de pouvoir éluant 1496
émanation *f* du radium 2926
émanations *fpl* 1206
embranchement *m* 423
émetteur *m* 1213
empoisonnement *m* du catalyseur 499
empreinte *f* digitale 1362
émulsion *f* 1216
émulsoïde *m* 1219
énantiotropie *f* 1223
enceinte *f* étanche 1636
encombrement *m* stérique 3397
énergie *f*
~ à seuil 3570
~ coulombienne 784
~ d'activation 63
~ d'activation apparente 203
~ d'activation réelle 3633
~ d'échange 1287
~ d'excitation 1296
~ d'ionisation 1832
~ de délocalisation 931
~ de désintégration radioactive 1233
~ de Fermi 1340
~ de la particule 2566
~ de liaison chimique 567
~ de liaison nucléaire 2406
~ de localisation 2054
~ de mésomérie 3038
~ de partié 2536
~ de réaction nucléaire 1232
~ de réseau 1975
~ de résonance 3038
~ de rotation 3077
~ de séparation 832, 3188
~ de stabilisation 833
~ de symétrie nucléaire 2436
~ de translation 3611
~ de vibration 3704

énergie
~ interne 1771
~ libre 1579
~ ~ à temperature constante 1579
~ liée 412
~ nucléaire 2408
~ nulle 3798
~ quantique 2837
~ superficielle 1421
enlèvement *m* 2664
énolisation *f* 1237
enrichissement *m* isotopique 1895
ensemble *m*
~ canonique de Gibbs 468
~ grand cannonique 1504
~ ~ ~ de Gibbs 1504
~ microcanonique de Gibbs 2220
~ représentatif 3374
~ statistique 3374
~ virtuel 3374
enthalpie *f* 1240
~ d'activation 1241
~ d'hydratation 1563
~ de réaction à l'état standard 3357
~ libre 1485
~ ~ à pression constante 1485
~ molaire 2258
~ standard de réaction 3357
entraînement *m* 3120, 3121
entraîneur *m* 491, 3119, 3301
anti-~ 1612
~ de rétention 1612
~ en retour 1612
~ isotopique 1893
~ nonisotopique 2392
entrode *f* 166
entropie *f* 1242
~ d'activation 1246
~ de mélange 1247
~ échangée 3459
~ spécifique molaire 2259
enzyme *m* 1249
épaisseur *f*
~ de l'atmosphère ionique 2928
~ de solution absorbante 3102
épimères *mpl* 1253
épimérisation *f* 1252
époxydation *f* 1256
équation *f*
~ BET 346
~ calorique d'état 463
~ chimique 574
~ d'Arrhenius 223
~ d'Einstein 1088
~ d'Einstein-Smoluchowski 1089
~ d'état 3527
~ d'état de Beattie-Bridgeman 324
~ d'état de Berthelot 336
~ d'état de Clausius 635
~ d'état de Dieterici 987
~ d'état de gaz parfait 1694
~ d'état de Van der Waals 3683
~ d'état réduite 2980
~ d'état virielle 3716
~ d'Hammett 1538
~ d'Ilkovič 1700
~ d'ondes contenant le temps 3574
~ de bilan pour l'impulsion dans les systèmes continus 300
~ de Boltzmann 388
~ de Born 401
~ de Bragg 420
~ de Brönsted-Bjerrum 437
~ de continuité 2126
~ de Gibbs 1482, 1488
~ de Henderson 1581
~ de Hugoniot 1647
~ de Kelvin 1917
~ de Laplace 2695
~ de Lewis et Sargent 2003
~ de Liouville 2035
~ de Lippmann 2037

équation
~ de Perrin 310
~ de Schrödinger 3575
~ de Szyszkowski 3492
~ de Van der Waals 3688
~ de vitesse 1928
~ de von Neumann 3737
~ fondamentale 1488
~ opératorielle de Schrödinger 3737
~ réduite de vitesse 1257
~ stœchiométrique 574
~ thermique d'état 3527
~ thermochimique 3532
équations *fpl* de Gibbs-Helmholtz 1486
équilibre *m* 3537
~ chimique 575
~ ~ instable 1320
~ de sédimentation 3154
~ électrochimique 2394
~ entre phases 2612
~ local 2053
~ mécanique 2157
~ membraneux 2173
~ physique 2612
~ radioactif 2882
~ ~ transitoire 3602
~ séculaire 3150
~ tautomérique 3500
~ thermique 3528
~ thermodynamique 3537
~ transitoire 3602
équivalent *m* 576
~ chimique 576
~ de dose 1051
~ en eau 1230
~-gramme 1502
erbium *m* 1264
erreur *f* 1512
~ absolue 6
~ accidentelle 2939
~ aléatoire 2939
~ ~ d'échantillonage 3103
~ d'analyse 164
~ d'indicateur 1710
~ d'une goutte 1065
~ de justesse 460
~ de parallaxe 2544
~ de sels 3098
~ de titrage 3585
~ personnelle 2464
~ pour-cent 2585
~ propre à la méthode 2202
~ relative 3011
~ standard de moyenne 3355
~ systématique 3490
espace *m*
~ des phases 2615
~ mort 883
~ vide 1422
espèce *f* chimique 588
espérance *f* mathématique 1305
essai *m*
~ à blanc 373
essaim *m* ionique 1801
estérification *f* 1271
transestérification 3599
estimateur *m* statistique 1273
estimation *f* 1272
~ ponctuelle 1272
étain *m* 3578
étalon *m*
~ extérieur 1314
~ externe 1313
~ intérieur 1779
~ interne 1778
~ primaire 2774
~ secondaire 3140, 3141
étape *f*
~ cinétique 1932
~ hydrodynamique 1661
~ initiale 1735

état *m*
~ amorphe des polymères 150
~ colloïdale 664
~ cristallin 837
~ critique 815
~ d'agrégat 3364
~ d'élasticité élevée 1601
~ d'oxydation 2522
~ d'oxydation du ion central 2526
~ de gaz 1462
~ de transition 58
~ de vibration-rotation 3711
~ dégénéré 904
~ excité 1301
~ fondamental 1515
~ gazeux 1462
~ instable 3672
~ liquide 2047
~ macroscopique 2090
~ mésomorphe 2183
~ métallique 2193
~ métastable 2198, 2201
~ microscopique 2228
~ nonstationnaire 2395
~ physique 3364
~ plasma 2681
~ plastique 2682, 2683
~ pseudo-cristallin 2812
~ quantique 2940
~ singulet 3226
~ solide 3251
~ stable 3348
~ standard 3360
~ stationnaire 3371
~ subcritique 2758
~ triplet 3629
~ vibronique 3711
~ vitreux 3728
états *mpl* correspondants 781
étendue *f* d'une série statistique 2944
éthynylation *f* 1276
étrangeté *f* 3410
europium *m* 1277
eutectique *m* 1278
~ binaire 350
~ solide 3248
eutectoïde *m* 1281
exaltation *f* 2469
excimères *mpl* 1294
excitation *f* 1295
~ à deux photons 3644
~ monoélectronique 2459
exciton *m* 1302
exclusion *f* d'ions 1809
explosion *f* 1306
~ en chaîne 527
~ thermique 3529
exposition *f* 1308
extinction *f*
~ de luminescence 2847
extraction *f* 1315

F

facteur *m*
~ d'analyse 165
~ d'enrichissement 1239
~ de Boltzmann 389
~ de compressibilité 694
~ de correction 444
~ de décomposition spectrale 1962
~ de décontamination 901
~ de dégénérescence 2
~ de dépolarisation 911
~ de fréquence 2761
~ de géométrie 1479
~ de l'entropie 1244
~ de Landé 1962
~ de Landé nucléaire 2414
~ de probabilité 3396
~ de qualité 2832

facteur
~ de saturation 3110
~ de séparation 3189
~ de transmission 3616
~ kT/h 3664
~ stérique 3396
facteurs mpl d'intensité 1752
faisceau m 323, 639
~ d'electrons 1137
~ de rayonnement 323
famille f
~ de l'actinium 53
~ de l'actino-uranium 53
~ de l'azote 2383
~ de l'oxygène 538
~ de l'uranium 3674
~ des éléments 1321, 1322
~ du carbone 482
~ du cuivre 775
~ du curium 857
~ du thorium 3563
~ du zinc 3801
~ principale 2118
~ secondaire 3430
familles fpl radioactives 2886
faraday m 1324
femtomètre m 1334
fer m 1846
radiofer m 2906
ferment m 1249
fermentation f
~ alcoolique 111
~ butyrique 453
~ cytrique 628
~ lactique 1957
~ oxydative 2527
fermions mpl 1341
fermium m 1345
ferrimagnétiques mpl 1347
ferrimagnétisme m 1346
ferrites fpl 1348
ferroélectrique m 1349
ferromagnétiques mpl 1351
antiferromagnétiques 195
ferromagnétisme m 1350
antiferromagnétisme 194
~ parasite 2533
fidélité f de la méthode 2756
filtrat m 1359
filtration f 1360
ultrafiltration 3649
filtre m
~ à échelons 3382
~ coloré 18
~ interférentiel 1756
~ neutre 2356
~ optique 2012
fin f de titrage 1226
fission f 2410
~ nucléaire 2410
~ spontanée 3337
photofission 2636
flambage m 2754, 2764
flexibilité f des macromolécules 1385
floculation f 1386
flottation f 1387
fluctuation f
~ absolue 3353
~ quadratique moyenne 2155
~ relative 3017
fluctuations fpl 1389
~ d'une grandeur 1389
~ en équilibre 1261
fluence f 1390
~ énergetique 1231
fluide m 1391
fluidité f 1392

fluor m 1400
fluoration f 1399
fluorescence f 1393
~ retardée 928
~ sensibilisée 3185
fluorimètre m 1398
fluorimétrie f 1394
flux m 1401
~ d'entropie 1245
~ d'induction magnétique 2101
~ de rayonnement 1403
~ magnétique 2101
flux mpl 3539
fonction f
~ d'acidité de Hammett 46
~ d'excitation 1297
~ d'onde 3741
~ de dissipation de l'énergie mécanique 1035
~ de distribution et position et impulsion de n
   corpuscules 1476
~ de distribution n-ple 1476
~ de distribution radiale 2856
~ de formation 1408
~ de Gibbs 1485
~ de Helmholtz 1579
~ de Massieu 2129
~ de Morse 2304
~ de partition de rotation 3073
~ de partition de translation 3612
~ de partition de vibration 3705
~ de partition électronique 1160
~ de partition interne 1773
~ de partition interne de la molécule 1773
~ de partition potentielle 2572
~ de Planck 2676
~ de répartition 2788
~ densité 2787
~ ~ de la fréquence 2787
~ ~ de répartition 2787
~ des probabilités totales 2788
~ propre 1085
~ radiale de distribution 2856
grande ~ de partition 1505
fonctions fpl
~ caractéristiques 545
~ d'état 3365
fond m
~ Compton 695
~ spectral 293
fondant m 1401
force f
~ de liaison 400
~ électromotrice 1133
~ ~ standard 3356
~ ionique 1824
forces fpl 3540
~ d'adsorption 95
~ de Coulomb 785
~ de dispersion 1024
~ de Van der Waals 3689
~ généralisées 3540
~ intermoléculaires 1763
~ nucléaires 2411
formation f d'éther-oxyde 1274
forme f
~ aci 50
~ bateau 379
~ cétonique 1923
~ chaise 537
~ cyclique 866
~ énolique 1236
~ érythro 1268
~ fondamentale 3659
~ ~ primitive 2776
~ méso 2179
~ polymorphe 2725
~ primitive 2776
~ thréo 3568
formes fpl de Kékulé 1915
formule f
~ brute 2277

formule
~ chimique 577
~ d'Arrhenius 223
~ de Breit-Wigner 431
~ de Clausius-Mossotti 636
~ de Debye et Hückel 888
~ de Freundlich 1427
~ de Hückel 1644
~ de Kirchhoff 1934
~ de Langmuir 1964
~ de Lorentz-Lorenz 2064
~ de Mayer 2145
~ de structure 3424
~ de Tafel 3495
~ de Van't Hoff 3694
~ développée 3424
~ électronique 1156
~ empirique 1215
~ en projection de E. Fischer 1370
~ massique de Weizsäcker 3751
~ projetée 2790
~ stérique 3389
formules *fpl* de Kékulé 1915
formylation *f* 1409
fraction *f*
~ d'échange 1288
~ en masse 2128
~ en volume 3732
~ molaire 2289
fragmentation *f* 1412
fragments *mpl* de fission 1374
francium *m* 1413
fréquence *f* 1425
~ caractéristique 543, 544
~ d'échange 1290
~ de Larmor 1968
~ de Raman 2934
~ moyenne de collision 660
~ relative 3012
frittage *m* 3227
~ du catalyseur 3228
front *m* Compton 696
frottement *m* intérieur 3720
fugacité *f* 3729
fusion *f* 2171
furet *m* 2849

G

gadolinium *m* 1443
gallium *m* 1444
gamma *m* 748
gaz *m* 1455
~ d'électrons 1152
~ électronique dégénéré 903
~ parfait 1693
~ porteur 494
~ réel 2971
gaz *mpl*
~ inertes 1727
~ nobles 1727
~ rares 1727
gazométrie *f* 1464
gel *m* 1471
coagel 642
hydrogel 1663
lyogel 2090
organogel 2489
xérogel 3770
gélatinisation *f* 1472
gelée *f* 2080
gélification *f* 1472
générateur *m*
~ de neutrons 2366
géochimie *f* 1477
géométrie *f*
~ de comptage 796
~ de l'absorption des rayonnements 1480

germanium *m* 1481
glucinium *m* 338
glucinium *m*, magnésium *m* et metaux *mpl* alcalino-terreux 339
gonflement *m* 3476
goniomètre *m* de Weissenberg 3750
grand potentiel *m* 1506
grand potentiel thermodynamique 1506
grande fonction *f* de partition 1505
grandeur *f* d'état 3366
grandeurs *fpl*
~ critiques 813
~ d'excès 1284
~ de mélange 1440
~ extensives 1309
~ macroscopiques 3365
~ molaires 2262
~ ~ partielles 2564
granulométrie *f* 2569
grappe *f* 3341
gravimétrie *f* 1507
thermogravimétrie 3547
gray *m* 1508
groupe *m* 167, 1516
~ à déplacer 1991
~ à substituer 1991
~ analytique 167
~ auxiliaire 3430
~ d'espace 3272
~ du bore 406
~ du carbone 482
~ du chrome 613
~ du fer 1352
~ du soufre 3444
~ du titane 3581
~ fonctionnel 1437
~ ponctuel 2693
~ pont 433
~ principal 2118
~ spatial 3272
second ~ fonctionnel 760
sous-~ 3430
groupement *m*
~ acide 43
~ bathochrome 319
~ fonctionnel 1437
~ polaire 2696
groupes *mpl* ponctuels cristallins 842

H

hadrons *mpl* 1523
hafnium *m* 1524
halogénation *f* 1535
halogènes *mpl* 1536
hauteur *f* du plateau théorique 1080
hélicité *f* 1576
hélium *m* 1577
hétéroatome *m* 1586
hétéroazéotrope *m* 1587
hétérolyse *f* 1595
hétéropolyacide *m* 1598
hétéropolymérisation *f* 1599
holmium *m* 1616
homoazéotrope *m* 1617
homogénéisation *f* des émulsions 1625
homologie *f* 1629
homologues *mpl* 1628
homolyse *f* 1630
homopolymètre *m* 1632
homopolymérisation *f* 1633
hybridation *f* des orbitales 1651

M

mécanisme *m*
~ A_{Ac}^1 138
~ A_{Ac}^2 139
~ A_{Ai}^1 140
~ B_{Ac}^2 377
~ B_{Ai}^1 378
~ d'échange homolytique 3241
~ de l'addition électrophile à la liaisons multiples 2159
~ de l'addition nucléophile au liaison éthylénique 2164
~ de l'addition nucléophile au carbonyle 2163
~ de l'addition radicalaire aux oléfines 2161
~ de l'élimination α 127
~ de l'élimination E_1 1208
~ de l'limination E_2 1209
~ de la polymérisation par les radicaux 2162
~ de la réaction $S_N 1$ 3236
~ de la réaction $S_N 2$ 3237
~ de la substitution électrophile aromatique 2160
~ de la substitution électrophile biparticulaire 3235
~ de la substitution électrophile $S_E 1$ 3234
~ de la substitution nucléophile aromatique 2165
~ de la substitution nucléophile biparticulaire 3237
~ de la substitution nucléophile biparticulaire avec transposition allylique 3238
~ de la substitution nucléophile interne 3239
~ de la substitution nucléophile monomoléculaire 3236
~ de la substitution radicalaire 3241
~ de réaction 2960
~ des réactions radiochimiques 2158
~ push-pull 2821
~ $S_E 2$ 3235
~ $S_N 2'$ 3238
~ $S_N i$ 3239
~ $S_N i'$ 3240
mécanochimie *f* 2166
médiane *f* 2167
mélange *m*
~ d'isotopes 2244
~ eutectique 1278
~ idéal 1695
~ ~ des gaz parfaits 2587
~ inactif dédoublable 2852
~ racémique 2852
membrane *f* échangeuse d'ions 1808
mendélévium *m* 2175
ménisque *m* 2176
mer *m* 739
mercure *m* 2177
mésomérie *f* 2182
méson *m*
~ π 2670
~ K 1936
mésons *mpl* 2186
mésothorium *m*
~ I 2187
~ II 2188
mesure *f* absolue de l'activité 4
métal *m* carbonyle 2189
métalloïdes *mpl* 2194, 2393
métamagnétiques *mpl* 2196
métamagnétisme *m* 2197
métamérie *f* 1439
métaux *mpl* 2195
~ alcalino-terreux 118
~ alcalins 116
~ de la famille du chrome 613
~ de la famille du manganèse 2120
~ de la mine de platine 2687
~ des terres rares 1966

métaux
~ légers de la mine de platine 2013
~ lourds de la mine de platine 1573
méthode *f*
macrométhode 1503
~ à courant 1388
~ à étalon interne 2209
~ à secteur rotatif 3074
~ centigrammique 515
~ classique 1503
~ CLOA 1989
~ coulostatique 791
~ d'Arndt-Eistert 216
~ d'émanation 1205
~ d'Enskog et Chapman 540
~ d'intégration 2207
~ d'isolement 1872
~ d'isolement d'Ostwald 1872
~ d'opposition de la mesure de f.e.m. 2690
~ d'oscillation 2501
~ de Bodroux 380
~ de Bouveault 413
~ de Combes 675
~ de comparaison à série d'étalons 3724
~ de comparaison visuelle 3724
~ de correction géométrique 316
~ de courbe d'étalonnage 3352
~ de dilution isotopique 1888
~ de Hull-Debye-Scherrer 891
~ de l'étalonnage intern 1780
~ de l'évaluation du potentiel de jonction d'Henderson 1582
~ de la détermination des nombres de transport de Hittorf 1606
~ de la liaison de valence 3680
~ de la résonance 3039
~ de Paneth-Hevesy 2540
~ de Reppe 3028
~ de Riley 3063
~ de rotation 3071
~ de variation d'épaisseur de la solution 2203
~ de Zincke 3802
~ des battements 1591
~ des indicateurs isotopiques 1899
~ des indicateurs radioactifs 2890
~ des mesures comparatives 2204
~ des moindres carrés 2208
~ des orbitales moléculaires 2280
~ des perturbations 2606
~ des rayons moléculaires 2268
~ des solutions diluées 1008
~ des surfaces limites mobiles 2312
~ des variations 3702
~ différentielle 992
~ ~ de Van't Hoff 992
~ du calcul du potentiel de jonction de Planck 2677
~ du champ moléculaire self-consistent 1544
~ du crystal rotatif 3071
~ du déplacement des surfaces limites 2311
~ dynamique 1388
~ grammique 1503
~ mécanique d'oscillation du cristal 2501
~ ~ de rotation du cristal 3071
~ microgrammique 2224
~ milligrammique 2236
~ nanogrammique 2331
~ par dilution 1009
~ par voie humide 3755
~ picrogrammique 2665
~ statique 3367
microméthode 2236
semi-microméthode 515
submicrométhode 2331
subultramicrométhode 2665
ultramicrométhode 2224
ultraultramicrométhode 2331
méthodes *fpl*
~ d'amplification 158
~ de condensation 705
~ de dispersion 1026
~ de relaxation 3022
~ galvanostatiques 1447
~ potentiostatiques 2750

méthylation *f* 2210
aminométhylation 146
~ totale 1303
transméthylation 3614
micelle *f* 2212
microanalyse *f* 2216
semi-~ 3718
microanalyseur *m* à sonde électronique 1175
microcalorimètre *m* 2218
~ adiabatique construit par Swiętosławski et Dorabialska 1049
microcalorimétrie *f* 2219
microchimie *f* 2221
microcomposant *m* 2222
microméthode *f*
semi-~ 515
microphotomètre *m* 2226
microscope *m* polarisant 2705
microspectrophotométrie *f* 2229
migration *f* des ions 2234
mi-largeur *f* 1526
milieu *m* dispersif 1025
minéralisation *f* 2237
miscibilité *f*
~ illimitée 683
~ limitée 2562
mobilité *f* d'ion 1815
modèle *m*
~ à particules indépendantes 1707, 2460
~ d'Ising 1858
~ de faisceaux 640
~ de la goutte liquide 2043
~ de Nilsson 2377
~ de superconductivité 3448
~ des couches 3192
~ du gaz de Fermi 1339
~ nucléaire alpha 129
~ optique 2472
~ unifié 3658
modèles *mpl*
~ collectifs 657
~ du noyau 2248
~ nucléaires 2248
modérateur *m* 2249
modération *f* 3232
modificateur *m* d'une réaction en chaîne 2250
molalité *f* 2254
molarité *f* 2256
mole *f* 2266, 2267
molécularité *f* de la réaction 2278
molécule *f* 2288
macromolécule 2087
~ activée 59
~ cuspidale asymétrique 237
~ ~ sphérique 3319
~ ~ symétrique 3478
~ diatomique hétéronucléaire 1597
~ ~ homonucléaire 1631
~ dipolaire 2706
~ excitée 1299
~-gramme 2267
~ linéaire 2026
~ polaire 2706
~ sandwich 3105
molécules *fpl* isoélectroniques 1868
molybdène *m* 2290
moment *m*
deuxième ~ 3145
~ angulaire du noyau 2435
~ ~ orbital 174
~ ~ total 3588
~ cinétique orbital 174
~ ~ total 3588
~ d'un groupement 1517

moment
~ de liaison 392
~ de spin 3326
~ de transition 3606
~ dipolaire 1099
~ ~ induit 1716
~ ~ permanent 2598
~ électrique dipolaire 1099
~ magnétique 2104
~ ~ atomique 2105
~ ~ nucléaire 2418
~ orbital 2481
~ quadrupolaire 2828
~ ~ nucléaire 2426
second ~ 3145
moniteur *m* 2292
monochromateur *m* 2294
monocristal *m* 3222
monomère *m* 2297
monotropie *f* 2303
monovariant système *m* de phases 3662
mouillage *m* 3756
moulin *m* colloïdal 666
mousse *f* 1404
mouvement *m*
~ brownien 440
~ perpétuel de première espèce 2602
~ ~ de seconde espèce 2603
~ propre 294
moyenne *f*
~ arithmétique 214
~ d'ensemble 3373
~ d'une variable aléatoire 1305
~ géométrique 1478
~ temporelle 3573
multiplet *m* 301, 2321, 2322
~ de Balandin 301
multiplets *mpl* de charge 553
multiplication *f* due au gaz 1463
multiplicité *f* 2323, 2324
muon *m* 2313
mutarotation *f* 2329
mutation *f* radio-induite 2868

N

n-conductivité *f* 1140
néodyme *m* 2346
néon *m* 2347
néphélomètre *m* 2349
néphélométrie *f* 2350
neptunium *m* 2351
neutralisation *f* 2357
neutretto *m* 2326
neutrino *m* 2358
antineutrino 198 (*vide*)
~ ν_μ 2326
~ ν_e 1159
neutron *m* 2359
antineutron 199
neutrons *mpl*
~ de fission 1375
~ différés 929
~ épithermiques 1255
~ immédiats 2793
~ intermédiaires 1760
~ lents 3233
~ prompts 2793
~ rapides 1328
~ retardés 929
~ thermiques 3530
nickel *m* 2373
niobium *m* 2379
nitration *f* 2380
nitrosation *f* 2384

niveau *m*
~ accepteur 30
~ de Fermi 1340
~ de signification 3210
~ donneur 1047
~ énergétique du noyau 2409
niveaux *mpl*
~ de rotation de la molécule 2282
~ de vibration de la molécule 2287
~ électroniques de la molécule 2273
~ énergétiques de la molécule 2274
~ ~ de rotation de la molécule 2282
~ ~ de vibration de la molécule 2287
~ ~ électroniques de la molécule 2273
~ Zeeman 3791
niveleur *m*
solvant-~ 2001
nobélium *m* 2386
noircissement *m* 370
nombre *m*
~ atomique 250
~ ~ effectif 1077
~ baryonique 312
~ d'Avogadro 282
~ d'hydratation d'ion 1657
~ d'occupation 281, 2447
~ d'ondes 3742
~ d'or 1492
~ d'oxydation 2522
~ de constituants 2444
~ de constituants indépendants 2444
~ de coordination 765, 766
~ de coordination caractéristique 542
~ de degrés de liberté 2442
~ de masse 2130
~ de plateaux théoriques 3519
~ de solvatation d'ion 3261
~ de symétrie 3483
~ de transport d'ion 3600
~ de variables indépendantes 2445
~ de variables intensives indépendantes 2445
~ des réactions chimiques indépendantes 2443
~ effectif de magnétons de Bohr 1081
~ leptonique 1996
~ ~ électronique 1158
~ maximum de coordination 2139
~ moyen d'occupation 281
~ ~ de coordination 2009
~ muonique 2327
~ quantique azimutal 2482
~ ~ de rotation 3079
~ ~ de spin 2109, 3329
~ ~ de vibration 3706
~ ~ effectif 10ⁿ2
~ ~ magnétique 2107
~ ~ principal 2778
~ ~ secondaire 2482
nombres *mpl*
~ aléatoires 2940
~ magiques 2094
non-métaux *mpl* 2393
normalisation *f* de la fonction d'onde 2404
normalité *f* de solution 2400
noyau *m*
~ aromatique 219
~ atomique 249
~ bimagique 1064
~ carbocyclique 478
~ composé 692
~ excité 1300
~ hétérocyclique 1590
~ homocyclique 1619
~ impair-impair 2451
~ impair-pair 2450
~ isocyclique 1619
~ magique 2093
~ métastable 2199
~ pair-impair 1283
~ pair-pair 1282
noyaux *mpl*
~ condensés 710
~ miroirs 2238
n-semi-conducteur *m* 1150

nuage *m*
~ de charge 549
~ électronique 1139
nucléide *m* 2439
~ père 2555
~ radioactif 2912
~ stable 3347
radionucléide 2912 (*vide*)
nucléon *m* 2437
nuclide *m* 2439

O

obstacle *m* stérique 3398
occlusion *f* d'ions 1809
octet *m* électronique 1176
oligomère *m* 2454
oligomérisation *f* 2455
onde *f*
~ de choc 3198
~ de de Broglie 886
~ de détonation 954
~ élastique 1092
ondes *fpl* de spin 3334
opalescence *f* critique 810
opérateur *m*
~ d'Hamilton 1537
~ densité 3376
~ statistique 3376
opérateurs *mpl* de la mécanique quantique 2838
opérations *fpl* de symétrie 3484
or *m* 1491
orbitale *f* 2479
~ π 2672
~ σ 3309
~ antiliante 193
~ atomique 251
~ de Slater 3230
~ hybride 1652
~ hydrogénoïde 1671
~ liante 395
~ moléculaire 2279
~ ~ délocalisée 933
~ ~ localisée 2057
~ non liante 2389
orbite *f* de Bohr 382
ordre *m*
~ à longue échéance 2063
~ à parcour réduit 3200
~ apparent de la réaction 206
~ d'interférence 2484
~ de réflexion 2484
~ de la réaction 2485
~ des stabilités d'Irving-Williams 1856
~ global de la réaction 2485
~ partiel de la réaction 2486
organogel *m* 2489
organosol *m* 2492
orientation *f* de substitution aromatique 2494
orthogonalité *f* des fonctions d'onde 2498
oscillateur *m* 2503
~ anharmonique 176
~ de Planck 2678
~ harmonique 1543
oscillopolarographe *m* 2504
osmium *m* 2507
osmomètre *m* 2508
osmose *f* 2509
électro-~ 1181
ouverture *f* d'un cycle 3065
oxydant *m* 2520

oxydation *f* 2521
auto-~ 279
~ anodique 1126
~ biologique 355
~ d'Oppenauer 2465
photooxydation 2648
oxydimétrie *f* 2528
oxygène *m* 2530
~ lourd 1572
ozonation *f* 2532
ozonisation *f* 2532
ozonolyse *f* 2533

P

p-conducteur *m* 1143
p-semi-conducteur *m* 1143
paire *f*
~ d'électrons 1177
~ d'ions 1840
palladium *m* 2539
papier *m*
~ réactif 1711
~ universel 3666
paquet *m* d'ondes 3743
parachor *m* 2542
~ atomique 252
paramagnétiques *mpl* 2551
paramagnétisme *m* 2550
superparamagnétisme 3454
paramètre *m* 2552
~ de population 2552
~ du système 3366
~ fissible 1376
~ indépendant 3366
parcours *m*
libre ~ moyen 2149
~ de particule 2946
parité *f* 2556
paroi *f*
~ adiabatique 86
~ diathermane 970
~ semi-perméable 3179
particule *f* 1192
antiparticule 200
~ α 128
~ β 341
~ ionisante 1835
particules *fpl*
~ α 130
~ β 344
~ colloïdales 663
~ élémentaires 1194
~ étranges 3412
partie *f* de matériau 445
passivité *f* du métal 2575
pellicule *f*
~ superficielle 3469
~ surfactive 3464
peptisation *f* 2583
percée *f* 673
père *m* nucléaire 2555
période *f* 2590
~ d'induction 1722
~ de relaxation 3023
~ radioactive 1528
perméabilité *f* magnétique 2106
permittivité *f* 2599
~ complexe 687
~ du vide 2600
~ optique 2473
~ relative 3014
~ statique 3368
pertes *fpl* diélectriques 977
pesée *f* 3749

pH 2608
pH-mètre *m* 2621
pH-métrie *f* 2622
phase *f* 2609
~ critique 811
~ dispersée 1019
~ gazeuse 1461
~ liquide 2046
~ métastable 2200
~ mobile 2247
~ solide 3249
~ stationnaire 3370
phases *fpl*
~ condensées 709
~ de Hume-Rothery 1648
~ de Laves 1978
~ énantiotropes 1222
~ intermétalliques 1761
~ mésomorphes 2042
~ monotropes 2302
phénomène *m* de Barkhausen 308
phénomènes *mpl*
~ de surface 3470
~ dissipatifs 1034
phonochimie *f* 3267
phonon *m* 2623
phosphore *m* 2625
radiophosphore 2913
phosphorescence *f* 2624
phosphorylation *f* 2626
phosphorylisation *f* 2626
photo-conducteur *m* 2633
photo-ionisation *f* 2640
photochimie *f* 2630
photoconductivité *f* 2632
photodésintégration *f* 2647
photodissociation *f* 2643
photoélectron *m* 2635
photofission *f* 2636
photofraction *f* 2637
photographie *f*
Laue ~ 1977
photoisomérisation *f* 2641
photoluminescence *f* 2642
photolyse *f* 2643
~ éclair 1383
photomètre *m* 2644
microphotomètre 2226
~ de flamme 1380
spectrophotomètre 3310
photométrie *f* 2645
~ de flamme 1381
~ énergétique 2660
~ photographique 2638
~ visuelle 3725
spectrophotométrie 3312
photon *m* 2646
~ d'annihilation 183
photons *mpl* gamma en cascade 1453
photooxydation *f* 2648
photopic *m* 2649
photopolymérisation *f* 2652
photoproduction *f* de mésons 2184
photoréduction *f* 2653
photosynthèse *f* 2657
physique *f* chimique 581
physisorption *f* 2658
pic *m* 2579
photopic 2649
~ d'échappement 1269
~ d'énergie totale 2649
~ de deuxième échappement 1060
~ de fuite 1269
~ de fuite des photons 3773

pic
~ de premier échappement 3223
~ de résonance 3042
~ de rétrodiffusion 296
~ de somme 3446
~ photoélectrique 2649
pics *mpl* de paire 2537
piège *m* 3619
piégeage *m* des radicaux 3621
piézo-électricité *f* 753, 2669
piézo-électrique *m* 2668
pile *f* 1445
~ à combustibles 1435
~ à combustion 1435
~ chimique 568
~ Daniell 875
~ de concentration 698
~ de concentration à transport d'ions 700
~ de concentration sans transport d'ions 699
~ étalon 3350
~ ~ Weston 3754
~ Leclanché 1994
~ réversible 3051
~ thermoélectrique 1129
pion *m* 2670
pK 2674
plan *m*
~ de symétrie avec glissement 1490
~ réticulaire 1976
plaque *f* photographique 2639
plasma *m* 2679
~ chaud 1638
~ froid 656
plateau *m* du compteur 2684
platine *m* 2685
platinoïdes *mpl* 2687
plomb *m* 1990
actino-~ 52
~ de l'actinium 52
~ de l'uranium 2927
~ du thorium 3562
radioplomb 2925
urano-~ 2927
plutonium *m* 2688
pOH 2691
poids *m*
~ atomique 3008
~ moléculaire 3013
~ ~ moyen au nombre 2441
~ ~ moyen au poids 2124
~ relatif d'atome 3008
~ statistique 3380
point *m*
~ azéotropique 288
~ congruent 730
~ d'inflammation 1699
~ de Boyle 416
~ de charge zéro 3795
~ de cryohydrate 823
~ de Curie 854
~ de Curie paramagnétique 2547
~ de fusion incongruent 1704
~ de l'espace des phases 2614
~ de Néel 2333
~ de rosée 960
~ de sublimation 3432
~ distéctique 1038
~ équivalent 1262
~ eutectique 1279
~ hétéroazéotropique 1588
~ isoélectrique 1867
~ normal d'ébullition 2397
~ péritectique 2593
~ triple 3627
poison *m* du catalyseur 498
polarimètre *m* 2697
spectropolarimètre 3313
polarimétrie *f* 2698

polarisabilité *f* 2699
~ atomique 253
~ d'orientation 2495
~ électronique 1161
~ totale 3592
polarisabilités *fpl* principales 2777
polarisation *f* 978, 979
~ atomique 254, 255
~ circulaire de la lumière 622
~ d'activation 66, 3607
~ d'orientation 2496, 2497
~ de charge d'espace 3271
~ de concentration 702, 1004
~ de cristallisation 841
~ de la lumière 2700, 2701
~ de réaction 2963
~ de résistance 3034
~ diélectrique 978, 979
~ ~ complexe 684
~ électrochimique 1127
~ électrolytique 1127
~ électronique 1162, 1163
~ induite 1717
~ molaire 2261
~ ohmique 3034
~ rotatoire 3082
~ spécifique 3285
~ spontanée 3339
pseudo-~ de résistance 2452
polarité *f* de liaison 399
polarogramme *m* 2707
polarographe *m* 2708
oscillopolarographe 2504
polarographie *f* 2712
~ à impulsions 2317
~ à tension carrée 3344
~ à tension sinusoïdale surimposée 132
~ dérivée 993
~ oscillographique 2506
~ par redissolution anodique 137
polonium *m* 2713
polyacide *m* 2714
polyaddition *f* 2715
polycondensat *m* 706
polycondensation *f* 707
polydispersité *f* 2718, 2724
~ moléculaire 2724
polyélectrode *f* 2719
polyélectrolyte *m* 2720
polymère *m* 2721
bipolymère 359
copolymère 772 (*vide*)
~ atactique 238
~ cristallin 836
~ d'addition 79
~ greffé 1498
~ irrégulier 1851
~ isotactique 1883
~ linéaire 2027
~ multibranche 421
~ ramifié 421
~ régulier 3004
~ réticulé 2354
~ séquencé 375
~ stéréorégulier 3390
~ syndiotactique 3486
~ tactique 3494
prépolymère 2763
terpolymère 3512
polymérisation *f* 2722
copolymérisation 773
photopolymérisation 2652
~ en bloc 376
~ en émulsion 1218
~ en masse 446
~ greffée 1499
~ ionique 1816
~ par addition 80
~ radicalaire 1417
~ radicalique 1417
~ stéréospécifique 3393

recoupement *m* par bombardement 818
recouvrement *m* des orbitales 2518
recuit *m* 181
réducteur *m* 2987
réduction *f* 2988
photoréduction 2653
~ de Béchamp 325
~ de Birch 361
~ de l'activité du catalyseur 879
~ de Rosenmund 3069
~ électrolytique 1128
~ sélective 3162
~ selon Clemmensen 637
~ selon Meerwein-Ponndorf-Verley 2170
réductométrie *f* 2989
référence *f*
~ extérieure 1314
~ intérieure 1779
réflexion *f* diffuse 999
réfraction *f*
~ atomique 256
~ de lumière 2997
~ double 1061
~ ionique 1822
~ moléculaire 2263
~ spécifique 3288
réfractivité *f* moléculaire 2263
réfractomètre *m* 2999
réfractométrie *f* 3000
réfrigérant *m* du réacteur nucléaire 2433
refroidissement *m* 757
région *f*
~ de sous-exposition 3656
~ de surexposition 2717
~ infra-rouge 1731, 2232
~ ultraviolette 2337, 3652
~ visible 3722
règle *f*
~ d'Allinger 120
~ d'Auwers et Skita 3734
~ d'Hoffmann 1610
~ d'Hudson 1646
~ de Blanc 372
~ de Bredt 429
~ de Curtin et Hammett 861
~ de Djerassi et Klyne 285
~ de Freudenberg 1426
~ de Guldberg 1520
~ de Hückel 1645
~ de l'octet 1
~ de l'orientation stérique 806
~ de la force ionique de Lewis et Randall 2002
~ de Le Châtelier 1993
~ de Markowinkoff 2122
~ de Neumann-Kopp 1946
~ de Prelog 2762
~ de Saytzeff 3113
~ de Schmidt 3124
~ de superposition optique de Van't Hoff 3697
~ de Traube 3622
~ de Trouton 3632
~ de Van't Hoff 3696
~ de Walden 3740
~ des gaz rares 1
~ des octants 2449
~ des phases 1487
~ des trois sigmas 3567
règles *fpl*
~ de Hund 1649
~ de résonance 3083
~ de sélection 3158
règression *f* 3003
~ linéaire 2028
relargage *m* 3099
relation *f*
~ d'Eötvös 1251
~ d'incertitude 3654
~ de Brönsted 438
~ de Clapeyron 634
~ de Duhem-Margules 1070

relation
~ de Geiger-Nuttal 1470
~ de Gibbs-Duhem 1484
~ de Heisenberg 3654
~ de Latimer 1973
~ de Mayer 2145
relations *fpl*
~ de Gibbs-Helmholtz 1486
~ de Maxwell 2144
~ de réciprocité d'Onsager 2461
relaxation *f* 3020
~ chimique 584
~ de tension 3415
~ diélectrique 980
~ longitudinale 3325
~ quadrupolaire 2829
~ spin-réseau 3325
~ ~-spin 3332
~ transversale 3332
rem *m* 3024
rémanence *f* 105
rendement *m* 3785
~ chimique 2897
~ de fission 1378
~ du comptage 799
~ du détecteur 1787
~ en courant 860
~ ionique 1826
~ photochimique 2628
~ quantique de la luminescence 2843
~ radiochimique 2874
~ radiolytique 2874
renversement *m* 3173
~ d'une raie 3173
rep *m* 3026
répartition *f*
~ de Boltzmann 2142
~ de Bose-Einstein 408
~ de Fermi-Dirac 1337
~ de Maxwell-Boltzmann 2142
~ de Poisson 2694
~ du binôme 352
~ normale 2401
répétabilité *f* 3027
reproductibilité *f* 3031
réseau *m*
~ cristallin 835
~ de Bravais 426
super-~ 3458
résine *f*
~ amphotère 156
~ échangeuse 1807
résines *fpl* échangeuses d'électrons 1151
résistance *f* interne de cellule 1775
résolution *f* 3035
~ de pics 2580
résonance *f*
double ~ 1062
~ de Fermi 1342
~ magnétique 2108
~ ~ nucléaire 2419
~ paramagnétique électronique 1178
~ qadrupolaire nucléaire 2427
résonances *fpl* 3043
~ des baryons 314
~ des mésons 2185
retardation *f*
ion ~ 1842
rétention *f* 3046
~ de configuration 3047
réticulation *f* 819
retombée *f* radioactive 2883
retour *m*
~ externe 1312
~ interne 1776
rétrodiffusion *f* 295
révélation *f* 950
rhéologie *f* 3059
~ chimique 585

solvant
~ basique 2808
~ capable de différencier des acides ou des bases 996
~ capable de niveler des acides ou des bases 2001
~ différenciant 996
~ inert 208
~-niveleur 2001
~ proton-accepteur 2808
~ proton-donneur 2803
solvat *m* 3258
solvatation *f* 3260
~ d'ion 3262
solvatochromisme *m* 3264
solvolyse *f* 3265
somme *f*
~ d'états de translation 3612
~ d'états de vibration 3705
~ des états de rotation 3078
sorbant *m* 3268
sorption *f* 3269
soufre *m* 3443
radiosoufre 2920
source *f*
~ d'entropie 2052
~ d'excitation 1298
~ de rayonnement 2873
~ de référence 2994
~ interne 1777
~ ~ du rayonnement 1777
~ isotopique de neutrons 1897
~ non scellée 3670
~ radioactive 2887
~ scellée 3135
sous-couche *f* électronique 1168
sous-groupe *m* 3430
~-~ du scandium 3115
~-~-du vanadium 3676
~-~-du zinc 3801
spallation *f* 3274
spécifité *f* 3284
spectranalyse *f*
~ d'absorption 21
~ moléculaire 2284
spectre *m* 3317
~ atomique 259
~ continu 746
~ d'absorption 24
~ d'activation de fluorescence 1395
~ d'arc 212
~ d'émission 1212
~ d'étincelles 3275
~ d'inversion 1795
~ de bande 304
~ de bandes électroniques 1174
~ de fluorescence 1396
~ de masse 2135
~ de microondes 2231
~ de neutrons 2369
~ de Raman 2933
~ de résonance magnétique nucléaire 2422
~ de rotation 3080
~ de vibration 3709
~ de vibration-rotation 3708
~ du rayonnement 2429
~ du rayonnement nucléaire 2429
~ discontinu 1017
~ ~ des rayons X 548
~ électronique 1167
~ infra-rouge 1732
~ ionique 1844
~ moléculaire 2286
~ optique 2477
~ résolu dans le temps 3577
~ ultraviolet 3653
~ visible 3723
spectrofluorimètre *m* 3303
spectrofluoromètre *m* 3303
spectrofluorimétrie *f* 3304

spectrogramme *m* 3305
spectrographe *m* 3306
~ de masse 2132
spectromètre *m* 3308
~ à fluorescence de rayons X 1397
~ à rayons X 3783
~ à rayons gamma 1454
~ à résonance magnétique nucléaire 2420
~ à scintillation 3130
~ de Bragg 419
~ de masse 2133
~ gamma 1454
spectrométrie *f*
~ de fluorescence 1394
~ de fluorescence atomique 246
~ de masse 2134
~ Raman 2936
spectrophotomètre *m* 3310
~ à flamme 1382
~ bifaisceau 1057
~ enregistreur 2973
~ monofaisceau 3220
~ non enregistreur 2121
spectrophotométrie *f* 3312
~ d'absorption 22
~ d'absorption atomique 241
~ de flamme 159
~ généralisée 1475
spectropolarimètre *m* 3313
spectropolarimétrie *f* 3314
spectroscope *m* 3315
spectroscopie *f* 3316
~ atomique 258
~ cinétique 1931
~ d'absorption 23
~ d'éclair 1384
~ d'émission 1211
~ de microondes 2230
~ de plasma 2680
~ de résonance magnétique nucléaire 2421
~ électronique 1166
~ hertzienne 2918
~ moléculaire 2285
~ Mössbauer 2309
~ nucléaire 2434
~ par réflection 2995
~ Raman 2937
sphère *f*
~ d'activité 2151
~ de coordination 763
~ de solvatation 3263
spin *m* 3320
~ isobarique 1880
~ isotonique 1880
~ nucléaire 2435
spin-orbitale *f* 3327
spur *m* 3341
stalagmomètre *m* 3349
statistique *f* 3372
~ de Bose-Einstein 409
~ de Fermi-Dirac 1338
~ de Maxwell-Boltzmann 390
~ mathématique 3381
stéréochimie *f* 3387
stéréo-isomérie *f* 3388
stœchiométrie *f* 3405
stripage *m* 3418
strontium *m* 3423
~ radiostrontium 2919
structure *f*
~ cristalline 847
~ en domaines 1044
~ fine 1361
~ hyperfine 1685, 1686
~ magnétique 2110
~ mosaïque 2306
superstructure 3458
structures *fpl*
~ de résonance 3044

INDEKS POLSKI

A

absorbancja f 10
absorbowalność f 25
absorpcja f 12
~ dielektryczna 975
~ promieniowania jonizującego 19
~ światła 20
absorpcjometr m 2631
absorpcyjność f 25
acetoliza f 34
acetylowanie n 35
achiralność f 37
acydoliza f 48
acydymetria f 44
acylacja f 75
acylowanie n 75
addend m 2006
addycja f 77
~ dipolarna 1,3 1011
~ Michaela 2213
addytywność f 82
adhezja f 83
adiabata f 84
adsorbat m 88
adsorbent m 89
adsorpcja f 90
~ aktywowana 57
~ chemiczna 592
~ fizyczna 2658
~ hydrolityczna 1676
~ jednocząsteczkowa 2296
~ jednowarstwowa 2296
~ monomolekularna 2296
~ selektywna 3159
~ siłami van der Waalsa 2658
~ wewnętrzna 1766
~ wielocząsteczkowa 2318
~ wielowarstwowa 2318
~ wymienna 1285
~ zlokalizowana 2055
adsorpcyjność f 102
adsorptyw m 88
aerozol m 103
agregacja f 109
agregaty mpl Chamié 539
ajnsztajn m 1087
akcelerator m 551
~ liniowy elektronów 2023
akceptor m 28, 29
~ elektronów 1135
~ reakcji sprzężonej 28
~ wolnych rodników 3118
aktor m 74
~ reakcji sprzężonych 74
aktyn m 51
~ D 52
aktynon m 55
aktynoołów m 52
aktynouran m 56

aktynowce mpl 54
aktywacja f 60
~ cząsteczki 65
~ elektronowa 64
~ fotojądrowa 1449
~ neutronami rezonansowymi 3040
~ neutronowa 2361
~ radiacyjna 2867
~ rezonansowa 3040
aktywator m katalizatora 500
aktywność f 70, 71
~ bezwzględna 2
~ ciśnieniowa 3729
~ katalityczna 497
~ katalizatora 497
~ nasycenia 3109
~ optyczna 2466
~ powierzchniowa 3465
~ promieniotwórcza 70
~ stężeniowa 71
~ termodynamiczna 71
~ właściwa 3277
~ wzbudzona 1714
akumulator m 31
akwametria f 209
akwokompleks m 210
albedo n 110
alkacymetria f 41
alkaliczność f 317
alkalimetria f 117
alkilacja f 119
alkilowanie n 119
alkoholiza f 113
alkoholometria f 121
alotropia f 112
ameryk m 141
amfolit m 154
amfoteryczność f 157
α-amidoalkilowanie n 143
aminoliza f 145
aminometylowanie n 146
aminowanie n 144
amminokompleks m 147
amokompleks m 147
amonoliza f 148
amperometria f 152
amplituda f prawdopodobieństwa 3741
anaforeza f 171
analiza f
~ absorpcyjna atomowa 241
~ ~ beta 342
~ ~ gamma 1452
~ ~ neutronowa 2360
~ aktywacyjna 61
~ ~ destrukcyjna 948
~ ~ instrumentalna 2391
~ ~ neutronowa 2362
~ ~ niedestrukcyjna 2391
~ ~ z zastosowaniem cząstek naładowanych 552
~ amperometryczna 152
~ całkowita 681

analiza
~ chemiczna 565
~ chromatograficzna 607
~ chromatopolarograficzna 611
~ chronopotencjometryczna 618
~ czołowa 1433
~ elektrochemiczna 1102
~ elektrograficzna 1117
~ elektrograwimetryczna 1118
~ elementarna 1191
~ fluorescencyjna 1394
~ ~ atomowa 246
~ ~ rentgenowska 3780
~ fluorymetryczna 1394
~ fotoaktywacyjna 2627
~ fotometryczna płomieniowa 1381
~ gazometryczna 1464
~ gazomiernicza 1464
~ gazowa 1456
~ grawimetryczna 1507
~ ilościowa 2833
~ instrumentalna 1744
~ interferometryczna 1759
~ izotopowa 1892
~ jakościowa 2830
~ kolorymetryczna 668
~ konduktometryczna 717
~ konformacyjna 728
~ konwencjonalna 752
~ kroplowa 3340
~ kupelacyjna 850
~ luminescencyjna 2074
~ mechaniczna 2569
~ metodą rozcieńczenia 1888
~ ~ wstecznego rozpraszania cząstek β 343
~ miareczkowa 3587
~ nefelometryczna 2350
~ objętościowa 3587
~ organoleptyczna 2491
~ polarograficzna 2709
~ polarymetryczna 2698
~ potencjometryczna 2747
~ radiochemiczna 2895
~ radiometryczna 2909
~ ramanowska 2936
~ refleksometryczna 2995
~ refraktometryczna 3000
~ rentgenofluorescencyjna 3780
~ rentgenowska absorpcyjna 3773
~ ~ dyfrakcyjna 3776
~ ~ spektralna 3732
~ sedymentacyjna 3153
~ sitowa 3204
~ spektralna 3298
~ ~ absorpcyjna 21
~ ~ atomowa 257
~ ~ cząsteczkowa 2234
~ emisyjna 1210
~ ~ masowa 2131
~ spektrochemiczna 3299
~ spektrofotometryczna 161
~ ~ płomieniowa 159
~ ~ precyzyjna 2757
~ spektrograficzna 3307
~ spektrometryczna 3309
~ ~ masowa 2131
~ spektropolarymetryczna 3314
~ spektroskopowa 3298, 3726
~ śladowa 3593
~ termiczna 3521
~ ~ różnicowa 995
~ termograwimetryczna 3547
~ ~ różnicowa 995
~ termomagnetyczna 3550
~ turbidymetryczna 3641
~ umowna 752
~ w skali makro 2084
~ w skali mikro 2216
~ w skali półmikro 3178
~ w szybkozmiennym polu elektrycznym 2505
~ wagowa 1507
~ wariancji 160
~ według grup funkcyjnych 1438
~ widmowa 3298
makroanaliza 2084
mikroanaliza 2216

analogi mpl dienowe 984
anihilacja f 182
~ tryplet-tryplet 3630
anion m 177
~ enolanowy 1235
anionit m 178
anionotropia f 179
anizotropia f 180
~ diamagnetyczna 963
~ magnetyczna 2096
anniling m 181
anoda f 185
anolit m 187
anomery mpl 188
ansa-związki mpl 189
antagonizm m jonowy 190
antyauksochrom m 191
antycząstka f 200
antyelektron m 2735
antyferromagnetyki mpl 195
antyferromagnetyzm m 194
antymateria f 196
antymery mpl 1221
antymon m 197
antyneutrino n 198
~ elektronowe 1153
~ mionowe 2325
antyneutron m 199
antypody pl optyczne 1221
antyproton m 201
argon m 213
aromatyzacja f 221
arsen m 226
arsonowanie n 227
arylowanie n 228
~ Meerweina 2169
asocjacja f 231
asocjat m 229
astat m 234
astrochemia f 782
atmosfera f jonowa 1810
atom m 240
~ centralny 516
~ donorowy 759
~ koordynujący 759
~ ligandowy 759
~ mostkowy 432
~ spiro 3336
~ węgla asymetryczny 235
~ zjednoczony 3660
atomy mpl
~ epitermiczne 1254
~ gorące 1635
~ odrzutu 1635
~ znaczone 1952
atropoizomeria f 264
auksochrom m 280
autoabsorpcja f 3167
~ linii 3168
autojonizacja f 274, 275
autokataliza f 267
autokatalizator m 268
autokompleks m 270
autooksydacja f 279
autoprotoliza f 275
autoradiografia f 277
autoradiogram m 276
autoradioliza f 278
azeotrop m 286
~ dodatni 2732
~ ujemny 2340

cząsteczka
~ typu bąka niesymetrycznego 237
~ ~ ~ sferycznego 3319
~ ~ ~ symetrycznego 3478
~ wzbudzona 1299
cząsteczki *fpl* izoelektronowe 1868
cząsteczkowość *f* reakcji 2278
cząstka *f*
~ chemiczna 1192
~ α 128
~ β 341
~ jonizująca 1835
cząstki *fpl*
~ dziwne 3412
~ elementarne 1194
~ koloidalne 663
częstość *f* 1425, 3012
~ charakterystyczna 543, 544
~ ~ Debye'a 543
~ grupowa 544
~ Larmora 1968
~ larmorowska 1968
~ ramanowska 2934
~ wymiany chemicznej 1290
~ względna 3012
częstotliwość *f* 1425
czułość *f* 3180
czynnik *m*
~ azeotropowy 287
~ Boltzmanna 389
~ chelatujący 563
~ elektrofilowy 1182
~ *g* jądra 2414
~ giromagnetyczny 1522
~ indukcji 1721
~ Landégo 1962
~ nukleofilowy 2438
~ rozszczepienia spektroskopowego g_J 1962
czynność *f* optyczna 2466
czystość *f*
~ jądrowa 2425
~ odczynnika 2970
~ optyczna 2474
~ promieniotwórcza 2885
~ radiochemiczna 2896

D

daltonidy *mpl* 3404
dawka *f*
~ całkowita 1749
~ dopuszczalna największa 2140
~ ekspozycyjna 1308
~ głęboka 944
~ głębokościowa 944
~ pochłonięta 11
~ promieniowania 2864
~ równoważna 1051
~ śmiertelna 1999
~ ~ medianowa 2168
~ ~ średnia 2168
~ zaabsorbowana 11
dawkomierz *m* 1052, 1055
~ filmowy 1358
~ fotograficzny 1358
~ fotometryczny 1358
~ Frickego 1429
~ indywidualny 2604
~ luminescencyjny 2075
~ osobisty 2604
~ progowy 3569
~ termoluminescencyjny 3549
debaj *m* 894
debajogram *m* 2751
decyklizacja *f* 3065
defekt *m* masy 2127
defekty *mpl*
~ chemiczne 1701

defekty
~ elektryczne 1155
~ Frenkla 1424
~ liniowe 2030
~ punktowe 2692
~ Schottky'ego 3127
~ sieci krystalicznej 1974
~ sieciowe 1974
deficyt *m*
~ elektronowy 1144
~ masy 2127
deformacja *f* jąder atomowych 902
degeneracja *f* 906
~ gazu doskonałego kwantowa 1459
~ poziomu energetycznego 906
degradacja *f* 907, 908
~ amidów Hofmanna 1609
~ ~ von Brauna 3736
~ Barbiera i Wielanda 306
~ Barbiera, Locquina i Wielanda 306
~ Emdego 1207
~ Hofmanna 1607
~ polimeru 908
~ ~ radiacyjna 2862
~ Ruffa i Fentona 3086
~ von Brauna 3736
~ Weermana 3748
~ Wohla i Zempléna 3762
dehalogenacja *f* 922
dehydratacja *f* 923
dehydrocyklizacja *f* 924
dehydrogenacja *f* 925
dehydrohalogenacja *f* 926
dejonizacja *f* 927
dekarboksylacja *f* 896
dekarbonylacja *f* 895
dekontaminacja *f* 900
demetylowanie *n* 935
demineralizacja *f* 927
denaturacja *f* białek 937
depolaryzator *m* 941
depolimeryzacja *f* 942
derywatografia *f* 945
desmotropia *f* 946
desorpcja *f* 947
destylacja *f* nośnikowa 492
desulfuracja *f* 949
detekcja *f* promieniowania jądrowego 951
detektor *m*
~ całkowy 1748
~ germanowo-litowy 2050
~ jonizacyjny 1831
~ neutronów 2364
~ ~ rezonansowy 3041
~ półprzewodnikowy 3176
~ promieniowania 2863
~ różnicowy 990
~ termokonduktometryczny 3523
detonacja *f* 953
deuter *m* 956
deuteron *m* 957
deuterowanie *n* 955
dezaktywacja *f*
~ cząsteczki 881
~ kontaktu 879
~ stanu wzbudzonego 880
dezalkilowanie *n* 884
dezaminowanie *n* 885
diagram *m*
~ fazowy 2611
~ Lauego 1977
~ molekularny 2272
~ proszkowy 2751
~ warstwicowy 3072
diagramy *mpl* Feynmana 1353
dializa *f* 962

dializator *m* Grahama 1500
diamagnetyki *mpl* 964
diamagnetyzm *m* 967
diastereoizomeria *f* 968
diastereoizomery *mpl* 969
dichroizm *m* 972
~ kołowy 621
dielektryk *m* 974
dimeryzacja *f* 1010
dipol *m*
~ elektryczny 1098
~ indukowany 1715
~ magnetyczny 2098
~ trwały 2597
długość *f*
~ fali de Broglie'a termiczna 3524
~ łańcucha 532
~ ~ kinetycznego 531
~ ~ polimeru 532
~ ~ reakcji 531
~ wiązania 396
dokładność *f* 32
~ metody 33
domeny *fpl* 1043
domieszki *fpl* 1701
donor *m* 1046
~ elektronów 1147
doza *f* promieniowania 2864
dozymetr *m* 1052, 1055
dozymetria *f* 1056
~ chemiczna 571
~ promieniowania jonizującego 1056
drgania *npl* 3710
~ charakterystyczne 1519
~ deformacyjne 328
~ grup atomów 1519
~ ~ funkcyjnych 1519
~ normalne 2405
~ rozciągające 3417
~ walencyjne 3417
~ zdegenerowane 905
~ zginające 328
drobina *f* 2288
droga *f*
~ reakcji 2962
~ swobodna średnia 2149
dublet *m*
~ elektronowy 1177
~ widmowy 1063
duchy *mpl* (*w widmie*) 3293
dwójłomność *f* 362
dwuazowanie *n* 971
dwubarwność *f* 972
dwurodniki *mpl* 360
dyfrakcja *f*
~ atomów i cząsteczek 997
~ elektronów 1146
~ neutronów 2365
~ promieni rentgenowskich 3775
~ światła 2010
dyfraktogram *m*
~ kryształu wahanego 2502
~ obrotowy 3072
~ proszkowy 2751
~ warstwicowy 3072
dyfraktometr *m* rentgenowski 3777
dyfuzja *f* 1000
~ powierzchniowa 3468
dylatancja *f* 1007
dyslokacje *fpl* 2030
~ brzegowe 1076
~ krawędziowe 1076
~ śrubowe 3134
dysmutacja *f* 1029
~ rodników 1030

dysocjacja *f* 1036
~ elektrolityczna 1122
~ fotochemiczna 2643
~ pod wpływem pola elektrycznego 1037
~ termiczna 3526
dyspergowanie *n* 1018
dyspersja *f*
~ dielektryczna 976
~ liniowa 2022
~ odwrotna 2975
~ przewodnictwa 1027
~ rotacyjna 2475
~ skręcalności optycznej 2475
~ statystyczna 1021
~ światła 2011
dyspersoid *m* 1020
dysproporcjonowanie *n* 1029
~ rodników 1030
dysproz *m* 1072
dyssypacja *f* 1032
dystektyk *m* 1038
dystrybuanta *f* zmiennej losowej 2788
dziura *f*
~ dodatnia 1613
~ elektronowa 1613
dziwność *f* 3410

E

ebuliometria *f* 1073
echo *n* spinowe 3323
efekt *m*
~ asymetrii 3021
~ Augera 265
~ Bakera i Nathana 1682
~ Barkhausena 308
~ Bartona 2061
~ batochromowy 320
~ białkowy 2800
~ chelatowy 559
~ Comptona 697
~ Cottona 621
~ ~ i Moutona 783
~ dalekiego zasięgu 2061
~ Debye'a i Falkenhagena 1027
~ de Haasa i van Alphena 921
~ Dopplera 1048
~ Dorna 3155
~ Dufoura 1069
~ dysocjacyjny pola elektrycznego 1037
~ dyspersyjny 1023
~ elektroforetyczny 1184
~ elektrokapilarny 1101
~ elektromeryczny 1132
~ elektrowiskozowy 1188
~ fazowy 1083
~ fotoelektryczny wewnętrzny 2632
~ ~ zewnętrzny 2634
~ Francka i Rabinovitcha 457
~ Halla 1533
~ Hedvalla 1575
~ hypochromowy 1688
~ hypsochromowy 1690
~ ±I 1723
~ indukcyjny 1720, 1723
~ izotopowy kinetyczny 1829
~ Jahna i Tellera 1904, 1905
~ Joule'a i Thomsona 1911
~ Kerra 1921
~ klatkowy 457
~ ±M 2181
~ magnetokaloryczny 2113
~ mezomeryczny 2181
~ mocy dawki 1054
~ Mössbauera 2308
~ napięcia B 297
~ następczy 105
~ natężenia pola elektrycznego 1357
~ ochronny 2799
~ odrzutu 3491

emiter *m* 1213
emulsja *f* 1216
emulsoid *m* 1219
enancjomery *mpl* 1221
enancjotropia *f* 1223
endoosmoza *f* 1181
energia *f*
~ aktywacji 63
~ ~ pozorna 203
~ ~ rzeczywista 3633
~ cząstki 2566
~ delokalizacji 931
~ Fermiego 1340
~ Gibbsa 1485
~ jądrowa 2408
~ jonizacji 1832
~ kulombowska 784
~ kwantu promieniowania 2837
~ lokalizacji 2054
~ mezomerii 3038
~ oscylacyjna 3704
~ powierzchniowa 1421
~ powinowactwa elektronowego 1136
~ progowa 3570
~ przemiany promieniotwórczej 1233
~ reakcji jądrowej 1232
~ rezonansu 3038
~ rotacyjna 3077
~ rozszczepienia polem krystalicznym ligandów 832
~ ~ ~ ligandów 832
~ separacji 3180
~ sieci krystalicznej 1975
~ sparowania 2536
~ stabilizacji 2038
~ ~ polem krystalicznym ligandów 833
~ ~ ~ ligandów 833
~ swobodna 1579
~ ~ Gibbsa 1485
~ ~ Helmholtza 1579
~ symetrii jądra atomowego 2436
~ translacyjna 3611
~ wewnętrzna 1771
~ wiązania chemicznego 567
~ ~ jądra 2406
~ wymiany 1237
~ wzbudzenia 1296
~ zerowa 3798
~ związana 412
enolizacja *f* 1237
entalpia *f* 1240
~ aktywacji 1241
~ molowa 2258
~ reakcji chemicznej standardowa 3357
~ swobodna 1485
entropia *f* 1242
~ aktywacji 1246
~ mieszania 1247
~ molowa 2259
~ przekazana 3459
enzym *m* 1249
epimery *mpl* 1253
epimeryzacja *f* 1252
epoksydowanie *n* 1256
erb *m* 1264
estryfikacja *f* 1271
estymacja *f* 1272
~ punktowa 1272
estymator *m* 1273
eteryfikacja *f* 1274
etynylowanie *n* 1276
europ *m* 1277
eutektoid *m* 1281
eutektyk *m* 1278
~ ciekły 1278
~ dwuskładnikowy 350
~ podwójny 350
~ stały 3248

F

fala *f*
~ de Broglie'a 886
~ detonacyjna 954
~ materii 886
~ polarograficzna 2711
~ sprężysta 1092
~ uderzeniowa 3198
fale *fpl* spinowe 3334
faraday *m* 1324
faza *f* 2609, 2614
~ ciekła 2046
~ gazowa 1461
~ krytyczna 811
~ metastabilna 2200
~ nieruchoma 3370
~ rozpraszająca 1025
~ rozproszona 1019
~ ruchoma 2247
~ stała 3249
fazy *fpl*
~ enancjotropowe 1222
~ Hume-Rothery'ego 1648
~ intermetaliczne 1761
~ Lavesa 1973
~ mezomorficzne 2042
~ międzymetaliczne 1761
~ monotropowe 2302
~ pośrednie 1648
~ skondensowane 709
femtometr *m* 1334
ferm *m* 1345
ferment *m* 1249
fermentacja *f*
~ cytrynowa 628
~ masłowa 453
~ mlekowa 1957
~ tlenowa 2527
fermiony *mpl* 1341
ferrimagnetyki *mpl* 1347
ferrimagnetyzm *m* 1346
ferroelektryk *m* 1349
ferromagnetyki *mpl* 1351
ferromagnetyzm *m* 1350
~ pasożytniczy 2553
ferryty *mpl* 1348
filtr *m* 2012
~ absorpcyjny 18
~ interferencyjny 1756
~ neutralny 2356
~ optyczny 2012
~ stopniowy 3382
~ szary 2356
~ świetlny 2012
fizyka *f* chemiczna 581
fizykochemia *f* 2659
fizysorpcja *f* 2658
flokulacja *f* 1386
flotacja *f* 1387
fluencja *f* 1390
~ cząstek 1390
~ energii 1231
fluktuacja *f*
~ bezwzględna 3353
~ kwadratowa średnia 2155
~ względna 3017
fluktuacje *fpl*
~ równowagowe 1261
~ statystyczne 1389
~ termodynamiczne 1261
fluor *m* 1400
fluorescencja *f* 1393
~ opóźniona 928
~ rentgenowska 3779
~ sensybilizowana 3185
fluorowanie *n* 1399

granice *fpl*
~ samozapłonu 2019
~ wybuchowości 1307
~ wybuchu 1307
~ zapalności 2020
grawimetria *f* 1507
grej *m* 1508
grono *n* 639
grubość *f* warstwy absorbującej 3102
grupa *f* 1516
~ analityczna 167
~ antyauksochromowa 191
~ auksochromowa 280
~ batochromowa 319
~ blokująca 2796, 3398
~ chromoforowa 615
~ dodatkowa 3430
~ funkcyjna 1437
~ główna 2118
~ hipsochromowa 1689
~ koordynująca 760
~ kwasowa 43
~ mostkowa 433
~ opuszczająca 1991
~ osłaniająca 2796
~ poboczna 3430
~ polarna 2696
~ przestrzenna 3272
~ punktowa 2699
~ skoordynowana 2006
~ solotwórcza 43
~ zabezpieczająca 2796
grupy *fpl* punktowe krystalograficzne 842

H

hadrony *mpl* 1523
hafn *m* 1524
halogenokompleks *m* 1532
halogeny *mpl* 1536
hamiltonian *m* 1024
hamowanie *n* jonów 1842
hel *m* 1577
helowce *mpl* 1727
heteroatom *m* 1586
heteroazeotrop *m* 1587
heteroliza *f* 1595
heteropolikwas *m* 1598
heteropolimeryzacja *f* 1599
hiperkoniugacja *f* 1682
hiperładunek *m* 1681
hiperon *m*
~ Λ 1960
~ Ξ 3771
~ Σ 3208
~ Ω⁻ 2456
~ kaskadowy 3771
hiperony *mpl* 1687
hipoteza *f* zerowa 2440
hipsochrom *m* 1689
histereza *f* 1691
~ adsorpcji 96
~ adsorpcyjna 96
~ magnetyczna 2102
holm *m* 1616
homoazeotrop *m* 1617
homogenizacja *f* emulsji 1625
homoliza *f* 1630
homologi *mpl* 1628
homologia *f* 1629
homopolimer *m* 1632
homopolimeryzacja *f* 1633

hybryd *m* 1652
~ digonalny 1006
~ liniowy 1006
~ sp 1006
~ sp² 3623
~ sp³ 3515
~ tetraedryczny 3515
~ trygonalny 3623
hybrydyzacja *f* orbitali 1651
hydratacja *f* 1654, 1655
~ jonu 1658
hydroborowanie *n* 1660
hydroformylowanie *n* 1662
hydrogenizacja *f* 1665
hydrogenoliza *f* 1672
hydroksokompleks *m* 1678
hydroksylowanie *n* 1679
~ Prévosta 2766
hydroliza *f* 1673, 1674
hydrozol *m* 1677
hydrożel *m* 1663

I

identyfikacja *f* 1696
iloczyn *m*
~ jonowy 1817
~ ~ rozpuszczalnika 1818
~ ~ wody 1819
~ rozpuszczalności 3255
ilość *f*
~ materii 151
~ substancji 151
ind *m* 1712
indeksy *mpl*
~ reaktywności 2968
~ ~ dynamiczne 1071
~ ~ statyczne 3369
indukcja *f*
~ chemiczna 578
~ jądrowa 2415
~ magnetyczna 2103
induktor *m* 1724
~ reakcji sprzężonej 1724
indykator *m* 1708
indywiduum *n* chemiczne 588, 3434
inhibitor *m* 1733
~ reakcji łańcuchowej 1734
inicjator *m* reakcji łańcuchowej 529
inicjowanie *n* łańcucha reakcji 528
interferometr *m*
~ Fabry'ego i Perota 1319
~ optyczny 2470
inwersja *f*
~ cukru 1791
~ konfiguracji 1794
~ pierścienia 756
~ Waldena 3739
iryd *m* 1845
iterb *m* 3787
itr *m* 3788
izentropa *f* 84
izobara *f* 1860
~ reakcji 3694
~ van't Hoffa 3694
izobary *fpl* 1862
izochora *f* 1863
~ reakcji 3695
~ van't Hoffa 3695
izodimorfizm *m* 1865
izodoza *f* 1866
izolator *m* 1099
izomer *m*
~ *cis* 623, 624
~ *trans* 3603, 3604

izomeria *f* 1874
~ *cis-trans* 625, 626
~ elektronowa 1157
~ endo-egzo 1224
~ funkcyjna 1439
~ geometryczna 625, 626
~ grup funkcyjnych 1439
~ hydratacyjna 1656
~ jądrowa 2416
~ jonizacyjna 1833
~ koordynacyjna 764
~ łańcuchowa 530
~ optyczna 2471
~ podstawienia 2731
~ polimeryzacyjna 2723
~ położenia 2731
~ pozycji koordynacyjnych 767
~ przestrzenna 3387
~ rozmieszczenia 767
~ solna 2034
~ strukturalna 2034, 3425
~ syn-anti 3485
~ wiązania 2034
~ Z-E 625
izomery *mpl* 1876
~ jądrowe 2417
~ zwierciadlane 1221
izomeryzacja *f* 1875
izomorfizm *m* 1877
izopolikwas *m* 1878
izospin *m* 1880
izostera *f* adsorpcji 98
izoterma *f* 1884
~ adsorpcji 99
~ ~ BET 347
~ ~ Freundlicha 1428
~ ~ Langmuira 1965
~ ~ liniowa 2025
~ BET 347
~ Freundlicha 1428
~ Langmuira 1965
~ liniowa 2025
~ krytyczna 808
~ podziału 1040
~ reakcji 2959
~ van der Waalsa 3690
~ van't Hoffa 2959
~ wymiany 1805
izotony *mpl* 1887
izotop *m*
~ długożyciowy 2062
~ krótkożyciowy 3199
izotopy *mpl* 1891
izotropia *f* 1900

J

jakość *f* promieniowania 2872
jama *f* potencjału 2745
jaskrawość *f* 2072
jasność *f* 2072
jądra *npl* zwierciadlane 2238
jądro *n* 249
~ atomowe 249
~ izomeryczne 2199
~ magiczne 2093
~ metastabilne 2199
~ metatrwałe 2199
~ nieparzysto-nieparzyste 2451
~ nieparzysto-parzyste 2450
~ parzysto-nieparzyste 1283
~ parzysto-parzyste 1282
~ podwójnie magiczne 1064
~ wzbudzone 1300
~ złożone 692
jednostka *f*
~ konfiguracyjna enancjomeryczna 1220
~ konstytucyjna 739
~ monomeryczna 2298

jednostka
~ podstawowa konfiguracyjna 723
~ ~ konstytucyjna 737
~ powtarzalna konstytucyjna 737
~ ~ przestrzennie 3385
~ przestrzennie podstawowa 3385
jednostki *fpl* atomowe 261
jod *m* 1798
~-131 2905
jodowanie *n* 1797
jon *m* 1799, 1827
~ centralny 518
~ dwubiegunowy 155
~ fenoniowy 2620
~ karboniowy 477
~ kompleksowy 636
~ liatowy 2079
~ lionowy 2081
~ macierzysty 2554
~ obojnaczy 155
~ wodoropodobny 1670
jonit *m* 1806
~ amfoteryczny 156
~ chelatujący 3281
~ ciekły 2044
~ makroporowaty 2088
~ nieorganiczny 1741
~ organiczny 1807
~ selektywny 3281
~ specyficzny 3281
jonizacja *f* 1828
~ pierwotna 2769
~ właściwa 3282
~ wtórna 3137
jonoforeza *f* 1839
jonorodnik *m* 1841
jony *mpl* nieklasyczne 2390

K

kadm *m* 455
kajzer *m* 1912
kaliforn *m* 461
kaloria *f* 1901
~ międzynarodowa 1901
~ piętnastostopniowa 464
~ termochemiczna 3536
~ termotechniczna 1901
kalorymetr *m* 465
~ adiabatyczny 85
~ izotermiczny 1885
~ lodowy Bunsena 447
mikrokalorymetr 2218
kalorymetria *f* 466
kalorymetry *mpl* bliźniacze 3642
kanał *m*
~ reakcji jądrowej 2431
~ reaktora doświadczalny 322
kaon *m* 1936
kapilara *f* Ługgina 2071
kapilarność *f* 469
karben *m* 475
karbenoid *m* 476
karboanion *m* 474
karbokation *m* 477
karboksylowanie *n* 486
karboksymetylowanie *n* 487
karbonylek *m* metalu 2189
karbonylowanie *n* 485
karta *f* kontrolna 749
kaseta *f* poczty pneumatycznej 2849
kaskada *f* kwantów 1453
kataforeza *f* 505

konformacja
~ częściowo ekliptyczna 2561
~ ekliptyczna 1075
~ kopertowa 1248
~ koronowa 821
~ krzesłowa 537
~ łódkowa 379
~ s-*cis* i s-*trans* 3132
~ skośna 1468
~ ±syn-klinalna 1463
~ ±syn-periplanarna 1075
~ zgiętego arkusza 1248
konformery *mpl* 729
konglomerat *m* 2851
kontakt *m* 3247
kontaminacja *f* 744
kontener *m* 743
kontinuum *n* komptonowskie 695
kontynuator *m* łańcucha 526
konwencja *f*
~ E-Z 1318
~ R-S 3083
konwersja *f*
~ interkombinacyjna 1784
~ pierścienia 756
~ wewnętrzna 1767, 1768
konwerter *m* strumienia (neutronów) 1402
koordynacja *f* 761
koplanarność *f* 771
kopolimer *m* 772
~ blokowy 375
~ przemienny 131
~ szczepiony 1498
kopolimeryzacja *f* 773
koprecypitacja *f* 776
korelacja *f* 779
~ elektronów 1142
~ konfiguracji 780
kosmochemia *f* 782
kowalencyjność *f* 803
krawędź *f* komptonowska 696
kreacja *f*
~ pary 2538
kręt *m* 174
~ całkowity 3588
kriohydrat *m* 822
kriometria *f* 824
kriozol *m* 826
krotność *f* produkcji wielorodnej 2324
kruszenie *n* 3274
krypton *m* 1950
kryptonat *m* promieniotwórczy 2894
krystalit *m* 838
krystalizacja *f* 839
~ z fazy gazowej 840
krystalochemia *f* 829
krystaloid *m* 844
krystaloluminescencja *f* 845
kryształ *m* 827
~ doskonały 2586
kryształy *mpl* atomowe 245
~ ciekłe 2042
~ cząsteczkowe 2271
~ jonowe 1814
~ kowalencyjne 245
~ mieszane 3250
~ ~ pseudoracemiczne 1041
~ molekularne 2271
~ mozaikowe 2305
~ walencyjne 245
~ z wiązaniem wodorowym 1668
~ zonalne 3804
kryterium *n* magnetyczne 2097
krzem *m* 3211

krzepnięcie *n* 1423
krzywa *f*
~ absorpcji 16, 17
~ analityczna 163
~ Bragga 417
~ charakterystyczna 1217
~ dyspersji 1022
~ elektrokapilarna 1100
~ elucji 1203
~ energii potencjalnej 2741
~ kalibrowania 1217
~ kinetyczna 1929
~ miareczkowania 3584
~ Morse'a 2304
~ polarograficzna 2707
~ przebicia 428
~ równowagi fazowej 2613
~ spektrofotometryczna 16
~ tworzenia 1408
~ wzorcowa 163
~ zaczernienia (emulsji fotograficznej) 1217
ksenon *m* 3769
kserożel *m* 3770
kulometr *m* 787
kulometria *f* 789
~ przy stałym potencjale 790
~ przy stałym prądzie 736
kupelacja *f* 850
kuweta *f* 13
kwadrupol *m* 2827
kwant *m*
~ γ 1450
~ anihilacyjny 183
~ działania 2675
~ promieniowania elektromagnetycznego 2646
kwantowanie *n* 2835
~ przestrzenne 3273
kwanty *mpl* γ 1451
kwantyzacja *f* 2835
kwarki *fpl* 2844
kwas *m* 38
kwasowość *f* 45
~ zasady 47

L

laboratorium *n* gorące 1637
lampa *f*
~ rentgenowska 3784
~ z katodą wnękową 1615
lantan *m* 1967
lantanowce *mpl* 1966
lauegram *m* 1977
lepkosprężystość *f* 3717
lepkościomierz *m* 3718
lepkość *f* 3720, 3721
~ anormalna 3426
~ bezwzględna 3721
~ dynamiczna 3721
~ graniczna 1789
~ kinematyczna 1926
~ newtonowska 2372
~ normalna 2372
~ strukturalna 3426
~ właściwa 3291
~ względna 3019
~ zredukowana 2985
lepton *m* 2313
leptony *mpl* 1998
lewoskrętność *f* 1959
liczba *f*
~ atomowa 250
~ ~ efektywna 1077
~ Avogadry 282
~ barionowa 312
~ falowa 3742
~ hydratacji jonu 1657

liczba
~ koordynacyjna 765, 766
~ ~ charakterystyczna 542
~ ~ maksymalna 2139
~ kwantowa efektywna 1082
~ ~ główna 2778
~ ~ magnetyczna 2107
~ ~ ~ spinowa 2109
~ ~ orbitalna 2482
~ ~ oscylacji 3706
~ ~ poboczna 2482
~ ~ rotacji 3079
~ ~ spinowa 3329
~ lepkościowa 2985
~ ~ graniczna 1789
~ leptonowa 1996
~ ~ elektronowa 1158
~ ~ mionowa 2327
~ ligandowa 2009
~ masowa 2130
~ niezależnych reakcji chemicznych 2443
~ obsadzenia 2447
~ obsadzenia średnia 281
~ porządkowa 250
~ postępu reakcji 1310
~ półek teoretycznych 3519
~ przenoszenia jonu 3600
~ rubinowa 3085
~ składników 2444
~ ~ niezależnych 2444
~ solwatacji jonu 3261
~ stopni swobody 2442, 2445
~ symetrii 3493
~ ~ cząsteczki 3483
~ termodynamicznych stopni swobody 2445
~ utlenienia 2522
~ zderzeń 660
~ złota 1492
liczby fpl
~ losowe 2940
~ magiczne 2094
licznik m 2860
~ 2π 792
~ 4π 793
~ cieczowy 2041
~ Czerenkowa 521
~ G-M 1469
~ ~ i Müllera 1469
~ neutronów 2363
~ okienkowy 3760
~ promieniowania 2860
~ proporcjonalny 2794
~ scyntylacyjny 3129
~ studzienkowy 3752
~ wnekowy 3752
ligand m 2006
~ chelatowy 560
~ jednodonorowy 3657
~ jednofunkcyjny 3657
~ jednomiejscowy 3657
~ wielodonorowy 2317
~ wielofunkcyjny 2317
~ wielokleszczowy 2317
likwidus m 2048
linia f
~ analityczna 168
~ antystokesowska 202
~ atomowa 263
~ jonowa 1837
~ ramanowska 2935
~ rayleighowska 2950
~ spektralna 3294
~ stokesowska 3407
~ widmowa 3294
linie fpl
~ homologiczne 1626
~ ostatnie 2932
liożel m 2080
lit m 2049
litowce mpl 116
lorens m 1988
lotność f 3729

luka f 3677
~ elektronowa 1613
luminancja f 2072
luminescencja f 2073
chemiluminescencja 591
krystaloluminescencja 845
~ krystalizacji 845
~ radiacyjna 2907
termoluminescencja 3548
luminofory mpl 2076
lutet m 2078
ładunek m przestrzenny 3270
łańcuch m
~ boczny 3201
~ główny 2117
~ kinetyczny 533
~ normalny 2398
~ prosty 2398
~ reakcji 533
~ rozgałęziony 1407
~ węglowy 481

M

macierz f gęstości 3376
magnetochemia f 2114
magneton m
~ Bohra 381
~ jądrowy 2423
magnetostrykcja f 2115
magnetyzacja f 2112
magnez m 2095
magnony mpl 2116
makroanaliza f 2084
makrocząsteczka f 2087
makrodrobina f 2087
makrometoda f 1503
makromolekula f 2087
makroskładnik m 2085
maksima fpl polarograficzne 2710
mangan m 2119
manganowce mpl 2120
masa f
~ atomowa 3008
~ ~ względna 3003
~ cząsteczkowa liczbowo średnia 2441
~ ~ wagowo średnia 2124
~ ~ (względna) 3013
~ kontaktowa 3247
~ krytyczna 809
~ molowa 2260
~ właściwa 938
~ zredukowana 2981
maszyna f cieplna 1551
materia f 3433
materiał m rozszczepialny 1372
~ wzbogacony 1238
matryca f 2137
mechanika f
~ falowa 2841
~ kwantowa 2841
~ statystyczna 3375
mechanizm m
~ $A_{Ac}1$ 138
~ $A_{Ac}2$ 139
~ $A_{Al}1$ 140
~ $B_{Ac}2$ 377
~ $B_{Al}1$ 378
~ chemicznych reakcji radiacyjnych 2158
~ dwucząsteczkowego podstawienia elektrofilowego 3235
~ ~ ~ nukleofilowego 3237
~ E_1 1208
~ E_2 1209

miareczkowanie
~ spektrofotometryczne 3311
~ strąceniowe 2755
~ termometryczne 3551
~ turbidymetryczne 1596
~ wizualne 3727
micela *f* 2212
miedziowce *mpl* 775
miedź *f* 774
mieszalność *f*
~ nieograniczona 683
~ ograniczona 2562
mieszanina *f*
~ azeotropowa 286
~ doskonała 1695
~ eutektyczna 1278
~ gazowa doskonała 2587
~ heteroazeotropowa 1587
~ homoazeotropowa 1617
~ idealna 1695
~ racemiczna 2851
międzywęźle *n* 1782
migracja *f* jonów 2234
mikroanaliza *f* 2216
mikrochemia *f* 2221
mikrofotometr *m* 2226
mikrokalorymetr *m* 2218
~ Świętosławskiego i Dorabialskiej 1049
mikrokalorymetria *f* 2219
mikrometoda *f* 2236
mikroskładnik *m* 2222
mikroskop *m* polaryzacyjny 2705
mikrosonda *f* elektronowa 1175
mikrospektrofotometria *f* 2229
mineralizacja *f* 2237
mion *m* μ 2313
młyn *m* koloidalny 666
mnożnik *m* analityczny 165
moc *f*
~ dawki 1053
~ jonowa 1824
~ kermy 1919
~ wiązania 400
model *m*
~ alfowy 129
~ cząstek niezależnych 1707, 2460
~ gazu Fermiego 1339
~ gronowy 640
~ Isinga 1858
~ klastrowy 640
~ kroplowy 2043
~ nadprzewodnikowy 3448
~ Nilssona 2377
~ optyczny 2472
~ powłokowy 3192
~ radiolizy dyfuzyjny 1003
~ skorelowanych par 3448
~ sztywnych kul 1542
~ uogólniony 3658
modele *mpl* jądra atomowego 2248
~ kolektywne 657
moderator *m* (reaktorowy) 2249
modyfikator *m* reakcji łańcuchowej 2250
mol *m* 2266, 2267
molalność *f* (roztworu) 2254
molarność *f* (roztworu) 2256
molekuła *f* 2288
molibden *m* 2290
molowość *f* (roztworu) 2256
moment *m*
~ dipolowy 1099
~ ~ elektryczny 1099
~ ~ grupy 1517
~ ~ indukowany 1716
~ ~ trwały 2598
~ ~ wiązania 392

moment
~ drugi 3145
~ drugiego rzędu 3145
~ grupy 1517
~ kwadrupolowy 2828
~ ~ elektryczny 2828
~ ~ jądra 2426
~ magnetyczny 2104
~ ~ atomu 2105
~ ~ efektywny 1081
~ ~ jądra 2418
~ orbitalny atomu 2481
~ pędu całkowity 3588
~ ~ jądra 2435
~ ~ orbitalny 174
~ ~ spinowy 3320
~ przejścia 3606
~ spinowy atomu 3326
~ wiązania 392
monitor *m* 2292
monochromator *m* 2294
monokryształ *m* 3222
monomer *m* 2297
monotropia *f* 2303
mostek *m*
~ elektrolityczny 3097
~ wodorowy 1667
multiplet *m* 2321
~ Bałandina 301
~ widmowy 2322
multipletowość *f* (termu) 2323
multiplety *mpl*
~ izospinowe 553
~ ładunkowe 553
mutacja *f* radiacyjna 2868
mutarotacja *f* 2329

N

naczynko *n* 13
~ pomiarowe 13
nadciekłość *f* 3451
nadfiolet *m* 2337, 3652
~ bliski 2337
~ daleki 1327
~ próżniowy 3673
~ Schumanna 3678
nadnapięcie *n* 2519
nadparamagnetyzm *m* 3454
nadprzewodnictwo *n* 3447
nadstruktura *f* 3458
nadwymiana *f* 3449
nakładanie *n* się orbitali 2518
namagnesowanie *n* 2112
~ nasycenia 3111
~ samorzutne 3338
~ spontaniczne 3338
~ szczatkowe 3033
napięcie *n*
~ Galvaniego 1446
~ I 1781
~ międzyfazowe 1755
~ Pitzera 2673
~ powierzchniowe 3473
~ rozkładowe 899
~ Volty 742
naprężenie *n* Pitzera 2673
napromienianie *n* 1848
~ impulsowe 2816
nasycenie *n* dielektryczne 982
natężenie *n*
~ magnetyzacji 2112
~ pola magnetycznego 2100
~ światła 2077
nefelometr *m* 2349
nefelometria *f* 2350

rozpad
~ β 340
~ heterolityczny 1595
~ homolityczny 1630
~ ketonowy 1924
~ kwasowy 42
~ promieniotwórczy 2880
~ rozgałęziony 423
~ złożony 423

rozpowszechnienie *n* izotopu 26

rozpraszanie *n*
~ cząstek 3116
~ elastyczne 1091
~ klasyczne 2951
~ koherentne 653
~ komptonowskie 697
~ nieelastyczne 1725
~ niekoherentne 1703
~ niespójne 1703
~ niesprężyste 1725
~ promieniowania 3117
~ ramanowskie 2933
~ rayleighowskie 2951
~ spójne 653
~ sprężyste 1091
~ wymienne 1293

rozproszenie *n* wsteczne 295
~ zwrotne 295

rozpuszczalnik *m*
~ amfiprotonowy 153
~ amfoteryczny 153
~ aprotonowy 208
~ kwaśny 2803
~ niwelujący 2001
~ obojętny 208
~ protonoakceptorowy 2808
~ protonodonorowy 2803
~ protonofilowy 2808
~ protonogenowy 2803
~ różnicujący 996
~ wyrównujący 2001
~ zasadowy 2808

rozpuszczalność *f* 3253

rozrzut *m* 1021

rozsprzężenie *n* spinowe 3321

rozstęp *m* 2944

rozszczepienie *n* 2410
~ jądra atomowego 2410
~ nadsubtelne 1684
~ samorzutne 3337

roztwór *m* 3257
~ atermiczny 239
~ buforowy pH 443
~ doskonały 1695
~ idealny 1695
~ koloidalny 3245
~ mianowany 3582
~ nasycony 3108
~ niedoskonały 2972
~ nieskończenie rozcieńczony 1692
~ porównawczy 2993
~ prawidłowy 3005
~ przesycony 3456
~ regularny 3005
~ rozcieńczony idealny 1692
~ rzeczywisty 2972
~ stały 3250
~ ~ addycyjny 1783
~ ~ ciągły 745
~ ~ międzywęzłowy 1783
~ ~ nieuporządkowany 2942
~ ~ racemiczny 1041
~ ~ substytucyjny 3437
~ ~ subtrakcyjny 3441
~ ~ śródwęzłowy 1783
~ ~ uporządkowany 2483
~ właściwy 3636
~ wzorcowy 3359

rozwijanie *n*
~ chromatogramu 958
~ łańcucha reakcji 534

rój *m* jonowy 1801

równania *npl*
~ fenomenologiczne 2619
~ Gibbsa i Helmholtza 1486

równanie *n*
~ adiabaty uderzeniowej 1647
~ adsorpcji Gibbsa 1482
~ Arrheniusa 223
~ BET 346
~ Boltzmanna 388
~ Borna 401
~ Braggów 420
~ Brönsteda i Bjerruma 437
~ Brunauera, Emmetta i Tellera 346
~ chemiczne 574
~ Clapeyrona 1694
~ Clausiusa i Clapeyrona 634
~ Clausiusa i Mossottiego 636
~ Debye'a, Hückela i Onsagera graniczne 887
~ Debye'a i Hückela 888
~ Debye'a i Hückela graniczne 890
~ Duhema i Margulesa 1070
~ Freundlicha 1427
~ Gibbsa 1488
~ Gibbsa i Duhema 1484
~ Hammetta 1533
~ Hendersona 1581
~ Hückla 1644
~ Hugoniota 1647
~ Ilkoviča 1700
~ izobary van't Hoffa 3694
~ izochory van't Hoffa 3695
~ Kelvina 1917
~ kinetyczne 1928
~ zredukowane 1257
~ ~ ~ Zawidzkiego 1257
~ Kirchhoffa 1934
~ ~, Rankina i Dupré 1934
~ Langmuira 1964
~ Latimera 1973
~ Liouville'a 2035
~ Lipmanna 2037
~ Lorentza i Lorenza 2064
~ Poissona 2695
~ reakcji 574
~ Schrödingera nie zawierające czasu 3575
~ ~ zawierające czas 3574
~ stanu Beattiego i Bridgemana 324
~ ~ ~ Berthelota 336
~ ~ ~ Clausiusa 635
~ ~ ~ Dietericiego 987
~ ~ ~ gazu doskonałego 1694
~ ~ ~ kaloryczne 463
~ ~ ~ termiczne 3527
~ ~ ~ termodynamiczne 463
~ ~ ~ van der Waalsa 3688
~ ~ ~ wirialne 3716
~ ~ ~ zredukowane 2980
~ stechiometryczne 574
~ szybkości reakcji 1928
~ Szyszkowskiego 3492
~ termochemiczne 3532
~ van der Waalsa 3688
~ von Neumanna 3737

równowaga *f*
~ chemiczna 575
~ ~ nietrwała 1320
~ ~ pozorna 1320
~ Donnana 2173
~ elektrochemiczna 2394
~ fazowa 2612
~ mechaniczna 2157
~ membranowa 2173
~ międzyfazowa 2612
~ promieniotwórcza 2882
~ ~ przejściowa 3602
~ ~ trwała 3150
~ przejściowa 3602
~ przeponowa 2173
~ radiacyjna 2865
~ reakcji 575
~ ~ trwała 575
~ sedymentacyjna 3154
~ tautomeryczna 3500
~ termiczna 3528
~ termodynamiczna 3537
~ ~ lokalna 2053

Ż

РУССКИЙ УКАЗАТЕЛЬ

А

абсолютная активность *f* 2
абсолютная конфигурация *f* 3
абсолютная ошибка *f* 6
абсолютная скорость *f* движения иона 1815
абсолютная скорость *f* реакции 7
абсолютная температура *f* 3546
абсолютная флуктуация *f* 3353
абсолютно неустойчивое состояние *n* 3672
абсолютно чёрное тело *n* 369
абсолютный нуль *m* 3797
абсолютный нуль *m* температуры 3797
абсолютный показатель *m* преломления 9
абсорбционная спектроскопия *f* 23
абсорбционная спектрофотометрия *f* 22
абсорбционный рентгеновский анализ *m* 3773
абсорбционный светофильтр *m* 18
абсорбционный спектральный анализ *m* 21
абсорбция *f* 12
диэлектрическая ∼ 975
автокатализ *m* 267
автокатализатор *m* 268
автокаталитическая реакция *f* 269
автокомплекс *m* 270
автоокисление *n* 279
автопротолиз *m* 275
авторадиолиз *m* 278
агент *m*
азеотропный ∼ 287
циклообразующий ∼ 563
агрегат *m*
кристаллический ∼ 2716
поликристаллический ∼ 2716
агрегатное состояние *n* 3364
агрегаты *mpl* Шамье 539
агрегация *f* 109
адгезия *f* 83
адденд *m* 2006
однокоординационный ∼ 3657
полидентатный ∼ 2317
аддитивность *f* 82
аддитивные свойства *f* 81
адиабата *f* 84
∼ Гюгоньо 1647
обратимая ∼ 84
адиабатический калориметр *m* 85
адиабатический процесс *m* 87
адиабатная диафрагма *f* 86
адиабатная оболочка *f* 86
адиабатный процесс *m* 87
адроны *mpl* 1523
адсорбат *m* 88

адсорбент *m* 89
адсорбтив *m* 88
адсорбционная ёмкость *f* 101
адсорбционная плёнка *f* 3464
адсорбционная способность *f* 102
адсорбционная теория *f* двойного электрического слоя Штерна 3399
адсорбционная хроматография *f* 92
адсорбционное уравнение *n* Гиббса 1482
адсорбционные силы *fpl* 95
адсорбционный индикатор *m* 97
адсорбционный потенциал *m* 100
адсорбционный ток *m* 94
адсорбционный центр *m* 91
адсорбция *f* 90
активированная ∼ 57
внутренняя ∼ 1766
гидролитическая ∼ 1676
локализованная ∼ 2055
мономолекулярная ∼ 2296
обменная ∼ 1285
полимолекулярная ∼ 2318
селективная ∼ 3159
физическая ∼ 2658
химическая ∼ 592
азеотроп *m*
гетероазеотроп 1587
гомоазеотроп 1617
азеотропия *f* 291
азеотропная система *f* 290
азеотропная смесь *f* 286
азеотропная точка *f* 288
азеотропный агент *m* 287
азеотропный предел *m* 289
азосочетание *n* 801
азот *m* 2382
акваметрия *f* 209
аквокомплекс *m* 210
аккумулятор *m* 31
аксиальная связь *f* 284
аксиальное правило *n* галогенкетонов 285
активатор *m* 500
активационная поляризация *f* 66
активационный анализ *m* 61
активационный анализ *m* с применением заряженных частиц 552
активация *f* 60
∼ быстрыми электронами 64
∼ молекулы 65
∼ произведенная быстрыми электронами 64
∼ резонансными нейтронами 3040
нейтронная ∼ 2361
радиационная ∼ 2867
резонансная ∼ 3040
фотоактивация 1449
активированная адсорбция *f* 57
активированная молекула *f* 59

активированный комплекс m 58
активная молекула f 59
активная поверхность f 69
активная поверхность f адсорбента 69
активная частица f 526
активное давление n 3729
активное столкновение n 1078
активное столкновение n молекул 1078
активность f 70, 71
абсолютная ~ 2
~ катализатора 497
~ насыщения 3109
индуцированная ~ 1714
каталитическая ~ 497
наведённая ~ 1714
оптическая ~ 2466
поверхностная ~ 3465
термодинамическая ~ 71
удельная ~ 3277
активные центры mpl 67
активный комплекс m 58
активный осадок m 68
актиниды mpl 54
актиний m 51
~ D 52
радиоактиний 2878
актиноиды mpl 54
актинон m 55
актиноуран m 56
актор m 74
~ сопряжённых реакций 74
акцептор m 28, 29, 3118
~ сопряжённой реакции 28
~ электронов 1135
акцепторный уровень m 30
алкалиметрия f 117
алкалиметрия f и ацидиметрия f 41
алкилирование n 119
алкоголиз m 113
алкоголиметрия f 112
алкогольное брожение n 111
аллильная перегруппировка f 125
аллотропия f 121
альбедо n 110
альдегидо-кетонная перегруппировка
  f 114
альдольная конденсация f 115
альфа-частичная модель f 129
альфа-частичная модель f ядра 129
алюминий m 133
амальгамная полярография f 137
америций m 141
α-амидоалкилирование n 143
аминирование n 144
аминолиз m 145
аминометилирование n 146
амминокомплекс m 147
аммонолиз m 148
аморфное строение n полимеров 150
аморфное тело n 149
амперометрическое титрование n 152
амперометрическое титрование n с двумя
  поляризованными электродами 348
амперометрия f 152
хроноамперометрия 616
ампульный источник m нейтронов 1897
амфион m 155
амфипротный растворитель m 153

амфотерность f 157
амфотерный ионит m 156
амфотерный растворитель m 153
амфотерный электролит m 154
анализ m
абсорбционный рентгеновский ~ 3773
абсорбционный спектральный ~ 21
активационный ~ 61
активационный ~ с применением
  заряженных частиц 552
~ веществ по отражению β-частиц 343
~ изотопов 1892
~ методом изотопного разбавления 1888
~ мокрым путём 3755
~ по вторичным рентгеновским спектрам
  3780
~ по спектрам комбинационного рассеяния
  2936
~ по спектрофотометрии пламени 159
~ по фотометрии пламени 1381
атомный абсорбционный спектральный ~ 241
атомный спектральный ~ 257
атомный флуоресцентный ~ 246
весовой ~ 1507
визуальный спектральный ~ 3726
газовый ~ 1456
гамма-активационный ~ 2627
гранулометрический ~ 2569
дериватографический ~ 945
дисперсионный ~ 160
дифференциальный спектрофотометрический
  ~ 994
дифференциальный термический ~ 995
изотопный ~ 1892
инструментальный ~ 1744
инструментальный активационный ~ 2391
интерферометрический ~ 1759
капельный ~ 3340
качественный ~ 2830
количественный ~ 2833
колориметрический ~ 668
конформационный ~ 728
люминесцентный ~ 2074
макроанализ 2084
масс-спектральный ~ 2131
масс-спектрометрический ~ 2131
механический ~ 2569
микроанализ 2216
молекулярный спектральный ~ 2284
нейтронноактивационный ~ 2362
нейтронный активационный ~ 2362
объёмный ~ 3587
органолептический ~ 2491
полный ~ 681
полумикроанализ 3178
полярографический ~ 2709
проявительный ~ 1202
радиометрический ~ 2909
радиохимический ~ 2895
рентгеновский дифракционный ~ 3776
рентгеновский спектральный ~ 3782
рентгеноспектральный ~ 3782
седиментационный ~ 3152
седиментометрический ~ 3152
ситовый ~ 3204
спектральный ~ 3298
спектрографический ~ 3307
спектрометрический ~ 3309
спектрополяриметрический ~ 3314
спектрофотометрический ~ 161
спектрохимический ~ 3299
термический ~ 3521
термомагнитный ~ 3550
титраметрический ~ 3587
флуоресцентный рентгеновский ~ 3780
фотоактивационный ~ 2627
фронтальный ~ 1433
функциональный ~ 1438
химический ~ 565
хроматографический ~ 607
хроматополярографический ~ 611
хронопотенциометрический ~ 618
электроанализ 1118
электровесовой ~ 1118
электрографический ~ 1117

Б

баланс m
тепловой ~ 1546
энергетический ~ ядерной реакции 2432
банановая связь f 329
барий m 307
барионное число n 312
барионные резонансы mpl 314
барионный заряд m 312
барн m 309
барьер m
~ вращения 3075
~ центробежной энергии 520
потенциальный ~ 2739, 2740
потенциальный ~ реакции 2740
потенциальный кулоновский ~ 786
энергетический ~ 2740
батохромная группа f 319
батохромное смещение n 320
безвариантная система f 1790
безвариантная термодинамическая
    система f 1790
безызлучательный переход m 2870
безызлучательный процесс m 2870
бекерель m 327
белковая ошибка f 2800
белое рентгеновское излучение n 3774
бензидиновая перегруппировка f 330
бензильная перегруппировка f 331
бензоилирование n 333
бензоиновая конденсация f 332
бериллий m 338
беркелий m 335
бертоллиды mpl 2396
бесконечно разбавленный раствор m 1692
бесконечно тонкий слой m 1729
бетатрон m 345
бимолекулярная реакция f 349
бимолекулярный нуклеофильный
    механизм m замещения 3237
бинарная система f 351
биноминальное распределение n 352
биокатализатор m 353
биологическая защита f 356
биологическая химия f 354
биологическое окисление n 355
биолюминесценция f 357
биосинтез m 358
биохимия f 354
биполимер m 359
1,3-биполярное присоединение n 1011
бирадикалы mpl 360
биуретовая реакция f 365
благородные газы mpl 1727
ближний порядок m 3200
ближняя инфракрасная область f 2336
ближняя ультрафиолетовая область f
    2337
блок m 374
блоксополимер m 375
блоксополимеризация f 376
блочная полимеризация f 446
бозоны mpl 410
боковая цепь f 3201
большая статистическая сумма f 1505
большая сумма f состояний 1505

большое каноническое множество n
    Гиббса 1504
большое каноническое распределение n
    Гиббса 1504
большой канонический ансамбль m 1504
большой канонический ансамбль m
    Гиббса 1504
большой термодинамический потенциал
    m 1506
бомба f
атомная ~ 242
водородная ~ 1666
калориметрическая ~ 391
ядерная ~ 242
бор m 405
боровская орбита f 382
боровский радиус m 384
брожение n
алкогольное ~ 111
лимоннокислое ~ 628, 1957
маслянокислое ~ 453
молочнокислое ~ 1957
бром m 436
бромирование n 435
броуновское движение n 440
бумага f
реактивная ~ 1711
универсальная реактивная ~ 3666
бумажная хроматография f 2541
буфер m
спектроскопический ~ 3300
буферный раствор m 443
быстрая коагуляция f 2948
быстрые нейтроны mpl 1328
бэр m 3024

В

вакантный узел m 3677
вакансия f 3677
вакуумная ультрафиолетовая область
    f 3678
валентная зона f 3679
валентная оболочка f 3683
валентность f 3684
~ иона 1825
главная ~ 2775
отрицательная ~ 2342
побочная ~ 3142
положительная ~ 2734
электровалентность 1187
электрохимическая ~ 2522
валентные колебания npl 3417
валентные кристаллы mpl 245
валентные электроны mpl 2468, 3682
вальденовское обращение n 3739
ван-дер-ваальсов радиус m 3691
ван-дер-ваальсова связь f 3687
ван-дер-ваальсовы силы fpl 3689
ванадий m 3685
ванна f 379
вариантность f системы 2445
вариационный метод m 3702
вековое равновесие n 3150
вектор m
волновой ~ 3744
величина f
~ R_f 3091
~ R_M 3092

изохорный процесс *m* 1864
изоциклическое кольцо *n* 1619
изоэлектрическая точка *f* 1867
изоэлектронные молекулы *fpl* 1868
изоэнтропийный процесс *m* 1857
изоэнтропический процесс *m* 1857
илид *m* 3786
импульсная спектроскопия *f* 1384
импульсное облучение *n* 2816
импульсный радиолиз *m* 2818
импульсный реактор *m* 2815
импульсный фотолиз *m* 1383
инвариантная система *f* 1790
инвариантная фазовая система *f* 1790
инверсионная линия *f* 1793
инверсионная ось *f* 1792
инверсионная ось *f* симметрии 1792
инверсионный спектр *m* 1795
инверсия *f*
~ сахаров 1791
~ цикла 756
ингибитор *m* 1733
~ цепной реакции 1734
индексы *mpl*
динамические ~ реакционной способности 1071
~ реакционной способности 2968
миллеровские ~ 2235
статические ~ реакционной способности 3369
индивидуальная проба *f* 3661
индивидуальный дозиметр *m* 2604
индивидуум *m* 588
химический ~ 3434
индий *m* 1712
индикатор *m* 1708
адсорбционный ~ 97
внешний ~ 1311
внутренний ~ 1772
изотопный ~ 1898
pH-~ 40
кислотно-основный ~ 40
кислотно-щелочной ~ 40
комбинированный ~ 3665
радиоактивный ~ 2889
универсальный ~ 3665
индикаторная ошибка *f* 1710
индикаторный электрод *m* 1709
индикация *f* 950
индофениновая реакция *f* 1713
индуктор *m* 1724
~ сопряжённой реакции 1724
индукционный эффект *m* 1720, 1723
индукция *f*
магнитная ~ 2103
химическая ~ 578
ядерная ~ 2415
индуцированная активность *f* 1714
индуцированная поляризация *f* 1717
индуцированный диполь *m* 1715
индуцированный дипольный момент *m* 1716
индуцируемые реакции *fpl* 800
инертные газы *mpl* 1727
инертный комплекс *m* 1726
инициатор *m* цепной реакции 529
инициирование *n* 528
инконгруэнтная точка *f* 1704
инструментальная ошибка *f* 1745
инструментальный активационный анализ *m* 2391
инструментальный анализ *m* 1744

интеграл *m*
~ взаимодействия 2572
~ столкновений 659
конфигурационный ~ 2572
интегральная доза *f* 1749
интегральная ёмкость *f* двойного слоя 1747
интегральная теплота *f* адсорбции 1750
интегральная теплота *f* набухания 1751
интегральная теплота *f* разведения 1557
интегральная теплота *f* растворения 1568
интегральный детектор *m* 1748
интегральный метод *m* (определения порядка реакции) 2207
интенсивные величины *fpl* 1752
интенсивные параметры *mpl* 1752
интенсивные признаки *mpl* 1752
интеркомбинационная конверсия *f* 1784
интеркомбинационный переход *m* 1753
интерметаллические фазы *fpl* 1761
интерференционный светофильтр *m* 1756
интерферометр *m* 2470
~ Фабри-Перо 1319
оптический ~ 2470
интерферометрический анализ *m* 1759
инфракрасная область *f* 1731, 2232
инфракрасный спектр *m* 1732
иод *m* 1798
~-131 2905
радиоиод 2905
иодирование *n* 1797
ион *m* 1799
амфион 155
водородоподобный ~ 1670
~ карбония 477
~-радикал 1841
карбониевый ~ 477
комплексный ~ 686
лиат-~ 2079
лионий-~ 2081
феноний-~ 2620
центральный ~ 518
ион-молекулярная реакция *f* 1838
ионизационная изомерия *f* 1833
ионизационная камера *f* 1829
ионизационный детектор *m* 1831
ионизация *f* 1828
вторичная ~ 3137
первичная ~ 2769
преионизация 274
удельная ~ 3282
фотоионизация 2640
ионизирующая частица *f* 1835
ионизирующее излучение *n* 1836
ионий *m* 1827
ионит *m* 1806
амфотерный ~ 156
жидкий ~ 2044
крупнорешётчатый ~ 2088
неорганический ~ 1740
органический ~ 1807
селективный ~ 3281
ионитовая диафрагма *f* 1808
ионная атмосфера *f* 1810
ионная линия *f* 1837
ионная пара *f* 1840
ионная полимеризация *f* 1816
ионная реакция *f* 1821
ионная рефракция *f* 1822
ионная связь *f* 1811
ионная сила *f* 1824
ионная электропроводность *f* 1812

колебательный спектр *m* 3709
количественный анализ *m* 2833
количественный состав *m* 2834
количество *n* вещества 151
коллективные модели *fpl* 657
коллектор *m* фракций 1411
коллигативные свойства *npl* 658
коллоид *m* 662
защитный ~ 2798
лиофильный ~ 2082
лиофобный ~ 2083
макромолекулярный ~ 2270
необратимый ~ 1853
обратимый ~ 3053
полуколлоид 232
семиколлоид 232
коллоидная мельница *f* 666
коллоидная система *f* 662
коллоидная химия *f* 665
коллоидное состояние *n* 664
коллоидные частицы *fpl* 663
коллоиды *mpl*
радиоколлоиды 2901
колонка *f*
ионообменная ~ 1804
капиллярная ~ 470
~ Голея 470
хроматографическая ~ 609
колонна *f*
тепловая ~ 3522
колориметр *m* 667
фотоколориметр 2631
колориметрический анализ *m* 668
колориметрическое титрование *n* 669
колориметрия *f* 670
кольца *npl* Лизеганга 2005
кольцо *n*
гидроароматическое ~ 1659
гомоциклическое ~ 1619
изоциклическое ~ 1619
карбоциклическое ~ 478
хелатное ~ 561
кольчато-цепная таутомерия *f* 3064
комбинационное рассеяние *n* света 2933
комбинированный индикатор *m* 3665
компаратор *m* 679
компенсационный метод *m* измерения
Э.Д.С. 2690
комплекс *m* 763
автокомплекс 270
аквокомплекс 210
активированный ~ 58
активный ~ 58
амминокомплекс 147
внешнеорбитальный ~ 2515
внутреннеорбитальный ~ 1738
высокоспиновый ~ 3324
галогенокомплекс 1532
гидроксокомплекс 1678
донорно-акцепторный ~ 1045
инертный ~ 1726
π-~ 2666
σ-~ 3206
~ внешних орбиталей 2515
~ внутренних орбиталей 1738
~ проникновения 2582
координационно насыщенный ~ 769
лабильный ~ 1955
малоустойчивый ~ 3671
многоядерный ~ 2727
неравновесный ~ 2582
нестойкий ~ 3671
нискоспиновый ~ 2069
нормальный ~ 2399
одноядерный ~ 2301
оксокомплекс 2529
октаэдрический ~ 2448
плоский квадратный ~ 3343

комплекс
прочный ~ 3346
равновесный ~ 2399
смешанный ~ 2243
тетраэдрический ~ 3516
циклический ~ 557
комплексная диэлектрическая
поляризация *f* 684
комплексная диэлектрическая
проницаемость *f* 68
комплексная соль *f* 689
комплексное соединение *n* 763
комплексный ион *m* 686
комплексометрическое титрование *n* 685
комплексометрия *f* 685
комплексообразователь *m* 519
компонент *m*
макрокомпонент 2085
микрокомпонент 2222
независимый ~ фазовой системы 1706
комптон-эффект *m* 697
комптоновский край *m* 696
комптоновское распределение *n* 695
комптоновское рассеяние *n* 697
конверсия *f*
внутренняя ~ 1767, 1768
интеркомбинационная ~ 1784
конгломерат *m* 2851
конгруэнтная точка *f* 730
конденсационные методы *mpl* 705
конденсация *f* 703, 704
альдольная ~ 115
ацилоиновая ~ 76
бензоиновая ~ 332
капиллярная ~ 471
~ Бозе-Эйнштейна 407
~ Клайзена 629
~ Клайзена-Шмидта 631
~ Меервейна 2169
~ Пехмана 2581
~ Ульмана 3648
~ Штоббе 3402
ретроградная ~ 3049
сложноэфирная ~ 629
конденсированная система *f* 711
конденсированные фазы *fpl* 709
конденсированные циклы *mpl* 710
конденсированный двойной слой *m* 678
кондуктометрическое титрование *n* 717
кондуктометрия *f* 718
конечная точка *f* титрования 1226
конечный продукт *m* реакции 2964
конкурирующая ядерная реакция *f* 1757,
2768
консекутивные реакции *fpl* 734
константа *f*
~ гидролиза 1675
~ Керра 1920
~ Кюри 452
~ нестойкости 1743
~ равновесия 1260
~ связи 802
~ скорости каталитической реакции 502
~ скорости реакции 2966
~ спин-спиновой связи J 3331
~ Фарадея 1324
~ экранирования 3193
~ химического равновесия 1260
~ электролитической диссоциации 1123
~ электролитической диссоциации
классическая 1123
~ электролитической диссоциации
термодинамическая 1124
~ ячейки 514
криоскопическая ~ 825
общая ~ нестойкости 2516

методы
конденсационные ~ 705
~ анализа по поглощению γ-излучения 1452
~ диспергирования 1026
потенциостатические ~ 2750
релаксационные ~ 3022
pH-метр *m* 2621
pH-метрия *f* 2622
механизм *m*
бимолекулярный нуклеофильный ~ замещения 3237
~ A_{Ac} 1 138
~ A_{Ac} 2 139
~ A_{Al} 1 140
~ B_{Ac} 2 377
~ B_{Al} 1 378
~ бимолекулярного отщепления 1209
~ бимолекулярного электрофильного замещения 3235
~ внутримолекулярного нуклеофильного замещения S_Ni 3239
~ гомолитического замещения 3241
~ S_E 1-замещения 3234
~ S_E 2-замещения 3235
~ S_N 1-замещения 3236
~ S_N 2-замещения 3237
~ S_N i'-замещения 3240
~ мономолекулярного отщепления E_1 1208
~ нуклеофильного замещения в бензольном кольце 2165
S_N 2' ~ нуклеофильного замещения включающий аллильную перегруппировку 3238
~ нуклеофильного присоединения к этиленовой связи 2164
~ E_2-отщепления 1209
~ присоединения нуклеофильных реагентов к карбонильной группе 2163
~ радиационно-химических реакций 2158
~ радикального присоединения к кратным углеродным связям 2161
~ реакции 2960
~ электрофильного ароматического замещения 2160
~ электрофильного присоединения к кратным углеродным связям 2159
~ α-элиминирования 127
пуш-пульный ~ 2821
радикальный ~ замещения 3241
ударно-тянущий механизм *m* 2821
механика *f*
квантовая ~ 2841
статистическая ~ 3375
механический анализ *m* 2569
механическое равновесие *n* 2157
механохимия *f* 2166
меченое соединение *n* 1953
меченые атомы *mpl* 1952
мигающий реактор *m* 2815
миграционная поляризация *f* 3271
миграционный ток *m* 2233
миграция *f* ионов 2234
микроанализ *m* 2216
микроволновая спектроскопия *f* 2230
микроволновой спектр *m* 2231
микрогетерогенный катализ *m* 2225
микрокалориметр *m* 2218
~ Свентославского и Дорабяльской 1049
микрокалориметрия *f* 2219
микроканонический ансамбль *m* 2220
микроканоническое распределение *n* 2220
микрокомпонент *m* 2222
микрокристаллоскопическая реакция *f* 2223

микрометод *m* 2236
микроскоп *m*
поляризационный ~ 2705
ультрамикроскоп 3650
микроскопическое сечение *n* 2227
микроскопическое состояние *n* 2228
микроскопическое эффективное сечение *n* 2227
микрофотометр *m* 2226
микрохимия *f* 2221
миллеровские индексы *mpl* 2235
минерализация *f* 2237
мицелла *f* 2212
мишень *f* 3498
толстая ~ 3554
тонкая ~ 3557
многокомпонентная система *f* 2316
многообразие *n* 2726
многофазная система *f* 1594
многоядерный комплекс *m* 2727
множественное рождение *n* 2320
множественность *f* рождения 2324
множитель *m*
аналитический ~ 165
~ kT/h 3664
предэкспоненциальный ~ 2761
стерический ~ 3396
энтропийный ~ 1244
модели *fpl*
коллективные ~ 657
ядерные ~ 2248
модель *f*
альфа-частичная ~ 129
альфа-частичная ~ ядра 129
гидродинамическая ~ 2043
диффузионная ~ радиолиза 1003
капельная ~ 2043
~ ассоциаций 640
~ газа Ферми 1339
~ Изинга 1858
~ независимых частиц 1707
~ Нильссона 2377
~ парных корреляций 3448
обобщенная ~ 3658
оболочечная ~ 3192
оптическая ~ 2472
сверхтекучая ~ 3448
модификатор *m* цепной реакции 2250
мозаичная структура *f* 2306
мозаичные кристаллы *mpl* 2305
молекула *f* 2288
активированная ~ 59
активная ~ 59
возбуждённая ~ 1299
гетеронуклеарная двухатомная ~ 1597
гомонуклеарная двухатомная ~ 1631
грамм-молекула *f* 2267
дипольная ~ 2706
линейная ~ 2026
макромолекула 2087
~ типа асимметричного волчка 237
~ типа симметричного волчка 3478
~ типа сферического волчка 3319
полярная ~ 2706
сверхвозбуждённая ~ 3450
молекулы *fpl*
изоэлектронные ~ 1868
молекулярная диаграмма *f* 2272
молекулярная орбиталь *f* 2279
молекулярная поляризация *f* 2261
молекулярная рефракция *f* 2263
молекулярная связь *f* 3687
молекулярная спектроскопия *f* 2285
молекулярная формула *f* 2277
молекулярная электропроводность *f* 2257

состояние
неустойчивое ~ 3672
основное ~ 1515
переходное ~ 59
пластическое ~ 2682
псевдокристаллическое ~ 2812
синглетное ~ 3226
сингулетное ~ 3226
~ плазмы 2681
стабильное ~ 3348
стандартное ~ 3360
стационарное ~ 3371
стеклообразное ~ 3728
твёрдое ~ 3251
термодинамически равновесное ~ 3537
триплетное ~ 3629
установившееся ~ 3371
устойчивое ~ 3348
состояния npl
соответственные ~ 781
сохранение n конфигурации 3047
спаренные электроны mpl 2535
спекание n 3227
~ катализатора 3228
спектр m 3317
атомный ~ 259
видимый ~ 3723
вращательный ~ 3080
дискретный ~ 1017
дуговой ~ 212
инверсионный ~ 1795
инфракрасный ~ 1732
ионный ~ 1844
искровой ~ 3275
колебательно-вращательный ~ 3708
колебательный ~ 3709
масс-~ 2135
микроволновой ~ 2231
молекулярный ~ 2286
оптический ~ 2477
полосатый ~ 304
рентгеновский характеристический ~ 548
~ испускания 1212
~ комбинационного рассеяния 2938
~ нейтронов 2369
~ поглощения 24
~ разрешённый по времени 3577
~ флуоресценции 1396
~ ядерного магнитного резонанса 2422
сплошной ~ 746
ультрафиолетовый ~ 3653
электронно-колебательно-вращательный ~ 1174
электронный ~ 1167
энергетический ~ 2429
энергетический ~ излучения 2429
спектральная кривая f 16
спектральная линия f 3294
спектральная сенсибилизация f 2476
спектральная серия f 3296
спектрально чистый реактив m 3295
спектральный анализ m 3298
спектральный дублет m 1063
спектральный мультиплет m 2322
спектральный терм m 3297
спектральный фон m 293
спектрограмма f 3305
спектрограф m 3306
масс-~ 2132
спектрографический анализ m 3307
спектрометр m 1454, 3308
гамма-~ 1454
масс-~ 2133
рентгеновский ~ 3783
рентгеновский флуоресцентный ~ 1397
~ Брэгга 419
~ ЯМР 2420
сцинтилляционный ~ 3130
спектрометрический анализ m 161, 3309
спектрометрия f
атомная флуоресцентная ~ 246

спектрометр
масс-~ 2134
спектрополяриметр m 3313
спектрополяриметрический анализ m 3314
спектрополяриметрия f 3314
спектропроектор m 3318
двойной ~ 1059
спектроскоп m 3315
спектроскопический буфер m 3300
спектроскопический носитель m 3301
спектроскопия f 3316
абсорбционная ~ 23
атомная ~ 258
диэлектрическая ~ 983
импульсная ~ 1384
мёссбауэровская ~ 2309
микроволновая ~ 2230
молекулярная ~ 2285
отражательная ~ 2995
радиоспектроскопия 2918
~ комбинационного рассеяния 2937
~ Мёссбауэра 2309
~ плазмы 2680
~ ядерного магнитного резонанса 2421
электронная ~ 1166
эмиссионная ~ 1211
ядерная ~ 2434
ЯМР-~ 2421
спектрофлуориметр m 3303
спектрофлуориметрия f 3304
спектрофлуорометр m 3303
спектрофлуорометрия f 3304
спектрофотометр m 1382, 3310
двухлучевой ~ 1057
нерегистрирующий ~ 2121
однолучевой ~ 3220
пламенный ~ 1382
регистрирующий ~ 2978
спектрофотометрический анализ m 161
спектрофотометрия f 3312
абсорбционная ~ 22
атомная абсорбционная ~ 241
кинетическая ~ 1931
~ пламени 159
ультрамикроспектрофотометрия 3651
спектрохимический анализ m 3299
спектрохимический ряд m 3302
спектрохимический ряд m Цушиды 3302
специфическая реакция f 3286
специфический катализ m 3276
специфический кислотно-основной катализ m 3276
специфический реагент m 3287
специфичная реакция f 3286
специфичность f
~ катализатора 3283
~ реагента 3284
спин m 3320
изобарический ~ 1880
изоспин 1880
изотопический ~ 1880
~ ядра 2435
спин-орбиталь m 3327
спин-орбитальная связь f 3328
спин-орбитальное взаимодействие n 3328
спин-решёточная релаксация f 3325
спин-спиновая релаксация f 3332
спин-спиновая связь f 3330
спин-спиновое взаимодействие n 3330
спиновая плотность f 3322
спиновая температура f 3333
спиновое квантовое число n 3329
спиновое эхо n 3323